PHYSIOLOGY OF BEHAVIOR

PHYSIOLOGY OF BEHAVIOR

Second Edition

Neil R. Carlson

University of Massachusetts

Allyn and Bacon, Inc.
Boston   London   Sydney   Toronto

This book is dedicated to the people closest to me:
my wife, Mary; my children, Kerstin and Paul;
and my parents, Alice and Fritz.

Production Editor: David Dahlbacka/Margaret Pinette
Designer: Armen Kojoyian
Series Editor: Bill Barke

Library of Congress Cataloging in Publication Data

Carlson, Neil R   1942–
  Physiology of behavior.

  Bibliography: p.
  Includes index.
  1. Psychology, Physiological. I. Title. [DNLM:
1. Behavior—Physiology. 2. Nervous system—Physiology.
3. Psychophysiology. WL 102 C284p]
QP360.C35 1981   152   80-26868
ISBN 0-205-07262-3
ISBN 0-205-07291-7 (international student ed.)

Printed in the United States of America.

10 9 8 7 6   85 84 83

CONTENTS

PREFACE

I am delighted to be writing the preface to the second edition of *Physiology of Behavior*. The preface is the last part of a book to be written; the manuscript has been sent to the compositor, and soon I shall be seeing the galley proofs. Now I have some time to reflect about what I have written.

Although only four years separated the writing of the first and second editions, the field of physiological psychology changed much in this interval of time. New experimental methods such as amino acid autoradiography and retrograde transport of horseradish peroxidase have made it possible to determine neural pathways that eluded degenerating-axon techniques. Radioactive 2-DG has been used to label neurons that are more metabolically active than their neighbors; thus, it is possible to label neurons that respond to a particular stimulus, or become active during a particular behavior. A quick-freeze technique has been developed that catches synaptic vesicles in the process of exocytosis. Progesterone receptors can now be assayed. CAT scans are making the task of the neuropsychologist and clinical neurologist much easier; now the investigator can estimate the location and extent of the lesion while the patient is still alive. New techniques have been developed to study the contribution of gastric receptors to the regulation of food intake. The discovery of the role of the suprachiasmatic nucleus has given a fresh impetus to the field of sleep and biological rhythms. More and more peptides are found to serve as neurotransmitters and/or neuromodulators; it seems hard to believe that the endogenous opiates were discovered in 1975.

I have not altered the basic outline of the book. As before, the foundations—biology, neurophysiology, neuropharmacology, neuroanatomy, and techniques of neuroscience research—are covered in the first seven chapters. However, the text is considerably revised. Many students and faculty members have written to me to make suggestions, a good number of which I used in the revision. In addition, my own students made suggestions directly, by proposing changes, or indirectly, by demonstrating a difficulty in understanding what I thought was clearly stated. I think those sections are more clear in the revision.

Since the second edition covers new research, and since I did not want to make the book much longer than it was, I had to drop some material. Obviously, I omitted mention of work that has been shown to be incorrect, except where it was necessary to clear up common misconceptions. In addition, interest and research activity has declined in some fields of inquiry; these, too, I dropped. I added a chapter on the physiology of mental disorders. In order to make room for this new chapter, I devoted two chapters, rather than three, to learning and memory. I hope that this revision will make it possible for more instructors to fit these two chapters into their syllabus, even if they include the new chapter on mental disorders.

Artists from the medical illustration department at the Boston University School of Medicine have prepared many new figures, and have revised many of the ones that appeared in the first edition. A glance through the book will show you how skillful and knowledgeable these people are. Physiological psychology is a topic that must be explained by illustrations just as much as by words. The new figures should make the book much easier to understand, as well as improve its appearance.

I did not change my general approach to writing, or to the study of physiological psychology. To describe these, I shall freely plagiarize the preface to my first edition. As my colleagues know, a competent physiological psychologist must know something of anatomy, physiology, biochemistry, pharmacology, and endocrinology—and have a good understanding of behavioral processes. Depending upon one's particular area of study, special competence in some of the biological subfields is necessary. Also, one cannot perform research without having the appropriate skills—for example, in surgery, histology, electronics, or computer programming.

The breadth of the field we call physiological psychology, and the extent to which this field relies on other disciplines, makes its study a rather daunting prospect for most students. Similarly, a prospective author is daunted by the task of writing a textbook that will provide students with a solid introduction to the field. If ambiguities or contradictory results are glossed over, students are misled, and an injustice is done to the many careful researchers who are trying to unravel the subtle complexities of the physiology of behavior. On the other hand, a disservice is done to students, and to the field, by becoming enmeshed in the details of research that does not reveal anything about *process*, or the underlying physiological mechanisms that determine an organism's behavior.

This text is designed for serious students who are willing to work. In return for their effort, I have endeavored to write a book that will provide a solid foundation for further study in the area. Those students who will not take any subsequent courses in this or related areas should receive the satisfaction of a much better understanding of their own behavior. Also, they will have a greater appreciation for the forthcoming advances in medical practices related to various disorders that affect a person's perceptual processes or behavior. It was my wish, in writing this text, that a student who carefully reads this book will henceforth perceive human behavior in a new light.

Besides the effort to avoid oversimplifications or excessive attention to details that might produce unnecessary confusion, I had several other goals in mind as I first wrote this book, and as I prepared this revision. First, I attempted to describe the physiology of *human* behavior as much as possible. A rat is a good subject for the study of the physiological control of food intake. However, if we wish to understand the anatomy and physiology of human memory, we must study the human brain itself, or at least the brains of higher primates. We humans are able to learn and integrate information across sensory modalities in ways that most other animals cannot.

A second principle that had guided my writing is to leave no steps out of logical discourse. My teaching experience has taught me that an entire lecture can be wasted if the students do not understand all of the "obvious" conclusions of a given experiment before the next one is described. The problem for the instructor is that the puzzlement usually leads to feverish notetaking ("I'll get all the facts down so that I can figure out what they mean later"). And a roomful of busy, attentive students tends, unfortunately, to reinforce the lecturer for what he or she is doing. I am sure all of my colleagues have been stopped cold by a question from a student that shows a lack of understanding of details long since passed, and, even more alarming, quizzical looks on the other students' faces, confirming that they, too, have the same question. Painful experiences such as these have taught me to examine all the logical steps between the discussion of one experiment and the next and to make sure that they are all explicitly stated. A textbook writer should address him/herself to students, and not to colleagues who are already acquainted with much of what he or she will say.

A third principle, which is closely related to the second, is to provide the student with the background necessary for the understanding of a given set of experiments. In this regard, students are taught the nature of postsynaptic potentials and the decision-making process of integration, the mechanism of the liberation of transmitter substance, the biosynthesis of the neurotransmitters, and principles of neuropharmacology so that they can understand the use of drugs in the treatment of schizophrenia and the affective disorders. I have been careful to provide enough biological background for the various topics in this book so that a course in zoology or physiology need not be a prerequisite to understanding the content. However, students with such a background will find the reading easier, and they will probably get more out of it.

In this book I have emphasized the fact that physiological psychology is the study of the *physiological*, not merely *neural*, bases of behavior. We must not restrict our view of the organism to those parts above the neck; the rest of the body does more than move the head around. For example, unless we understand how the kidney works and how blood volume is monitored and maintained, we cannot understand the physiology of thirst, no matter how many brain lesions we produce, or how many neurons we record from, or how many areas we electrically stimulate.

So far as writing style goes, I have tried to be as clear and precise as possible. If I have succeeded in what I want to accomplish, then the reader will find that I have no "style" at all. My thoughts about the use of language in writing are expressed better than I can by Graves and Hodge (1943): ". . . good English is a matter not merely of grammar and syntax and vocabulary, but also of sense: the structure of the sentences must hold together logically . . . even phrases which can be justified both grammatically and from the point of view of sense may give his reader a wrong first impression, or check his reading speed, tempting him to skip." "Readability" is not to be determined by counting syllables in words or words in sentences; it is a function of the clarity of thought and expression.

Although I must take all the blame for the shortcomings of this book, I want to thank students and colleagues for helping me by sending reprints of their work, by reading drafts of chapters, by suggesting topics that I should cover, by sending photographs that have been reproduced in this book, and by pointing out deficiencies in the first edition. I want to express special thanks to Frédéric Bremer, Ronald Clavier, J. A. Deutsch, John Donahoe, Robert Feldman, Kay Fite, Mark Friedman, Albert Galaburda, John Heuser, David Hubel, Ivan Hunter-Duvar, Alan Kim Johnson, John Johnson, Andrew Kertesz, John Liebeskind, Ralph Lydic, Bruce McCutcheon, Bruce McEwen, Jerrold Meyer, John Moore, Martin Moore-Ede, Antonio Nuñez, Roger Oesterreich, Ada Olins, Mauricio Russek, William Schwartz, S. Murray Sherman, Jerome Siegel, D. Nico Spinelli, Edward Stricker, and George Wade.

Several of my colleagues reviewed the manuscript of this book and made suggestions for improving the final draft. I want to thank James Brown, San Diego State University; Leo M. Chalupa, University of California—Davis; Dale McAdam, University of Rochester; James C. Mitchell, Kansas State University; Edward J. Ray, University of Illinois—Urbana; John Santelli, Fairleigh Dickinson University; and William Wheeler, University of South Florida.

Leanna Standish, Department of Psychology, Smith College, has prepared a student workbook to accompany this text. We all know how important active participation is in the learning process, and Leanna's workbook provides an excellent framework to guide the student's study behavior. My own students found the first edition of the workbook invaluable; the improvements she made in the second edition should make it even more useful.

I also want to thank the people at Allyn and Bacon: the psychology editor, Bill Barke, who provided assistance, support and encouragement; the developmental editor, Allen Workman, who made many valuable suggestions about the book's organization and style; the production editors, David Dahlbacka and Margaret Pinette, who checked my grammar and made sure that I really said what I hoped to, and who guided the book through all the stages of production; and, last but not least, medical illustrators Mark Lefkowitz and Marcia Williams and graphic illustrators Terry Buzzee, Jo Ellen Murphy, and Alice Vickery, whose superb work speaks for itself. I also want to thank Beth Powell, who cheerfully made many trips to the library to get books and make copies of journal articles for me. My family deserves special thanks for the encouragement and moral support they gave me. Without Mary's help with typing and checking of references I would still be working on this book. And again, I want to thank the University of Massachusetts for supporting me on a sabbatical leave, during which I wrote the first edition of the book, and the University of Victoria, British Columbia, for generously providing me the office space and use of its library during my stay there.

I was delighted to hear from many students and colleagues who read the first edition of my book, and I hope that the dialogue will be continued. I hope very much that you will write to me and tell me what you like and dislike about it. Please write to me at the Department of Psychology, Tobin Hall, University of Massachusetts, Amherst, Massachusetts 01003. I also hope that you will write to discuss issues that I have raised in the book or issues that I have not raised, but should have. I enjoyed writing and revising this book even more than I thought I would; the pleasure will be extended by receiving correspondence from you.

1

Introduction

Physiological psychologists, like other psychologists, try to explain behavior. This book will introduce you to what has been learned about the physiology of behavior, and, perhaps more importantly, will describe the methods that are used to investigate this problem. I hope that I will not discourage you unduly if I say that much of what we "know" about the causes of behavior is less than complete. The study of the physiology of behavior is very much an ongoing quest for answers. This book is much more a description of that quest than a list of answers.

The study of the physiology of behavior is (I think) such an intrinsically interesting pursuit that it does not require a pep talk, so I will not deliver one here. Physiological psychology does not have to be made to sound interesting—it *is* interesting. So instead of trying to convince you that what you are about to read is fascinating stuff, I shall try to accomplish two other purposes in this introduction. First, there are some issues that are so much a part of the scientific life of a physiological psychologist that he or she does not think about them very much. These issues do not lend themselves to discussion in any of the chapters that deal with specific topics that are the subjects of research effort. Second, there are some things I should like to say about the organization of the book that you might find helpful in your studies.

EXPLANATION IN PHYSIOLOGICAL PSYCHOLOGY

First, a word about *explanation*. What does a scientist do when he or she "explains" something? Scientific explanation comes in two forms: generalization and reduction. Most psychologists deal with *generalization*. They explain particular instances of behavior as examples of general laws. For instance, most psychologists would explain a pathologically strong fear of dogs as an example of classical conditioning. Presumably, the person was frightened earlier in life by a dog. An unpleasant stimulus was paired with the sight of the animal, (perhaps the person was knocked down by an exuberant dog or even attacked by a vicious one) and the subsequent sight of dogs evokes the earlier response—fear.

Most physiologists deal with *reduction*. Phenomena are explained in terms of simpler phenomena. For example, the movement of a muscle is explained in terms of changes in the membrane of muscle cells, entry of particular chemicals, and interactions between protein molecules within these cells. A molecular biologist would "explain" these events in terms of forces that bind various molecules together, and cause various parts of these molecules to be attracted to one another. In turn, the job of atomic physicists is to describe matter and energy themselves, and account for the various forces that are found in nature. Practitioners of each branch of science call on more elementary generalizations to explain the phenomena that they are interested in.

The task of physiological psychology is to "explain" behavior in physiological terms. Like other scientists, physiological psychologists believe that all natural phenomena—including human behavior—are subject to the laws of physics. Thus, the laws of behavior can be reduced to descriptions of physiological processes. No consideration has to be given to concepts like free will. (More about this later.)

How does one study the physiology of behavior? Physiological psychologists cannot simply be reductionists. It is not enough to observe behaviors and correlate them with physiological events that occur at the same time. Identical behaviors, under different conditions, may occur for different reasons, and thus be initiated by different physiological mechanisms. This means that we must understand "psychologically" why a particular behavior occurs before we can understand what physiological events made it occur.

Let me give a concrete example. Mice, like many other mammals, often build nests. Behavioral observations show that mice will build nests under two conditions: when the air temperature is low, and when the animal is pregnant. A nonpregnant mouse will not build a nest if the weather is warm, whereas a pregnant mouse will build one regardless of what the temperature is. The same behavior occurs for different reasons. Thus, it should not be a surprise that these behaviors are initiated by different physiological mechanisms. Nest building can be studied as a behavior related to the process of temperature regulation or it can be studied in the context of parental behavior.

Sometimes physiological mechanisms tell us something about psychological processes; physiological psychologists do not always work from the top down. This is particularly true of complex phenomena like language, memory, and mood, which are poorly understood psychologically. Let me describe a few examples of physiological data suggesting psychological explanations. (The examples I give will be sketchy, but they will be described in more detail in later chapters.)

Damage to a specific part of the brain (which can occur when a blood clot blocks a cerebral artery) can cause very specific impairments in a person's language abilities. The nature of these impairments suggests how these abilities are organized. For example, damage to a brain region that is important in analyzing speech sounds also produces deficits in spelling, which suggests that the ability to recognize a spoken word and the ability to spell it call upon related brain mechanisms. Damage to another region can produce extreme difficulty in sounding out unfamiliar words that are being read without impairing the person's ability to read words with which he or she is already familiar. This suggests that reading comprehension can take two routes: one is related to speech sounds, and the other is primarily visual, bypassing any sort of acoustic representation.

Physiological research has also contributed to our understanding of mental disorders. There is a severe disturbance in mood known as manic-depressive illness, which is characterized by alternations between crushing mental depression and mania—a compulsive, hyperactive sort of cheerless "happiness." Many clinical psychologists and psychiatrists believed that depression was the primary symptom, and the mania was a defensive reaction to it. The obsessive happiness was a preventive measure to forestall the very painful depression. However, it was found that a common mineral, lithium, could be used to treat manic-depressive illness. The drug appeared to exert its therapeutic effect by preventing the manic phase of the cycle; when administered during the depressive phase it had no effect. However, once the mania was abolished by the lithium, the depressive part of the cycle never occurred again. Thus, depression appears to be a reaction to mania rather than the reverse.

In actual practice, the research efforts of physiological psychologists involve both forms of explanation—generalizations and reduction. Ideas for experiments are stimulated by the investigator's knowledge of psychological generalizations about behavior and information about physiological mechanisms. This means that a good physiological psychologist must be a good psychologist and a good physiologist.

BEHAVIOR AND EVOLUTION

In order to understand the workings of a complex piece of machinery, it helps to know what its function is. This is just as true for a living organism as it is for a mechanical device. However, there is an important difference between machines and organisms; machines have inventors who had a purpose in mind when they designed them, whereas organisms are the result of a long series of accidents. Thus, strictly speaking, we cannot say that any physiological mechanisms of living organisms have a purpose. We can only say that the mechanisms have proved to be useful.

The cornerstone of evolution is the principle of natural selection. Briefly, here is how the process operates for sexually-reproducing multicellular animals. Every organism consists of a very large number of cells, each of which contains chromosomes. Chromosomes are very large, complex molecules that contain the codes for producing the proteins that the cell needs to grow and perform its functions. In essence, the chromosomes contain the blueprints for the construction (that is, embryological development) of a particular member of a particular species. If the plans are altered, a different organism is produced.

The plans do get altered; mutations occur from time to time. Mutations are accidental changes in the chromosomes of sperms or eggs that develop into new organisms. For instance, a cosmic ray might strike a chromosome in a cell of a parental testis or ovary, thus affecting the offspring. Most mutations are deleterious: the offspring either fails to survive, or it survives with some sort of deficit. However, a small percentage of mutations are beneficial, and confer a *selective* advantage to the organism that possesses them. That is, the animal is more likely to live long enough to reproduce, and hence to pass on its chromosomes (with their alteration) to its own offspring. Many different kinds of traits can confer a selective advantage. A few examples are resistance to a particular disease, the ability to digest new kinds of foods, more effective weapons for defense or for procurement of prey, or even a more attractive appearance to members of the opposite sex. (One must reproduce in order to pass on one's chromosomes.)

Naturally, the traits that can be altered by mutations are physical ones; chromosomes make proteins, which affect the structure and chemistry of cells. But the *effects* of these physical alterations can be seen in an animal's behavior; thus the process of natural selection can act upon behavior. For example, if a particular mutation results in changes in the brain that cause the animal to stop moving and freeze when it perceives a novel stimulus, that animal is more likely to escape undetected when a predator passes nearby. This means that the animal is more likely to survive to produce offspring, and pass its genes on to future generations.

We can see that the process of natural selection tries out a variety of alterations in the structure of organisms. If a particular alteration produces a behavior that helps the animal to survive, it is likely that this change will be passed on to future generations. Thus, the behavioral repertoire of a species is gradually built up by making changes in its physical structure.

An understanding of the principle of natural selection plays some role in the thinking of every person who undertakes research in physiological psychology. Some people explicitly consider the genetic mechanisms of various behaviors and the physiological processes that these behaviors depend upon. Others are concerned with comparative aspects of behavior and its physiological basis; they compare the nervous systems of animals up and down the phylogenetic scale in order to make hypotheses about the evolution of brain structure and the corresponding behavioral capacities that emerged. But even if they are not directly involved with the problem of evolution, the principle of natural selection guides the thinking of all physiological psychologists. We ask ourselves what the selective advantage of a particular trait might be. We think about how nature might have used a physiological mechanism that already existed to perform more complex functions in more complex organisms. When we entertain hypotheses, we ask ourselves whether a particular explanation makes sense in an evolutionary perspective.

i.e.) the history of the evolution of a species

CONSCIOUSNESS AND FREE WILL

I wish that I could tell you what consciousness, or self-awareness, is. Although we have made a great deal of progress in our study of the physiology of *behavior*, consciousness is another matter. We know that it is altered by changes in the structure or chemistry of the brain, and thus, we conclude that consciousness is a physiological function, just like behavior. We can even speculate about the selective value of self-awareness: consciousness and the ability to communicate seem to go hand in hand; species like ours, with complex social structures and an enormous capacity for learning, are well served by the ability to express our intentions to each other. This communication makes large-scale cooperation possible and permits us to establish customs and laws of behavior. Perhaps the ability to make plans, and communicate these plans to others, is what was selected for in the evolution of consciousness. Even if these speculations are true (and I think that they are very reasonable), we have no idea at all how a complex assemblage of neurons can be aware of itself. What properties of the circuits of neurons in my brain allow me to know who I am? I cannot imagine how we shall answer this question.

The question of human consciousness suggests another issue: determinism and free will. Self-awareness seems to bring with it a feeling of control; most people feel as if their minds make their brains do what they do. But physiological methods are limited to matter and energy—things that can be measured by physical means. If a nonmaterial mind controls what the brain does, then our methods will never succeed in understanding the causes of behavior. Therefore, in our research, we act like determinists. That is, we assume that behavior will be explained (at least in principle) down to the last detail when we completely understand physiology. My qualifying phrase "at least in principle" must not be taken lightly. Even if someday we discover all there is to be known about behavior, we will not be able to predict a particular person's behavior on a particular occasion. In order to apply the physical laws governing behavior, we would have to know *everything* that is presently going on in the person's body in order to predict what he or she will do next.

From our present perspective, this knowledge would seem to be impossible to obtain, which means that for all practical purposes, physiological psychology will never take all the mystery out of an individual's behavior. Even physical scientists have to confront the difference between understanding general laws and predicting the behavior of complex systems. Although it is impossible to predict where a feather would fall if it were dropped from a tall building during a windstorm, no one would assert that the landing place is affected by free will on the part of the feather. A determinist would maintain that free will in humans is a myth that is perpetuated for two reasons: the complexity of the human brain (which is far greater than the forces that act on a falling feather), and our own consciousness, which makes us feel that our mind controls our body, rather than the reverse.

I am sure that many of you do not agree with the determinist position; you may feel that you are in control of your own behavior, and you can point out, correctly, that I cannot prove otherwise. However, the issue is a philosophical (and religious) one that can be divorced from scientific investigation of the physiology of behavior. If a person wants to believe in his or her own free will, that is fine, so

long as he or she acts like a determinist in the laboratory. We must
limit the scope of our hypothesis to the methods of investigation that
we have at hand. Since our techniques are physical, our explanations
must be physical, also. If organisms do have nonphysical minds or
souls that control their behavior, the methods of physiological psychology will never detect them. Nonphysical entities cannot be investigated by physical means. Thus, we must assume that the behavior of our subjects is determined by their physiology.

The Unity of Consciousness

Human awareness brings with it a feeling of unity of consciousness;
each of us is a single individual, with a store of memories, needs,
hopes, and feelings. Our personalities may have different aspects,
and our tastes may change, but to each of us our mind appears to be
a single entity. When we observe an object we do not experience the
feel of it, the sound of it, and the sight of it as belonging to different
objects. Neither do we experience our various sense modalities as
providing separate awarenesses; we perceive them as different inputs to a single consciousness. And yet, as we shall see later (in
Chapter 19), this does not appear to be the case for most animals
other than ourselves. Their minds (if the term *mind* can be applied
to animals other than humans) are probably *not* unitary, since present evidence suggests that experiences gained by means of one sense
modality do not easily modify experiences gained by means of another sense modality. A human would have no trouble recognizing,
by touch, objects previously seen but not felt. However, this is not
generally true for other animals.

Information about the environment is conveyed to the brain by
sensory nerves, and it is analyzed there by various sensory systems.
These systems occupy relatively distinct parts of the brain. This separation means, for example, that the identification of patterns of
sound as words is accomplished by regions of the brain that are
distinct from those that interpret a visual pattern as a representation
of the person who is speaking these words. Of course, the physiological mechanisms that mediate these functions must communicate
and cooperate with each other. The integration of information that
is received by different sensory modalities appears to be accomplished by means of physical interconnections between brain regions
that perform these analyses. In effect, these interconnections unify
separate aspects of our consciousness.

Two Minds in One: The Effects of Surgical
Division of the Brain

Different aspects of consciousness depend on specific regions of the
brain, and integration of these separate aspects of awareness de-

pends on connections between these specific regions. These facts are illustrated very forcefully by the consequences of brain damage in humans. For example, accidental brain damage can impair a person's ability to read, although he or she is not blind. However, the person can still write. This means that the person can write, but cannot read what he or she has just written. Damage to another part of the brain makes it difficult or impossible for a person to repeat the words that he or she has just heard, but the person is still able to understand what has been said. The patient can *answer* questions, but cannot *repeat* them. Our own introspection into the workings of our mind would never predict such phenomena as these; it appears to each of us that we first hear the words, then identify them, then figure out what is being asked, then find the answer to the question, and finally say the answer. How can we possibly answer a question without knowing what the words of the question are? (I will leave you in suspense until Chapter 19, when these matters will be discussed in detail.)

Another division of consciousness, which I will describe in more detail here, was discovered by use of a surgical procedure that was devised to alleviate the symptoms of severe epilepsy. The most important connection between the right and left halves of the brain is provided by the *corpus callosum*. Besides conveying information, the corpus callosum occasionally transmits a violent storm of neural impulses from one side of the brain to the other in people with certain forms of epilepsy. A number of years ago it was discovered that severing the corpus callosum surgically would provide a great measure of relief to people whose epilepsy could not be treated with medication (Bogen, Fisher, and Vogel, 1965). (See **FIGURE 1.1**.)

One might think that such an operation would disrupt a person's behavior in a clearly observable manner. Instead, these people appear quite normal, at least to the casual observer. Careful study is needed to reveal the fact that the cerebral hemispheres (right and left halves of the brain), no longer able to exchange information, can independently perceive, think, act, and remember. In fact, as we shall see, the hemispheres can even engage in conflict.

Although the right and left sides of the brain operate independently after the corpus callosum has been severed, the left hemisphere, because of its unique capabilities, dominates the person's behavior. Mechanisms that are located in the left hemisphere are responsible for the comprehension and production of speech. For example, damage to one region of the left hemisphere will impair a person's ability to comprehend the meaning of speech. Damage to another region will prevent normal speech, although the patient will still be able to understand spoken words. Destruction of the same regions of the right hemisphere will not produce these disabilities; people with damage to the right hemisphere can converse normally. (Contrary to popular belief, speech mechanisms are usually located in the left hemisphere whether a person is left- or right-handed. The

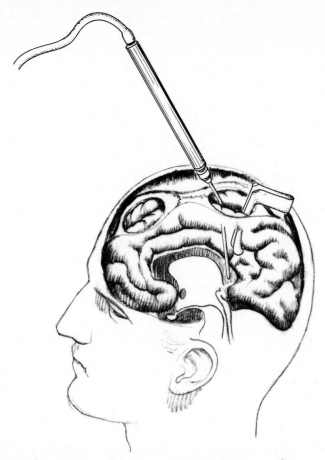

FIGURE 1.1   Schematic view of the surgical separation of the cerebral hemisphere produced by cutting the corpus callosum. (From Gazzaniga, M. S., *Fundamentals of Psychology*. New York: Academic Press, 1973.)

right hemisphere is dominant for speech in only a minority of the population.)

The fact that a person's verbal activities are normally controlled by one hemisphere accounts for the relatively minor change that is seen in these activities after the corpus callosum is severed. This normal hemispheric speech dominance also accounts for the fact that a person with a bisected brain does not give the impression of possessing two independently operating hemispheres, since only one of them (the left) controls speech. The right hemisphere is not able to express itself verbally; it is mute.

Besides being mute, the right hemisphere is also an extremely poor reader. Because of this fact, one of the first things a person with a split brain learns is to hold a book with the right hand while reading. The muscles of the arms and legs are largely controlled by the opposite side of the brain. (The nerve fibers conveying the impulses from brain to muscle cross to the other side in lower parts of the

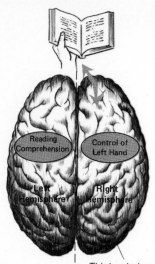

This hemisphere
cannot understand
what is being read.

B R A I N

FIGURE 1.2 The right
hemisphere is not interested in
a book it cannot read.

brain and in the spinal cord and are thus not affected when the corpus callosum is severed.) This arrangement means that the right hemisphere controls the left hand. Since the right hemisphere is not able to comprehend the meaning of the words on a page that is being read by the left hemisphere, it will find reading to be a boring task. Therefore, a book held in the left hand (which is controlled by the right hemisphere) will quickly be put down. (See **FIGURE 1.2.**) Since the right hemisphere is able to enjoy and appreciate nonverbal visual stimuli, the left hand quite readily holds up interesting pictures so that they may be examined. Patients with a split brain are struck by the independence of the left side of their body; they (that is, their verbal left hemispheres) note that the left arm seems to have a mind of its own.

Since verbal communication with a person whose brain has been bisected means communication with the left hemisphere, an observer tends not to detect the presence of the right hemisphere, except in clear-cut cases where the left hand attempts to perform a task that contradicts the person's verbal behavior. However, in the laboratory, sensory information can be presented independently to the left and right hemispheres. These studies are described by Gazzaniga and LeDoux (1978). For example, a blindfolded patient can easily give the name of a simple object felt by the right hand. However, if an object is placed in the left hand, the patient cannot verbally identify it. (Sensory information from each hand is sent to the opposite hemisphere.) If the object is then placed among several others and the blindfold is removed, the patient is able to point to the appropriate object (with the left hand, of course). Thus, the right hemisphere is capable of perceiving and remembering, and it can also initiate hand and arm movements that demonstrate the perception and the memory.

Just as a given hemisphere controls, and receives sensory information from, the limbs on the other side of the body, so visual stimuli that are located to the left and right of the *fixation point* (the place at which the eyes are looking) are transmitted to the contralateral (opposite-sided) hemisphere. If we look straight ahead, the left hemisphere sees objects to our right, and the right one sees objects to our left. Thus, the cerebral hemispheres of a patient with a split brain can be shown different visual stimuli. For example, if the word *fork* is shown just to the left hemisphere, the person can say "fork" or can select a fork from other objects with the right hand (but not the left hand, of course). If the words are simple enough, the right hemisphere, with its rudimentary language ability, can read words and select the appropriate objects. However, if the patient is asked to repeat the word, he or she reports seeing nothing.

COMMUNICATION BETWEEN TWO MINDS IN ONE HEAD. Although the hemispheres of a patient with a split brain can no longer communicate by means of the corpus callosum, they can sometimes convey messages to each other by different means. The

10

right hemisphere hears whatever is spoken by the left hemisphere, of course, and understands much of what is being said. (The right hemisphere is therefore much more aware of the separate existence of another entity in the person's body.) In certain situations the right hemisphere can pass information on to the left. If a patient is confronted with a simple test that requires yes or no answers based upon visual stimuli, correct answers are often obtained even when the stimuli are shown to the right hemisphere. For example, the right hemisphere might be shown a patch of red. The patient is asked, "Was that red?" Sometimes the patient says yes, and waits for the next item. At other times the patient first says no, shakes his or her head, and, frowning, says: "That's wrong. I meant yes." Here is why. The right hemisphere perceives the stimulus. If it hears the left hemisphere give an incorrect response, it initiates a headshake and a frown, which tells the left hemisphere that the answer was incorrect, so the left hemisphere changes its answer.

It is quite clear that a person with a bisected brain possesses two minds. It would surely be quibbling to assert that there is no "mind" in the right hemisphere because it cannot talk to us. Each hemisphere can perceive and react to its own sensory input, and thus it appears likely that each hemisphere has its own consciousness. The left hemisphere can answer yes to the question "Are you conscious and aware of yourself?" but the right hemisphere cannot. The fact that the nonverbal hemisphere can appreciate the sight of an interesting picture (as shown by the fact that the patient will hold the picture in the left hand) argues for an independent consciousness. Even more compelling is the fact that the right brain can apparently become impatient with the ineptness of the left brain on some tasks. The right hemisphere is much better at the visual perception of three-dimensional objects, and can draw better pictures or arrange patterns of blocks much better than the left hemisphere can. On occasions where the right hand has been required to perform one of these tasks, the left hand has been observed to push the clumsily-performing right hand away and take over the task.

The point of this discussion is that mind is a phenomenon produced by a functioning brain. The fact that disconnecting the hemispheres gives rise to two distinct minds—with different capacities, memories, and (probably) personalities—provides, I believe, the most persuasive proof that the unity of our conscious awareness is a product of the interconnections of various regions of the brain.

THE BRAIN: ORGAN OF PERCEPTION, DECISION, AND CONTROL

The brain receives information, makes decisions, and produces effects. People are most familiar with input of a sensory nature: vision, audition, touch. We are also familiar with the brain's output through the nerves that control the muscles with which we move our bodies. But the brain also receives information other than that transmitted through the sensory nerves, and it controls more than the skeletal muscles.

FIGURE 1.3 schematically classifies the communication channels of the brain into inputs and outputs and further divides each of these classifications according to the medium that is used for the transmissions—neural and nonneural. (See **Figure 1.3.**) The familiar neural (sensory) inputs will be discussed in Chapters 8 and 9. The motor outputs (neural and nonneural) will be discussed in a general manner in Chapter 10. Besides controlling the skeletal muscles, neural outputs of the brain also control muscles such as those of the gut, the heart, the urinary bladder, and the iris of the eye. The brain also controls the hormonal output of some glands by means of neural connections. For example, the *adrenal medulla* produces *epinephrine* (adrenaline) in response to neural commands from the brain. Finally, the brain produces and secretes hormones of its own, which control the hormone output of the *anterior pituitary gland*, which is attached to the base of the brain. In turn, the hormones of the anterior pituitary gland affect many physiological systems of the body. Growth of

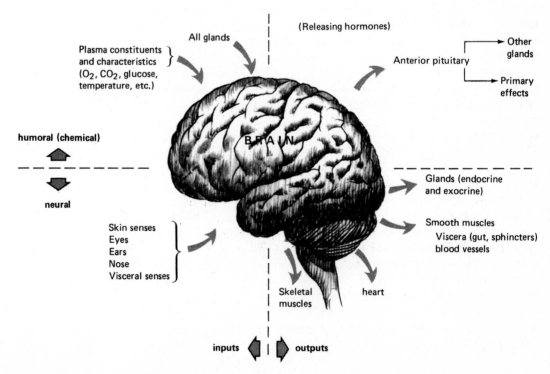

FIGURE 1.3 Schematic representation of the communication channels between the brain and the rest of the body.

the bones, sexual maturation, and regulation of the body's mineral balance are a few of the phenomena that are controlled by the brain through its influence on the hormonal output of this gland. Specific examples of the role of the nonneural outputs of the brain will be discussed in Chapter 11 (sexual development and behavior), Chapter 12 (hunger), and Chapter 16 (emotion).

The final quadrant of Figure 1.3 represents the communication channel that is probably the least familiar: nonneural input to the brain. The brain (and, therefore, the animal's behavior) is affected by many hormones, as will be discussed in Chapters 11, 12, 13, and 16. The brain also responds to other nonneural inputs, such as body temperature, the amount of oxygen carried by the blood, and the concentration of the blood plasma. Some of the nonneural, nonhormonal inputs to the brain will be considered in the chapters on hunger and thirst (Chapters 12 and 13).

Before the physiology of behavior can be discussed, it will be necessary to have a basic understanding of the way the brain processes information. Therefore, Chapters 2, 3, 4, and 5 are concerned with those elements of the brain that transmit and process information: the neurons. These chapters explain how neurons communicate and make decisions, and how information is transmitted into, out of, and within the brain. The chemistry of neural communication is described, providing a basis for understanding how drugs affect behavior. Chapter 6 outlines the anatomy of the nervous system and introduces a few facts about the functions of some of its structures. In Chapter 7 the research methods of the neuroscientist are described and discussed, along with some of the limitations of these methods.

Chapters 8 and 9 deal with sensory processes. Chapter 8 describes the way in which environmental events are translated into neural activity, and how the information thus represented is transmitted to the brain. Chapter 9 discusses the nature of the neural representation and the means by which it is analyzed by the brain. The motor (output) systems of the brain are described in Chapter 10.

Chapters 11 through 20 outline the progress that has been made by physiological psychologists and other neuroscientists in understanding the physiology of sexual behavior, hunger, thirst, sleep, emotional behavior, reinforcement, pain, learning, memory, and mental disorders. In these chapters I have attempted to select the most important experiments and to integrate the information that I present. By being as concrete and explicit as possible in my conclusions I hope to establish a conceptual framework that will facilitate your efforts to learn more about the physiology of behavior. New facts, standing in isolation, are difficult to learn and to remember. (You will certainly agree with this statement as you acquire the new vocabulary that is presented in the early chapters of this book. Nothing is more difficult to learn than a set of new names.) However, once a conceptual framework is established, it is quite easy to digest new information. In some instances I have made logical jumps (where experimental data are not available) in order to tie some facts together. Subsequent research may prove the tentative conclusions to be incorrect, but your grasp of the conceptual framework will make it easy for you to correct the erroneous conclusions. Without the framework you would probably forget much of the information learned in a course in physiological psychology, and the results of the subsequent research might not mean anything at all to you.

SOME MECHANICAL DETAILS

A few words about the format of the book: I have tried to integrate text and illustrations as closely as possible. In my experience, one of the most annoying aspects of reading some books is not knowing when to look at an illustration. When reading complicated material, I have found that sometimes I look at the figure too soon, before I have read enough to understand it, or that I look at it too late and realize that I could have made more sense out of the text if I had just looked at the figure sooner. Furthermore, after looking at the illustration, I often find it difficult to return to the place where I stopped reading. Therefore, in this book you will find the figure references in boldface (like this: **FIGURE 5.6**), which means, "stop reading, and look at the figure." I have placed these references in the locations I think will be optimal. If you look away from the text then, you will be assured that you will not be interrupting a line of reasoning in a crucial place and will not have to reread several sentences to get going again. Furthermore, the boldface will make it easier for you to find your place again. You will find sections like this: "Figure 6.1 shows an alligator and a human. The alligator is certainly laid out in a linear fashion—we can draw a straight line that starts between its eyes and continues down the center of its spinal cord. (See **FIGURE 6.1.**)" This particular example is a trivial one and will give you no problems no matter when you look at the figure. But in other cases the material is more complex, and you will have less trouble if you know what to look for before you stop reading and examine the illustration.

You will notice that some words in the text are *italicized*, while others are printed in ***bold italics***. Italics mean one of two things: the word is being stressed for emphasis, and is not a new term, or I am pointing out a new term that I do not think is necessary for you to learn. On the other hand, a word in bold italics is a new term that you should probably try to learn. Most of the bold italicized terms in the text are part of the vocabulary of the physiological psychologist. Often, they will be used again in a later chapter. As an aid to your studying, I have included a list of newly introduced terms at the end of each chapter, along with the page number where the term was first used. Pronunciations of terms that might not be obvious are also given there.

There is also a glossary at the end of the text. If a term was used only once, and is not one that you will encounter later in the text, it is probably not included in the glossary. Its definition will be found where it is first introduced in the text. But if, in a later chapter, you encounter a scientific term that is not in bold italics, it was probably introduced earlier. If you need to refresh your memory, the easiest place to find a definition is in the glossary. In addition, a rather comprehensive index at the end of the book provides a list of terms and topics, with page references.

I hope that in reading this book you will come not only to learn more about the brain but also to appreciate it for the marvelous instrument it is. The brain is wonderfully complex, and perhaps the most remarkable thing of all is that we are able to use it to understand it.

SUGGESTED READINGS

GAZZANIGA, M. S. AND LEDOUX, J. E. *The Integrated Mind.* New York: Plenum Press, 1978. A fascinating account of the effects of the split-brain operation.

2

The Cells of the Nervous System

The body is made of living cells, and the nervous system is no exception to this rule. This chapter describes the cells of the nervous system and the special functions they perform.

CELLS OF THE CENTRAL NERVOUS SYSTEM

Neurons

The nerve cell, or neuron, is the information processing and transmitting element of both the central nervous system (brain and spinal cord) and the peripheral nervous system (nerves and other structures outside the central nervous system). Neurons can receive information about the environment. They can "argue" and reach decisions. They can produce muscular movements and glandular secretions. They can store information and retrieve it—even years later.

Before describing the particular characteristics of neurons, I will describe the structures and properties these cells have in common with other cells of the body. Not all structures will be mentioned; I will discuss only those that serve roles related to the unique characteristics of neurons.

STRUCTURES OF CELLS.   Figure 2.1 illustrates a typical animal cell. (See **FIGURE 2.1.**) The *membrane* defines the boundary of the cell. It consists of a double layer of lipid (fatlike) molecules in which float various structures made of protein molecules. The membrane is an exceedingly complex structure; it is far more than a bag holding in the contents of the cell. It is an active part of the cell, keeping in some substances and keeping out others. It even uses up energy by actively extruding some substances and pulling others in. The membrane contains specialized molecules that detect substances outside the cell (such as hormones) and pass information about the presence of these substances to the interior of the cell. The membrane of the neuron is especially important in the transmission of information, and its characteristics will be discussed in more detail later.

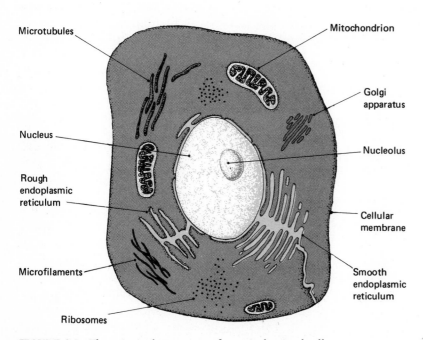

Microtubules

Mitochondrion

Golgi apparatus

Nucleus

Nucleolus

Rough endoplasmic reticulum

Cellular membrane

Microfilaments

Smooth endoplasmic reticulum

Ribosomes

FIGURE 2.1   The principal structures of a typical animal cell.

The bulk of the cell consists of **cytoplasm.** Cytoplasm is complex and varies considerably across types of cells, but it can be most easily characterized as a jellylike, semiliquid substance, filling the space outlined by the membrane. Cytoplasm is not static and inert; it streams and flows.

Mitochondria are important in the energy cycle of the cell; many of the biochemical steps involved in the extraction of energy from the breakdown of glucose take place on the *cristae* of mitochondria. Highly active cells, then, contain a large proportion of mitochondria. These structures are shaped like oval beads and are formed of a double membrane. The inner membrane is wrinkled, and the wrinkles make up a set of shelves (cristae) that fill the inside of the bead.

Endoplasmic reticulum appears in two forms: rough and smooth. Both types consist of membrane and are continuous with the outer membrane of the cell. It is thought that endoplasmic reticulum is concerned with transport of substances around the cytoplasm and provides channels for segregation of various molecules involved in different cellular processes. Rough endoplasmic reticulum may be differentiated from the smooth type in that it contains **ribosomes.** These structures, which are attached to the endoplasmic reticulum, are the sites for protein synthesis in the cell. The protein produced by the ribosomes attached to rough endoplasmic reticulum is destined to be transported out of the cell. For example, the hormone insulin (a protein) is manufactured there in certain cells of the pancreas. Unattached ribosomes are also distributed around the cytoplasm; the unattached variety is thought to produce protein for use within the cell.

Active not inert

The **Golgi apparatus** also consists of membrane. This structure serves as a wrapping or packaging agent. For example, secretory cells wrap their product in a membrane produced by the Golgi apparatus. When the cell secretes its products, the container migrates to the outer membrane of the cell, fuses with it, and bursts, spilling the product into the extracellular fluid. As we shall see, neurons communicate with each other by secreting chemicals by these means.

The **nucleus** of the cell is round or oval and is covered by the nuclear membrane. The **nucleolus** and the **chromosomes** reside within this membrane. The nucleolus is concerned with the manufacture of ribosomes. The chromosomes contain the genetic information of the organism. They consist of long strands of **deoxyribonucleic acid** (DNA). When active, portions of the chromosomes (genes) cause production of another complex molecule, **messenger ribonucleic acid** (mRNA). Messenger RNA leaves the nuclear membrane and attaches to ribosomes, where it causes the production of a particular protein. (See **FIGURE 2.2.**) Protein synthesis is an important function of cells, and this topic will be examined in greater detail later (in Chapter 18). Proteins serve as **enzymes,** as well as providing structure, and enzymes direct the chemical processes of a cell by selectively facilitating specific chemical reactions.

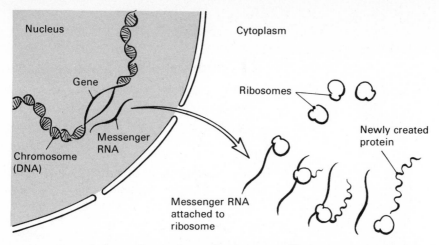

Arranged throughout the cell are **microfilaments** and **microtubules.** These long, slender, hairlike structures serve as a matrix, or framework, into which are imbedded the various components of the cytoplasm. They also assist in transporting molecules and structures from place to place within the cell, and they apparently play an essential role in the communication of information from neuron to neuron.

The nerve cell has a particular specialty—information transmission and decision making. Some neurons (sensory neurons) have an additional talent: they can detect changes in the environment and transmit information about these changes to other neurons. Another kind of neuron produces hormones or hormonelike substances; these **neurosecretory cells** can release chemical substances that may affect other cells within the brain or in other organs.

STRUCTURES OF NEURONS.   Neurons come in many shapes and varieties, according to the specialized job they perform. They usually have, in one form or another, the following four structures or regions: (1) cell body, or **soma;** (2) **dendrites;** (3) **axon;** and (4) **terminal buttons.**

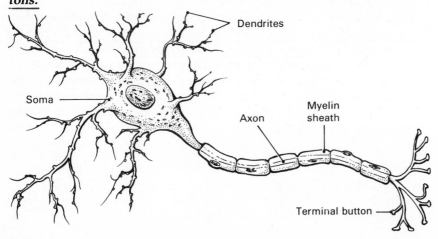

FIGURE 2.3 The principal structures or regions of a multipolar neuron.

19

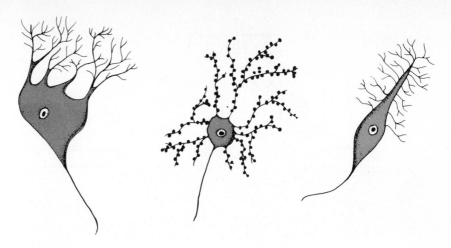

FIGURE 2.4 A sample of the variety of dendritic shapes found in various types of neurons.

Cell body (or Soma). The cell body (or soma) contains the nucleus and much of the machinery that provides for the life processes of the cell. (See **FIGURE 2.3** on the preceding page.) Its shape varies considerably in different kinds of neurons.

Dendrites. Dendron is the Greek word for tree, and the dendrites of the neuron look very much like trees. (See **FIGURE 2.3.**) These dendrites vary in shape even more widely than do real trees; a glance at Figure 2.4 will give you an idea of the many forms they may take. (See **FIGURE 2.4.**) Neurons "converse" with one another, and dendrites, along with the membrane of the soma, receive these neural messages. The messages that pass from neuron to neuron are transmitted across the **synapse,** a junction between terminal buttons (described below) of the sending cell and a portion of the somatic or dendritic membrane of the receiving cell.

Synapses on dendrites of many neurons occur not on the branches or twigs but on little buds known as **dendritic spines.** Figure 2.5 illustrates a terminal button of the sending cell and a dendritic spine of the receiving cell. (See **FIGURE 2.5.**) Communication at a synapse is one way; a terminal button transmits messages to the receiving cell but does not receive messages from it.

Synapses occur not only between terminal button and dendritic membrane, but also between terminal button and somatic membrane. In most cases there are no spines on the membrane of the soma; the terminal buttons just meet the smooth surface of the membrane. (See **FIGURE 2.6.**)

Axon. The dendritic and somatic membranes receive messages from other cells. The information is processed (Chapters 3 and 4 describe this processing) and the resulting message, if any, is passed down the axon to another set of cells—the cells that *this* neuron transmits messages to. (See **FIGURE 2.7.**)

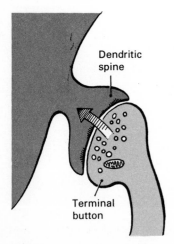

Dendritic spine

Terminal button

FIGURE 2.5 A synapse of a terminal button on a dendritic spine.

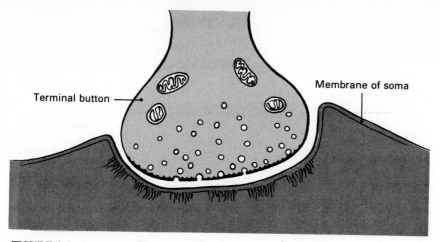

FIGURE 2.6   A synapse of a terminal button on somatic membrane.

The axon is a long, slender tube. It carries information away from the cell body down to the terminal buttons. The message is electrical in nature, but it is not carried down the axon the way a message travels down a telephone wire. Just how it is transmitted will be described in Chapter 3.

Axons and their branches come in different shapes, as do dendrites. There are three principal types of neurons, classified accord-

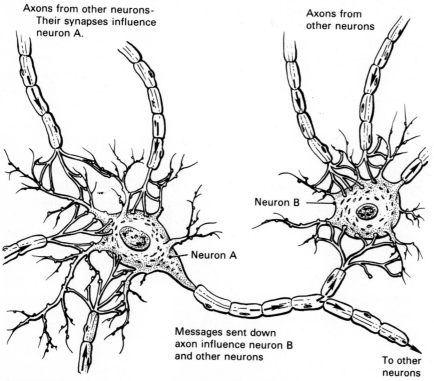

FIGURE 2.7   An overview of the synaptic connections between neurons.

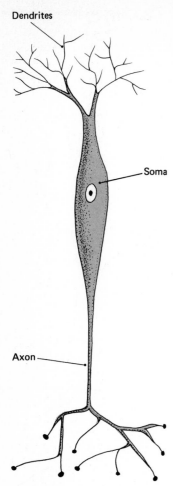

Dendrites

Soma

Axon

FIGURE 2.8 A bipolar neuron.

ing to the way their axons and dendrites leave the soma.

The neuron depicted in Figures 2.3 and 2.7 is the most common type found in the central nervous system; it is **multipolar.** That is, the somatic membrane gives rise to one axon, but to the trunks of many dendritic trees.

Bipolar neurons give rise to one axon and one dendrite, at opposite ends of the soma. (See **FIGURE 2.8**.) These neurons are usually sensory in nature; that is, they are located out and away from the central nervous system. They convey visual, auditory, and vestibular (balance, from the inner ears) information in mammals. The dendrites of these cells receive information directly from special *receptor cells* (see Chapter 8) about what is going on in the environment, and their axons pass this information along to other neurons back in the central nervous system (usually referred to as the *CNS*.)

The third type of nerve cell is the **unipolar** neuron. It has only one stalk leaving the soma; this stalk then divides into two branches a short distance from the soma. (See **FIGURE 2.9**.) Both ends of these branches arborize (divide up like the branches of a tree—for some reason they arborize in Latin rather than dendrotize in Greek). The unipolar cell, like the bipolar cell, transmits information from the environment into the CNS. The distal arborizations (those away from the CNS) are dendrites; the proximal arborizations (those toward—actually, within—the CNS) end in terminal buttons. These unipolar neurons receive sensory information either from specialized receptor cells, as bipolar neurons do, or directly from the environment, thus serving as receptor cells themselves. The message does not pass across the somatic membrane of unipolar neurons, as it does in multipolar and bipolar neurons. The soma, in the case of unipolar neurons, serves principally as the factory that supplies the axon with what it needs to be alive and functioning. The somas of all types of neurons serve this role also, of course, but the somatic membranes of multipolar and bipolar neurons additionally participate in transmission of information and in the decision-making process.

Terminal buttons.   The axon divides and branches a number of times. At the end of each of the twigs is found a little knob called the terminal button. These buttons have a very special function; when a message is passed down the axon, the terminals of the transmitting cell secrete a chemical called a **transmitter substance.** This chemical (there are a variety of different transmitter substances in the CNS) is detected by the receiving cell and its presence produces an effect there. The effect becomes a factor that is considered in the decision-making process of the receiving cell, and, depending on the decision that is made, information will (or will not) be sent down the axon of *that* cell. In Chapter 4 the detailed structure of terminal buttons will be described, along with the rest of the structures that make up synapses.

Glia

Neurons constitute only about half of the volume of the central nervous system. The rest consists of a variety of supporting cells, the most important of which is the **neuroglia,** or "nerve glue." In a very real sense, glial cells (as they are usually called) glue the CNS together. But they do much more than that. Neurons lead a very sheltered existence; they are physically and chemically buffered from the rest of the body by the glial cells. These cells surround neurons and hold them in place; they supply the neurons with various chemicals, including some that they need in order to exchange messages with other neurons; they insulate neurons from one another so that neural messages do not get scrambled; they regulate the chemical composition of the extracellular fluid, and they even act as housekeepers, destroying and removing the carcasses of neurons that are killed by injury or that die as a result of old age.

There are several types of glial cells, each of which plays a special role in the central nervous system.

ASTROCYTES (ASTROGLIA). **Astrocyte** means "star cell," and this name accurately describes the shape of these cells. Astrocytes are rather large, as glia go, and provide physical and nutritional support to neurons. Together with microglia, they also clean up debris within the brain. Finally, they chemically buffer the extracellular fluid; this function will be described further in Chapter 3.

Transport of chemicals. Some of the astrocyte's processes (the arms of the star) are wrapped around blood vessels; other processes are wrapped around parts of neurons, so that the somatic and dendritic membrane of neurons is largely surrounded by astrocytes. (See **FIGURE 2.10.**) Many years ago, Camillo Golgi (discoverer of the

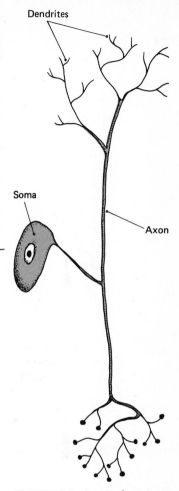

FIGURE 2.9 A unipolar neuron.

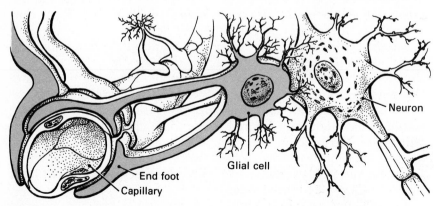

FIGURE 2.10 Structure and location of astrocytes. (Adapted from Kuffler, S. W., and Nicholls, J. G. *From Neuron to Brain*. Sunderland, Mass.: Sinauer Associates, 1976.)

Golgi apparatus) suggested that astrocytes supplied neurons with nutrients from the capillaries, and disposed of their waste products (Golgi, 1903). Nutrients were thought to pass from capillary to the cytoplasm of the astrocytes, and then through the cytoplasm to the neurons. Waste products followed the opposite route. This hypothesis is very plausible, as a glance at Figure 2.10 will show. (See **FIGURE 2.10**.) Many people have accepted this hypothesis as fact. However, the evidence that is available so far does not appear to warrant such a conclusion (Kuffler and Nichols, 1976). The Golgi hypothesis may be true, but it is still a hypothesis.

There is good evidence from the peripheral nervous system that satellite cells *do* transport substances to neurons, even if we do not know whether all nutrients reach neurons by means of astrocytes. Lasek, Gainer, and Przybylski (1974) found that in the peripheral nervous system of a squid, neurons take up proteins that are synthesized in adjacent satellite cells. Furthermore, some neurons lose their ability to produce the transmitter substance if the surrounding satellite cells are destroyed (Patterson and Chun, 1974).

In the central nervous system, astrocytes undoubtedly contribute much to the biochemical processes that take place in neurons, but the precise nature of their role is still a mystery.

Support. Besides having some kind of role in transporting chemicals to neurons, astrocytes serve as the matrix that holds neurons in place. They also surround and isolate synapses, preventing transmitter substances from leaving the immediate vicinity. Thus, astrocytes provide each synapse with an isolation booth, keeping the neurons' conversations private.

Housekeeping. Neurons occasionally die of old age (that means we do not know exactly why they die). Others are killed by head injury, cerebrovascular accident (blood clot or rupturing of a blood vessel), or disease. Certain kinds of astrocytes (along with the microglia) then take up the task of cleaning away the debris. These cells are able to travel around the CNS; they extend and retract their processes (*pseudopodia*, or "false feet") and glide about the way amoebas do. They come in contact with pieces of debris, pushing themselves against and finally engulfing and digesting portions of the dead neurons. We call this process **phagocytosis,** and cells capable of doing this are called **phagocytes** (eating cells). If there is a considerable amount of injured tissue to be cleaned up, astrocytes will divide and produce enough new cells to do the task. Once the dead tissue is broken down, a framework of glial cells will be left to fill in the vacant area, and a specialized kind of astrocyte will form scar tissue, walling off the area.

MICROGLIA. **Microglia** are smaller than the other types of glia. They serve as phagocytes, along with the astrocytes.

OLIGODENDROGLIA. **Oligodendroglia** are residents of the CNS,
and their principal function is to provide the **myelin sheath,** which
insulates most axons from one another. (Some axons are not myelinated and lack this sheath.) Myelin is made of lipid (70–80 percent)
and protein (20–30 percent). It is produced by the oligodendroglia in
the form of a tube surrounding the axon. This tube does not form a
continuous sheath, but consists of a series of segments, with a small
portion of uncoated axon between the segments. This bare portion
of axon is called a **node of Ranvier,** after its discoverer. The myelinated axon, then, resembles a string of elongated beads. (Actually, the
beads are very much elongated, their length being approximately 80
times their width.)

A given oligodendroglial cell produces only one segment of myelin covering a given axon, but it may provide one segment for each
of several different axons. During development of the CNS, oligodendroglia form processes shaped something like canoe paddles.
Each of these paddle-shaped processes then wraps many times
around a segment of an axon and, while doing so, produces layers of
myelin. Each paddle, then, becomes a segment of an axon's myelin
sheath. (See **FIGURE 2.11.**)

Unmyelinated axons of the CNS are not actually naked; they are

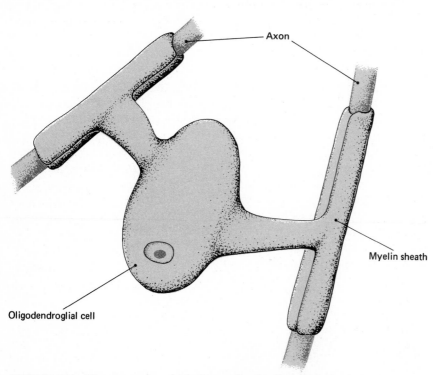

FIGURE 2.11   The process by which the myelin sheath is formed on axons
of the central nervous system by the oligodendroglia.

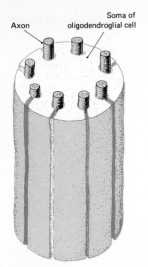

Axon   Soma of
        oligodendroglial cell

FIGURE 2.12 Support is provided for unmyelinated axons of the central nervous system by the oligodendroglia.

also covered by oligodendroglia. However, in this case the glial cells do not manufacture myelin. The axons are covered in a different way, also. Instead of being wrapped in segments by the processes of the oligodendroglia, unmyelinated axons pass right through the cell bodies of the glial cells. (See **FIGURE 2.12.**) Each oligodendroglial cell provides support for several axons.

Neurons and glia are the two principal types of cells in the CNS. A few other kinds of cells also reside there—**ependymal** cells that line the hollow, fluid-filled **ventricles** in the brain, cells that make up the blood vessels that serve the brain, and, of course, the red and white blood cells within the blood vessels. There are also cells that form a set of membranes that surround the brain and spinal cord. This lining (the **meninges**) will be discussed in Chapter 6.

CELLS OF THE PERIPHERAL NERVOUS SYSTEM

There are three principal types of cells in the peripheral nervous system: neurons (which have just been described), satellite cells, and Schwann cells.

Satellite Cells

The supportive function served by glia in the CNS is served by their counterparts in the peripheral nervous system (PNS), the **satellite cells.** These cells provide physical support to the cell bodies of neurons located outside the central nervous system. Groups of neural cell bodies (excluding those found in sense organs) are located in three places in the PNS: (1) in the **ganglia** of **spinal nerves,** (2) in the ganglia of **cranial nerves,** and (3) in the ganglia of the **autonomic nervous system.** (Chapter 6 contains a description of these ganglia.) Ganglia are groups of nerve cells bodies that are located outside the CNS. They are covered with connective tissue and form discrete nodules. The neurons packed within these nodules are held in place by satellite cells.

Schwann Cells

Some kinds of oligodendroglia produce myelin in the CNS, and other kinds merely envelop a number of axons. In the peripheral nervous system the **Schwann cells** provide the same functions. Most axons in the PNS are myelinated. The myelin sheath occurs in segments, as it does in the CNS, and each segment consists of a single Schwann cell, wrapped many times around the axon. In the CNS the oligo-dendroglia grow a number of paddle-shaped processes that wrap around a number of axons. In the PNS a Schwann cell provides myelin for only one axon, and the entire cell—not merely a cellular process—surrounds the axon.

A rather crude way of visualizing the relationship of Schwann cells to an axon is to picture a series of fried eggs (the yolks of the eggs represent the nuclei of the Schwann cells) wrapped around a rope. (I find this analogy helpful, if unappetizing.) Figure 2.13 schematizes the process by which a Schwann cell wraps around an axon. (See **FIGURE 2.13**.)

Schwann cells also differ from their CNS counterparts, the oli-

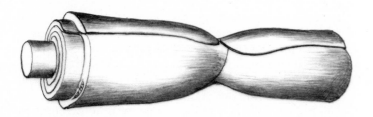

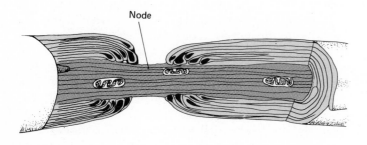

Node

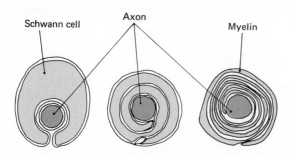

Schwann cell   Axon   Myelin

FIGURE 2.13   The process by which the myelin sheath is formed on axons of the peripheral nervous system by the Schwann cells.

godendroglia, in an important way. If damage occurs to a peripheral nerve (consisting of a bundle of many myelinated axons, all covered in a sheath of tough, elastic connective tissue), the Schwann cells aid in the digestion of the dead and dying axons. Then the Schwann cells arrange themselves in a series of cylinders, which act as guides for regrowth of the axons. The distal portions of the severed axons die, but the stump of each severed axon grows sprouts, which then grow in all directions. If one of these sprouts encounters a cylinder provided by a Schwann cell, that sprout will quickly grow through the tube (at a rate of up to 3 or 4 mm/day), while the other, non-productive sprouts of that axon wither away. If the cut ends of the nerve are still located close enough to each other, the axons will reinnervate the muscles and sense organs they serve.

On the other hand, if the cut ends of the nerve are displaced, or if a section of the nerve is damaged beyond repair, the axons will not be able to find their way to the original sites of innervation. In such cases, neurosurgeons can sew the cut ends of the nerves together, if not too much of the nerve has been damaged. (Nerves are flexible and can be stretched a bit.) If too long a section has been lost, and if the nerve was an important one (controlling hand muscles, for example), a piece of nerve of about the same size can be taken from another part of the body. Nerves overlap quite a bit in the area of tissue they innervate. Neurosurgeons have no trouble finding a branch of a nerve that serves a large leg muscle, for example, that the patient can lose without ill effect. The piece of nerve from the leg can then be grafted to the damaged nerve in the arm. The surgeon works under a special microscope using the very precise technique of *microsurgery*. The axons in the excised and transplanted piece of nerve die away, of course, but the tubes produced by the Schwann cells then guide the sprouts of the damaged nerve and help them find their way back to the hand muscles.

Unfortunately, the glial cells of the CNS are not so cooperative as the supporting cells of the PNS. If axons in the brain or spinal cord are damaged, new sprouts will form, as in the PNS. However, the budding axons encounter scar tissue produced by the astrocytes, and they cannot penetrate this barrier. Even if they could get through, the axons would not reestablish their original connections without guidance similar to that provided by the Schwann cells of the PNS. Thus, the difference in the regenerative properties of the CNS and PNS results from differences in the characteristics of the supporting cells, not from differences in the neurons.

BLOOD-BRAIN BARRIER

The final topic to be discussed in this chapter is the **blood-brain barrier.** If a dye such as trypan blue is injected into the bloodstream, all tissues except the brain and spinal cord will be tinted blue. However, if this same dye is injected into the ventricles of the brain, the blue color will spread throughout the CNS. This demonstrates the fact that a barrier exists between the blood and the extracellular fluid of the brain.

This blood-brain barrier is selectively permeable. Some substances can cross (permeate) this barrier; others cannot. There appear to be two components to the blood-brain barrier: the endothelial cells that line the walls of the capillaries, and the astrocytes that surround them. In most of the body, the endothelial lining of the capillaries is perforated with small holes, and this permits the free exchange of most substances between blood plasma and extracellular fluid. The capillaries of the brain lack these perforations, and hence many substances cannot get out of the capillaries. Others can get past the endothelial lining but are blocked by the astrocytes.

The blood-brain barrier is not uniform throughout the nervous system. In several places the barrier is more permeable, allowing certain substances excluded elsewhere to cross relatively freely. For example, the *area postrema* of the lower portion of the brain is involved in the initiation of vomiting. The blood-brain barrier is somewhat weaker there, and hence this region is more sensitive to toxic substances in the blood. A poison that enters the circulatory system from the stomach can thus stimulate this area to initiate vomiting. If the organism is lucky, the poison can be expelled from the stomach before it causes too much damage.

It has been noted that the blood-brain barrier works, to a certain extent, both ways. (That is, there is also a *brain-blood barrier*.) Proteins present in the extracellular fluid of the brain cannot enter the blood supply, and hence are not recognized by the **immune system** as belonging to the body of the organism. Indeed, if the protein constituent of CNS myelin is injected into an animal's blood supply, antibodies are produced that destroy tissue in the central nervous system. The disease thus produced is called **experimental allergic encephalomyelitis** and resembles the disease multiple sclerosis (Einstein, 1972).

It is thought that multiple sclerosis results from virus-produced damage to the blood-brain barrier, which allows myelin protein to enter the blood supply. This invasion of a "foreign protein" mobilizes the immune system against CNS myelin. With the insulation gone, messages being carried by the axons are no longer kept separate, and the scrambling of messages results in sensory disorders and loss of muscular control. As the disease progresses, the axons themselves are destroyed.

KEY TERMS

astrocyte (*ASS tro site*) p. 23
autonomic nervous system p. 26

axon p. 19
bipolar neuron p. 22

SUGGESTED READINGS

DYSON, R. D. *Cell Biology: A Molecular Approach.* Boston: Allyn and Bacon, 1978. Information about cellular machinery in general (not specifically cells found in the nervous system) can be found in a cell biology text such as this one.

KUFFLER, S. W. AND NICHOLLS, J. G. *From Neuron to Brain: A Cellular Approach to the Function of the Nervous System.* Sunderland, Mass.: Sinauer Associates, 1976. An excellent and comprehensive description of the material contained in this chapter and in the next two chapters.

3

Membrane Potentials and the Transmission of Information

Chapter 2 described various types of cells that make up the nervous system, and noted that neurons constitute the decision-making elements of the CNS. The decision they make—based on the information that is received from the terminal buttons of other cells—is whether or not the axon will send a message down to its terminal buttons. If this message is sent, the terminal buttons then "talk" to the cells on which they synapse.

The present chapter considers the nature of the message carried by the axon and the way in which this message is initiated and transmitted. The message is electrical in nature, but the axon does not carry it the way a wire transmits electrical current. Instead, the message is transmitted by means of complex changes in the membrane of the axon, which result in exchanges of various chemical constituents of the extracellular fluid and the fluid within the axon. These exchanges produce electrical currents.

Forces That Move Molecules

To understand the entire process, it is necessary first to examine some of the forces involved in propagation of information along the axon. I will describe a series of simple experiments that will make it easier for me to explain some much more complex experiments later.

DIFFUSION.   When a spoonful of sugar is carefully poured into a container of water, it settles to the bottom. After a time, the sugar dissolves, but it remains close to the bottom of the container. After a much longer time (probably several days), the molecules of sugar distribute themselves evenly throughout the water, even if no one stirs the liquid. The process whereby molecules distribute themselves evenly throughout the medium in which they are dissolved is called **diffusion.** The same process occurs with mixture of gases, and even (on a tremendously longer time scale) with solids.

Diffusion is characterized by movement of molecules down a **concentration gradient.** Once the sugar dissolves, it is located in highest concentration at the bottom of the container. The concentration of sugar molecules is the lowest at the top. Thus, there is a gradient of concentration of sugar molecules, which goes from highest (at the bottom) to lowest (at the top).

When there are no forces or barriers to prevent it, molecules diffuse down their concentration gradient; that is, they travel from regions of high concentration to regions of low concentration. Molecules are constantly in motion, and their rate of movement is proportional to the temperature. Only at absolute zero (0° Kelvin = − 273.1° C = −459.6° F) do molecules cease their random movement. As they move about, they bump into one another, and the colliding molecules veer off in different directions. The result of these collisions, in our example, is to produce a net movement of water molecules downward and of sugar molecules upward. (Just as there is a concentration gradient for sugar molecules, so there is one for water molecules; initially, water is most concentrated at the top of the container, and least concentrated at the bottom.)

Imagine an experiment that could be performed in the absence of gravity (in a manned satellite orbiting the earth, for example). This experiment uses a glass container with a barrier down the middle, separating the vessel into two equal chambers. Suppose nylon mesh is used as a "barrier." A 10-percent sugar solution is poured into the left side and a 5-percent sugar solution into the right. Since the nylon mesh barrier is **permeable** to both sugar and water molecules (i.e., it lets them both pass through freely), the container soon has a 7.5-percent sugar on both sides. This is merely another example of diffusion. (See **FIGURE 3.1.**)

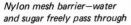

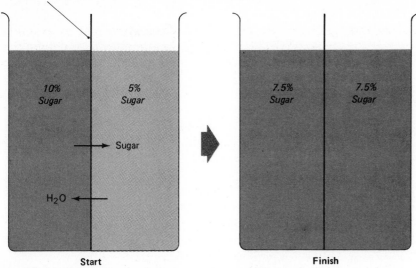

FIGURE 3.1   Both water and sugar molecules readily diffuse through the
nylon mesh barrier, moving down their concentration gradients.

OSMOSIS.   Suppose, however, that we replace the barrier with a
thin piece of uncoated cellophane. (This experiment will not work
with regular cellophane, which is coated to make it waterproof.)
Uncoated cellophane is not waterproof; it has pores large enough to
allow water molecules to pass through. However, the pores are too
small to permit passage of sugar molecules. If we once more pour a
10-percent sugar solution into the left compartment and a 5-percent
solution into the right one we have again created concentration gradients for sugar and water molecules. But, since sugar molecules
cannot pass through the cellophane membrane, there will be no
movement of sugar.

However, the water can get through the barrier, and after a period of time there will be a 7.5-percent sugar solution on both sides
of the barrier. (See **FIGURE 3.2** on the following page.) Another way
of describing the process is to say that it started with a 95-percent
"water solution" on the right and a 90-percent solution on the left.
Water moved down its concentration gradient, reducing the concentration of water molecules on the right and increasing the concentration on the left until the concentrations on both sides were equal.

Note that although the concentrations are equal now, the volumes are not. The levels of liquid on each side of the cellophane are
different. (See **FIGURE 3.2.**) That disparity causes no trouble in the
gravity-free environment where this experiment is being conducted,
but things would not be so simple on earth, as will soon be apparent.

The phenomenon just described is called _**osmosis**_; it is defined
as diffusion through a semipermeable membrane.

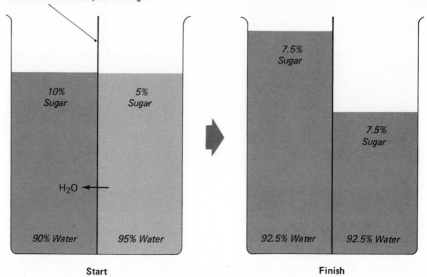

FIGURE 3.2   In the absence of gravity, water diffuses through the semipermeable membrane, moving down its concentration gradient. Osmotic equilibrium is thus achieved.

HYDROSTATIC PRESSURE.   Imagine performing another experiment with the container, again using the cellophane barrier. However, this time we shall conduct the experiment on earth. We shall fill both chambers with pure water: 100 ml in the left one and 50 ml in the right. Soon the right-hand container gains water and the left one loses it, until each side contains 75 ml. (See **FIGURE 3.3**.) Because we used pure water, osmosis is not observed. Instead, the water volumes become equal because of the effects of hydrostatic pressure, supplied by the force of gravity. We all know that water runs down hill, so no one should be surprised by the results of this experiment.

DYNAMIC EQUILIBRIUM: PUTTING THE FORCES TOGETHER. So far we have seen examples of two different forces (osmotic pressure and hydrostatic pressure) that bring a system from an initial state of **disequilibrium** to one of **equilibrium.** (*Libra* means balance, so a system is at equilibrium when there is an equal balance of forces.)

    To complicate things just a bit, imagine performing the osmosis experiment on earth. We place equal volumes of sugar solution in the container, a 10-percent sugar solution in the left and a 5-percent solution in the right. Initially, only one force is present—osmotic pressure, causing water molecules to move from right to left. How-

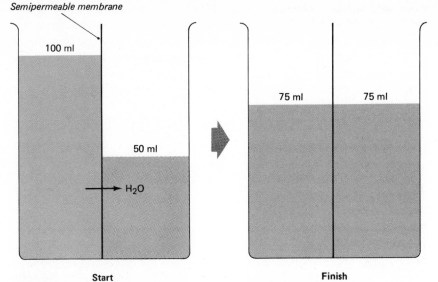

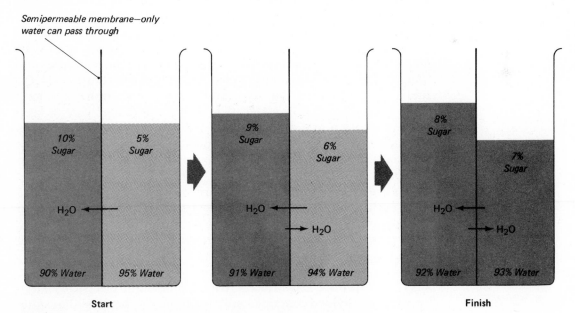

FIGURE 3.3   In the presence of gravity, water is forced by gravity across
the semipermeable membrane until the levels of liquid on both sides of
the container is equal.

FIGURE 3.4   Water moves from right to left, down its osmotic gradient,
but as water accumulates on the left side of the container, hydrostatic
pressure begins to cause a movement of water toward the right.
Eventually a point of equilibrium is reached, and there is no net
movement of water.

ever, as soon as a difference in the height of the solution on each side begins to occur, hydrostatic pressure starts pushing water from left to right. Eventually the forces reach a standoff, so that the movement of molecules going left and right is equal. (See **FIGURE 3.4** on the preceding page.) Equilibrium is reached, but it is the result of two equal and opposing disequilibriums; there is neither hydrostatic nor osmotic equilibrium.

DEVELOPMENT OF THE MEMBRANE POTENTIAL

The phenomena described in the previous section and the concept of the equilibrium of opposing forces can be related to the way neurons work. The membrane of the axon, like the uncoated cellophane barrier, is semipermeable, and it is part of a system that is normally at equilibrium. However, the extracellular and intracellular fluids are more complicated than sugar solutions are, and it is not necessary to consider hydrostatic pressure in describing the transmission of information by the axon. ***Electrostatic pressure*** will take its place.

ELECTROSTATIC PRESSURE. Many substances split into two parts, each with an opposing electrical charge, when the substance is dissolved in water. We call substances with this property ***electrolytes***; the charged particles into which they decompose are called ***ions.*** Ions have either a positive charge (***cation***) or a negative charge (***anion***). For example, many molecules of sodium chloride (NaCl, table salt) split into sodium cations (Na^+) and chloride anions (Cl^-) when dissolved in water. (I find that the easiest way to keep the terms cation and anion straight is to think of the cation's plus sign as a cross, and remember the superstition of a black cat crossing your path. Stupid little mnemonic devices like this are the easiest to remember.)

As everyone has undoubtedly learned, particles with like charges repel each other (+ repels + and − repels −), but particles with different charges are attracted to each other (+ and − attract). The force exerted by this attraction or repulsion provides electrostatic pressure.

A HYPOTHETICAL EXAMPLE. We can produce an electrical charge by using our divided container and a solution of a chemical that splits into ions. We will put a concentrated solution of potassium acetate in the left compartment and a dilute solution on the right. If the membrane were permeable to both potassium (K^+) and acetate (AC^-) ions, the result would be a migration of both ions (and of water, also) down their concentration gradients. If only water could get through the membrane, we would essentially be repeating the last experiment we conducted with the sugar solutions.

Suppose, however, that the K^+ ions could pass through the pores in the membrane but that the acetate ions could not, because of their greater size. For the sake of simplicity, let us assume that we have started with 6 molecules of potassium acetate on the left and 2 molecules

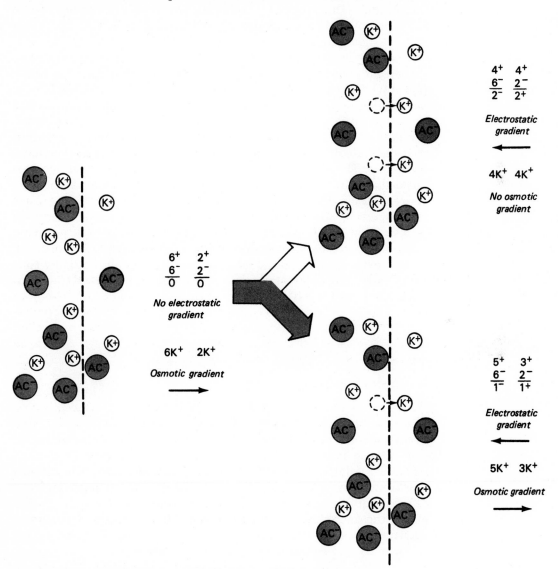

FIGURE 3.5 Development of a membrane potential (hypothetical). The starting point is shown on the left. The upper right shows how an electrostatic charge would develop if potassium ions were allowed to move down their concentration gradient until osmotic equilibrium was achieved. This result could never occur. The lower right portion of the illustration shows how the osmotic and electrostatic pressures balance each other, just as hydrostatic and electrostatic pressures balanced in the example shown in Figure 3.4.

on the right. We will assume that all the molecules dissociate (split into ions), giving the result seen in **FIGURE 3.5, LEFT** on the preceding page. Since there is an equal number of anions and cations on each side of the membrane, there is no net electrical charge.

Osmotic equilibrium would demand an equal number of K^+ ions on each side of the membrane (remember, AC^- ions cannot cross the membrane). This would be accomplished by the movement of two K^+ ions down their concentration gradient, from left to right. This state is shown in the upper right drawing of Figure 3.5. However, if you count the ions, you will see that there are two extra anions on the left and two extra cations on the right. (See **FIGURE 3.5, UPPER RIGHT.**) The two K^+ ions that just crossed to the right would now be repelled by the excess positive charge on that side and attracted by the negative charge on the left. In other words, the **electrostatic gradient** for these ions goes from right to left. This means, of course, that the state of affairs illustrated in the upper right drawing would not occur. The electrostatic pressure that builds up as the K^+ ions go to the right opposes further migration of these ions down their concentration gradient. Just as there was a standoff in the previous experiment between osmotic and hydrostatic pressure, so there will be a standoff in this case between osmotic and electrostatic pressure. The actual outcome is illustrated in Figure 3.5, lower right. The drawing shows seven K^+ ions on the left, and five of them on the right. Thus the osmotic and electrostatic pressures balance each other. (See **FIGURE 3.5, LOWER RIGHT.**)

The Nernst Equation: What Causes the Axonal Membrane Potential?

Since the amount of osmotic pressure that is exerted across a membrane depends upon the relative concentrations of the solutions on each side, it is possible to calculate this pressure merely by measuring the relative concentrations. Furthermore, since the electrostatic and osmotic forces balance perfectly once the system has reached equilibrium, we then know the value of the electrical charge across the membrane. (If you know the value of one force, you know the value of an opposing force that balances it. See **FIGURE 3.6.**)

The Nernst equation expresses this relationship:

$$V = k \log ([X_L]/[X_R])$$

where V = voltage across the membrane (in millivolts, or 1/1000 volt)

k = a constant that depends on such experimental conditions as temperature (its sign depends on the charge of the ion in question)

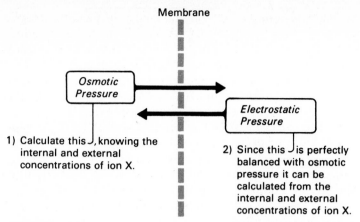

FIGURE 3.6   A schematic representation of the rationale for the Nernst equation.

$[X_L]$ and $[X_R]$ = the concentrations of the ion in question on the left
and right of the membrane

If the ion is equally distributed on both sides of the membrane, $[X_L]/[X_R] = 1$, and since the logarithm of 1 is 0, the membrane voltage = 0. As the disparity between the concentrations of the ion increases, so does the membrane voltage.

The Nernst equation, then, allows us to answer this question: How much charge must there be across the membrane to counteract the osmotic pressure that results from the observed ionic imbalance?

Now let us look at the axon. The membrane of the axon is selectively permeable, allowing some but not all ions to cross; thus, there is a charge across the axonal membrane (membrane potential). Axons that are found in mammalian nervous systems are difficult to study; they are small, not easily isolated, and impossible to separate from their associated oligodendroglia or Schwann cells. Fortunately, nature has provided the squid with an axon of enormous diameter (relatively speaking, that is). The giant squid axon (this is a giant axon of a squid, not the axon of a giant squid) is about 0.5 mm in diameter. That might not sound huge, but it is hundreds of times greater than the diameter of the largest mammalian axon.

The cations that are most important to axonal transmission of information are those of sodium (Na^+) and potassium (K^+). (The single plus sign indicates a single unit of positive charge.) The most important anions are those of chloride (Cl^-) and various protein anions (usually symbolized by A^-). The concentrations of these ions within the cytoplasm (more properly called axoplasm) of the giant squid axon and the concentration of these ions in the extracellular fluid (seawater) are given below, in millimoles/liter. (I should note that you cannot predict the membrane potential by adding up the values, since many of the protein anions have two or three negative charges.)

Ion	External Concentration	Internal Concentration
Na^+	460	50
K^+	10	400
Cl^-	560	40
A^-	0	345

If we apply the Nernst equation to the various ions, we can see whether it is possible to predict the membrane potential from the distribution of ions on each side of the membrane.

Before doing any calculations, we should first examine the limitations of the Nernst equation. In order for it to give valid results, several conditions must be met. In other words, assumptions about the system are made that must be true if the equation is to work.

These assumptions are as follows:

1. *There are no forces other than electrostatic pressure counterbalancing the osmotic pressure.* If there are other forces, then osmotic and electrostatic pressures need not be equal, and knowing the osmotic pressure will not tell us the value of the electrostatic pressure.
2. *The ion in question is free to traverse the membrane.* If the ion cannot cross the barrier, unequal concentrations do not have to be opposed by electrostatic pressure; the membrane barrier is sufficient by itself. If the barrier were of glass, for example, the two solutions would not interact at all.
3. *The system has reached equilibrium.* The Nernst equation applies only after enough time has elapsed for all the forces to be balanced.

It is important to note that the form of the Nernst equation used here is simplified a bit and will work only for ions with a single positive or negative charge. Also, the sign of the constant k depends on whether we examine an anion or a cation.

Let's calculate what the membrane potential should be, given the extracellular and intracellular concentrations of the chloride ion.

The Nernst equation, again, is as follows:

$$V = k \log ([X_o]/[X_i])$$

where $[X_o]$ and $[X_i]$ refer to the concentrations of ion X outside and inside the axon.

At 20° C, and for an anion, $k = -58$, so we can solve the equation for Cl^- as follows:

$$V = k \log ([Cl^-_o]/[Cl^-_i])$$
$$V = -58 \log (560/40)$$
$$= -58 \log (14)$$
$$= -58(1.1461)$$
$$= -66.5 \text{ mV}$$

The Nernst equation predicts that the inside of the axon should be charged negatively, compared with the outside. The value of this charge should be 66.5 mV. Let me go through the logic of this calculation.

Consider the concentration of Cl^- ions. There are many more of them outside than in. This means that there is a considerable amount of osmotic pressure that tends to force Cl^- ions inward. Since the system is at equilibrium, there must be an electrostatic force that tends to push Cl^- ions *out of* the axon, exactly balancing the inward-pushing osmotic pressure. What kind of charge would push negatively charged Cl^- ions out of the axon? Since like charges repel each other, the force must be exerted by a negative charge on the inside. Thus, the inside must be negatively charged (and, of course, the outside must be positively charged). The exact value of the charge is equal to the osmotic pressure, which is calculated as $-58 \log (560/40)$, or -66.5 mV.

Is this clear? Look at the Nernst equation again and see if it makes sense now. If it still does not, read the preceding paragraph again.

We need not be content with calculating the membrane potential; we can measure it directly with a voltmeter by placing one wire inside the axon and another into the seawater bathing the axon. (See **FIGURE 3.7.**)

Measurements like this have shown that the membrane potential is -70 mV, which is very close to the voltage we predicted from the distribution of Cl^- using the Nernst equation. (Convention has the membrane potential expressed as inside relative to outside. A negative membrane potential means that the inside is charged negatively, the outside positively.)

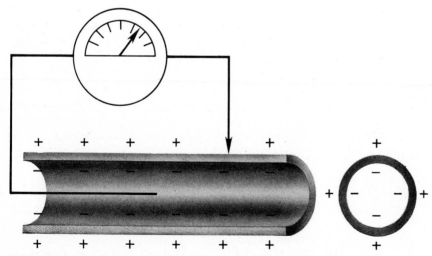

FIGURE 3.7 A schematic drawing of the method for recording the membrane potential of an axon.

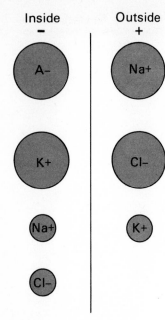

Inside
−

Outside
+

FIGURE 3.8 The relative concentration of some of the most important ions inside and outside the axon.

The equation can be tried with the other ions. Before we even bother with A⁻ (the protein anions located exclusively within the cell) I will tell you that the membrane is impermeable to these large molecules; therefore, for this ion, assumption 2 of the Nernst equation is violated. The membrane, and not electrostatic pressure, holds the A⁻ ions in, so the Nernst equation should not be used here. You can see that osmotic pressure would tend to push A⁻ out, and so would electrostatic pressure; clearly, these forces do not oppose in the case of A⁻. (See **FIGURE 3.8**.)

Now let's try potassium. At 20°C, and for a cation, $k = 58$, so the equation is as follows:

$$V = k \log ([K^+_o]/K^+_i])$$
$$= 58 \log (10/400)$$
$$= 58 \log (0.025)$$
$$= 58(-1.6)$$
$$= -93 \text{ mV}$$

The calculated voltage is somewhat greater than the charge that is actually observed (-70 mV). Either there is a certain amount of measurement error or something is wrong. The reason for the discrepancy will soon be apparent.

The calculations for the sodium ion are as follows:

$$V = k \log ([Na^+_o]/[Na^+_i])$$
$$= 58 \log (460/50)$$
$$= 58 \log (9.2)$$
$$= 58(0.96)$$
$$= +56 \text{ mV}$$

This voltage ($+56$ mV) does not come anywhere near the actual membrane potential (-70 mV). Something is clearly wrong.

The measurements of ionic concentration and of the membrane potential are accurate. The Nernst equation is not in error; it has a solid foundation in the laws of thermodynamics. Therefore, one or more of its assumptions must have been violated.

As we shall see, assumption 1 is violated; electrostatic pressure and osmotic pressure are *not* the only forces that act on Na⁺. As Figure 3.8 indicates, both osmotic pressure *and* electrostatic pressure tend to push Na⁺ into the cell. And yet, it stays out. Thus, there must be something that opposes both of these forces. (See **FIGURE 3.8**.)

The Sodium-Potassium Pump

If there were an absolute barrier to Na⁺ (as there is for A⁻), it would never enter the axon. However, a simple experiment shows that this is not the case. If a giant squid axon is placed in a dish of seawater containing some radioactive Na⁺, is allowed to sit a while, and then is removed and washed off, the axon is found to be radioactive. This means that Na⁺ from the seawater got into the axon. Furthermore, if we analyze the axoplasm for Na^+, we find that its concentration is unchanged. This indicates that as much sodium left the axon as came in.

It's easy to see how Na^+ got into the axon—it was pushed through a slightly leaky barrier by electrostatic and osmotic pressure. But what force pushed an equal number of sodium ions back outside?

Studies have shown that there is a pump in the membrane: the ***sodium-potassium pump.*** This mechanism forces sodium out of the cell and forces potassium in. Since the membrane resistance to Na^+ is high, most of it stays out. Since the membrane resistance to K^+ is low (about 1/100 that of Na^+), only a little extra potassium remains inside the cell; most goes back outside.

This pump uses energy; up to 40 percent of the neuron's energy expenditure goes into the sodium-potassium pump. If the axon is poisoned with DNP (dinitrophenol), a chemical that halts biochemical processes that are necessary for energy production in a cell, the pump stops, and Na^+ leaks into the cell. Also, if the temperature is varied, the pump speeds up or slows down, thus altering the concentration of Na^+ within the axon. Therefore, there is good evidence that a metabolically active sodium-potassium pump exists in the cell membrane. We do not know precisely how the pump operates, but a schematic illustration of one hypothetical model is shown in figure 3.9. (See **FIGURE 3.9.**)

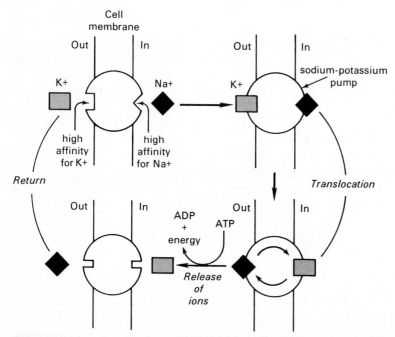

FIGURE 3.9   A schematic representation of the sodium-potassium pump. (Adapted from Dyson, R. D., *Cell Biology: A Molecular Approach.* Boston: Allyn and Bacon, 1978.)

The effect of the sodium-potassium pump, then, is to produce a tremendous osmotic and electrostatic gradient for sodium; if the membrane barrier for sodium were to break down, Na^+ ions would rush into the axon. The pump also produces a slight excess of K^+ ions inside. The process is rather like trying to fill a leaky bucket, since most of the K^+ ions that are brought into the cell are able to get back outside. However, there is some membrane resistance to the flow of K^+, so the pump is able to build up a slight osmotic gradient for potassium, beyond the value that balances the electrostatic force of the membrane potential. If the barrier to K^+ were to fall completely, there would be a net movement of these ions out of the cell, and the membrane would become more charged than it is (there would be even more cations on the outside, and fewer inside).

We have seen that, because of the selective permeability of the membrane and because of the presence of the sodium-potassium pump, the membrane is electrically polarized, or charged. The outside is charged positively, and the inside is charged negatively. A decrease in the resistance of the membrane to sodium would allow a great number of Na^+ ions to enter the cell (lowering the membrane potential, whereas a decrease in the resistance of the membrane to potassium would *raise* the membrane potential, since some of the excess K^+ ions would leave the cell).

THE ACTION POTENTIAL

Now that we understand the production of the membrane potential, we can examine its role in the transmission of information in the nervous system. As we shall see, changes in membrane permeability to Na^+ and K^+ produce the electrical message that is carried down the axon.

Measurement of Changes in the Axonal Membrane Potential

First we shall examine the nature of the message. To do this, we place an isolated squid axon in a dish of seawater and use an oscilloscope to record electrical events from the axon. A more detailed description of this device can be found in Chapter 7, but it will be sufficient to say here that an oscilloscope records voltages, as does a voltmeter, but also produces a record of changing voltages, graphing them as a function of time. For example, Figure 3.10 shows a hypothetical curve drawn by an oscilloscope. This is not a graph of a real phenomenon, but only a curve for illustrative purposes. We can see that the voltage being recorded started at -100 mV, changed rapidly to − 50 mV (in about 5 milliseconds, of 5/1000 of a second), and then more slowly changed back to − 100 mV. (See **FIGURE 3.10.**)

We will attach the oscilloscope to the axon via two wires (more properly called *electrodes*), one recording the voltage on the outside, and one inserted into the axoplasm. Since the resting potential of the axon is around − 70 mV, the graph obtained (when the axon is not disturbed) is a straight horizontal line at − 70 mV. In addition, we will use another device—a shocker that will allow us to pass electrical current through the axonal membrane and thus artificially alter the membrane potential. (See **FIGURE 3.11.**) The shocker can pass positive or negative current through the electrode on the outside of the axon, thus raising the membrane potential in that region (making it more positive outside) or lowering it (making it less positive). Positive shocks, which increase the membrane potential, produce *hyperpolarization* (more polarization). Negative shocks, which decrease the membrane potential, produce *hypopolarization* (less polarization). To be consistent with other writers, I shall use the term *depolarization* instead of hypopolarization. This term is not really correct (depolarization literally means removal of polarization), but it is the term most generally used.

Triggering the Action Potential

Figure 3.12 shows a graph drawn by an oscilloscope monitoring the effects of a series of brief depolarizing and hyperpolarizing shocks to the axon. The graphs of the effects of these separate shocks are su-

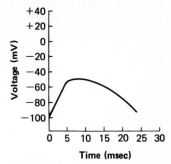

FIGURE 3.10 An example of a curve that could be traced by an oscilloscope.

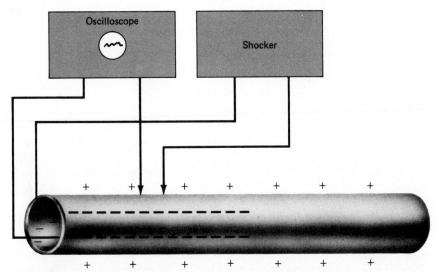

FIGURE 3.11 A schematic representation of the means by which an axon can be shocked while its membrane potential is being recorded.

perimposed on the same drawing so that we can compare them. The shock intensity is labelled in arbitrary units from 1 to 5, 5 representing the highest intensity, + representing hyperpolarizing pulses, and − representing depolarizing pulses. (See **FIGURE 3.12.**).

As you can see, successively stronger hyperpolarizations (+ shocks) produce successively larger disturbances in the membrane potential. The changes in the membrane potential last considerably longer than the very brief shocks. The reasons for this phenomenon (having to do with the resistive and capacitive properties of the membrane and axoplasm) need not concern us here. Figure 3.12 also shows the effects of the depolarizing shocks. The results in this case are different; the effects of the low-intensity shocks (− 1 to − 3) mirror those of the hyperpolarizing shocks, but once the membrane potential reaches a certain point (approximately − 65 mV for most axons), it suddenly breaks away and reverses itself (so that the outside becomes *negative*). Then the potential quickly returns to normal. The whole process takes about 2 milliseconds. (See **FIGURE 3.12.**)

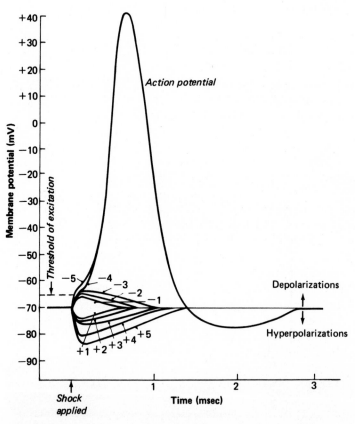

FIGURE 3.12 A schematic drawing of the results that would be seen on the oscilloscope screen if various shocks were delivered to the axon pictured in Figure 3.11.

This phenomenon, a very rapid reversal of the membrane potential, is called the ***action potential.*** It constitutes the message carried by the axon. The voltage level at which the membrane "breaks away" with an action potential is called the ***threshold of excitation.*** This process is what the decision made by the neuron is all about: whether or not to produce an action potential. The decision depends on whether the threshold of excitation is reached. Furthermore, the action potential is the event that initiates the release of transmitter substance from the terminal buttons, thus making them "talk" to the receiving cells.

Ionic Events during the Action Potential

Experiments have shown that the action potential results from a transient drop in the membrane resistance to Na^+ (allowing these ions to rush into the cell), immediately followed by a transient drop in the membrane resistance to K^+ (allowing these ions to rush out of the cell). It is time now to examine the evidence for these events.

EVIDENCE FOR THE INFLUX OF SODIUM.   We have seen that an axon will become radioactive after being placed in a dish of seawater containing radioactive sodium. The rate of uptake of radioactivity is a measure of the amount of sodium that gets through the membrane in a given period of time. If this axon is now repeatedly stimulated to produce action potentials, we find that it becomes even more radioactive; the electrical activity of the axon has resulted in an increased influx of sodium. However, the concentration of sodium in the axoplasm has still not changed. The sodium-potassium pump must have become more active in order to counteract the increased influx of sodium.

Perhaps the following analogy will help make clear why the radioactivity of the axon increases even though the intracellular Na^+ concentration remains the same. Suppose we have two beakers of water, a large one and a small one. If we repeatedly pour a tiny amount of water from the large one into the small one, and then pour the same amount back, we do not increase the volume of water in the small beaker, but we gradually introduce more and more molecules of water from the large beaker into the small one. If we started with ink in the large beaker, the contents of the small beaker gradually become darker as we exchange a little bit of their contents. Similarly, radioactive sodium from the seawater mixes with the sodium in the axoplasm even though increased activity of the sodium-potassium pump keeps the amount of intracellular sodium at a constant level.

It appears, then, that sodium enters the axon during action potential, and excess intracellular sodium is then removed by the sodium-potassium pump. A very clever set of experiments by Hodgkin, Huxley, and Katz (see Katz, 1966) revealed the time course of the sodium influx (and of the potassium efflux, or outflow, that immediately follows it) during the action potential.

THE SEQUENCE OF IONIC FLUXES.   The process of ionic flow occurs in the following sequence:

1. As soon as the threshold of excitation is reached, the membrane barrier to sodium drops (actually, to ⅛ the normal value), and Na⁺ comes rushing in, down its electrostatic and osmotic gradients. This produces a rapid change in the membrane potential, from -70 mV to $+50$ mV.

2. The membrane now drops its resistance to potassium. This resistance was not extremely high to begin with, but now it is quite low. Since, at the peak of the action potential, the inside of the axon is now positively charged, K⁺ ions are driven *out* of the cell, down the osmotic and electrostatic gradients. This efflux of cations causes the membrane potential to go back down again. As a matter of fact, it overshoots the usual value of the resting potential (-70 mV) and only gradually returns to normal. (See **FIGURE 3.13.**) If you will recall, the Nernst equation indicated that the membrane potential should be at -93 mV to account for the observed internal and external concentrations of K⁺, thus showing that the sodium-potassium pump kept some excess potassium within the cell. With the drop in the membrane resistance to K⁺, this excess potassium can leak out and make the outside even more positive—the membrane becomes more polarized than usual.

3. The membrane once again becomes resistant to the flow of sodium; the whole process of sodium influx lasts only about one millisecond.

4. Finally, the K⁺ resistance of the membrane goes back up to its normal level. The sodium-potassium pump removes the Na⁺ ions that leaked in and retrieves the K⁺ ions that leaked out.

How much ionic flow is there? When I say "sodium rushed in," I do not mean that the entire axoplasm becomes flooded with sodium. Because the drop in membrane resistance to sodium is so brief, and because diffusion over any appreciable distance takes some time, not too many Na⁺ molecules flow in, and not too many K⁺ molecules flow out. At the peak of the action potential, a very thin layer of fluid immediately inside the axon is full of Na⁺ ions that have just arrived; this is, indeed, enough to reverse the membrane potential. However, not enough time has elapsed for these ions to fill the entire axon. Before this event can take place, the Na⁺ barrier goes up again, and K⁺ starts flowing out. Experiments have shown that an action potential temporarily increases the number of Na⁺ ions inside the giant squid axon by 1/300,000. The concentration just inside the membrane is high, but the total number of ions entering the cell is very small, relative to the number already there. On a short-term basis the sodium-potassium pump is not very impor-

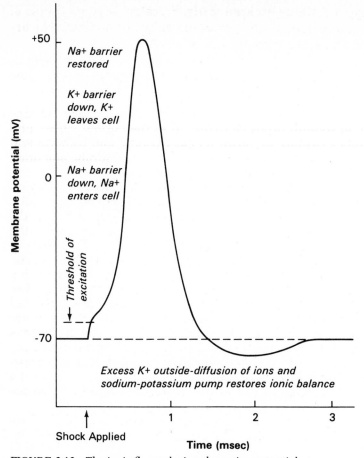

Membrane potential (mV)

+50

*Na+ barrier
restored*

*K+ barrier
down, K+
leaves cell*

*Na+ barrier
down, Na+
enters cell*

0

*Threshold of
excitation*

-70

*Excess K+ outside-diffusion of ions and
sodium-potassium pump restores ionic balance*

1   2   3

↑
Shock Applied

Time (msec)

FIGURE 3.13   The ionic fluxes during the action potential.

tant. The few Na^+ ions that leak in diffuse into the rest of the axo-plasm, and the slight increase in Na^+ concentration is hardly no-ticeable. On a long-term basis, of course, the sodium-potassium pump is necessary, but it would take many action potentials to cause any serious gain of Na^+ within the cell.

The situation is different for the smallest unmyelinated axons in the mammalian CNS (approximately 0.1 micrometer, or 1/10,000 mm in diameter). These axons increase their intracellular Na^+ con-centration by 10 percent when the axon fires, so to these fibers the sodium-potassium pump is important even on a short-term basis.

So far we have seen what produces the membrane potential and how changes in this potential, initiated by depolarization of the membrane to around -65 mV (the threshold of excitation), result in an action potential. Now let us examine the characteristics of the membrane that are responsible for selective permeability, and for the changes in permeability that produce the action potential.

**Membrane Changes Responsible
for the Ionic Fluxes**

Although sodium molecules are smaller than potassium molecules, when these substances are dissolved in water their ions attract a number of molecules of water. These water molecules attach to the ions, forming a sort of bubble surrounding each of them. Sodium ions attract more water molecules, and hence the hydrated ("watered") Na^+ ions are larger than the hydrated K^+ ions. It has been suggested that the membrane is perforated by small pores. These hypothetical pores could account for the membrane's selective permeability; the smaller hydrated K^+ ions would pass through the pores more easily than the larger hydrated Na^+ ions.

If the existence of pores in the membrane can account for selective permeability, how can one account for the temporary decrease in permeabilities to Na^+ and K^+ during the action potential? It cannot just be that the pores open up more, or else the membrane resistances to K^+ and Na^+ would drop simultaneously. Remember, hydrated Na^+ ions are the larger of the two, and if the pores opened up enough to let these ions through, membrane resistance to the smaller K^+ ions would also have to fall. Instead, the membrane resistance to Na^+ falls first and is followed by a fall in the K^+ resistance. Various theories have been proposed to account for this selective permeability, but no one can say with any assurance what the actual mechanism might be.

One hypothesis is rather intriguing, however, since it accounts for two phenomena—the selective fall in Na^+ resistance and the transitory nature of this fall. The drop in the membrane resistance to sodium is triggered by depolarization of the membrane, but if the Na^+ resistance were merely a function of the membrane potential, then there would be no way to explain why the Na^+ resistance reasserts itself while the membrane is still depolarized. The hypothesis suggests that there is a finite number of **sodium carriers** in the membrane. When triggered by depolarization of the membrane, these carriers transport sodium across the membrane. The sodium carriers remain on the inside of the membrane until the resting potential is restored. Then they move back to the outside of the membrane, ready to transport another cargo of Na^+ into the axoplasm. (See **FIGURE 3.14.**) This hypothesis has yet to be proved, but it does account for the selectivity of the fall in the Na^+ resistance of the membrane and also for its transitory nature; there are only so many carriers, and they are not reset until the resting potential is restored.

We need not hypothesize carriers for the fall in K^+ resistance that follows the fall in Na^+ resistance. If the pores in the membrane open up a little wider, they will allow K^+ to pass through more freely without allowing the larger hydrated Na^+ ions through.

I would like to emphasize the following point again: the exist-

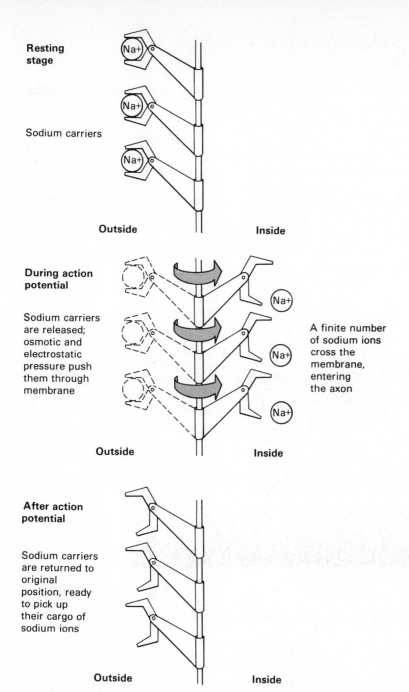

Resting stage

Sodium carriers

Outside   Inside

During action potential

Sodium carriers are released; osmotic and electrostatic pressure push them through membrane

A finite number of sodium ions cross the membrane, entering the axon

Outside   Inside

After action potential

Sodium carriers are returned to original position, ready to pick up their cargo of sodium ions

Outside   Inside

FIGURE 3.14   A schematic representation of hypothetical sodium carriers.

ence of pores and Na$^+$ carriers is hypothetical, and that means un-proved. This may be how the membrane works, but much study and experimentation will be needed to elucidate these mechanisms.

PROPAGATION OF THE ACTION POTENTIAL

Conduction by Cable Properties

Now that we have a basic understanding of the membrane potential and production of the action potential, we can discuss the final topic of this chapter—movement of the message down the axon, or *propagation of the action potential*. To study this phenomenon, we shall again make use of the giant squid axon. A shocker will be attached to one electrode in the axoplasm and another on the outside of the membrane, at one end of the axon. An oscilloscope will be used to record membrane potentials. It will also be attached to two electrodes, one in the axoplasm and the other touching the outside of the axonal membrane. This external electrode is movable, so that recordings can be made from any location along the surface of the axon. (See **FIGURE 3.15**.)

A subthreshold, depolarizing shock (too small to produce an action potential) is produced at one end of an axon, and the results are recorded at a point near the site of stimulation. This procedure is repeated, recording farther and farther down the axon. These shocks produce disturbances in the axon that diminish as the recording electrode moves farther away from the point of stimulation. (See **FIGURE 3.16**.) Note, however, that there is no noticeable delay in the onset of this disturbance; transmission is almost instantaneous. (See **FIGURE 3.16**.) This means that transmission of the sub-threshold depolarization is *passive*. The axon is acting like a cable,

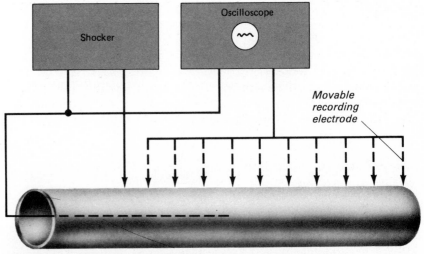

FIGURE 3.15 A schematic representation of the procedure by which shocks can be delivered to one end of the axon while recording the membrane potential various distances away.

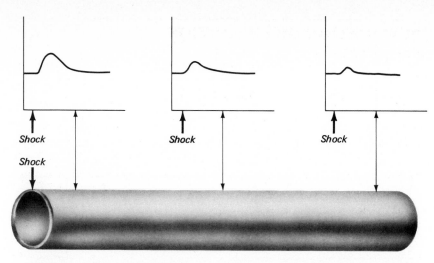

FIGURE 3.16   The results obtained when a subthreshold shock is applied
to the axon. This demonstrates decremental conduction.

carrying along the current started at one end. This property of the
axon follows laws, discovered in the nineteenth century, concerning
conduction of electricity along submarine telegraph cables laid along
the ocean floor. Submarine cables degrade an electrical signal because of leakage of the insulator, resistance of the wire, and capac-
itance between the wire and the conductor (seawater) surrounding
it. If you put a large pulse in at one end, you get, at the other end,
a much smaller signal. Furthermore, the crispness of the pulse is
lost; you get a slowly changing potential similar in shape to the
disturbances we recorded from the axon. (Remember, the signal applied to the axon was a brief, sharp pulse.) We say that the transmission of subthreshold depolarizations follows the laws describing
the **cable properties** of the axon. The axon is not doing anything
different from what we would expect from a submarine cable. (Of
course, the submarine cable has a much lower resistance and transmits this kind of electrical disturbance much farther than an axon
could, but both of these conductors follow the same laws.) Since
hyperpolarizations never elicit action potentials, these disturbances
are also transmitted via the passive cable properties of an axon.

All-or-None Conduction
of the Action Potential

Now let's produce a number of suprathreshold (above-threshold) depolarizing shocks to the end of the axon, and record at successively
greater distances from the site of stimulation. This time we shall
record an action potential, with a peak of around +50 mV, at each
point along the axon. However, conduction of the action potential is

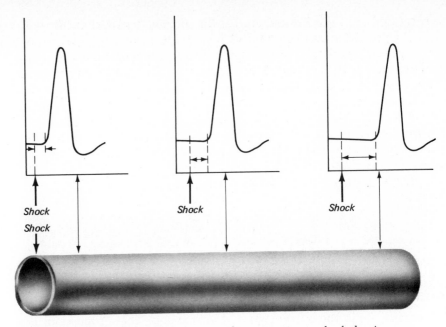

FIGURE 3.17   The results obtained when the axon is given a shock that is above the threshold of excitation. This demonstrates the nondecremental conduction of an action potential.

much slower than conduction of electrical disturbances that are transmitted via the passive cable properties of the axon. The action potential arrives later and later as we move the recording electrode away from the stimulating electrode. (See **FIGURE 3.17.**)

By doing these experiments, we have established <u>a basic law of</u> <u>axonal transmission</u>: the ***all-or-none law.*** <u>This law states that you</u> <u>get an action potential or you do not; once it has been triggered, it</u> <u>is transmitted down the axon to its end.</u> It always remains the same size, without growing or diminishing. I should note here that the axon will transmit an action potential in either direction, or even in both directions if you start one in the middle of its length. However, since action potentials are started, in intact organisms, at one end only, the axon exclusively carries one-way traffic.

How is the action potential propagated? Since we know how the action potential is triggered, and since we know about the passive cable properties of axons, it will be easy to explain this phenomenon. Consider, for sake of simplicity, that the axon is divided into segments. (See **FIGURE 3.18.**) We shall stimulate the segment at the left end, triggering an action potential there. This potential will spread, via passive cable properties, to the next segment. It may decline a bit in size, but it will still be large enough to depolarize the next segment above the threshold of excitation, so an action potential is triggered there. The action potential then is transmitted via passive cable properties of the axon to the next segment, and . . . well, I think you get the picture (**FIGURE 3.18**).

The naked axon is not really divided into segments; propagation of the action potential is a smooth, continuous process. If you think of the axon as being divided into infinitesimally small segments, you can describe a continuous process.

CONDUCTION IN MYELINATED AXONS. You will remember from the last chapter (I hope) that most axons are myelinated; segments of them are covered by a myelin sheath produced by the CNS oligodendroglia or the PNS Schwann cells. These segments are separated by portions of naked axon, the nodes of Ranvier. Myelinated axons really *are* segmented, and they transmit the action potential in the way I just described for the hypothetically segmented axon. Let us shock one end of a myelinated axon and then record the membrane potential at various places underneath the Schwann cell covering one segment. We shall record with extremely thin electrodes that can pierce the Schwann cell without significant damage (see Chapter 7 for a description of these electrodes). The results are shown

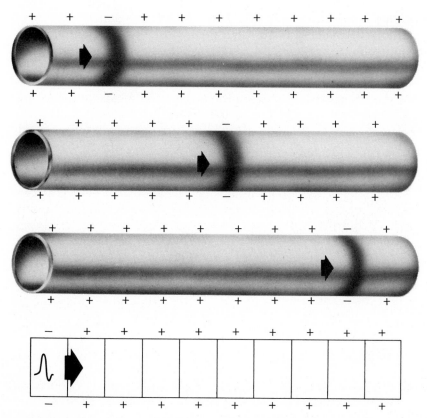

FIGURE 3.18 Propagation of an action potential down an unmyelinated axon.

in Figure 3.19. (See **FIGURE 3.19.**) Note that the "action potential" gets smaller as it passes along the membrane wrapped by the Schwann cell. Why did I write "action potential" rather than action potential? Because the disturbance only looks like an action potential—it really is not one. If it were, we'd have to repeal the all-or-none law.

We do not have to repeal the law, however. The Schwann cell (and the oligodendroglia of the CNS) wraps very tightly around the axon, leaving no measureable extracellular fluid around it except at the nodes of Ranvier, where the axon is naked. If there is no extracellular fluid, there can be no inward flow of Na^+ when the resistance of the membrane to Na^+ drops. There is no extracellular sodium to get into the cell. Then how does the "action potential" travel along the axonal membrane covered by the myelin sheath? You guessed it—cable properties. The axon passively transmits the electrical disturbance resulting from the action potential down to the next node of Ranvier. The disturbance gets smaller, but it is still large enough at the next node to trigger an action potential there. The action potential gets retriggered, or repeated, at each node of Ranvier, and it is passed, via the cable properties of the axon, to the next node.

What advantage is there to this kind of axon? There are two: **Saltatory conduction** (from the Latin *saltare*, "to dance"), jumping from node to node, is faster than the continuous conduction down an unmyelinated axon. The fastest myelinated axon can transmit action potentials at a speed of 100 meters per second (that is around 224 miles per hour). This increased speed of conduction occurs because the transmission between the nodes is by means of the axon's

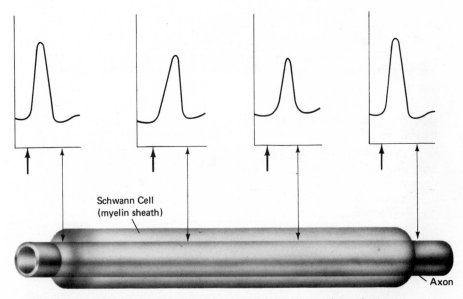

FIGURE 3.19   Propagation of an action potential down a myelinated axon.

cable properties, and you will recall that this conduction is extremely fast. A delay is introduced at each node, of course, as the action potential is regenerated.

The second advantage to saltatory conduction is an economic one. Energy must be expended by the sodium-potassium pump to get rid of the excess sodium ions that leak into the axon during the action potential. This pump is given work to do all along an unmyelinated axon, because sodium leaks in everywhere. Since sodium can leak into a myelinated axon only at the nodes of Ranvier, much less gets in, and consequently much less has to be pumped out again. Therefore, a myelinated fiber does not need to expend as much energy to maintain its sodium balance.

Regulation of Extracellular Fluid during Neural Activity

The total amount of extracellular fluid in the body is vast, but in the central nervous system the neurons and glia take up most of the space. Extracellular fluid is restricted to small channels between the cells. Therefore, if a group of axons in one region begins to fire at a high rate, the concentration of K^+ ions in the extracellular fluid in that region will begin to increase. The Na^+/K^+ pump will not be able to keep up with the neural firing. An increase in the extracellular concentration of K^+ makes the axon more excitable; an action potential can be triggered more easily. (The explanation for this phenomenon would take too long to be given here.)

In Chapter 2 I outlined the various supportive functions of astrocytes and mentioned that one of them is to aid in the regulation of ionic concentrations of the extracellular fluid. Studies have shown that glia will temporarily take up the excess K^+ ions that are released into the extracellular fluid, and thus prevent the axons from becoming too excitable. In fact, there appears to be a $Na^+–K^+$ pump in the membrane of glial cells whose activity depends upon the amount of K^+ in the extracellular fluid. If the concentration of extracellular K^+ increases, the pump works harder, storing the excess K^+ inside the glial cells (Henn, Haljämae, and Hamberger, 1972).

The importance of the electrolyte-regulating function of glia is shown in the disease called epilepsy. This disorder is characterized by periodic bouts of uncontrolled firing of neurons, especially in the cerebral cortex (a layer of neural tissue covering the largest portion of the brain). It is thought, by some investigators, that epilepsy results from a deficient potassium uptake mechanism in glial cells. The increased extracellular potassium increases the excitability of cerebral neurons (Tower, 1960). The disorder is treated with anticonvulsant drugs that depress neural conduction, probably by interfering with sodium conductivity of the membrane.

In this chapter, we have seen how the membrane produces a resting potential, and how changes in its Na$^+$ and K$^+$ permeability result in an action potential. Finally, we have seen how the axon transmits this action potential down to the end. In the next chapter we will see how the neuron decides whether to "fire" the axon, and we will investigate the nature of the message transmitted by terminal buttons at the synapse.

KEY TERMS

SUGGESTED READINGS

ECCLES, J. C. *The Understanding of the Brain.* New York: McGraw-Hill, 1973.

KUFFLER, S. W. and NICHOLLS, J. G. *From Neuron to Brain: A Cellular Approach to the Function of the Nervous System.* Sunderland, Mass.: Sinauer Associates, 1976. The book by Kuffler and Nicholls is probably the single best source for further information about material covered in Chapters 2–4. The book by Eccles is also excellent, but is less detailed.

4

Neural Communication and the Decision-Making Process

In Chapter 3 we learned how the characteristics of the axonal membrane result in a resting membrane potential and how a small decrease in this potential results in an action potential that is propagated down the axon. In this chapter we shall study (1) *synaptic transmission*, the way in which terminal buttons send their messages across the synapse to the receiving cells; (2) *postsynaptic potentials*, the response of the receiving cells to the transmitter substance that is released by the terminal buttons of the transmitting cells; and (3) the process of *integration*, by which a neuron decides whether or not to send an action potential down its axon. We shall also see how (4) *presynaptic inhibition* reduces the amount of transmitter substance that is released by a terminal button, and hence diminishes the message sent to the receiving cell.

SYNAPTIC TRANSMISSION

As we learned in Chapter 2, neurons "talk to" other neurons by means of synapses, and the medium used for these one-sided conversations is the diffusion of transmitter substance across the gap between the terminal button and the membrane of the receiving cell. When a terminal button "talks," it causes a brief alteration in the membrane of the receiving cell. The net effect of these disturbances in the resting potential, occurring at the many synapses on the receiving cell, is what determines whether the cell fires its axon and talks to *its* receiving cells.

The topic of synaptic transmission will be divided into two parts: a description of the structure of synapses, and a discussion of the production, release, and deactivation of transmitter substances.

Structure of Synapses

Synapses are junctions between the terminal buttons at the ends of the axonal branches of one cell and (usually) somatic or dendritic membrane of another. Since a message is transmitted only one way, it makes sense to use different terms in referring to the two membranes on each side of the synapse; the membrane of the transmitting cell is the *presynaptic membrane,* and that of the receiving cell is the *postsynaptic membrane.* As shown in Figure 4.1, these membranes are separated by a small gap, variable from synapse to synapse, but usually around 200 Å wide. (See **FIGURE 4.1.**)

The space separating the presynaptic and postsynaptic membranes is referred to as the *synaptic cleft.* This region contains extracellular fluid, through which the transmitter substance diffuses. For many years it was thought that the synaptic cleft was an open, fluid-filled space, but in recent years some photographs have shown the presence of numerous filaments joining the presynaptic and postsynaptic membranes. The function of this so-called *synaptic web* is not clear. It does appear to hold the presynaptic and postsynaptic membranes together. Brain tissue can be broken apart and homogenized (run through a blender), and synaptic knobs can be separated out by means of differential centrifugation. (This process allows us to isolate particles on the basis of their specific gravity; if a container filled with a suspension of heavy and light particles is spun in a centrifuge, the heavy particles will migrate toward the bottom. The same phenomenon drives the mercury to the bottom of a clinical thermometer as you "shake it," or swing it through an arc.) The terminal buttons of brain tissue subjected to such treatment appear in *synaptosomes* ("synapse bodies"), which usually include a piece of the postsynaptic membrane, suggesting that the synaptic web holds the membranes together. There is no evidence, however, that

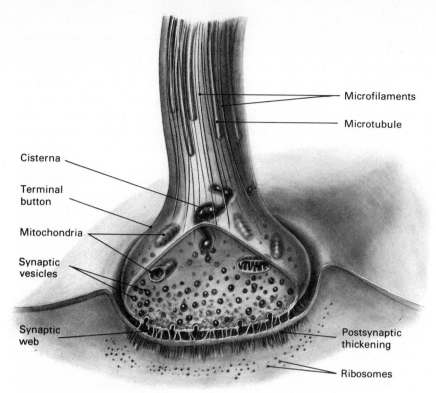

Microfilaments

Microtubule

Cisterna

Terminal
button

Mitochondria

Synaptic
vesicles

Synaptic
web

Postsynaptic
thickening

Ribosomes

FIGURE 4.1   Details of a synapse.

the synaptic web plays any role in the process of synaptic transmission.

As you may have noticed in Figure 4.1, there are two prominent structures in the cytoplasm of the terminal button—mitochondria and **synaptic vesicles.** You will recall from Chapter 2 that many of the biochemical steps in the extraction of energy from glucose are located on the cristae of the mitochondria; hence, their presence near the terminal button suggests the presence of processes there that require energy.

Synaptic vesicles are small rounded objects; some are spheres and others are oval. They appear to be packages (the term *vesicle* means "little bladder") that contain transmitter substance. They are produced locally, in the cisternae. (See **FIGURE 4.1.**) Actually, as we shall see a little later, the cisternae are recycling plants rather than manufacturing plants. These structures do not make synaptic vesicles out of raw materials, but out of the membrane of old vesicles that have expelled their contents into the synaptic cleft.

Synaptic vesicles are also produced in the soma of the neuron. If a nerve is *ligated* (that just means "tied") with a bit of thread and later examined under the electron microscope, a collection of vesicles—along with a lot of other material—will be found on the **proximal** (closer to soma) side of the obstruction. These vesicles are presumably produced in the Golgi apparatus, the packaging plant of the cell, and are sent down to the terminals via **axoplasmic transport,** an active process involving movement of substances along microtubules that run the length of the axon.

Ribosomes are found around the cytoplasm of the receiving cell, near the postsynaptic membrane, but they are not found in the terminal buttons. Since proteins are made at the ribosomes, the fact that ribosomes are found near the postsynaptic membrane implies that protein synthesis is important for some aspect of the process of receiving messages from terminal buttons.

The microtubules and microfilaments that are found running down the length of the axon also travel through the terminal buttons. Just as axoplasmic transport is accomplished by these thin fibers, so they are thought to transport substances within the terminal buttons.

The postsynaptic membrane under the terminal button is somewhat thicker than the membrane elsewhere. Some synapses are characterized by thicker postsynaptic membranes than others, but it is not known whether these differences have any functional significance.

Mechanisms of Synaptic Transmission

Transmitter substance is produced in the terminal buttons. Raw materials are sent down the axoplasm from the soma and are used to make transmitter substance. These materials may also be recycled; the breakdown products of the transmitter substance usually reenter the terminal button to be used again. In many cases, as we shall see, the transmitter substance itself is retrieved intact and used again.

RELEASE OF TRANSMITTER SUBSTANCE.   The release of transmitter substance, and the recycling of the container, is a fascinating process. When the axon fires (propagates an action potential), a number of synaptic vesicles migrate to the presynaptic membrane, adhere to it, and then rupture, spilling their contents into the synaptic cleft. (See **FIGURE 4.2.**)

The evidence that synaptic vesicles contain transmitter substance and that they spill their contents into the synaptic cleft is quite strong. If an isolated frog nerve-muscle pair is poisoned with venom of the black widow spider, the muscle will contract vigorously for several minutes because the venom causes release of transmitter substance from the terminal buttons that synapse on the muscle fiber. (These synapses are actually called **neuromuscular junctions.**) If the terminal buttons are then examined under the microscope, synaptic vesicles are found to be lacking (Clark, Hurlbut, and Mauro, 1972). The terminal buttons also appear to be larger, as though ad-

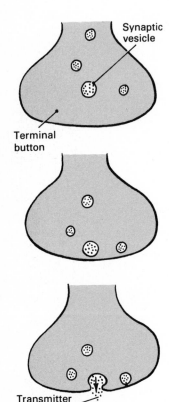

Synaptic vesicle

Terminal button

Transmitter substance

FIGURE 4.2   The process by which synaptic vesicles release the transmitter substance into the synaptic cleft.

ditions (from the expended vesicles) had been made to the membrane.

Even more direct evidence concerning the release of transmitter substance has been obtained. The release of transmitter substance is a very rapid event, taking only a few milliseconds to occur. A clever experiment by Heuser and his colleagues (Heuser, 1977; Heuser, Reese, Dennis, Jan, Jan, and Evans, 1979) was able to stop the action so that the details could be studied. The nerve attached to an isolated frog muscle was electrically stimulated, and then the muscle was dropped against a block of pure copper that had been cooled to 4° K (approximately −453° F). The outer layer of tissue became frozen in 2 milliseconds or less. The ice held the components of the terminal buttons in place until they could be chemically stabilized and examined with an electron microscope.

Figure 4.3 shows the vesicles in cross-section; note the vesicles that appear to be fused with the presynaptic membrane. (See **FIGURE 4.3, TOP,** on the following page.) The lower two pictures show an end view of the surface of the terminal button, as the postsynaptic membrane would see it. The middle picture shows a presynaptic membrane that was frozen 3 msec after the nerve was shocked, which is too soon for the synaptic vesicles to burst open. (See **FIGURE 4.3, MIDDLE.**) The bottom picture shows a presynaptic membrane that was frozen 5 msec after stimulation; note the vesicles, caught in the act of emptying their contents into the synaptic cleft. (See **FIGURE 4.3, BOTTOM.**)

RECYCLING OF THE VESICULAR MEMBRANE. A study by Heuser and Reese (1973) has outlined the means by which the membrane used in synaptic vesicles appears to be recycled. As the synaptic vesicles fuse with the presynaptic membrane and burst open, the terminal button gets larger. At the point of junction between the axon and the terminal button, little buds of membrane pinch off into the cytoplasm (a process called *pinocytosis*) and migrate to the cisternae, where they fuse into a large, irregularly shaped structure. Then, pieces of membrane are broken off the cisternae and get filled with molecules of transmitter substance. (See **FIGURE 4.4,** on page 65.)

Heuser and Reese performed an experiment using *horseradish peroxidase.* This substance, with such an improbable name, is very useful to neurochemists and neuroanatomists. It can be placed somewhere in the body and found later when the tissue of the animal is examined microscopically—there are specific stains that indicate its presence and whereabouts. Heuser and Reese placed a frog muscle in a solution of horseradish peroxidase, and the nerve that was attached to it was electrically stimulated, causing the axons to fire. Shortly thereafter, they removed the muscle, stained it, and looked for the presence of horseradish peroxidase in the terminal buttons that synapse on the muscle. They found the peroxidase inside small

FIGURE 4.3 Photomicrographs of the fusion of synaptic vesicles with the membrane of terminal buttons that synapse with frog muscle. *Top*: A cross-section through a terminal button. *Middle and bottom*: Views of the surface of terminal buttons. The dimples are synaptic vesicles that are in the process of fusing with the membrane (see text). (From Heuser, J. E. In *Society for Neuroscience Symposia, Volume II*, edited by W. M. Cowan and J. A. Ferrendelli. Bethesda, Md.: Society for Neuroscience, 1977; and in Heuser, J. E., Reese, T. S., Dennis, M. J., Jan, Y., Jan, L., and Evans, L. *Journal of Cell Biology*, 1979, *81*, 275–300.)

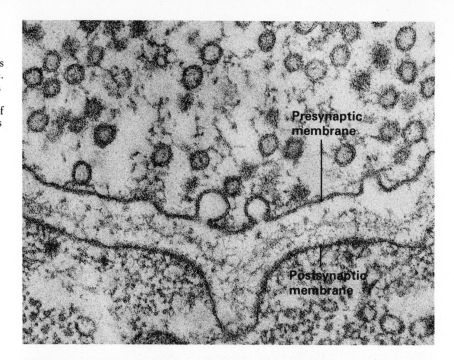

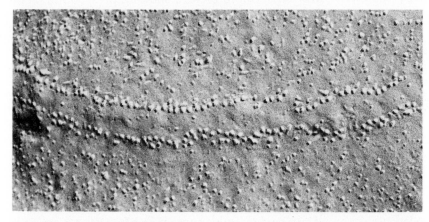

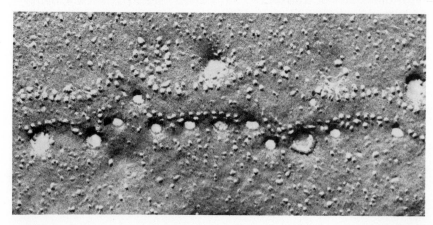

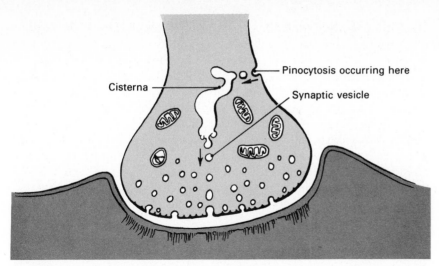

Cisterna

Pinocytosis occurring here

Synaptic vesicle

FIGURE 4.4   Recycling of the vesicular membrane.

vesicles in the cytoplasm near the junction of the axon and terminal
button. If they waited a longer time before they killed and stained
the tissue, they found the peroxidase in the cisternae. If they waited
even longer, they found it in the synaptic vesicles themselves.

We have seen the cycle followed by the synaptic vesicles, but
what force guides and moves them around? It has been found that
the membrane of synaptic vesicles is coated with a protein called
stenin, and the presynaptic membrane with a protein called **neurin.**
The microtubules that run through the cytoplasm are also coated
with a substance like neurin, and they appear to be able to interact
with the stenin on the membrane of the vesicles. It has been suggested by Berl, Puszkin, and Nicklas (1973) that an interaction between the microtubules and the stenin that coats the vesicles causes
a ratchetlike action that propels the vesicles to the presynaptic membrane.

Once the vesicles are attached to the presynaptic membrane, the
neurin on this membrane and the stenin on the vesicles interact in
a similar way, rupturing the vesicles. Neurin and stenin appear to
be very similar to actin and myosin, two proteins that are found in
muscle. These substances interact in a ratchetlike arrangement, and
provide the motive force of muscles.

As we shall see in Chapter 10, actin and myosin are stimulated
to exert the force in a muscular contraction by the entry of calcium
into the cell. Similarly, the synaptic vesicles migrate to the presynaptic membrane and rupture only after calcium enters the terminals.
(If the extracellular fluid is artificially depleted of calcium, the synapses can no longer function.) The entry of calcium into the terminal
buttons (perhaps by means of temporary opening of "calcium pores")
appears to be the event that initiates synaptic release of transmitter
substance (Miledi, 1973; Blaustein, Kendrick, Fried, and Ratzlaff,
1977).

Although the vescicular hypothesis of neurotransmitter release is accepted by almost all neuroscientists, there is some evidence that suggests that transmitter substances can sometimes be released directly from the cytoplasm of terminal buttons. In fact, some investigators argue that release of neurotransmitter through vescicular migration occurs only after the cytoplasmic pool of transmitter substance is depleted. The issue is very controversial, as you may see if you examine the April, May, and November, 1979, issues of *Trends in NeuroSciences*. If you think that scientists are impartial and dispassionate about interpretations of research findings, you will be disabused of this notion as you read the various articles and letters to the editor that are contained in the issues; the debate is lively, indeed.

TERMINATION OF THE POSTSYNAPTIC POTENTIAL We have examined the production and release of transmitter substance. I have already mentioned that the postsynaptic membrane is depolarized or hyperpolarized (depending on the nature of the synapse) by the transmitter substance. However, the change in the membrane potential is brief. If the presence of transmitter substance alters the membrane potential, what restores the situation to normal so quickly?

There are two answers, because there are two processes, *deactivation* of the transmitter substance and its *re-uptake* by the terminal button. Deactivation is accomplished by an enzyme that destroys the transmitter molecule. For example, transmission at the neuromuscular junction and at some neuron-neuron synapses is mediated by a chemical called **acetylcholine** (abbreviated ACh). Release of this chemical into the synaptic cleft causes a change in the postsynaptic membrane potential. This change is short-lived because the postsynaptic membrane contains an enzyme called **acetylcholinesterase** (mercifully referred to as AChE). AChE destroys ACh by cleaving it into *choline* and the *acetate* ion. Neither of these substances affects the electrical potential of the postsynaptic membrane, so as the ACh molecules are destroyed, the resting potential is restored. Acetylcholinesterase is an extremely energetic ACh destroyer; one molecule of AChE will chop apart more than 64,000 molecules of acetylcholine each second!

Re-uptake is the other process that keeps the effect of the transmitter substance brief. (Re-uptake seems to me to be redundant—uptake would say it just as well—but re-uptake has become the standard term.) This process is nothing more than an extremely rapid removal of transmitter substance from the synaptic cleft by the terminal button. Exactly how the transmitter substance gets back we do not know. The transmitter substance does *not* return in the vesicles that get pinched off the membrane at the shoulder of the terminal button. The mechanism that retrieves the transmitter substance is probably similar to the sodium-potassium pump.

The postsynaptic membrane of synapses using this method need not contain an enzyme to inactivate the transmitter; the synaptic vesicles of the terminal button squirt a little out, and the presynaptic membrane takes it back in, giving the postsynaptic membrane only a brief exposure to the transmitter substance.

POSTSYNAPTIC POTENTIALS

The brief depolarization or hyperpolarization produced in the postsynaptic membrane by the action of the transmitter substance is called a postsynaptic potential. Since depolarization brings the membrane of the cell closer to the threshold of excitation (the point at which an action potential will be elicited in the axon), and thus raises the probability of axonal firing, depolarizing effects of synaptic transmission are referred to as *excitatory postsynaptic potentials* (*EPSPs*). Hyperpolarizing postsynaptic potentials move the membrane potential away from the threshold of excitation and are hence referred to as *inhibitory postsynaptic potentials* (*IPSPs*). In a later section (Integration) we shall see how these effects, occurring at synapses on the somatic and dendritic membrane, determine whether or not the neuron fires, but in this section we shall concern ourselves with the mechanism whereby postsynaptic potentials are produced.

Transmitter substance has an effect on the membrane potential only at the synapses; if a little bit is injected into the extracellular fluid on a synapse-free region of a neuron, no postsynaptic potential is recorded. It appears that there must be special molecules attached to the postsynaptic membrane. These molecules (*receptor sites*) and the transmitter substance have an affinity for each other; when a molecule of the transmitter substance diffuses across the synaptic cleft, it meets one of these specialized molecules and attaches to it. Once it does, changes in permeability of the postsynaptic membranes are somehow initiated, and the result is a flow of ions through the membrane. This process is schematized in Figure 4.5. It is as if molecules of the transmitter substance act as keys to unlock doors covering pores in the membrane. (See **FIGURE 4.5**.)

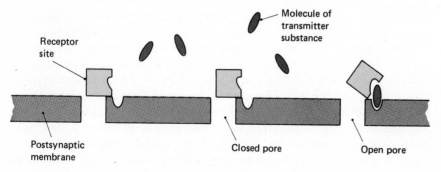

FIGURE 4.5 A possible way in which molecules of transmitter substance can alter the permeability of the postsynaptic membrane. (Adapted from Eccles, J. C., *The Understanding of the Brain*. New York: McGraw-Hill, 1973.)

The illustration does not pretend to show how the mechanism actually works, since that is not known. It just serves as an easy way to picture how transmitter substances might increase membrane permeability.

Since the postsynaptic potentials can be either depolarizing (excitatory) or hyperpolarizing (inhibitory), the alterations in membrane permeability must be specific to particular species of ions. Let's review the ionic balance across the membrane before we examine the nature of these changes in permeability.

There is a large electrostatic and osmotic gradient for sodium; this ion would rush in, depolarizing the membrane, if it were allowed to. There is a small osmotic gradient for potassium; it would flow out (hyperpolarizing the membrane) if it could. Osmotic and electrostatic pressures balance very nicely for the chloride ion. The membrane provides a bit of resistance to the flow of Cl^- ions, but since there is no chloride pump, the distribution of chloride is able to balance itself across the membrane. A decrease in the membrane resistance to chloride alone, then, would have no effect on the resting membrane potential. The protein anions within the cytoplasm are enormous in size, relative to the potassium, chloride, and sodium ions; there are no pores in the membrane wide enough to allow these ions to leak out.

Ionic Movements during EPSPs

Let us consider EPSPs first. They are depolarizations, so a positive ion must leak in, or perhaps a negative ion leaks out. (See **FIGURE 4.6.**) The only negative ion that could leak out would be chloride, but since the distribution of chloride is balanced at the resting potential, we have to reject this possibility; therefore, positive ions must leak in. The most likely candidate is sodium. If the membrane resistance to sodium were to drop, Na^+ would rush in (this happens, of course, in the action potential) and depolarization would occur. Experiments have shown this to be the case.

In fact, experiments have shown that the pores that open up in response to liberation of an excitatory transmitter substance allow two ions to flow through the membrane—sodium and potassium. Sodium is, of course, the most important one; an influx of Na^+ depolarizes the cell. As this occurs, however, it drives some K^+ out. Without this efflux of K^+ the EPSP would be even larger. One would think that if the pores were large enough to admit sodium, certainly chloride ions should get through also; after all, Cl^- ions are even a bit smaller than K^+ ions. However, it has been shown experimentally that chloride ions do not cross the membrane during the EPSP.

The most likely description of the pores is as follows: they are large enough to admit Na^+ and K^+ ions, but the walls of the pore contain a negative charge that repels the Cl^- ion. These pores are

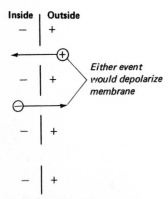

Inside | Outside

Either event would depolarize membrane

FIGURE 4.6   The ionic movements could depolarize the membrane.

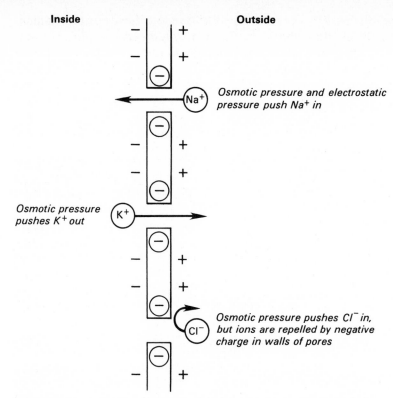

Inside   **Outside**

Osmotic pressure and electrostatic pressure push Na+ in

Osmotic pressure pushes K+ out

Osmotic pressure pushes Cl⁻ in, but ions are repelled by negative charge in walls of pores

FIGURE 4.7   An EPSP results from an influx of sodium and an efflux of potassium.

shown schematically in Figure 4.7. Sodium ions are shown entering the cell through some pores, while potassium ions leave via some others. Chloride ions are repelled by the negative charge and hence cannot enter the channels. (See **FIGURE 4.7.**)

Ionic Movements during IPSPs

IPSPs are easier to describe. These potentials result from the temporary opening of pores large enough to admit potassium or chloride ions, but still too small to admit sodium ions. When these pores are opened by molecules of the transmitter substance, K^+ ions leave the cell, down their concentration gradient. This migration of ions hyperpolarizes the membrane potential and begins to attract some Cl^- ions out of the cell. If it were not for the mobility of these chloride ions through the membrane, the IPSP would be even larger than is actually observed.

    Thus, to account for the hyperpolarizing effects of the inhibitory transmitter substance, we only have to assume that the pores that open up are just large enough to admit potassium and chloride ions, but not quite large enough to let sodium ions get through.

INTEGRATION

The only reason for studying postsynaptic potentials in such detail is that these potentials determine whether or not a cell sends a message down its axon. To illustrate integration, or the decision-making process of neurons, we can perform an experiment similar to one we did in Chapter 3.

A giant squid axon is isolated in a dish of seawater and a recording electrode is placed on the membrane. Electrodes from two shockers are attached to the end of the axon; one delivers brief positive shocks to the outside of the membrane (hyperpolarizations) and the other delivers brief negative shocks (depolarizations). (See **FIGURE 4.8**.)

We know from Chapter 3 that if the depolarizing shocks cause the membrane of the axon to reach the threshold of excitation, an action potential will be elicited. Subthreshold depolarizations, and all hyperpolarizations, will produce **graded potentials** that will be propagated down the axon according to its passive cable properties. (This sort of transmission is usually referred to as **decremental conduction,** as opposed to the nondecremental conduction of the action potential, which is transmitted without a decrease in size.)

The depolarizing shocker (the excitatory shocker) is set at a level

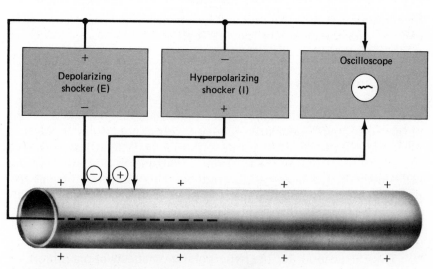

FIGURE 4.8 A schematic representation of the means by which an axon can be stimulated with both negative and positive shocks.

just barely sufficient to trigger an action potential. We will find that, if we deliver a pulse from the hyperpolarizing shocker (inhibitory shocker) just before delivering an excitatory shock, we prevent the occurrence of the action potential. (That is to say, we *inhibit* the action potential.) The inhibitory pulse increases the membrane potential, so that the effects of the excitatory pulse do not make it to the threshold of excitation. (See **FIGURE 4.9.**) Thus we see how excitatory and inhibitory changes in the membrane potential can cancel each other.

Let us examine another phenomenon. This time we administer a small, subthreshold, excitatory pulse. Before the effects of this one are over, we shock the membrane again, and again, until eventually the axon fires. (See **FIGURE 4.10.**) The effects of closely spaced subthreshold excitations are cumulative; a number of small excitations can achieve the same result as a single large one. This process is called **temporal summation,** that is, the addition of the effects of small excitations across time.

Let us demonstrate one more phenomenon. If we simultaneously deliver excitatory shocks from two electrodes placed on the axonal membrane, the effects of these shocks are added together. Therefore, two subthreshold shocks can summate and trigger an action potential. We call the addition of potential changes from shocks applied to various parts of the membrane **spatial summation.** (See **FIGURE 4.11,** on the following page.)

Of course, there are no electrodes on the axons of neurons in the intact animal. There are, instead, terminal buttons arranged on the somatic and dendritic membrane of these cells, capable of producing EPSPs and IPSPs. These postsynaptic potentials summate, in a process referred to as **integration.** (Integration simply means addition.)

The membrane of dendrites and soma is not capable of producing an action potential; only an axon can do that. The membrane potential at the **axon hillock** is what determines whether the axon fires. This region, the junction between soma and axon, is capable of producing an action potential, and if it fires, an impulse is propagated down the axon. EPSPs and IPSPs occurring at the synapses on the soma and dendrites are transmitted decrementally, according to the passive cable properties of the neural membrane. These postsynaptic potentials are integrated, and whenever the net result (at the axon hillock) exceeds the threshold of excitation, the axon fires. If many excitatory synapses on the cell fire at a high rate, the axon of the cell will also fire at a high rate. If inhibitory synapses now begin firing, there will be a fall in the rate of production of action potentials. The rate of a cell's firing, then, depends on the relative numbers of EPSPs and IPSPs occurring on its dendritic and somatic membrane. In Chapter 10 we shall see how choices are made by the process of integration. Conflicts between very simple competing behaviors (reflexes) are resolved in favor of the one that is most able to excite the appropriate motor neurons and most able to inhibit the ones that produce the competing behaviors.

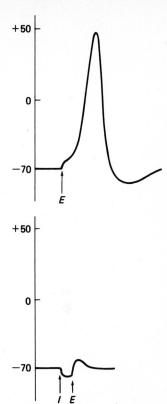

FIGURE 4.9 A depolarizing shock that normally produces an action potential can be inhibited by a previously produced hyperpolarizing shock.

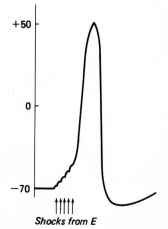

FIGURE 4.10 Temporal summation can occur if subthreshold depolarizations are presented in rapid succession.

Excitatory shocks
administered simultaneously

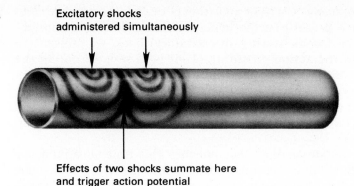

Effects of two shocks summate here
and trigger action potential

FIGURE 4.11   Spatial summation can occur if subthreshold
depolarizations are presented simultaneously at more than one location.

PRESYNAPTIC INHIBITION

So far I have been discussing only postsynaptic excitation or inhibition, occurring at **axosomatic** or **axodendritic** synapses. There is
another kind of synapse (**axoaxonic**), which consists of a junction
between a terminal button and an axon of another cell. I have left
discussion of this type of synapse until last, because it does not contribute to a process of integration, as do the axosomatic and axodendritic synapses. Instead, axoaxonic synapses alter the amount of
transmitter substance liberated by the terminal buttons of the axon
they synapse with (but only if that axon itself fires).

The release of transmitter substance by a terminal button is
initiated by an action potential. Experiments have shown that the
amount released (and hence the size of the resulting PSP) depends
on the magnitude of the change in the membrane potential. Normally this is a constant amount: from -70 mV to $+50$ mV, or a total
change of 120 mV. Furthermore, the transition in potential must be
rapid. A slow change in the membrane potential will not release
transmitter substance.

Here, then, is how presynaptic inhibition works. The axon of cell
A fires, and its terminal button liberates transmitter substance,
which depolarizes the membrane of axon B. (See **FIGURE 4.12**.)
Please note that I said it produces depolarization, which at axosomatic or axodendritic synapses would be called an EPSP. This depolarization does not bring the membrane of axon B to its threshold
of excitation and is therefore conducted to the terminal button via
passive cable properties of the axon. Since the potential change is
relatively slow no transmitter substance is released by the terminal
button of axon B. If an action potential is now sent down axon B,
the membrane potential will quickly shoot up to $+50$ mV and then

Somatic membrane

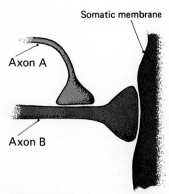

Axon A

Axon B

FIGURE 4.12   An axoaxonic
synapse.

drop back down. Since transmitter substance is liberated in proportion to the change in the membrane potential, less than the normal amount will be released. The action potential is occurring in a less-than-normally polarized terminal, and thus the total change in the membrane potential will be smaller than usual. (See **FIGURE 4.13.**)

Axoaxonic synapses can obviously have an effect only if the axon being inhibited is actually firing. These synapses might be useful in the following way: presynaptic inhibition on a particular input to a cell will reduce the responsiveness of that cell to only that kind of input, but the cell will respond just as readily to its other inputs. (See **FIGURE 4.14.**)

This chapter has described how cells communicate and how their one-way conversations determine the rate of axonal firing of the receiving cell. In the next chapter we shall examine the nature of the various transmitter substances and the effects of various drugs and biochemical agents on the functions of axonal transmission and synaptic communication.

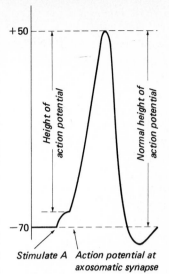

Stimulate A   Action potential at axosomatic synapse

FIGURE 4.13   Presynaptic inhibition. The action potential is recorded from the terminal button of axon B, shown in Figure 4.12. The small depolarization that precedes the action potential is produced by the activity of axon A.

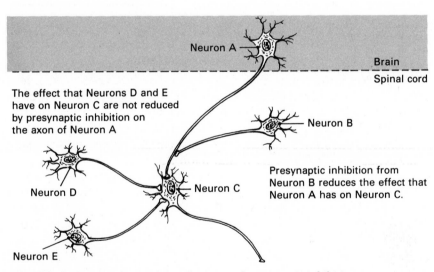

FIGURE 4.14   An example of the advantage of presynaptic inhibition.

KEY TERMS

acetylcholine (*a set ul KO leen*) p. 66
acetylcholinesterase (*a set ul ko lin ES ter ace*)
   p. 66

axon hillock p. 71
axoaxonic synapse p. 72
axodendritic synapse p. 72

SUGGESTED READINGS

DUNN, A. J., and BONDY, S. C. *Functional Chemistry of the Brain.* Flushing, N.Y.: SP Books, 1974. An excellent source of information about the mechanisms of synaptic transmission: release of transmitter substance, recycling of the vesicular membrane, etc.

ECCLES, J. C. *The Understanding of the Brain.* New York: McGraw-Hill, 1973.

KUFFLER, S. W. and NICHOLLS, J. G. *From Neuron to Brain: A Cellular Approach to the Function of the Nervous System.* Sunderland, Mass.: Sinauer Associates, 1976. These two books have already been mentioned. The one by Kuffler and Nicholls is the more detailed of the two.

5

Biochemistry and Pharmacology of Synaptic Transmission

As we have seen in Chapters 3 and 4, the brain processes and transmits information through the movements of chemicals: synaptic transmission is accomplished by the extrusion of transmitter substance from terminal buttons and its acceptance by receptor sites. This chapter will describe in more detail the biochemical processes involved in the synthesis, storage, release, and deactivation of transmitter substances, and it will show how various drugs interact with these processes.

THE VARIETY OF TRANSMITTER SUBSTANCES

As we saw in Chapter 4, transmitter substances have two general effects on the postsynaptic membrane: depolarization (EPSP) or hyperpolarization (IPSP). One might suspect, then, that there would be two kinds of transmitter substances, inhibitory and excitatory. There are, instead, many more kinds of transmitter substances. Some are exclusively (so far as we know) excitatory or inhibitory, and some may produce either excitation or inhibition, depending on the nature of the postsynaptic receptor sites.

It is extremely difficult to prove that a given chemical serves as a transmitter substance, especially in the central nervous system. Table 5.1 provides a list of compounds that are suspected of being transmitter substances, along with their distribution in the nervous system and their hypothesized effects on postsynaptic cells. (See **TABLE 5.1.**)

It is important to note that *neural* inhibition (that is, an inhibi-

TABLE 5.1 Probable transmitter substances

Probable transmitter substance	*Location*	*Hypothesized Effect*
Acetylcholine (ACh)	Brain, spinal cord, autonomic ganglia, target organs of the parasympathetic nervous system	Excitation in brain and autonomic ganglia, excitation or inhibition in target organs
Norepinephrine (NE)	Brain, spinal cord (see Figure 15.12), target organs of sympathetic nervous system	Inhibition in brain, excitation or inhibition in target organs
Dopamine (DA)	Brain (see Figure 15.11)	Inhibition
Serotonin (5-hydroxytryptamine, or 5-HT)	Brain, spinal cord	Inhibition
Gamma-amino-butyric acid (GABA)	Brain (especially cerebral and cerebellar cortex), spinal cord	Inhibition
Glycine	Spinal cord interneurons	Inhibition
Glutamic acid	Brain, spinal sensory neurons	Excitation
Aspartic acid	Spinal cord interneurons, brain (?)	Excitation
Substance P	Brain, spinal sensory neurons (pain)	Excitation (and inhibition?)
Histamine, taurine, other amino acids, oxytocin, endogenous opiates, many others	Various regions of brain, spinal cord, and peripheral nervous system	(?)

tory postsynaptic potential) does not always produce *behavioral* inhibition. For example, let us suppose that a group of neurons inhibit a particular movement. If *these* neurons are now inhibited, they will no longer be able to suppress the behavior. The net effect of inhibition of the inhibitory neurons is excitatory on the behavior. Of course, the same goes for neural excitation, in just the opposite way. Neural *excitation* of neurons that *inhibit* a behavior will have a suppressive effect on that behavior.

One might reasonably wonder why nature saw fit to provide the nervous system with so many transmitters (and I might note that most neurochemists suspect that further research will greatly expand the list). There would appear to be at least three explanations for the diversity of transmitter substances: "historical accident," different types of postsynaptic potentials, and biochemical separation of functional systems.

1. *Historical accident.* Different functions of the nervous system developed at different times in our evolutionary history; earlier-evolving functions had a smaller set of biochemical mechanisms to choose from, whereas functions that developed later could make use of a wider range of mechanisms that had evolved for other purposes. For example, there are two distinct components of the mammalian visual system. One component is made up of small, unmyelinated axons. This system is presumably of earlier evolutionary origin, since it anatomically resembles the visual system of lower vertebrates, and it is involved more with visual reflexes than with visual discrimination. The "newer" component consists of larger, myelinated fibers. Presumably, the older component evolved before the evolutionary development of large myelinated axons; functional systems were constructed out of the building blocks that were available at the time.

   Similarly, we may make some guesses concerning the evolutionary "age" of various transmitter substances. **Glutamic acid** and **gamma-amino butyric acid (GABA)** are thought to be the oldest. They have direct excitatory (glutamic acid) and inhibitory (GABA) effects on axons, which suggests that they had a general modulating role even before the evolutionary development of specific receptor molecules. Cells have evolved a biochemical pathway called the *GABA shunt*. In the extraction of energy from glucose, the cell performs a complex series of biochemical reactions constituting the **Krebs citric acid cycle.** The GABA shunt consists of a series of reactions that take a metabolic "shortcut" across this cycle and, in so doing, extract less energy from glucose. However, these reactions produce the compounds glutamic acid and GABA, which might then be used as neurotransmitters.

   I should point out that we actually do not—and cannot—know when the various transmitter substances were evolved, not having been present at the time. It is interesting to make these speculations, however.

2. *Different types of postsynaptic potentials.* I have not discussed the time course of postsynaptic potentials, but have only described the direction of the change in the electrical charge across the neural membrane. Postsynaptic potentials can vary in duration from a few milliseconds up to a second or more. The duration of a PSP is, of course, a function of many things: amount of transmitter substance liberated, type and number of receptor sites, amount of deactivating enzyme present at the synapse, and rate of re-uptake at the terminal button, to name a few. Studies have shown that the time course of alteration of the postsynaptic membrane potential is also a function of the transmitter substance; some generally produce short-lived PSPs, whereas others produce a more prolonged effect. The nervous system undoubtedly uses these fast-acting and slow-acting synapses in different ways.

It has also been suggested that, besides triggering the PSPs themselves, transmitter substances might produce biochemical changes in the postsynaptic cells, causing long-term alterations in cell functioning. This is a very plausible hypothesis. As we shall see, some transmitter substances do more than cause conformational changes in the membrane that alter its permeability to the various ions. Some transmitter substances produce their effects by triggering biochemical reactions within the cytoplasm of the postsynaptic cell. The biochemical changes within the cytoplasm might very well produce more (and longer-term) effects besides alterations in membrane permeability. As you might imagine, much attention has been paid to the possibility that these hypothetical changes are involved in the memory process. This hypothesis will be discussed in Chapter 18.

3. *Biochemical separation of functional systems.* To a certain extent, particular behaviors (like sleep, sexual behavior, and feeding) are facilitated or inhibited by neurons that use particular transmitter substances. This means that the behaviors can be inhibited or facilitated by **neuromodulators** that interact with the neurons that control them. Some neuromodulators (including hormones such as testosterone and estrogen) are produced outside the nervous system. They are brought to the CNS in the blood. Other neuromodulators are produced by neurosecretory cells within the brain itself. A good example is the set of opiate-like substances that the brain produces. These substances (called endogenous opiates or **endorphins**) interact with various neural systems and diminish the perception of pain. Opiates like morphine and heroin produce their effects because they stimulate neural receptors for the endorphins.

Some neuromodulators stimulate specialized receptors that are not associated with synapses. Others appear to diffuse into the synaptic cleft, where they interact with the postsynaptic receptors. The biochemical separation of functional systems makes it possible for neuromodulators of this second type to affect a particular brain function. For example, suppose that a particular function (sleep, let's say) is regulated by a system of neurons that use a particular transmitter substance. A neuromodulator that stimulates the postsynaptic receptors for this transmitter substance can facilitate or inhibit these neurons, and thus play a part in regulating sleep. The reason that some drugs can affect behaviors in particular ways is probably that they facilitate or inhibit one or more of these functional systems.

PHARMACOLOGY OF SYNAPSES

Before I launch into a discussion of the drugs that affect the various kinds of synapses, it would be worthwhile to discuss the general ways in which drugs can facilitate or interfere with synaptic transmission. Drugs that block or inhibit the postsynaptic effects are called **antagonists;** drugs that facilitate them are called **agonists.** (The Greek word *agon* means "contest.") Figure 5.1 illustrates the ways in which drugs can act as agonists or antagonists. (Glance at **FIGURE 5.1.**)

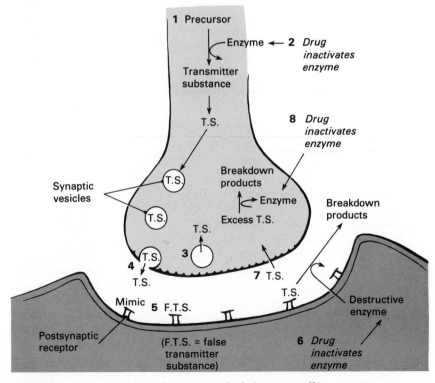

FIGURE 5.1   An overview of the ways in which drugs can affect synaptic transmission.

First, the transmitter substance must be synthesized from its precursors. The synthesis usually takes several steps, and usually occurs in the cell body. The transmitter substance is then placed in vesicles that are transported down the axon to the terminals. In some cases, additional amounts of transmitter substance are resynthesized out of breakdown products in the terminal buttons. Sometimes, the amount of transmitter substance that is produced is limited by the amount of precursor that is available, and thus, administration of the precursor results in increased production of the transmitter substance. In these cases, artificially-administered precursor serves as an agonistic drug. (See step 1, **FIGURE 5.1.**)

The various steps of the synthesis of transmitter substances are controlled by *enzymes.* Enzymes consist of special protein molecules that act as catalysts; that is, they cause a chemical reaction to take place without becoming a part of the final product themselves (note that all enzymes' names end in -*ase*). Since cells contain the constituents that are needed to synthesize nearly anything, the compounds they actually do produce depend mainly upon the particular enzymes that are present. Furthermore, there are enzymes that break molecules apart as well as put them together, so the enzymes present in a particular region of a cell determine which molecules are permitted to remain intact. In the case of the reversible reaction below, the relative concentrations of enzymes X and Y determine whether the complex substance AB, or its constituents, will predominate in the cell. Enzyme X makes A and B join together, while enzyme Y splits AB apart. (Energy may also be required to make the reactions proceed.)

$$A + B \underset{Y}{\overset{X}{\rightleftharpoons}} AB$$

The synthesis of a transmitter substance is thus controlled by the appropriate enzymes. Therefore, if a drug inactivates one of these enzymes, it will prevent the transmitter substance from being produced. This makes the drug an antagonist. (See step 2, **FIGURE 5.1.**)

Transmitter substances are stored in synaptic vesicles, which are transported to the presynaptic membrane, where the chemicals are released. Some drugs make the membrane of particular kinds of synaptic vesicles become "leaky." The transmitter substance escapes, and hence nothing is released when the vesicles rupture along the presynaptic membrane. These drugs serve as antagonists. (See step 3, **FIGURE 5.1.**)

Some drugs act as antagonists by preventing the release of transmitter substances from the terminal button. Others act as agonists by stimulating release. (See step 4, **FIGURE 5.1.**)

Once the transmitter substance is released, it must stimulate the postsynaptic receptors. Some drugs bind with and activate the receptors directly, mimicking the effects of the transmitter substance. (Naturally, this means that they are agonists.) Other drugs also bind with the postsynaptic receptors but do *not* activate them. Since they occupy the receptors, they prevent the transmitter substance from exerting its effect, and hence act as antagonists. These drugs are called **false transmitters.** (An analogy to their method of operation would be plugging a keyhole with putty, which would prevent the key from opening the lock. See step 5, **FIGURE 5.1.**)

The next step after stimulation of the postsynaptic receptors is enzymatic deactivation or re-uptake of the transmitter substance, which terminates the postsynaptic potential. First, let us consider enzymatic deactivation (such as the destruction of ACh by AChE). A drug that deactivates the destructive enzyme allows the transmitter substance to remain intact for a longer time. Thus, it serves as an agonist. (See step 6, **FIGURE 5.1.**) The postsynaptic activity of transmitter substances other than ACh is terminated by active uptake mechanisms in the terminal button. Drugs that block or retard the re-uptake process serve as agonists, since the transmitter substance remains on the synaptic cleft for a longer time, producing a prolonged postsynaptic potential. On the other hand, drugs that increase the rate of re-uptake serve as antagonists, since the transmitter substance is removed at a faster-than-normal rate. (See step 7, **FIGURE 5.1.**)

Besides being present in the postsynaptic membrane, destructive enzymes are also present within the cytoplasm of the terminal button. The apparent function of these enzymes is to destroy any extra transmitter substance that is produced. Drugs that inactivate these enzymes allow excess transmitter substance to be produced and stored, and hence more will be released when the axon fires. Thus, the drugs act as agonists. (See step 8, **FIGURE 5.1.**)

There is still another way that drugs can affect synaptic activity, but I shall have to postpone discussion of this effect until I present the role of the "second messenger" in neural transmission.

This section has covered a lot of information. Since the descriptions that follow are built on the general principles that have been outlined here, you might find it worthwhile to read this section again.

TRANSMITTER SUBSTANCES

Having examined some reasons put forth for the variety of transmitter substances that are present in the nervous system, and the ways that drugs can facilitate or impair synaptic transmission, let us turn our attention to the nature and mode of action of the transmitters themselves. We know more about **acetylcholine** and **norepinephrine** than any other transmitters because one or the other of these two substances occurs exclusively in various parts of the pe-

ripheral nervous system. The accessibility of these peripheral neurons, and the fact that they are segregated as to type of transmitter,
has made them the easiest to study. Conclusive evidence concerning
transmitter substances residing within the CNS is much more difficult to come by, since the neurons not only are harder to get at, but
they are also intermingled with neurons of other types.

Acetylcholine

DISTRIBUTION.   In vertebrates, acetylcholine (ACh) is the substance that is liberated at synapses on skeletal muscles (neuromuscular junctions). ACh is also the transmitter substance in the ganglia
of the autonomic nervous system. Ganglia are collections of cell bodies and nerve terminals that are located outside the CNS. The autonomic nervous system (described in more detail in Chapter 6) is the
part of the peripheral nervous system that is concerned with "vegetative functions"—control of the digestive system, heart rate, blood
pressure, etc. Axons from ***preganglionic neurons*** whose cell bodies
are located within the CNS enter the peripheral ganglia, where they
synapse with the ***postganglionic neurons,*** whose axons then proceed
to the target organ they innervate (heart, blood vessel, intestine,
sphincter, etc.). Acetylcholine is used to transmit EPSPs from preganglionic to postganglionic neurons. The postganglionic neurons
then affect their target organs by secreting either acetylcholine or
norephinephrine. Figure 5.2 schematizes this relationship. (See **FIG-
URE 5.2.**)

SYNTHESIS AND DEACTIVATION.   Acetylcholine is produced by
means of the following reaction:

$$\text{acetyl CoA} + \text{choline} \xrightarrow{\text{choline acetylase}} \text{ACh} + \text{CoA}$$

Coenzyme A (CoA) is a complex molecule, consisting in part of the
vitamin pantothenic acid. CoA is a ubiquitously useful substance,
taking part in many reactions in the body. All substances that enter
the Krebs citric acid cycle, where they are metabolized to provide
energy, are first converted to acetate and joined with CoA to form
acetyl CoA. Similarly, in the synthesis of acetylcholine, CoA acts as
a carrier for the acetate ion, which gets attached to choline to make
acetylcholine (ACh). Acetyl CoA, then, is coenzyme A with its attached acetate ion. Choline, a substance derived from the breakdown
of lipids (fatty substances), is taken into the neuron from general
circulation. In the presence of the enzyme ***choline acetylase*** the acetate ion is transferred from the CoA molecule to the choline molecule,
yielding a molecule of ACh and one of plain old CoA.
    When a substance is produced within a cell, its rate of synthesis
is limited by some factors that prevent an excessive amount of pro-

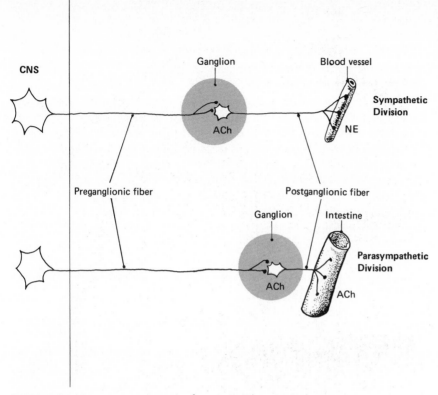

FIGURE 5.2   Neurotransmitters in the autonomic nervous system.

duction. In the terminal button of an acetylcholine-secreting neuron, there is a more-than-sufficient quantity of acetyl CoA to combine with available choline, and adequate supplies of choline acetylase are present to catalyze the reaction. The rate of reaction, then, is limited by the amount of available choline. This means that, should a molecule of choline come by, it will be grabbed and will get an acetate ion clapped onto it.

It makes a great deal of sense to make choline the rate-limiting substance in the synthesis of acetylcholine. When ACh is liberated into the synaptic cleft, it is quickly split apart and thus destroyed by acetylcholinesterase (AChE) of the postsynaptic cell, as we saw in Chapter 4. The choline, or most of it, is then taken up by the terminal button, to be used again in new molecules of ACh. An overabundance of acetyl CoA and choline acetylase means that the rate of ACh synthesis will depend mostly on the rate of the re-uptake of choline, subsequent to its use as part of a molecule of transmitter substance.

Sometimes excess ACh is produced—more than can be stored in the synaptic vesicles. For this reason, AChE is also present in the cytoplasm of the terminal buttons. This AChE cannot destroy the ACh that is stored in the vesicles, but it can—and does—destroy any transmitter substance produced by the cell that exceeds the storage capacity of the synaptic vesicles.

TYPES OF CHOLINERGIC RECEPTORS.   In general, acetylcholine is an excitatory transmitter substance, but it can also have inhibitory effects; the nature of the PSPs produced appear to depend upon the type of receptor molecule at the postsynaptic membrane. In the mammalian nervous system there appear to be two types of receptors (both mediating EPSPs), **muscarinic** and **nicotinic.** These receptors are so named because they are stimulated by the drugs *muscarine* or *nicotine*. The *cholinergic* receptors (those stimulated by ACh) of the autonomic ganglia and those of the neuromuscular junction are nicotinic. The cholinergic receptors of the target organs of the autonomic nervous system are muscarinic. Both types of cholinergic receptors are found in the CNS. The spinal cord contains nicotinic receptors. These receptors also predominate in the brain, but some muscarinic receptors are also found there. It is not known why two types of cholinergic receptors exist; perhaps they respond to different kinds of neuromodulators.

DRUGS THAT AFFECT CHOLINERGIC SYNAPSES.   In order to outline the various kinds of drugs that might affect cholinergic synapses, let us examine the steps in the production and use of ACh. As we have seen, ACh is produced from choline and acetate; its synthesis may be indirectly inhibited by the drug **hemicholinium,** which prevents the transport of choline across the membrane of the terminal button. Since ACh is resynthesized from the choline that is taken up by the terminals, administration of hemicholinium will cause the rapid depletion of ACh in the nerve terminals. This makes hemicholinium a potent cholinergic antagonist.

Once ACh is produced and stored in synaptic vesicles, it is normally released into the synaptic cleft upon stimulation of the presynaptic axon. The very poisonous drug **botulinum toxin** (present in improperly preserved food that is infected with the microorganism *Clostridium botulinum*) prevents release of ACh, and thus selectively shuts off cholinergic synapses. Venom of the black widow spider, on the other hand, causes the continuous release of ACh into the synaptic cleft, acting as a potent agonist.

Once ACh is liberated into the synaptic cleft, it must interact with functioning cholinergic receptors in order to produce a PSP. False transmitters can prevent ACh from activating the receptors; these substances occupy the receptor sites and prevent ACh from attaching to them, but they do not trigger the membrane events that normally produce PSPs. Since the false transmitters are not destroyed by AChE, they remain in the receptor sites for a long time. Muscarinic synapses are blocked by plant alkaloids such as **atropine** (named after Atropos, one of the three Fates of Greek mythology, who cut the thread of life). Nicotinic synapses are blocked by **d-**

tubocurarine, the active ingredient of **curare,** a potent paralytic agent discovered by South American Indians. This drug paralyzes skeletal muscles by blocking nicotinic receptors of the membrane of the muscle fibers. Curare does not affect the brain, since it cannot cross the blood-brain barrier. If artificial respiration is not provided, the fully conscious recipient of this drug dies of suffocation, not being able to breathe.

There are also excitatory compounds that mimic the effects of ACh by attaching to cholinergic receptors and inducing postsynaptic potentials; examples of this kind of drug, muscarine and nicotine, have already been given.

Finally, after triggering the postsynaptic events that produce a PSP, ACh is normally deactivated by AChE (acetylcholinesterase). AChE can be temporarily inhibited by drugs such as **eserine,** or permanently inhibited by diisopropylfluorophosphate (happily called **DFP**). These drugs, because they deactivate AChE, greatly potentiate the effects of cholinergic activity by prolonging the PSPs produced by ACh. Thus, they are very effective agonists. Many organophosphate insecticides work in this way; hence, the antidote for these poisons is atropine, which binds with and blocks cholinergic receptors, thus compensating for the excessively long PSPs produced by liberation of ACh at the synapse.

Figure 5.3 schematizes the sites of drug action at a cholinergic synapse. (See **FIGURE 5.3.**)

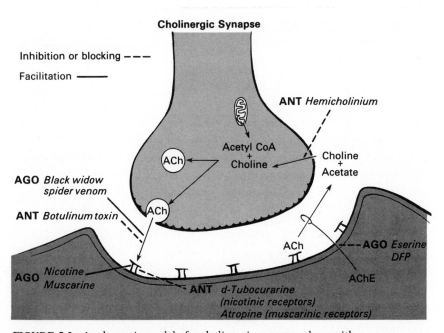

FIGURE 5.3   A schematic model of a cholinergic synapse, along with some drugs that can affect it.

Norepinephrine

Norepinephrine, dopamine, and serotonin are three neurotransmitters that belong to a family of compounds called *monoamines.* Because of the structural similarities of these substances, some drugs affect the activity of all of them, to some degree. The first two, norepinephrine and dopamine, belong to a subclass of monoamines called *catecholamines.* (See **TABLE 5.2.**)

DISTRIBUTION.   Because of its presence in some postganglionic terminals of the autonomic nervous system (Figure 5.2), norepinephrine has received much experimental attention, along with ACh. I should note now that *adrenalin* and *epinephrine* (and, of course, *noradrenalin* and noradrenaline) are synonymous. Epinephrine (adrenalin) is produced by the medulla (central core) of the adrenal glands, small endocrine glands located above the kidneys. *Ad renal* means "toward kidney" in Latin. In Greek, one would say *epi nephros* (upon the kidney); hence the term epinephrine. The latter term has been adopted by pharmacologists (probably because the word Adrenalin has been appropriated by a drug company as a proprietary name); therefore, to be consistent with general usage I'll call the transmitter substance norepinephrine (or more often, simply *NE*). Unfortunately, everyone still uses "noradrenergic" for the adjective form; "norepinephrinergic" just never caught on.

Norepinephrine generally (but not always) appears to be inhibitory in effect on neurons of the CNS. At the target organs of the sympathetic nervous system it is usually excitatory. Falck, Hillarp, Thieme, and Torp (1962) discovered that when fish brain tissue was exposed to dry formaldehyde gas, noradrenergic neurons would fluoresce a bright yellow color when the tissue was examined under ultraviolet light. The development of this histofluorescence technique has made it possible to trace the precise circuitry of noradrenergic neurons in the brain (Figure 15.12).

DRUGS THAT AFFECT NORADRENERGIC SYNAPSES.   Noradrenergic synapses may be affected by drugs in a variety of ways. The amino acid *tyrosine* serves as the precursor for NE. Tyrosine is transformed into L-**DOPA** (L-*3,4-dihydroxyphenylalanine*) by action of the enzyme *tyrosine hydroxylase* (a hydroxyl group is added). L-

TABLE 5.2   Classification of the monoamine transmitter substances

Monoamines	
Catecholamines	*Indolamines*
Norepinephrine	Serotonin (5-HT)
Dopamine	

DOPA then loses a carboxyl group through the activity of **DOPA decarboxylase** and becomes **dopamine.** Finally, dopamine gains a hydroxyl group via **dopamine-β-hydroxylase** and becomes norepinephrine. These reactions are shown in **FIGURE 5.4.**

Most transmitter substances are synthesized in the cell body, packaged in synaptic vesicles, and transported down the axon to the terminal buttons. However, in the case of norepinephrine, the final biosynthetic step is omitted until the vesicles reach the synapse. The vesicles are filled with dopamine, and are transported to the terminal buttons. There, the dopamine is converted to norepinephrine through the action of dopamine-β-hydroxylase. If the organism is treated with an inhibitor of this enzyme (such as **FLA-63**), the contents of the synaptic vesicles will not be converted to norepinephrine, and the noradrenergic terminals will secrete dopamine instead of norepinephrine.

The synthesis of NE may be blocked by administration of **α-methyl-para-tyrosine (AMPT)**. This drug inhibits tyrosine hydroxylase and prevents the synthesis of DOPA, and thus of dopamine and NE. (See **FIGURE 5.4.**) Drugs that block the effects of other relevant biosynthetic enzymes have also been found, but AMPT is most often used experimentally.

Reserpine, a drug used extensively as a hypotensive (blood pressure-reducing) agent, decreases activity at NE synapses by making the vesicular membrane become "leaky," so that the transmitter substance cannot be held inside. It escapes and is destroyed by MAO and COMT, which will be discussed later. (The effect is not specific for noradrenergic synapses; reserpine also impairs vesicular storage of dopamine and serotonin.) On the other hand, **amphetamine** stimulates the release of NE (and dopamine, as well) into the synaptic cleft, and also retards the re-uptake of these transmitter substances. These effects make amphetamine a potent agonist for the catecholamines.

We saw that there are two types of cholinergic receptors. Similarly, there are two types of noradrenergic receptors, **alpha** and **beta.** Alpha receptors mediate smooth muscle contraction (such as that of the blood vessels, thus raising blood pressure when stimulated), whereas beta receptors increase heart muscle contractions and relax smooth muscles of the bronchi and intestines. Thus, beta adrenergic agonists are used in treatment of the symptoms of asthma. Both alpha and beta receptors are stimulated by **isoproterenol,** but this drug most effectively stimulates beta receptors. Beta receptors are selectively blocked by **propranolol** whereas **phentolamine** antagonizes alpha receptors.

The postsynaptic effects of NE are normally terminated by re-uptake of the transmitter by the terminals, rather than by enzymatic deactivation (as is the case with ACh). Therefore, drugs that affect the rate of re-uptake alter the properties of noradrenergic synapses. **Desipramine** acts as a potent inhibitor of re-uptake of NE; hence,

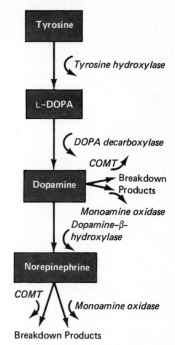

FIGURE 5.4 Biosynthesis of catecholamines.

the PSPs produced at noradrenergic synapses are considerably facilitated. Desipramine, imipramine, and other drugs with similar effects are often referred to as ***tricyclic antidepressants***. (*Tricyclic* refers to the molecular shape of the drug.) They have been used in the treatment of chronic depression, which appears to result from decreased activity in noradrenergic (and/or serotonergic) neurons. In fact, depression is a side effect that sometimes accompanies the use of such hypotensives as reserpine. (The story is more complicated than this, as we shall see in Chapter 20, where I shall describe the biochemistry of depression in more detail.)

Monoamine oxidase (MAO) and ***catechol-O-methyltransferase (COMT)*** are enzymes that break down dopamine and norepinephrine, as well as other biologically active amines. It is thought that MAO is used by catecholamine-producing cells to regulate the amounts of transmitter substance that is produced. The drug ***pargyline*** acts as an inhibitor of MAO; its administration therefore increases production and liberation of the monoamines. Like imipramine, pargyline has been successfully used in the treatment of depression, but is not used much now because it produces undesirable side effects.

The discovery of a substance with a lethal affinity for noradrenergic (and also dopaminergic) neurons has provided neuroscientists with a very useful tool. The drug ***6-hydroxydopamine (6-HD)*** is taken up selectively by terminal buttons of these cells, and once the drug reaches a sufficient cytoplasmic concentration, it kills the cells. Although 6-HD crosses the blood-brain barrier quite slowly and is therefore not very useful in systemic administration, it can be injected directly into specific regions of the brain, where it selectively destroys the catecholaminergic cells. By observing the behavior of animals that now lack these neurons, one can attempt to infer their functions.

Figure 5.5 schematizes the sites of drug action at a noradrenergic synapse. (See **FIGURE 5.5.**)

Dopamine

Like NE, dopamine appears to produce inhibitory postsynaptic potentials. Development of a histofluorescence technique that is specific for dopamine has made it possible to diagram its distribution in the CNS. (Figure 15.11) We have already seen the biosynthetic pathway for ***dopamine (DA)*** in Figure 5.4; this transmitter is the immediate precursor of norepinephrine. Like NE, its synthesis is inhibited by AMPT; it is impossible to block dopamine synthesis without also affecting NE. Many of the drugs that affect NE synapses similarly affect dopaminergic synapses: reserpine prevents storage of the transmitter in the synaptic vesicles, and amphetamine facilitates its release into the synaptic cleft.

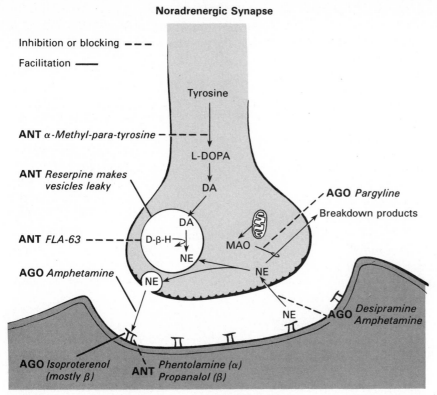

Noradrenergic Synapse

Inhibition or blocking – – –

Facilitation ———

Tyrosine

ANT α-*Methyl-para-tyrosine*

L-DOPA

DA

ANT *Reserpine makes vesicles leaky*

AGO *Pargyline*

Breakdown products

DA

ANT *FLA-63* D-β-H

MAO

NE

AGO *Amphetamine*

NE

NE

NE

AGO *Desipramine Amphetamine*

AGO *Isoproterenol (mostly β)*

ANT *Phentolamine (α) Propanalol (β)*

FIGURE 5.5   A schematic model of a noradrenergic synapse, along with some drugs that can affect it.

Like other monoamines, dopamine is destroyed by monoamine oxidase. However, recent studies have found that a specific form of MAO, type B, appears to be more specific for dopamine than for the other monoamines. Therefore, ***deprenyl,*** a drug that deactivates MAO type B, serves as a dopaminergic agonist. Amphetamine, which facilitates the release of the monoamines, also retards the re-uptake of dopamine (and to a lesser extent, norepinephrine as well). Thus, it is a very potent dopaminergic agonist.

Some drugs have been found that selectively affect DA terminals. ***Gamma-hydroxybutyrate*** inhibits the release of dopamine, and hence serves as an antagonist. ***Apomorphine,*** an agonist, stimulates dopamine receptors.

Parkinson's disease, which is characterized by tremors and progressive rigidity of the limbs, appears to result from degeneration of dopaminergic neurons in brain structures involved in movement—specifically, a pathway from the substantia nigra to the caudate nucleus (see Chapter 10). The substantia nigra is so called because it is naturally stained black with melanin, the substance that gives color to skin. This compound is produced by the breakdown of dopamine.

The DA neurons in this pathway normally inhibit cholinergic neurons whose activity cause muscle movements. Hence, the release of the cholinergic neurons from inhibition results in the patient's rigidity. Administration of L-DOPA increases the amount of dopamine that is synthesized in the patient's brain. This facilitates the activity of the remaining dopaminergic neurons, and thus alleviates symptoms of the disease.

Dopamine has been implicated as a transmitter that might be involved in schizophrenia, a serious mental disorder (psychosis) characterized by disruption of normal, logical thought processes. ***Chlorpromazine*** and related compounds have been found to have a profound antipsychotic effect; independent of sedative effects, these drugs alleviate the symptoms of schizophrenia—enough to have dramatically reduced the hospital population of schizophrenics. The most important neuropharmacological effect of the antipsychotic drugs seems to be blockage of dopaminergic receptors in the brain. If ***apomorphine,*** which directly stimulates DA receptors, is given to schizophrenic patients whose symptoms have successfully been brought under control by one of the antipsychotic drugs, their symptoms return. As you might predict from the previous paragraph, a

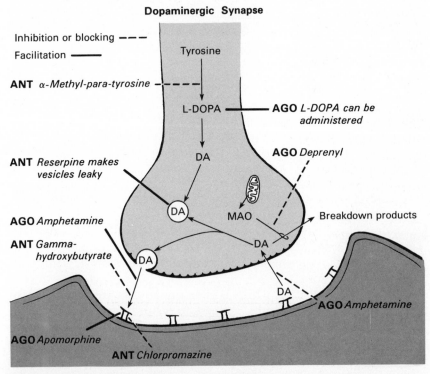

FIGURE 5.6 A schematic model of a dopaminergic synapse, along with some drugs that can affect it.

frequent side effect of the antipsychotic drugs is development of symptoms like those of Parkinson's disease. Fortunately, drugs are available that combine the antidopaminergic effect with an anticholinergic effect (thus suppressing activity of cholinergic neurons no longer inhibited by the DA cells). These drugs produce minimal parkinsonian symptoms. The role of dopamine in schizophrenia is described in fuller detail in Chapter 20.

Figure 5.6 schematizes the sites of drug action at a dopaminergic synapse. (See **FIGURE 5.6.**)

Serotonin

The inhibitory transmitter **serotonin** (also called **5-hydroxytryptamine**, or **5-HT**) has received a considerable amount of experimental attention. It is produced by means of a biosynthetic pathway starting with the amino acid **tryptophan.** Tryptophan receives a hydroxyl group via the enzyme **tryptophan hydroxylase** and becomes **5-hydroxytryptophan (5-HTP).** The enzyme **5-HTP decarboxylase** removes a carboxyl group from 5-HTP, and the result is 5-HT (serotonin). Figure 5.7 illustrates these reactions. (See **FIGURE 5.7.**)

The regional distribution of 5-HT in the CNS has been studied with the advent of specific staining techniques. Serotonergic neurons in various hindbrain locations send axons to the spinal cord and into forebrain regions.

There are a number of pharmacological agents that affect serotonergic neurons. **Para-chlorophenylalanine (PCPA)** inhibits the enzyme tryptophan hydroxylase, thus blocking the synthesis of 5-HT. Reserpine prevents vesicular storage of 5-HT (as well as norepinephrine and dopamine); thus it deactivates these synapses.

Several drugs, such as cinnanserin and methysergide, act as potent blockers of 5-HT receptors in peripheral tissue, but studies have failed to identify any agents that definitely block 5-HT receptors in the central nervous system. The hallucinogenic drug **lysergic acid diethylamide (LSD)** *increases* the activity of neurons in the brain that are normally *inhibited* by serotonergic neurons; thus it appears to be a serotonergic antagonist. Studies have shown that the effects are not due to blockade of postsynaptic 5-HT receptors. Instead, LSD appears to interfere with a regulatory mechanism that normally control the activity of serotonergic neurons; it "fools" the system in such a way as to suppress the release of serotonin.

Although the search for serotonin receptor blockers that act in the central nervous system has been unsuccessful so far, a drug has been found that directly stimulates central 5-HT receptors: **quipazine**.

Amitryptyline blocks the re-uptake of 5-HT (which normally terminates the PSP), and thus serves as an agonist. Amitryptyline, like desipramine, is used to treat psychotic depression. This fact suggests

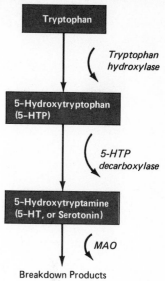

FIGURE 5.7   Biosynthesis of serotonin (5-HT).

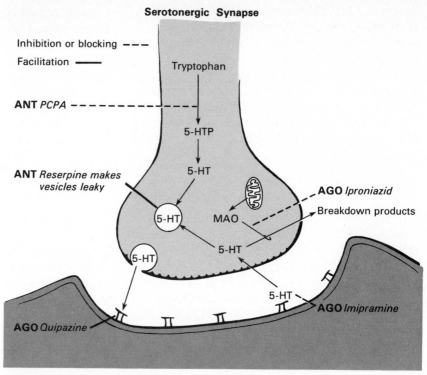

Serotonergic Synapse

Inhibition or blocking ---
Facilitation ———

ANT *PCPA* - - - - - -

Tryptophan

5-HTP

5-HT

ANT *Reserpine makes vesicles leaky*

5-HT

5-HT

MAO

AGO *Iproniazid*
Breakdown products

5-HT

AGO *Quipazine*

5-HT

AGO *Imipramine*

FIGURE 5.8   A schematic model of a serotonergic synapse, along with some drugs that can affect it.

that 5-HT, as well as norepinephrine, may play a role in this disorder. (This topic, too, will be covered in Chapter 20.)

Serotonergic synapses can also be facilitated with *iproniazid,* which inhibits MAO, the enzyme that destroys excess 5-HT in the terminal buttons. Another strategy that has been used to increase the amount of brain 5-HT has been to inject the animal with the 5-HT precursor, 5-HTP. Adequate supplies of 5-HTP decarboxylase, which convert the 5-HTP to 5-HT, are present in the brain.

Recent research on the role of serotonin-containing neurons in the CNS has been facilitated by the discovery of the substance *5,6-dihydroxytryptamine (5,6-DHT).* This drug produces effects like that of 6-HD on NE and DA cells; it is selectively taken up by serotonergic cells and then it subsequently kills them. The sites of drug action at a serotonergic synapse are schematized in **FIGURE 5.8.**

Glutamic acid

Glutamic acid is found throughout the brain. Its precursor, alpha-ketoglutaric acid, is available in abundant quantities from the Krebs citric acid cycle. Chinese food also provides a lot of glutamic acid; MSG (monosodium glutamate) is the sodium salt of glutamic acid. Most investigators believe that it is a transmitter substance—indeed, it is probably the principal excitatory neurotransmitter in the brain—but evidence is still not conclusive. It appears that the post-synaptic effects of glutamic acid are terminated by re-uptake.

So far, only two drugs are known to specifically affect glutamic acid synapses: (1) an inhibitor, *glutamic acid diethylester,* which blocks glutamic acid receptors, and (2) a facilitator, *glutamic acid dimethylester,* which blocks re-uptake of glutamic acid by the terminal buttons.

GABA

Gamma-amino butyric acid (GABA) is produced from glutamic acid by the action of *GAD (glutamic acid decarboxylase).* GABA appears to be inhibitory, and it appears to have a widespread distribution throughout the gray matter (cellular areas) of the brain. GABA is also found in the dorsal horn of the spinal cord (see Chapter 6).

Relatively few drugs have been discovered that specifically affect GABA. *Tetanus toxin* inhibits release of GABA into the synaptic cleft. *Picrotoxin* and (more specifically) *bicuculline* antagonize the effects of GABA on the receptor sites, whereas *muscimol* (a hallucinogen) stimulates these receptors. The drug *2-hydroxy GABA* retards the re-uptake of GABA, and thus serves as an agonist.

Excess GABA is destroyed by an enzyme called *GABA trans-aminase (GABA-T),* the way MAO gets rid of excess DA, NE, and 5-HT. A drug called *n-propylacetic acid* deactivates GABA-T, and hence acts as an agonist. Since GABA seems to be the principal inhibitory transmitter substance in the brain, GABA agonists have an inhibitory effect upon neural firing. Indeed, *n*-propylacetic acid has been used to treat epilepsy.

GABA has been implicated in a serious hereditary neurological disorder, Huntington's chorea. This disease is characterized by involuntary movements, depression, progressive mental deterioration, and, ultimately, death. It has been suggested (Perry, Hansen, and Kloster, 1973) that the disease results from degeneration of GABA cells in the basal ganglia (brain structures concerned with motor control—see Chapters 6 and 10).

Figure 5.9 schematizes the sites of drug action at a gabaminergic synapse. (See **FIGURE 5.9** on the following page.)

Glycine

The amino acid *glycine* appears to be the inhibitory neurotransmitter in the spinal cord and lower portions of the brain. Little is known

Gabaminergic Synapse

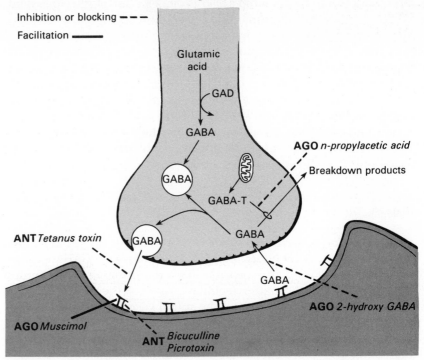

Inhibition or blocking ━ ━ ━

Facilitation ━━━━━

Glutamic acid

GAD

GABA

GABA

AGO *n-propylacetic acid*

Breakdown products

GABA-T

ANT *Tetanus toxin*

GABA

GABA

GABA

AGO *2-hydroxy GABA*

AGO *Muscimol*

ANT *Bicuculline Picrotoxin*

FIGURE 5.9   A schematic model of a gabaminergic synapse, along with some drugs that can affect it.

about its biosynthetic pathway; there are several possible routes, but not enough is known to decide how neurons produce glycine. Only two drugs affect the hypothesized glycine synapses with any degree of specificity; tetanus toxin inhibits glycine release (along with that of GABA), whereas the poison **strychnine,** by blocking glycine receptors (which mediate inhibition), acts as a CNS stimulant.

Other Suspected Transmitters

Many other substances are suspected to serve as neurotransmitters, including amino acids such as **taurine, aspartic acid,** and **serine;** polypeptides such as the endorphins, **substance P** and **oxytocin;** and amines such as **histamine.** In fact, there is good evidence that substance P is the transmitter substance secreted by neurons that bring information related to pain into the central nervous system, and that an endogenous opiate serves as a neurotransmitter, producing presynaptic inhibition on the terminal buttons that secrete substance P. In addition, histamine has been implicated in psychotic depression. More research will be needed to confirm the status of these substances as neurotransmitters. The final list will undoubtedly contain a large number of compounds.

MECHANISMS OF POSTSYNAPTIC ACTIVITY

The exact way in which transmitter substances alter the ionic permeability of postsynaptic neurons is not known. Some transmitters (e.g., ACh) appear to have a relatively direct effect on receptor molecules that determine the permeability characteristics of the membrane "pores." The mode of action of some other transmitters is much less direct. They involve activation of the so-called "second messenger"—substances known as **cyclic nucleotides.**

Situated in the postsynaptic membrane of DA and NE synapses (and probably in many others, also), are molecules of **adenyl cyclase,** attached to receptor molecules. When stimulated by the neurotransmitter, adenyl cyclase becomes active and produces, in the cytoplasm of the postsynaptic cell, **cyclic AMP** (cyclic adenosine monophosphate) from **ATP (adenosine triphosphate).** (In some cases, cyclic GMP is produced from GTP, but I shall stick with cyclic AMP for simplicity's sake.)

The cyclic AMP then acts as a cofactor for enzymes called **kinases;** these enzymes, becoming active in the presence of cyclic AMP, cause **phosphorylation** (addition of a phosphate group) of proteins in the postsynaptic membrane. Phosphorylation changes the physical configuration (bending and folding) of a protein. This process, then, could alter the shape of proteins responsible for the permeability properties of the membrane. Figure 5.10 illustrates this process. (See **FIGURE 5.10.**)

Phosphorylation of proteins associated with the nucleus can also alter regulation of protein synthesis of the cell. (In fact, cyclic AMP-

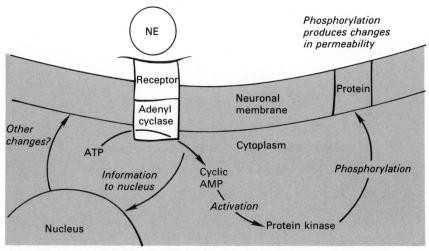

FIGURE 5.10   A schematic model of a postsynaptic membrane containing receptors that activate adenyl cyclase.

stimulated protein synthesis is the means by which some hormones, such as insulin, affect their target cells.) Thus, in investigations of the memory process, neuroscientists have recently given much attention to processes mediated by cyclic AMP.

As you know, the postsynaptic effects are terminated by reuptake and (in the case of ACh), enzymatic deactivation. Some mechanism must also be present to terminate the effects of cyclic AMP. This is accomplished by an enzyme called ***phosphodiesterase.*** Thus, a phosphodiesterase inhibitor would act as an agonist at synapses whose second messenger is cyclic AMP. The most commonly-administered phosphodiesterase inhibitor is **caffeine,** which exerts its stimulating effect by prolonging the life of cyclic AMP, the "second messenger" at many synapses.

In this chapter I have described the substances that are thought to serve as neurotransmitters, and I have described the ways that drugs can selectively affect the production and activity of these substances. In later chapters we shall see how neuroscientists have used these pharmacological agents to try to discover how the brain controls behavior.

KEY TERMS

acetylcholine p. 81

adenosine triphosphate (ATP) (*a DEN o seen*) p. 96

adenyl cyclase (*a DEEN ul SY klase*) p. 95

adrenalin p. 86

agonist p. 79

alpha adrenergic receptor p. 87

alpha methylparatyrosine (AMPT) p. 87

amitryptyline (*am a TRIP ta leen*) p. 91

amphetamine p. 87

antagonist p. 79

apomorphine p. 89

aspartic acid (*ass PAR tik*) p. 94

atropine p. 84

beta adrenergic receptor p. 87

bicuculline (*by KEW kew leen*) p. 93

botulinum toxin p. 84

caffeine p. 96

catecholamine p. 86

catechol-O-methyl transferase (COMT) p. 88

chlorpromazine p. 90

choline acetylase (*a SEET a lase*) p. 82

coenzyme A (CoA) p. 82

curare (*kew RAHR ee*) p. 85

cyclic AMP p. 96

cyclic nucleotide p. 95

deprenyl (*DEP ren ill*) p. 89

desipramine (*dez IP ra meen*) p. 87

5,6-dihydroxytryptamine (5,6-DHT) p. 92

DFP (diisopropylfluorophosphate) (*dy EYE so et cetera*) p. 85

DOPA decarboxylase p. 87

dopamine (DA) p. 87

dopamine-β-hydroxylase p. 87

endorphin p. 78

enzyme p. 80

epinephrine (*ep i NEF rin*) p. 86

eserine (*ESS er in*) p. 85

false transmitter p. 81

FLA-63 p. 87

GABA transaminase (GABA-T) p. 93

gamma-amino butyric acid (GABA) (*bew TEER ik*) p. 77

gamma hydroxybutyrate (*BEW ta rate*) p. 89

glutamic acid (*glu TAM ik*) p. 77

glutamic acid decarboxylase p. 93

glutamic acid diethylester (*dy eth ul ES ter*) p. 93

SUGGESTED READINGS

DUNN, A. J., AND BONDY, S. C. *Functional Chemistry of the Brain.* Flushing, N.Y.: SP Books, 1974. An excellent source of information about the mechanisms of synaptic transmission: release of transmitter substance, recycling of the vesicular membrane, etc.

COOPER, J. R., BLOOM, F. E., AND ROTH, R. H. *The Biochemical Basis of Neuropharmacology.* New York: Oxford University Press, 1978.

SIEGEL, G. J., ALBERS, R. W., KATZMAN, R., AND AGRANOFF, B. W. (Editors) *Basic Neurochemistry.* Boston: Little, Brown, 1976. All three books are excellent. The volume by Siegel et al. is more advanced. It is more of a handbook than a text, containing 36 chapters written by various experts in the field.

6

Introduction to the Structure of the Nervous System

So far, we have studied the structure and function of neurons and their supporting cells. We have seen how neurons communicate and reach decisions. Now it is time to begin discussion of some functional systems of the brain; that means it will be necessary, first, to become acquainted with the general structure of the nervous system.

In this chapter I shall make no attempt to present a course in neuroanatomy. I would rather present less material and have you learn nearly all of it than present a lot of material and have you try to figure out what is important enough to remember. I shall try to keep the number of terms introduced here to a minimum (as you will see, the minimum is still a rather large number). In later chapters dealing with specific functional systems, I shall present more information concerning the relevant anatomy; you should have no trouble incorporating the new information into the general framework you will receive in this chapter. This scheme will permit me to distribute the anatomical details a bit more (minimizing the probability of neuroanatomical information overload) and will allow me to discuss structure and function together. Anatomy is always more interesting (and easier to learn) when it is presented in a functional context.

Introduction to the
Structure of the
Nervous System

The nervous system consists of the brain and spinal cord, which make up the central nervous system (CNS), and the cranial nerves, spinal nerves, and peripheral ganglia, which constitute the peripheral nervous system (PNS). The central nervous system is encased in bone; the brain is covered by the skull, and the spinal cord resides within the *vertebral column.*

DIRECTIONS AND PLANES OF SECTION

When describing topographical features (hills, roads, rivers, etc.), we need to use terms denoting directions (e.g., the road goes north, and then turns to the east as it climbs the hill). Similarly, in describing structures of the nervous system and their interconnecting pathways, we must have available a set of words that define geographical relationships. We could use the terms *in front of, above, to the side of,* etc., but then different terms would have to be used in describing the same set of neural structures of animals like humans, who walk erect and whose spinal cord is oriented at right angles to the ground, and animals like the rat, whose spinal cord is normally parallel to the ground. To make it easier to describe anatomical directions, a standard set of terms has been adopted.

Directions in the nervous system are normally described relative to the *neuraxis,* an imaginary line drawn through the spinal cord up to the front of the brain. For simplicity's sake, let us consider an animal with a straight neuraxis. Figure 6.1 shows an alligator and a human. This alligator is certainly laid out in a linear fashion—we can draw a straight line that starts between its eyes and continues down the center of its spinal cord. (See **FIGURE 6.1.**) The front end is **_anterior,_** and the tail is **_posterior._** The terms *rostral* (toward the beak) and *caudal* (toward the tail) are also employed; I shall use these latter terms more often. The top of the head and the back are part of the *dorsal* surface, while the *ventral* surface faces the ground. (The spinal cord, then, is located in a position dorsal to the abdominal surface and ventral to the surface of the animal's back.) These directions are somewhat more complicated in the human, since the neuraxis takes a 90-degree bend. The front views of the alligator and the human illustrate the terms *lateral* and *medial,* toward the side and toward the midline, respectively. (See **FIGURE 6.1.**) When describing brain structures, the terms *superior* and *inferior* are often used. The superior structure is above (dorsal to) the inferior one.

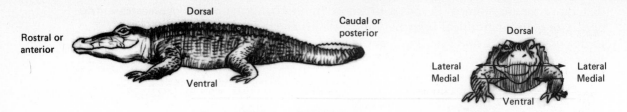

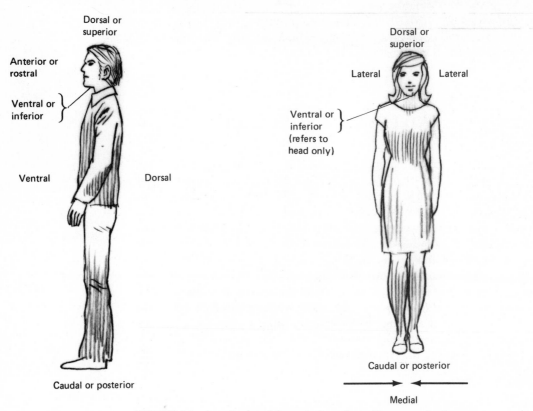

FIGURE 6.1 A sagittal and frontal view of an alligator and a human, with the terms denoting anatomical directions.

To see what is in the nervous system, we have to cut it open; to be able to convey information about what we find, we slice it in a standard way. Figure 6.2 shows the nervous system of an alligator and that of a human. Again, let us consider the alligator first. We can slice the nervous system in three ways: (1) *transversely*, like a salami, giving us **cross sections**, or **transverse sections** or **frontal sections** or **coronal sections;** (2) parallel to the ground, giving us **horizontal sections;** and (3) perpendicular to the ground and parallel to the neuraxis, giving us **sagittal sections.** It is unfortunate that so many terms are used to refer to the first plane of section; I shall try to use only two in the text—*frontal sections* through the brain and *cross sections* through the spinal cord. (See **FIGURE 6.2.**)

**Introduction to the
Structure of the
Nervous System**

CNS of Alligator

Transverse (frontal, coronal) plane (cross sections lie in this plane)

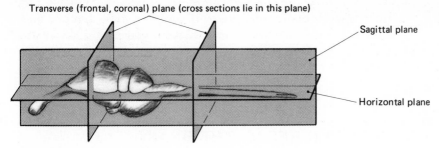

Sagittal plane

Horizontal plane

CNS of Human

Transverse plane

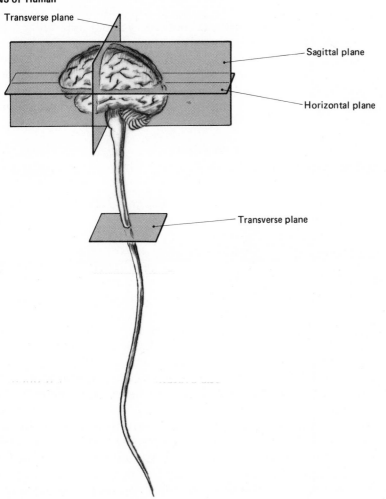

Sagittal plane

Horizontal plane

Transverse plane

FIGURE 6.2 Planes of section as they relate to the central nervous system
of an alligator and a human.

I should mention one section that you will encounter often in this text—the ***midsagittal section.*** If you sliced the brain down the middle, dividing it into its two symmetrical halves, you would have cut it through its *midsagittal plane.* If you then looked at one half of the brain with its cut surface toward you, you would be getting a *midsagittal view.* You will see many of these views of the brain in this text.

GROSS FEATURES OF THE BRAIN

Figure 6.3 illustrates the relationship of the brain and spinal cord to the head and neck of a human. Do not worry about unfamiliar labels on this figure; these structures will be described later. (See **FIGURE 6.3.**) The brain is a large mass of neurons, glia, and supporting cells. It is the most protected organ of the body, encased in a tough, bony skull and floating in a pool of *cerebrospinal fluid.* The brain receives a copious supply of blood and is chemically guarded by the blood-brain barrier.

Blood Supply

The brain receives approximately 20 percent of the blood flow from the heart, and it receives this blood flow continuously. Other parts of the body (e.g., skeletal muscle, digestive system) receive varying quantities of blood, depending on their needs, relative to those of other regions. <u>The brain, however, always receives its share.</u> The brain cannot store its fuel (primarily glucose and ketone bodies), nor can it temporarily extract energy without oxygen as the muscles can; therefore, a consistent blood supply is essential. A one-second interruption of the blood flow to the brain uses up much of the dissolved oxygen; a six-second interruption produces unconsciousness. Permanent damage occurs within a few minutes.

Circulation of blood in the body proceeds from arteries to arterioles to capillaries; the capillaries then drain into venules, which collect and become veins. The veins travel back to the heart, where the process begins again. Regional blood flow is controlled by smooth muscles in the walls of arteries and arterioles. Contraction of circular fibers constricts the arterioles and restricts blood flow; contraction of longitudinal fibers dilates the arterioles. The smooth muscles of the arterioles are controlled by nerve endings and by levels of various hormones. However, the arterioles of the brain are not very responsive to changes in the general physiological state of the organism. Instead, increases in local levels of carbon dioxide cause dilation of the brain's blood vessels. Blood flow to various

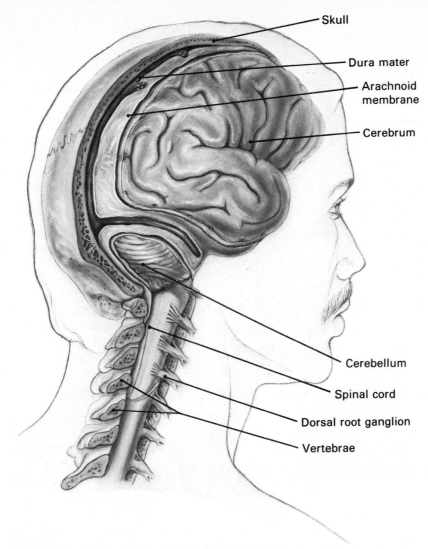

Introduction to the
Structure of the
Nervous System

FIGURE 6.3   The relationship of the brain and spinal cord to the head
and neck.

regions is thus regulated by local demand; an increased rate of metabolism produces excess carbon dioxide, which results in a corresponding increase in regional blood flow.

Figure 6.4 shows a bottom view of the brain and its major arterial supply. (The spinal cord has been cut off, as have the left half of the cerebellum and the left temporal lobe.) Two major sets of arteries serve the brain: the **vertebral arteries**, which serve the caudal portion of the brain, and the **internal carotid arteries**, which serve the rostral portions. (See **FIGURE 6.4** on the following page.) You can see that the blood supply is rather peculiar; major arteries

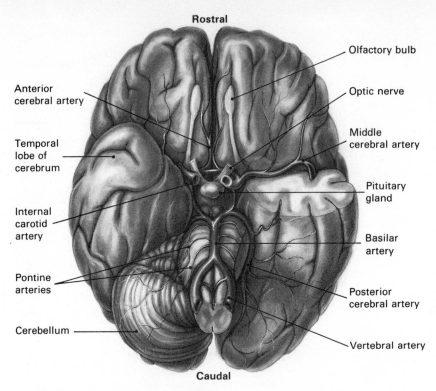

Rostral

Olfactory bulb

Anterior
cerebral artery

Optic nerve

Middle
cerebral artery

Temporal
lobe of
cerebrum

Pituitary
gland

Internal
carotid
artery

Basilar
artery

Pontine
arteries

Posterior
cerebral artery

Cerebellum

Vertebral artery

Caudal

FIGURE 6.4   Arterial supply to the brain.

join together and then separate again. Normally, there is little mixing of blood from the rostral and caudal arterial supplies, or, in the case of the rostral supply, that of the right and left sides of the brain. However, if a blood vessel becomes occluded blood flow can follow alternative routes, reducing the probability of loss of blood supply and subsequent destruction of brain tissue.

Venous drainage of the brain is shown in Figure 6.5. Major veins, like major arteries, are interconnected, so that blood in some veins can flow in either direction (shown by double-ended arrows), depending on intracerebral pressures in various parts of the brain (See **FIGURE 6.5.**)

Meninges

The entire nervous system—brain, spinal cord, cranial and spinal nerves, and autonomic ganglia—is covered by tough connective tissue. The protective sheaths around the brain and spinal cord are referred to as the **_meninges._** The meninges (singular *meninx*) consist of three layers, which are shown in Figure 6.6. The outer layer is thick, tough, and unstretchable; its name, **_dura mater,_** means "hard mother." The middle layer of the meninges, the **arachnoid,** gets its name from the weblike appearance of the **_arachnoid trabec-_**

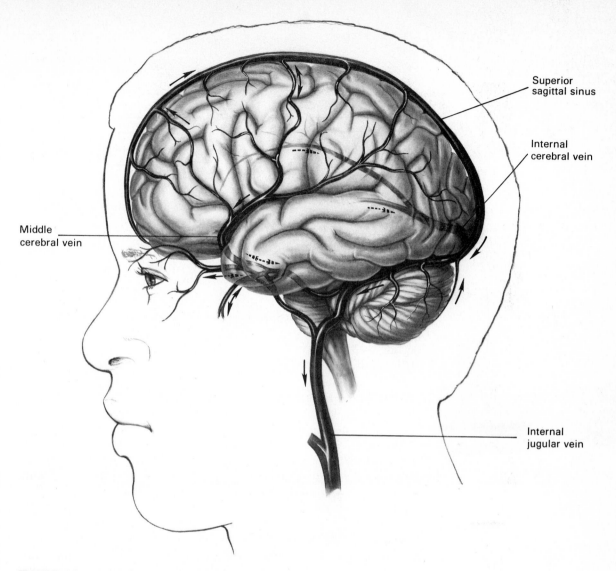

Superior
sagittal sinus

Internal
cerebral vein

Middle
cerebral vein

Internal
jugular vein

FIGURE 6.5   Venous drainage of the brain.

ulae that protrude from it (*arakhnē*—"spider"). The arachnoid is soft and spongy, and lies beneath the dura mater. Closely attached to the brain and spinal cord, and following every surface convolution, is the ***pia mater*** ("pious mother"). The smaller surface blood vessels of the brain and spinal cord are contained within this layer. Between the pia mater and arachnoid is a gap called the ***subarachnoid space***. This space is filled with cerebrospinal fluid (CSF), and through it pass large blood vessels. (See **FIGURE 6.6** on the following page.)

The peripheral nervous system is covered with two layers of meninges. The middle layer (arachnoid), with its associated pool of CSF, covers only the brain and spinal cord. Outside the CNS the

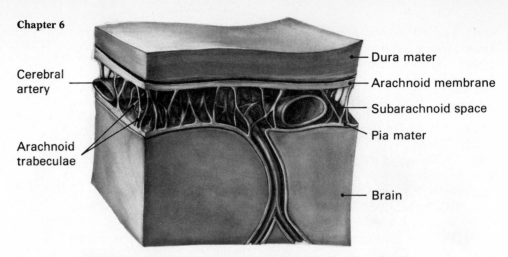

FIGURE 6.6 A schematic drawing of the arachnoid, subarachnoid space, and pia mater.

outer and inner layers (dura mater and pia mater) fuse and form a sheath covering the spinal and cranial nerves and the autonomic ganglia.

In the first edition of this book I said that I did not know why the outer and inner layers of the meninges were referred to as "mothers." I received a letter from medical historians at the Department of Anatomy at UCLA that explained the name. (Sometimes it pays to proclaim one's ignorance.) A tenth century Persian physician, Ali ibn Abbas, used the Arabic term *al umm* to refer to the meninges. The term literally means "mother," but was used to designate any swaddling material, since Arabic lacked a specific term for membrane. The tough outer membrane was called *al umm al djafiya* and the soft inner one was called *al umm al riqiqa*. When the writings of Ali ibn Abbas were translated into Latin during the eleventh century, the translator, who was probably not familiar with the structure of the meninges, made a literal translation of *al umm*. He referred to the membranes as the "hard mother" and the "pious mother," rather than using a more appropriate Latin word. And why was the inner meninx called "pious"? The Arabic *riqiqa* means "tender," which could refer to its delicate nature or the way in which it holds fast to the surface of the brain. ("Tender" and "tenacious" have the same root.) Thus, a better translation would have been "tenacious membrane" rather than "pious mother."

Ventricular System and the Production of CSF

The brain is very soft and jellylike. The considerable weight of a human brain (approximately 1400 gm), along with its delicate con-

Introduction to the
Structure of the
Nervous System

struction, necessitates that it be protected from shock. A human brain cannot even support its own weight well; it is extremely difficult to remove and handle a fresh brain from a recently deceased human without damaging it.

Fortunately, the intact brain within a living human is well protected. It floats in the bath of cerebrospinal fluid contained within the subarachnoid space. Since the brain is completely immersed in liquid, its net weight is reduced to approximately 80 gm, and pressure on the base of the brain is therefore considerably diminished. The cerebrospinal fluid surrounding the brain and spinal cord also reduces the shock to the CNS that would be caused by sudden head movement. Painful headaches accompany head movements after clinical removal of the cerebrospinal fluid prior to special X-ray tests; these headaches attest to the value of CSF as a shock-absorbing medium. (Fresh CSF is subsequently produced, and the headaches disappear.)

The brain contains a series of hollow, interconnected chambers that are filled with cerebrospinal fluid. These chambers are connected with the subarachnoid space by means of small openings (*foramina*), and they are also continuous with the narrow, tubelike *central canal* of the spinal cord. Figure 6.7 shows a human brain with the ventricular system shaded in. (See **FIGURE 6.7** on the following page.) The largest chamber is the paired set of *lateral ventricles.* The lateral ventricles are connected by the **foramen of Monro** to the **third ventricle.** The third ventricle is located at the midline of the brain; it is a single structure, and its walls divide the local region of the brain into symmetrical halves. A bridge of neural tissue (**massa intermedia**) crosses through the middle of the third ventricle. The **cerebral aqueduct,** a long tube, connects the third ventricle to the **fourth ventricle,** which, at its caudal end, connects with the central canal of the spinal cord. (There is no first or second ventricle.) (See **FIGURE 6.7.**)

Cerebrospinal fluid is manufactured by a special vascular structure called the **choroid plexus,** which protrudes into each of the ventricles and produces CSF from blood plasma. CSF is continuously produced; the total volume is approximately 125 ml, and the half-life of CSF (the time it takes for half of the CSF present in the ventricular system to be replaced by fresh fluid) is about three hours, so several times this amount is produced by the choroid plexus each day. The continuous production of CSF means that there must be a mechanism for its removal; Figure 6.8 illustrates the production, circulation, and reabsorption of cerebrospinal fluid. (See **FIGURE 6.8** on page 109.)

The illustration shows a slightly rotated midsagittal view of the central nervous system, so the lateral ventricles, located on each side of the brain, cannot be shown. (To visualize the lateral ventricles, refer to Figure 6.7. (See **FIGURES 6.7** and **6.8.**) CSF is produced by the choroid plexus of the lateral ventricles, and it flows through the

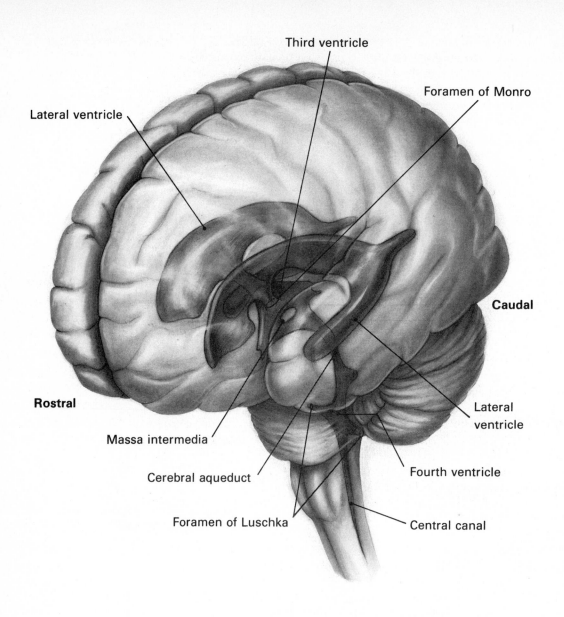

Third ventricle

Foramen of Monro

Lateral ventricle

Caudal

Rostral

Lateral ventricle

Massa intermedia

Cerebral aqueduct

Fourth ventricle

Foramen of Luschka

Central canal

FIGURE 6.7   The ventricular system of the brain.

foramen of Monro into the third ventricle. Additional CSF is produced in this ventricle and then flows through the cerebral aqueduct to the fourth ventricle, where still more CSF is produced. CSF leaves the fourth ventricle via the **foramen of Magendie** and the **foramina of Luschka** and collects in the subarachnoid space surrounding the brain. All CSF thus flows into the subarachnoid space around the

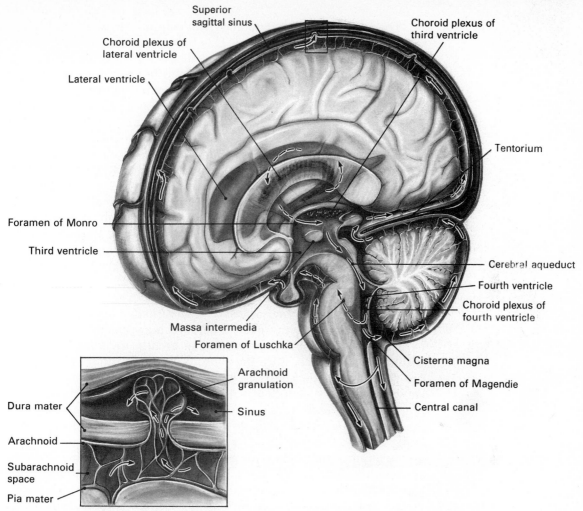

Superior sagittal sinus

Choroid plexus of lateral ventricle

Lateral ventricle

Choroid plexus of third ventricle

Tentorium

Foramen of Monro

Third ventricle

Cerebral aqueduct

Fourth ventricle

Choroid plexus of fourth ventricle

Massa intermedia

Foramen of Luschka

Cisterna magna

Foramen of Magendie

Central canal

Arachnoid granulation

Sinus

Dura mater

Arachnoid

Subarachnoid space

Pia mater

FIGURE 6.8   Circulation of cerebrospinal fluid through the ventricular system.

central nervous system where it is reabsorbed into the blood supply through the **arachnoid granulations.** The arachnoid granulations protrude into the **superior sagittal sinus,** which eventually drains into the veins serving the brain. (See **INSERT, FIGURE 6.8.**)

Figure 6.8 also illustrates a fold of dura mater, the **tentorium,** which separates the cerebellum from the overlying cerebrum. (See **FIGURE 6.8.**) This tough sheet of connective tissue extends to the top of the brainstem, leaving an opening (**tentorial notch**) through which the brainstem passes. Prizefighters (unless they are extremely good and quit early enough) usually receive a good many blows to the head. This repeated jarring of the brain often causes bruising of the brainstem against the edge of the tentorium; this assault is one of the causes of the "punch-drunk" syndrome.

The meninges, skull, and vertebral column encase the CNS in a rigid container of fixed volume. This fact means that any growth in the mass of the brain must result in displacement of the fluid contents of the container. Hence, growth of a brain tumor, depending on its location, will often deform the walls of the ventricular system, as the invading mass takes up volume previously occupied by CSF. (It is quite fortunate that these hollow ventricles exist; the only other fluid-filled spaces are the blood vessels, which would be constricted by a growing tumor if there were no ventricles in the brain.)

In order to diagnose and locate brain tumors, clinicians examine X-ray photographs of the ventricular system. By comparing the **ventriculogram** thus obtained with normal ones, a radiologist can determine the location and extent of tumors within the brain. In the past, air or a radiopaque dye was injected to make the ventricular system visible. However, a recently developed technique (the CAT scan, which will be described in Chapter 7) makes it possible to photograph the ventricular system without having to inject air or chemicals.

Occasionally the flow of CSF is interrupted at some point in its route of passage. For example, the cerebral aqueduct may be blocked by a tumor. This occlusion results in greatly increased pressure within the ventricles, since the choroid plexus continues to produce CSF. The walls of the ventricles then expand and produce a condition known as **hydrocephalus** (literally, "water-head"). If the obstruction remains, and if nothing is done to reverse the increased intracerebral pressure, blood vessels will be occluded, and permanent—perhaps fatal—brain damage will occur. The optic nerves, serving the eyes, are covered with meningeal layers, including the flexible but unstretchable dura mater. Increased intracerebral pressure is thus transmitted through the contents of the optic nerve the way water pressure can be transmitted through a hose. The pressure will cause the *optic disk* (attachment of the optic nerve to the eye) to bulge forward into the fluid cavity of the eye. The state of the optic disk, which can be seen at the back of the eye by looking through the pupil, is used as a clinical sign in the diagnosis of hydrocephalus.

ANATOMICAL SUBDIVISIONS
OF THE BRAIN

The brain is usually divided into **forebrain, midbrain,** and **hindbrain.** These major divisions are further subdivided; Table 6.1 presents these subdivisions, along with the principal structures found in each region. (See **TABLE 6.1**.)

Table 6.1 Anatomical subdivisions of the brain

Major division	Subdivision	Principal structures
Forebrain	Telencephalon	Cerebral cortex
		Basal ganglia
		Limbic system
	Diencephalon	Thalamus
		Hypothalamus
Midbrain	Mesencephalon	Tectum
		Tegmentum
Hindbrain	Metencephalon	Cerebellum
		Pons
	Myelencephalon	Medulla oblongata

Telencephalon

The "end brain" includes the **cerebral cortex,** covering the surface of the cerebral hemispheres, the **basal ganglia,** and the **limbic system.** The latter two sets of structures are located, principally, within the deep or **subcortical** portions of the brain.

CEREBRAL CORTEX.   Cortex means "bark," and the cerebral cortex surrounds the cerebral hemispheres like the bark of a tree. In humans the cortex is greatly convoluted; these convolutions, consisting of **sulci** (small grooves), **fissures** (large grooves), and **gyri** (bulges between adjacent sulci or fissures), greatly enlarge the surface area of the cortex, as compared with a similarly sized smooth brain. The amount of cerebral cortex, relative to the size of the rest of the brain, correlates well with phylogenetic development; higher animals have more cortex.

The surface of the cerebral hemispheres is divided into five lobes. Four of these lobes (**frontal, parietal, temporal,** and **occipital**) are visible on the lateral surface and are shown in Figure 6.9. (See **FIGURE 6.9** on the following page.)

The midsagittal view in Figure 6.10 shows the **limbic lobe** and lets us see the inner surface of the other four lobes (See **FIGURE 6.10** on the following page.)

The **corpus callosum** is the largest **commissure** in the brain; it consists of white matter (myelinated axons) that connect the two hemispheres of the brain. It unites **homotopic,** or geographically similar regions. (*Homo* = "same"; *topos* = "place.") In order to prepare a midsagittal view of the brain, one must slice through the middle of the corpus callosum. (See **FIGURE 6.10.**).

The actual appearance of part of the corpus callosum is shown in Figure 6.11. Some of the cortex has been dissected away to reveal bundles of fibers that interconnect homotopic region. A fine sheet of gray matter (nerve cell bodies) called the **indusium griseum** ("gray

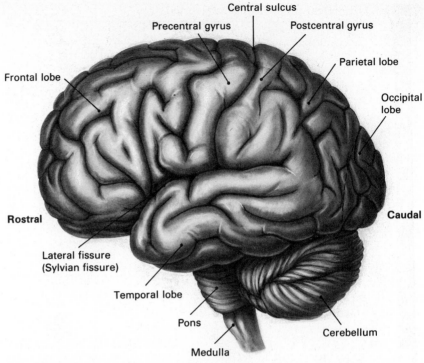

FIGURE 6.9   A lateral view of the human brain.

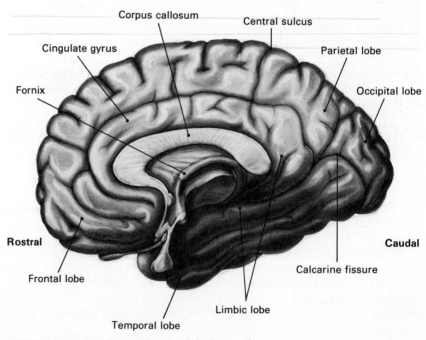

FIGURE 6.10   A midsagittal view of the human brain.

**Introduction to the
Structure of the
Nervous System**

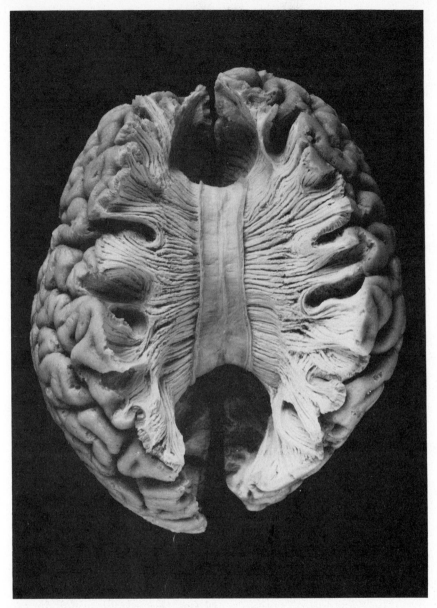

FIGURE 6.11 A dorsal view of a human brain, showing the appearance of
the corpus callosum. (From Gluhbegovic, N., and Williams, T. H. *The
Human Brain: A Photographic Atlas*. Hagerstown, Md.: Harper & Row, 1980.)

tunic") covers the medial part of the corpus callosum. (See **FIGURE
6.11.**)

The frontal, parietal, temporal, and occipital lobes consist of
neocortex—the most recently evolved neural tissue. The limbic lobe
consists of limbic cortex (often called paleocortex) and is a part of
the limbic system, which includes a number of interconnected subcortical brain structures.

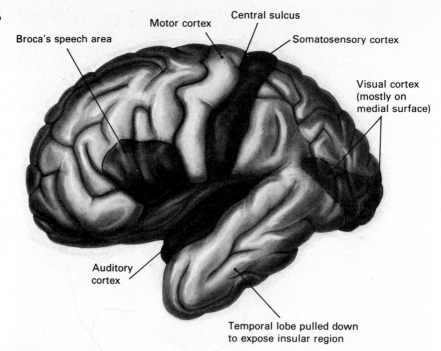

FIGURE 6.12 A schematic lateral view of the human brain showing some of the important cortical areas. Only the more prominent sulci and gyri are shown.

There are several neocortical regions that have special functions. Some of these areas are shown in Figure 6.12. **Sensory areas** receive primary sensory information from the receptors for audition, vision, somatosenses (touch, pressure, temperature, etc.), and taste. **Motor cortex** contains neurons that participate in the control of movement. Other regions are associated with higher functions; for example, damage to **Broca's speech area** in the frontal lobe leads to severe difficulty in speaking. (See **FIGURE 6.12.**)

As we shall see in later chapters, the terms *motor, sensory,* and *association* are too simple. Motor cortex (often referred to today as motor-sensory cortex) contains neurons that receive direct inputs from the ascending sensory system. Somatosensory cortex, on the other hand, contains neurons that join the descending motor fiber pathways (as does much of association cortex).

LIMBIC SYSTEM. The limbic system consists of a set of interconnected structures including limbic cortex, **hippocampal formation** (usually referred to as **hippocampus**), **amygdaloid complex** (usually called **amygdala**), **septum, anterior thalamus**, and **mammillary body.** (The latter two structures are part of the diencephalon, not the telencephalon, but they are generally considered to be part of

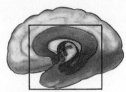

**Introduction to the
Structure of the
Nervous System**

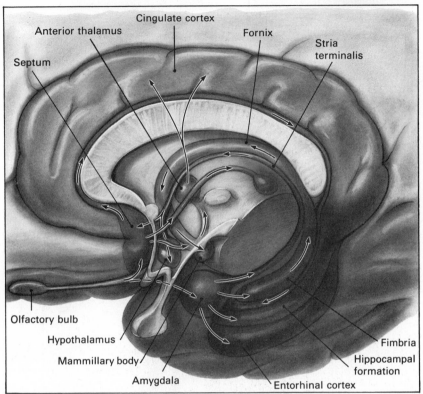

FIGURE 6.13   A very schematic, simplified representation of the limbic
system.

the limbic system.) Figure 6.13 illustrates these structures and their
interconnections. (See **FIGURE 6.13.**) The limbic system is involved
in emotional behavior, motivation, and (perhaps) memory; it will be
discussed in more detail in later chapters.

BASAL GANGLIA.   The basal ganglia (*amygdala, globus pallidus,
caudate nucleus,* and *putamen*) are shown in Figure 6.14. (See **FIG-
URE 6.14** on the following page.) The basal ganglia are concerned
with motor control and constitute a major portion of the extrapyramidal motor system (described in more detail in Chapter 10). For
example, Parkinson's disease is characterized by degeneration of dopaminergic cells that send axons to the caudate nucleus.

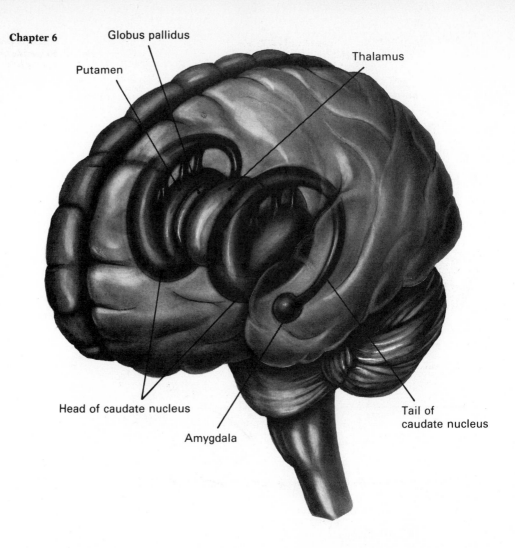

FIGURE 6.14   A schematic representation of the basal ganglia of a human brain.

The basal ganglia also play a role in emotional behavior; destruction of portions of the amygdala (which shares membership in the limbic system and basal ganglia) often produces taming of normally aggressive animals. Recently, neurosurgeons have surgically removed amygdalas of pathologically violent humans in an attempt to control their aggressive outbursts. (These operations will be described in Chapter 16.)

Diencephalon

The diencephalon is the most rostral portion of the brainstem. (See **FIGURE 6.15.**) The two most important structures of the "inter-

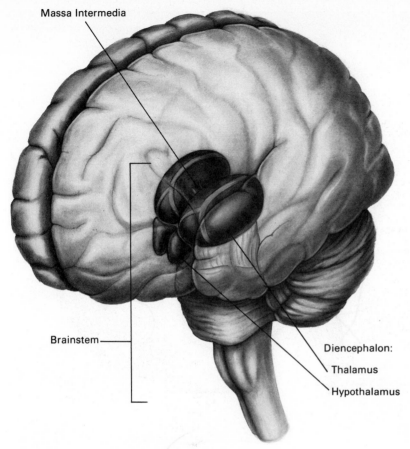

Massa Intermedia

Brainstem

Diencephalon:

Thalamus

Hypothalamus

FIGURE 6.15   A schematic representation of the human brainstem and diencephalon.

brain" are the thalamus and hypothalamus.

THALAMUS.   The thalamus is located in the dorsal part of the diencephalon. It is a large, two-lobed structure, whose sides are connected by the massa intermedia, which pierces the middle of the third ventricle. (See **FIGURE 6.15.**) (The massa intermedia is probably not a very important structure, since the brains of some apparently normal people have been observed to lack this bridge of tissue.)

Most neural input to the cerebral cortex is received from the thalamus; indeed, much of the cortical surface can be divided into regions receiving *projections* from specific parts of the thalamus. (***Projection fibers*** are those sets of axons, from cell bodies located in one region of the brain, that synapse on other neurons located within another specific region.) Figure 6.16 shows the appearance of some of the projection fibers from thalamus to cortex. Even a gross dis-

FIGURE 6.16 A lateral view of a partially-dissected human brain, showing projection fibers from thalamus to neocortex. (From Gluhbegovic, N., and Williams, T. H. *The Human Brain: A Photographic Atlas.* Hagerstown, Md.: Harper & Row, 1980.)

section such as this reveals how massive the thalamo-cortical projection is. (See **FIGURE 6.16.**)

The thalamus can be divided into a number of **nuclei** which are discrete clumpings of a large number of neurons of similar shape. Some of these nuclei (**sensory relay nuclei**) receive sensory information from terminals of incoming axons. The neurons in these nuclei then relay the sensory information to specific sensory projection areas of the cortex. For example, the **lateral geniculate nucleus** projects to visual cortex, the **medial geniculate nucleus** projects to auditory cortex, and the **ventral posterior nucleus** projects to somatosensory cortex. (See **FIGURE 6.17.**)

Other thalamic nuclei project to specific regions of cortex, but they do not relay primary sensory information. For example, the **ventrolateral nucleus** projects to motor cortex, the **dorsomedial nu-**

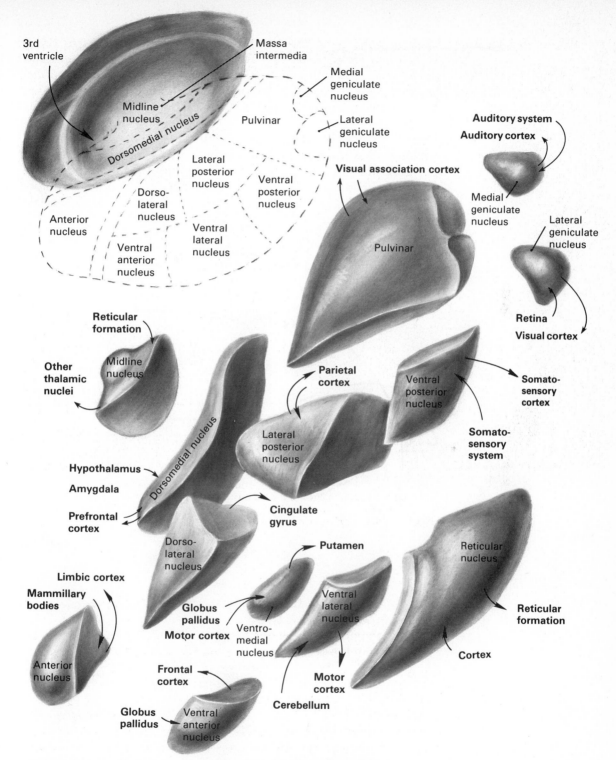

FIGURE 6.17 A schematic representation of the thalamus, along with some of the principal inputs and outputs of the various thalamic regions. Besides these inputs and outputs, there are many interconnections among the various thalamic regions.

cleus projects to prefrontal cortex, the ***pulvinar*** projects to visual association areas in parietal and temporal cortex, and the ***anterior nuclei*** project to limbic cortex. Other thalamic nuclei (e.g., ***midline*** and ***reticular*** nuclei) project diffusely to widespread regions of cortex and to other thalamic nuclei. (See **FIGURE 6.17.**)

HYPOTHALAMUS.   The hypothalamus lies at the base of the brain. Although it is a relatively small structure, it is of considerable importance in control of the autonomic nervous system, in reflex integration, in control of the endocrine system, and in organization of behaviors that are related to survival of the species. (One often refers

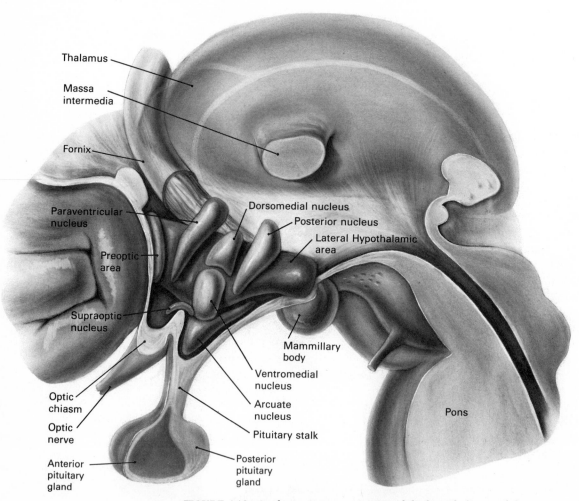

FIGURE 6.18   A schematic representation of the hypothalamus of the human brain.

to the hypothalamus as controlling the four Fs: fighting, feeding, fleeing, and mating.)

The hypothalamus is situated on both sides of the inferior portion of the third ventricle. As its name implies, this structure is located beneath the thalamus (not immediately beneath it; the **sub-thalamus**—concerned with motor control—is located between the thalamus and the hypothalamus). The hypothalamus is a very complex structure. It contains many nuclei and fiber tracts; Figure 6.18 indicates its location and size. Note that pituitary gland is attached to the base of the hypothalamus via the **pituitary stalk.** (See **FIGURE 6.18**.) The role of the hypothalamus in control of the four Fs will be considered in several chapters later in the book (Chapters 11, 12, 13, 16), along with more detailed descriptions of its anatomy.

Mesencephalon

The mesencephalon, or midbrain, is the portion of the brainstem just caudal to the diencephalon. It is usually divided into parts: the dorsal **tectum** and the **ventral tegmentum.**

TECTUM.   The principal structures of the tectum ("roof") are the **superior colliculi** and **inferior colliculi** (collectively referred to as the **corpora quadrigemina,** or "bodies of four twins"). Figure 6.19 illustrates a dorsal and ventral view of the brainstem, with the overlying cerebrum and cerebellum removed. The four bumps on the dorsal surface are the superior and inferior colliculi (a colliculus is a "small hill"). (See **FIGURE 6.19** on the following page.) The inferior colliculi are a part of the auditory system. As you will see in Chapter 9, all fibers conveying auditory information synapse in or pass through the inferior colliculi on the way to the medial geniculate nucleus and auditory cortex. The superior colliculi are part of the visual system. In mammals, however, they are not the most important part. Mammals have evolved an addition to the collicular system (the sole visual system in lower vertebrates): the direct retino-geniculate-visual cortex system. However, the superior colliculi have been left with a role in vision—principally in visual reflexes and reactions to moving stimuli, although recent evidence suggests that the perceptual capacities of the tectum have been underestimated.

TEGMENTUM.   The tegmentum consists principally of the rostral end of the **reticular formation,** several nuclei controlling eye movements, and the **red nucleus** and **substantia nigra** (parts of the extra-pyramidal motor system).

RETICULAR FORMATION.   The reticular formation is a large structure consisting of many nuclei (over ninety in all). It is also characterized by a diffuse, interconnected network of neurons with

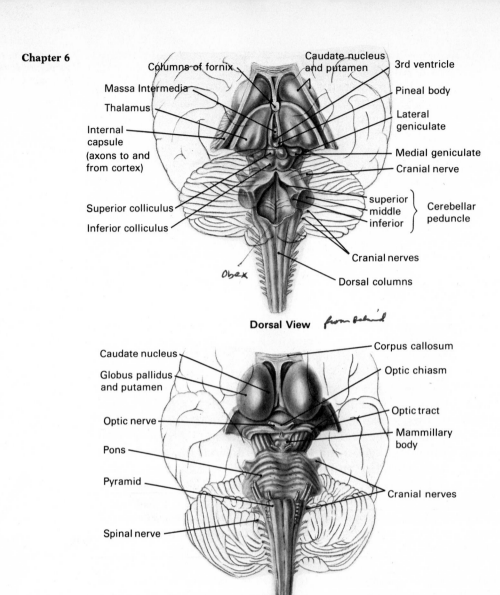

Caudate nucleus
and putamen

3rd ventricle

Columns of fornix

Massa Intermedia

Pineal body

Thalamus

Lateral
geniculate

Internal
capsule
(axons to and
from cortex)

Medial geniculate

Cranial nerve

Superior colliculus

superior
middle
inferior

Cerebellar
peduncle

Inferior colliculus

Cranial nerves

Obex

Dorsal columns

Dorsal View *from behind*

Caudate nucleus

Corpus callosum

Globus pallidus
and putamen

Optic chiasm

Optic nerve

Optic tract

Mammillary
body

Pons

Pyramid

Cranial nerves

Spinal nerve

Ventral View *from in front*

FIGURE 6.19   Two views of the human brainstem.

complex dendritic and axonal processes. (Indeed, *reticulum* means
"little net"; early anatomists were struck by the netlike appearance
of the reticular formation.) The reticular formation occupies the central core of the brainstem, from the lower border of the medulla to
the upper border of the midbrain. (Most neuroscientists consider the
reticular and midline nuclei of the thalamus to be a functional extension of the reticular formation.) The reticular formation receives
sensory information by means of various pathways, and projects fi-

**Introduction to the
Structure of the
Nervous System**

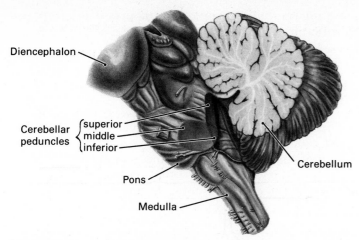

Diencephalon

Cerebellar
peduncles {superior
middle
inferior

Cerebellum

Pons

Medulla

FIGURE 6.20   A schematic representation of the human brainstem,
showing how the cerebellum connects with this region.

bers to cortex, thalamus, and spinal cord. It plays a role in sleep and
arousal, selective attention, muscle tonus, and control of various vital reflexes. Some of these functions will be described more fully in
later chapters.

CEREBELLUM.   The *cerebellum* ("little brain") is like a miniature
version of the cerebrum. It is covered by *cerebellar cortex* and has
a set of *deep cerebellar nuclei* that project to cerebellar cortex just
as the thalamic nuclei project to cerebral cortex. Figure 6.20 shows
the brainstem with the cerebellum dissected away on one side, to
illustrate the superior, middle, and inferior *cerebellar peduncles,*
bundles of white matter that connect the cerebellum to the brainstem. (See **FIGURE 6.20.**)
   Without the cerebellum it would be impossible to stand, walk,
or perform any coordinated movements. (A virtuoso pianist or other
performing musician probably owes much to his or her cerebellum.)
The cerebellum receives visual, auditory, vestibular, and somatosensory information, and it also receives information about individual
muscular movements being directed by the brain. The cerebellum
integrates this information and modifies the motor outflow, exerting
a coordinating and smoothing effect on the movements. Cerebellar
damage results in jerky, poorly coordinated, exaggerated movements; extensive cerebellar damage makes it impossible even to
stand.

PONS.   The *pons,* a rather large bulge in the brainstem, lies immediately ventral to the cerebellum. Refer back to **FIGURE 6.20.**
The pons contains, in its core, a portion of the reticular formation,

including some nuclei that appear to be important in sleep and arousal. Several cranial nerve nuclei, serving the head and facial region, are located in the pons.

Myelencephalon

MEDULLA. The *medulla oblongata* (usually just called the medulla) is the most caudal portion of the brainstem; its lower border is the rostral end of the spinal cord. The medulla contains part of the reticular formation, including nuclei that control vital functions such as regulation of the cardiovascular system, respiration, and skeletal muscle tonus. Several cranial nerve nuclei are also located there.

SPINAL CORD

The spinal cord is a roughly cylindrical structure, containing a bulge in its lower middle region and gradually tapering to a point in the lower back region. The principal function of the spinal cord is to distribute motor fibers to the effectors of the body (glands and muscles) and to collect somatosensory information to be passed on to the brain. The spinal cord also has a certain degree of autonomy from the brain; various reflexive control circuits (some of which are described in Chapter 10) are located there.

The spinal cord is protected by the vertebral column, which is composed of twenty-four individual vertebrae of the *cervical, thoracic* and *lumbar* regions, and the fused vertebrae making up the *sacral* and *coccygeal* portions of the column. Figure 6.21 illustrates the divisions and structures of the spinal cord and vertebral column. Note that the spinal cord is only about ⅔ as long as the vertebral column; the rest of the space is filled by a mass of *spinal roots* composing the *cauda equina* ("mare's tail"). (See **FIGURE 6.21.**)

Early in embryological development the vertebral column and spinal cord are of the same length. As development progresses, the vertebral column grows faster than does the spinal cord. This differential growth rate causes the spinal roots to be displaced downward; the most caudal roots travel the farthest before they emerge through the intervertebral foramina and thus compose the cauda equina. To produce the *caudal block* sometimes used in childbirth, a local anesthetic can be injected into the CSF contained within the dural sac surrounding the cauda equina.

Small bundles of fibers (*fila*) emerge from the spinal cord in two straight lines along its dorsolateral and ventrolateral surfaces. Groups of these fila fuse together and become the thirty-one paired

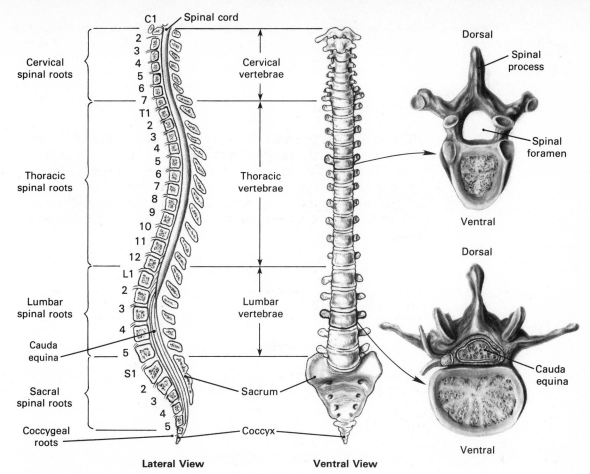

Lateral View　　　**Ventral View**

FIGURE 6.21　The human spinal column, with details showing the anatomy of the vertebrae, and the relationship between spinal cord and spinal column.

sets of ***dorsal*** and ***ventral roots.*** These roots, in turn, join together as they pass through openings between the vertebrae (***intervertebral foramina***) and become the thirty-one pairs of spinal nerves. Figure 6.22 illustrates a cross section of the spinal column taken between two adjacent vertebrae, showing the junction of the dorsal and ventral roots in the intervertebral foramina. (See **FIGURE 6.22** on the following page.)

The spinal cord, like the brain, consists of white matter and gray matter. Unlike the brain, its white matter (consisting of ascending and descending bundles of myelinated axons) is on the outside; the gray matter (mostly cell bodies and short, unmyelinated axons) is on the inside. Figure 6.23 shows a cross section through the spinal cord. Ascending tracts are shown in dark gray; descending tracts, in light

Dorsal *towards back*

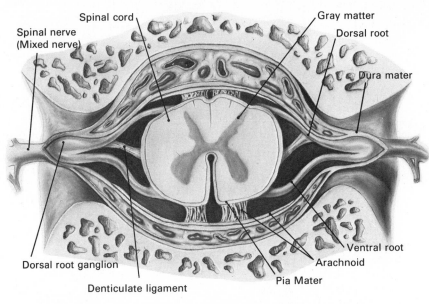

FIGURE 6.22   A detailed view of a section through a vertebra, showing the spinal cord, dorsal and ventral roots, and spinal nerves.

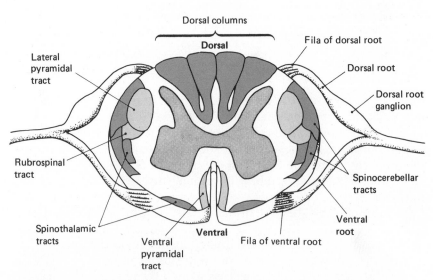

FIGURE 6.23   A schematic cross section through the spinal cord, showing the principal ascending (sensory) and descending (motor) tracts.

gray. (See **FIGURE 6.23.**) The ascending (sensory) tracts will be described in a section on somatosenses in Chapter 9. Similarly, the descending (motor) tracts will be described in Chapter 10.

PERIPHERAL NERVOUS SYSTEM

The brain and spinal cord communicate with skeletal muscles and sensory systems of the rest of the body via the cranial and spinal nerves. Glands, smooth muscles, and cardiac muscle are regulated by the autonomic nervous system, which consists of offshoots of spinal and cranial nerves and the associated autonomic ganglia.

Spinal Nerves

The spinal nerves begin at the junction of the dorsal and ventral roots of the spinal cord. The nerves leave the vertebral column and travel, branching repeatedly as they go, to the muscles or sensory receptors they innervate. Branches of spinal nerves often follow blood vessels, especially those branches that innervate skeletal muscles. Figure 6.24 shows a dorsal view of a human with a few branches of the spinal nerves. Note that the spinal nerves of the thoracic region follow spaces between the ribs. These intercostal nerves (*costa* means "rib") are paralleled by the intercostal arteries and veins. (See **FIGURE 6.24** on the following page.) Note also that some of the spinal nerves fuse together and then divide again; the fusions are referred to as *plexuses* (braids).

Now let us consider the pathways by which sensory information enters the spinal cord and motor information leaves it. The cell bodies of all nerve fibers that bring sensory information into the brain and spinal cord are located outside the central nervous system. (The sole exception is the visual system; the retina of the eye is considered to be a part of the brain.) The cell bodies that give rise to the axons afferent to the spinal cord reside in the *dorsal root ganglia,* rounded swellings of the dorsal root. The afferent neurons are of the unipolar type (described in Chapter 2). The axonal stalk divides close to the cell body; it sends one limb into the spinal cord and the other limb out to the sensory organ. Note that the dorsal root is completely sensory in nature; all its fibers are afferent to the spinal cord. (Refer back to **FIGURE 6.22.**)

Cell bodies that give rise to the ventral root are located within the gray matter of the spinal cord. The axons of these multipolar neurons leave the spinal cord via a ventral root, which joins a dorsal root to make a spinal nerve (often referred to as a *mixed nerve* because it carries both sensory and motor fibers).

127

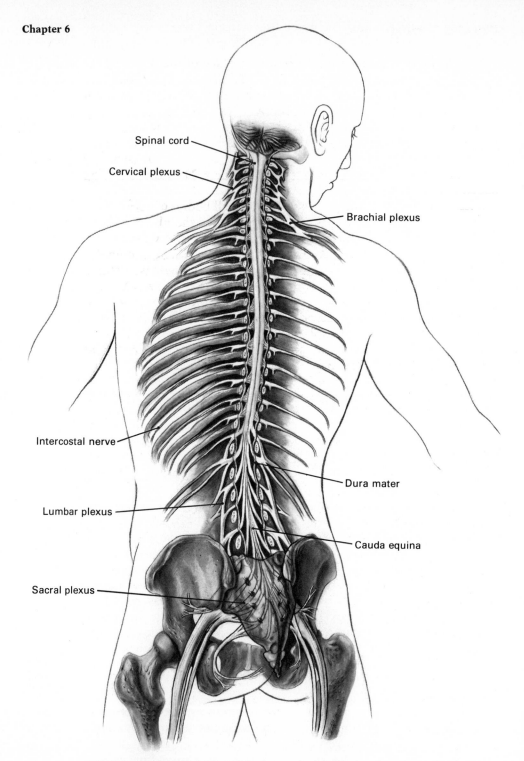

Spinal cord

Cervical plexus

Brachial plexus

Intercostal nerve

Dura mater

Lumbar plexus

Cauda equina

Sacral plexus

FIGURE 6.24   A dorsal view of the human body, showing the routes traveled by the principal spinal nerves.

Cranial Nerves

Twelve pairs of nerves leave the ventral surface of the brain. These cranial nerves serve sensory and motor functions of the head and neck region. (One of them, the *tenth*, or **vagus nerve,** serves autonomic functions of organs in the thoracic and abdominal cavities.) Figure 6.25 presents a view of the base of the brain and illustrates the cranial nerves and the structures they serve. Note that efferent (motor) fibers are drawn as solid lines and that afferent (sensory) fibers are drawn as dashed lines. (See **FIGURE 6.25** on the following page.)

As I mentioned in the previous section, cell bodies of nerve fibers afferent to the brain and spinal cord (except for the visual system) are located outside the CNS. Somatosensory information (and also gustation, or taste) is received, via the cranial nerves, from unipolar neurons whose cell bodies reside in **cranial nerve ganglia** (structures similar to the dorsal root ganglia of the spinal cord). (See **FIGURE 6.25.**) Auditory, vestibular, and visual information is received via fibers of bipolar neurons (described in Chapter 2). Olfactory information is received via a complex system; the olfactory bulbs, at the ends of the olfactory nerves, contain a considerable amount of neural circuitry. The cell bodies that give rise to the olfactory nerve fibers are of the multipolar type. The various sensory receptors will be described in Chapter 8; the corresponding neural pathways will be found in Chapter 9.

AUTONOMIC NERVOUS SYSTEM

The autonomic nervous system (ANS) is concerned with regulation of smooth muscle, cardiac muscle, and glands. Smooth muscle is found in the skin (associated with hair follicles), in blood vessels, in the eye (controlling pupil size and accommodation of the lens), and in the wall and sphincters of the gut, gall bladder, and urinary bladder. Merely describing the organs innervated by the autonomic nervous system suggests the function of this system: regulation of "vegetative processes" in the body. The ANS consists of two anatomically separate systems, the *sympathetic* and *parasympathetic* divisions.

Sympathetic Division of the ANS

The sympathetic division is most active when **catabolic** processes are required (i.e., those involved with expenditure of energy from reserves that are stored in the body). For example, increased blood flow to skeletal muscles, secretion of epinephrine (resulting in in-

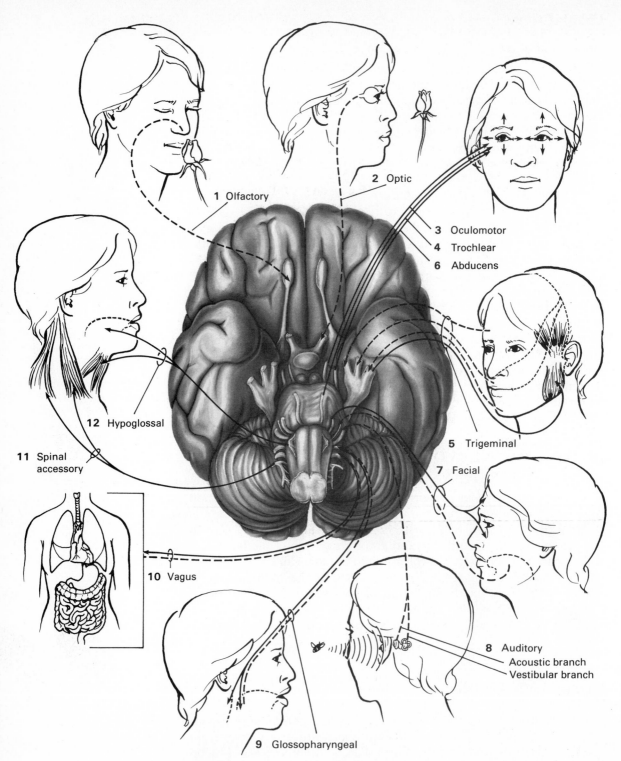

1 Olfactory

2 Optic

3 Oculomotor
4 Trochlear
6 Abducens

5 Trigeminal

7 Facial

8 Auditory
Acoustic branch
Vestibular branch

9 Glossopharyngeal

10 Vagus

11 Spinal
accessory

12 Hypoglossal

FIGURE 6.25 The locations and functions of the cranial nerves.

creased heart rate and a rise in blood sugar level), and piloerection (erection of fur in mammals who have it and production of "goose bumps" in humans) are some effects that are mediated by the sympathetic nervous system during excitement. The cell bodies of sympathetic motor neurons are located in the intermediate horn of the gray matter of the thoracic and lumbar regions of the spinal cord. The fibers of these neurons exit via the ventral roots. After joining the spinal nerves, the fibers branch off and pass into *spinal sympathetic ganglia.* Figure 6.26 shows the relationship of these ganglia to the spinal cord. Note that the various spinal sympathetic ganglia are connected to the neighboring ganglia above and below, and thus form the *sympathetic chain.* (See **FIGURE 6.26** on the following page.)

All sympathetic motor fibers enter the ganglia of the sympathetic chain, but not all of them synapse there. Some leave the ganglia and travel to one of the *sympathetic prevertebral ganglia,* which are located away from the spinal cord, among the internal organs. All sympathetic motor fibers synapse on neurons located in the ganglia. The neurons upon which they synapse are called *postganglionic neurons.* In turn, the postganglionic neurons send axons to the target organs, such as intestine, stomach, kidney, or sweat glands. (See **FIGURE 6.26.**)

All synapses within the sympathetic ganglia are cholinergic; the terminals on the target organs, belonging to the postganglionic fibers, are noradrenergic. The exception to this rule is provided by the sweat glands, which are innervated by cholinergic terminals. The medulla of the adrenal gland is innervated directly by preganglionic sympathetic fibers (whose terminals are cholinergic). The secretory cells of the adrenal medulla may be thought of as the postganglionic cells; they are, of course, adrenergic, secreting epinephrine and norepinephrine. (See **FIGURE 6.26.**)

Parasympathetic Division of the ANS

The parasympathetic division of the autonomic nervous system is concerned with effects supporting *anabolic* activities (those concerned with increases in the body's supply of stored energy). The anabolic and catabolic processes of the body together make up its *metabolism.* Such effects as salivation, gastric and intestinal motility, secretion of digestive juices, and increased blood flow to the gastrointestinal system are mediated by the parasympathetic division of the ANS.

Cell bodies that give rise to preganglionic nerve fibers are located in two regions: the nuclei of the cranial nerves and the intermediate horn of the gray matter in the sacral region of the spinal cord. Thus, the parasympathetic division of the ANS has often been referred to as the *craniosacral* system. Parasympathetic ganglia are located in the immediate vicinity of the target organs; the postgan-

Introduction to the
Structure of the
Nervous System

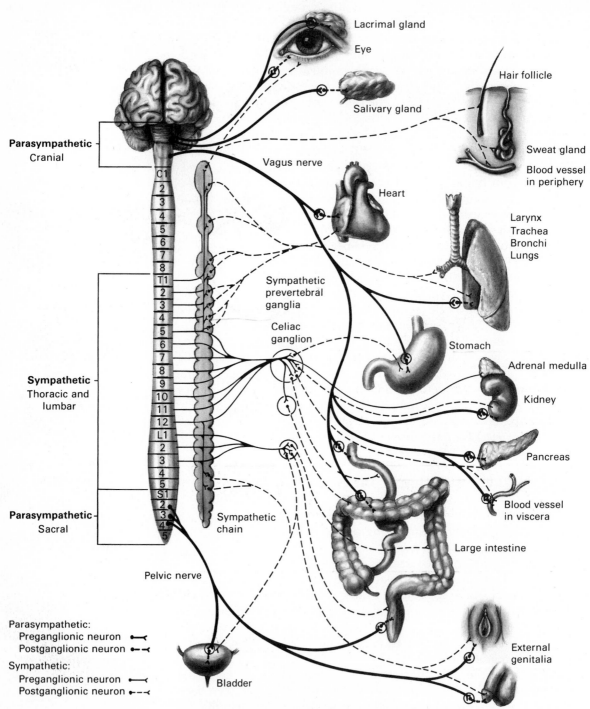

Parasympathetic
Cranial

Sympathetic
Thoracic and
lumbar

Parasympathetic
Sacral

Lacrimal gland

Eye

Hair follicle

Salivary gland

Sweat gland

Vagus nerve

Blood vessel
in periphery

Heart

Larynx
Trachea
Bronchi
Lungs

Sympathetic
prevertebral
ganglia

Celiac
ganglion

Stomach

Adrenal medulla

Kidney

Pancreas

Blood vessel
in viscera

Sympathetic
chain

Large intestine

Pelvic nerve

External
genitalia

Parasympathetic:
  Preganglionic neuron
  Postganglionic neuron
Sympathetic:
  Preganglionic neuron
  Postganglionic neuron

Bladder

FIGURE 6.26 A schematic representation of the autonomic nervous
system and the target organs it serves.

132

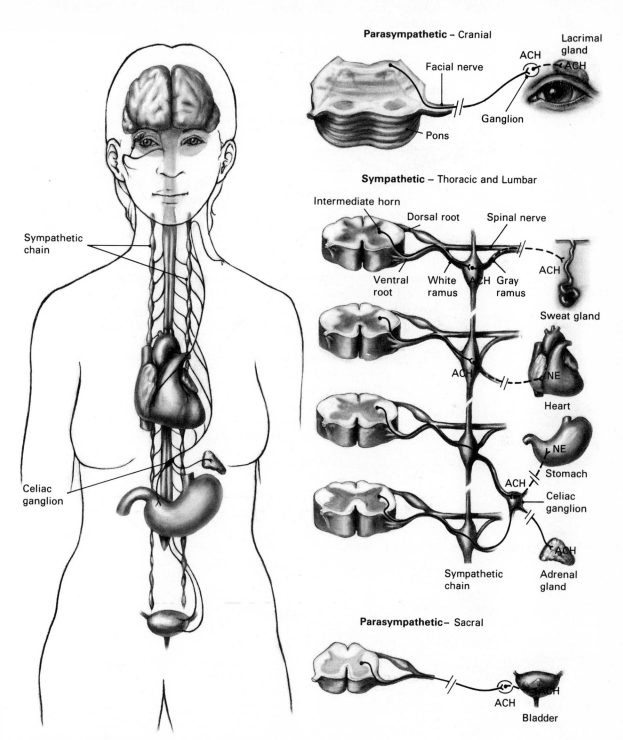

Parasympathetic – Cranial

Facial nerve
ACH
Lacrimal gland
ACH
Ganglion
Pons

Sympathetic – Thoracic and Lumbar

Intermediate horn
Dorsal root
Spinal nerve
Ventral root
White ramus
ACH
Gray ramus
ACH
Sweat gland

NE
Heart

NE
Stomach
ACH
Celiac ganglion

Sympathetic chain

ACH
Adrenal gland

Sympathetic chain

Celiac ganglion

Parasympathetic – Sacral

ACH
ACH
Bladder

FIGURE 6.27   Preganglionic and postganglionic fibers of the two branches of the autonomic nervous system.

glionic fibers are therefore relatively short. (See **FIGURE 6.27.**) The terminals of both preganglionic and postganglionic neurons are cholinergic.

Division of the autonomic nervous system into sympathetic and parasympathetic can be done easily on the basis of anatomy, but I should emphasize that these systems function together in a cooperative fashion. For example, if a man sits resting on a hot day, digesting his meal, parasympathetic activity increases the blood flow to the gastrointestinal system, stimulates motility of the stomach and gut, and increases secretion of digestive juices. At the same time, sympathetic fibers produce sweating. Perhaps the man and a particularly attractive female friend catch sight of each other; sympathetic activity dilates their pupils and increases their heart rate and rate of respiration. The man's parasympathetic activity results (perhaps, depending on the nature of their prior associations) in penile erection. We can see, from this example, that the two branches of the autonomic nervous system have very specific effects, and one branch can dominate in a given organ while the other branch exerts its effect in a different organ. (There *are* general effects, however. If the people in my example subsequently engage in sexual behavior, the man's parasympathetic activity, which is producing his erection and continuing to aid in the digestion of his food, will suddenly be replaced by massive sympathetic activity—with accompanying loss of erection, cold clammy sweat, cessation of the digestive process, rapid heart rate, shallow, rapid breathing, and secretion of epinephrine— should one of their spouses surprise them.)

KEY TERMS

amygdala (*a MIG da la*) p. 114
amygdaloid complex p. 114
anabolism (*an AB a liz'm*) p. 131
anterior p. 99
anterior nuclei (of thalamus) p. 118
anterior thalamus p. 114
arachnoid (*a RAK noyd*) p. 104
arachnoid granulation p. 109
arachnoid trabecula (*tra BEK u la*) p. 105
Broca's speech area p. 114
basal ganglia p. 111
catabolism (*ka TAB a liz'm*) p. 129
cauda equina (*KAW da ee KWINE a*) p. 124
caudal p. 99
caudate nucleus p. 115
central canal p. 107.
cerebellar cortex p. 123

cerebellar peduncles p. 123
cerebellum p. 123
cerebral aqueduct (*sur EE brul*) p. 107
cerebral cortex p. 111
cerebrospinal fluid (*sur EE bro SPY nul*) p. 102
cervical region (of spinal cord) p. 124
choroid plexus (*CORE oyd*) p. 107
coccygeal region (of spinal cord) (*cox i GEE ul*) p. 124
commissure (*KAHM i shure*) p. 111
coronal section p. 100
corpora quadrigemina p. 121
corpus callosum p. 111
cranial nerve ganglion p. 129
craniosacral system (*kray nee o SAY krul*) p. 131
cross section p. 100
deep cerebellar nuclei p. 123

SUGGESTED READINGS

CARPENTER, M. B. *Core Text of Neuroanatomy.* Baltimore: Williams & Wilkins, 1972.

NETTER, F. H. *The Ciba Collection of Medical Illustrations.* Vol. 1, *Nervous System.* Summit, N.J.: Ciba Pharmaceutical Products Co., 1953.

NOBACK, C R., and DEMAREST, R. J. *The Nervous System: Introduction and Review.* New York: McGraw-Hill, 1972.

The nervous system is exceedingly intricate; I am sure you do not have to be told that after reading this chapter. The nervous system is also beautiful when it is rendered by a good artist. The drawings in these three books are indeed beautiful—especially the full-color plates in Netter's volume. The books by Carpenter and by Noback and Demarest are more informative, however.

CROSBY, E. C., HUMPHREY, T., and LAUER, E. W. *Correlative Anatomy of the Nervous System.* New York: Macmillan, 1962.

TRUEX, R. C., and CARPENTER, M. B. *Human Neuroanatomy,* 6th edition. Baltimore: Williams & Wilkins, 1969.

The first three books are useful for learning neuroanatomy; these latter two are for reference once you have studied one of the books from the first list. The amount of detail contained in the book by Crosby, Humphrey, and Lauer and in the one by Truex and Carpenter makes them heavy going for the beginner.

GLUHBEGOVIC, N., and WILLIAMS, T. H. *The Human Brain: A Photographic Guide.* New York: Harper & Row, 1980.

This volume presents a series of beautiful photographs of human brains, whole and dissected. Since most of us do not have the dissecting skills that Drs. Gluhbegovic and Williams do, studying this book is probably *better* than dissecting a brain by yourself.

7

Research Methods of Physiological Psychology

Study of the biological basis of behavior involves the efforts of scientists in many disciplines: physiology, neuroanatomy, biochemistry, psychology, endocrinology, and histology, to name but a few. To pursue a project of research in physiological psychology requires competence in many experimental techniques. Since contradictory results are often obtained by the use of different procedures, one must be very familiar with the advantages and limitations of the various methods employed, so that the truth can be arrived at. Scientific investigation entails a process of asking questions of nature. The method that is used frames the question. Often, we receive a puzzling answer, only to realize later that we were not asking the question we thought we were.

The experimental methods used by physiological psychologists include neuroanatomical techniques, ablation, electrical recording, electrical stimulation, and chemical techniques. The boundaries of these categories are not distinct; for example, electrical stimulation and recording techniques can be used to obtain anatomical information.

NEUROANATOMICAL TECHNIQUES

Histological Procedures

The gross anatomy of the brain was described long ago, and everything that could be identified was given a name. Early anatomists named most brain structures according to their similarity to commonplace objects: amygdala, or "almond-shaped object"; hippocampus, or "seahorse"; genu, or "knee"; cortex, or "bark"; pons, or "bridge"; uncus, or "hook"—to give a few examples. More recent names tend to be duller: for example, the *nucleus ventralis medialis thalami, pars magnocellularis* (this clever name means "large-celled part of the ventral medial nucleus of the thalamus").

Detailed anatomical information about the brain has been obtained through use of various histological, or tissue-preparing, techniques. As we have seen, the brain consists of many billions of neurons and glial cells, the nerve cells forming distinct nuclei and fiber bundles. We cannot possibly see any cellular detail by gross examination of the brain. Even a microscope is useless without **fixation** and **staining** of the neural tissue.

FIXATION.   If we hope to study the tissue in the form it had at the time of the organism's death, we must destroy the **autolytic** (literally, "self-dissolving") **enzymes,** which will otherwise turn the tissue into amorphous mush. The tissue must also be preserved to prevent its decomposition by bacteria or molds. To achieve both of these objectives, the neural tissue is placed in a **fixative.** The most commonly used fixative is **formalin,** an aqueous solution of formaldehyde, a gas. Formalin halts autolysis, hardens the very soft and fragile brain, and kills any microorganisms that might destroy it. There are other fixatives, but formalin preserves and hardens well, penetrates the tissue rapidly, and is very cheap and easy to obtain, so this substance is almost always the one chosen to fix brain tissue.

Often the fixation process is given a head start by **perfusion** of the cerebral vascular system with the fixative. Almost always, the brain is at least perfused with a saline solution. The following procedure is used: The animal whose brain is to be studied is killed with an overdose of a general anesthetic (usually ether or sodium pentobarbital). The thoracic cavity is opened, and the *right atrium* (one of the chambers of the heart) is cut open. A needle is inserted into the heart, or a cannula (small metal tube) is inserted into the *aorta,* the large artery leaving the heart, and the animal's blood is replaced with a salt solution; the blood leaves via the cut in the atrium. (The animal's brain is perfused because better histological results are obtained when there is no blood present in the tissue.) Often the vascular system is then perfused with formalin, in order to speed up the process of fixation. The head is removed, the skull is opened, and the brain is removed and placed in a jar containing the fixative. (Brain removal—or perhaps I should say removal of an *intact* brain—takes a bit of practice. The brain is exceedingly delicate.)

Once the brain has been fixed, it is necessary to slice it into thin sections and to stain various cellular structures in order to see anatomical details. Some procedures require that the tissue be stained before being sliced, but the techniques that are most commonly used by physiological psychologists call for sectioning first, and then staining. Therefore, I shall describe the procedures in that order.

SECTIONING.　In order to slice (or *section*) neural tissue, a **microtome** is commonly used. A microtome (literally, "that which slices small") is an instrument capable of slicing tissue into very thin sections. Sections prepared for examination under a light microscope are typically 10 μm to 80 μm in thickness; those prepared for the electron microscope are generally cut at less than 1 μm. (A **micrometer,** abbreviated μm, is 1/1000 of a millimeter.) Electron microscopy will be considered separately in a later section.

A microtome must contain three parts: a knife, a platform on which to mount the tissue, and a mechanism that advances the knife

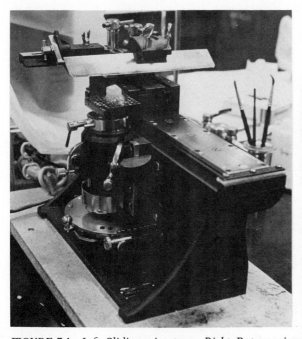

FIGURE 7.1　*Left*: Sliding microtome. *Right:* Rotary microtome.

(or the platform) the correct amount after each slice, so that another section can be cut. Figure 7.1 shows two commonly used microtomes. The one on the left is a sliding microtome. The knife holder slides forward on an oiled rail and takes a section off the top of the tissue mounted on the platform. The platform automatically rises by a predetermined amount as the knife and holder are pushed back. (See **FIGURE 7.1, LEFT** on the previous page.) To operate the rotary microtome shown on the right, you simply turn the wheel. The tissue platform moves up and down relative to the vertically mounted knife. The tissue is cut as it descends, and the platform is automatically advanced on the upstroke, moving the tissue into position for the next slice. (See **FIGURE 7.1, RIGHT.**)

Slicing brain tissue is not quite so simple as it might at first appear. As I mentioned, raw neural tissue is very soft. Fixation by immersion in formalin will harden the brain (to the texture of cheese, but without its characteristic stickiness). Either of two techniques can be used to make the tissue hard enough to cut thinly: freezing or embedding.

Freezing is conceptually the easier. The tissue can be chilled with blasts of compressed carbon dioxide, with dry ice (solid carbon dioxide), or with various kinds of refrigeration units. A particular type of microtome (a cryostat) operates within the confines of a freezer. Often the brain is first soaked in a sucrose (table sugar) solution; this procedure minimizes tissue damage by preventing the formation of large ice crystals as the brain freezes. The temperature of the brain must be carefully regulated; if the block is too cold, the tissue will shatter into little fragments. If it is too warm, a layer of tissue will be torn off rather than sliced off. Frozen sections can be cut with either sliding or rotary microtomes.

The brain can be embedded in materials that are of sliceable consistency at room temperature. (That has to be pretty consistent, for some substances. The laboratory can be brought to a halt if the air conditioning fails on a hot day.) The usual materials for embedding are paraffin and nitrocellulose. Paraffin comes in various grades, according to the room temperature at which it can best be sliced. The brain is first soaked in a solvent for paraffin (such as xylene) and then soaked in successively stronger solutions of paraffin (in an oven kept at a temperature just above the melting point for paraffin). Then the brain is placed in a small container of liquid paraffin, which is allowed to cool and harden. The entire block is sliced, the paraffin providing physical support for the tissue. A rotary microtome is used for cutting paraffin-embedded sections. It is fascinating to watch the cutting of these sections; as the knife passes through the block, it warms it slightly so that the sections get glued together, end to end. (See **FIGURE 7.2.**) It is immensely gratifying, after having spent a lot of time preparing the tissue, to see the sections emerge in a beautiful ribbon.

The brain can also be embedded in nitrocellulose. This com-

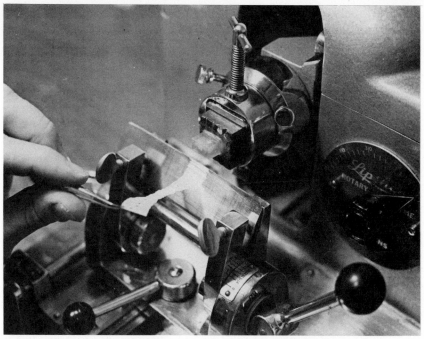

FIGURE 7.2 A "ribbon" being formed as a paraffin-embedded brain is sectioned on a rotary microtome.

pound, produced from the reaction of nitric acid with cellulose, is sometimes referred to as gun cotton. (Cotton can serve as the source of cellulose fibers.) In past years, histological technicians had to make sure they kept their gun cotton wet, because once it dried out, the vibration of even a footstep in the lab might cause the nitrocellulose to blow up. Today we usually purchase a commercial product that will not explode upon impact.

The brain is infiltrated with successively more concentrated solutions of nitrocellulose. The embedding medium is hardened into a block by the action of chloroform vapor. The block, along with its enclosed brain tissue, attains the consistency of soft cheese rind (again, without any stickiness). The block is usually cut on a sliding microtome.

Plastics can be used for embedding tissue, but this procedure is reserved for preparation of sections for the electron microscope.

After the tissue is cut, it is usually mounted on glass microscope slides with an agent such as albumin (protein extracted from egg whites). The slide is dried and heated, and the albumin becomes insoluble, cementing the tissue sections to the glass. The tissue can then be stained, the entire slide being put into the various chemical solutions. (Some staining procedures require that the individual sections be stained first, and then mounted onto slides—a much more tedious process.) The stained and mounted sections are then covered with a mounting medium and a very thin glass coverslip is placed over the sections. The mounting medium (which is very thick and resinous) gradually dries out, thus keeping the coverslip in position.

STAINING.   If you looked at an unstained section of brain tissue under a microscope, you would be able to see the outlines of some large cellular masses and the more prominent fiber bundles, especially if the tissue were kept wet. However, no fine detail would be revealed. The study of microscopic neuroanatomy requires special histological stains. There are basically three types of stains used for neural tissue: those which reveal cell bodies by interacting with contents of the cytoplasm, those which selectively color myelin sheaths, and those which stain the cell membrane (entire cell, or just the axons).

Cell-body stains.   Cell-body stains give color to the **Nissl substance** contained in the cytoplasm, thus outlining cell bodies. In the late nineteenth century, Franz Nissl, a German neurologist, discovered that methylene blue, an aniline dye (derived from the distillation of coal tar), would stain cell bodies of brain tissue. The material that takes up the dye consists of RNA, DNA, and nucleoproteins, located in the nucleus and scattered, in the form of granules, in the cytoplasm. Many dyes can be used to stain cell bodies, but the most frequently used is **cresyl violet.** The dyes, by the way, were not developed for histological purposes; they were manufactured for use in dyeing cloth.

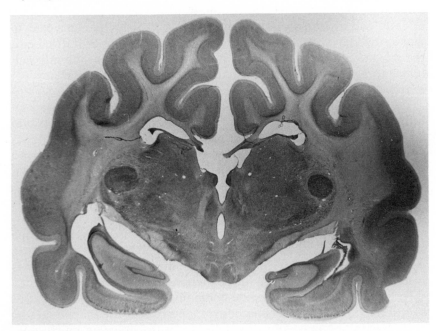

FIGURE 7.3   A frontal section of a cat brain, stained with a Nissl stain (cresyl violet). (Histological material courtesy of Mary Carlson.)

The discovery of cell-body stains (also called Nissl stains) made it possible to identify nuclear masses in the brain. Figure 7.3 shows a frontal section of a cat brain stained with a Nissl stain (cresyl violet). Note that it is possible to observe fiber bundles by their lighter appearance; they do not take up the stain. (See **FIGURE 7.3.**) The stain is not selective for *neural* cell bodies. All cells are stained, neurons and glia alike. It is up to the investigators to determine which is which—by size, shape, and location.

Myelin Stains. **Myelin stains** color myelin sheaths. These stains (such as the Weil method, which employs **hematoxylin,** a dye extracted from logwood) make it possible to identify fiber bundles. (What is light in Figure 7.3 is dark in Figure 7.4.) However, pathways of single fibers cannot be traced. There is simply too much intermingling of the individual fibers. (See **FIGURE 7.4.**)

Membrane stains. **Membrane stains** contain salts of various heavy metals (silver, uranium, osmium, etc.) that interact with the somatic, dendritic, and axonal membrane. The **Golgi-Cox stain** (which uses silver) is, for some mysterious reason, highly selective, staining only a fraction of the neurons in a given region. (Why this happens is not known, and the selectivity undoubtedly gives a biased view of the neurons that populate a given region.) The selective staining makes it possible, however, to observe the processes and arborizations of individual neurons and to trace details of synaptic inter-

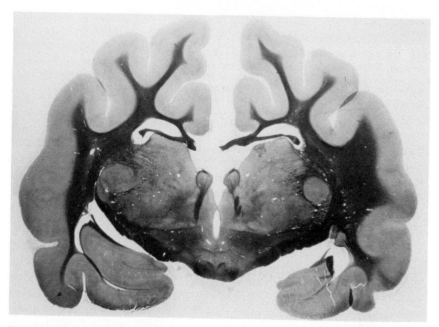

FIGURE 7.4   A frontal section of a cat brain, stained with a myelin stain (Weil method). (Histological material courtesy of Mary Carlson.)

FIGURE 7.5 A section of cortex of a cat brain, stained by the Golgi-Cox method, as modified by D. N. Spinelli and J. K. Lane. Unlike previous versions of the Golgi-Cox method, this one stains neural processes right to the outer edge of the cortex. (Histological material courtesy of D. N. Spinelli and J. K. Lane.)

connections. Figure 7.5 shows the appearance of individual neurons of the cerebral cortex stained by a new modification of the Golgi-Cox stain. Note the individual neurons and their interconnecting processes. The large cells in the center are oligodendroglia, providing myelin sheaths for the bundles of fibers running horizontally. (See **FIGURE 7.5.**)

Degenerating-axon stains. A special variant of the cell membrane stain was devised by Walle Nauta and Paul Gygax in the middle of this century. The Nauta-Gygax stain (which uses silver) identifies axons that are dying and are in the process of being destroyed by phagocytes. This stain might seem too esoteric to be of any practical use, but it has contributed immeasurably to our knowledge of the interconnections of various neural structures. However, as we shall see, new techniques are replacing the Nauta-Gygax method.

Tracing Neural Pathways

The central nervous system contains many billions of neurons, most of which are gathered together in thousands of discrete nuclei. These nuclei are interconnected by incredibly complex systems of axons. The problem of the neuroanatomist is to trace these connections: which nuclei are connected to which others, and what route is taken by the interconnecting fibers? The problem cannot be answered by means of histological procedures that stain all neurons, such as cell-body, membrane, or myelin stains. Close observation of a brain that has been prepared by these means reveals a tangled mass of neurons. Something must be done to make the connections that are being investigated stand out from all of the others.

DEGENERATION STUDIES. The first method to identify individual neural connections used the Nauta-Gygax degenerating axon stain. If a cell body is destroyed, or if the axon is cut, the distal portion of the axon quickly dies and disintegrates. This process is called *anterograde degeneration.* (See **FIGURE 7.6.**) A degenerating axon stain will identify these dying axons as trails of black droplets. (See **FIGURE 7.7.**)

The fact that only degenerating axons will be stained means that the Nauta-Gygax technique can be used to label one specific set of axons. The procedure works like this: A specific region of the brain (for example, a particular nucleus) is destroyed by means of stereotaxic surgery (which will be described later in this chapter). The animal is permitted to live for a few days, so that the axons that are attached to neurons in the destroyed area will have time to begin degenerating. Next, the animal is killed, the brain is removed and sliced, and the sections are prepared with the Nauta-Gygax stain. Only the degenerating axons will absorb the silver, and thus be stained, which means that the neuroanatomist can trace the fibers that leave the area that was surgically destroyed.

This technique has been extremely valuable, and has supplied the bulk of our knowledge of the circuitry of the brain. When it was developed, it revitalized the field of neuroanatomical research, which was at a dead end with the techniques that were then available. Now a new set of techniques has again revolutionized neuroanatomical investigation.

AMINO ACID AUTORADIOGRAPHY. Amino acids are used by neurons for construction of protein. These substances enter the cell from the cytoplasm by means of special uptake mechanisms in the membrane. Once they are within the cell, the amino acids are incorporated into proteins, which are then transported to places where they are needed, including the terminal buttons of the axon. As we saw in Chapter 2, there is a special means of transportation, axoplasmic flow, that delivers substances—including proteins—through the axon to the terminal buttons. Thus, some of the amino acids that are

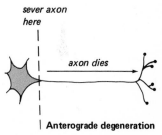

FIGURE 7.6 Anterograde degeneration.

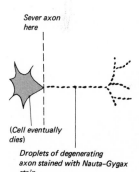

FIGURE 7.7 A schematic representation of the appearance of a degenerating axon as shown by the Nauta-Gygax stain.

145

present in the extracellular fluid that surrounds the cell body of a neuron will subsequently be found in the terminal buttons of that neuron.

The technique of **amino acid autoradiography** takes advantage of the axoplasmic transport of proteins. Radioactive amino acids are injected into a particular region of the brain. The animal is permitted to live for a day or two, so that the radioactive amino acids will be incorporated into proteins and transported through the axons of the neurons that reside in the region that received the injection. Next, the animal is killed, its brain is removed and sliced, and the sections are placed on microscope slides. The slides are taken into a darkroom, where they are coated with a photographic emulsion (the substance found on photographic film). After a wait of several weeks, the slides, with their coatings of photographic emulsion, are developed, just as film is developed. The radioactive proteins show them-

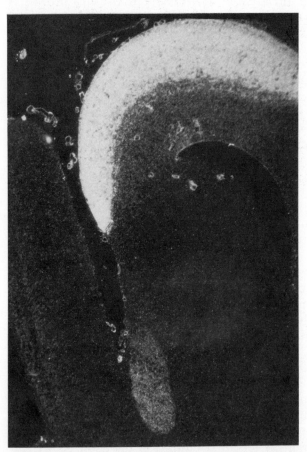

FIGURE 7.8 Amino acid autoradiography. The regions of brightness reveal the radioactive proline that was originally injected into the frog's eye. (Histological material courtesy of K. V. Fite.)

selves as black spots in the developed emulsion; the radioactivity exposes the emulsion, just as X-rays or light will do. (The term autoradiography can roughly be translated as "writing with one's own radiation.")

Figure 7.8 illustrates the actual appearance of the grains of exposed silver in the photographic emulsion. A radioactive amino acid (proline) was injected into a frog's eye, where it was taken up by neural cell bodies in the retina, incorporated into protein, and transported through the optic nerve to the terminal buttons, which are located in the optic tectum (large crescent-shaped area at the top). The radioactive protein was released by the terminal buttons, taken up by postsynaptic neurons in the optic tectum, and carried by the axons of these neurons to the nucleus isthmi, located toward the bottom of the photograph. The brain section was photographed by means of dark-field illumination, which makes the silver grains scatter a beam of light, and thus look white. (See **FIGURE 7.8.**) The technique has revealed that neurons in the retina of the frog send axons to the optic tectum, and that the neurons with which they synapse send axons to nucleus isthmi.

HORSERADISH PEROXIDASE STUDIES.   Horseradish peroxidase is a rather unlikely name for a substance that is used in neuroanatomical research. (Believe it or not, there is also a turnip peroxidase.) Horseradish peroxidase (generally referred to as HRP) is an enzyme—a protein that is capable of splitting certain peroxide molecules, turning them into insoluble salts. We already saw its use (in Chapter 4) as a tracer that demonstrated recycling of the membrane of terminal buttons during the release of transmitter substances.

When HRP is injected into the brain, it is taken up, by means of active transport mechanisms, into axons and terminal buttons. (Why it is taken up is not known.) It is subsequently transported by retrograde axoplasmic flow—that is, flow directed back toward the cell soma. Thus, the HRP reaches the cell bodies that send axons into the region of the brain that received the injection.

The technique works like this: HRP is injected into the part of the brain that is under investigation. After a survival time of a day or two, the animal is killed, the brain is sliced, and the sections are soaked in a sequence of chemical baths that visibly mark the location of the HRP, which has been carried back to the cell bodies. Thus, the HRP technique permits the identification of neurons that project axons *to* a particular region. Figure 7.9 illustrates the appearance of cells that have been labeled with HRP; the oculomotor (third cranial) nerve was soaked with horseradish peroxidase. The peroxidase was carried back, by retrograde axoplasmic flow, to the motor neurons that give rise to the axons within this nerve. (See **FIGURE 7.9** on the following page.)

The HRP technique, in conjunction with the radioactive amino acid technique, which identifies the pathways of axons that project

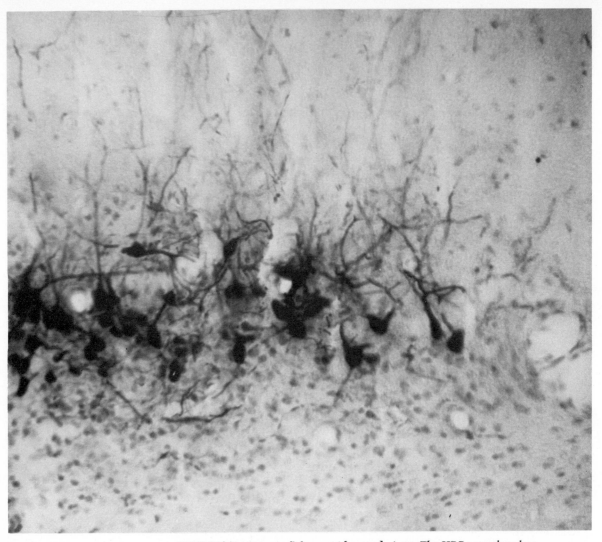

FIGURE 7.9 Horseradish peroxidase technique. The HRP was placed on the third cranial nerve, and was transported back to the cell bodies whose axons travel through this nerve. (Histological material courtesy of K. V. Fite.)

from a particular region, permits the interconnections of the brain to be studied with a high degree of accuracy and sensitivity. Furthermore, under some circumstances HRP will enter cell bodies, and will subsequently be transported by means of anterograde axoplasmic flow. Thus, it can be used to provide results similar to those of the radioactive amino acid technique.

Study of the Living Brain

Advances in X-ray techniques and computers have led to the development of a method of studying the anatomy of the living brain.

As we saw in Chapter 6, the ventricular system can be seen by displacing the cerebrospinal fluid with air, and then making X-rays of the head. However, this invasive procedure is no longer needed, and information about much more than the shape of the ventricular system can be obtained.

G. N. Hounsfeld and a British electronics firm, EMI, developed a technique called *computerized axial tomography.* This procedure, usually referred to as a *CAT scan,* works as follows: The patient's head is placed in a large ring that somewhat resembles a doughnut. The ring contains an X-ray tube and, directly opposite it (on the other side of the patient's head), an X-ray detector. The X-ray beam is passed through the patient's head, and the amount of radioactivity that gets through the patient's head is measured by the detector. The X-ray tube and detector are then moved around the ring by a small amount, and the transmission of radioactivity is measured again. The process is repeated until the brain has been scanned throughout 360 degrees. The computer takes the information and plots a two-dimensional picture of a horizontal section of the brain. The patient's head is then moved up or down through the ring, and a scan is taken of another section of the brain. The procedure is shown schematically in Figure 7.10, along with two CAT scans, one from a normal brain (right), and one from a patient with a brain lesion (left). (See **FIGURE 7.10.**)

Computerized axial tomography has been used extensively in the diagnosis of various brain pathologies, including tumors, blood clots, hydrocephalus, and degenerative diseases such as multiple sclerosis. The benefits to the patient are obvious; a CAT scan can often tell the physician whether brain surgery is necessary. The technique is also of considerable importance to neuropsychologists, who try to infer brain functions by studying the behavioral capacities of people who have sustained brain damage by disease or physical injury. The CAT scan makes it possible for them to obtain the approximate location of the lesion.

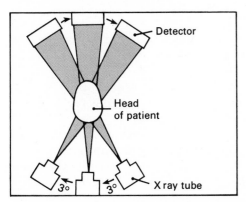

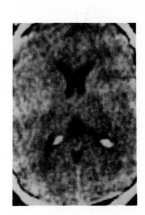

FIGURE 7.10   A schematic explanation of computerized axial tomography (CAT scan). The left scan shows a large lesion in the posterior part of the left hemisphere, which caused a severe speech defect in the patient. The right scan shows a normal brain. (Scans courtesy of A. Kertesz.)

An even more exciting technique is currently being developed: *positron emission transverse tomography,* or **PETT.** This technique will permit the investigator to measure biochemical changes in specific regions of the human brain. The patient receives a radioactive substance that takes part in the biochemical process of the brain. The radioactive molecules emit a particle called a positron, which affects the passage of X-rays from a scanning device. Thus, the computer can determine which regions of the brain have taken up the radioactive substance, and inferences can be made about the biochemical processes that are taking place. The scan even produces a motion picture; the biochemical changes can be observed over a course of time. These techniques were not even dreamt of a few years ago; who knows what the next decade will bring?

Electron Microscopy

The light microscope is limited in its ability to resolve extremely small details. Because of the nature of light itself, magnification of more than approximately 1500 times does not add any detail. In order to see such small anatomical structures as synaptic vesicles and details of cell organelles it is necessary to use an electron microscope. A beam of electrons, smaller than the photons that constitute visible light, is passed through the tissue to be examined. (The tissue must first be coated with a substance that produces detailed variations in the resistance to the passage of electrons, much like staining for light microscopy, causing various portions of the tissue to absorb light.) A shadow of the tissue is then cast upon a sheet of photographic film, which is exposed by the electrons. Electron micrographs produced in this way can provide information about structural details on the order of a few angstrom units.

Stereotaxic Surgery

Many procedures used in neuroscience research require the placement of an object such as a wire or the tip of a metal cannula in a particular part of the brain. For example, one might want to inject a chemical into the brain, which would require the insertion of a cannula, or one might want to destroy a particular region of the brain, which would require the insertion of a metal electrode through which destructive electrical current may be passed. If one were simply to slice the brain open to get to the appropriate part, great damage would be done. Stereotaxic surgery permits the insertion of an object into the depths of the brain without serious damage to overlying tissue.

Stereotaxis literally means "solid arrangement"; more specifically, it refers to the ability to locate objects in space. A *stereotaxic apparatus* permits the investigator to locate brain structures that are hidden from view. This device contains a holder that fixes the animal's head in a standard position, and a carrier that moves an electrode or a cannula through measured distances in all three axes of space. However, in order to perform stereotaxic surgery one must first study a *stereotaxic atlas,* so I shall describe the atlas first.

THE STEREOTAXIC ATLAS.   No two brains of animals of a given species are completely identical, but there is enough similarity between different individuals to predict the location of a given brain structure, relative to external features of the head. For instance, a given thalamic nucleus of a rat might be so many millimeters ventral, anterior, and lateral to a point formed by the junction of several bones of the skull. Figure 7.11 shows two views of a rat skull: a

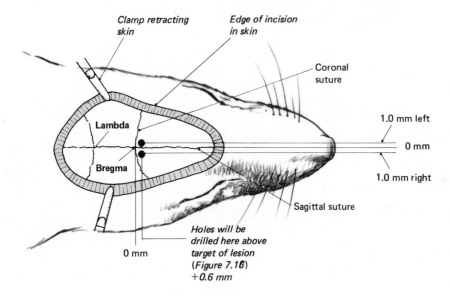

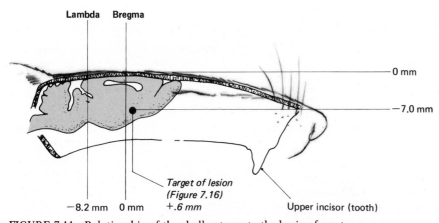

FIGURE 7.11   Relationship of the skull sutures to the brain of a rat.

drawing of the dorsal surface and, beneath it, a midsagittal view. (See **FIGURE 7.11** on the preceding page.) The junction of the coronal and sagittal sutures (seams between adjacent bones of the skull) is labeled **bregma**. If the animal's skull is oriented as shown in the illustration, a given region of the brain occupies a fairly constant location in space, relative to bregma. Not all atlases use bregma as the reference point, but this reference is the easiest to describe.

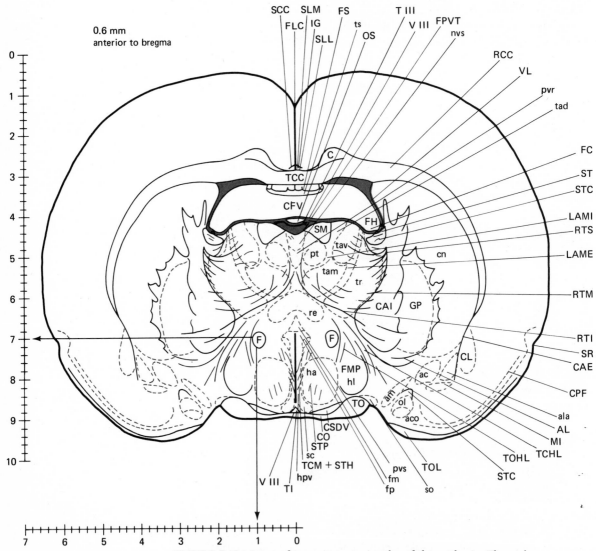

FIGURE 7.12 A page from a stereotaxic atlas of the rat brain. The scale on the side and bottom of the figure has been modified in order to be consistent with my description of bregma as a reference point. (From König, J. F. R., and Klippel, R. A., *The Rat Brain: A Stereotaxic Atlas of the Forebrain and Lower Parts of the Brain Stem.* Copyright 1963 by the Williams & Wilkins Co., Baltimore.)

A stereotaxic atlas contains pages that correspond to frontal sections taken at various distances anterior and posterior to bregma. For example, the page shown in Figure 7.12 is identified as being 0.6 mm anterior to bregma. If we wanted to place the tip of a wire in the structure labelled *F* (the fornix), we would have to drill a hole through the skull 0.6 mm anterior to bregma (because the structure shows up on the 0.6 mm page) and 1.0 mm lateral to the midline. (See **FIGURE 7.11** and **FIGURE 7.12**.) The electrode would be lowered through the hole until the tip was 7.0 mm lower than the skull height at bregma. (See **FIGURE 7.12**.) Thus, by finding a neural structure (which we cannot see in our animal) on one of the pages of a stereotaxis atlas, we can determine the structure's location relative to bregma (which we can see). I should note that, because of variations in different strains and ages of animals, the atlas gives only an approximate location. It is always necessary to try out a new set of coordinates, perform histology on the animal's brain to see where you really placed the lesion, correct the numbers, and try again.

There are human stereotaxic atlases, by the way. Usually multiple landmarks are used, including X-ray ventriculograms (see Chapter 6), and the location of the wire (or other device) inserted into the brain is verified by taking X-rays before producing a brain lesion.

THE STEREOTAXIC APPARATUS. The principle for operation of the stereotaxic apparatus is quite simple. There is a headholder that orients the animal's skull in the proper direction, a holder for the electrode, and a calibrated mechanism that moves the electrode holder in measured distances along the three axes: anterior-posterior, dorsal-ventral, and lateral-medial. Figure 7.13 illustrates a stereotaxic apparatus designed for small animals; various headholders can be used to outfit this device for such diverse species as rats, mice, hamsters, pigeons, and turtles. (See **FIGURE 7.13** on page 154.)

Once the stereotaxic coordinates have been obtained, the animal is anesthetized and placed in the apparatus, and the scalp is cut open. The skull is exposed and the canulla or electrode is placed with its tip on bregma. The location of bregma is measured along each of the axes, and the cannula or electrode is moved the proper distance along the anterior-posterior and lateral-medial axes to a point just above the target. A hole is drilled through the skull just below the cannula or electrode, and the device is lowered into the brain by the proper amount. Now, the tip of the cannula or electrode is in its proper position in the brain.

The next step depends upon what is to be accomplished by the surgery. In some cases, the cannula or electrode is chronically implanted with acrylic plastic, a procedure that will be described later. In other cases, a chemical is infused into the brain through the cannula, or an electrical current is passed through the electrode in order to destroy brain tissue. In either of these cases, the device is subse-

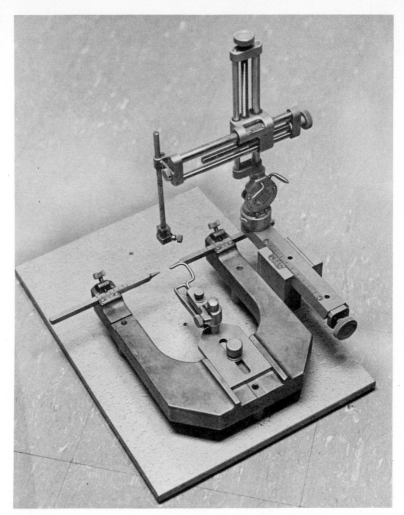

FIGURE 7.13   A stereotaxic apparatus for performing brain surgery on rats.

quently removed from the brain. In all cases, when surgery is complete, the wound is sewed together and the animal is taken out of the stereotaxic apparatus and allowed to recover from the anesthetic.

LESION PRODUCTION/BEHAVIORAL EVALUATION

Behavioral Evaluation

One of the most important methods of inquiry into brain functions requires the destruction of neural tissue in order to determine what function that tissue performs. <u>The rationale for these studies is that the function of an area of the brain is inferred from the behavioral capacity that is missing from the animal's repertoire after destruction of that area.</u> To take an obvious example: if an animal can no longer see after the destruction of a particular part of the brain, the destroyed area was probably involved in vision.

However, we must be very careful in interpreting the effects of brain lesions. For example, how did we ascertain that the operated animal was blind? Did it bump into objects, or fail to run through a maze toward the light that signals the location of food, or no longer constrict its pupils to light? An animal could bump into objects because of deficits in motor coordination, or it could have lost memory for a maze problem, or it could see quite well but could have lost visual reflexes. The experimenter must be clever enough to ask the right question, especially when studying complex processes such as hunger, attention, or memory. Even when studying simpler processes, people can be fooled. For years it was thought that the albino laboratory rat was blind. (It isn't.) Think about it: how would you test to see whether a rat could see? Remember, they have vibrissae (whiskers) that can be used to detect a wall before bumping into it, or the edge of a table before walking off it. The animals can also follow odor trails around the room.

<u>The interpretation of lesion studies is also complicated by the fact that all regions of the brain are interconnected, and no one part does any one thing.</u> When a nucleus is destroyed, the lesion may also sever axons passing through the area. If a structure normally inhibits another, the observed changes in behavior might really be a function of disinhibition of that second structure. <u>Yet another complication was pointed out by Illis (1963), who noted that brain lesions produced temporary degenerative changes in synapses in the vicinity of the lesion.</u> These effects lasted for at least two weeks. Very often we see a partial recovery of function some time after production of a brain lesion. It is impossible to say whether this recovery results from a "taking over" of the function of the damaged structure by some other brain region or from repair of the temporarily injured synapses. In later chapters we shall see numerous examples of the difficulty of deducing the role of a brain region from the behavior of an animal lacking that region. I should note that, <u>these problems notwithstanding, the lesion technique is the most frequently used experimental method of the physiological psychologist.</u>

I should also note that histological evaluation must be made of each animal's brain lesion. Brain lesions often miss the mark, and it is necessary to verify the precise location of the lesions after testing the animals behaviorally. Nothing gladdens an experimenter's heart more than finding that the one animal in the study whose behavior was not consistent with that of the other animals (or with the experimenter's hypothesis) actually had a brain lesion in the wrong location. Now this animal's data can be thrown out! (All too often, though, it is one of the "good" animals that has the bad lesion.)

Production of Brain Lesions

A lesion literally refers to a wound or injury, and when a physiological psychologist destroys part of the brain, he or she usually refers to the damage as a brain lesion. It is very easy to destroy parts of the dorsal surface of the cortex; the animal is anesthetized, the scalp is cut, part of the skull is removed, and the cortex is brought into view. Almost always, a suction device is used to aspirate the brain tissue. The dura mater is cut away and a glass pipette is placed on the surface of the brain. A vacuum pump attached to the pipette is used to suck up the brain as if it were jelly. The pipette also removes blood, which is highly toxic to the brain when it gets outside the blood vessels. With practice, it is quite easy to aspirate away the cortical gray matter, stopping at the underlying layer of white matter, which has a much tougher consistency. A *cautery* (instrument with a heated point) can also be used to destroy regions of cortex, but the extent of the damage is more difficult to control.

Subcortical brain lesions are usually produced by passing current through a wire of stainless steel or platinum that is electrically insulated except for a portion of the tip. The wire is stereotaxically guided so that the end of it is in the appropriate location, and the lesion maker is turned on. Two kinds of electrical current can be used. Direct current (d.c.) produces lesions by initiating chemical reactions whose products destroy the cells in the vicinity of the electrode tip. Radiofrequency (RF) lesion makers produce alternating current of a very high frequency. This current does not stimulate neural tissue, nor does it produce chemical reactions. It destroys cells with the heat that is produced by the passage of the current

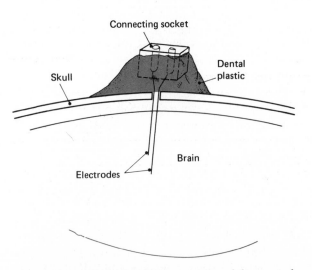

FIGURE 7.14 A schematic representation of the means by which electrodes can be chronically implanted in an animal's brain.

through the tissue, which offers electrical resistance. Radiofrequency lesions have a distinct advantage—no metal ions are left in the damaged tissue. In contrast, when d.c. lesions are produced, some of the electrode is left behind in the brain because ions of metal are carried away by the electrical current. It would appear to be more prudent to make radiofrequency lesions and thus avoid possible long-term effects produced by metal deposition.

Even more specific methods of lesion production are available. As we saw in Chapter 5, there are drugs that are selectively taken up by serotonergic or catecholaminergic neurons, and which subsequently kill these cells. These drugs can be injected directly into particular regions of the brain, thus killing specific populations of neurons that use one of these transmitter substances.

RECORDINGS OF THE BRAIN'S ELECTRICAL ACTIVITY

Rationale

Axons produce action potentials, and terminal buttons elicit post-synaptic potentials. These electrical events can be recorded (as we have already seen), and perhaps regional changes in electrical activity might indicate the functions of the area being recorded from, during, for example, stimulus presentations, or decision making, or motor activities. The rationale seems sound, but, as we shall see, electrical recordings of neural activity, especially of those electrical events that might be correlated with complex behaviors, are very difficult to interpret.

Recordings can be made chronically, over an extended period of time after the animal recovers from surgery, or for a relatively short period of time, during which the animal is kept anesthetized. Acute recordings, made while the animal is anesthetized, are usually restricted to studies of sensory pathways or (in conjunction with electrical stimulation of the brain) to investigations of anatomical pathways in the brain (to be described later.) Acute recordings seldom involve behavioral observations, since the behavioral capacity of an anesthetized animal is, at best, limited.

Chronic electrodes can be implanted in the brain with the aid of a stereotaxic apparatus, and an electrical socket, attached to the electrode, can be cemented to the animal's skull (with an acrylic plastic that is normally used for making dental plates). Then, after recovery from surgery, the animal can be "plugged in" to the recording system. (See **FIGURE 7.14**.)

Electrical recordings can be taken through very small electrodes that detect the activity of one (or just a few) neurons, or large electrodes that respond to the electrical activity of large populations of neurons can be used. The electrical signal detected by the electrode is quite small and must be amplified. A biological amplifier works just like the amplifiers in a stereo system, converting the weak signals recorded at the brain (similar in magnitude to the signals from a phonograph cartridge) into stronger ones, large enough to be displayed on the appropriate device. (The analogy with the sound system even applies here; recordings of the activity of single neurons are often played through loudspeakers, as we shall see.) Output devices vary considerably. Their basic purpose is to convert the raw data (amplified electrical signals from the brain) into a form we can perceive—usually a visual display.

Electrodes

MICROELECTRODES.   *Microelectrodes* have a very fine tip, small enough to be able to record the electrical activity of individual neurons. (We usually refer to this process as *single-unit* recording, a unit referring to a single neuron.) Microelectrodes can be constructed of metal or of glass tubes. Metal electrodes are sharpened by electrolytic etching in an acid solution. Current is passed through a fine wire as it is moved in and out of the solution. The tip erodes away, leaving a very fine, sharp point. The wire (usually of tungsten or stainless steel) is then insulated with a special varnish. The very end of the tip is so sharp that it does not retain insulation and thus can record electrical signals.

Electrodes can also be constructed of fine glass tubes. Glass has an interesting property: if a hollow tube of glass is heated until it is soft, and if the ends are pulled apart, the softened glass will stretch into a very fine filament. This filament, no matter how thin, will still have a hole running through it. To construct glass microelectrodes, one heats the middle of a length of capillary tubing (glass with an outside diameter of approximately 1 mm) and then pulls the ends sharply apart. The glass tube is drawn out finer and finer, until the tube snaps apart. The result is two *micropipettes,* as shown in **FIGURE 7.15.** (These devices are usually produced with the aid of a special machine, called a micropipette puller.) Glass will not conduct electricity, so the micropipette is filled with a conducting liquid, such as a solution of potassium chloride. Glass micropipettes were used to provide the data concerning axonal conduction and synaptic transmission, described in earlier chapters of this book.

FIGURE 7.15   The production of two micropipettes from a single piece of glass tubing.

MACROELECTRODES.   *Macroelectrodes,* which record the activity of a very large number of neurons, are not nearly so difficult to make or to use. They can be constructed from a variety of materials: stainless steel or platinum wires, insulated except for the tip (these can be inserted into the brain or placed on top of it); small balls of metal attached to wires, which can be placed on the exposed cortex; screws that can be driven into holes in the skull; flat disks of silver or gold that can be attached to the scalp with an electrically conductive paste (used for recordings taken from the human head); or thin, uninsulated platinum wires that can be inserted into the scalp (again, for human recordings).

Macroelectrodes do not detect the activity of individual neurons; rather, the records obtained with these devices represent the slow potentials (EPSPs and IPSPs) of many thousands—or millions—of cells in the area of the electrode. Recordings taken from the scalp, especially, represent the electrical activity of an enormous number of neurons.

Amplifiers

The main function of amplifiers used for electrical recording is to increase the amplitude of the signal obtained from the electrode. Another function is to filter the input, that is, to amplify only a selected range of frequencies. For example, if we wanted to record unit activity with a microelectrode, we would set the filtration controls on the amplifier so that only high frequencies would be amplified. Thus, we would see the occurrence of action potentials, but low-frequency signals would not be seen. On the other hand, we would want to amplify only lower frequencies (approximately 2–50 Hz) when recording the summed slow potentials from a larger area of the brain. Figure 7.16 illustrates this filtration process—a composite signal, containing unit activity along with slow potentials, is shown on the left, with the selectively filtered signals shown on the right. (See **FIGURE 7.16.**) (A microelectrode is capable of recording slow potentials as well as units. If we were interested only in slow potentials, however, we would make a macroelectrode rather than take the trouble to make a microelectrode.)

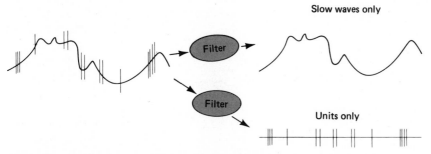

FIGURE 7.16 The separation of a composite signal, containing both slow waves and action potentials, into two separate signals.

Output Devices

We obviously cannot directly observe the electrical activity from an amplifier; just as a sound system is useless without speakers or headphones, so a biological amplifier is useless without an output device. I shall describe three output devices: oscilloscopes, ink-writing oscillographs, and computers.

OSCILLOSCOPE.   In Chapter 3 I described the basic principle of an oscilloscope—the plotting of electrical potentials as a function of time. The display unit of an oscilloscope is demonstrated in **FIGURE 7.17.** This cathode-ray tube contains an electron gun, which emits a focused stream of electrons toward the face of the tube. A special surface on the inside of the glass converts some of energy of the electrons into a visible spot of light. Electrons are negatively charged and are thus attracted to positively charged objects and repelled by negatively charged ones. The plates arranged above and below the electron beam, and on each side, can be electrically charged, thus directing the beam to various places on the face of the tube. The dot can thus be moved independently by the horizontal and vertical deflection plates. (See **FIGURE 7.17.**) (If you have ever played with a video game, or used an Etch-a-Sketch, you will know what I mean. One knob can move the dot up and down while the other moves it left or right—these movements are independent of each other.)

The horizontal deflection plates are usually attached to a timing circuit, which sweeps the beam from left to right at a constant speed. Simultaneously, the output of the biological amplifier moves the beam up and down. We thus obtain a graph of electrical activity as a function of time.

To illustrate the use of an oscilloscope for the recording of unit activity, consider the experiment outlined in **FIGURE 7.18.** A light

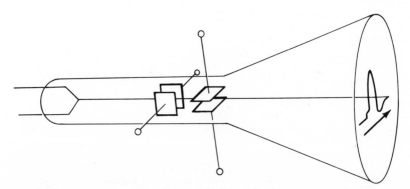

FIGURE 7.17   A schematic and simplified representation of the cathode-ray tube from an oscilloscope.

is flashed in front of the cat, and, at the same time, the dot on the oscilloscope is started across the screen. (We say that we "trigger the sweep.") Let's say that we are recording from a cell in visual cortex that responds to a light flash by giving a burst of action potentials. The record of this event will be seen on the face of the oscilloscope as vertical lines superimposed on a horizontal one (if the beam is moved slowly). If the beam is moved rapidly, we will see the shapes of the individual action potentials. (See **FIGURE 7.18.**) You might wonder how the illustrated display can be seen—after all, the display consists of a moving dot, and I have shown a continuous line. Most oscilloscope screens exhibit *persistence*, however. As the dot moves, it leaves a trace of its pathway behind, which slowly fades away.

When recording single-unit activity, the investigator usually also attaches the output of the amplifier to a loudspeaker. When a microelectrode is lowered through the brain, it is necesary to stop a bit before the optimal recording point, to allow the brain tissue, which has been pushed down a little by the progress of the electrode, to spring back up. If the electrode goes down too far, there is a very high probability that the cell will become injured or killed by the electrode as the tissue moves up. Therefore, it is necessary to detect the firing of a cell as soon as possible, so that the progress of the

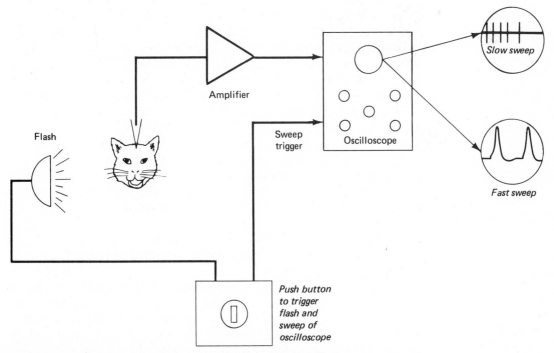

FIGURE 7.18   A schematic representation of the means by which the responses of single units to a flash of light can be recorded.

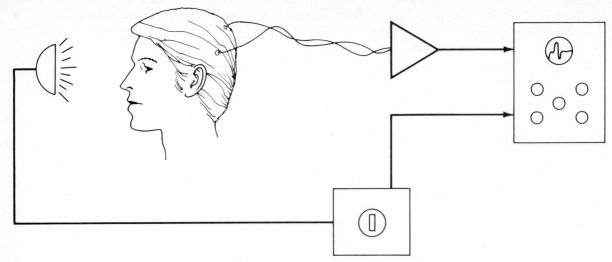

FIGURE 7.19 A schematic representation of the means by which evoked potentials in response to a flash of light can be recorded from the scalp of a human head.

electrode can be stopped in time. Our ears are much more efficient than our eyes in extracting the faint signal of a firing neuron from the random background noise. You can hear the ticking, snapping sound of unit activity from the speaker long before its presence can be visually detected on the face of the oscilloscope. Our auditory system really does a superb job of extracting signal from noise.

Oscilloscopes are also ideal for the display of **evoked potentials.** When a stimulus is presented to an organism, a series of electrical events is initiated at the receptor organ. This activity is conducted into the brain and propagates through various neural pathways in the brain. If an electrode is placed in or near these pathways, it is possible to record the electrical activity that is evoked by the stimulus. For example, one might place a scalp electrode on the back of a person's head and present a flash of light, while simultaneously triggering the sweep of an oscilloscope. Figure 7.19 shows such an experiment, along with the evoked potential from visual cortex, recorded through the skull and scalp. (See **FIGURE 7.19.**)

INK-WRITING OSCILLOGRAPH.   Oscilloscopes are most useful in the display of phasic activity, such as an evoked potential, which occurs during a relatively brief period of time. If neural activity is continuously recorded and displayed on an oscilloscope screen, it will be seen as a series of successive sweeps of the beam, thus presenting a rather confusing picture. A much better device for such a purpose is the ink-writing oscillograph (often called a polygraph).

The time base of the polygraph is provided by a mechanism that moves a very long strip of paper past a series of pens. The pens are,

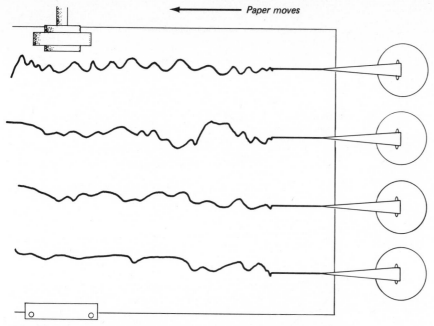

Paper moves

FIGURE 7.20   A record from an ink-writing oscillograph (polygraph).

essentially, the pointers of large voltmeters, moving up and down in response to the electrical signal sent to them by the biological amplifiers. Figure 7.20 illustrates a record of electrical activity recorded from electrodes attached to various locations on a person's scalp. (See **FIGURE 7.20.**) Such records are called ***electroencephalograms (EEGs),*** or "writings of electricity from the head." They can be used to diagnose epilepsy or brain tumors, or to study various stages of sleep and wakefulness, which are associated with characteristic patterns of electrical activity. This last phenomenon will be described in more detail in Chapter 14. I should note that many modern polygraphs do not use ink; they print electrostatically, or with heated pens on specially treated paper. Nevertheless, their principle of operation is the same.

COMPUTER.   Finally, a computer can be used as an output device. A computer can convert the analog signal (one that can continuously vary, like the EEG) received from the biological amplifier into a series of numbers (digital values). Figure 7.21 illustrates how a series of digital values can represent an evoked potential. Each point represents the voltage of the continuous analog signal at an instant of time. (Each point is one millisecond apart.) The values were stored in a computer and were later displayed on the screen of an oscilloscope. (See **FIGURE 7.21** on the following page.)

A computer can do more than display the data, furthermore; it can perform many kinds of analyses. For example, it can compute

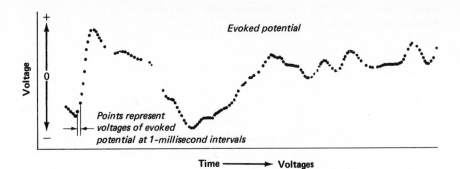

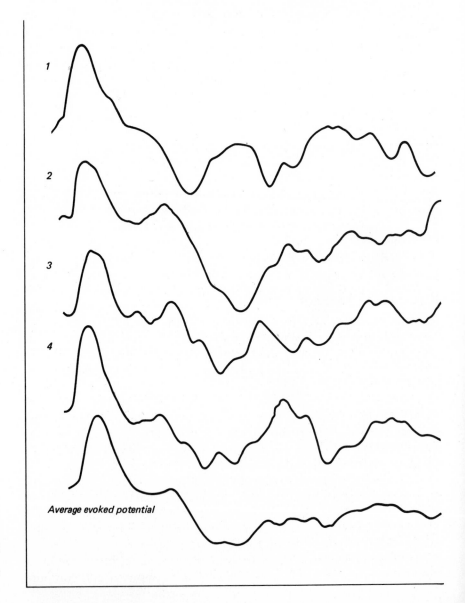

FIGURE 7.21 An evoked potential as represented by a display of points digitized by and stored in a computer.

FIGURE 7.22 Evoked potentials in response to a flash of light, recorded from rat visual cortex. The top curves are individual evoked potentials; the bottom curve represents the average of the individual curves.

the *latencies* (time intervals) between stimulus and the onset of an evoked potential, or it can count the number of action potentials elicited by a stimulus. One of the most common functions of a computer is the averaging of a series of individual evoked potentials.

Figure 7.22 shows a series of evoked potentials recorded from rat visual cortex, in response to a flash of light. Notice that these evoked potentials generally resemble one another, but do not look identical. (See **FIGURE 7.22**.) The individual waves were digitized by a computer, the series of numbers thus obtained were added together, and the total was divided by the number of sweeps (four sweeps, in this case). The resulting numbers describe the average wave, shown at the bottom. The individual variations of the single sweeps, which represent "noise," or other brain activity not related to the light flash, are "averaged out" by this process. Usually, many more than four waves are averaged together, and the result is even smoother than is shown in this example. (See **FIGURE 7.22**.)

The averaging technique is extremely useful in extracting a signal from the ongoing electrical activity of the brain. The average of several hundred sweeps will often reveal that a brain structure is receiving previously unsuspected sensory information. There are dangers, however. Figure 7.23 shows a series of (imaginary) evoked potentials, each of which has a single, sharp-peaked wave occurring at various times. Note, however, that the average curve has a broad, gentle wave and thus resembles none of the individual evoked potentials. The averaging process gives a very poor representation of discrete events that have any variability in time of occurrence. (See **FIGURE 7.23**.) One must guard against such possibilities by examining individual waves, as well as the averages.

Behavioral Studies

The usefulness of electrical recording in the study of sensory pathways should be obvious. In these studies we know when the electrical signals are elicited at the receptors, and we look for subsequent neural events. We can even study motor systems in a similar manner. The recorded electrical activity is stored (on magnetic tape, or in a computer memory) and when a movement is made the records are averaged *backward* in time from the onset of the movement, to look for potentials that were present before the movement. Presumably, the potentials represent neural events that were instrumental in producing the movement.

Other studies might involve the correlation of continuously recorded electrical activity in various brain structures with the ongoing behavior of the animal. Stages of sleep and activity, feeding, and other "spontaneous behaviors" can be observed, or the animals can be trained in some task or subjected to diverse situations in order to observe the response of various brain structures. If we find that a

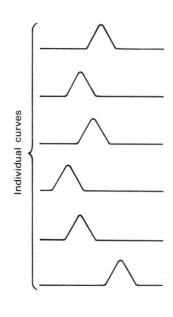

Individual curves

Average curve

FIGURE 7.23   A fallacy introduced by the averaging process. The average curve at the bottom contains a broad, low rise, whereas the individual curves each contain a sharp peak, which occurs at variable times.

brain region becomes more active after an animal is deprived of access to water, perhaps that region has something to do with mechanisms of thirst. The problems associated with use of electrical recordings will be discussed in later chapters dealing with particular physiological and behavioral processes.

ELECTRICAL STIMULATION OF THE BRAIN

Neural activity can be elicited by electrical stimulation. In Chapter 3, I described the way in which action potentials could be produced by delivering electrical pulses to an axon. Various neural structures can also be stimulated through electrodes inserted into the brain; the stimulation experiment, like the recording experiment, can be either acute or chronic.

Identification of Neural Connections

One of the most clear-cut uses of electrical stimulation is demonstrated in neuroanatomical studies. If stimulation delivered through an electrode in structure A produces an evoked potential recorded from an electrode in structure B, then the structures must be connected. Details of their interconnections—directness of the pathway, diffuseness of connections, etc.—can also be inferred from the record obtained (See **FIGURE 7.24.**)

Electrical Stimulation during Neurosurgery

One of the more interesting uses of electrical stimulation of the brain was developed by Wilder Penfield (see Penfield and Jasper, 1954) in the treatment of *focal epilepsy.* This disease is characterized by a localized region of neural tissue that periodically irritates surrounding areas, triggering an epileptic seizure (wild, sustained firing of cerebral neurons, resulting in some degree of behavioral disruption, often including convulsions). If severe cases of focal epilepsy do not respond to medication, surgical excision of the focus is necessary. But how can the focus be identified?

Prior to the onset of a seizure, many patients report the experience of an *aura.* Auras are diverse in nature; some people describe odors, some say that everything appears bright and shimmering, and others report a sense of fear and dread. Since the auras probably resulted from stimulation of the brain tissue surrounding the focus, Penfield reasoned that he might identify this area by electrically

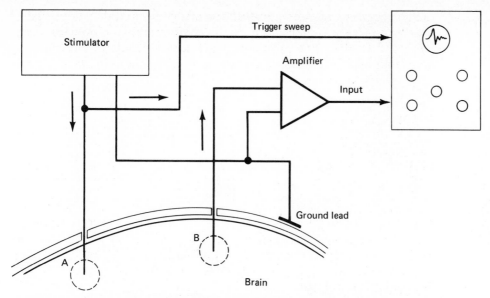

FIGURE 7.24 A schematic representation of the means by which one area of the brain can be stimulated while recording in another area. This procedure permits inferences to be made concerning the nature of the neural connections between these areas.

stimulating various parts of the brain in the unanesthetized patient. The electrode locations that would produce the aura would identify the focus.

Patients undergoing open-head surgery first have their heads shaved. Then a local anesthetic is administered to the scalp along the line that will be followed by the incision. The scalp is cut all the way around, and the skull under the cut is sawed through. The top of the skull can then be removed (like the top of a soft-boiled egg, if you'll excuse the analogy). The dura mater is then cut and folded back, and the naked brain is thus available to the surgeon.

Penfield touched the tip of a metal electrode to various parts of the brain and noted the effects on the patient's behavior. For example, stimulation of motor cortex produced movement, and stimulation of auditory cortex elicited reports of the presence of buzzing noises. If all went well, an aura was produced, and the appropriate region was removed. The dura was sewn back together and the top of the skull was replaced.

Besides giving the patients relief from their epileptic attacks, the procedure provided Penfield with a lot of interesting data. As he stimulated various parts of the brain, he noted the effect and placed a sterile piece of paper, on which was written a number, on the point stimulated. After the operation he could compare the recorded notes with a photograph of the patient's brain, showing the location of the points of stimulation. (See **FIGURE 7.25** on the following page.)

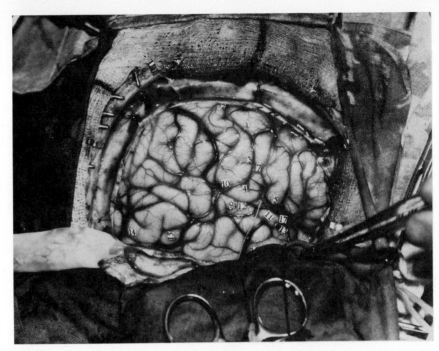

FIGURE 7.25 The appearance of the cortical surface of a conscious patient whose brain has been stimulated. The points of stimulation are indicated by the numbered tags that have been placed there by the surgeon. (From Case, M. M. in Wilder Penfield, *The Mystery of the Mind: A Critical Study of Consciousness and the Human Brain*. With Discussions by William Feindel, Charles Hendel, and Charles Symonds. [Copyright © 1975 by Princeton University Press; Princeton Paperback, 1978] Fig. 4, p. 24. Reprinted by permission of Princeton University Press.)

Behavioral Studies

Stimulation of the brain of an unanesthetized, freely moving animal often produces behavioral changes. For example, hypothalamic stimulation can elicit such behaviors as feeding, drinking, grooming, attack, or escape, which suggests that the hypothalamus is involved in the control of these behaviors. Stimulation of the caudate nucleus often halts ongoing behavior, suggesting that this structure is involved in motor inhibition. Brain stimulation can serve as a signal for a learned task or can even serve as a rewarding or punishing event. These latter two phenomena are described in Chapter 17.

There are problems in the interpretation of the results of electrical stimulation of the brain. An electrical stimulus (usually a series of pulses) can never duplicate the natural neural processes that go on in the brain. The normal interplay of spatial and temporal patterns of excitation and inhibition is destroyed by the artificial stimulation of an area. It would almost be like attaching ropes to the arms of the members of an orchestra and then shaking all the ropes simultaneously to see what they can play. (Sometimes, in fact, local stimulation is used to produce a "temporary lesion"—the region is put out of commission by the meaningless artificial stimulation.) The surprising thing is that stimulation so often *does* produce orderly changes in behavior.

CHEMICAL TECHNIQUES

A growing number of investigations into the physiological bases of behavior use various chemical techniques. The importance of these techniques will be made obvious to you as you read the rest of this book. In this section I will only mention some of the basic procedures.

Identification of Transmitter Substances

A few structures (e.g., autonomic ganglia) contain only one transmitter substance, which makes its identification relatively easy (with the stress on *relatively*, not *easy*). In the brain, identifying the transmitter substance is more difficult. The principal methods used to identify a neurotransmitter are the use of staining techniques for the transmitter, recording of postsynaptic potentials after **iontophoretic** application of the transmitter into the vicinity of the cell, and localization of transmitter-destroying enzymes.

As we saw in Chapter 5, exposure of fresh brain sections to formaldehyde gas results in chemical reactions that cause monoamine-containing cells (i.e., those containing dopamine, norepinephrine, or serotonin) to fluoresce a bright green or yellow color when the tissue is examined under a special microscope that uses ultraviolet light. Anatomical pathways of neurons containing dopamine, norepinephrine, and serotonin have thus been discovered. Confirmation of these results, and the tentative identification of other transmitters whose presence cannot be visualized by histofluorescence techniques has been obtained through use of iontophoresis. A **double-barreled micropipette** is prepared, one pipette inside the other. The longer (inner) pipette is inserted into a neuron, and records are made of its membrane potential. The shorter pipette is filled with an ionized form of the neurotransmitter to be tested. Very small amounts of the neurotransmitter can be ejected from the pipette by passing electrical current through it; the charged molecules are carried with the electrical current out through the tip of the micropipette. (See **FIGURE 7.26** on page 170.) If applications of this neurotransmitter, and not of others, produce PSPs, it suggests that the postsynaptic membrane contains receptors for this substance. (By inference, the local

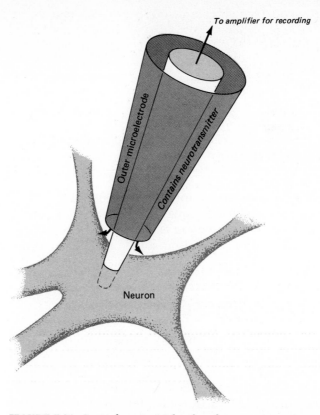

To amplifier for recording

Outer microelectrode

Contains neurotransmitter

Neuron

FIGURE 7.26   Iontophoresis. Molecules of a neurotransmitter are carried out of the outer pipette by an electric current. Intracellular recordings are made by means of the inner pipette to determine whether the neuron responds to the neurotransmitter.

terminal buttons also use this chemical to transmit synaptic information.)

Sometimes neurotransmitters are located indirectly. Acetylcholine is produced by means of choline acetylase and deactivated by acetylcholinesterase. Although ACh cannot be identified by staining, AChE and choline acetylase can. The presence of these enzymes, then, suggests that the terminals synapsing there secrete this neurotransmitter.

Stimulation or Inhibition of Particular Transmitters

In Chapter 5 I described various chemicals that mimic a particular transmitter substance, inhibit or facilitate its production and/or release, prevent its release, block the postsynaptic receptors, or prevent the destruction or re-uptake of the transmitter. Thus, these substances can be used to observe the behavioral effects of stimulation or inhibition of a particular neurotransmitter. For example, depletion of serotonin by PCPA produces insomnia (at least temporarily), and thus suggests the involvement of this transmitter in sleep.

Particular kinds of synapses can be inhibited or stimulated by injection of various pharmacological agents directly into parts of the brain. This can even be done chronically; a *cannula* (small metal tube) can be placed in an animal's brain, and a fitting, attached to the cannula, can be cemented to the skull. At a later date a flexible tube can be connected to the fitting, and a chemical can be injected into the brain. (See **FIGURE 7.27.**) Other methods have been devised to permit the introduction of a powdered chemical into a specific brain region.

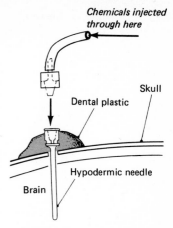

FIGURE 7.27　A chronic intracranial cannula. Chemicals can be infused into the brain through this device.

Radioactive Tracers

Radioactive tracers are radioactive chemicals that become incorporated into chemical processes within cells. They provide the investigator with a labeled substance whose location can be followed by various means. We saw one of the most important uses of radioactive tracers in the radioactive amino acid technique, which uses anterograde axoplasmic flow to trace the pathways followed by axons of neurons that reside in some particular region of the brain. I will describe some basics here, with particular procedures to follow in later chapters.

DETERMINATION OF RATE OF CHEMICAL REACTIONS.　Radioactive chemicals can be used to determine the rate of incorporation of a given substance in various chemical reactions. For example, if we wanted to determine the relative rates of protein synthesis in various neural structures, we could inject an animal with a measured amount of radioactive amino acids, which are used as building blocks for protein. After waiting for a period of time, we would kill the animal and dissect the brain. The amount of radioactivity in the protein that was extracted from a particular portion of brain tissue would tell us how much of that protein had been constructed since the radioactive amino acid had been administered. "Old" protein would not be radioactive; only "new" protein would. The amount of radioactivity in a given amount of protein would tell us how much of that protein was "new."

DETERMINATION OF SITE OF ACTION OF A BEHAVIORALLY EFFECTIVE CHEMICAL.　Suppose that we were interested in finding out where estrogen (a female sex hormone) is taking up in the brain. The fact that estrogen can affect behavior suggests that some brain cells selectively take up this hormone. We could inject an animal with radioactive estrogen, wait a while, kill the animal, slice the brain, place the sections on microscope slides, and prepare au-

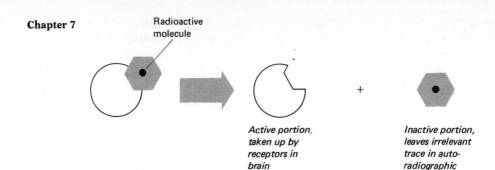

FIGURE 7.28   A possible source of error in autoradiographic tracer studies.

toradiographs by coating the sections with a photographic emulsion. (The technique of preparing autoradiographs would be the same as the one I described in the section on neuroanatomical techniques.) If the estrogen was taken up selectively by any neurons, its radioactivity would cause black spots in the developed photographic emulsion.

There are two restrictions in the use of autoradiography. The compound must be available in radioactive form, and the substance must produce its effect in an intact form, and not break down or be converted into other substances. Very misleading results might be obtained if the substance breaks down and only one of the resulting products produces the change in behavior. In particular, if the "behaviorally inert" portion of the molecule retains the radioactive atom, the results of autoradiography will be meaningless. (See **FIGURE 7.28.**) The problem is that all too often the identity of the effective molecule is not known.

This chapter has barely skimmed the surface of the techniques that are available to students of brain functions. However, you have been introduced to the major types of research methods that are used, and you should have no trouble understanding the particular experiments that I shall describe in later chapters of this book. I hope that you have also learned enough so that you will be able to understand the rationale behind most experimental procedures you might read about in scientific journals or other books.

KEY TERMS

amino acid autoradiography p. 146
anterograde degeneration p. 145
autolytic enzyme (*aw toh LIT ik*) p. 138
autoradiography p. 146

bregma p. 152
cautery p. 156
computerized axial tomography (CAT scan) p. 149
cresyl violet p. 142

SUGGESTED READINGS

SKINNER, J. E. *Neuroscience: A Laboratory Manual*. Philadelphia, Saunders, 1971.

WEBSTER, W. G. *Principles of Research Methodology in Physiological Psychology*. New York: Harper & Row, 1975.

These are "how to do it" manuals designed for laboratory courses in physiological psychology. Webster's book emphasizes "experiments" that can be done as laboratory projects, whereas Skinner's book emphasizes technique. Skinner's book also includes a valuable stereotaxic atlas of the rat brain and excellent guides for the dissection of cow and sheep brains. Both books contain a list of references pertaining to particular investigative techniques.

SLOTNICK, B. M., AND LEONARD, C. M. *A Stereotaxic Atlas of the Albino Mouse Forebrain*. Rockville, Md.: Public Health Service, 1975. (U.S. Government Printing Office Stock Number 017-024-00491-0.)

KÖNIG, J. F. R., AND KLIPPEL, R. A. *The Rat Brain: A Stereotaxic Atlas of the Forebrain and Lower Parts of the Brain Stem*. Baltimore: Williams & Wilkins, 1963.

PELLIGRINO, L. J., AND CUSHMAN, A. J. *A Stereotaxic Atlas of the Rat Brain*. New York: Appleton-Century-Crofts, 1967.

SNIDER, R. S., AND NIEMER, W. T. *A Stereotaxic Atlas of the Cat Brain*. Chicago: University of Chicago Press, 1961.

These stereotaxic atlases (arranged in order of brain size) are the standard references for investigators who use these species. Of the two rat brain atlases, König and Klippel's is more useful for studying neuronatomy, whereas Pelligrino and Cushman's atlas is more useful as a guide for stereotaxic surgery.

8

Receptor Organs and the Transduction of Sensory Information

In order for us to experience the world, there must be changes in patterns of neural activity in our brains that correspond to physical events in the environment. The real world surrounds us, but our perception of it takes place within our head. The topic of this chapter will be the process by which environmental change affects neural firing (**sensory transduction**); in this chapter and in Chapter 9 we shall study the subsequent anatomical pathways followed by sensory information and the nature of **sensory coding** in the brain.

We receive information about the environment from our sensory receptors. Stimuli impinge on the receptors and, through various processes, alter their electrical characteristics. These electrical changes (called either receptor potentials or generator potentials—the distinction will be made shortly) modify the pattern of firing in axons leading into the CNS. In this chapter I shall describe the anatomy and location of the various receptors, and shall summarize what is known or hypothesized about the nature of sensory transduction.

RECEPTOR AND GENERATOR POTENTIALS There are two basic types of receptor cells. One kind has an axon and communicates with other neurons by means of normal synaptic transmission. Sensory events affecting these cells produce **generator potentials.** These potentials are similar in function to postsynaptic potentials; they raise or lower the probability that the axon of the sensory neuron will fire (or, more descriptively, they raise or lower the axon's rate of firing). The other type of receptor cell does not have an axon and does not produce action potentials. This receptor sustains slow potential changes (**receptor potentials**) when stimulated by sensory events. Receptor potentials are transmitted, either electrically or chemically, to neurons with axons capable of producing action potentials. The receptor potential thus alters the rate of firing of these latter neurons. The basic distinction, then, is between receptor cells that have axons (and are true neurons) and those that do not. Figure 8.1 illustrates the process of transduction of sensory information into neural firing. (See **FIGURE 8.1.**)

THE SENSORY MODALITIES. It is often said that there are five senses: sight, hearing, smell, taste, and touch. Actually, we have more than five senses. The problem is that people do not agree on just how many more. Certainly, sensory physiologists agree that we should add the vestibular senses; the inner ear provides us with information about orientation and angular acceleration (changes in speed of rotation) of the head, as well as auditory information. Vestibular information is very important; we use it to maintain our balance. However, we are not "aware" of it as we are of vision or olfaction, for example, so it is easy to see why vestibular sensation was left off the list (which was drawn up long before there were any scientific investigations on sensory systems). Furthermore, receptors for vestibular information are not visible unless one dissects the region of the inner ear, and the receptors are always functioning; we cannot turn them off the way we can close our eyes. If something is constantly present, we tend not to notice it.

The sense of "touch" also includes several kinds of information. The **somatosenses** (a much better word than touch) include sensitivity to pressure, touch, warmth, cold, skin vibration, limb position and movement, and pain. There is no disagreement about the fact that we can detect these stimuli; the issue is whether or not they are detected by separate senses. As we shall see in the discussion of somatosenses in this chapter, we do not even know which receptors are responsible for some of these sensations.

The rest of this chapter will be divided into seven sections, covering the transduction of visual, auditory, vestibular, cutaneous, kinesthetic/organic, gustatory, and olfactory information. Each section will describe (1) the nature of the stimulus that excites the receptors; (2) anatomy of the receptor organ, including a discussion of ways in which the stimulus is modified by the sensory apparatus or organ; (3) anatomy of the receptor cells; (4) the process of transduction of

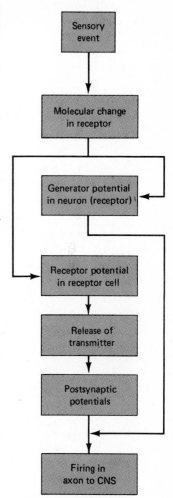

FIGURE 8.1 A schematic representation of the way in which physical stimuli are transduced into neural activity.

physical energy into patterns of neural activity; (5) the route followed by the sensory information through the peripheral nervous system; and (6) the role of efferents from the CNS to the sense organ and its receptors.

VISION

The Stimulus

As we all know, we must have light in order to make use of our eyes. Light, for humans, is a narrow band of the electromagnetic spec-

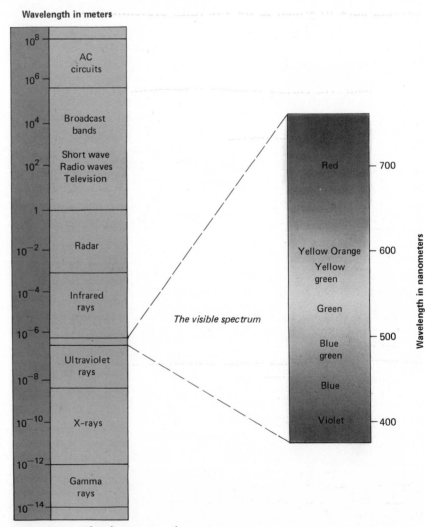

Wavelength in meters

FIGURE 8.2   The electromagnetic spectrum.

trum. Electromagnetic radiation with a wavelength of between 380 and 760 nm (nanometer, or one-billionth of a meter) is visible to us. (See **FIGURE 8.2.**) Other animals can sense different ranges of electromagnetic radiation. A rattlesnake, for example, can use infrared radiation in detecting its prey; it can thus locate warm-blooded animals in the dark. The range of wavelengths we call light is that part of a continuum which we humans can see.

Anatomy of the Eye

The eyes are suspended in the **orbits.** They are moved by six muscles attached to the outer coat of the eye (**sclera**). Normally, we cannot look behind our eyeballs and see these muscles because of the presence of the **conjunctiva.** These mucous membranes line the eyelid and fold back to attach to the eye (thus preventing a contact lens that has slipped off the cornea from "falling behind the eye"). Figure 8.3 illustrates the external and internal anatomy of the eye. (See **FIGURE 8.3** on the next page.)

The outer layer of the eye, the sclera, is opaque, not permitting entry of light. The **cornea,** however, is transparent and admits light. The amount of light entering the eye is regulated by the size of the pupil, formed by the opening in the iris, which consists of a ring of muscles situated behind the cornea. The iris contains two bands of muscles, the dilator (whose contraction enlarges the pupil) and the sphincter (whose contraction reduces it). The sphincter is innervated by cholinergic fibers of the parasympathetic nervous system; cholinergic blockers (for example, belladonna alkaloids such as atropine) thus produce pupillary dilation by relaxing the sphincter of the iris. In fact, belladonna received its name from this effect. Belladonna means "beautiful lady" and was used in ancient times to enhance a woman's sex appeal. (Dilated pupils often indicate interest, and there is, to almost any man, nothing more attractive than a woman who finds him interesting.)

The lens is situated immediately behind the iris. It consists of a series of transparent (naturally) onionlike layers. Its shape can be altered by contraction of the **ciliary muscles.** Normally, because of the tension of elastic fibers that suspend the lens, the lens is relatively flat (and thus is focussed on distant objects). When the ciliary muscles contract, tension is taken off these fibers, and the lens springs back to its normally rounded shape. Therefore, the lens can focus images of near or distant objects on the **retina,** the light-receptive surface on the back of the eye.

Light then passes through the fluid-filled **posterior chamber** of the eye and falls on the retina. In the retina are located the receptor cells, the **rods** and **cones.** Cones, which mediate color vision and provide vision of the highest acuity (ability to detect fine details), are most densely packed at the back of the eye and thin out toward

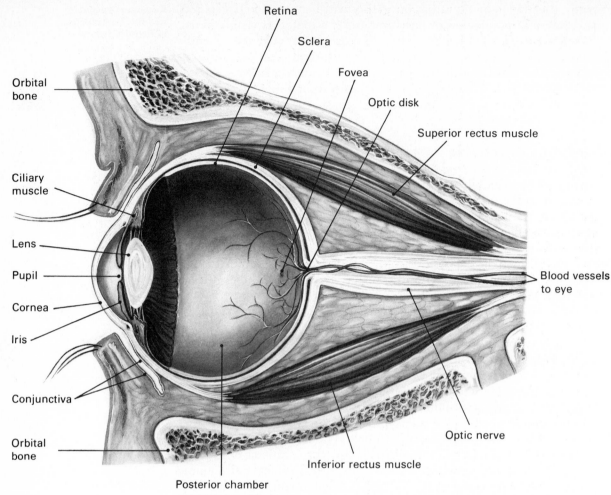

Retina
Sclera
Fovea
Optic disk
Superior rectus muscle

Orbital
bone

Ciliary
muscle

Lens

Pupil

Cornea

Iris

Conjunctiva

Orbital
bone

Blood vessels
to eye

Optic nerve

Inferior rectus muscle

Posterior chamber

FIGURE 8.3   The human eye.

the periphery. The **fovea,** or central region of the retina, which mediates our most detailed and accurate vision, contains only cones. Rods, on the other hand, are most numerous in the periphery. Rods are incapable of detecting colors, but are more sensitive to light. In a very dimly lighted room we use our rod vision; therefore, we are color blind, and we cannot see very well at the center of our visual field (because there are no rods in the fovea). You have probably noticed, while out on a dark night, that looking directly at a dim, distant light (that is, placing the image of the light on the fovea) causes it to disappear.

Another feature of the retina is the **_optic disk,_** where the axons conveying visual information gather together and leave the eye via the optic nerve. The optic disk produces a *blind spot* because no receptors are located there. We do not normally perceive our blind

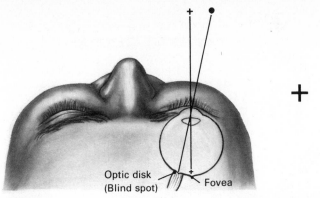

FIGURE 8.4   A test for the blind spot. With the left eye closed, look at the
+ with the right eye and move the page back and forth. At about 20 cm
the black circle disappears from the visual field because it falls on the
blind spot.

spots, but their presence can be demonstrated. If you have not found
your blind spot before, you might want to try the exercise described
in **FIGURE 8.4.**

Close examination of the retina shows that it consists of several
layers of neuron cell bodies, their axons and dendrites, and the receptors themselves. Figure 8.5 illustrates a schematic cross section
through the retina, which is usually divided into three main layers,
the photoreceptive layer, the bipolar cell layer, and the ganglion cell
layer. Note that the photoreceptors are at the *back* of the retina; light
must pass through the overlying layers to get to them. (See **FIGURE
8.5** on the following page.)

Anatomy of the Photoreceptors

Figure 8.6 shows a drawing, reconstructed from electron micrographs, of a single rod and cone of a frog. Note that each photoreceptor contains a layered outer segment, connected by a cilium to
the inner segments. The nucleus is located within the inner segment.
(See **FIGURE 8.6** on the following page.) Figure 8.7 shows, in greater
detail, the outer segments of the rods and cones. Although outer
segments of both receptors are layered (the layers are called ***lamellae***), there is a basic difference. Rods contain free-floating disks,
whereas the lamellae of the cones consist of one continuous folded
membrane. (See **FIGURE 8.7** on page 182.) According to studies by
Young (1970), new protein is perpetually produced by the receptor
cells to replace that which gets worn out. Rods continuously shed
old disks off the end, while new replacement disks are produced at
the base of the outer segment. However, cones do not produce new
folds. The tapered shape of the outer segment is apparently determined by development. The outer lamellae are produced early in

179

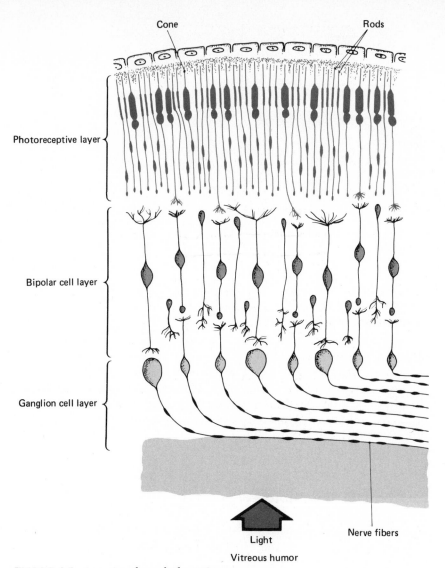

FIGURE 8.5   A section through the retina.

fetal life, whereas inner ones are produced later. As the cones grow, the segments they produce are correspondingly larger. At the base of the outer segments the cones produce new protein, which migrates outward and becomes incorporated into the lamellae.

Transduction of Visual Information

Light can be conceived of as electromagnetic radiation or as particles of energy (photons). This fact bothers many people who prefer a unified view of the world, but physicists have learned to live with

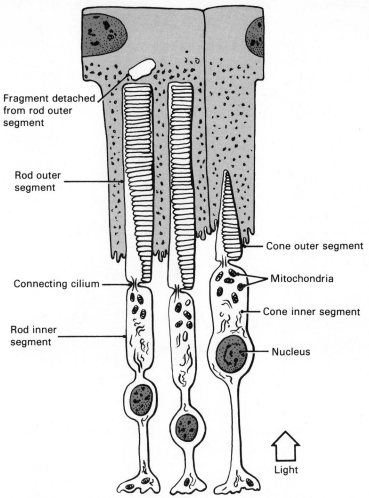

FIGURE 8.6 Photoreceptors. (Redrawn from Young, R. W., "Visual cells."
Copyright 1970 by Scientific American, Inc. All rights reserved.)

these contradictory theories—both of which appear to be correct.
The photon is the stimulus that excites a receptive cell of the retina,
and it appears that rods are able to detect the presence of only one
photon, the minimum quantity of light that can exist. Our photoreceptors, then, are exquisitely sensitive detectors of light.

The first step in the process of transduction of light is chemical,
and it involves a special chemical, or ***photopigment.*** Photopigments
consist of two parts, ***opsin*** (a protein) and ***retinal*** (a smaller molecule,
derived from vitamin A). There are several forms of opsin; let us
consider the photopigment of human rods, ***rhodopsin,*** which consists
of rod opsin plus retinal.

Retinal is synthesized from ***retinol*** (vitamin A). (This is why car-
rots, rich in retinol, are said to be good for your eyesight.) Retinal

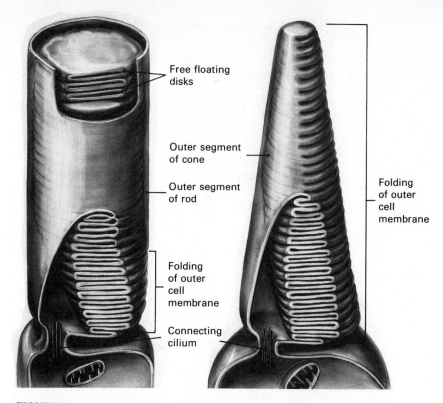

Free floating
disks

Outer segment
of cone

Outer segment
of rod

Folding
of outer
cell
membrane

Connecting
cilium

Folding
of outer
cell
membrane

FIGURE 8.7   Details of photoreceptors. (Redrawn from Young, R. W., "Visual cells." Copyright 1970 by Scientific American, Inc. All rights reserved.)

is a molecule with a long chain that is capable of bending at a specific point. The straight-chained form of retinal is called ***all-trans retinal;*** the form with a bend is called ***11-cis retinal.*** The bent form, 11-*cis* retinal, is the only naturally occurring form of retinal capable of attaching to rod opsin to form rhodopsin. The 11-*cis* form of retinal, moreover, is very unstable; it can exist only in the dark. When a molecule of rhodopsin is exposed to light (i.e., absorbs a photon), the bend in the retinal chain straightens out, and the retinal assumes the all-*trans* form. Since rod opsin cannot remain attached to all-*trans* retinal, the rhodopsin breaks into its two constituents. (See **FIGURE 8.8.**)

The splitting of the photopigment causes a sudden decrease in the sodium permeability of the outer membrane of the photoreceptor, which results in hyperpolarization. This constitutes the receptor potential. But since the receipt of a single photon by a photopigment molecule can produce a detectible receptor potential, there must be some intermediate messenger between the membrane of the lamellae and the outer membrane of the cell.

Indeed, it has been found that the bleaching of one rhodopsin molecule results in the disappearance of 50,000 molecules of cyclic

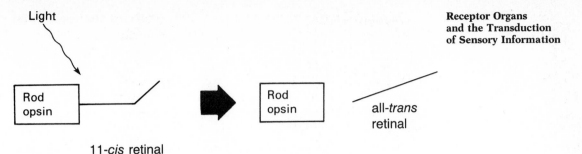

FIGURE 8.8 A schematic representation of the way in which rhodopsin is split by light.

GMP (Woodruff, Bownds, Green, Morrisey, and Shedlovsky, 1977; Woodruff and Bownds, 1979). It has been hypothesized that the fission of a molecule of photopigment activates molecules of phosphodiesterase, which, in turn, destroy molecules of cyclic GMP. The cyclic GMP normally holds open some sodium channels in the membrane, so when cyclic GMP disappears, so do the sodium channels. Thus, the sodium conductivity of the membrane falls, and it becomes hyperpolarized. (See **FIGURE 8.9.**)

The details of the process have not been worked out yet. In addition to alterations in cyclic GMP levels, the concentration of Ca^{++} also changes. (See Hubbell and Bownds, 1979, for a review.) The relationship between changes in cyclic GMP and calcium ions is not known.

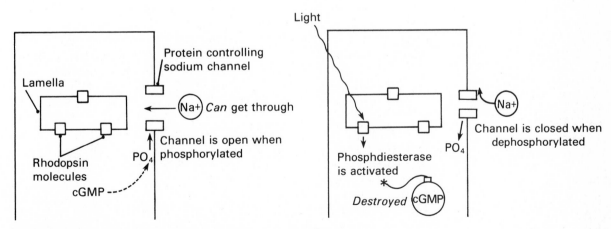

FIGURE 8.9 A schematic representation of the way in which receptor potentials might be produced in photoreceptors.

In most higher primates, four different opsins (rod opsin and three kinds of cone opsins) join with retinal to produce four different photopigments. Each of these compounds most readily absorbs light of a particular wavelength. A given cone contains only one of the three cone photopigments; the various cones are thus maximally sensitive to light of long, medium, or short wavelength. The visual system uses information from these three types of cones to produce color vision.

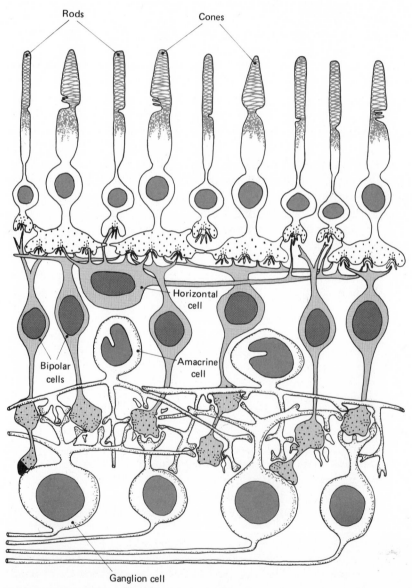

Rods

Cones

Horizontal cell

Bipolar cells

Amacrine cell

Ganglion cell

FIGURE 8.10   Details of retinal circuitry. (Redrawn by permission of the Royal Society and the authors from Dowling, J. E., and Boycott, B. B., *Proceedings of the Royal Society (London)*, 1966, Series B, *166*, 80–111.)

Origin of the Optic Nerve

The ganglion cell layer of the retina gives rise to fibers of the optic nerve (the second cranial nerve). Figure 8.10 is a schematic drawing of a primate retina. (See **FIGURE 8.10.**) The photoreceptors synapse with the **bipolar cells**—rods with rod bipolar cells and cones with midget bipolar or flat bipolar cells. The bipolar cells are not capable of producing action potentials; they produce graded potentials that travel across their membrane. They synapse with the ganglion cells. The **horizontal cells** receive synapses from a number of photoreceptors, and themselves synapse upon bipolar cells. The distal (away from photoreceptor) ends of the bipolar cells are interconnected via the **amacrine cells.** Finally, the **ganglion cells** send axons toward the optic disk and give rise to the optic nerve. (See **FIGURE 8.10.**)

Efferent Control from the CNS

As we have seen, the amount of light falling on the retina is controlled by the size of the pupillary aperture, which depends on the degree of contraction of the sphincter and dilator muscles of the iris. Pupillary size depends on two factors. The first factor is the degree of arousal of the organism; if sympathetic activity dominates, the pupils will dilate, whereas parasympathetic activity produces constriction. The second factor is the amount of light falling on the retina. Increases or decreases in the level of illumination produce corresponding pupillary constriction or dilation. Exposure to extremely bright light will produce reflexive squinting, which further reduces the size of the aperture.

Shape of the lens is also controlled by the brain, to focus the image of near or distant objects on the retina. This accommodation for distance is normally integrated with convergence of the eyes. When we look at a near object, the eyes turn inward so that the two images of the object fall on corresponding portions of the retinas. As we shall see later, this leads to stimulation of corresponding cells in visual cortex on both sides of the brain, producing a fused image. Convergence of the eyes and accommodation of the lens normally occur together, so that the object on which the eyes are focused is also the object on which the eyes converge. If you hold a pencil in front of you and focus on distant objects, you see two blurry pencils. If you then focus on the pencil, you get two blurry views of the background.

Control of eye movement is a very complicated process; as we study the process, we realize that it takes a very sophisticated computer to accomplish what our brain does in moving the eyes. When the brain commands skeletal muscles to pick up a weight, it relies on sensory feedback to determine whether the arms in fact moved; perhaps the weight is too heavy to lift. The eyes, on the other hand, are free to move, and when the brain sends signals to the six muscles controlling eye movement, it "assumes" that the eye does as it is told. Sensory organs within the eye muscles signal information concerning muscle length back to the brain (Davson, 1972), but it is not yet clear what the brain does with this information. The process by which the brain controls eye movements is schematized in Figure 8.11. (See **FIGURE 8.11.**)

The lack of feedback from the eye can be demonstrated easily. We all perceive the world to be fixed in one place; if we shift our gaze, we experience a constant environment being scanned by moving eyes. However, the retinas themselves do not have enough information to make this decision. A moving world and a moving eyeball produce the same changes in the retinal image. The brain has access to the commands it gave to the eye muscles, however, and if changes in retinal image correspond with changes in the commands to the eye muscles, we perceive the world as being stationary.

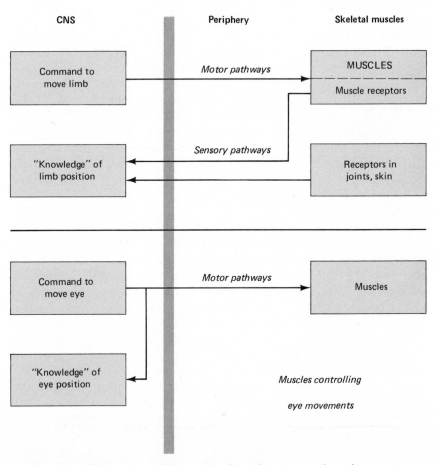

FIGURE 8.11 Interpretation of movement depends upon signals to the eye muscles rather than upon proprioceptive feedback.

Suppose the retinal image of the world changed while the commands to the eye muscles were constant: we would perceive movement. Try this experiment: Cover your left eye with your left hand. Look down a bit and to the left. Now touch the right side of the upper eyelid of your right eye and push against your eyeball. The eye will move, and as it does, it will appear to you that the world is also moving. The brain detected movement of the image on the retina, but there were no commands to move the muscles; ergo, the world moved. Fortunately, there is more to our brains than that. Other parts of the brain are carrying out the experiment, and they say, "No, the world just looks as though it is moving; the finger is moving the eye."

The opposite experiment has been performed—the eyes were held in place while the brain commanded their movement (Hammond, Merton, and Sutton, 1956). Merton's eye muscles were injected with curare (which, you will recall, blocks nicotinic cholinergic synapses, and hence paralyzes the muscles). Merton then "willed" an upward movement of his eyes. Signals went to the appropriate eye muscles. Other brain structures were informed that the eyes had moved (but, of course, they had not). The retinal image did not change, even though the eye muscles were commanded to move; therefore, the system concluded that the world had moved up. Every time the subject "willed" a movement of his eyes, he felt as if the whole world was moving, precisely following his gaze.

Besides CNS control of eye movement, pupil size, and accommodation of the lens, there appears to be a much more subtle control of retinal information. Spinelli, Pribram, and Weingarten (1965) found that auditory clicks produced electrical activity in the optic nerve. The responses ceased when the nerve was cut between the brain and the recording electrode. Thus, there appear to be efferent fibers from brain to retina. Cragg (1962) confirmed studies from the late nineteenth century that found nerve terminals, among the amacrine cells, at the ends of axons coming from the vicinity of the optic disk. These, presumably, are terminals of the efferent fibers whose presence was demonstrated by Spinelli and his coworkers. The role of these fibers in visual functioning is not known.

AUDITION

The Stimulus

We hear sounds, which are normally transmitted via rapid successive condensations and rarefactions of air. If an object vibrates at the proper frequency (between approximately 20 and 15,000 times per second), the pressure changes it induces in the air will, if the object is near enough, ultimately stimulate receptive cells in our ears. We can also stimulate these receptors by placing a vibrating object against the bones of the head, bypassing air conduction altogether.

Anatomy of the Ear

Figure 8.12 shows a section through the ear and auditory canal, and illustrates the apparatus of the middle and inner ear. (See **FIGURE 8.12.**) Sound is funnelled via the *pinna* (external ear) through the *external auditory canal* to the *tympanic membrane* (eardrum), which vibrates with the sound. We are not very good at moving our ears,

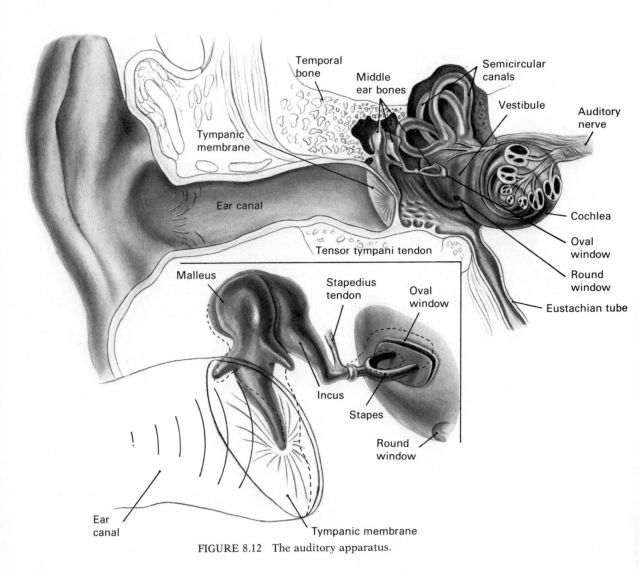

FIGURE 8.12   The auditory apparatus.

but by orienting our heads, we can modify the sound that finally reaches the receptors. A muscle in the tympanic membrane (**tensor tympani**) can alter the membrane's tension and thus control the amount of sound that is permitted to pass through to the middle ear.

The **ossicles,** bones of the middle ear, are set into vibration by the tympanic membrane. The **malleus** (hammer) connects with the tympanic membrane and transmits vibrations via the **incus** (anvil) and **stapes** (stirrup) to the **cochlea,** the inner ear structure containing the receptors. The baseplate of the stapes presses against the membrane behind the **oval window,** the opening in the bony process surrounding the cochlea. (See **FIGURE 8.12.**) The **stapedius muscle,** when contracted, directs the baseplate of the stapes away from its normal point of attachment to the oval window, and hence dampens the vibration passed on to the receptive cells.

The cochlea is filled with fluid; this means that sounds transmitted through the air must be transferred into a liquid medium. This process normally is very inefficient—99.9 percent of the energy of airborne sound would be reflected away if the air impinged directly against the oval window of the cochlea. (If you have ever swum underwater, you have probably noted how quiet it is there; most of the sound arising in the air is reflected off the surface of the water.) The chain of ossicles serves as an extremely efficient means of energy transmission. The bones provide mechanical advantage, the baseplate of the stapes making smaller, but more forceful, excursions against the oval window than the tympanic membrane makes against the malleus.

The name cochlea comes from the Greek word *kokhlos*, or land snail. It is indeed snail-shaped, consisting of two and three-quarters turns of a gradually tapering cylinder. The cochlea is divided longitudinally into three sections, as shown in **FIGURE 8.13** on the following page. The auditory receptors are called **hair cells,** and they are anchored, via rodlike **Deiters' cells,** to the **basilar membrane.** The cilia of the hair cells pass through the **reticular membrane** and their ends attach to the fairly rigid **tectorial membrane,** which projects overhead like a shelf. (The entire structure, including the basilar membrane, hair cells, and tectorial membrane, is referred to as the **organ of Corti.**) (See **FIGURE 8.13.**) Sonic vibrations cause movement of the basilar membrane relative to the tectorial membrane, and the resultant stretch exerted on the cilia of the hair cells produces the receptor potential.

If the cochlea were a closed system, no vibration would be transmitted via the oval window, since liquids are essentially incompressible. However, there is a membrane-covered opening, the **round window,** which allows the fluid contents of the cochlea to move back and forth. The baseplate of the stapes presses against the oval window, which increases the hydrostatic pressure within the *vestibule*, a chamber to which the cochlea is attached. The **scala vestibuli** (literally, the stairway of the vestibule) connects with the vestibule and

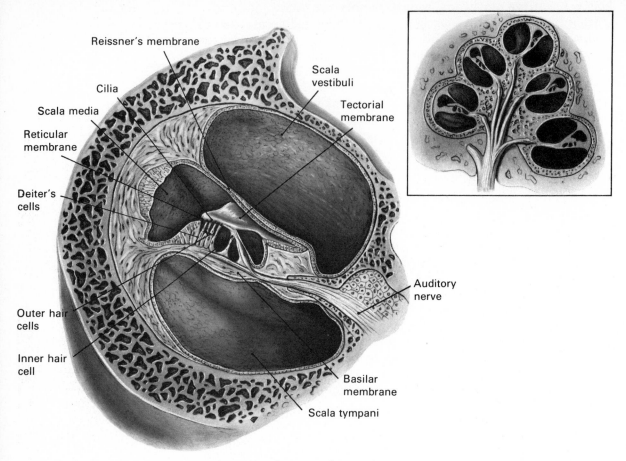

Reissner's membrane

Cilia

Scala media

Reticular membrane

Deiter's cells

Outer hair cells

Inner hair cell

Scala vestibuli

Tectorial membrane

Auditory nerve

Basilar membrane

Scala tympani

FIGURE 8.13   The organ of Corti.

conducts the pressure around the turns of the cochlea. Some portion of the basilar membrane vibrates with the sound waves, transmitting the waves of pressure changes into the **scala tympani.** Pressure changes in the scala tympani are transmitted to the membrane of the round window, which moves in and out in a manner opposite to movements of the oval window. Figure 8.14 shows the cochlea partially straightened out. A sound wave is deforming the basilar membrane. (See **FIGURE 8.14.**)

Anatomy of the Auditory Hair Cells

There are two types of auditory receptors, *inner* and *outer* hair cells, lying on the inside and outside of the cochlear coils. In the human cochlea, there are 3400 inner hair cells and 12,000 outer hair cells. Figure 8.15 illustrates the two types of cells and their supporting Deiters' cells. (See **FIGURE 8.15.**) These cells synapse with nerve

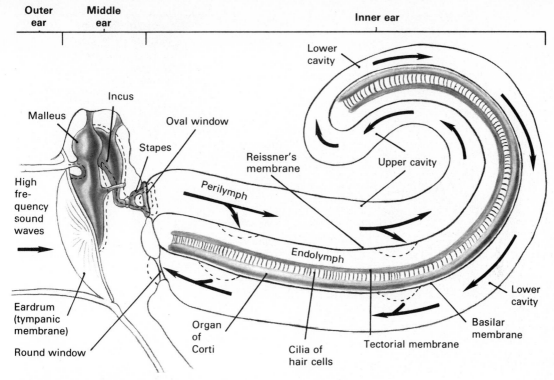

FIGURE 8.14   Sound waves transmitted through the oval window deform a portion of the basilar membrane.

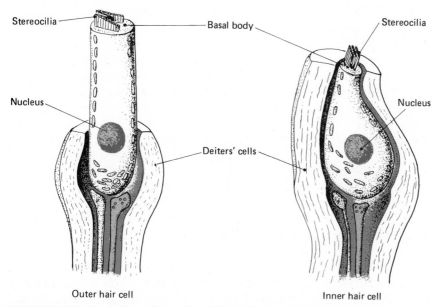

Outer hair cell

Inner hair cell

FIGURE 8.15   The auditory hair cells. (Adapted from Gulick, W. L., *Hearing: Physiology and Psychophysics.* New York: Oxford University Press, 1971.)

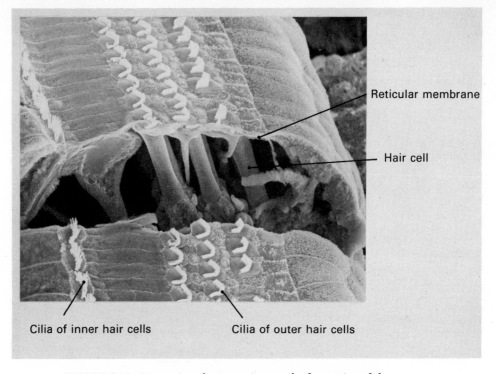

Reticular membrane

Hair cell

Cilia of inner hair cells

Cilia of outer hair cells

FIGURE 8.16 A scanning electron micrograph of a portion of the organ of Corti, showing the actual appearance of the cilia of the inner and outer hair cells. (Photomicrograph courtesy of I. Hunter-Duvar, The Hospital for Sick Children, Toronto, Ontario.)

terminals that pass through the neural channels. Figure 8.16 shows the actual appearance of the inner and outer hair cells and the reticular membrane in a photograph taken by means of a scanning electron microscope, which shows excellent three-dimensional detail. Note the three rows of outer hair cells on the right and the single row of inner hair cells on the left. (See **FIGURE 8.16.**)

Transduction of Auditory Information

Although the process by which the organ of Corti converts mechanical energy into neural activity has received intense study, we still do not know how this process occurs. George von Békésy, in a lifetime of brilliant studies on cochleas of various animals, from human cadavers to elephants, found that the vibratory energy exerted on the oval window resulted in deformations in the shape of the basilar membrane called "traveling waves." These deformations occur at different portions of the membrane, depending upon the frequency of the stimulus. The physics of the production of these waves is too complicated to be presented here (anyway, I take it on faith, myself). It suffices to say that, because of resonances of the cochlear spirals, and because of the physical properties of the basilar membrane, low-frequency sounds cause a maximum deformation at the apical end of the basilar membrane, whereas high-frequency sounds cause the end nearest the oval window to bend.

As the basilar membrane bends, it produces a shearing force on the stiff cilia of the hair cells, which is thus transmitted to the **cuticular plate** anchoring the base of the cilia to the top of the hair cells. Somehow, the shearing force produces a receptor potential. Many suggestions have been made concerning the means by which the shearing results in a receptor potential, but the evidence (which I shall now review) is still inconclusive.

In 1930, Wever and Bray discovered that if one amplified signals from a cat's **cochlear nerve,** and passed these signals through a loudspeaker, the cat's ear served as a very good microphone. One could talk into the cat's ear and hear one's voice over the loudspeaker. Wever and Bray were not recording nerve action potentials; these so-called **microphonics** could be recorded directly from the cochlea even in animals whose cochlear nerves had been previously cut and whose auditory fibers had subsequently degenerated. Thus, the electrical signal came from the organ of Corti itself. Once the blood supply to the inner ear was cut off, the **cochlear potential** (the more usual term for the cochlear microphonics) disappeared within a few hours.

It is generally (but not universally) assumed that the cochlear potential represents a summed recording of the receptor potentials of individual hair cells. Many different transducer mechanisms have been suggested; I will describe one of them (from Gulick, 1971), which I find appealing. Figure 8.17 shows a cross section through a coil of the cochlea. There are two (and probably three) kinds of liquid within the cochlea; the scala vestibuli and scala tympani contain **perilymph,** while the scala media contains **endolymph.** The region immediately surrounding the bodies of the hair cells contains **cortilymph,** thought to be different from the endolymph of the scala media (Angelborg and Engström, 1973). The interior of the hair cells is negative (by 20 to 70 mV) relative to the fluid that bathes them; furthermore, there is a potential difference between perilymph and endolymph, the endolymph being 50 to 80 mV positive to the perilymph. (See **FIGURE 8.17** on the following page.) Although there are large differences in the potassium ion content of perilymph and endolymph, it appears that this ionic difference is not the cause of the electrical potential difference; experimental manipulations of K^+ concentrations of the perilymph and endolymph have not led to corresponding changes in electrical charge (Eldredge and Miller, 1971).

Close microscopic examination of the hair cells shows that there is a bare spot in the cuticular plate; just below this bare spot is the

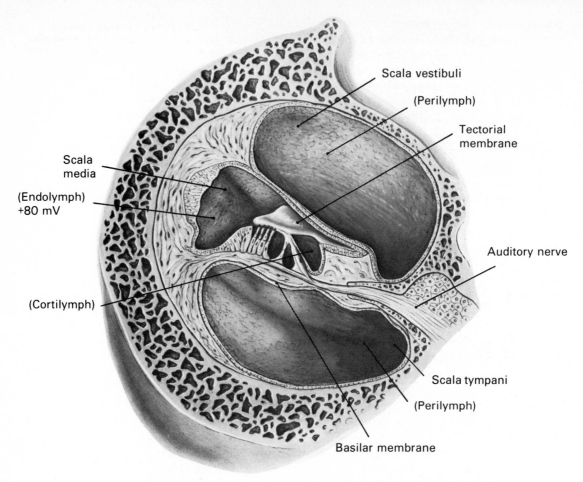

Scala vestibuli

(Perilymph)

Tectorial membrane

Scala media

(Endolymph) +80 mV

Auditory nerve

(Cortilymph)

Scala tympani

(Perilymph)

Basilar membrane

FIGURE 8.17   The electrical charges in the various regions of the cochlea.

basal body of the hair cell, and just above it is a small pore, approximately 1.5 μm in diameter (Hawkins, 1965). (See **FIGURE 8.18.**) A considerable amount of Golgi apparatus and a large number of mitochondria and granules are located around the basal body, which has led to the suggestion (Engström, Ades, and Hawkins, 1965) that the basal body might be the excitable portion of the hair cell. When the basilar membrane deforms and causes the cilia to be pulled, the cuticular plate moves relative to the basal body. This alters exposure of the membrane of the basal body to the endolymph that enters through the pore in the reticular membrane. Changes in ionic composition of the fluid bathing the basal body thus alter its membrane potential. (Alternatively, the basal body may be electrically stimulated by a flow of current through the pore and opening in the cuticular plate.) The potential changes are then propagated toward the base of the hair cell and alter flow of transmitter substance to the

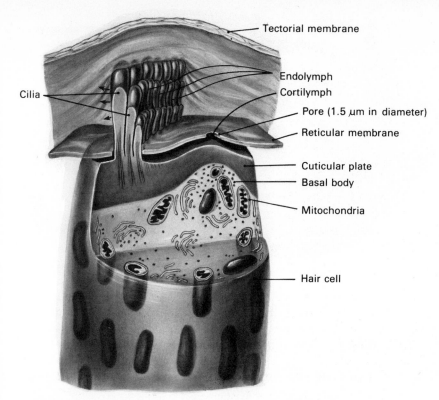

FIGURE 8.18   A shearing force on the cilia causes movement of the cuticular plate, and an ensuing movement of ions might be the event that produces the auditory generator potential.

dendrites of afferent neurons of the cochlea. (See **FIGURE 8.18.**)

Origin of the Auditory Nerve

The afferent fibers of the cochlear nerve (a branch of the **auditory nerve,** or eighth cranial nerve) are produced by bipolar cells that reside in the **spiral ganglion.** The reason for this ganglion's name is made clear in Figure 8.19, which shows the scala media (*not* the entire cochlea) with its associated neural structures. Note that the spiral ganglion is not one structure, but consists of numerous bunches of nerve fibers, each containing a small nodule. The cell bodies of the bipolar sensory neurons reside within these nodules. (See **FIGURE 8.19** on the following page.)

Figure 8.20 shows a diagram of the synaptic connections of inner and outer hair cells. Note that there are both afferent and efferent connections; the cochlear nerve contains both incoming and outgoing fibers. (See **FIGURE 8.20** on the following page.) The appearance of vesicles within the efferent nerve terminals, and within the cyto-

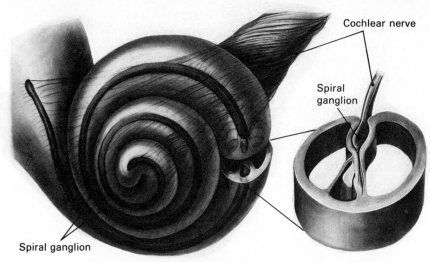

FIGURE 8.19   The spiral ganglion and the cochlear nerve.

plasm of the hair cells in the vicinity of the afferent nerve endings, suggests that information is transmitted chemically. The transmitter substance that conveys sensory information from the hair cells to the dendrites of the cochlear nerve neuron has not been identified (Fex, 1972); efferent transmission (which is inhibitory in nature) appears to be cholinergic (Fex, 1973).

Approximately 95 percent of the afferent cochlear nerve fibers synapse with the inner hair cells, on a one receptor–one neuron basis (Spoendlin, 1973). The other 5 percent of the sensory fibers synapse

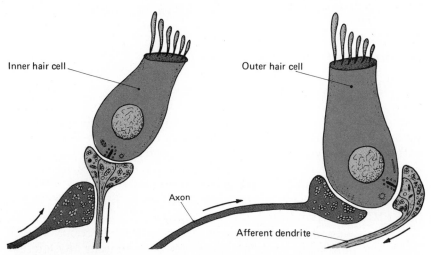

FIGURE 8.20   Details of the synaptic connections of the auditory hair cells. (Adapted from Spoendlin, H., The innervation of the cochlear receptor. In *Basic Mechanisms in Hearing*, edited by A. R. Møeller. New York: Academic Press, 1973.)

with the outer hair cells, on a ten receptors–one neuron basis. Thus, although the inner hair cells represent only 22 percent of the total number of receptive cells, they appear to be of primary importance in transmission of auditory information to the CNS.

Efferent Control from the CNS

In describing the muscles of the middle ear, the stapedius and tensor tympani, I also described their function in protective reflexes acting to reduce the probability of hair cell damage upon exposure to loud noise. This is about all the control we humans have over the nature of vibrations reaching the oval window (other than turning our head or putting our fingers in our ears). There is some evidence that the size of the external auditory canal and the elasticity of its walls can be altered slightly by the contraction of surrounding muscles, but the effects of these modifications are minor. Other mammals (e.g., cat, rabbit) have horn-shaped ears that can be independently turned toward the source of the sound.

As I noted in the previous section, the hair cells receive a considerable number of efferent terminals, which are inhibitory in effect. The efferent terminals synapse directly upon the outer hair cells and upon the dendrites to which the inner hair cells direct their sensory information. (Refer back to **FIGURE 8.20.**) The cell bodies of the efferent fibers are located in the *superior olivary nuclei* of the medulla. Approximately 500 efferent axons leave the nucleus on each side of the brain; 400 travel to the *contralateral* (other side) cochlea, and 100 go to the *ipsilateral* (same side) cochlea.

The precise role of these fibers is not known. Some investigators have suggested that they play a role in selective attention; others propose that they are involved in the sharpening of frequency-specific information from the hair cells.

VESTIBULAR SYSTEM

The Stimuli

The vestibular system has two components: the *vestibular sacs* and the *semicircular canals.* They represent two of the three components of the bony labyrinths. (We just studied the other component, the cochlea.) As we shall see, the vestibular sacs respond to the force of gravity and inform the brain about the head's orientation. The semicircular canals respond to angular acceleration; they detect changes in rotation of the head, but not steady rotation. They also respond (but rather weakly) to position or linear acceleration.

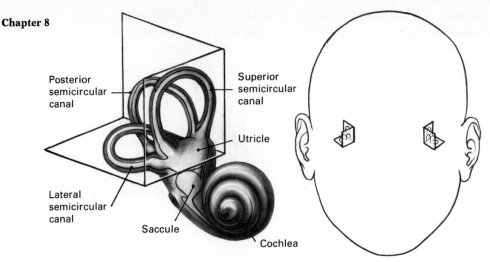

FIGURE 8.21   The bony labyrinths of the of the inner ear.

Anatomy of the Vestibular Apparatus

Figure 8.21 shows the bony labyrinths: the cochlea, the semicircular canals, and the two vestibular sacs—the **utricle** and the **saccule**. (See **FIGURE 8.21.**) The semicircular canals will be considered first.

The semicircular canals approximate the three major planes of the head: sagittal, transverse, and horizontal. Each canal responds maximally to angular acceleration in its plane. Sections through one semicircular canal are represented schematically in Figure 8.22. The enlargement (**ampulla**) contains the **crista**, the organ containing the sensory receptors. (See **FIGURE 8.22.**) The crista consists of a large

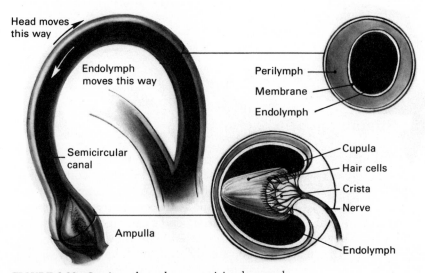

FIGURE 8.22   Sections through one semicircular canal.

number of hair cells, whose cilia are embedded in a gelatinous mass called the *cupula*. The semicircular canal consists of a membranous canal floating within a bony one; the membranous canal contains endolymph and floats within perilymph. The crista consists of a gelatinous mass that blocks part of the ampulla.

In order to explain the effects of angular acceleration on the semicircular canals, I shall first describe an "experiment." If we place a glass of water on the exact center of a turntable, and then start the turntable spinning, the water in the glass will, at first, remain stationary (the glass will be moving with respect to the water it contains). Eventually, however, the water will begin rotating with the container. If we then shut the turntable off, the water will continue spinning for a while because of its inertia.

The semicircular canals operate on the same principle. The endolymph within these canals, like the water in the glass, resists movement when the head begins to rotate. This inertial resistance pushes the endolymph against the cupula, causing it to bend until the fluid begins to move at the same speed as the head. If the head rotation is then stopped, the endolymph, still circulating through the canal, pushes the cupula the other way. Angular acceleration is thus translated into bending of the crista, which exerts a shearing force on the cilia of the hair cells. This process was directly observed by Steinhausen (1931), who injected a drop of oil in the partially dissected semicircular canal of a pike (a fish with a large and easily accessible vestibular apparatus). Figure 8.23 shows the effects of rotation of the semicircular canal in a clockwise direction; the endolymph resists the rotation and pushes the cupula to the right. (See **FIGURE 8.23.**)

The vestibular sacs (utricle and saccule) work very differently. These organs are roughly circular in shape, and each contains a

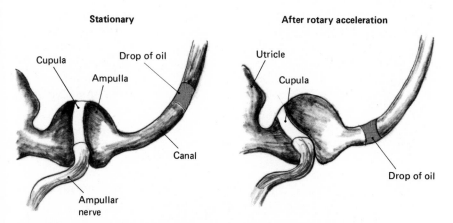

Stationary   **After rotary acceleration**

Cupula
Drop of oil
Ampulla
Utricle
Cupula
Canal
Drop of oil
Ampullar nerve

FIGURE 8.23   A diagram of Steinhausen's demonstration that movement of the endolymph causes displacement of the cupula. (Adapted from Dohlman, G., *Proceedings of the Royal Society of Medicine*, 1935, *28*, 1371–1380.)

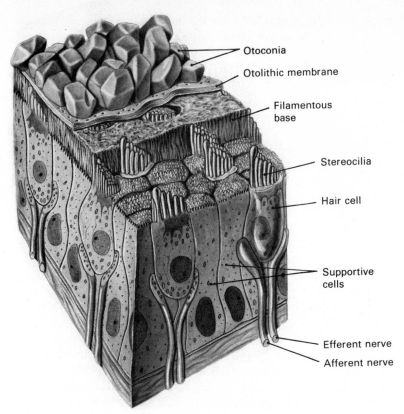

FIGURE 8.24 The receptive tissue of the utricle and saccule.

patch of receptive tissue (on the "floor" of the utricle and on the "wall" of the saccule, when the head is in an upright position). The receptive tissue, like that of the semicircular canals and cochlea, contains hair cells. The cilia of these receptors are embedded in an overlying gelatinous mass, which contains something rather unusual—*otoconia*, small crystals of calcium carbonate. (See **FIGURE 8.24.**) The weight of the crystals causes the gelatinous mass to shift in position as the orientation of the head changes. Thus, movement produces a shearing force on the cilia of the receptive hair cells.

Anatomy of the Receptor Cells

The hair cells of the semicircular canal and vestibular sacs are very similar in morphology. There are two types of cells, appearing in both organs, as shown in **FIGURE 8.25.** The type I hair cell is embedded in a dendritic process (called a calyx) similar in shape to an egg cup. Transmission appears to be chemically mediated across slight indentations, indicated by the small arrows in the drawing. (See **FIGURE 8.25.**) Efferent terminals (apparently cholinergic) synapse

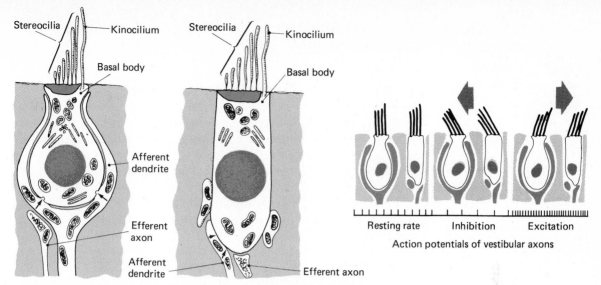

Stereocilia — Kinocilium

Basal body

Afferent dendrite

Efferent axon

Afferent dendrite

Stereocilia — Kinocilium

Basal body

Efferent axon

Resting rate | Inhibition | Excitation

Action potentials of vestibular axons

FIGURE 8.25   The two types of vestibular hair cells. (Adapted from Ades, H. W., and Engström, H., Form and innervation in the vestibular epithelia. In *The Role of the Vestibular Organs in the Exploration of Space*, edited by A. Graybill. U.S. Naval School of Medicine: NASA SP–77, 1965.)

on the outside of the calyx, but not on the type I hair cell itself. Type II hair cells are not surrounded by a calyx; they synapse with both afferent and efferent terminals. As the figure indicates, hair cells of both types may synapse with branches of the same dendrite. (See **FIGURE 8.25.**)

Each hair cell contains one long ***kinocilium*** and several ***stereocilia,*** which decrease in size away from the kinocilium. These cilia are rooted in a cuticular plate. A basal body underlies the kinocilium (indeed, the bare patch above the basal body of the auditory hair cell represents a vestigial kinocilium; during embryological development the auditory receptors possess kinocilia, which later degenerate, leaving the patch). It has been suggested (Flock, 1965; Wersäll, Flock, and Lundquist, 1965) that the orientation of the cilia gives the receptor maximal sensitivity to shearing force in one direction—namely, across the stereocilia, toward the kinocilium. (See **FIGURE 8.25.**)

The hair cells of the crista are all oriented in one direction, and they are thus sensitive to movement of the cupula in one direction. When head rotation causes the cupula to bend toward the utricle, the hair cells are stimulated, which produces an increased firing rate of associated afferent neurons in the *vestibular nerve*. Bending of the cupula in the opposite direction produces a slight decrease in firing rate. Thus, the semicircular canals of the right and left ear together provide information about the magnitude and direction of angular rotation of the head. (See **FIGURE 8.25.**)

The hair cells of the utricle and saccule are oriented in various

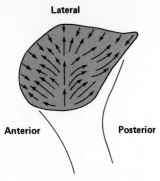

FIGURE 8.26 Hair cells in different regions of the utricle are sensitive to shearing forces in different directions. (Adapted from Flock, A., *Journal of Cell Biology*, 1964, 22, 413–431.)

directions; thus, different groups of hair cells signal different angles of head tilt. (See **FIGURE 8.26.**)

Transduction of Vestibular Information

As we saw in previous sections, the hair cells of the vestibular apparatus apparently produce a receptor potential in response to a shearing force across the cilia, and they pass this information on to the afferent neurons by means of chemical transmission. It is not known how the shearing force produces a receptor potential; it seems likely that the transduction mechanisms of vestibular and auditory hair cells are similar (one can also record microphonics in response to vestibular stimulation).

Origin of the Vestibular Nerve

The vestibular and cochlear nerves constitute the two branches of the eighth cranial nerve (auditory nerve). The bipolar cell bodies that give rise to the afferent fibers of the vestibular nerve are contained in the **vestibular ganglion,** which appears as a nodule on the vestibular nerve. Efferent fibers arise from cell bodies in the **fastigial nucleus** of the cerebellum and the **vestibular nuclei** of the medulla.

Efferent Control from the CNS

The efferent fibers of the vestibular nerve appear to exert an inhibitory effect on the firing rate of the afferent neurons. Activity of these efferents changes during tactile stimulation, somatic movement, and eye movements (Goldberg and Fernandez, 1975), but the role of modulation of receptor activity by the CNS is completely unknown.

SKIN SENSES

The somatosenses are usually divided into two groupings: (1) the skin senses (cutaneous senses) and (2) kinesthesia and organic sensitivity. I shall discuss the skin senses in this section.

The Stimuli

At least three qualities of sensation are received by the skin: pressure, temperature, and pain. This is the minimal number; many investigators class cold and warmth sensitivity separately and distinguish between touch and pressure, for example. The classical method for determining which sensory qualities are detected has been to make a 20 × 20 mm grid on the skin (with a rubber stamp) and test each square separately for sensitivity to various kinds of stimuli produced by pins, hot or cold probes, or fine hairs. Studies such as this indicate that not all squares are sensitive to all stimuli. One square might be sensitive to cold and pain, another to touch alone, another to warmth, touch, and pain, etc. The results suggest that there are independent systems mediating sensitivity to various stimuli. Furthermore, if the squares of skin are stimulated with a small electric spark, the same stimulus gives rise to different sensations in different squares (Bishop, 1943).

I shall divide cutaneous sensation into the three broadest categories—pressure, temperature, and pain—and shall discuss thermoreceptors, mechanoreceptors, and pain receptors (some call them *nociceptors*—"hurt" receptors).

The stimulus that excites thermoreceptors is obvious: temperature (chiefly changes in temperature). Pressure receptors are stimulated by mechanical deformation of the skin. Again, these receptors respond best to changes; mechanoreceptors quickly adapt to constant stimuli of moderate intensity. The stimulus that excites pain receptors has not yet been specified. Pain seems to be elicited by a variety of procedures that produce tissue damage. These procedures might produce a common effect, which then stimulates pain receptors in a similar way, or a variety of mechanisms might transduce the stimuli that cause pain. To further complicate the story, pain can be aroused via almost any sensory modality, if the stimulus is intense enough. This section, however, will deal only with cutaneous pain.

Anatomy of the Skin and Its Receptive Organs

The skin is a complex and vital organ of the body—one that we tend to take for granted. We cannot survive without it; extensive skin burns are fatal. Our cells, which must be bathed by a warm fluid, are protected from the hostile environment by the skin's outer layers. The skin participates in thermoregulation by producing sweat, thus cooling the body, or by restricting its circulation of blood, thus conserving heat. Its appearance varies widely across the body, from mucous membrane to hairy skin to the smooth, hairless skin of the palms and on the soles of the feet.

Skin consists of subcutaneous tissue, dermis, and epidermis, and contains various receptors scattered throughout these layers. Figures 8.27 and 8.28 show a section through hairy and *glabrous* skin

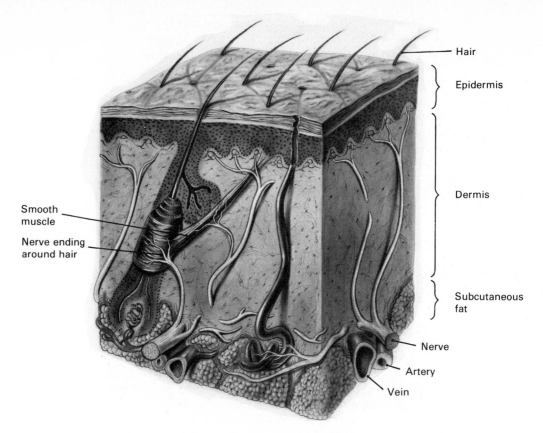

FIGURE 8.27   Schematic section through hairy skin.

(smooth, hairless skin, such as we have on our fingertips and palms). Hairy skin contains mostly unencapsulated (free) nerve endings. These nerve endings are found just below the surface of the skin, in a basketwork around the base of hair follicles, and around the emergence of hair shafts from the skin. (See **FIGURE 8.27.**)

Glabrous skin, however, contains both free nerve endings and axons that terminate within specialized end organs. Over the years, investigators have described a large number of cutaneous receptors, but subsequent research has shown many of them to be *artefacts* of the staining process used to reveal the microscopic structure of the skin. (Artefact means "made by art"—hence artificially introduced, not normally existing in the tissue.) Other specialized end organs have been shown to be variants of a single form, changing shape as a function of age. Sinclair (1967) suggests that there are really only four organized endings found in glabrous skin.

Pacinian corpuscles are the largest sensory end organs in the body. They are found in glabrous skin, external genitalia, mammary glands, and various internal organs. These receptors consist of series of onionlike layers wrapped around the terminal of a myelinated

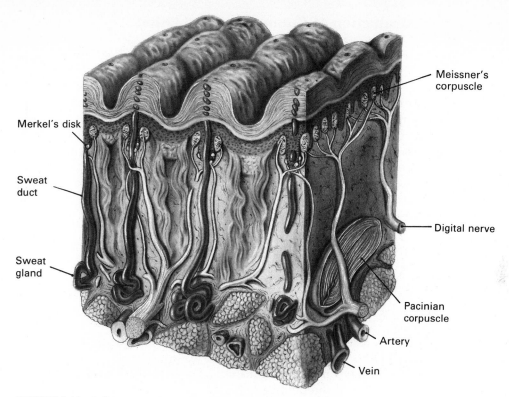

Merkel's disk

Sweat duct

Sweat gland

Meissner's corpuscle

Digital nerve

Pacinian corpuscle

Artery

Vein

FIGURE 8.28   Schematic section through glabrous (smooth) skin.

fiber. (See **FIGURE 8.28.**) They are especially sensitive to touch, their axons giving a burst of responses when the capsule is moved relative to the fiber. The inside of the corpuscle is filled with a viscous substance that offers some resistance to the movement of the nerve ending inside. This construction gives the Pacinian corpuscle peculiar response characteristics, as we shall see later.

Meissner's corpuscles are smaller than Pacinian corpuscles. They are found in the papillae, which are small elevations of the dermis that project up into the epidermis. These end organs are innervated by several axons, unlike the Pacinian corpuscles, each of which contains a single nerve fiber. (See **FIGURE 8.28.**)

Merkel's disks are found in the same locations as Meissner's corpuscles. The disks are single, flattened epithelial cells, each lying in proximity to the terminal branches of an axon. (See **FIGURE 8.28.**)

Krause end bulbs are found in *mucocutaneous zones*—the junctions between mucous membrane and dry skin, such as the edge of the lips, eyelids, glans penis, and clitoris. They consist of loops of unmyelinated fibers similar in appearance to balls of yarn. Each end bulb represents two to six fibers.

Transduction of Cutaneous Stimulation

Many unsuccessful attempts have been made to relate receptor type to sensory quality. It would seem logical to assign a particular function to a particular receptor, but there does not appear to be a consistent relationship between stimulus and receptive end organ. For example, hairy skin contains only free nerve endings and Iggo corpuscles, but it can detect the same sensory qualities as glabrous skin, with all of its specialized terminal structures. (The sensitivities of these areas differ considerably, of course.) Furthermore, affected areas of the skin of people afflicted with psoriasis (a hereditary skin disease) show gross changes in sensory innervation without accompanying changes in cutaneous sensitivity.

TEMPERATURE. Thermal receptors are difficult to study, since changes in temperature alter metabolic activity, and also rate of axonal firing, of a variety of cells. For example, a receptor that responds to pressure might produce varying amounts of activity in response to the same mechanical stimulus, depending upon the temperature. Nevertheless, investigators have recently identified cold receptors as being small myelinated axons that branch into unmyelinated fibers in the upper layer of the dermis (Hensel, 1974). Warmth receptors have not been anatomically identified, and so far, the transduction of temperature changes into rate of axonal firing has not been explained.

An ingenious experiment by Bazett, McGlone, Williams, and Lufkin (1932) showed long ago that receptors for warmth and cold lie at different depths in the skin. The investigators lifted the prepuce (foreskin) of uncircumsized males with dull fish hooks. They applied thermal stimuli on one side of the folded skin and recorded the rate at which the temperature changes were transmitted through the skin by placing small temperature sensors on the opposite side. They then correlated these observations with verbal reports of warmth and coolness. The investigators concluded that cold receptors were close to the skin and that warmth receptors were located deeper in the tissue. (This experiment shows the extremities to which scientists will go to obtain information—pun intended.)

PRESSURE. Pressure sensitivity appears to be related to movement of the skin. The skin is sensitive only to deformation, or bending, and not to pressure exerted evenly across its surface. For example, if you dipped your finger into a pool of mercury, pressure would be exerted on all portions of your skin below the surface of the liquid. Nevertheless, you would feel only a ring of sensation, at the air/mercury junction. This is the only part of the skin that is bent, and it is here that skin receptors are stimulated. (Since mercury is quite poisonous if it enters a break in the skin, please do not try this demonstration—just imagine it.)

Mechanical pressure appears to be transduced by a variety of receptors, both encapsulated and unencapsulated. The best-studied one is the Pacinian corpuscle. This organ responds to bending, relative to the axon that enters it. If the onionlike layers are dissected away (Loewenstein and Rathkamp, 1958), this receptor still responds to bending of the naked axon; thus, the transducer is the terminal itself. The generator potential that is produced is proportional to the degree of bending. If the threshold of excitation is exceeded, an action potential is produced at the first node of Ranvier. Loewenstein and Mendelson (1965) have shown that the layers of the corpuscle alter the mechanical characteristics of the organ, so that the axon responds briefly when the intact organ is bent and again when it is released.

We do not know how bending of the tip of the nerve ending in a Pacinian corpuscle produces a generator potential. Presumably, the bending changes membrane permeability, and the resulting flow of the ions across the membrane alters its potential. It is tempting to speculate that membrane "pores" are somehow enlarged as a result of this bending and stretching of the membrane. It is not known whether other mechanoreceptors operate on the principle followed by the ending within the Pacinian corpuscles. Most investigators assume that they do, and that the encapsulated endings serve only to modify the physical stimulus transduced by a portion of the neuron itself.

A special form of mechanoreception is produced by the bending of a hair. As was seen in Figure 8.27, there is a basketlike nerve ending around the base of the hair, and there are also free nerve endings near the location where the hair emerges from the skin. The basketlike endings seem to be less sensitive to hair movement. Stetson (1923) glued the hair to the skin at its point of emergence, so that movement of the hair stimulated only the deeper nerve endings. (To visualize this, picture the hair as an oar. The glue made the skin stiff and served as an oarlock; movement of the hair "rowed" the base of it through the underlying layers of skin.) Under these conditions, the hair could be moved quite a bit before giving rise to sensation. Conversely, an artificial hair glued to the skin worked as well as a real one (as far as tactile sensitivity goes), suggesting that normal sensitivity to touch in hairy areas is mediated by free nerve endings near the surface.

PAIN. The story of pain is quite different from that of temperature and pressure; the analysis of this sensation is extremely difficult. It is obvious that our awareness of pain and our emotional reaction to it depend on central factors. We can, for example, have a tooth removed painlessly while under hypnosis, which has no effect on stimulation of pain fibers. I shall deal with pain in greater detail in a later chapter on reward and punishment (Chapter 17); here I shall consider only the peripheral aspects of pain.

Most investigators identify pain reception with the networks of free nerve endings in the skin. Pain appears to be produced by a variety of procedures that cause tissue damage. Most investigators believe that the damage leads to an increase in some extracellular substance that stimulates pain fibers. Many substances can be injected into the skin to produce pain, and a variety of them have been proposed as candidates for the role of chemical mediator of pain. (The usual procedure for testing chemically mediated pain nowadays is to produce a blister with *cantharides*—Spanish fly—and then to pick off the top of the blister and treat the raw skin with the chemical to be tested.) A good relationship between intensity of pain and concentration of the potassium ion has been observed (Keele, 1966). This is of significance because tissue damage produces an extracellular increase in K^+ concentration. Other investigators have noted that pain is also produced by a low pH level (acidic solution), and by histamine, acetylcholine, and serotonin (Sinclair, 1967). The chemical mediator has not yet been identified; the nature of the substance, when identified, should suggest the means of sensory transduction. (There could, of course, be more than one mediator.)

Route of Somatosensory Fibers to the CNS

Somatosensory fibers enter the CNS via spinal and cranial (principally the *trigeminal*, or fifth) nerves. The cell bodies of these unipolar neurons are located in the dorsal root ganglia and cranial nerve ganglia.

Efferent Control from the CNS

There do not appear to be any efferent mechanisms that specifically alter responsiveness of cutaneous receptors. Changes in peripheral blood flow, sweating, and piloerection could, of course, alter the firing patterns of receptors to the appropriate stimuli, and touching and feeling an object obviously involve motor mechanisms as much as sensory mechanisms. However, there are no direct efferents to the receptors themselves, in contrast to what we have seen in the case of other sense modalities.

KINESTHESIA AND ORGANIC SENSITIVITY

We are aware of the position and movement of limbs of our body, and we certainly can detect when our intestines are swollen with gas, or when a kidney stone passes through a ureter. And we know when our bladders are full. **Kinesthesia** (literally, "movement sensation") generally refers to appreciation of both movement and position of the limbs, whereas **organic sensitivity** refers to feelings received from internal organs.

The Stimuli

Stretch receptors in skeletal muscles report changes in muscle length to the CNS, and stretch receptors in tendons measure the force being exerted by the muscles. Receptors within joints between adjacent bones respond to the magnitude and direction of limb movement. The muscle length detectors (sensory endings on the **intrafusal muscle fibers**) do not give rise to conscious sensations; their information is used in motor control systems. These receptors will be discussed separately in Chapter 10. Organic sensitivity is provided via receptors in the linings of muscles, outer layers of the gastrointestinal system and other internal organs, and linings of the abdominal and thoracic cavities. Many of these tissues are sensitive only to stretch and do not report sensations when cut, burned, or crushed. In addition, the stomach and esophagus are responsive to heat and cold and to some chemicals.

Anatomy of the Organs and Their Receptive Cells

A schematic view of a skeletal muscle is shown in **FIGURE 8.29** on the following page. Four kinds of information are received by muscle and tendon afferents.

1. The sensory endings on the intrafusal muscle fibers signal muscle length.
2. The sensory endings within the **Golgi tendon organ** at the muscle/tendon junction respond to tension exerted by the muscle on the tendon.
3. The membranous covering of the muscle (fascia) contains Pacinian corpuscles. These receptors apparently signal deep pressure exerted upon muscles, which can be felt even if the overlying cutaneous receptors are anesthetized or denervated.
4. Throughout the muscle and its overlying fascia are distributed free nerve endings, which generally follow the blood supply. These receptors presumably signal pain that accompanies prolonged exertion or muscle cramps.

The tissue that lines the joints contains free nerve endings and encapsulated receptors, such as Pacinian corpuscles. The encapsulated endings presumably mediate sensitivity to joint movement and position, while stimulation of the free nerve endings produces pain (such as that which accompanies arthritis).

Pacinian corpuscles and free nerve endings are also found in the

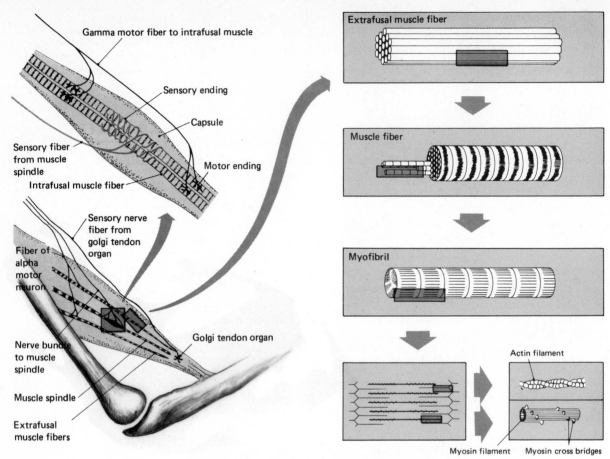

FIGURE 8.29 A skeletal muscle and its sense receptors. (Adapted from Bloom and Fawcett, *A Textbook of Histology*. Philadelphia: W. B. Saunders, 1968.)

outer layers of various internal organs and give rise to organic sensations.

Transduction of Kinesthetic and Organic Information

The mechanoreceptors and pain receptors are similar to those found in the skin; presumably, they transduce sensory information in similar ways.

Route of Kinesthetic and Organic Afferent Fibers to the CNS

The cell bodies of the receptors reside within the dorsal root ganglia or cranial nerve ganglia. Kinesthetic fibers are carried in the same nerves that convey motor fibers to the skeletal muscles. Organic sensitivity, however, is conveyed over fibers that travel with efferents of the autonomic nervous system and thus pass (without synapsing) through the autonomic ganglia on their way to the CNS. In general, pain is conveyed via afferents that accompany sympathetic fibers, whereas nonpainful stimuli are transmitted via nerves containing parasympathetic afferents.

Efferent Control from the CNS

The kinesthetic receptors described in this section respond to the effects of CNS motor outflow (by definition), and organic sensitivity is modified by activity of the gastrointestinal system, for example. There does not appear to be any direct control of these receptors by efferent fibers. The role of CNS feedback in the intrafusal muscle fiber system is described in Chapter 10.

GUSTATION

So far, we have been studying stimuli with physical energy: thermal, photic, or kinetic. The stimuli received by the last two senses to be studied, gustation and olfaction, do not, in any obvious way, transmit energy to the receptors.

The Stimuli

For a substance to be tasted, molecules of it must dissolve in the saliva and stimulate the taste receptors on the tongue. Tastes of different substances vary, but much less than we generally realize. There are only four qualities of taste: bitter, sour, sweet, and salty. Much of the flavor of good steak depends on its odor; to an *anosmic* person (lacking the sense of smell) or to a person whose nostrils are stopped up, an onion tastes like an apple, a steak like salty cardboard.

Anatomy of the Taste Buds
and Gustatory Cells

The tongue, palate, pharynx, and larynx contain approximately 10,000 taste buds. Most of these receptive organs are arranged around *papillae*, small protuberances of the tongue. Papillae are surrounded by moatlike trenches that serve to trap saliva. The taste buds (approximately 200 of them, for the larger papillae) surround

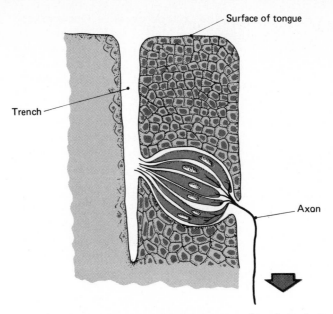

FIGURE 8.30   A taste bud. (Adapted from Woodworth, R. S., *Psychology*, 4th edition, New York: Holt, 1940.

the trenches, and their pores open into them. Figure 8.30 shows a cross section through a taste bud opening into a trench in the tongue. (See **FIGURE 8.30.**)

Individual papillae can be examined by placing a small glass tube over them and applying gentle suction, turning them inside out (von Békésy, 1964). Von Békésy found that electrical or chemical stimulation produced only a single sensation, and he concluded that each papilla is specific for one of the four taste qualities. However, Bealer and Smith (1975) later found that one-third of the papillae tested produced responses to stimuli of all four taste qualities.

Taste receptor cells are not neurons; they are specialized cells that synapse with dendrites of sensory neurons. The receptor cells possess hairlike processes that project through the pores of the taste bud into the trench adjacent to the papilla. It was previously thought that there were two types of cells in the taste buds, receptors and sustentacular (supporting) cells, but it has been shown that the various shapes of cells represent different portions of the life process of a receptor cell. Gustatory receptors have a life-span of only ten days. They quickly wear out, being directly exposed to a rather hostile environment. As they degenerate, they are replaced by newly developed cells; the dendrite of the sensory fiber is somehow passed on to the new cell (Beidler, 1970). The presence of vesicles within the cytoplasm of the receptor cell around the synaptic region suggests that transmission at this synapse is by chemical means.

Transduction of Gustatory Information

It seems most likely that transduction of taste is accomplished by means of a process similar to chemical transmission at synapses: some characteristic of the stimulus molecule is "recognized" by the receptor, and it produces changes in membrane permeability and subsequent receptor potentials. Attempts have been made to specify the molecular characteristics that stimulate different types of receptors, thus determining taste.

Acids, in general, are sour. The lower the pH, the more sour the taste. To taste salty, a substance must ionize. Salty substances contain such metallic cations as Na^+, K^+, and Li^+, with a halogen or other small anion (Cl^-, Br^-, SO_4^{--}, NO_3^-, etc.). As the salt molecules begin to get larger, they begin to taste bitter (e.g., sodium acetate). Bitter substances are difficult to characterize. They tend to be large molecules and to contain nitrogen. Various alkaloids (e.g., quinine) impart a bitter taste. Sweet substances tend to be moderately large, nonionizing, organic molecules. Like the salts, as they grow larger, they tend to elicit sensations of bitterness along with sweetness (e.g., saccharine). Dzendolet (1968) has suggested a common characteristic of sweet substances, a molecular arrangement that makes them hydrogen ion acceptors. Removal of hydrogen ions from the receptor sites presumably stimulates the sensation of sweetness. However, there is ample evidence that there is more than one kind of sweet receptor. For example, some molecules appear to taste sweet to some species, but not to other, and there are species differences in the effects of chemicals that block sweet receptors (Pfaffmann, Frank, and Norgren, 1979).

Obviously, since we do not understand precisely which molecular structures are associated with taste qualities, we are a long way from understanding the mechanism of transduction of gustatory stimuli.

Route of Gustatory Fibers to the Brain

The cell bodies that give rise to afferent fibers are located in the ganglia of the seventh (facial), ninth (glossopharyngeal), and tenth (vagus) nerves. Taste buds on the anterior two-thirds of the tongue synapse with fibers of the **chorda tympani,** a division of the facial nerve. (This nerve passes through the middle ear, and can be recorded from or stimulated electrically. Recordings have even been taken from this nerve during the course of human ear operations.) Fibers of the glossopharyngeal nerve serve the posterior third of the tongue, while branches of the vagus nerve innervate the pharynx and larynx.

Efferent Control from the CNS

Some investigators suspect that there might be efferent synapses on taste receptors, but these suggestions are quite tentative (Murray and Murray, 1970). There is no evidence for a mechanism of efferent control of taste receptors.

OLFACTION

The Stimulus

It is easy to give a superficial description of the stimulus for odor—molecules of those substances that are volatile (i.e., that evaporate at a reasonable temperature) and can dissolve in the mucus that coats the olfactory epithelium. A more specific description is a different matter. As we shall see, it is much more difficult to explain,

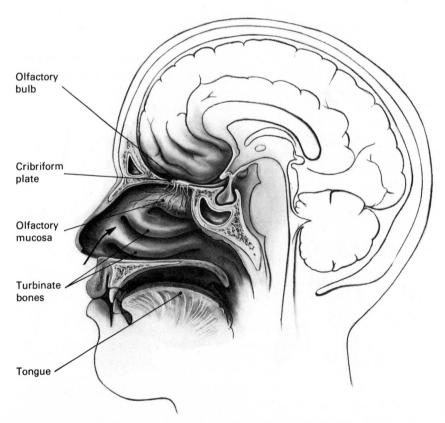

Olfactory
bulb

Cribriform
plate

Olfactory
mucosa

Turbinate
bones

Tongue

FIGURE 8.31 A schematic representation of the olfactory bulb and olfactory mucosa.

on a molecular basis, why different substances have different odors.

Anatomy of the Olfactory Apparatus

Our olfactory receptors reside within two patches of mucous membrane (**olfactory epithelium**), each having an area of about one square inch. The olfactory epithelium is located at the top of the nasal cavity as shown in **FIGURE 8.31.** Air entering the nostrils is swept upward (especially when we sniff at an odor) by action of the *turbinate bones* and reaches the sensory receptors.

The **olfactory bulb,** an enlargement on the end of the olfactory (first cranial) nerve, lies at the base of the brain, just above the bony **cribriform plate.** The olfactory receptors communicate with the olfactory bulbs via groups of axons that pass through the numerous small holes in the cribriform plate. (*Cribrum* = "sieve.") Afferent fibers from the trigeminal nerve also terminate here, in the form of free nerve endings, and presumably mediate pain in response to noxious chemical stimulation.

Anatomy of the Olfactory Receptors

Figure 8.32 illustrates a group of olfactory receptors, along with the sustentacular cells that support them. (See **FIGURE 8.32.**) The receptors are cell bodies of neurons, and they give rise to the axons

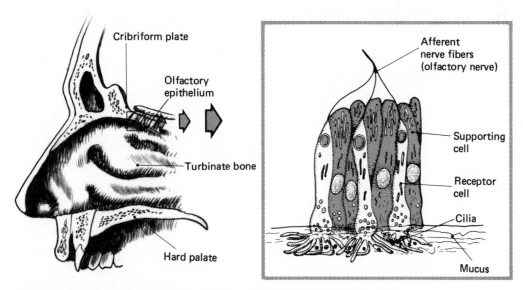

FIGURE 8.32 The olfactory receptors. (Adapted from de Lorenzo, A. J., Studies on the ultrastructure and histophysiology of all membranes, nerve fibers, and synaptic junctions in chemoreceptors. In *Olfaction and Taste*, edited by Y. Zotterman. New York: Macmillan, 1963.)

that pass through the cribriform plate to the olfactory bulb. The receptors possess numerous cilia, which project from the surface of the mucosa. It is assumed that primary reception of the odor molecules takes place on these cilia.

Transduction of Olfactory Information

The means by which odor molecules produce generator potentials is a complete mystery. Apparently, there are receptor sites on the cilia that are stimulated by odor molecules. Figure 8.33 shows the relative size of a cross section through an olfactory cilium and of a camphor molecule, indicated by a dot. We can see that the process of transduction must be quite subtle, involving molecular events. (See **FIGURE 8.33.**) Information is propagated down the cilia to the cell body, and ultimately to the axon, where generator potentials are translated into altered rates of firing. Information transfer down the cilia might be electrical, or it might even be mechanical. Cilia are known to be motile (e.g., the cilia of cells lining the respiratory tracts, the cilia and flagella of bacteria and sperm cells) and their motility and stiffness vary as a function of changes in ionic concentrations.

Cross section of olfactory cilium

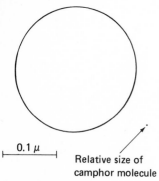

├─── 0.1 μ ───┤

Relative size of
camphor molecule

FIGURE 8.33 Relative sizes of the cross section through an olfactory cilium and a camphor molecule.

Origin of Olfactory Nerve Fibers

The axons of the olfactory receptors do not enter the olfactory nerve. Instead, they synapse with the dendrites of *mitral cells* in the olfactory bulbs, in the complex axonic and dendritic arborizations called *olfactory glomeruli*. The axons of the mitral cells then enter the olfactory nerve. Some axons synapse in the brain, whereas others cross the brain, enter the other olfactory nerve, and synapse in the contralateral olfactory bulb.

The olfactory bulbs are much more than bumps on the olfactory nerves. They contain a considerable amount of neutral circuitry and receive efferent fibers from the brain. A good deal of sensory integration undoubtedly takes place in the olfactory bulb. The nature of this integration, however, is not known.

Efferent Control from the CNS

Efferent fibers from several locations in the brain enter the olfactory bulbs. The synapses of these fibers appear to be inhibitory, but their role in the processing of olfactory information is a mystery.

The brain also controls the effects of olfactory stimuli in a more obvious way: we can sniff the air, maximizing the exposure of our olfactory epithelium to the odor molecules, or we can pinch our nostrils and breathe through the mouth, thus producing minimal olfactory stimulation.

KEY TERMS

SUGGESTED READINGS

GELDARD, F. A. *The Human Senses,* 2nd edition. New York: Wiley & Sons, 1972.

UTTAL, R. W. *The Psychobiology of Sensory Coding.* New York: Harper & Row, 1973.

These two references cover all the senses. Geldard's book contains more information about *psychophysics* (the study of the relationship between physical events and sensation), whereas Uttal emphasizes neural coding.

DAVSON, H. *The Physiology of the Eye,* 3rd edition. New York: Academic Press, 1972.

GULICK, W. L. *Hearing: Physiology and Psychophysics.* New York: Oxford University Press, 1971.

YOST, W. A., AND NIELSEN, D. W. *Fundamentals of Hearing.* New York: Holt, Rinehart and Winston, 1977.

9

Sensory Coding

In the previous chapter we saw how environmental events produce generator or receptor potentials, with subsequent changes in firing rates of afferent neurons of the spinal and cranial nerves. In this chapter I shall describe how various features of the environment— for example, shape, color, and brightness of visual forms, pitch and loudness of sounds—are coded into particular spatial and temporal patterns of neural firing, and how this coded information is transmitted to the cortex.

The concept of **sensory coding** deserves some discussion. A code consists of a set of rules whereby information may be transformed from one set of symbols into another. The different forms a message takes as it gets translated by various coding rules need not resemble each other. Spatial differences in the original messages may be represented by temporal differences, temporal differences by spatial ones. Graphs, such as the ones I used in Chapters 3 and 4, represent the use of spatial coding to portray a variety of dimensions; a graph of the action potential uses two spatial dimensions to represent (a) the electrical charge across the membrane and (b) time. So long as we know the rules of the transformations (and these are given by the labels on the axes), we can reconstruct the event represented by the graph.

To illustrate further: a written message might be given to the first of several people standing in a line. That person could read the message in Spanish to the second, who would say it in English to the third, who would convey it via semaphore signals to the fourth, who might tell the fifth person, "The message is Hamlet's soliloquy." Up to the fourth person, the representation of the message was transformed in reasonably simple ways. Transmission between the fourth and fifth person did not resemble previous transmissions; the coding used here relied on the fact that the fourth and fifth people were each familiar with Hamlet's soliloquy. As we shall see, it is possible that some elements of the sensory system "learn" frequently perceived stimulus complexes; the subsequent presence of these stimuli can then be represented more simply, by the activity of these elements.

At the level of the receptor, sensory events can be represented in a graded manner. For example, the cochlear microphonic encodes auditory information perfectly, with all its nuances. On the other hand, a single axon has a one-letter alphabet available to it. It can transmit an action potential or it can remain silent. In order to transmit information more complicated than presence or absence of a given stimulus, the axon must use the only available dimension, time. For instance, intensity of stimulation can be encoded by rate of neural firing; the more intense the stimulus is, the faster the axon fires.

When we consider a large number of neurons, we can see another way in which information can be encoded. Stimulus intensity, transformed into a temporal code (rate of firing) by a single neuron, can be represented spatially by a number of neurons. For example, if receptors were individually tuned to respond selectively to particular intensities of the stimulus, we would be able to determine the magnitude of the sensory event by noting the location of the active receptor. The nervous system thus has available to it only these two basic formats in which to represent information: **spatial coding** and **temporal coding.** Spatial coding is often referred to as *line-specific coding;* activity of a given line (chain of neurons from the receptor organ to the brain) represents a particular intensity, quality, or location of the stimulus.

Temporal codes can be much more complicated than rate. Any complex message capable of being put into words can be transmitted via Morse code, for instance. And most computers communicate with remote instruments such as teletypewriters and display consoles, and even other computers, by means of temporal pulse codes transmitted on a single line. These mechanical devices use patterns of pulses, rather than mere rate, to represent information. The pulses represent successive **bits** (*bi*nary dig*its*) of information, which in turn represent numbers or letters. For example, if the letter B is represented by the bit pattern 00110001, we could just as well signal the zeros and ones with the absence and presence of electrical pulses at the appropriate times. The pattern 00110001 would be transmitted as a short pause (two units long), two pulses, a longer pause (three units long), and another pulse. Of course, the receiving device would have to know when the message started, and it would need some kind of clock to measure the interval between pulses to determine how many missing pulses (zeros) there were.

The nervous system might use temporal codes of similar complexity to represent sensory information. And spatial codes could be much more complex than "which neuron is firing?"; the stimulus could be represented by patterns of activity across many thousands of neurons. Perkel and Bullock (1968) list a number of ways in which the nervous system could code sensory information. They call these *"candidate neural codes."* If we find that some complex aspect of neuronal firing is related to a stimulus dimension, we have not necessarily identified a neural code. The relationship could be an **epiphenomenon,** spuriously introduced by the process of producing the real code.

The only way to find out whether a candidate code is actually one used by the nervous system is to see whether it conveys information that ultimately affects the behavior of the organism. The following analogy should make this point clear. It has been observed that porpoises and other marine mammals can produce a huge repertoire of complex sounds. Are these sounds used for communication? Do they encode information? Studies performed on a single porpoise could not answer these questions; they could do no more than establish "candidate codes." We might find that particular sounds were associated with feeding, or play, or frustration, etc. But unless we established that these sounds subsequently affected the behavior of other porpoises, we would not have shown them to be a means of communication. Similarly, what a given neuron, or group of neurons, says in response to stimulus changes does not provide enough information to establish a neural code. We must also show that other neurons are "listening" to that message. As you might imagine, very few neural codes have been established in such a way.

Another problem we encounter when we consider sensory coding is the identification of the ultimate destination of the information. Where does the message go? For simple organisms, or for simple reflex activity in more complex organisms, we can follow the message right out to the effectors. The stimulus elicits activity in a chain of neurons beginning at a set of receptors and ending at a set of effectors. Very often, however, the stimulus produces no immediate effect on behavior, but instead results in some unobservable internal changes that manifest themselves in behavior much later. Obviously, we have now entered the domain of memory. But where, anatomically, did we enter that domain? And what neurons are responsible for our *experience* of this stimulus? Most scientists investigating the transmission of sensory information would consider their task complete when they had described the nature of the message as it appears in sensory cortex. As Uttal puts it:

> ... there really is no requirement for any decoding of the message at its destination. We certainly do not know what aspects of neural action are identifiable ... with experience, but it seems almost certain that all that is required is that there *exist* some representation of the pattern at some appropriately high level of the nervous system. (Uttal, 1973, p. 208)

Uttal's position seems reasonable. Obviously, the human nervous system gives rise to conscious experience, but since we have no idea how to explain this very private phenomenon, we must, by default, identify experience of a stimulus with its cortical representation. We must be careful not to seek for a decoder that looks at this representation and interprets the pattern. If we do so, we commit the error of looking for a **homunculus,** a "little man" who resides in our heads, looking at, and interpreting, the activity of cortical neurons the way we might look at a display panel of some piece of complex machinery. Experience must be recognized as an emergent property of the nervous system. Arrival of coded sensory information—somewhere in the brain—*is* the experience itself.

With this discussion concluded, let us move to something more concrete—a description of the ascending sensory pathways and a summary of the sensory codes that have been identified so far. Most of these codes express relationships between stimulus parameters and neural events; therefore, they must still be regarded as candidate codes.

VISION

Spatial Representation in the Ascending Visual Pathway

The best word to describe the first level of analysis of the retinal image would be mosaic. Literally, *a mosaic* is a picture consisting of a large number of discrete elements—bits of glass or ceramic, for example. The lens of the eye casts an image of the environment on the retinal photoreceptors, and these receptors each code their small portion of the image by responding at a rate that is related to local photic intensity. The photoreceptors are not connected to the ganglion cells (whose axons transmit visual information over the optic nerves) on a one-to-one basis. At the periphery of the retina, many individual receptors (mostly rods) converge on a single ganglion cell. Foveal vision is more direct, with approximately equal numbers of ganglion cells and cones. These receptor-to-axon relationships accord very well with the fact that our foveal (central) vision is most acute, and our peripheral vision much less precise. In a sense, the pieces constituting the matrix get larger as one goes from fovea to periphery, and the image transmitted to the brain becomes correspondingly cruder.

The retina also contains a considerable amount of circuitry that encodes the visual information in a more complex way. The ganglion cells do more than signal local light intensity, as the photoreceptors do. The activity of these cells is modified by complex spatial and temporal aspects of the visual stimulus. Before we examine the nature of this level of coding, however, we should become acquainted with the anatomy of the ascending visual system.

The route from optic nerve to primary visual cortex is the pathway that is easiest to describe of all the sense modalities. The axons of the retinal ganglion cells ascend via the optic nerves (which become the optic tracts as they enter the brain) to the dorsal lateral geniculate nucleus of the thalamus. The terminals of these axons

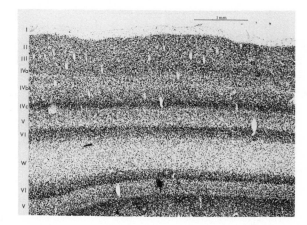

FIGURE 9.1 A photomicrograph of a small section of striate cortex (left) and a cross-section taken through the entire striate cortex of a rhesus macaque monkey (right). The ends of striate cortex are shown by arrows. (From Hubel, D. H., and Wiesel, T. N. Functional architecture of macaque monkey visual cortex. *Proceedings of the Royal Society of London, B.* 1977, *198*, 1–59.)

synapse on cells in this nucleus, which in turn send their axons via the *optic radiations* to primary visual cortex, that region surrounding the *calcarine fissure* (*calcarine,* "spur-shaped") at the most posterior region of the cerebrum. This area is often called *striate cortex* because of the dark layer (*striation*) of cells it contains. (See **FIGURE 9.1** on preceding page.)

Ignoring for a moment the complexities of coding that take place in the retina, we find that the visual system maintains the spatial code seen on the retinal mosaic all the way up to visual cortex. There is a *retinotopic representation* on the cortex. That is, a given region of the retina excites a given set of cells in visual cortex, and adjacent retinal regions excite adjacent cortical areas. This topographic representation appears to be a real sensory code. If a two-dimensional array of electrodes is placed over human visual cortex, the person will report "seeing" geometric shapes that correspond to the pattern of electrodes stimulated (Dobelle, Mladejovsky, and Girvin, 1974). This study was carried out on peripherally blinded people to ascertain the feasibility of providing visual prostheses. (A *prosthesis,* literally, "addition," is an artificial device made to take the place of a missing or damaged part of the body.) Unfortunately, long-term electrical stimulation results in tissue damage and this fact rules out the use of such synthetic replacement parts in the immediate future. Some other way will have to be found to stimulate cortical neurons. Further evidence for the reality of the spatial code on the surface of the cortex comes from the fact that damage to restricted regions of visual cortex produces specific blindness for corresponding portions of the visual receptive field. (It is interesting to note that the brain "fills in" these *scotomas* the way it fills in the blind spot caused by the exit of the optic nerve from the eye. People are often not aware of small scotomas until these blind spots are carefully tested for.)

The retinal surface is not represented on visual cortex in a linear fashion; the picture is much distorted. It is as if you printed a picture on a sheet of rubber and stretched it in various directions. The center of the sheet is stretched the most; foveal vision, with its great acuity, takes up approximately 25 percent of visual cortex.

Let us now examine the ascending visual system in more detail. Figure 9.2 illustrates a diagrammatic view of the human brain as observed from below. The optic nerves join together at the base of the brain to form the *optic chiasm* (*khiasma,* "cross"). There, axons from ganglion cells serving the inner halves of the retina (the nasal sides) cross through the chiasm and ascend to the *dorsal lateral geniculate nucleus* of the opposite side of the brain. (See **FIGURE 9.2.**) The lens inverts the image of the world projected on the retina (and similarly reverses left and right). Therefore, the *decussation* (crossing) of axons representing the nasal retinal fields results in a separate representation of each half of the visual field, divided vertically, in the dorsal lateral geniculate nucleus of the opposite side of the brain.

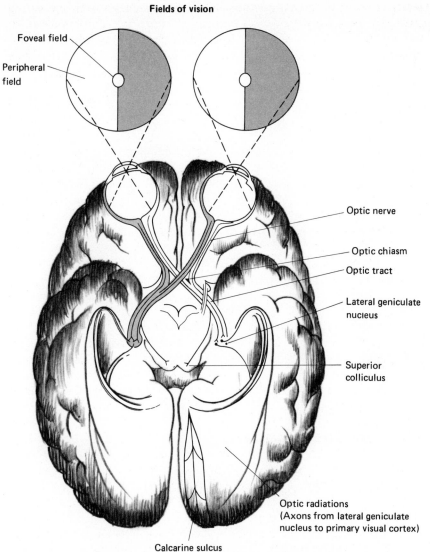

Fields of vision

Foveal field

Peripheral field

Optic nerve

Optic chiasm

Optic tract

Lateral geniculate nucleus

Superior colliculus

Optic radiations
(Axons from lateral geniculate nucleus to primary visual cortex)

Calcarine sulcus

FIGURE 9.2   The primary visual pathways. (Adapted with permission from McGraw-Hill Book Co. from *The Human Nervous System*, 2nd edition, by Noback and Demarest. Copyright 1975, by McGraw-Hill, Inc.)

(See **FIGURE 9.2.**) Each dorsal lateral geniculate nucleus then projects in a straightforward manner to the ipsilateral visual cortex; as a matter of fact, the lateral geniculate nuclei are the only subcortical afferents of primary visual cortex of primates. Since there is a considerable amount of overlap in the visual fields of our eyes, this means that many cortical regions receive information about the same point in the visual field from both eyes.

Figure 9.3 shows the actual appearance of the base of the brain,

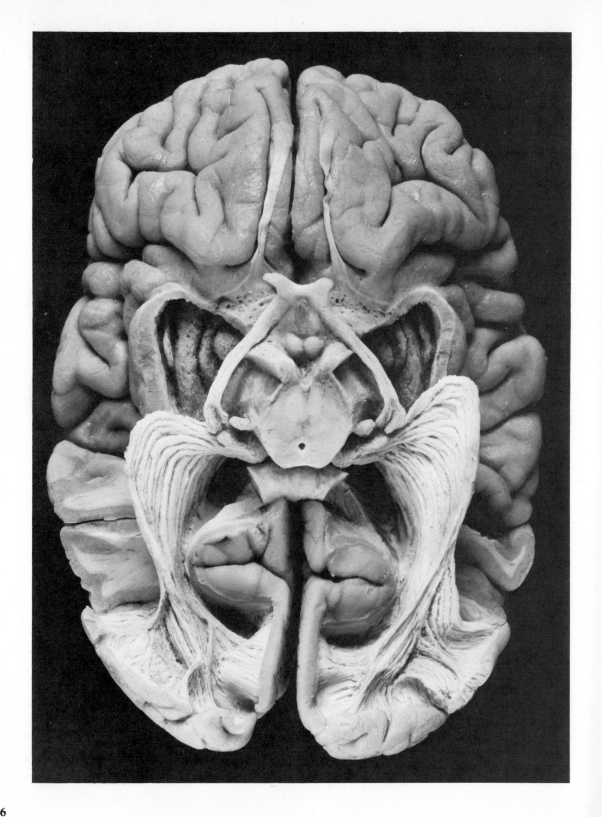

with neural tissue dissected away so that the optic radiations can be seen. Note the heavy projection to the upper lip of the calcarine sulcus. (See **FIGURE 9.3.**)

Besides the primary retino-geniculo-striate pathway, there are five other pathways taken by fibers from the retina.

1. **Suprachiasmatic nucleus.** This region of the hypothalamus controls various behaviors and physiological processes that vary across the day/night cycle. Input from the retina synchronizes its activity. (This structure will be discussed in Chapter 15.)
2. **Accessory optic nuclei.** These nuclei, located in the brainstem, play a role in coordinating eye movements that compensate for head movements, thus keeping the eyes "on track" (Ito, 1977). This system also involves the floccular region of the cerebellum.
3. **Pretectum.** This pathway terminates near the superior colliculus, and plays a role in control of pupillary size (Sprague, Berlucchi, and Rizzolatti, 1973).
4. **Superior colliculus.** The superior colliculus plays a role in attention to visual stimuli and control of eye movements. This structure sends fibers to various areas of visual cortex (not to primary visual cortex, however). The superior colliculus also receives afferents from most areas of visual cortex.
5. **Ventral lateral geniculate nucleus.** The dorsal lateral geniculate nucleus is the principal relay station between retina and striate cortex. The ventral lateral geniculate nucleus also receives direct visual input, but relays it only to subcortical structures: pretectum, superior colliculus, pontine nuclei, and suprachiasmatic nucleus. Its function is not known.

Coding of Intensity

The eye is a remarkably sensitive organ, responding to an incredible range of stimulus intensity. The smallest stimulus that can be detected is much less than one-millionth the intensity of the brightest light to which the eye can be exposed without damage. Obviously, then, this range of brightness cannot be faithfully represented by neural firing rate; a neuron cannot vary its firing rate by a factor of more than a million. The upper limit of most neurons is less than 1000 impulses per second. One-millionth that rate would be one action potential every 1⅔ minutes—a rate so slow as to be meaningless. So how does the visual system encode this tremendous range of intensity?

It does so in several ways. First, photic intensity is represented

---

FIGURE 9.3 A photograph of the base of the brain, with tissue dissected away so that the projections between dosal lateral geniculate nucleus and primary visual cortex are visible. (From Gluhbegovic, N., and Williams, T. H. *The Human Brain: A Photographic Atlas.* Hagerstown, Md.: Harper & Row, 1980.)

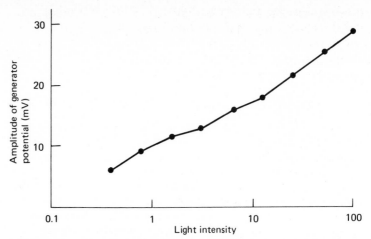

FIGURE 9.4   The relationship between light intensity and the generator potential in the eye of the horseshoe crab. (Adapted from Fuortes, M. G. F., *American Journal of Ophthalmology*, 1958, *46*, 210–223.)

in a *nonlinear* fashion by firing rate. Figure 9.4 illustrates the relationship between light intensity and the amplitude of the generator potential that is produced by the photoreceptive cells in the eye of the *Limulus* (horseshoe crab). You can see that the curve in this figure is not at all linear: changes in intensity of the light stimulus at the upper end of the scale produce hardly any alterations in the amplitude of the generator potential. The receptors are much more sensitive to changes at the lower end of the brightness scale. (Note that the horizontal axis is not linear. See **FIGURE 9.4**.) This nonlinear responsiveness represents a compression of information. A large range of stimulus intensity is represented by a smaller range in amplitude of the generator potential.

A second method for extending the range of intensity that can be represented by the primate eye (and in the eye of most other mammals) is provided by the existence of two types of receptors. Rods can detect light at levels of brightness that are too dim for cones to respond. Intensity is thus represented by the firing rate of two populations of receptors. For the visual system, then, spatial (line-specific) coding, as well as temporal coding (rate), is used to represent intensity.

Finally, some of the information about the range of intensity is just not needed. It is much more important for an animal to detect *changes* in brightness (which represent potentially significant changes in the world) or *differences* in the brightness of various portions of the visual field (which provide the basis of form perception). In fact, as one passes from the photoreceptors to the ganglion cell layer of the retina, one finds that few cells respond in a simple way to the amount of light falling on a small portion of the retina. A small percentage of cells respond in a simple fashion to overall brightness, but most cells respond to particular *features* of the retinal image (DeValois, 1965). Psychologically, absolute brightness (except at the extremes of the range) is not an important variable. We tend to adapt to the overall level of illumination and judge various portions of the visual field as to their relative brightness. A white piece of paper seen in dim light appears to us to be brighter than a piece of gray paper seen in brighter light, even though the intensity of light being reflected from the gray paper might be greater. We should not be surprised that most cells of our visual system generally exhibit a similar disregard for overall brightness.

Receptive Fields and Feature Detection

I have described the fact that spatial relations of the retinal mosaic are maintained in the lateral geniculate nucleus and visual cortex. The representation is not in the form of simple dots. Research has shown that individual cells respond to more complex features of the retinal image. The portion of the visual field to which a given cell responds is defined as the cell's ***receptive field.*** The identification of a cell's receptive field is accomplished by a procedure called mapping. The animal is anesthetized and a screen is placed in front of it. Recordings of action potentials are taken from single neurons while a small spot is moved around on the screen. (Sometimes a small spot of light is shone directly on the retina.) The receptive field is defined as the area of the screen (or retina) in which the stimulus elicits a response from the cell. Response merely means change in rate of firing; information is coded just as well by a decreased rate of firing as it is by an increased rate.

RECEPTIVE FIELDS IN THE RETINA. Kuffler (1953), recording from ganglion cells in the retina of the cat, first discovered the basic type of receptive field that has been shown to exist in the mammalian retina. He found that the receptive field consists of a roughtly circular center, surrounded by a ring. In his experiments, cells responded in opposite manner to the two regions. A spot of light presented to the central field produced a burst of unit activity. When the spot was presented to the surrounding field, no response was detected, but the cell fired vigorously for a while when the spot of light was turned off. The cell thus responded in a center-on, surround-off manner. Kuffler also identified cells that give a contrary center-off, surround-on response. Subsequent investigators (e. g., Rodieck and Stone, 1965) found that simultaneous presentation of a stimulus to both center and surround gives no response; the cells, therefore, serve to compare the brightness of the center spot with its surround, giving the greatest response when the contrast is maximal.

Not all retinal ganglion cells showed this cancellation. Those that did show a null position (i.e., center + surround stimulation counterbalancing each other) were called **X cells;** those that did not were called **Y cells.** Subsequent studies have shown that there appear to be morphological differences in these two types of ganglion cells (Cleland, Levick, and Wässle, 1975). In addition, whereas the dorsal lateral geniculate nucleus receives input from both X and Y cells, only Y cells project to the superior colliculus. X cells also have smaller receptive fields than Y cells do, and they respond more vigorously to stationary stimuli; Y cells tend to respond best to change. Since more Y cells are found in the peripheral retina, and since they project to superior colliculus as well as to visual cortex, perhaps they are important in directing eye movements toward changing or moving stimuli at the periphery of vision. Finally, only X cells convey color-coded information (Schiller and Malpeli, 1977).

Coding of Features in Visual Cortex

It is a difficult and tedious process to investigate feature detection that is performed by cells of visual cortex. Once a microelectrode has been inserted into the brain, there is a finite amount of time in which to search for the optimal stimulus. Slight movement of the brain tissue relative to the electrode can cause the neuron that is being recorded from to become lost, and physical contact between electrode and cell can kill the neuron. Therefore, one has to start hunting quickly for the "best" stimulus. The fact that it is impossible to try out every possible shape, moving in all possible directions within all parts of the visual field, means that the best stimulus found for that cell is merely the best one that has been tried. The stimulus finally chosen might bear a very poor resemblance to the real "best stimulus."

Bearing these limitations in mind, let us examine the pioneering (and continuing) work of David Hubel and Torsten Wiesel of Harvard University. It is largely because of the work of Hubel and Wiesel that we know more about the structure of primary visual cortex and the response characteristics of its cells than any other region of cortex. (See Hubel and Wiesel, 1977; 1979). I must warn you now that although their anatomical evidence represents the "state of the art," and is very impressive, many investigators support an alternative to their model of feature detection and information processing in the mammalian visual system. I will present their model first, and then I will discuss some contrary evidence.

RECEPTIVE FIELDS OF CORTICAL NEURONS. As we saw, retinal ganglion cells have circular receptive fields, divided into a center and a contrasting surround. Neurons in the dorsal lateral geniculate nucleus have similar response patterns; therefore, neurons in layer IV of striate cortex, which receive input from the dorsal lateral geniculate nucleus, respond in a similar manner.

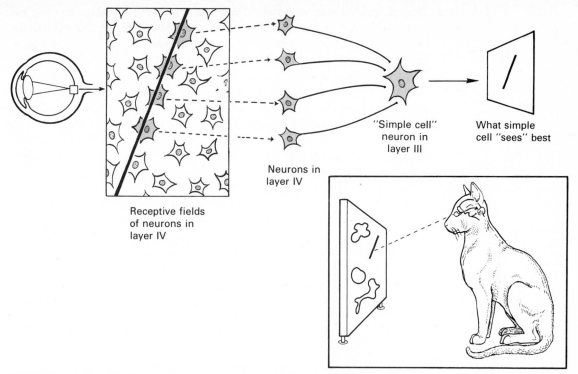

FIGURE 9.5   A schematic representation of Hubel and Wiesel's model of
the neural circuitry that determines what a simple cell "sees."

Neurons in layer IV project to the layers just above (Lund, 1973).
Neurons in these layers, like those in layer IV, have restricted receptive fields, in which the stimulus must be placed in order to produce
a response. However, the best stimulus, for these neurons, is not a
spot of light; it is a line, with a particular *orientation*. That is, some
neurons respond best to a vertical line, some to a horizontal line,
and some to a line oriented somewhere in between. The selectivity
of different neurons varies, but in general, their response falls off
when the line tilts by more than 10 degrees. (For comparison's sake,
the angle between the hands of a clock at one o'clock are 30 degrees
apart.)

Hubel and Wiesel speculate that these cells, which they call **simple cells,** receive input from a number of individual layer IV cells
whose receptive fields are arranged in a straight line. If the stimulus
is a line that excites the centers of the receptive fields of all the layer
IV cells that project to a given simple cell, that cell will be maximally
stimulated. (See **FIGURE 9.5.**)

Complex cells are found in other layers of striate cortex, both
above and below layer IV. These cells, like simple cells, respond to
segments of lines with a particular orientation. However, the line
can be moved around somewhat and still continue to excite the cell.

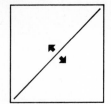

Response increases when line is moved in direction of either arrow.

FIGURE 9.6 The best stimulus for a complex cell is a line of a particular orientation, moving perpendicular to its orientation.

For example, suppose that a particular complex cell is stimulated best by a 45 degree line, slanted from lower left to upper right. A simple cell that is stimulated by a 45 degree line will respond to it only if the line is placed in one particular location. In contrast, a complex cell will respond to a 45 degree line anywhere in its receptive field, which is much larger than that of a simple cell. Moreover, it will probably respond even more if the line is moved perpendicular to its orientation. (See **FIGURE 9.6.**) Hubel and Wiesel suggest that a complex cell receives input from a number of simple cells, all of which respond to line segments with the same orientation, but within different regions of the visual field. (See **FIGURE 9.7.**)

DISTRIBUTION OF RECEPTIVE FIELDS. Hubel and Wiesel stress that the anatomy and response patterns of neurons show that information processing in striate cortex is strictly a local matter. A particular region of cortex receives information from a fairly restricted area of the retina; no neurons there receive information from both the center and the periphery of the visual field. Thus, although primary visual cortex analyzes the visual scene into line segments of various orientations, a composite picture cannot be "put together"

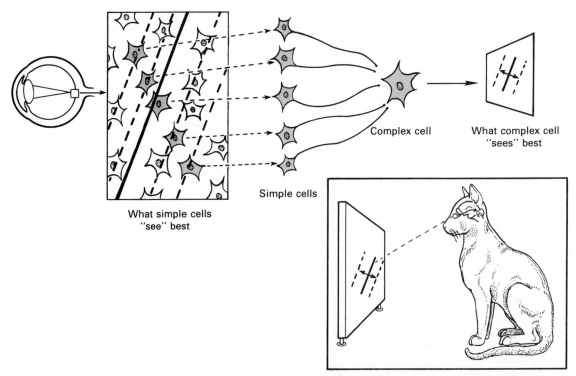

FIGURE 9.7 A schematic representation of Hubel and Wiesel's model of the neural circuitry that determines what a complex cell "sees."

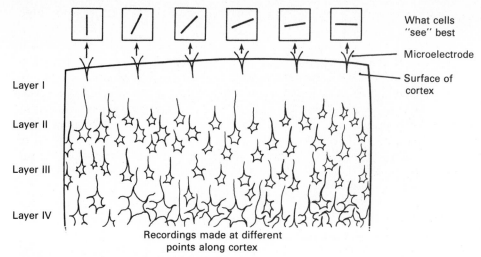

Layer I

Layer II

Layer III

Layer IV

What cells "see" best

Microelectrode

Surface of cortex

Recordings made at different points along cortex

FIGURE 9.8   As the recording electrode is moved across the surface of visual cortex, the orientation of the "best stimulus" for complex cells rotates.

there. Presumably, this function takes place in visual association cortex.

Hubel and Wiesel found that striate cortex is arranged in columns of cells with functions similar to their neighbors. That is, if a cell responded best to a particular stimulus, so did a cell immediately above or below it. As the recording electrode moved laterally, across the surface of the cortex, the orientation of the lines the cells responded to gradually changed. As the electrode moved across the cortex, the preferred orientation of the lines rotated in a clockwise or counterclockwise direction; each 25 μm of movement produced a corresponding rotation of 10 degrees in the preferred stimulus. (See **FIGURE 9.8.**)

An ingenious study (Hubel, Wiesel, and Stryker, 1978) demonstrated the pattern of preference for lines of a particular orientation throughout visual cortex. The investigators injected a monkey with radioactive *2-deoxyglucose (2-DG)*, and then presented a pattern of vertical stripes, moving back and forth from left to right. The injected chemical resembles normal glucose, and hence is taken up in greatest quantities by neurons that have the highest metabolic rate. However, unlike glucose, 2-DG cannot be metabolized, and, in addition, cannot leave the cell once it enters. Thus, the procedure labels metabolically active cells with radioactivity.

Since the retina was stimulated with a pattern of vertical lines, those cortical neurons that respond best to lines of this orientation would be firing at the highest rate, and would hence be the most active metabolically. Therefore, they should also be the most radioactive. The monkey's brain was sliced, and the sections were examined by means of autoradiography.

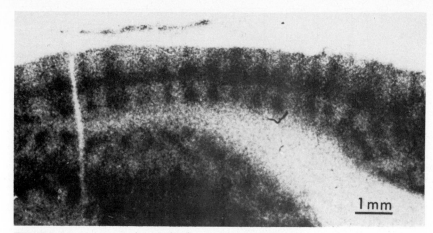

FIGURE 9.9   An autoradiograph of a cross-section through striate cortex of a rhesus monkey that had been given an injection of ^{3}H-2-deoxyglucose and exposed to a moving pattern of vertical stripes. (From Hubel, D. H., Wiesel, T. N., and Stryker, M. P. Anatomical demonstration of orientation columns in macaque monkey. *Journal of Comparative Neurology*, 1978, *177*, 361–380.)

Figure 9.9 shows a slice through striate cortex taken perpendicular to the surface. Note that columns of neurons, approximately 0.5 mm apart, were labeled with radiography, indicating that they responded best to vertically oriented lines. In contrast, neurons in layer IV were uniformly dark, because they responded to any stimulus within the center of their receptive field. (See **FIGURE 9.9.**) The distribution of these columns is in the form of isolated patches scattered throughout striate cortex. Figure 9.10 shows two slices taken almost parallel to the surface of the cortex. (See **FIGURE 9.10.**)

OCULAR DOMINANCE AND THE PERCEPTION OF DEPTH.   We have two eyes, and most of their visual fields overlap. As we saw earlier, the right halves of the retinas (representing the left visual field) project to the right hemisphere, and the left halves (representing the right visual field) project to the left hemisphere. How is the information from each eye combined within primary visual cortex?

Hubel and Wiesel found that all simple cells responded to stimuli presented to one eye or the other. None of these cells responded to both eyes. Some complex cells responded only to stimuli presented to one eye while others responded to stimuli presented to either eye. Most of the cells that responded to either eye showed a preference for one, responding more vigorously to stimuli presented to that eye. Eye preference, like preference for lines of a particular orientation, was found to be consistent within a vertical column of neurons, but varied as the electrode moved laterally along the surface of the cortex. Thus, if recordings started in a region of cortex that contained neurons that responded exclusively to stimuli presented to the left

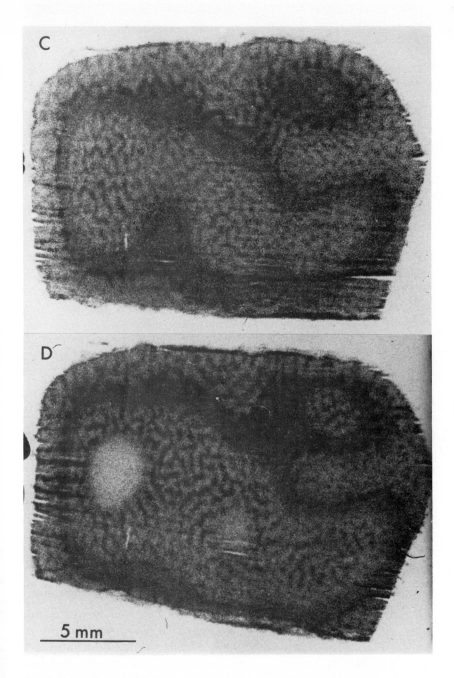

FIGURE 9.10 An autoradiograph of two sections taken nearly parallel to the surface of striate cortex of a rhesus monkey that had been given an injection of ³H-deoxyglucose and exposed to a moving pattern of vertical stripes. (From Hubel, D. H., Wiesel, T. N., and Stryker, M. P. Anatomical demonstration of orientation columns in macaque monkey. *Journal of Comparative Neurology*, 1978, *177*, 361–380.)

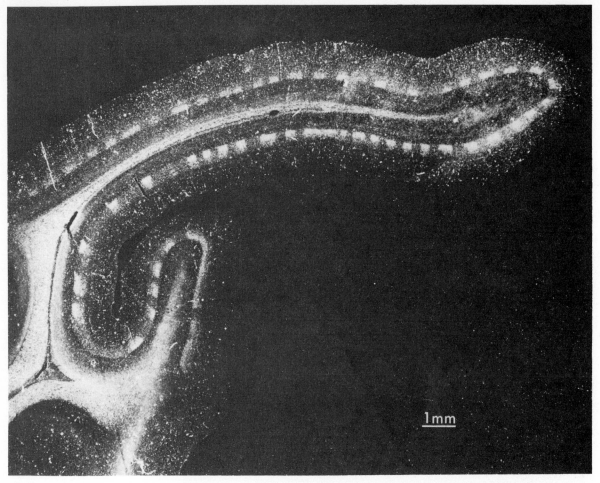

FIGURE 9.11 Dark-field autoradiograph (radioactivity is shown by light areas) of a cross-section through striate cortex of a rhesus macaque monkey that had received an injection of radioactive amino acid in the eye. (From Hubel, D. H., and Wiesel, T. N. Functional architecture of macaque monkey visual cortex. *Proceedings of the Royal Society of London, B.* 1977, *198*, 1–59.)

eye, more and more responsiveness to the right eye would be recorded as the electrode was moved away from the starting point. Finally, a region of exclusive "right eye" cells would be encountered.

Hubel, Wiesel, and Stryker (1978) injected one eye of a monkey with a radioactive amino acid. Anterograde axoplasmic flow carried the amino acid to the dorsal lateral geniculate. There, a small amount was released by the terminals and was transported into cell bodies that project to striate cortex. A very sensitive autoradiographic technique was used to determine the location of ocular dominance columns: that is, the cortical regions that received input from the left eye.

Figure 9.11 shows the results in a slice through striate cortex. Dark and light stripes confirm the results of recording studies—alternating regions of cortex are dominated by input from one eye or the other. (See **FIGURE 9.11**.)

As we saw earlier, orientation preferences are shown by isolated patches, or islands, of cortical tissue. In contrast, ocular dominance occurs in interleaved "stripes" of tissue, shown in a slice taken almost parallel to the surface of the cortex. (See **FIGURE 9.12** on next page.)

Not all neurons in striate cortex follow the scheme that has just been outlined. A small proportion of them are *strictly* binocular; that is, they will respond only when both eyes are simultaneously stimulated. These cells will not respond to stimulation of one eye alone (Poggio and Fischer, 1977; Clarke and Whitteridge, 1978). They generally respond best when there is a stimulus in almost, but *not quite,* the same location of the visual field of both eyes; thus they respond to **retinal disparity,** or a stimulus that is slightly out of register. Therefore, these cells would appear to be important in depth perception.

In fact, it appears that there are two classes of cells that respond to retinal disparity. If you focus your eyes so that they converge on a particular point in space, all points that lie the same distance from your eye as the fixation point are said to be on the **fixation plane.** (Of course, the "plane" is not really flat; it is spherical in shape, with your head at the center of the sphere.) One class of cells is excited by visual stimuli that are located in front of the fixation plane (that is, nearer to the observer) but are inhibited by stimuli located behind it (farther away). The other class responds in just the opposite way— inhibition in response to the nearer stimuli and excitation in response to the distant ones (Poggio and Fischer, 1977). The biological significance of these cells receives support from the report (Richards, 1977) that there are two classes of people who have difficulty in judging the distance of objects: some of them misjudge objects in front of the fixation plane (closer to them), while others have difficulty with objects that are behind it. It is possible that the people with this affliction lack one or the other of the two classes of cells that detect retinal disparity.

FEATURE DETECTION: EDGES OR SPATIAL FREQUENCIES? Hubel and Wiesel's model of feature detection asserts that the retina, dorsal lateral geniculate nucleus, and layer IV of primary visual cortex see the world in terms of spots. This information is relayed to neurons (simple cells) in other layers of visual cortex, where it is combined into information about lines of a particular orientation. In turn, these cells send information to other neurons (complex cells), which detect the presence of lines of a particular orientation that are located anywhere within a particular region of the visual field. This model makes two major assumptions: (1) Feature detection in the visual system is based on lines, which presumably represent

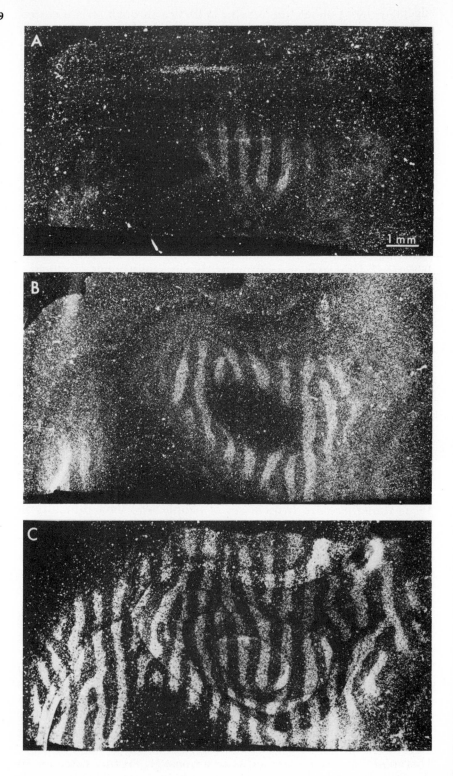

edges of objects in the real world. (2) Analysis is *serial,* or *sequential;* that is, visual information is transmitted from one group of neurons to the next, with more and more complex analyses being performed. However, as we shall see, there is evidence that contradicts these two assumptions.

In this section I shall discuss research that relates to assumption 1—that the visual system deals with edges. It appears that the best stimulus for most neurons in primary visual cortex is not an edge or a sharp line; it is a ***sine-wave grating*** (DeValois, Albrecht, and Thorell, 1978). A sine-wave grating looks rather like a series of fuzzy, unfocused, parallel bars. Figure 9.13 illustrates a sine-wave grating. A sine-wave grating is designated by its ***spatial frequency.*** We are accustomed to frequencies (for example, of sound waves or radio waves) being expressed in terms of time (cycles per second). However, spatial frequencies are expressed in terms of distance—for example, cycles per centimeter. Since the image of a stimulus varies in size according to how close it is to the eye, ***visual angle*** is generally used instead of the physical distance between adjacent cycles. Thus, the spatial frequency of a visual stimulus is measured in cycles per degree of visual angle. (See **FIGURE 9.13.**)

Hubel and Wiesel's model has the advantage of simplicity; it is easy to conceive of the visual system using information about edges—areas of sharp contrast in brightness—to construct a comprehensive image of the visual world. How can information about spatial frequencies be used to analyze visual information? Unfortunately, although we have good experimental evidence that suggests that the visual system does analyze information in terms of spatial frequency, we do not have adequate, comprehensive models to explain how it might do so. The best I can do is to give you some pieces of the picture. (The pun was not planned, but I like it.)

Consider the photograph that is shown in Figure 9.14. If we take a thin horizontal slice through it, we get a line that varies in brightness, much like one of the lines that constitute a television picture. (See **FIGURE 9.14** on following page.) If we measure the brightness of all points on this line, we can draw a graph of changes in brightness along the length of it. (See **FIGURE 9.14.**)

The waveform that is shown in Figure 9.14 is a complex one. However, mathematicians have a technique called ***Fourier analysis*** that permits complex functions like these to be described in terms of simple sine waves. It is possible to calculate a series of pure sine waves of different frequencies that can be added together to produce the complex one that is shown in the figure. Perhaps the visual system performs a Fourier analysis. If there are different classes of neu-

Actual appearance of sine-wave grating

When viewer is farther from stimulus, visual angle between cycles is smaller, therefore grating has more cycles per degree of visual angle.

FIGURE 9.13 Sine-wave gratings and visual angle. The actual appearance of a sine-wave grating is shown at the top. A schematic explanation of the concepts of visual angle and spatial frequency is shown below. V-shaped wedges are drawn between the stimulus and eye, with the open end on adjacent dark portions of the sine-wave grating, and the point on the viewer's eye. The *visual angle* between adjacent dark portions of the stimulus is smaller when the observer is farther from the stimulus. (Photograph courtesy of S. Murray Sherman.)

FIGURE 9.12 Autoradiographs of nearly-horizontal sections through striate cortex of a rhesus macaque monkey that had received an injection of radioactive amino acid in the eye. (From Hubel, D. H., and Wiesel, T. N. Functional architecture of macaque monkey visual cortex. *Proceedings of the Royal Society of London,* B. 1977, *198,* 1–59.)

FIGURE 9.14 A graph of the variation in brightness of a slice of a photograph. The waveform that is shown could be analyzed for its content of various pure sine-wave frequencies by Fourier analysis. Presumably, the visual system performs a similar analysis.

rons in the visual system that respond best to stimuli of particular spatial frequencies, then a slice of a visual scene (such as you see in Figure 9.14) might be analyzed in terms of different amounts of activity in each of these classes.

Of course, the process gets much more complicated when you consider a complete two-dimensional scene. Presumably, different neurons respond to different slices of the scene, sliced in many directions. In addition, you will remember that Hubel and Wiesel demonstrated that the information that is received by individual cortical neurons comes from a limited portion of the visual field; therefore, individual neurons do not respond to the spatial frequency of an entire slice, but only a limited length of it. If a Fourier analysis is performed by the brain, it must add together all these small pieces of information.

What about evidence? Many investigators (see DeValois and DeValois, 1980, for a review) have shown that neurons in visual cortex respond best to stimuli of a particular spatial frequency. Moreover, a wide range of frequencies is represented by different neurons. Therefore, the raw information for a Fourier analysis—information about a wide range of spatial frequencies—is present in the visual system. In addition, nonphysiological perceptual studies strongly suggest that the brain uses information about spatial frequency to analyze the visual world. These studies (pioneered by Blakemore and Campbell, 1969) make use of the phenomenon of *adaptation.*

When a particular stimulus is presented for a prolonged time, the sensory system becomes less sensitive to the stimulus; weak values of this stimulus, which were previously detectable, cannot be perceived any longer. When people stare at sine-wave gratings of particular frequencies they undergo a temporary loss of sensitivity to that frequency, but not to others. This adaptation is not a matter of fatigue in particular photoreceptors; the subjects do not have to stare at one point, but can let their eyes roam around the display, which means that a moving stimulus is presented to the retina. In addition, adaptation is specific to the orientation of the gratings. For example, if the subjects stare at a pattern of vertically oriented gratings and are then tested with a pattern of horizontally oriented gratings of the same spatial frequency, no loss in sensitivity is seen. This finding implies that different elements of the visual system look at different "slices" of the visual scene. These elements seem to be located in the cortex, since exposure of one eye to a particular spatial frequency causes a loss of sensitivity to patterns with the same frequency (and same orientation) that are presented to the opposite eye. If the analysis were performed within the retina itself, then adaptation of one eye should not affect the other. (See Sekuler, 1974, for a review of these studies.)

Low spatial frequencies normally provide us with information about objects in the visual world, while high frequencies provide us with supplementary information about texture. An image that is deficient in high-frequency information looks somewhat blurry and out of focus, like the image that is seen by a slightly nearsighted person who is not wearing corrective lenses. This image still provides much information about forms and objects in the environment. However, removal of low-frequency information is devastating; images are very difficult to perceive. The importance of low-frequency information was illustrated very nicely by Harmon and Julesz (1973). Figure 9.15 (left) shows a picture that has been chopped into a series of squares. (See **FIGURE 9.15, LEFT** on following page.) Although it contains low-frequency information, it is difficult to recognize because there is a considerable amount of high-frequency "noise" that is provided by the edges of the composite squares. (Fourier analysis shows that an abrupt edge contains much high-frequency information; in this case, the information is meaningless, since it is not related to the content of the figure.) The same figure, with high-frequency information removed, is shown on the right. As you can see, this figure is much clearer, even though the low-frequency information is exactly the same in each case. (See **FIGURE 9.15.**)

The physiological evidence for neurons that respond selectively

FIGURE 9.15 Both pictures contain the same amount of low-frequency information, but extraneous high-frequencies have been filtered from the picture on the right. (From Harmon, L. D., and Julesz, B. Masking in visual recognition: Effects of two-dimensional filtered noise, *Science*, 1973, *180*, 1194–1197. Copyright 1973 by the American Association for the Advancement of Science.)

to particular spatial frequencies is very good; so is the behavioral evidence concerning the importance of spatial frequencies in perception. What we do not know is how the information from individual frequency-sensitive neurons is pooled to form images of the visual world. Understanding this process will not be a simple task, as you can surely appreciate.

What of the data obtained by Hubel and Wiesel? If we assume that the edge detectors that they found actually respond best to sine-wave gratings rather than bars, their results are easily accomodated. Perhaps their observations that different portions of cortex respond to lines of particular orientations can be modified to say "different gratings of particular orientations."

Serial or Parallel Processing?

Hubel and Wiesel's model of information processing by the visual system is a serial one; information at one level is passed on to more complex cells at the next level, which, in turn, pass on the results of their analysis to cells that perform even more complex analyses. However, there is good physiological evidence for parallel processing in the visual system. For example, Stone (1972) showed that simple cells and complex cells both receive direct input from the dorsal lateral geniculate nucleus. This means that at least some of the information that is received by complex cells is not first funneled through simple cells. In addition, the latency of responding is shorter for complex cells; if the information processing were sequential, one would expect the complex cells to respond later than the simple ones. Finally, the regions of visual cortex that receive input from peripheral parts of the retina contain relatively more complex cells than regions of visual cortex that receive information from the fovea. Since the fovea clearly provides more detailed visual information than the periphery does, this finding does not support the concept of a hierarchy of information processing, from simple cell to complex cell.

An alternative to the sequential hypothesis is that both kinds of cells receive input from the dorsolateral geniculate nucleus, but from different classes of neurons. In fact, it has been suggested that simple cells might receive information from X cells of the geniculate, while complex cells respond to geniculate Y cells. Of course, the apparent existence of parallel processing does not rule out the possibility that serial (sequential) processing could also take place. Indeed, it seems almost certain that a good deal of serial processing goes on as information is relayed from striate cortex to areas of visual association cortex.

The Plasticity of the Visual System

In 1971, Hirsch and Spinelli reported an astonishing discovery—visual experience could modify the receptive fields of neurons in the visual cortex of a cat. Hirsch and Spinelli raised kittens in the dark from birth. At three weeks of age the kittens were fitted with a special pair of goggles that presented a separate visual pattern to each eye. One eye was presented with a view of three horizonal bars; the other eye saw three vertical bars. A photograph of one of these kittens is shown in Figure 9.16. (The kittens spent only a part of each day wearing the goggles; most of their time was spent—goggle-less—in the dark. See **FIGURE 9.16** on following page.) When the cats were ten to twelve weeks of age, the receptive fields of cortical neurons were determined by a computer technique.

The investigators placed a round black spot on a white screen and moved the spot by means of a magnet attached to a computer-driven device behind the screen. The spot could be moved throughout the entire area of the screen by means of a series of fifty sequential horizontal or vertical sweeps. (See **FIGURE 9.17** on following page.) The stimulus spot was moved in a series of small discrete steps (fifty per line), and the computer recorded the number of action potentials produced while the stimulus was in each of the 2500 positions. The receptive field of the neuron recorded from was displayed by the computer on the face of an oscilloscope. Each of the stimulus positions was represented by a corresponding location on the oscilloscope screen; if the number of action potentials recorded

FIGURE 9.16   A kitten wearing the training goggles from one of Spinelli's experiments. The horizontal and vertical stripes on the outside of the goggles are for identification purposes only; the actual stimuli are contained within the goggles and are illuminated by transparent openings at the sides (out of which the kitten cannot see). The cardboard cone prevents the kitten from dislodging the goggles.

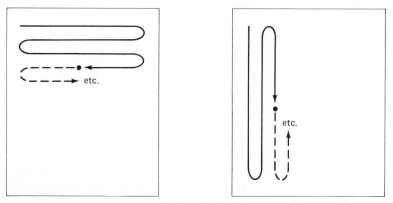

FIGURE 9.17   The horizontal and vertical scanning procedure used by Spinelli and his coworkers.

from the cell exceeded some criterion number while the stimulus was in a given position, a dot was displayed in the corresponding position on the oscilloscope.

Figure 9.18 presents some plots of receptive fields; note the vertically oriented and horizontally oriented visual fields. (See **FIGURE 9.18.**) It is interesting that the specifically oriented receptive fields were produced by stimulation of the eye that had been exposed to bars of the appropriate orientation. That is, receptive fields of neurons responding to the horizontally stimulated eye were found to be horizontal in their orientation; similarly, the vertically stimulated eye produced vertical receptive fields. Also, contrary to what is seen in normally reared kittens, the authors found no cells with binocular receptive fields. (The training procedure prevented stimuli from ever being binocularly perceived.)

Beyond Striate Cortex

The primate brain contains many different areas of cortex that have visual functions. All of occipital cortex, and some temporal cortex as well, serves vision. Neuroanatomists disagree about the precise number of anatomically-distinct areas of visual cortex, but it is probably a dozen or so (Van Essen, 1979). As we saw, primary visual cortex performs a local analysis of the visual world, in terms of rather restricted regions of light and dark. Information from both eyes is mixed, but also on a strictly local basis. Therefore, visual perception—that is, analysis of form and depth—must take place elsewhere.

It is widely believed that there are hierarchies of analysis from striate cortex to successive areas of visual association cortex. In support of this hypothesis, localized damage to striate cortex produces blindness in localized parts of the visual field. In contrast, the visual deficits that are produced by localized damage to visual association cortex are not limited to specific parts of the visual field.

A considerable amount of work has been done on *inferotemporal cortex* of the rhesus monkey. Klüver and Bucy (1939) observed an interesting change in the behavior of monkeys whose temporal lobes had been excised. The operated animals became hypersexual (which is saying a lot, for a monkey), less aggressive, and considerably less able to discriminate ordinary objects visually. The visual deficit (called "psychic blindness" by Klüver and Bucy) was characterized by a disruption of visual recognition and was later shown to be a result of cortical damage. (The changes in emotional behavior appeared to be produced by removal of subcortical structures.) The monkeys exhibited a considerable amount of "oral investigation"; they mouthed objects in their environment. Apparently, this increased oral behavior was secondary to the visual deficit. When the monkey was given a tray containing a mixture of food (raisins, peanuts, etc.) along with hardware of similar size (nuts and bolts), it

FIGURE 9.18 The receptive fields of single cortical neurons recorded from a cat that had worn goggles such as those shown in Figure 9.16. (From Hirsch, H. V. B., and Spinelli, D. N., *Experimental Brain Research*, 1971, *12*, 509–527.)

picked up each object and placed it in its mouth. If the object was a piece of hardware, the monkey returned it to the tray; if it was food, the monkey ate it. The monkeys could otherwise see quite well. They could get around the environment without bumping into things and could accurately pick up small objects with their fingers.

Recording studies have confirmed the importance of temporal cortex in the complex analysis of visual information. Cells there tend to respond to complex visual stimuli. Gross, Rocha-Miranda, and Bender (1972) found a particularly interesting cell, which gave its best response to a drawing of a monkey's hand; other stimuli produced a response whose magnitude depended on the degree of similarity (as judged by the experimenters) to a monkey's hand. It would be impossible, without presenting all possible stimuli, to demonstrate that a neuron in temporal cortex served as a "hand detector," but the results of investigations like this one suggest that temporal cortex (which receives visual information relayed from visual cortex) participates higher-level visual analysis. This region of cortex will be discussed again in Chapter 19; as we shall see, it appears to play a special role in visual memory.

Visual association cortex receives information from the superior colliculus as well as from striate cortex. Although the information from striate cortex is undoubtedly the most important for visual perception, the collicular input also appears to play a role. Weiskrantz, Warrington, Sanders, and Marshall (1974) tested people who had lesions of primary visual cortex. These subjects had scotomas, or specific blind regions, corresponding to the cortical damage. The investigators found that these people could detect differences between various stimuli (such as *X* and *O*) presented to their blind region when they were forced to make a choice among alternatives. They reported no "awareness" of stimuli presented there, but were nevertheless able to respond correctly.

Color Vision

Various theories of color vision have been proposed for many years—long before it was possible to disprove or validate them by physiological means. In 1807, Thomas Young proposed that color vision could be accounted for by the presence of three visual receptors, each sensitive to a single color. This theory was suggested by the fact that, for the human observer, any color can be reproduced by mixing various quantities of three colors judiciously selected from different points along the spectrum. (Actually, if you accept some restrictions, you can choose any three colors, so long as any one cannot be produced by mixing the other two. We may ignore that nicety, since, as we shall see, nature chose three colors judiciously.)

I want to emphasize the point that I am referring to *color* mixing, not pigment mixing. In combining pigments, we find that yellow and blue paint mix to make green. Color mixing refers to the addition of two or more light sources. If we shine a beam of red light and a beam of bluish green light together on a white screen, we will see yellow light. If we mix yellow and blue light, we get white light.

The concept of primary colors has been with us for a long time, and it appears to have some psychological validity. Humans have long regarded yellow, blue, red, and green as primary colors. All other colors can be described as mixtures of these primaries. One can speak of a bluish green or yellowish green, and orange appears to have both red and yellow qualities. Purple resembles both red and blue. But we would never describe yellow as anything but yellow; a slightly longer wavelength starts looking reddish, while a slightly shorter wavelength starts looking greenish. Similarly, we see green, blue, and red as primary. The psychological reality of these four colors has suggested that representation of these colors in the visual system provides the primary information for the perception of color.

COLOR CODING IN THE RETINA. Two theories of color vision predominated prior to the definitive physiological studies. The earliest theory maintained that there were three kinds of color receptors, responding maximally to red, blue, or green light. Other theories, taking into account the primariness of yellow, suggested that there were instead two kinds of color receptors, each of which represented a pair of complementary colors in opponent fashion. The two receptors (red-green and blue-yellow) would produce excitation in response to one of the colors, and inhibition in response to the other. Both types of theory also suggested that there was another class of receptors (rods) responding only to light intensity, in a color-blind manner.

Physiological investigations of retinal photoreceptors in higher primates have ruled in favor of the three-cone theory. Study has been made of the absorption characteristics of single cones isolated from the primate retina. The results show that a given receptor preferentially absorbs light of one of three wavelengths, giving strong evidence for the analysis of color by three different kinds of color receptors. (See **FIGURE 9.19** on next page.) Even more direct proof has been obtained from electrophysiological investigations. Recordings of generator potentials have been taken from single cones in the retina of the carp (a fish) in response to brief pulses of light of varying wavelength. The receptors fell into three types, as represented in **FIGURE 9.20** on page 249.

A fascinating phenomenon occurs in the neural circuitry of the retina between the cones and ganglion cells; the three-color code gets translated into an opponent-color system. Daw (1968) found that most receptive fields of color-sensitive ganglion cells are arranged in a center-surround fashion. For example, the cell might give an on-response to red and an off-response to green in the center, and the opposite set of responses in the surround. Furthermore, Gouras

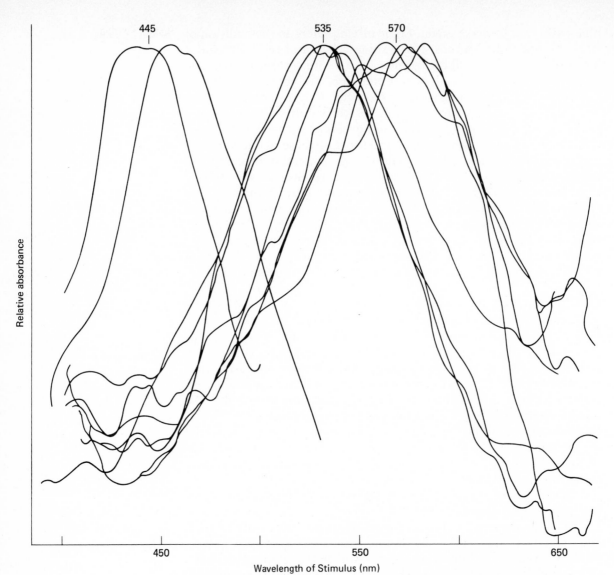

FIGURE 9.19   Absorption characteristics of single cones isolated from the
primate retina. (From Marks, W. B., Dobelle, W. H., and MacNichol, E. F.,
Science, 13 March 1964, *143*, 1181–1183. Copyright 1964 by the American
Association for the Advancement of Science.)

(1968) has found a few ganglion cells that respond best to only one
color. These cells also act as center-surround contrast detectors for
different intensities of light of that color.

COLOR CODING IN THE LATERAL GENICULATE NUCLEUS.   At
the level of the lateral geniculate nucleus, primate color vision appears to be entirely coded by cells that respond in opponent fashion
to complimentary colors. DeValois, Abramov, and Jacobs (1966), re-

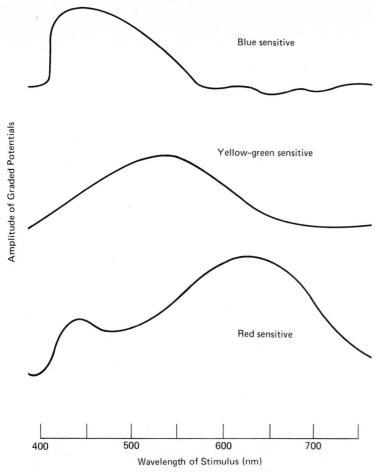

FIGURE 9.20 Generator potentials from single color-sensitive cones in the retina of the carp. (Redrawn from Tomita, T., Kaneko, A., Murakami, M., and Pautler, E., *Vision Research*, 1967, 7, 519–537.)

cording action potentials from lateral geniculate neurons, found two major types of opponent cells: red-green detectors and blue-yellow detectors. Each of these categories could be subdivided into two more categories, depending on the type of response seen. Letting *G, R, Y,* and *B* stand for green, red, yellow, and blue, and letting + and − represent excitation and inhibition, there were four types of opponent cells: $+R -G$, $-R +G$, $+Y -B$, and $-Y +B$. These four types of opponent cells are shown in Figure 9.21, which represents frequency of unit firing as a function of wavelength of the stimulus. (See **FIGURE 9.21** on following page.)

COLOR CODING IN CORTEX. The nature of sensory coding for color in the visual cortex is uncertain. Investigators have found cells that selectively respond to one of three colors, so there seems to be a reemergence of the three-color code seen at the level of the receptor, along with cells showing opponent-type responses (Motokawa, Taira, and Okuda, 1962). Andersen, Buchmann, and Lennox-Buchthal (1962) found cortical cells that responded best to light of particular wavelengths, but the best wavelengths were spread across the entire spectrum. The cells did not cluster into three groups. According to

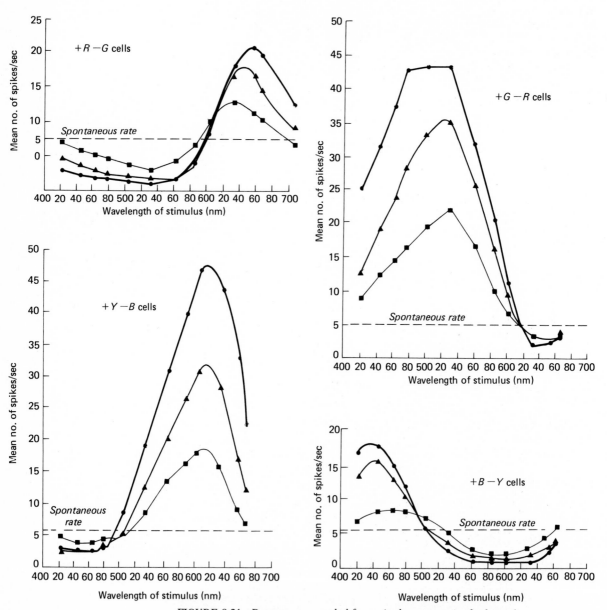

FIGURE 9.21 Responses recorded from single neurons in the lateral geniculate nucleus. These neurons appear to encode color in an opponent process. (From DeValois, R. L., Abramov, I., and Jacobs, G. H., *Journal of the Optical Society of America*, 1966, *56*, 966–977.)

Gouras and Krüger (1975), the study of color coding in cortical neurons is complicated by the fact that a given cell responds to form as
well as to color, and only careful quantititative study will identify
the chromatic information conveyed by these "multiduty detectors."

AUDITION

We have seen that the visual system maintains a spatial representation of the retinal surface, with additional levels of feature extraction. As we shall see, points on the basilar membrane are similarly
represented by locations on the surface of auditory cortex. Let us
first examine the anatomy of ascending auditory information.

The Ascending Auditory Pathways

The anatomy of the auditory system is more complicated than that
of the visual system. Rather than give a detailed verbal description
of the pathways, I shall refer you to **FIGURE 9.22.** Note that fibers
enter the cochlear nuclei of the medulla, and most of them cross to

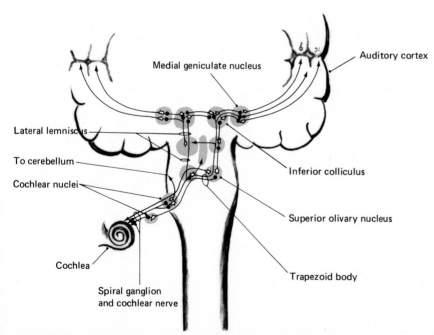

FIGURE 9.22   The pathway of the auditory system. (Adapted from
Noback, C. R., *The Human Nervous System*. New York: McGraw-Hill, 1967.
Copyright © 1967, by McGraw-Hill, Inc.)

the other side of the brain at the level of the pons. From there they go through the **lateral lemniscus** (a large bundle of axons) to the inferior colliculus, medial geniculate nucleus, and, finally, to auditory cortex. As you can see, there are many synapses along the way to complicate the story. And auditory information is relayed to the cerebellum and reticular formation as well. (See **FIGURE 9.22.**)

If we unrolled the basilar membrane into a flat strip and followed afferent fibers serving successive points along its length, we would reach successive points along the surface of auditory cortex, the basal end being represented most medially, the apical end most laterally. Since, as we shall see, various parts of the basilar membrane respond best to particular frequencies of sound, this point-to-point relationship between cortex and basilar membrane is referred to as **tonotopic representation** (*tonos*, "tone"; *topos*, "place").

Frequency Detection

CODING BY LOCATION ON THE BASILAR MEMBRANE. The psychological dimension of *pitch* best corresponds to the physical dimension of frequency of the sound stimulus. Pitch can be affected somewhat by other factors (loud tones have a slightly higher pitch than softer ones of the same frequency), but for the purposes of our discussion I shall equate pitch with frequency. The work of von Békésy has shown us that, because of the mechanical construction of the cochlea and basilar membrane, there is a relationship between the location of maximum deformation of the basilar membrane and frequency of the stimulus. Figure 9.23 illustrates the amount of de-

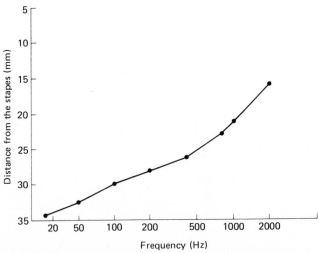

FIGURE 9.23   Encoding of pitch by location of maximum deformation of the basilar membrane. (From von Békésy, G., *Journal of the Acoustical Society of America*, 1949, *21*, 233–245.)

formation along the length of the basilar membrane produced by stimulation with tones of various frequencies. Note that higher frequencies produce more displacement at the basal end of the membrane (closest to the stapes). (See **FIGURE 9.23**.)

This spatial coding of frequency on the basilar membrane is more than just a candidate code. The antibiotic drugs *kanamycin* and *neomycin* produce degeneration of the auditory hair cells (and also, incidentally, vestibular hair cells). Damage to auditory hair cells begins at the basal end of the cochlea and progresses toward the apical end; this can be verified by killing experimental animals after treating them with the antibiotic for varying amounts of time. Longer exposures to the drug are associated with increased progress of hair cell damage down the basilar membrane. The progressive death of hair cells very nicely parallels a progressive hearing loss; the highest frequencies are the first to go, and the lowest are the last (Stebbins, Miller, Johnsson, and Hawkins, 1969).

Evidence from a variety of experiments has shown that although the basilar membrane codes for frequency along its length, the coding is not very specific. A given frequency causes a large region of the basilar membrane to be deformed. However, people can detect changes in frequency of only 2 or 3 Hz. It is difficult to understand how such precise pitch determinations can be made by an organ with such a broad tuning characteristic. Recordings made from single auditory nerve fibers also show a very precise degree of frequency tuning—better than would be predicted by the broad area of the basilar membrane that is deformed by a particular frequency. Figure 9.24 shows the frequency response of sixteen auditory nerve fibers. Note that each fiber is represented by a V-shaped curve. (See **FIGURE 9.24**.) The data were collected as follows: a fiber was located with a microelectrode, and tones of various frequencies and inten-

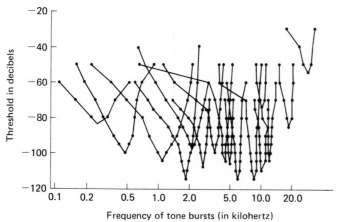

FIGURE 9.24 "Tuning curves" of single auditory nerve fibers. (From Kiang, N. Y.-S, *Acta oto-laryngologica*, 1965, 59, 186–200.)

sities were presented to the ear. For each cell, points were plotted that corresponded to the least intense tone that gave a response at a given frequency. The V shapes indicate that as the intensity goes up (as shown by points higher on the vertical axis), a given fiber responds to a wider range of frequencies. At low intensity levels, the frequency specificity of a given fiber is very good.

Some process that sharpens the frequency tuning characteristics of the auditory nerve fibers must take place on the basilar membrane. A mechanism that might account for this process is *lateral inhibition*. This concept has gained wide acceptance as an explanation for contrast enhancement, and it has been directly observed in the compound eye of the horseshoe crab (Hartline and Ratliff, 1958). If a single *ommatidium* (individual element of the compound eye)— let us call it O_1—is stimulated with light, the surrounding ommatidia become less sensitive to light; they are inhibited by O_1. The inhibition is propagated in all directions away from O_1, the closest receptors being inhibited the most. The amount of inhibition produced by an ommatidium, furthermore, is proportional to the amount of light falling on it. Let us see how this process sharpens responsiveness to contrast. First have a glance at **FIGURE 9.25.** If we allow 100 units of light to fall on O_1 and 70 units on O_2 (located to the right of O_1), the cells will mutually inhibit each other. O_1, however, being stimulated by a greater amount of light, will inhibit O_2 more than this

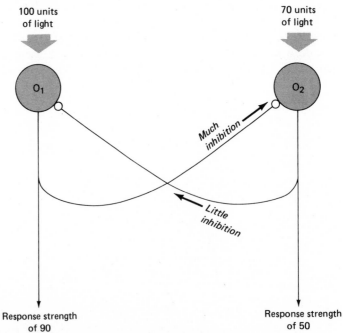

FIGURE 9.25   A schematic representation of the process of lateral inhibition.

cell will inhibit O_1. As the responsiveness of O_2 to light decreases, this ommatidium will consequently inhibit O_1 still less. O_1 will respond more and inhibit O_2 even more. Eventually an equilibrium will be achieved, with O_2 responding at a rate far below that of O_1. (The entire process actually takes place very rapidly.) (See **FIGURE 9.25.**) The net effect of the process of lateral inhibition in the eye of *Limulus* is to exaggerate the response of the receptor getting the most light, relative to the response of its neighbors.

Lateral inhibition has been suggested as a mechanism that would increase the sharpness of frequency tuning of the hair cells along the basilar membrane. The cells receiving the greatest stimulation would inhibit their neighbors more than they would be inhibited by their neighbors, so only the most-stimulated units would respond. Physiological evidence for such a tuning mechanism has not yet been obtained.

Frequency discrimination appears to be as sharp at the level of the auditory nerve as it is anywhere in the auditory system. Single-unit recordings have provided no evidence for any additional tuning mechanisms at higher levels; Figure 9.26 illustrates frequency-intensity responses of a number of units recorded from the cochlear nerve, trapezoid body, inferior colliculus, and medial geniculate nucleus. (See **FIGURE 9.26.**)

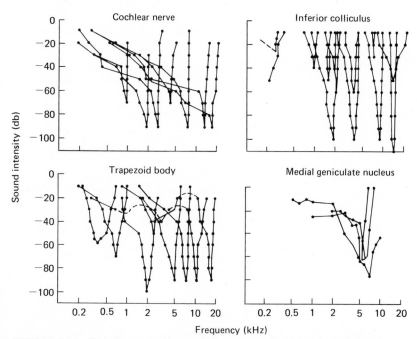

FIGURE 9.26 "Tuning curves" recorded at various levels of the auditory pathways. (From Katsuki, Y., Neural mechanism of auditory sensation in cats. In Rosenblith, W. A., *Sensory Communication.* Copyright 1961 by MIT Press, Cambridge, Mass.)

TEMPORAL CODING OF PITCH. We have seen that frequency is nicely coded for in a spatial manner on the basilar membrane. However, the lowest frequencies do not appear to be accounted for in this manner. Kiang (1965), whose data were presented in Figure 9.24, was unable to find any cells responding best to frequencies of less than 200 Hz. How, then, can low frequencies be distinguished? It appears that lower frequencies are encoded by neural firing that is synchronized to the movements of the apical end of the basilar membrane. The neurons fire in time with the sonic vibrations.

Good evidence for frequency coding by synchronized firing of the auditory hair cells is provided by a study by Miller and Taylor (1948). These investigators presented **white noise** (sound containing all frequencies, similar to the hissing sound you hear between FM radio stations, or the sound you hear on a vacant TV channel) to human observers. When they rapidly switched the white noise on and off, the observers reported that they heard a tone corresponding to the frequency of pulsation. The white noise, containing all frequencies, stimulated the entire length of the basilar membrane, so the frequency that was detected could not be coded by place. The auditory system, therefore, can detect pitch coded by synchronous firing of auditory nerve fibers.

Auditory nerve fibers also show a phenomenon called *phase locking*, even to rather high (5000 Hz) frequencies. Phase locking describes the tendency for a cell to fire only at a particular portion of the repetitive cycle of vibration of the basilar membrane. This phenomenon is illustrated in Figure 9.27. Vibration of the basilar membrane is represented by the wavy line at the top. The output of a cell that perfectly follows every wave is shown on line B. (See **FIGURE 9.27.**) Lines C and D show how other cells might not respond at the frequency of the stimulus all the time, but when they fire they tend to fire during the same portion of the cycle. (See **FIGURE 9.27.**) This phenomenon is demonstrated by the auditory system, as was shown by Rose, Brugge, Anderson, and Hind (1967). The graphs shown in Figure 9.28 are *frequency histograms*. Note the regularity of the distributions. (See **FIGURE 9.28.**) The horizontal axes represent time between two successive action potentials. To construct such plots, we find an active fiber and turn on the stimulus. We start a clock as soon as an action potential is recorded, and stop it when we detect another one. We then note the time, reset the clock, and see how long it takes for another action potential to occur. (Of course, being poor mortals, we engage the services of a high-speed computer to do the timing and recording for us.) After we have recorded and timed for a while we will have a series of numbers—inter-spike intervals. We note that there seem to be clusters of numbers; for a cell stimulated with a 1000-Hz tone, the inter-spike intervals tend to be 1, 2, 3, 4 (etc.) milliseconds. We find very few at 1.5 or 2.5, etc. When we plot the number of times we observe a given interval, we see these clusters of times in the figure. Note, especially, the graph for the 1000-

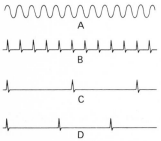

FIGURE 9.27   A schematic representation of the process of phase locking.

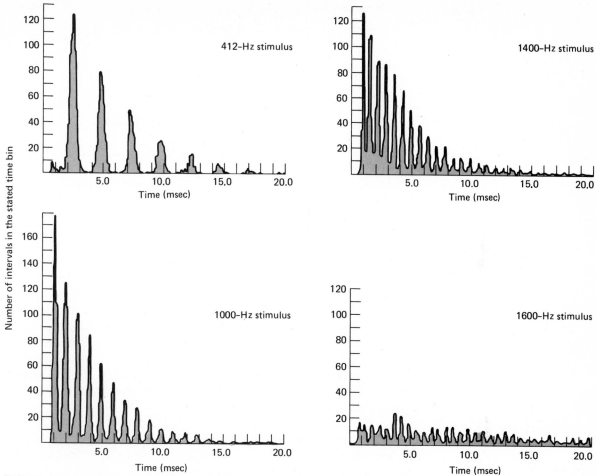

FIGURE 9.28 Evidence that individual auditory fibers show phase locking to pure tones. (From Rose, J. E., Brugge, J. F., Anderson, D. J., and Hind, J. E., *Journal of Neurophysiology*, 1967, *30*, 769–793.)

Hz tone, with its regular clusters at even 1-millisecond intervals. (See **FIGURE 9.29** on following page.) The cells clearly "lock on" to a portion of the wave of vibration, even if they do not always fire at the same rate.

Earlier theories of frequency coding suggested that even very high frequencies could be encoded by a **volley principle** (Wever, 1949). These theories suggested, in the absence of physiological data, that cells would phase lock to the stimulus in the manner that was demonstrated by Rose and his colleagues. Figure 9.29 illustrates how the firing of a number of fibers, phase-locked to the stimulus but firing at a much lower rate, could be "summed" to reproduce the frequency of the stimulus. (See **FIGURE 9.29.**) However, the bottom line, entitled "Fibers *a–e* combined," is rather bothersome. Wever's model does not explain *how* the information can be combined.

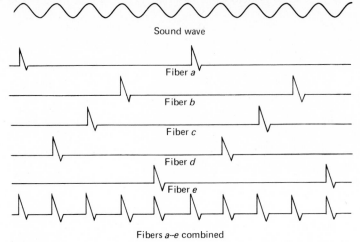

FIGURE 9.29 Wever's volley theory. But what combines fibers *a–e*? (Adapted from Wever, E. G., *Theory of Hearing*. New York: John Wiley & Sons, 1949.)

Clearly, there is no cell that can fire at this rate; after all, the volley theory was devised in order to get around the limited rate of neural firing.

The volley principle is not accepted today; the coding of higher frequencies is accounted for nicely by the location of maximum deformation of the basilar membrane. Why, then, do neurons in the auditory system show phase locking to the frequency of auditory stimuli? As we shall see in a later section, it appears quite likely that this phenomenon is used by the auditory system to localize the source of a sound.

Coding of Intensity

I noted that the visual system compresses the intensity range of light. This compression occurs even more in the auditory system. Wilska (1935) glued a small wooden rod to a volunteer's tympanic membrane (temporarily, of course). He made the rod vibrate longitudinally by means of an electromagnetic coil that could be energized with alternating current. The frequency and intensity of the current could be varied, which consequently changed the loudness and pitch of the stimulus to the subject. Wilska observed the rod under a microscope and measured the distance it moved in vibration. This movement was related to the amount of electrical current used, so that he could calculate the extremely minute vibrations that were too small to detect under the microscope. The astonishing result was that, in order for the subject to detect a sound, the eardrum need be vibrated a distance of less than the diameter of a hydrogen atom!

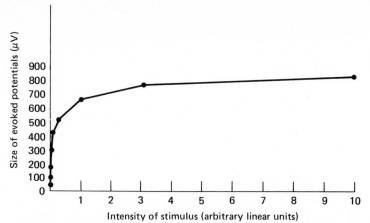

FIGURE 9.30   The relationship between size of evoked potentials recorded from the cochlear nucleus and intensity of the stimulus. (Data, redrawn on a linear scale, are from Saunders, J. C., Cochlear nucleus and auditory cortex correlates of a click stimulus-intensity discrimination in cats. *Journal of Comparative and Physiological Psychology*, 1970, 72, 8–16. Copyright 1970 by the American Psychological Association.)

This means that, in very quiet environments, a young, healthy ear is limited in its ability to detect sounds in the air by the masking noise of blood rushing through the cranial blood vessels, not by the sensitivity of the auditory system itself. A more recent study using modern measuring instruments (Tonndorf and Khanna, 1968) has shown that Wilska's measurements might even be on the conservative side.

The range of sounds to which the ear can respond is (I hesitate to say it) somewhere on the order of 100 trillion to one (Uttal, 1973). Such a phenomenal range is subject to great compression. Figure 9.30 shows the relationship between intensity of the stimulus and the size of the evoked potentials recorded at the cochlear nucleus. (See **FIGURE 9.30.**)

CODING OF INTENSITY IN THE BRAIN.   Coding of intensity seems to be rather constant throughout the auditory system. Response curves to various click intensities recorded at the inferior colliculus, medial geniculate nucleus, white matter containing axons to cortex, and auditory cortex are presented in Figure 9.31. Note that the curves appear quite similar. (The data are expressed as the percentage of the maximum amplitude of the evoked potential that was recorded at each point, so that the shapes of the functions can be compared.) (See **FIGURE 9.31** on the following page.)

I noted earlier that frequency is spatially coded on the surface of auditory cortex. Tunturi (1952) found that intensity is also spatially coded at this level—by vertical distance from the cortical surface. Tunturi placed electrodes at various locations on and in a dog's

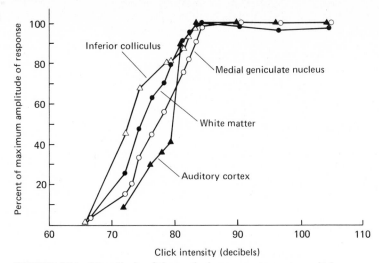

FIGURE 9.31 Magnitude of the evoked response to various click intensities recorded at different locations in the auditory system. (From Etholm, B., *Acta oto-laryngolgica*, 1969, *67*, 319–325.)

auditory cortex, and recorded evoked potentials to sound. He then stimulated the animal's ear with various frequencies and intensities. His plots of the most effective stimuli are shown in Figure 9.32. Note that the surface was most effectively stimulated by tones of lower intensity, whereas deeper locations required louder tones. (See **FIGURE 9.32.**) Low-frequency tones (below 100 Hz) are not represented; detection of these frequencies is presumably mediated by subcortical regions of the brain.

Auditory cortex appears to be unnecessary for frequency or intensity discriminations. If auditory cortex (primary and various secondary auditory areas surrounding it) is removed, a cat can still discriminate among tones of differing frequencies. What is lost is the ability to discriminate more complex characteristics of auditory information. For example, the cats cannot discriminate among different three-note "tunes" (Diamond and Neff, 1957).

Feature Detection in the Auditory System

So far I have discussed coding of only pitch and loudness. However the auditory system responds to other qualities of sonic stimuli besides these two. For example, our ears are very good at locating the source of sound in a lateral dimension. We cannot tell whether a sound is in back or in front of us, but we are very good at determining whether it is to the right or left of us. (To discriminate front from back, we merely turn our heads, transforming the discrimination into a left-right decision.) There are two separate physiological mechanisms used to detect sound sources; we use *phase differences*

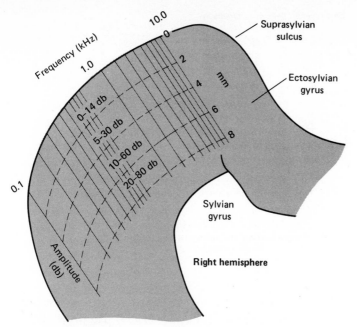

FIGURE 9.32   Coding of intensity and frequency of auditory stimuli by
location in auditory cortex. (From Tunturi, A. R., *American Journal of
Physiology*, 1952, *168*, 712–727.)

for low frequencies (less than approximately 3000 Hz) and intensity
differences for higher frequencies. Stevens and Newman (1936) found
that localization is worst at approximately 3000 Hz, presumably because both mechanisms are rather inefficient at that frequency.

LOCALIZATION BY MEANS OF PHASE DIFFERENCES.   Phase
differences refer to the simultaneous arrival, at each ear, of different
portions of the oscillating sound wave. If we assume a speed of 700
miles per hour for the propagation of sound through the air (the
actual value depends on temperature, barometric pressure, and humidity), adjacent cycles of a 1000-Hz tone are 12.3 inches apart.
Thus, if there are auditory neurons that are phase-locked to the stimulus (and we have seen that there are), a tone presented by a sound
source closer to one ear than the other would produce an asynchrony
in the firing patterns of the two ears. Comparing the auditory neurons of the left and right ears, we would find that impulses occurred
at slightly different times.

   The effects of a single click are easier to visualize than those of
a continuous tone. If the sound pressure wave from a click reaches
one ear sooner, it initiates action potentials before they can occur in
the other auditory system. Studies have shown that the human ear
can detect these differences down to a fraction of a millisecond (Wallach, Newman, and Rosenzweig, 1949).

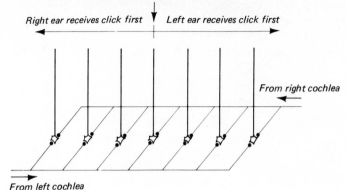

FIGURE 9.33   Licklider's model of the means by which the auditory system can detect which ear receives the sound first. (Adapted from Roederer, J. G., *Introduction to the Physics and Psychophysics of Music.* New York: Springer-Verlag, 1973.)

An ingenious neural model to account for such fine temporal discriminations was presented by Licklider (1959). He proposed the model shown in Figure 9.33. If we assume that the neurons that receive synaptic input will fire only if the incoming signals are nearly simultaneous, then the middle one of these cells will fire only if the sound (a click, for simplicity) reaches the ears simultaneously; the two incoming impulses will travel the same distance and converge on the cell in the middle. (See **FIGURE 9.33.**) If the stimulus reaches the right ear first, a cell toward the left will fire, because the signal on the right-ear line gets a head start on the left-ear line, and the two signals converge toward the left. (See **FIGURE 9.33.**) The model is attractive because it is so simple, but there is no evidence that the auditory system actually contains such a circuit.

LOCALIZATION BY MEANS OF INTENSITY DIFFERENCES.   The other localization mechanism relies upon the differences in intensity of sound pressure received by the ears. At high frequencies it is more difficult to detect phase differences in the stimulus, but intensity differences become greater. Low frequencies pass more easily through objects than do high frequencies; you can easily demonstrate this by comparing the frequencies of outside sounds you hear when a window is open with those you hear when the window is shut. Shrill noises become quieter with the window closed, but the rumble of trucks is much less affected. Similarly, if a high-frequency sound is presented from the right, the head casts a "sonic shadow" on the left ear, producing an intensity difference.

ECHO SUPPRESSION.   Whenever we are confronted with a sound within a building, or even outside (unless we are standing on a featureless plain), we are confronted with multiple echoes. However, unless the echo is very pronounced and of sufficiently long delay, we hear only one sound, the first to reach our ears. Etholm (1969) has observed a physiological mechanism that might possibly account for this "echo suppression." He found that if two closely spaced clicks were presented to the ear, the evoked response to the second would be considerably diminished. Thus, the first sharp, transient sound to arrive appears to suppress sounds that immediately follow it. A steady sound (much less important than transients for such things as speech detection) presumably would not produce such an effect.

DETECTION OF OTHER FEATURES.   The auditory system also detects other features, but the mechanisms are not nearly so well defined. Various studies (Whitfield and Evans, 1965; Goldstein, Hall, and Butterfield, 1968) have found cells responding only to onset or offset of a sound (or both), and to changes in pitch or intensity (sometimes only to changes in one direction). These results accord with the fact that changing stimuli attract our attention better than do steady ones. Data are scanty so far, so we have no real conception of the coding mechanism used, or even of precisely what features are coded.

VESTIBULAR SYSTEM

Vestibular stimulation does not produce any readily definable sensation; certain low-frequency stimulation of the vestibular sacs can produce nausea, and stimulation of the semicircular canals can produce dizziness and rhythmic eye movements *(nystagmus)*. However, there is no primary sensation, as there is for audition and vision, for example. Therefore, discussion of the vestibular system will be restricted to a description of its neural pathways and the results of vestibular stimulation.

THE VESTIBULAR PATHWAYS.   Figure 9.34 illustrates the afferent and efferent fibers of the vestibular system. (See **FIGURE 9.34** on following page.) Most of the afferent fibers of the vestibular nerve synapse within the four vestibular nuclei, but some fibers travel to the cerebellum. Neurons of the vestibular nuclei send their axons to the cerebellum, spinal cord, medulla, and pons. (See **FIGURE 9.34**.) There also appear to be vestibular projections to temporal cortex, but the precise pathways have not been determined. Most investigators believe that the cortical projections are responsible for feelings of dizziness, and that the activity of projections to the lower brainstem can produce nausea and vomiting. Projections to nuclei controlling neck muscles are clearly involved in maintaining an upright position of the head.

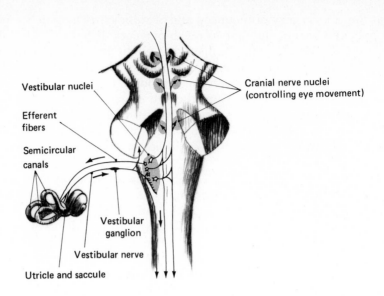

Vestibular nuclei

Efferent
fibers

Semicircular
canals

Vestibular
ganglion

Vestibular nerve

Utricle and saccule

Cranial nerve nuclei
(controlling eye movement)

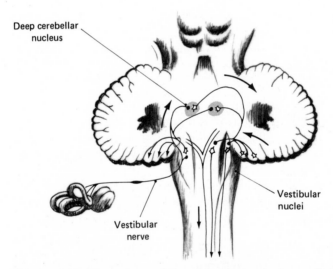

Deep cerebellar
nucleus

Vestibular
nuclei

Vestibular
nerve

FIGURE 9.34 The pathways followed by neurons in the vestibular
system. (Adapted with permission from McGraw-Hill Book Co. from *The
Human Nervous System*, 2nd edition, by Noback and Demarest. Copyright
1975, by McGraw-Hill, Inc.)

Perhaps the most interesting connections are those to the cranial
nerve nuclei (third, fourth, and sixth) controlling the eye muscles. As
we walk or—especially—run, the head is jarred quite a bit. The vestibular system exerts direct control on eye movement, to compensate
for the sudden head movements. This process maintains a fairly
steady retinal image. Test this reflex yourself: look at a distant object
and hit yourself on the side of the head. Note that your image of the

world jumps a bit, but not too much. People who have suffered vestibular damage, and who lack this reflex, have great difficulty seeing anything while walking or running. Everything becomes a blur of movement.

SOMATOSENSES

In the last chapter I classified the somatosenses into two categories: (1) cutaneous sensitivty and (2) kinesthesia and organic sensitivity. In this chapter I shall consider only cutaneous sensitivity; a discussion of kinesthesia will be left for Chapter 10. Too little is known about coding of organic sensitivity to warrant discussion here.

Cutaneous Pathways

There are several distinct neural pathways followed by cutaneous sensory information. Two will be described here. One (the **lemniscal system**) conveys precisely localized information from touch receptors. The other (**spinothalamic system**) carries pain and temperature sensation, which is less precisely localized.

LEMNISCAL PATHWAYS.   The lemniscal pathways are outlined in **FIGURE 9.35** on following page. Almost all of the fibers convey information to the contralateral hemisphere. The first-order neuron has a very long axon. In the case of spinal afferents, the axon of the unipolar cell (whose soma is located in a dorsal root ganglion) ascends through the dorsal columns to the cuneate or gracile nucleus of the medulla. (The destination depends on the location of the receptor—upper regions of the body send fibers to the cuneate nucleus, lower regions to the gracile nucleus.) (See **FIGURE 9.35.**) These axons are the longest nerve fibers of the body; a touch receptor in the big toe is on one end of a single axon stretching all the way to the medulla. Fibers of the second-order neurons decussate (cross to the side of the brain) and travel via the **medial lemniscus** (a large fiber bundle) to the ventral posterior nuclei of the thalamus. The neurons synapse there, and third-order neurons project to somatosensory cortex (postcentral gyrus of the parietal lobe).

Most touch receptors rostral to the ears send information via the trigeminal, facial and vagus nerves. We shall consider the major pathway, from the trigeminal nerve. This pathway is almost a replica of the dorsal column pathway. Most second-order neurons decussate and travel via the *trigeminal lemniscus* (parallel to the medial lemniscus) to the ventral posterior nuclei. The third-order fibers project to somatosensory cortex. (See **FIGURE 9.35.**)

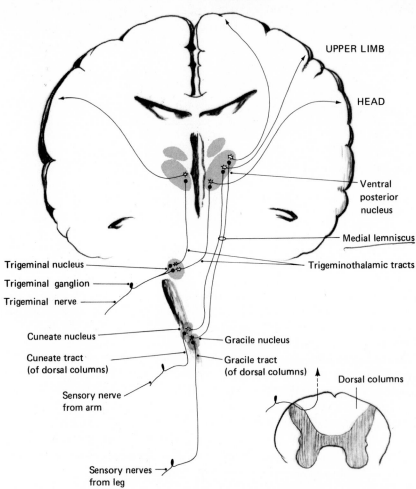

FIGURE 9.35   The pathways followed by neurons that mediate fine touch and pressure, and feedback from the muscles, and joint receptors. (Adapted from Noback, C. R., and Demarest, R. J., *The Nervous System: Introduction and Review*. New York: McGraw-Hill, 1972.)

SPINOTHALAMIC AND RETICULOTHALAMIC PATHWAYS.   Figure 9.36 illustrates the spinothalamic and reticulothalamic systems, which carry information about temperature and pain. (See **FIGURE 9.36.**) (Pain will be discussed separately in Chapter 17.) The afferent fibers synapse as soon as they enter the central nervous system, either in the dorsal horn of the spinal cord or in the ***trigeminal nucleus.***

Second-order neurons decussate immediately; some ascend via the spinothalamic (or trigeminothalamic) tracts directly to the ventral posterior nuclei of the thalamus. Others follow a diffuse, poly-

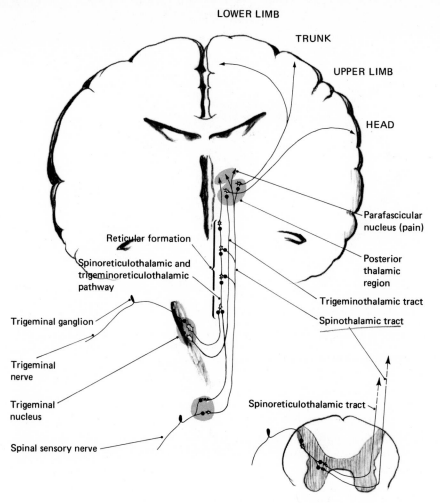

FIGURE 9.36   The pathways followed by neurons that mediate diffuse touch and pressure, temperature, and pain. (Adapted from Noback, C. R., and Demarest, R. J., *The Nervous System: Introduction and Review.* New York: McGraw-Hill, 1972.)

synaptic pathway through the reticular formation. Note that the spinothalamic tract also gives off collaterals to the reticular formation as it passes by. (See **FIGURE 9.36.**) From the thalamus, most third-order fibers conveying pain and temperature information project to *secondary somatosensory cortex.*

SOMATOSENSORY CORTEX.   Figure 9.37 shows a lateral view of a monkey brain, with a drawing vaguely resembling two monkeys. (See **FIGURE 9.37** on following page.) These drawings (animunculi—they would be called homunculi if drawn on a human brain) roughly

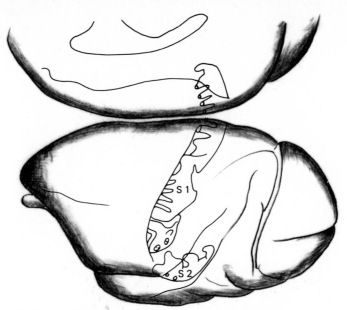

FIGURE 9.37   Sensory animunculi, indicating the regions of the body that project to particular areas of somatosensory cortex. (Adapted from Woolsey, C. N., Organization of somatic sensory and motor areas of the cerebral cortex. In *Biological and Biochemical Bases of Behavior,* edited by H. F. Harlow and C. N. Woolsey. Madison: University of Wisconsin Press, 1958.)

indicate which parts of the body project to which areas of primary and secondary somatosensory cortex. Note that an inordinate amount of cortical tissue is given to representation of fingers and lips, corresponding to the greater tactile sensitivity of these regions. (See **FIGURE 9.37.**)

Neural Coding of Somatosensory Information

CODING OF TACTILE STIMULI.   Coding of location of a stimulus on the body surface is accomplished by means of spatial coding—the *somatotopic representation* illustrated in the previous figure. Single units of the somatosensory system can be recorded from, and their receptive fields can be determined by stimulating the skin with the appropriate stimulus and noting the size and location of the area from which responses are elicited. In general, the larger myelinated fibers of the lemniscal system, serving touch and fine pressure, respond to a relatively small area of skin. Unmyelinated fibers and the smaller myelinated fibers have larger receptive fields and are part of the spinothalamic system, responding to temperature changes or pain-eliciting stimuli. The receptive fields, when measured by the response of cortical neurons, appear to have characteristics similar to retinal ganglion cells; there is a central region of skin that pro-

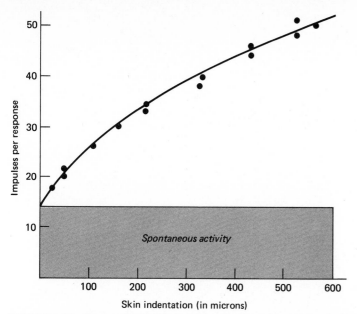

FIGURE 9.38 Response of a single neuron in ventral posterior thalamus as a function of skin indentation produced by a small probe. The neuron is responding to information from a single touch receptor in hairy skin of a monkey. (Redrawn from Mountcastle, V. B., The problem of sensing and the neural coding of sensory events. In *The Neurosciences*, edited by G. Quarton, T. Melnechuk, and F. O. Schmitt. New York: Rockefeller University Press, 1967.)

duces excitation, and a surrounding region that produces inhibition (Mountcastle and Powell, 1959). This phenomenon is presumably produced by lateral inhibitory mechanisms within the central nervous system, and it supposedly increases fineness of localization.

The range of stimulus energy transduced by the cutaneous receptors is a very small fraction of the range detected by the auditory and visual systems. Furthermore, the absolute level of stimulus intensity required to stimulate the cutaneous receptors is on the order of millions of times higher. Compression of the range of stimulus intensity occurs at the level of some cutaneous receptors; others transduce energy linearly. Figure 9.38 illustrates the response of a neuron in the ventral posterior area of the thalamus as a function of stimulus intensity (amount of skin indentation produced by a small probe) applied to a single touch receptor in hairy skin of a monkey. Note that changes in stimulus intensity produce less change in response at higher levels than at lower levels. (See **FIGURE 9.38**.) However, receptors in glabrous skin appear to transduce tactile stimuli in a linear manner, as is shown in Figure 9.39. In this experiment, recordings were taken directly from a myelinated axon serving tactile receptors in glabrous skin at the base of a monkey's thumb. (See **FIGURE 9.39** on following page.)

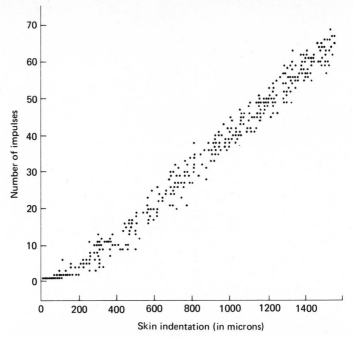

FIGURE 9.39 Response of a single axon served by pressure-sensitive receptors in the base of the thumb of a monkey. (Redrawn from Mountcastle, V. B., The problem of sensing and the neural coding of sensory events. In *The Neurosciences*, edited by G. Quarton, T. Melnechuk, and F. O. Schmitt. New York: Rockefeller University Press, 1967.)

In all cases, intensity appears to be coded in a linear fashion past the level of the receptor. Mountcastle (1967) summarized a number of experiments that showed a linear relationship between firing rate of the peripheral afferent fiber and all subsequent levels of the somatosensory system. Even estimations of stimulus intensity by humans were related to receptor stimulation in a linear fashion. Figure 9.40 shows perceived intensity as a function of stimulus intensity (amount of skin deformation from a small probe placed on the distal pad—fingerprint portion—of the middle finger). Note that the relationship is a nearly straight line. (See **FIGURE 9.40.**)

ADAPTATION. It has been known for a long time that a moderate, constant stimulus applied to the skin fails to produce any sensation after it has been present for a while. We not only ignore the pressure of a wristwatch, but we cannot feel it at all if we keep our arm still (assuming that the band is not painfully tight). Physiological studies have shown that the reason for the lack of sensation is absence of receptor firing; the receptors adapt to a constant stimulus.

This adaptation is not a result of any "fatigue" of physical or chemical processes within the receptor. Adaptation can be explained as a function of the mechanical construction of the receptors and

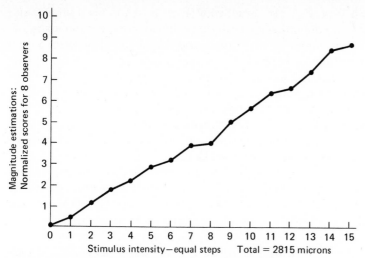

FIGURE 9.40 Estimated magnitude of intensity of tactile stimuli presented to the finger of human subjects. (Redrawn from Mountcastle, V. B., The problem of sensing and the neural coding of sensory events. In *Neurosciences*, edited by G. Quarton, T. Melnechuk, and F. O. Schmitt. New York: Rockefeller University Press, 1967.)

their relationship to skin and (in some cases) end organs. As we saw in Chapter 8, the fibers of Pacinian corpuscles respond once when the receptor is bent and again when it is released, because of the way the nerve ending "floats" within the viscous interior of the corpuscle. Therefore, these receptors adapt almost immediately to a constant stimulus. Most other fibers adapt less quickly. Nafe and Wagoner (1941) recorded the sensations reported by human subjects as a stimulus weight gradually moved downward as it deformed the skin. Pressure was reported until the weight finally stopped moving. When the weight was increased, pressure was reported until downward movement stopped again. Pressure sensations were also briefly recorded when the weight was removed while the surface of the skin regained its normal shape.

RESPONSIVENESS TO MOVING STIMULI. A moderate, constant, nondamaging stimulus is rarely of any importance to an organism, so this adaptation mechanism is useful. Our cutaneous senses are used much more often to analyze shapes and textures of stimulus objects moving with respect to the surface of the skin. Sometimes the object itself moves, but more often we do the moving ourselves. If I placed an object in your palm and asked you to keep your hand still, you would have a great deal of difficulty recognizing the object by touch alone. If you were permitted to move your hand, you would manipulate the object, letting its surface slide across your palm and the pads of your fingers. You would be able to describe its three-dimensional shape, hardness, texture, slipperiness, etc. (Obviously your motor system must cooperate, and you need kinesthetic sensation from your muscles and joints, besides the cutaneous information.) If you squeeze the object and feel a lot of well-localized pressure in return, it is hard. If you feel a less intense, more diffuse pressure in return, it is soft. If it produces vibrations as it moves over the ridges on your fingers, it is rough. If very little effort is needed to move the object while pressing it against your skin, it is slippery. If it does not produce vibrations as it moves across your skin, but moves in a jerky fashion, and if it takes effort to remove your fingers from its surface, it is sticky.

Mountcastle (1967) describes the results of experiments he and his colleagues performed in the investigation of movement-sensitive receptors. The investigators found that the receptors in glabrous skin could be divided into two types according to their sensitivity to vibration. Figure 9.41 illustrates the minimum amplitude on a sine-wave stimulus (vibrating the surface of the skin) necessary to produce a sensation in human subjects. As the frequency increases from 1 to 40 Hz, the stimulus amplitude needed for sensation steadily decreases. After 40 Hz, the decrease follows a very different function.

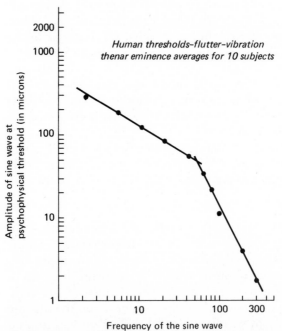

FIGURE 9.41   Amplitude of the sine-wave stimulus at psychophysical threshold (the point at which it could barely be felt) as a function of frequency of the stimulus. (Redrawn from Mountcastle, V. B., The problem of sensing and the neural coding of sensory events. In *The Neurosciences*, edited by G. Quarton, T. Melnechuk, and F. O. Schmitt. New York: Rockefeller University Press, 1967.)

(See **FIGURE 9.41** on previous page.) These results strongly suggest that two types of receptors respond to different frequency ranges. Physiological studies by Mountcastle and his colleagues indicated that low-frequency detection was performed by quickly adapting mechanoreceptors in the surface of the skin. Higher frequencies were detected by Pacinian corpuscles located in deeper tissue.

These investigators also studied the nature of neural coding for frequency. The signal from the low-frequency receptors was represented faithfully up to the level of the cortex. However, high-frequency signals from the Pacinian corpuscles, while still represented in the thalamus and thalamocortical fibers, were not observed in somatosensory cortex. Uttal (1973) suggests that it is likely that frequency, represented by rate of unit firing in the thalamus, becomes coded in a spatial manner at the level of the cortex. Frequency of vibration, perhaps, is coded by firing of different sets of cortical neurons.

DETECTION OF COOLNESS AND WARMTH.   Thermal sensation is a very complicated process, and not much is known about coding of temperature at higher levels of the CNS. Coolness and warmth, as we saw in the last chapter, are detected by different receptors. The responses of these two types of receptors combine to provide thermal sensations. There is a temperature level that, for a particular region of skin, will produce a sensation of temperature neutrality—neither warmth nor coolness is reported. This neutral point is not an absolute value but depends on the prior history of thermal stimulation of that area. If the temperature of a region of skin is raised by a few degrees, the initial feeling of warmth is replaced by one of neutrality. If the skin temperature is lowered to its initial value, it now feels cool. Thus, increases in temperature lower the sensitivity of warmth receptors and raise sensitivity of cold receptors. The converse holds for decreases in skin temperature. This adaptation to ambient temperature can be easily demonstrated by placing one hand in a bucket of warm water and the other in a bucket of cool water until some adaptation has taken place. Simultaneous immersion of both hands in water of intermediate temperature leads to a peculiar sensation: the water feels warm to one hand and cool to the other.

The necessity for the central nervous system to combine information from warmth and cold receptors is further underscored by data of Hensel and Kenshalo, illustrated in Figure 9.42. Average firing rates of warmth- and cold-sensitive fibers is shown as a function of temperature. Note that each of the curves successively rises and falls with an increase in temperature. (See **FIGURE 9.42.**) Thus, firing rate of either system does not reliably encode temperature; cold receptors, for example, fire at the same rate at 17° C as they do at 34° C. (See **FIGURE 9.42.**) Thus, the brain must use information from both populations of receptors in determining temperature (Hensel, 1974).

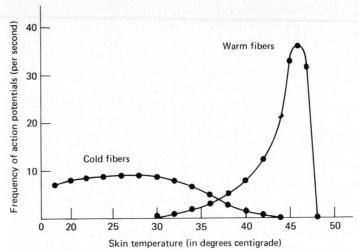

FIGURE 9.42 Since a given neuron responds at the same rate to two different temperatures, our perception of temperature depends on input from more than one type of neuron. (From Hensel, H., and Kenshalo, D. R., Warm receptors in the nasal region of cats. *Journal of Physiology (London)*, 1969, *204*, 99–112. Reprinted by permission of Cambridge University Press.)

GUSTATION

The Neural Pathway for Taste

Gustatory information is received from the anterior part of the tongue by the chorda tympani, which is part of the seventh cranial nerve. Most studies of gustatory function have used this branch, since the nerve is conveniently available, running immediately under the eardrum. Taste receptors in the posterior tongue send information through the lingual branch of the ninth cranial nerve, while the vagus (tenth) nerve receives information from receptors of the palate and epiglottis.

The first relay station for taste is the **nucleus of the solitary tract**, located in the medulla. (As we shall see in Chapter 15, this region also plays a role in sleep.) The taste-sensitive neurons of this nucleus send their axons a short distance forward, to the **parabrachial nucleus** of the pons. The details of this second relay station have been determined only for the rat, but it is believed that a similar area can be found in primates (Pfaffman, Frank, and Norgren, 1979). The pontine neurons then project to the **thalamic taste area**, situated at the most medial part of the ventral posterior group of thalamic nuclei (somatosensory relay nuclei). Projections are also sent to the subthalamus, lateral hypothalamus, and various parts of the limbic system. It is thought that the hypothalamic pathway may be important

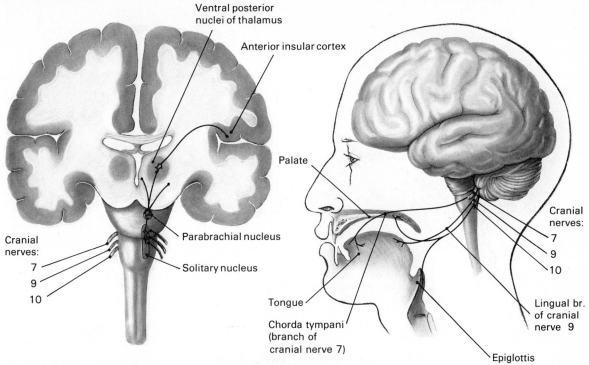

FIGURE 9.43 Neural pathways of the gustatory system.

in mediating the reinforcing effects of sweet and (under some circumstances) salty tastes. (The role of the lateral hypothalamus in reinforcement will be discussed in Chapter 17.)

The thalamic taste area projects to three areas in the primate brain: two regions of ventral somatosensory cortex, and a region of **anterior insular cortex,** a part of frontal cortex that is normally hidden under the anterior end of the temporal lobe. Unlike most other sense modalities, taste is ipsilaterally represented in the brain. (See **FIGURE 9.43.**)

Neural Coding of Taste

Almost all fibers in the chorda tympani nerve respond to more than one taste quality, and many respond to changes in temperature, as well. However, most show a preference for one of the four qualities (sweet, salty, sour, and bitter). Figure 9.44 shows the average responses of fibers in the rat chorda tympani and glossopharyngeal nerve to sucrose (S), NaCl (N), HCl (H), quinine (Q), and water (W), as recorded by Nowlis and Frank (1977). These investigators stimulated three different kinds of taste buds, and found the same general types of responses from each of them. (See **FIGURE 9.44** on the following page.)

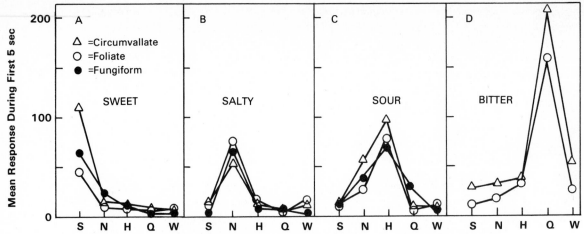

FIGURE 9.44   Mean number of responses recorded from axons in the chorda tympani and glossopharyngeal nerves during the first five seconds after the application of sweet, salty, sour, and bitter stimuli. (From Nowlis, G. H., and Frank, M. Qualities in hamster taste: behavioral and neural evidence. In *Olfaction and Taste 6*, edited by J. LeMagnen and P. MacLeod. London, Washington, D.C.: Information Retrieval, 1977.)

Similar kinds of response patterns are found in the nucleus of the solitary tract (Doetsch and Erickson, 1970). If anything, these neurons are even more broadly tuned. Funakoshi and Ninomiya (1977) reported data from cortical taste neurons in dogs and rats. Of 60 neurons recorded from, more than one-third responded to only one of the five taste stimuli that were used, while another third responded to only two of them. Ten cells were excited by one stimulus and inhibited by another. Thus, the response of the cortical neurons appears to be more sharply tuned than neurons in the periphery.

OLFACTION

The Neural Pathways for Odor

The complexity of the olfactory bulb, and the interconnections of these structures via fibers passing through the anterior commissure, was discussed in Chapter 8. The mitral cells in the olfactory bulb give rise to the afferent fibers of the olfactory nerve.

These axons enter the *lateral olfactory tract,* which projects to various parts of the limbic system—principally, the amygdala. They also project to orbitofrontal neocortex and to limbic cortex. Unlike all other sense modalities, olfactory information is not relayed from thalamus to cortex. Indeed, the reverse occurs; neurons in olfactory cortex project to the thalamic taste area. The fact that flavor is a composite of taste and smell may be related to this covergence of information. The hypothalamus also receives a considerable amount of olfactory information, which is probably important for the acceptance or rejection of food, and for the olfactory control of reproductive processes that is seen in many species of mammals.

Coding of Odor Quality

Although there have been many attempts to identify odor primaries corresponding to the sweet, sour, bitter, and salty qualities of taste, we still cannot say with any certainty how odor quality is encoded. This lack of knowledge about coding at the peripheral level certainly hampers research on central coding of olfactory information.

The concept of primary olfactory qualities has been prevalent for quite some time—partly because of the fact that it is easier to conceive of the coding of the overwhelming number of discriminable odors by a few primary dimensions than by a myriad of different types of receptors. After all, new substances with new odors are synthesized every year, and it would be unreasonable to expect that we evolved specific receptors for all the odors yet to be experienced.

Two other pieces of information suggest that odors may be sorted out according to some classification scheme. Some people have very specific anosmias; they cannot detect certain odors. This fact would suggest that there are various receptor types, and that these people lack one or more kinds of these specific receptors. The second piece of data is that we humans seem to be able to reach some agreements about the similarities of odors. Classifications such as fruity, pinelike, and musky make sense to most of us, and we are willing to say that there is more similarity between the odors of pine oil and cedar oil than between skunk and the smell of limes. More weight is given to these similarities by the fact that adaptation to one odor (i.e., loss of sensitivity after constant stimulation) produces a corresponding adaptation to similar odors (Moncrieff, 1956).

STEREOSPECIFIC THEORY.   How does one begin to construct a set of possible odor primaries in order to develop a theory of olfaction? Amoore tried to do so by means of a literature search. He looked through the chemical literature and noted that the terms used by chemists to describe the odor of a compound could be divided into seven classifications: camphoraceous, ethereal, floral, musky, pepperminty, pungent, and putrid. Amoore went on to examine the three-dimensional structures of the various molecules, to see if there could be anyway to classify them according to shape. His conclusion was that the "primary odors" could be characterized by seven different molecular configurations, recognized by receptive sites of sim-

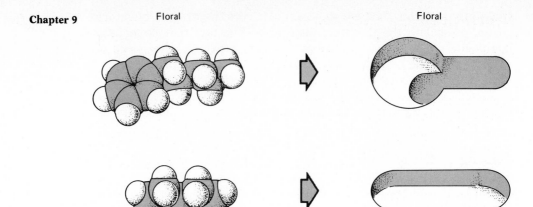

Floral   Floral

Ethereal   Ethereal

FIGURE 9.45   Examples of the molecular shape of two odorous substances, along with their hypothesized receptors. (Adapted from Amoore, J. E., *Molecular Basis of Odor*. Springfield, Ill.: Charles C Thomas, 1970.)

ilar shapes, two of which are shown in Figure 9.45. (See **FIGURE 9.45.**)

Amoore constructed plastic models of the molecules and receptors. Then he attempted, by several means, to correlate the "goodness-of-fit" of the molecular models and receptor models with observers' judgments of the similarities of the odors the models represented. In other words, were odors whose molecules fit the receptor model well judged more similar than odors whose molecules fit the receptor model poorly? Unfortunately, the relationship was not particularly good. Amoore then abandoned his model of the hypothesized receptor sites and concentrated on the molecular models themselves. He had much more success with this procedure. Three-dimensional similarity (judged by a shape-recognizing computer that "looked at" the molecular models with a television camera) correlated very well with observers' ratings of odor similarity. Thus, there appears to be supporting evidence for the suggestion that there are primary odors, and that the molecular configuration determines a substance's smell. We still do not know what the nature of the receptors might be. (See Amoore, 1970, for a review of his studies.)

The Amoore theory has not identified nature of the olfactory receptors: nor, for that matter, has any other theory. Various investigators have stressed other characteristics of odorous molecules besides their shape. For example, Wright has noted some correspondence between the odors of molecules and their patterns of absorption of infrared radiation (Wright, 1974). How this may ultimately relate to the transduction of olfactory information is still a mystery.

Unfortunately, recordings taken from single olfactory receptors have not helped to identify "odor primaries." In a discussion of their attempt to classify the response characteristics of individual olfactory fibers, Lettvin and Gesteland (1965) reported that they were unable to find any specific classes of fibers that coded for odor qualities. A given fiber would be found to respond differentially to various odor molecules, being excited by some, inhibited by others, and unaffected by still others. Its neighbor would have a different, idiosyncratic response. Coding of olfactory information into "odor primaries," if they indeed exist, must result from a complex interaction of various receptors.

Recordings have also been made in the central nervous system. At the level of the amygdala, neurons continue to exhibit a nonselective response pattern. Cain and Bindra (1972) recorded responses to eight different odors; all amygdala neurons responded to more than two of them. However, Tanabe, Iino, Oeshima, and Tagaki (1974, 1975) recorded from 40 cells in orbitofrontal olfactory cortex. Half of these cells responded to only one of the eight odors that the experimenters used, while decreasing numbers of cells responded to two, three, or four kinds of odors. Thus, neurons in olfactory cortex perform some sort of analysis of incoming information, sorting the olfactory stimuli into different categories. We do not know how this analysis is performed, or what the classification scheme is.

KEY TERMS

accessory optic nuclei p. 227

adaptation p. 241

anterior insular cortex p. 275

bit p. 221

calcarine fissure (KAL ka rin) p. 224

complex cell p. 231

decussation p. 224

2-deoxyglucose (2-DG) p. 233

dorsal lateral geniculate nucleus p. 224

epiphenomenon p. 221

fixation plane p. 237

Fourier analysis (FOO ree ay) p. 239

homunculus (ho MUNG kew luss) p. 222

inferotemporal cortex p. 245

lateral lemniscus (lem NISS kuss) p. 252

lateral olfactory tract p. 276

lemniscal system p. 265

medial lemniscus p. 265

nucleus of the solitary tract p. 274

ocular dominance p. 234

optic chiasm (KY az'm) p. 224

optic radiations p. 224

parabrachial nucleus (pair a BRAY kee ul) p. 274

phase difference p. 260

pretectum p. 227

prosthesis p. 224

receptive field p. 229

retinal disparity p. 237

retinotopic representation p. 224

scotoma p. 224

secondary somatosensory cortex p. 267

sensory coding p. 219

simple cell p. 231

SUGGESTED READINGS

GELDARD, F. A. *The Human Senses*, 2nd edition: New York: Wiley & Sons, 1972.

MASTERTON, R. B. (Ed.) *Handbook of Behavioral Neurobiology. Volume 1: Sensory Integration.* New York: Plenum, 1978.

UTTAL, R. W. *The Psychobiology of Sensory Coding.* New York: Harper & Row, 1972.

HUBEL, D. H. AND WIESEL, T. N. Brain mechanisms of vision. *Scientific American*, September, 1979.

The books by Geldard and Uttal have already been recommended. The article by Hubel and Wiesel is a very clear summary of their research. The book by Masterton is excellent; I recommend it highly. It contains chapters written by various experts on all of the sense modalities. You will also find articles in almost every issue of *Annual Review of Psychology* and *Annual Review of Neuroscience* on topics related to sensory processing.

10

Glands, Muscles, and the Control of Movement

So far, much has been said about neural communication and the transduction of sensory information and its transmission into the brain. However, all these processes would be useless without some means of interacting with the environment. There is no selective advantage in having sensory systems unless the animal can utilize the information thus provided by taking some physical action. To interact with the environment we need **effectors.** The name effector was appropriately chosen, since it refers to cells that are located at the distal end of peripheral efferent nerve fibers, which produce physical effects as a result of neural stimulation. We possess two types of effectors: muscle fibers and secretory cells. In this chapter I shall describe the principles of glandular secretion and muscular contraction. I shall then discuss the control of these effects, first on a simpler level by reflexive mechanisms and then on a more complex, integrative level by the motor systems of the brain.

GLANDS

Almost all glands, both *endocrine* and *exocrine,* are controlled by the brain. There are a few exceptions; for example, the parathyroids, which regulate calcium metabolism, respond directly to the calcium level of the blood. However, most other glands either receive direct neural control or are controlled by a series of events initiated by the *hypothalamic hormones.*

Exocrine Glands

Exocrine (literally, "outside-secreting") glands are those that have a duct through which the gland's products are secreted. For example, the lachrymal glands secrete tears through a duct to the inner surface of the eyelid; the liver and pancreas secrete digestive juices into the intestine; the seminal vesicles and prostate secrete fluids into the male genitourinary system; and the sweat glands and sebaceous glands secrete sweat and oil to the surface of the skin. All of the innervated exocrine glands are controlled by the autonomic nervous system. For example, the salivary glands receive postganglionic fibers of both the parasympathetic and the sympathetic divisions. The parasympathetic division, active during such anabolic processes as digestion, produces a copious secretion of thin, watery saliva. Stimulation of the sympathetic division also results in secretion, but of a smaller quantity of thick, viscous saliva. (This is why the inside of your mouth feels sticky during periods of fright or extreme excitement.)

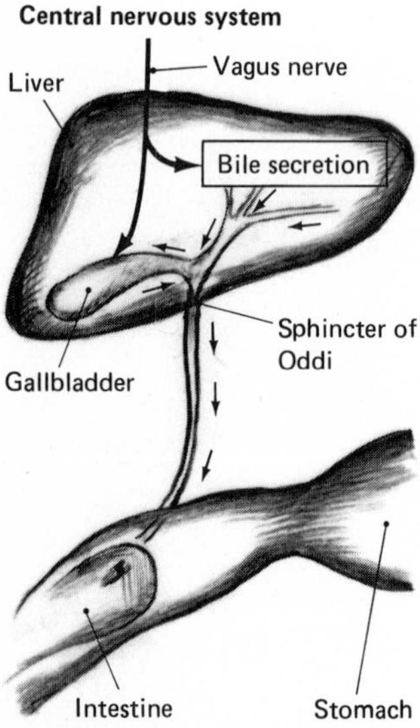

Some exocrine glands store their secretions and release them in response to neural stimulation. For example, the liver secretes bile, the amount of secretion being under control of both neural and hormonal factors. The bile is stored in the gallbladder and is retained there for so long as the sphincter of Oddi is constricted. A sphincter is a sort of valve that controls the passage of substances through a duct. It is made of smooth muscle, to be described later. Upon neural stimulation received via the vagus nerve, this sphincter relaxes and muscles in the wall of the gallbladder contract, thus allowing the stored bile to be secreted into the digestive system. (See **FIGURE 10.1.**)

FIGURE 10.1 Example of an exocrine gland controlled by the central nervous system.

Endocrine Glands

Endocrine (or "inside-secreting") glands secrete their products into the extracellular fluid surrounding capillaries; thus, the hormones they produce enter the blood. These glands do not store hormones in containers like the gallbladder; instead, the hormones are stored inside vesicles in the cytoplasm of hormone-producing cells. When the glands are stimulated to secrete their hormone, they do so by means of a process very similar to the liberation of transmitter substance by nerve terminals. Usually, the stimuli that trigger the release of the hormone also increase its rate of production.

As we saw in Chapter 1, the brain communicates with the effectors of the body by neural and nonneural means. Much of the endocrine system is controlled by the hypothalamic hormones produced in the brain. There is a small but very crucial vascular system (the **hypothalamic-hypophyseal portal system**) that interconnects the hypothalamus and **anterior pituitary gland** *(adenohypophysis)*. Arterioles of the hypothalamus branch into capillaries and then drain into small veins that travel to the anterior pituitary, where they branch into another set of capillaries. Therefore, substances that enter the hypothalamic capillaries of this system travel directly to the anterior pituitary before they are diluted in the large volume of blood in the vascular system. (See **FIGURE 10.2.**) There is also a small portal system going in the opposite direction, from anterior pituitary to hypothalamus. This system undoubtedly provides the

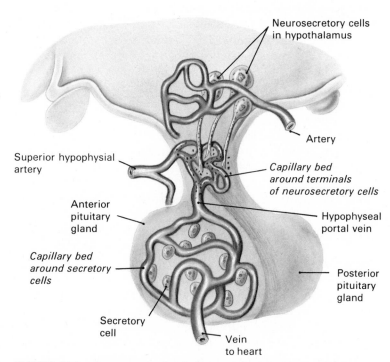

Neurosecretory cells in hypothalamus

Artery

Superior hypophysial artery

Capillary bed around terminals of neurosecretory cells

Anterior pituitary gland

Hypophyseal portal vein

Capillary bed around secretory cells

Posterior pituitary gland

Secretory cell

Vein to heart

FIGURE 10.2   The pituitary gland, showing the portal blood supply. Hypothalamic hormones are released by the neurosecretory cells in the capillaries of the hypothalamus, near its junction with the pituitary stalk.

hypothalamus with information concerning the release of pituitary hormones.

The hypothalamus produces a number of releasing and inhibiting hormones. These substances are secreted by neurosecretory cells into the extracellular fluid around the hypothalamic capillaries of the portal system. (See **FIGURE 10.2.**)

The hypothalamic hormones regulate the synthesis and release of the anterior pituitary hormones. Most of the anterior pituitary hormones (the *trophic hormones*) produce indirect effects; they stimulate the production and release of hormones in other endocrine glands. Two of them, prolactin and somatotrophic hormone (growth hormone), produce direct effects. The behavioral effects of many of these hormones will be discussed in later chapters.

The anterior pituitary hormones are polypeptides, consisting of chains of amino acids. They stimulate their *target cells* by interacting with receptors on the surface of these cells, in a manner very similar to the interaction between transmitter substance and receptor site. The acceptance of a hormone molecule by a receptor on the target cell initiates the appropriate changes in the cell. Most of these changes appear to be mediated by the adenyl cyclase-cyclic AMP process, which also mediates the postsynaptic effects of some neurotransmitters (Chapter 5).

Steroid hormones, produced by the adrenal cortex and gonads, are composed of very small fat soluble molecules that have no difficulty in entering the target cells. They attach to substances in the nucleus and direct the machinery of the cell to alter its protein production.

The hypothalamus also produces the hormones of the *posterior pituitary gland (neurohypophysis)*. These hormones (oxytocin, which stimulates ejection of milk and uterine contractions at the time of childbirth, and antidiuretic hormone, which regulates urine output by the kidney) are produced by cell bodies in the hypothalamus. The hormones travel in vesicles down through the axoplasm, where they collect in the terminals within the posterior pituitary. When an axon fires, the hormone contained within the terminals is liberated and enters the circulatory system.

The adrenal medulla, also controlled neurally, closely resembles a sympathetic ganglion. It is innervated by preganglionic fibers, and its secretory cells are analogous to postganglionic sympathetic neurons. These cells secrete epinephrine and a little norepinephrine when they are neurally stimulated. Secretions of the adrenal medulla function chiefly as an adjunct to the direct neural effects of sympathetic activity; for example, epinephrine increases heart rate and constricts peripheral blood vessels. This gland also stimulates a function that cannot be mediated neurally—an increase in the conversion of glycogen ("animal starch") into glucose within skeletal muscle cells, which increases the energy available to them.

MUSCLES

Mammals have three types of muscles: skeletal muscle (often called striated muscle because of its bands and stripes), cardiac muscle, and smooth muscle (so called because it lacks striations).

SKELETAL MUSCLE.   *Skeletal muscles* are usually attached to bone at each end and move the bones relative to each other. (Exceptions include eye muscles and some abdominal muscles, attached to bone at one end only.) Muscles are fastened to bones via tendons, which are very strong bands of connective tissue. Several different classes of movement can be accomplished by the skeletal muscles, but I shall refer principally to only two of them: *flexion* and *extension.* Contraction of a flexor muscle produces flexion, the drawing in of a limb. The opposite movement is produced by extensor muscles. These are the so-called "antigravity muscles"—the ones we use to stand up. Picture a four-legged animal. Lifting of a paw would be described as flexion; putting it back down would be extension. I should note that we sometimes talk about "flexing" our muscles. This is an incorrect use of the term. Muscles *contract*; limbs *flex*. When body builders show off their arm muscles, they are simultaneously contracting the flexors and extensors of that limb.

SMOOTH MUSCLE.   There are two types of *smooth muscle,* both of which are controlled by the autonomic nervous system. *Multiunit smooth muscles* are found in larger arteries, around hair follicles (where they produce piloerection), and in the eye (controlling lens adjustment and pupillary dilation). This type of smooth muscle is normally inactive, but it will contract in response to neural stimulation or to certain hormones. In contrast, *single-unit smooth muscles* normally contract in a rhythmic fashion. Some of these cells spontaneously produce *pacemaker potentials* (we could regard them as self-initiated EPSPs). These slow potentials elicit action potentials, which are propagated by adjacent smooth muscle fibers, resulting in a wave of muscular contraction. The efferent nerve supply (and various hormones) can modulate the rhythmic rate, increasing or decreasing it, instead of eliciting the individual contractions. Single-unit smooth muscles are found chiefly in the gastrointestinal system, uterus, and small blood vessels.

CARDIAC MUSCLE.   Finally, there is cardiac muscle, the location of which is specified by its name. This type of muscle looks somewhat like striated muscle, but acts like single-unit smooth muscle. The heart beats regularly, even if it is denervated. Neural activity and, again, certain hormones serve to modulate heart rate. A group of cells in the *pacemaker* of the heart are rhythmically active and initiate contractions of cardiac muscle.

Chapter 10 Anatomy of Skeletal Muscle

The detailed structure of a skeletal muscle is shown in **FIGURE 10.3.** As you can see from this illustration (which is highly schematic and omits many details for clarity), there are two types of muscle fibers, served by three kinds of nerve endings. The **extrafusal muscle fibers** are served by axons of the **alpha motor neurons.** These fibers are responsible for the force produced by a muscle when it contracts. The **intrafusal muscle fibers** are served by two axons, one afferent and one efferent. The afferent ending is found in the central region (capsule) of the intrafusal muscle fiber. This ending is a mechano-receptor, sensitive to forces applied to the ends of the intrafusal muscle fiber. There is more than one type of afferent ending, but for simplicity only one kind is shown here. The efferent to this muscle fiber can cause the fiber to contract; however, this contraction con-

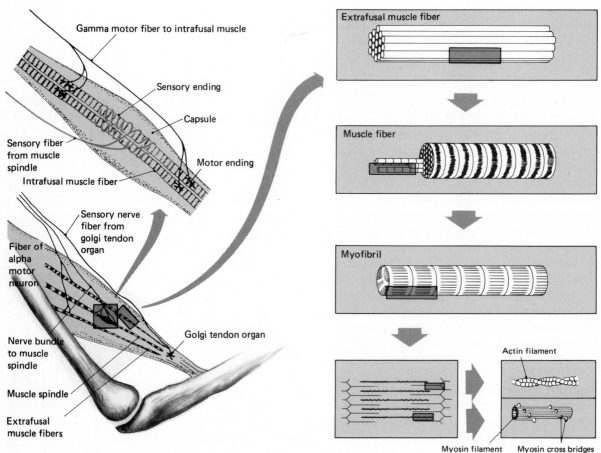

FIGURE 10.3   Anatomy of striated (skeletal) muscle. (Adapted from Bloom and Faucett, *A Textbook of Histology.* Philadelphia: W. B. Saunders, 1968.)

tributes an insignificant amount of mechanical force to the muscle as a whole. As we shall see, the only function of contraction of the intrafusal muscle fiber, is to modify the sensitivity of the fiber's afferent ending to stretch.

Note that a single myelinated axon of an alpha motor neuron serves several muscle fibers. (See **FIGURE 10.3.**) In primates, the number of muscle fibers served by a single axon varies considerably, from as few as a dozen in muscles controlling the fingers or eyes to many hundreds in large muscles of the leg. An alpha motor neuron, its axon, and associated extrafusal muscle fibers constitute an entity known as a *motor unit.*

A single muscle fiber can, in turn, be divided into a number of *myofibrils,* which consist of overlapping strands of *actin* and *myosin.* Note the small protrusions on the myosin filaments; these structures *(myosin cross bridges)* are the motile elements that interact with the actin filaments and produce muscular contractions. (See **FIGURE 10.3.**)

Neuromuscular Junctions

The synapse between an efferent nerve terminal and the membrane of a muscle fiber is called a *neuromuscular junction.* The nerve terminals synapse on *motor end plates,* located in grooves along the surface of the muscle fibers. When an axon fires, acetylcholine is liberated by the terminals and produces a depolarization of the postsynaptic membrane (*end plate potential,* or *EPP*). The end plate potential is much larger than a corresponding EPSP in the central nervous system; an EPP *always* causes muscle fiber to fire. The membrane of the muscle fiber propagates an action potential along its length, thus inducing a contraction, or *twitch,* of the muscle fiber.

The twitch of a muscle fiber is just as much an all-or-none phenomenon as the action potential of the axon. Thus, elicitation of an action potential in an alpha motor neuron *guarantees* the contraction of the muscle fibers that are a part of this motor unit. The last place where decisions can be made in the nervous system is at the level of the motor neuron. There, excitatory and inhibitory nerve terminals can produce their conflicting effects, but nothing can prevent the contraction of all the muscle fibers of a motor unit once the motor neuron fires. Sherrington, the great pioneering neurophysiologist, thus referred to the motor neuron of the spinal cord and cranial nerve nuclei as the site of the *final common pathway* of the decision-making process.

As the action potential, triggered by the EPP produced by release of ACh from the nerve terminal, propagates along the muscle fiber, it increases membrane permeability to Na^+ and K^+, a process associated with transmission of a nerve impulse. In addition, the membrane becomes more permeable to the calcium ion, and the entry of this ion activates the myosin cross bridges, thus causing the muscle twitch. Active membrane transport (i.e., a **calcium pump**) gets rid of the intracellular calcium and terminates the contraction.

Physical Basis of Muscular Contraction

The entry of Ca^{++} into the cytoplasm of a muscle fiber triggers a series of events that results in movement of the myosin cross bridges. These protrusions alternately attach to the actin strands, bend in one direction, detach themselves, bend back, reattach to the actin, etc. The cross bridges thus "row" along the actin filaments. Figure 10.4 illustrates the "rowing" sequence, and it shows how this sequence results in shortening of the muscle fiber. (See **FIGURE 10.4.**)

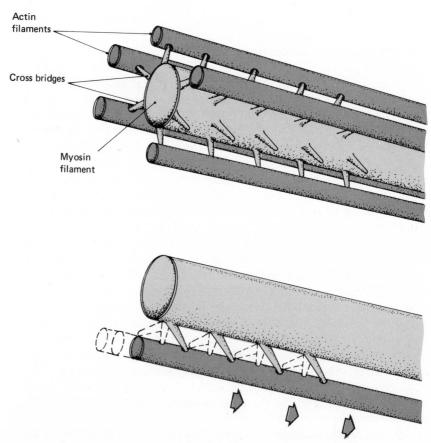

Actin filaments

Cross bridges

Myosin filament

FIGURE 10.4 The mechanism by which muscles contract. The myosin cross bridges perform "rowing" movements, which cause the actin and myosin filaments to move relative to each other. (Adapted from Anthony, C. P., and Kolthoff, N. J., *Textbook of Anatomy and Physiology*, 8th edition. St. Louis: C. V. Mosby, 1971.)

A single impulse of a motor neuron will produce a single twitch of a muscle fiber. The physical effects of the twitch will last considerably longer than will the action potential because of the elasticity of the muscle and the time required to rid the cell of calcium. Figure 10.5 shows how the physical effects of a series of action potentials can overlap, causing a sustained contraction by the muscle fiber. A single motor unit in a leg muscle of a cat can raise a 100 g weight, which attests to the remarkable strength of the contractile mechanism. (See **FIGURE 10.5.**)

As you know quite well, muscular contraction is not an all-or-none phenomenon, as are the twitches of the constituent muscle fibers. Obviously, strength of muscular contraction is determined by the average rate of firing of the various motor units. If, at a given moment, many units are firing, the contraction will be forceful. If few are firing, the contraction will be weak.

Sensory Feedback from the Muscles

The intrafusal muscle fibers contain sensory endings sensitive to stretch. These fibers (they are often referred to as **muscle spindles**) are arranged in parallel with the extrafusal muscle fibers. Therefore, they are stretched when the muscle lengthens and are relaxed when it shortens. Thus, even though these afferents are *stretch receptors*, they are actually *muscle-length detectors*. As we shall see, this distinction is important. Stretch receptors are also located within the tendons; this system is referred to as the **Golgi tendon organ.** The Golgi tendon organ contains receptors that detect stretch exerted by the muscle, via its tendons, on the bones to which the muscle is attached. The stretch receptors of the Golgi tendon organ encode degree of stretch by rate of firing. It does not matter how long the muscle is—only how hard it is pulling. The receptors of the muscle spindle, on the other hand, detect muscle length, not tension.

Figure 10.6 shows the response of afferents of the muscle spindles and Golgi tendon organ to various types of movements. Figure 10.6A shows the effects of passive lengthening of the muscle, the kind of movement that would be seen, for example, if your forearm, held in a completely relaxed fashion, were slowly lowered by someone who was supporting it. The rate of firing of one type of muscle spindle afferent increases (MS_1), while the activity of the afferent of the Golgi tendon organ (GTO) remains unchanged. (See **FIGURE 10.6A** on following page.) Figure 10.6B shows the same results if the arm were dropped quickly; note that this time MS_2 (the second type of muscle spindle afferent) fires a rapid burst of impulses. This fiber, then, signals rapid changes in muscle length. (See **FIGURE 10.6B.**) Figure 10.6C shows what would happen if a weight were suddenly dropped into your hand while your forearm was held parallel to the ground. MS_1 and MS_2 (especially MS_2, which responds to rate of change in

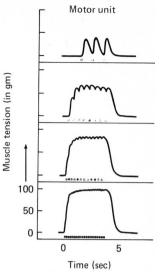

FIGURE 10.5   A rapid succession of action potentials can cause a muscle fiber to produce a sustained contraction. Each dot represents an individual action potential. (Adapted from Devanandan, M. S., Eccles, R. M., and Westerman, R. A., *Journal of Physiology (London)*, 1965, *178*, 359–367.)

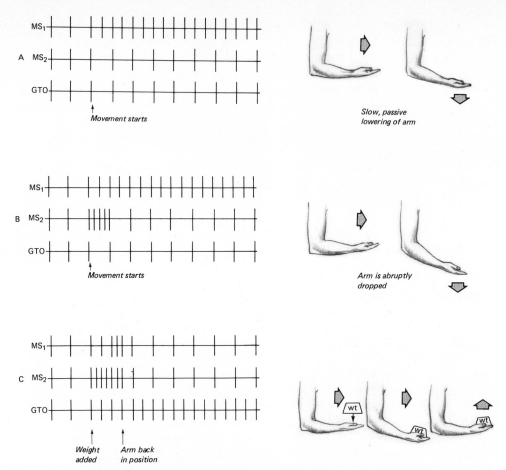

FIGURE 10.6   Effects of arm movements on the firing of muscle and tendon afferents. (A) Slow passive extension of the arm. (B) Rapid extension of the arm. (C) Addition of a weight to an arm held in a horizontal position. MS_1 and MS_2 are two types of muscle spindles; GTO is an afferent fiber from the Golgi tendon organ.

muscle length) will briefly fire, because your arm will lower briefly and then come back to the original position. GTO, monitoring strength of contraction, fires in proportion to the stress on the muscle, so it continues to fire, even after the original position of the arm is restored. (See **FIGURE 10.6C.**) I might note that, because of Archimedes' principle, which describes how force can be increased or decreased by means of levers, your biceps muscle must exert a force of 280 lb to support a weight of 40 lb carried in your hand. (See **FIGURE 10.7.**)

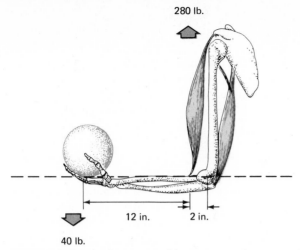

280 lb.

12 in.   2 in.

40 lb.

FIGURE 10.7   The force exerted by most muscles is considerably greater
than the weight supported by the limb.

REFLEX CONTROL OF MOVEMENT

The Monosynaptic Stretch Reflex

It is easy to demonstrate the activity of the simplest functional
neural pathway in the body. Sit on a surface high enough to allow
your legs to dangle freely and have someone lightly tap your patellar
tendon just below the kneecap. In response to this light tap your leg
will kick forward. (I am sure few of you will bother with this demonstration, since you are quite familiar with it; most physical examinations include a test of this patellar reflex.) The time interval
between the tendon tap and the start of the leg extension is about
50 msec. That interval is too short for the involvement of the brain—
it would take considerably longer for sensory information to be relayed to the brain and motor information to be relayed back. For
example, if a person is asked to respond as fast as possible to a light
flash by producing a muscular movement, the interval between the
stimulus and the start of the movement will be several times greater
than the time required for the patellar reflex. The patellar reflex
occurs in response to a brief, quick stretch to the muscle (the tendon
serves only to transmit the stretch), which causes sensory information to be sent to the spinal cord. Almost immediately, motor impulses are "reflected" back to the muscle—hence the term *reflex*.

Obviously, the patellar reflex as such has no utility; no selective
advantage is bestowed upon animals that kick a limb when a tendon

is tapped. However, if a more natural stimulus is applied, the utility of this mechanism becomes apparent. Figure 10.8 reproduces part of Figure 10.6, showing the effects of placing a weight in a person's hand. However, this time I have included a piece of the spinal cord, with its roots, to show the neural circuit that composes the ***monosynaptic stretch reflex.*** First follow the circuit: starting at the muscle spindle, afferent impulses follow the fiber to the gray matter of the spinal cord. The terminals synapse on an alpha motor neuron that innervates the extrafusal muscle fibers of the same muscle. Only one synapse is encountered along the route from receptor to effector—hence the term *monosynaptic.* (See **FIGURE 10.8.**)

Now consider the sequence of events. The alpha motor neurons (and their associated muscle fibers) must fire at some constant rate to keep the limb in a constant position, as shown in the left-hand figure. When the weight is increased, the forearm begins to move down. This movement lengthens the muscle and increases the firing rate of the muscle spindle afferents. Since the afferent fibers synapse on the alpha motor neurons (and produce EPSPs), the firing rate of the motor neurons increases. Hence, the muscle contracts and pulls the weight up. (See **FIGURE 10.8.**)

The monosynaptic stretch reflex is probably most important in

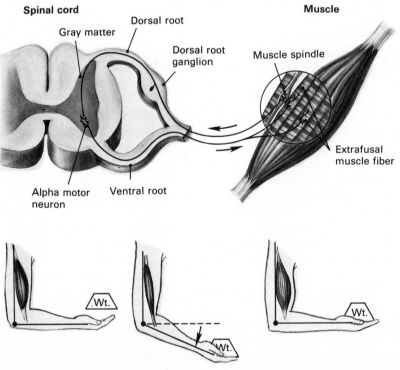

FIGURE 10.8   The monosynaptic stretch reflex.

initiating the movement to restore the limb to the original position. The brain then uses information it receives from the muscle spindles to set a new firing level of the motor neurons, keeping the limb in the correct position. If the spindle afferent did not synapse with the alpha motor neuron, but only sent information to the brain, there would be a great lag between increased weight and the start of the muscle activity to restore limb position. The weight would probably be dropped. The reflex works in the opposite direction, also. When we pick up a weight that is lighter than we expect, the sudden muscle shortening quickly reduces activity of the spindle afferent and removes excitatory activity from the alpha motor neurons, slowing the rate of contraction. Otherwise, we would probably throw the weight into the air.

Another very significant role played by the monosynaptic stretch reflex is its control of posture. In order to stand, we humans must keep our center of gravity above our feet, or we will fall. As we stand, we tend to oscillate back and forth, and from side to side. Our vestibular sacs and our visual system play a very significant role in the maintenance of posture. However, these systems are aided by the activity of the monosynaptic stretch reflex. For example, consider what happens when a person begins to lean forward. The large calf muscle (gastrocnemius) is stretched, and this stretching elicits compensatory muscular contraction that pushes the toes down, thus restoring upright posture. (See **FIGURE 10.9.**)

Polysynaptic Reflex Pathways

Before I begin to discuss some more complicated reflexes, I should mention the fact that the simple circuit diagrams used here (including the one you just looked at in Figure 10.8) are quite fallacious. Many neurons participate in even the simplest reflex, but, for simplicity's sake, only a single chain of neurons is drawn. You should bear in mind that each neuron shown in the diagrams represents hundreds or thousands of neurons, each axon synapsing on many neurons, and each neuron receiving synapses from many different axons. The multiple branching of axons represents *divergence* of information and the multiple input on a single neuron represents *convergence* of information. These processes are shown in **FIGURE 10.10** on the next page. The diagram of a reflex would be awfully untidy if I tried to represent the amount of divergence and convergence that really occurs.

As we previously saw, the afferent fibers from the Golgi tendon organs serve as detectors of muscle stretch. There are two populations of GTO afferents, with different sensitivities to stretch. The more sensitive afferents tell the brain how hard the muscle is pulling. The less sensitive ones have an additional function. Their terminals synapse on spinal cord *interneurons* (neurons that reside entirely

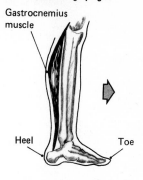

Standing upright

Gastrocnemius muscle

Heel

Toe

Leaning forward

Muscle lengthens, muscle spindles fire, alpha motor neurons are stimulated, muscle contracts

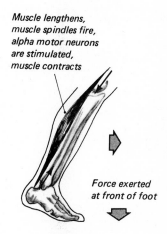

Force exerted at front of foot

Upright posture restored

FIGURE 10.9 The role of the monosynaptic stretch reflex in postural control.

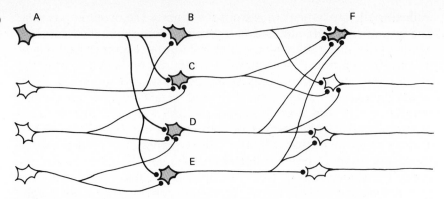

FIGURE 10.10   Examples of divergence (neuron A synapses with neurons B, C, D, and E) and convergence (neurons B, C, D, and E synapse with neuron F).

within the gray matter of the spinal cord and serve to interconnect other spinal neurons). These interneurons synapse, in turn, on the alpha motor neurons serving the same muscle. The interneurons liberate glycine, and hence produce IPSPs on the motor neurons. (See **FIGURE 10.11**.) The function of this reflex pathway is to decrease the strength of muscular contraction when there is danger of damage to the tendons or bones to which the muscles are attached. Weight lifters can lift heavier weights if their tendon organs are deactivated with injections of a local anesthetic, but they run the risk of pulling the tendon away from the bone, or even breaking the bone.

The discovery of the inhibitory tendon organ reflex provided the first real evidence of neural inhibition, long before the synaptic mechanisms were understood. A **_decerebrate_** cat (one whose brain stem has been cut through) exhibits a phenomenon known as **_decerebrate rigidity._** This rigidity results from excitation originating in the caudal reticular formation, which greatly facilitates all stretch

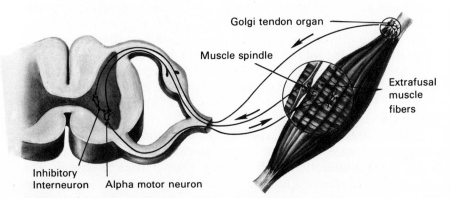

FIGURE 10.11   Input from the Golgi tendon organ can cause IPSPs to occur on the alpha motor neuron.

reflexes, especially of extensor muscles. (In a later section we shall see how this facilitation is accomplished.) Rostral to the brainstem transection is an inhibitory region of the reticular formation, which normally counterbalances the excitatory one. The transection removes the inhibitory influence on the motor neuron in the spinal cord, leaving only the excitatory one. If you attempt to flex the outstretched leg of a decerebrate cat, you will meet with increasing resistance, which suddenly melts away, allowing the limb to flex. It almost feels as though you were closing the blade of a pocketknife—hence the term ***clasp-knife reflex.*** The sudden release is, of course, mediated by activation of the tendon organ reflex.

The monosynaptic stretch reflex is not quite so simple as it might first appear. Muscles are arranged in opposing pairs. The ***agonist*** moves the limb in the direction being studied, and since muscles cannot push back, the ***antagonist*** muscle is needed to move the limb back in the opposite direction. (*Agōn* means "contest"; hence the terms agonist and antagonist.) Consider this fact, then: when a stretch reflex is elicited in the agonist, it contracts quickly, thus causing the antagonist to lengthen. It would appear, then, that the antagonist is presented with a stimulus that should elicit *its* stretch reflex. And yet the antagonist relaxes instead. Let's see why.

Afferents of the muscle spindles, besides sending terminals to the alpha motor neuron and to the brain, also synapse on inhibitory interneurons, which, in turn, synapse on alpha motor neurons serving the antagonistic muscle. (See **FIGURE 10.12.**) This means that a stretch reflex excites the agonist and inhibits the antagonist, so that the limb can move in the direction controlled by the stimulated muscle.

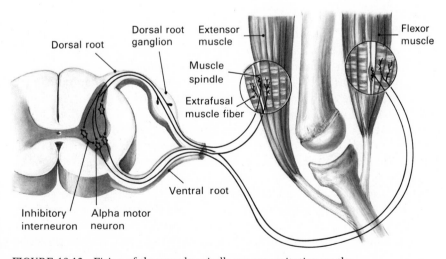

FIGURE 10.12   Firing of the muscle spindle causes excitation on the alpha motor neuron of the agonist, and inhibition on the antagonist.

These simple reflexes are only a small fraction of the many that have been discovered so far. For example, reflex flexion of one forelimb produces extension of the other one, and even involves the hindlimbs. But let us now examine the role of the fourth nerve fiber to the muscle—the efferent to the intrafusal muscle fiber.

The Gamma Motor System

The muscle spindles are very sensitive to changes in muscle length; they will increase their rate of firing when the muscle is lengthened by a very small amount. The interesting thing is that this detection mechanism is adjustable. Remember that the ends of the intrafusal muscle fiber can be contracted by activity of the associated efferent fiber; rate of firing determines the degree of contraction. When the muscle spindles are relaxed, they are relatively insensitive to stretch. However, if they are being stimulated at a high rate by their efferents, they are *very* sensitive to changes in muscle length. This property of adjustable sensitivity simplifies the role of the brain in controlling movements.

We already saw that the spindle afferents help maintain limb position even when the load carried by the limb is altered. Efferent control of the muscle spindles permits these muscle-length detectors to assist in changes in limb position, as well. Consider a single muscle spindle. When its efferent fiber is completely silent, the spindle is completely relaxed and extended. As the firing rate of the fiber increases, the spindle gets shorter and shorter. If, simultaneously, the rest of the entire muscle also gets shorter, there will be no stretch on the central region that contains the sensory endings, and the afferent fiber will not respond. However, if the muscle spindle contracts faster than does the muscle as a whole, there will be a considerable amount of afferent activity.

The motor system makes use of this phenomenon in the following way: When commands from the brain are issued to move a limb, both the alpha motor neurons and the *gamma motor neurons* (the cell bodies that activate the conduction of the muscle spindles) are activated. The alpha motor neurons start the muscle contracting. If there is little resistance, both the extrafusal and the intrafusal fibers will contract at approximately the same rate, and little activity will be seen from the spindle afferents. However, if resistance is met, these fibers will fire and thus cause reflexive strengthening of the contraction. The brain thus makes use of the gamma motor system in moving the limbs. By establishing a rate of firing in the *gamma motor system*, the brain determines the length of the muscle spindles and thus, indirectly, the length of the entire muscle.

It was formerly thought that only the gamma motor neurons were activated to initiate movements, and that the alpha motor neurons were stimulated solely by the spindle afferents. However, Vallbö

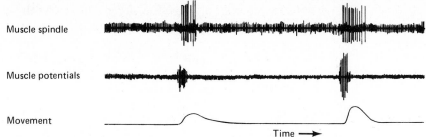

Muscle spindle

Muscle potentials

Movement

Time ——→

FIGURE 10.13   Evidence that the muscle begins moving before the muscle spindle begins firing proves that the alpha motor neurons directly initiate the movement. (From Valbo, Å. B. Muscle spindle response at the outset of isometric voluntary contractions in man: Time difference between fusiomotor and skeletomotor effects. *Journal of Physiology (London)*, 1971, *218*, 405–431. Used by permission of Cambridge University Press.)

(1971) put small electrodes into his own peripheral nerves and found that contraction of the muscle (as shown by its electrical activity, recorded in the **electromyogram**) always preceded activity of the spindle afferent. Thus, the alpha motor neurons must have been activated directly by the brain, because the movement started before the afferent impulses were observed. (See **FIGURE 10.13.**)

The level of activity of the gamma motor system is largely responsible for *muscle tone*. Even when we are resting, there is a certain amount of muscular activity. (This phenomenon is probably necessary for the health of our muscles. If a peripheral nerve is cut, the muscle it serves will atrophy—wither away. Direct electrical stimulation of the muscle prevents this effect.) The rate of firing of the gamma motor neurons largely determines the degree of muscle tone. You can excite the gamma motor system yourself, and demonstrate its effects, by performing the **Jendrassic maneuver.** Have a friend test your patellar reflex and note its strength. Now clasp your fingers together in front of you and pull your hands in opposite directions, forcefully. While you are doing this, have your friend again test your patellar reflex. It should be more vigorous now. The increased gamma motor activity to your arms "spills over" to the gamma motor neurons serving your leg muscles, in an example of divergence of information. The increased gamma motor activity enhances your patellar reflex.

Recurrent Inhibition

Figure 10.14 shows a more detailed view of an alpha motor neuron. Before its axon leaves the gray matter of the spinal cord, it sends off a collateral fiber. This fiber (called a **recurrent collateral**) branches, and the terminals synapse on inhibitory interneurons. These inter-

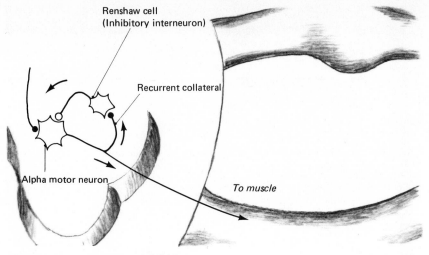

FIGURE 10.14   Recurrent inhibition.

neurons (called **Renshaw cells**) synapse on the same alpha motor neurons whose recurrent collateral stimulated them. (See **FIGURE 10.14.**) Thus, after an alpha motor neuron fires, it receives some self-initiated inhibition. (The IPSPs produced by Renshaw cells are quite long-lived—up to ½ sec.) This means that once the alpha motor neuron fires, it is difficult to get it to fire again for a period of time.

This phenomenon of **recurrent inhibition** accomplishes a useful purpose: the rotation of effort among the various motor units of a muscle. A weak muscular contraction requires a low rate of activity of the muscle fibers; during any given time interval, few of the fibers contract. This same effect could be accomplished by having a few fibers fire rapidly, or by having all fibers fire slowly. If the same small number of motor units fired repeatedly, however, they would quickly become fatigued. Since each motor unit follows a cycle of sensitivity—inhibition, followed by gradual recovery—a steady low level of excitation of all the neurons will cause each one to fire only in the sensitive portion of its cycle. Thus, the muscular effort is distributed among all the motor units, giving each one a period of rest. A more forceful contraction is produced by increased stimulation of the motor neurons, causing them to fire earlier in their cycle of sensitivity.

Complex Reflex Mechanisms

Reflex mechanisms can take care of many complex functions. Even if the spinal cord is severed from the brain, a female dog can become pregnant, carry her litter to term, and deliver the pups. In males, penile erection and ejaculation can be stimulated even after the

FIGURE 10.15   The tonic neck reflexes cause the cat's body to respond to movements of the head. (Adapted from Elliot, H. C., *Textbook of Neuroanatomy*, Philadelphia: J. B. Lippincott, 1963.)

spinal cord is cut. Humans with spinal cord damage have thus become fathers by means of artificial insemination.

The brain, as well as the spinal cord, participates in reflexes. A very high level of reflex integration can be demonstrated in a very nimble animal, the cat. If a cat's head is pushed down, the animal will flex its forelimbs and extend its hindlimbs. If its head is pushed up the opposite movements occur. (See **FIGURE 10.15.**) The cat's body posture will also be changed when the head turns right or left, or rotates. These **tonic neck reflexes** combine with vestibular reflexes to produce the ***righting reflex.*** Figure 10.16 illustrates what happens when a cat is held upside down and dropped. First the head begins to return to its normal orientation (vestibular reflexes) and then the body follows the head (tonic neck reflexes). The mechanisms go together very nicely so that the cat invariably lands on its feet. (See **FIGURE 10.16.**)

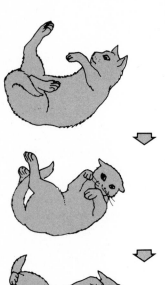

Many afferents of muscle spindles and cutaneous receptors travel to motor cortex, and it is apparent that they can initiate or facilitate reflex motor activity via this pathway. For example, Asanuma and Rosén (1972) and Rosén and Asanuma (1972) stimulated and recorded from discrete regions in motor cortex of monkeys. They found a reciprocal afferent and efferent organization in these cortical regions. For example, if stimulation of a locus on motor cortex produced thumb flexion, the same area was maximally sensitive to cutaneous stimulation of the ball of the thumb. Loci that produced thumb extension when stimulated appeared to be most responsive to touch along the back of the thumb. This organization, therefore, would appear to reinforce a movement that resulted in the touching of an object, and might, as Eccles (1973) notes, be responsible for a baby's grasp when an object is placed in its hand.

Vestibular reflexes (mediated by cerebellum) cause head to assume normal orientation with respect to ground

CENTRAL MOTOR CONTROL

So far I have been describing "automatic" motor control—something that is "involuntary," for the most part. I put these words in quotes because, although we all realize that there is a difference between such "involuntary" movements as the patellar reflex and such "voluntary" movements as I am making in moving my pen to write this, there is no precise way to define either of these terms. The word voluntary comes from the Latin *voluntas*—"will." If we take a mechanistic view of the body, we must reject the notion of will; the body is a machine, and all effects must have causes. If we postulate free will, then there would be effects not determined by causes. But we cannot throw away the concept of "voluntary." We can get rid of the term by substituting some euphemism or other, but until we know much more than we do about the mechanisms of the brain, we are going to be stuck with that fuzzily defined word.

Motor cortex can be defined in various ways. Anatomically, it

Tonic neck reflexes cause body to follow head

FIGURE 10.16   *The righting reflex.* Vestibular reflexes cause the cat's head to right first, and tonic neck reflexes cause the body to follow the cat's head. (Adapted from Marey, M., *Comptes Rendus des Seances de L'Academie des Sciences*, 1894, *119*, 714–721.)

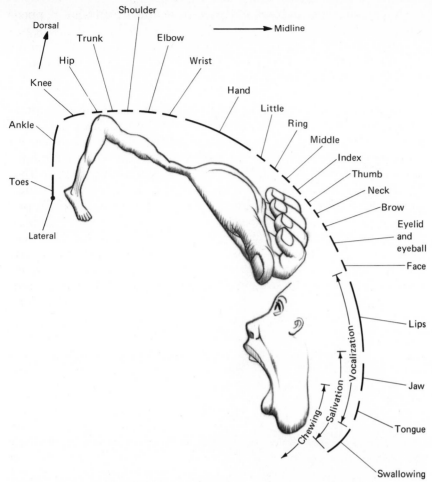

FIGURE 10.17 A motor homunculus. Stimulation of various regions of
motor cortex causes movement in muscles of various parts of the body.
(Adapted from Penfield, W., and Rasmussen, T., *The Cerebral Cortex of Man.*
New York: Macmillan, 1950.)

can be shown that some areas contribute to descending motor systems. Damage to motor functions can be observed after destruction
of various cortical regions. Or, electrodes can be placed on various
parts of cortex to determine which areas produce movements when
electrically stimulated. The "classical" motor cortex is the precentral gyrus, but movements can be elicited from many other cortical
regions. The "motor homunculus" of Figure 10.17 is drawn over
"primary motor cortex" and represents the regions of the body that
move in response to local electrical stimulation. (See **FIGURE 10.17.**)
Just as cutaneous receptors from the fingers and lips project to a
disproportionately large area of somatosensory cortex, a correspondingly large area of motor cortex elicits movements in our most pre-

cisely controlled structures—fingers, lips, tongue, and vocal apparatus.

The Pyramidal Motor System

The *pyramidal motor system* generally receives a disproportionate amount of attention in discussions of motor control mechanisms in mammals, considering its importance. The pyramidal motor system consists of a long, monosynaptic pathway from cortex to motor neurons (or interneurons) of the cranial nerve nuclei and ventral horn of the spinal cord. In humans, most of the fibers decussate at the level of the medulla; a variable number of fibers (averaging 20 percent) descend through the ipsilateral spinal cord. However, most of these ipsilateral fibers eventually decussate in the spinal cord near their level of termination. Collateral fibers are also sent to ventrolateral thalamus, red nucleus, pontine nuclei, inferior olive, and reticular formation. The system is schematically illustrated in **FIG-**

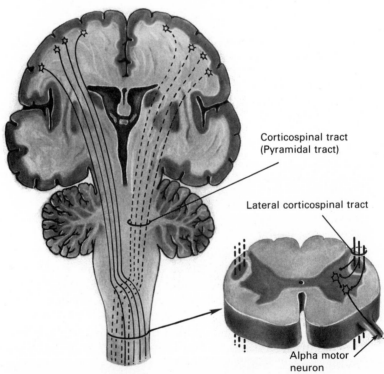

Corticospinal tract
(Pyramidal tract)

Lateral corticospinal tract

Alpha motor
neuron

FIGURE 10.18 Pathway of the pyramidal motor system. Note the complex interconnections with various subcortical motor nuclei. (Adapted with permission from McGraw-Hill Book Co. from *The Human Nervous System*, 2nd edition, by Noback and Demarest. Copyright 1975, by McGraw-Hill, Inc.)

It should be noted that, although the precentral gyrus is usually called "primary motor cortex," its cells contribute only about one-third of the descending fibers. The other fibers arise from cells of the postcentral gyrus (the somatosensory area), and various regions of frontal and temporal cortex.

Just as we do not know the location of the brain regions that "experience" stimuli, we do not know where "voluntary" movements originate. (We could say that they originate at the sense receptors, because our behavior is a function of sensory stimuli, in both the recent and distant past, but that would be begging the question.) It is clear that "voluntary" movements are *not* initiated in the precentral gyrus. Electrical stimulation of this area initiates irresistible movements in human patients, but the patients invariably report that the stimulation elicited the movement itself, and not a desire to move. We do not know the starting point of motor systems any more than we know the ending point of sensory systems. There is some evidence that the "associational" areas of cortex participate in the initiation of movements. Damage to these areas in humans leads to movements that are well integrated but robotlike in character, lacking purpose or direction (Crosby, Humphrey and Lauer, 1962). Furthermore, electrical potentials (so-called "readiness potentials") can be recorded from widespread areas of cortex some time before a movement begins. However, these results do not rule out the possible participation of subcortical areas of the brain in the control of these "voluntary" movements. As we shall see, subcortical control appears to be quite important.

INITIATION VERSUS CONTROL OF MOVEMENT: THE APRAXIAS.   The evidence that most directly implicates association cortex in the initiation of voluntary movements is provided by study of the *apraxias*—inabilities to perform certain types of movements after lesions of cerebral cortex. As Geschwind (1975) points out, some of these disorders can be viewed as deficits of understanding.

In order for a movement to be performed in response to spoken commands, the sounds must be analyzed for meaning and the results of this analysis must be forwarded to brain mechanisms involved in producing the appropriate movements. Speech is analyzed for meaning by a region of auditory association cortex located on the left temporal lobe. (This region will be discussed in more detail in Chapter 19.)

Certain brain lesions (produced by tumors or by cerebrovascular accident) can disrupt connections between this auditory association cortex and motor cortex of the frontal lobe. Therefore, motor mechanisms controlled by the frontal lobes cannot be activated by speech. However, the temporal lobes contain many motor neurons that project to the extrapyramidal motor system—especially the parts controlling postural changes and movements of the proximal musculature, as opposed to the distal musculature of the forearms and hands.

A patient with such lesions, asked to "assume a boxer's stance," readily does so. When asked to "jab with your hand," he or she remains immobile. Similarly, such a patient will not wave a hand in response to verbal commands, but will readily nod his or her head if asked to do so.

This syndrome can best be explained in the following way: The meaning of the verbal message is decoded by auditory association cortex of the left temporal lobe. If the motor neurons of the temporal lobe are able to initiate the commanded movement, the patient responds appropriately. If the movement requires the participation of frontal cortical mechanisms, the patient cannot respond. The fact that he is able to make movements of the distal musculature is shown by the fact that he can wave back if someone waves at him, and that he can handle objects appropriately if they are given to him. Therefore, this form of apraxia does not appear to result from damage to motor mechanisms, but rather to result from interrupted communication between mechanisms that initiate movements and those that control them.

THE PYRAMIDAL SYSTEM IN EVOLUTIONARY PERSPECTIVE. The pyramidal motor system is not essential to movement—not even to many well-integrated, skilled behaviors. In humans, the "pyramidal tract syndrome"—that is, the deficits following damage to these structures—characteristically includes **paresis** (partial paralysis), spasticity, depression of cutaneous reflexes, and exaggeration of stretch reflexes. However, it appears that much of the deficit that is seen actually results from damage to other motor systems adjacent to the pyramidal system. Instead, the principal motor effects appear to be decreased muscle strength (particularly in the flexors of the hands), clumsiness in movement of the hands and fingers, and increased reaction time in response to visual stimuli (Brooks and Stoney, 1971). Clearly, an individual could survive quite well without this system. This is not to say that the pyramidal motor system is not needed by our species; to the contrary, our evolution was shaped by the fact that we, as a group, are able to perform precise, fine movements with our hands and fingers. A clumsy individual human will survive, but our species would not have attained its present status if all humans had been clumsy.

It is commonly said that the pyramidal system is a recent evolutionary development and reaches its ultimate development in *Homo sapiens*. (We humans tend to think that *everything* reaches its ultimate development in our species.) However, Towe (1973) notes that analogous systems can be found in birds, and he argues that the pyramidal system was first seen in some common ancestor to mammals and birds and is thus very old, phylogenetically. He also notes that humans do not possess an inordinate number of pyramidal tract fibers, relative to body size; as a matter of fact, a mouse has fifty times more pyramidal tract fibers per kilogram of body weight than does a human. And some individual chimpanzees and seals have a greater absolute number of pyramidal fibers than do some individual humans, even though the humans outweigh them. As Towe puts it, ". . . from the present perspective on his pyramidal tract, man shows up as just another mammal."

In an analysis of the comparative anatomy of the pyramidal tract, Heffner and Masterton (1975) concluded that two measurements were best correlated with a species' digital dexterity: depth of penetration of pyramidal fibers down the spinal cord (caudally), and depth of penetration of these fibers into the spinal gray matter (ventrally). Other measures, such as number or sizes of fibers in the pyramidal tract, bore poor relationship to dexterity. It is fairly easy to interpret the significance of the correlation between dexterity and depth of penetration (ventrally) into the gray matter of the spinal cord. The farther these fibers penetrate, the closer their terminal buttons are to the spinal motor neurons controlling the muscles. More dorsal terminations represent indirect control of the motor neurons through interneurons. It is not clear why penetration into caudal regions of the spinal cord is related to dexterity; perhaps, as Heffner and Masterton suggest, dexterity of the hand and finger muscles requires better control over competing influences, arising from lower levels of the spinal cord, on the motor neurons controlling these muscles. In any event, their data seem to confirm the importance of the pyramidal system in control of fine movements of the fingers and hands.

ACTIVITY OF PYRAMIDAL TRACT NEURONS DURING MOVEMENT. Evarts (1965, 1968b, 1969) has recorded the activity of single neurons in the precentral gyrus of monkeys trained to move a lever back and forth by means of wrist flexions and extensions. Movements made in the correct amount of time resulted in the delivery of a bit of grape juice, a favored beverage for monkeys. The force needed to move the lever could be controlled by the experimenter. Figure 10.19 shows the experimental preparation as well as the relationship between lever movement and the firing of a cortical neuron. Note that the firing of this neuron is nicely related to the movement, the rate increasing during flexion. (See **FIGURE 10.19** on next page.) Evarts also found that the firing rate of pyramidal tract neurons was generally related to the force of a movement, but not to its extent.

THE PYRAMIDAL MOTOR SYSTEM AND VOLUNTARY MOVEMENT. We cannot conclude that the relationship between cell discharge and movement observed by Evarts proves that the cell thus initiates the movement (a point made by Evarts himself). Fetz (1973) found that neural activity of most cells he recorded in motor cortex correlated well with contraction of proximal limb muscles (closest

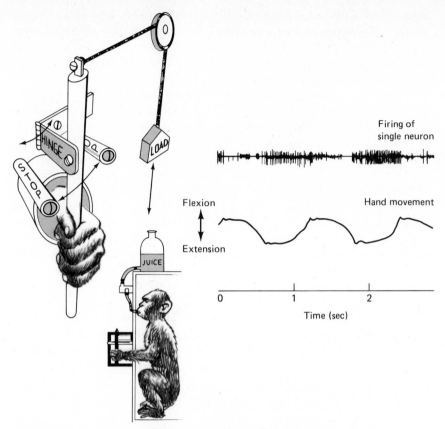

FIGURE 10.19 The relationship between firing of single neurons in motor cortex and hand movements. The single unit records are redrawn from the original data and are therefore only approximate representations.
(Redrawn from Evarts, E. V., Relation of pyramidal tract activity to force exerted during voluntary movement. *Journal of Neurophysiology*, 1968, *31*, 14–27.)

to the trunk). One might assume that electrical stimulation of the same sites would stimulate these muscles. However, such stimulation resulted in contraction of the *distal* limb muscles. Furthermore, stimulation of the appropriate sites in human precentral gyrus will produce facial movements, but Ward, Ojeman, and Calvin (1973) were unable to find many cells that fired in relation to "voluntary" movements of the same muscles.

Another study also suggests that the participation of the pyramidal system is of minor importance in the initiation of movement (Towe and Zimmerman, 1973). Muscular contractions can easily be elicited through cortical stimulation. Towe and Zimmerman found that transection of the pyramidal tracts had no effect on the threshold (i.e., weakest electrical shock capable of producing movements) or latency of movement (time between shock and onset of movement). And long ago, Lloyd (1941) showed that stimulation of the pyramidal tract produced movement only after a series of successive shocks, which contrasts with the fact that a single shock to cortex is sufficient to produce movement. Thus, cortically elicited movement does not appear to be mediated primarily via the corticospinal tract of the pyramidal motor system.

THE PYRAMIDAL SYSTEM AND CONTROL OF TACTUALLY GUIDED MOVEMENT.   H. Kornhuber has described a proposed set of brain mechanisms that can account for the control of voluntary movements. (I must emphasize that the proposal attempts to account for *control* of voluntary movements, and not their *initiation*, which is an entirely different matter.) His explanation of control of rapid and slow "ramp movements" will be described in the next section.

Kornhuber (1974) suggests that the role of motor cortex and the pyramidal system is to regulate movements that require guidance from somatosensory information. Anatomically, the most common type of sensory input to neurons of motor cortex is from the somatosenses; indeed, somatosensory input is the only direct input to motor cortex. All others are via multisynaptic pathways (Pandya and Kuypers, 1969). Furthermore, finger movements, which require the highest degree of control from tactual feedback, are represented by the largest amount of cortical area (as we saw in Figure 10.17). It is important to bear in mind that the most important movements of our hands and fingers are those which manipulate objects; hence, we require accurate (and rapid) somatosensory feedback. Perhaps the most compelling evidence comes from lesion studies; animals lose their ability to react to somatosensory input, but not to input from other sense modalities, after destruction of motor cortex (Bard, 1938; Denny-Brown, 1960).

Evarts (1974) has provided electrophysiological evidence that nicely supports Kornhuber's suggestion that the neurons of motor cortex mediate tactual control of hand and finger movements. Using the apparatus shown in Figure 10.19, Evarts trained monkeys to produce a hand movement to a flash of a light or to a tactual stimulus delivered by means of the handle they were holding. Pyramidal tract neurons in motor cortex began firing 100 milliseconds after a visual stimulus, but responded in as brief an interval as 25 milliseconds to a tactual stimulus. Evarts's data provide further evidence for the preferential access of somatosensory input to these neurons, and thus lend support to Kornhuber's suggestion.

The Extrapyramidal Motor System

The term **extrapyramidal** was poorly chosen. It means "other than pyramidal" and suggests a subservient role. The pyramidal tracts are easily seen during gross examination of the brain and were hence described much earlier than were the complicated, diffuse, polysynaptic pathways of the extrapyramidal motor system.

Until fairly recently, much more was known about the anatomy of the extrapyramidal motor system than about its function. A supportive role was typically assigned to it; one often read that the pyramidal motor system initiates movements, while the extrapyramidal motor system smooths out the movements and produces postural adjustments that support these movements. Portions of the extrapyramidal system do indeed appear to be involved in the smoothing-out of movements and in postural adjustments. However, the fact that well-integrated "voluntary" movements can occur after pyramidal tract sections would suggest that the extrapyramidal system is intimately involved in the initiation of movements as well.

The extrapyramidal system includes cortex and subcortical structures from forebrain to hindbrain. The telencephalic subcortical structures include parts of the basal ganglia—the caudate nucleus and putamen together constituting the **neostriatum** ("new grooved structure") and the globus pallidus constituting the **paleostriatum** ("old grooved structure"). The amygdala, although anatomically a part of the basal ganglia, is not functionally a part of the extrapyramidal motor system. Other parts of this system include various thalamic nuclei (ventral lateral, ventral anterior, and midline nuclei), subthalamic nucleus, red nucleus and substantia nigra of the pons, and portions of the brainstem reticular formation. The cerebellum plays a crucial role in integration and control of this system.

BASAL GANGLIA.   The interconnections of the cortex and basal ganglia are very complex. Figure 10.20 illustrates a very schematic and simplified view of the interconnections. Note that there are opportunities for feedback and integration of information at all levels of the system. In particular, note the loop indicated by the shaded arrows, which will be discussed shortly. (See **FIGURE 10.20**.)

The basal ganglia appear to be principally related to facilitation and inhibition of motor sequences. For example, the globus pallidus appears to be facilitatory in nature. Destruction of this structure, which is the only part of the basal ganglia that directly projects to lower motor structures, results in a severe decrease in motor activities; the subject remains passive and immobile. On the other hand, stimulation of the globus pallidus facilitates reflex movements or those artificially elicited by cortical stimulation. The caudate nucleus, which receives motor information from cortex and feeds information back to cortex via the thalamic nuclei, appears to be involved in suppression of motor activity. Its stimulation will often cause an animal to cease its ongoing behavior, whereas destruction will lead to hyperactivity, such as the **obstinate progression** seen in cats with lesions of the caudate nucleus. A cat with such a lesion will pace incessantly, like a little robot. If it encounters a wall, it will continue to exhibit walking movements with its head against the

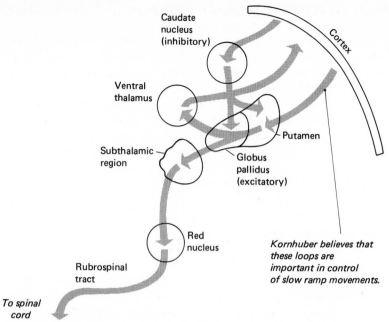

Caudate
nucleus
(inhibitory)

Cortex

Ventral
thalamus

Putamen

Subthalamic
region

Globus
pallidus
(excitatory)

Red
nucleus

*Kornhuber believes that
these loops are
important in control
of slow ramp movements.*

Rubrospinal
tract

*To spinal
cord*

FIGURE 10.20   Some interconnections of basal ganglia. Note the loop
that Kornhuber believes is involved in initiation of slow movements. Also
note that the red nucleus provides the major motor output of the basal
ganglia.

wall (if the floor is slippery enough to permit its paws to slide back-
wards). Complex movements like these do not appear to be "programmed" in the basal ganglia; the sequences of muscular movements involved in locomotion appear to be organized elsewhere.
However, the basal ganglia seem to have control over the starting
and stopping of these activities. Humans with damage to the basal
ganglia can usually walk, but they have difficulty in starting and
stopping the locomotor sequence.

Tonic motor activities are also controlled by the basal ganglia.
Parkinson's disease, which was discussed in Chapter 5, results from
damage to the dopaminergic neurons connecting the substantia nigra and neostriatum. The disease is characterized by muscular
tremor, as well as difficulty in the initiation of movements. Patients
with Parkinson's disease have difficulty in holding their hands still,
but once they begin to manipulate something, the tremor often disappears. I once heard of a very skillful surgeon who suffered from
this **resting tremor** characteristic of Parkinson's disease. When he
held his hands at his sides, they shook violently. However, once he
picked up a scalpel, he was as steady as a rock. Nevertheless, I shud-
der at the thought of a patient succumbing to the anesthetic just
after getting a glimpse of the trembling surgeon standing by the
table.

The **red nucleus** (also called *nucleus ruber,* for those who prefer the Latin) appears to be an important link in the flow of information from cortex to motor neuron. As we saw, section of the pyramidal tracts does not prevent "voluntary" movement of even the distal limbs. However, subsequent damage to the **corticorubrospinal system** (cortex to red nucleus to spinal cord) leads to a complete loss of movement of distal limb muscles. The two systems appear to cooperate, for the loss of movement is not seen after sections of the rubrospinal tract alone (Eldred and Buchwald, 1967).

There are also **reticulospinal** fibers; portions of the brainstem reticular formation play a role in control of the gamma motor system. The control appears to be mostly tonic (long-lasting), as opposed to phasic (brief). Activity of many reticulospinal fibers is modified by cutaneous pressure from all over the body, suggesting that this system plays a role in maintenance of posture (Wolstencroft, 1964). As we saw in a previous section, there appears to be a region that facilitates extensor reflexes (and produces some flexor inhibition) located in the caudal portion of the brainstem reticular formation. There is another, more rostral region, that inhibits these effects. Decerebrate rigidity results from the sole influence of the caudal, extensor-facilitating region. In humans, similar injury produces extension in the legs but flexion in the arms. Thus, we could really call the caudal region an antigravity-facilitating area. The flexors, rather than the extensors, of our arms are antigravity muscles, since we walk on two limbs. You have probably seen brain-damaged people who hold their hands, curled at the wrists and fingers, by their shoulders. This is a manifestation of flexor-facilitation—and probably a corresponding extensor-inhibition—of the muscles of the upper limbs.

THE BASAL GANGLIA AS A GENERATOR OF "SLOW RAMP" MOVEMENTS. Kornhuber (1974) points out that damage to the basal ganglia disrupts a patient's ability to perform slow, smooth movements. However, rapid movements made by the eyes are not impaired by such damage. (As we shall see, we cannot move our eyes slowly, unless we are tracking a slowly moving object.) The motor deficits that occur can be those of *release* (e.g., rigidity or uncontrollable writhing movements after damage to the caudate nucleus or putamen) or of *deficiency* (**akinesia,** or the inability to move, after damage to the globus pallidus or ventral thalamus). Therefore, he suggests that the basal ganglia participate in the control of these movements.

DeLong (1974) has obtained electrophysiological evidence in support of Kornhuber's suggestion. He recorded the activity of single neurons in the putamen during execution of both rapid and slow movements of a monkey's hand. A majority of the units preferentially responded before and during slow, rather than fast, movements.

CEREBELLUM. The cerebullum plays a very important role in motor control. It performs three major functions:

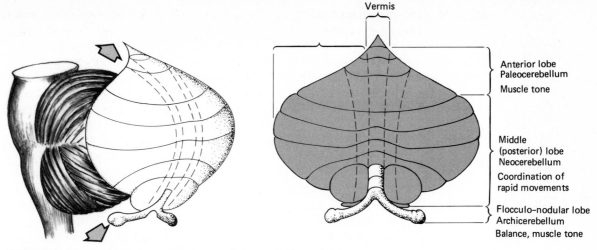

Vermis

Anterior lobe
Paleocerebellum

Muscle tone

Middle
(posterior) lobe
Neocerebellum

Coordination of
rapid movements

Flocculo–nodular lobe
Archicerebellum

Balance, muscle tone

FIGURE 10.21   A schematic representation of the cerebellar cortex.
(Adapted with permission from McGraw-Hill Book Co. from *The Human
Nervous System*, 2nd edition, by Noback and Demarest. Copyright 1975,
by McGraw-Hill, Inc.)

1. It interacts with the brainstem reticular formation in its control
   of the gamma motor system, receiving kinesthetic input and information from the vestibular nuclei.
2. It receives vestibular information and exerts control over the
   postural muscles through **cerebellovestibulospinal pathways.**
3. The newest part of the cerebellum (**neocerebellum**) is intimately
   involved in the execution of rapid, skilled movements.

Figure 10.21 shows a view of the surface of the cerebellum and illustrates the three regions associated with these three roles. (See
FIGURE 10.21.)

THE CEREBELLUM AS A PROGRAMMER OF RAPID MOVE-
MENTS.   The last of the three roles, control of rapid skilled movement, is the most interesting. Humans are able to perform motions
that are complexes of many individual joint movements without
"paying attention" to each component. For example, if you hold your
arm out straight in front of you, it is possible for you to move it
rapidly so that your hand describes a circle. Try this, and move your
arm as rapidly as you can. You will note that in doing so, you engage
not only the muscles of your arm, shoulder, and neck, but also those
of your trunk and (especially if you stand) your legs. A phenomenal
number of muscles are called into action, and at precisely the correct
time. Just considering the arm movement alone, various muscles
must begin and end their contractions at precisely the correct time
in order to produce a smooth motion. (After all, a single muscle
cannot produce a circular motion at the end of the arm.) I find that
I can perform this movement more than forty times in 10 seconds;
this means that it takes less than 250 msec for each circular motion.

The fastest pathway from muscle receptor to cortex to muscle would take on the order of 100 msec. This means that the movement cannot be controlled by sensory feedback: by the time the brain registered information about the whereabouts of the limb, and corrective information reached the muscles, almost one-half of the circle would have been completed. The fact that I had to practice a bit before I could even successfully count the movements suggests that feedback is not particularly prominent in this system.

Other examples of complex movements too fast for feedback correction abound. Almost any skilled **ballistic** (literally, "throwing") movement qualifies. Once you begin to swing a bat or a golf club, or throw a ball, you are "committed" to that movement. Midcourse corrections are exceedingly difficult. Instead, you observe the results and try again.

Figure 10.22 shows some of the inputs and outputs of the cerebellum. You will see that this structure not only receives information about what the muscles and limbs are doing, but also finds out what they are about to do, since it monitors information on its way down to the motor neurons. (See **FIGURE 10.22.**) Similarly, it sends fibers to various motor nuclei, thus modifying the commands received by the muscles. (See **FIGURE 10.22.**)

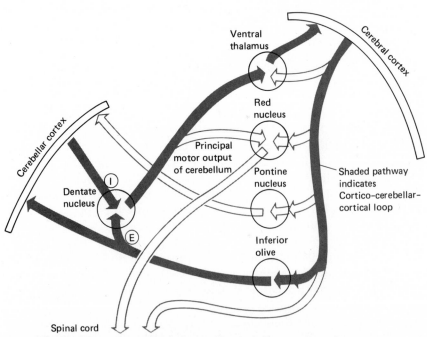

FIGURE 10.22   The relationship between the cerebellum and motor systems. Note the loop that Kornhuber believes is involved with control of rapidly executed movements. Many details are omitted, such as the sensory input to the cerebellum.

The nature of cerebellar control of skilled motor activity can best be seen by the effects of unilateral neocerebellar lesions in humans. People with such damage can produce skilled movements normally, on the unaffected side of the body. They can move the limbs on the affected side, but complex motions are made joint by joint. A patient once said, "The movements of my left arm are done subconsciously, but I have to think out each movement of the right (affected) arm. I come to a dead stop in turning and have to think before I start again" (Holmes, 1939).

Kornhuber has collected data that indicate that cerebellar cortex controls the initiation of rapid stepwise movements while the deep cerebellar nuclei are involved in stopping and holding after these movements. Control of eye movements is unique in that we are unable to move them slowly when scanning a stationary scene. Instead, our eyes make **saccades,** fast, jerky movements from place to place. The *duration,* and not the speed, of a saccade is what determines the size of an eye movement, since the motor neurons controlling the eye muscles fire at the maximal rate during a saccade, regardless of its size. Cerebellar damage disrupts control of these movements; the eyes continue to move in a stepwise manner, but a number of irregular movements are required before the eyes reach the target (Kornhuber, 1974). Thus, the deficit is one of timing of brief intervals. Smooth occular pursuit of a slowly moving object is not impaired by these lesions.

The same type of deficit can be seen during rapid limb movements; damage to cerebellar cortex impairs a patient's ability to aim a rapid hand movement at some point in space, although the movement can accurately be performed if it is done slowly.

In contrast to the timing function exhibited by cerebellar cortex, Kornhuber argues that the deep cerebellar nuclei are involved with stopping and holding. Lesions of the deep cerebellar nuclei result in an uncontrollable tremor when an attempt is made to fix the gaze; these tremors disappear when the eyes are closed (Aschoff, Conrad, and Kornhuber, 1970). Similarly, these lesions result in a rapid, oscillating tremor at the end of an aimed movement of the arm, although the ballistic movement itself is accurately performed (Kornhuber, 1974).

In summary, we have much to learn about where voluntary movements begin. However, a picture of the brain mechanisms that control fast and slow voluntary movements is beginning to emerge. The following movement summarizes Kornhuber's conclusions: Move your finger from a point a foot or so in front of your face up to the tip of your nose, as rapidly as you can. Cerebellar cortex is involved in timing the fast ballistic movement, whereas the deep cerebellar nuclei participate in stopping the motion. The basal ganglia then control a slow ramp motion that is terminated by tactual feedback from your finger touching your nose.

KEY TERMS

actin p. 287
agonist p. 295
akinesia (*ay kin EE zha*) p. 310
alpha motor neuron p. 286
antagonist p. 295
anterior pituitary gland p. 283
apraxia (*ay PRAKS ee uh*) p. 303
ballistic movement p. 312
calcium pump p. 288
cerebellovestibulospinal system p. 311
clasp-knife reflex p. 295
convergence p. 293
corticorubrospinal system p. 310
decerebrate (*dee SAIR a brit*) p. 294
decerebrate rigidity p. 294
divergence p. 293
effector p. 281
electromyogram (EMG) p. 297
endocrine gland p. 282
end plate potential (EPP) p. 287
exocrine gland p. 282
extension p. 285
extrafusal muscle fiber p. 286
extrapyramidal motor system p. 307
final common pathway p. 287
flexion p. 285
Golgi tendon organ (*GOAL jee*) p. 289
hypothalamic hormone p. 282
hypothalamic-hypophyseal portal system
    (*hy poff i SEE ul*) p. 283
interneuron p. 293
intrafusal muscle fiber p. 286

Jendrassic maneuver p. 297
monosynaptic stretch reflex p. 292
motor end plate p. 287
motor unit p. 287
multiunit smooth muscle p. 285
muscle spindle p. 289
myofibril (*my o FY brill*) p. 287
myosin (*MY o sin*) p. 287
neocerebellum p. 311
neostriatum p. 308
neuromuscular junction p. 287
obstinate progression p. 308
pacemaker potential p. 285
paleostriatum p. 308
paresis (*pa REE sus*) p. 304
posterior pituitary gland p. 284
pyramidal motor system p. 302
recurrent collateral p. 297
recurrent inhibition p. 298
red nucleus p. 309
Renshaw cell p. 298
resting tremor p. 309
reticulospinal system p. 310
righting reflex p. 300
saccade (*suh KADD*) p. 313
single-unit smooth muscle p. 285
skeletal muscle p. 285
smooth muscle p. 282
steroid hormone (*STEER oyd*) p. 284
target cell p. 284
tonic neck reflexes p. 300
trophic hormone (*TROW fik*) p. 284

SUGGESTED READINGS

KORNHUBER, H. H. Cerebral cortex, cerebellum, and basal ganglia: An introduction to their motor functions. In *The Neurosciences: Third Study Program*, edited by F. O. Schmitt and F. G. Worden. Cambridge, Mass.: MIT Press, 1974. I highly recommend this chapter of Kornhuber's. However, it is not particularly easy reading. I would suggest that you first read the *Scientific American* chapters before tackling this one.

LUCIANO, D. S., VANDER, A. J., AND SHERMAN, J. H. *Human Function and Structure*. New York: McGraw-Hill, 1978. This book contains excellent descriptions of muscles and glands (and a lot more). It is very well written, thorough, and up-to-date. The illustrations are beautiful.

EVARTS, E. B. Brain mechanisms of movement. *Scientific American*, September, 1979.

EVERT, J. P. The neural basis of visually guided behavior. *Scientific American*, March, 1974.

LLINÁS, R. R. The cortex of the cerebellum. *Scientific American*, January, 1975.

MERTON, P. A. How we control the contraction of our muscles. *Scientific American*, May, 1972.

11

Sexual Development and Behavior

The subject of this chapter is important to almost all of us. The topics of other chapters might (I hope) be interesting, but discussion of these topics is not capable of evoking the kinds of emotional reactions that may accompany discussions of sexual development and sexual behavior. We all have our individual beliefs concerning what constitutes appropriate and inappropriate sexual behavior and what behaviors and interests should or should not be associated with a person's gender. The discussion of sexual behavior in this chapter will be biological, not ethical, and I shall make no attempt to consider the varieties of human sexual practices, except as they relate to physiological mechanisms. For example, researchers have discovered no obvious neural or endocrinological mechanisms that might account for such phenomena as pederasty, voyeurism, or necrophilia. At the present time, such paraphilias (literally, "abnormal loves") can probably be accounted for better by general psychological theories than by specific physiological functions; these phenomena will not be discussed here. On the other hand, there may be some predisposing physiological bases for such psychosexual phenomena as homosexuality, and evidence concerning these phenomena will be reviewed.

All behavior is a function of physiology, but the biological-behavioral link is probably more obvious for sexual behavior than for any other. First of all, sexual behavior depends upon morphology; an individual's behavior is certainly controlled to a large degree by the possession of male or female genitals. Second, a person's sexual desire is affected by his or her own sex hormones. For example, a castrated man, lacking male sex hormones, will eventually lose interest in sexual activity unless he is given pills containing hormones similar to those formerly produced by his own testes. Of course, sexual behavior is also strongly influenced by learning; it is possible for a person to adopt a sex role different from the one normally associated with his or her hormones and sex organs.

Sex hormones have a dual role in the control of sexual behavior. They have an **organizational** effect, shaping the ultimate development of a person's sexual organs and brain. Exposure to certain sex hormones before the fetus is born will *organize* the developing cells so that they will later develop into male or female sex organs (or, if the hormonal exposure is abnormal, the cells may develop into something in between the male and female forms). Furthermore, the developing brain is affected by exposure to hormones before birth; a man's brain is probably not precisely the same as that of a woman. Because of the organizational effects of hormones, the physical (neural or genital) and hormonal determinants of sexual behavior go hand in hand (so to speak).

The second role of sex hormones is their **activational** effect. For example, male hormones are necessary if a man is to produce sperm and experience a normal sex drive. However, male hormones will not have the same effects on a woman. She cannot produce sperm, since she does not have testes. Neither will male sex hormones induce in her a desire to engage in sexual behavior with women. Since her body (including her brain) was *organized* as a female, the *activational* effects of sex hormones are different from those seen in a man.

In this chapter, then, I shall discuss sexual development and the organizational control exerted by hormones. I shall also consider the activational effects of hormones—the role these chemicals play in our day-to-day behavior. Finally, I shall discuss evidence concerning the neural bases of our sexual behavior, and the way in which hormones interact with these neural circuits.

SEXUAL DEVELOPMENT

Humans and other mammals are **sexually dimorphic**; we come in two forms, male and female. Besides the obvious differences in male and female genitalia, males are generally larger, broader in the shoulders, and narrower in the hips, and have varying amounts of facial and chest hair. Females have larger breasts.

Sexual development is a fascinating process. A person's gender is determined very early in life, when a pair of primitive organs develop as either ovaries or testes. That occurrence begins a chain of events that results in the development of a girl or boy who (if all continues to go well) becomes a woman or man. And the factor that causes the primitive organs to become ovaries or testes is the presence or absence of a single chromosome.

Determination of Sex

All cells of the body (other than *gametes*—sperms or ova) contain twenty-three pairs of chromosomes, including a pair of sex chromosomes. The genetic information that programs the development of a human is contained in the DNA constituting these chromosomes. (Our accomplishment of miniaturizing computer circuits on single silicon chips looks very primitive when we consider the fact that the blueprints for an entire human being are so small as to be invisible to the naked eye.) The identity of the sex chromosomes determines an individual's sex. There are two types of sex chromosomes, X and Y. All humans, male and female, possess at least one X chromosome. The additional possession of a Y chromosome causes the individual to develop into a male.

CELL DIVISION AND THE PRODUCTION OF GAMETES. It is possible to observe the twenty-three pairs of human chromosomes. Epithelial cells can be scraped from the mucous membrane on the inside of the cheek and then placed in a culture medium that supports their growth and division. During cell division (*mitosis*) the chromosomes must be duplicated so that the daughter cells (that is what they are called, regardless of the sex of the donor) each have the entire complement of genetic material. This process is shown schematically in Figure 11.1 (For simplicity's sake, only two pairs of chromosomes, rather than twenty-three, are shown.) (See **FIGURE 11.1.**) Once cellular division is established, the culture is treated with colchicine, a drug that halts the process in the phase shown in part C of Figure 11.1. The drug dissolves the spindle fibers that pull the chromosomes apart. (See **FIGURE 11.1C.**)

The dividing cells now contain a double set of twenty-three pairs of chromosomes in the nucleus, straightened out and easy to see. The cells are then squashed so as to flatten the chromosomes, and the genetic material is stained. Then the many cells are searched until one is found in which all the chromosomes can readily be seen. A photograph is taken, and the pictures of the chromosomes are cut out and rearranged according to size. Figure 11.2 illustrates a set of

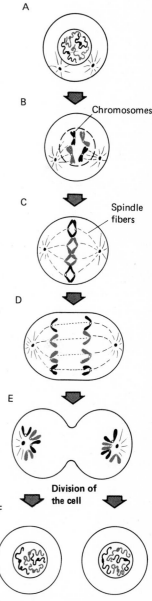

A

B

Chromosomes

C

Spindle fibers

D

E

Division of the cell

F

FIGURE 11.1 Mitosis, the process by which a somatic cell replicates itself. (From G. H. Orians, *The Study of Life.* Boston: Allyn and Bacon, Inc., 1973.)

317

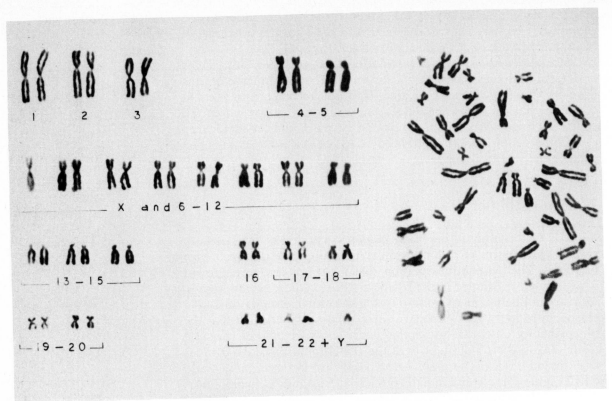

FIGURE 11.2   A karyotype of a cell whose division was arrested in
metaphase (phase C of Figure 11.1). (From Money, J., and Ehrhardt, A.,
Man & Woman, Boy & Girl. Copyright 1973 by The Johns Hopkins
University Press, Baltimore, Maryland. By permission.)

human chromosomes prepared in this way, before and after re-
arrangement. Remember, we can see twice as many chromosomes
as the cell normally contains, since the process of cell division was
arrested just before each member of a duplicated chromosome,
joined near the center, would normally separate and travel to each
of the daughter cells. (See **FIGURE 11.2.**) The *karyotype* (literally,
"nucleus mark") shown in this figure is that of a male, since we can
see a Y chromosome.

In the production of gametes (ova and sperms—*gamein* means
"to marry"), cells divide in a different way. *Meiosis* results in cells
that contain only one member of each of the 23 pairs of chromosomes. The development of a human begins when a single sperm and
ovum join, sharing their 23 single chromosomes to make up the full
complement of 23 pairs.

The particular member of a chromosome pair that goes to a
particular gamete appears to be determined by a random process.
If we tossed a coin twenty-three times, we would obtain one of
8,388,608 different sequences. Since the segregation of the chromo-

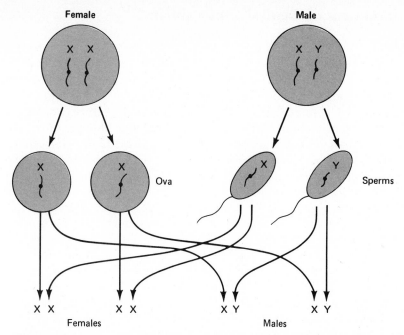

FIGURE 11.3   The sex of the offspring depends on whether the sperm cell carries an X or a Y chromosome.

some pairs is as random (apparently) as the coin toss, this means that a person can produce 8,388,608 different kinds of gametes. Since it takes the combination of two gametes to produce another human, a single couple could produce 8,388,608 × 8,388,608, or something like 70,368,774,177,664 different children. Genetically identical siblings could be conceived at different times, but the probability is not very high.

FERTILIZATION AND THE DETERMINATION OF SEX.   In the production of a new organism, a single sperm unites with a single ovum, and the genetic material contained in each of these gametes combines.

   If we consider only the sex chromosomes, there are two kinds of sperms, X-bearing and Y-bearing. Women, with their XX complement, produce only X-bearing ova. Thus, since men, with their XY complement, produce both X-bearing and Y-bearing sperms, the sex of the offspring is determined by the sex chromosome contained in the sperm that fertilizes the ovum. (See **FIGURE 11.3.**)

   Genetic sex, defined as female by the XX complement and as male by the XY complement, normally initiates a series of events that results in a reproductively competent organism, bearing the appropriate morphology and exhibiting the appropriate sexual behavior. Let us consider some of the details of the process.

TISSUE DIFFERENTIATION.   Specific locations (genes) on the chromosomes express themselves by means of a process that results in synthesis of protein. The proteins that are thus produced act as structural elements of the cell, or as enzymes, controlling biochemical processes. Since all the cells of the developing embryo contain the same genetic information, an additional mechanism must account for *tissue differentiation:* some cells become liver cells, others become bone cells, others become neurons. The expression of most genes in a given cell is repressed by special proteins. The interaction of these proteins with substances in the cytoplasm determines whether a given gene (or set of genes) will be active. Therefore, the factors that regulate the activity of the genes determine the identity and functions of a particular cell. As we shall see, the presence of particular hormones at critical stages of development can affect the process of differentiation, and hence the ultimate form the organism will take.

The process of tissue differentiation is still largely mysterious, although it is being studied very intensively. In the process of successive cells divisions starting with a single fertilized ovum, there must be some unequal divisions of the cytoplasmic contents, or an external stimulus (such as a hormone) must alter the contents of some, but not all, cells. If all the daughter cells of successive mitotic divisions received (and retained) the same cytoplasmic material, as well as the same genetic material, then how could the tissue differentiate? Instead of an embryo, there would be an amorphous mass of identical cells.

Once differentiation has begun, cells are influenced by their neighbors. At early stages of development it is possible to transplant a small piece of tissue to different parts of the organism and observe that the transplanted cells, and their daughter cells, will assume the form appropriate to their location. Thus, it is evident that cells can affect each other's protein synthesis; studies have shown that this interaction is transmitted by means of exchanges of secretions via the extracellular fluid and through contact among cells.

HORMONAL CONTROL OF SEXUAL DIFFERENTIATION.   Figure 11.4 shows the precursors of sex organs, both male and female. At this stage of development (around 7 to 8 weeks after the mother's last menstrual period), the fetus is bisexual. The precursor of the internal female sex organs is called the *Müllerian system;* the *Wolffian system* is the precursor of the male sex organs. A pair of sex glands, the *primordial* ("first begun") *gonads,* is capable of developing into either ovaries or testes. (See **FIGURE 11.4.**) At this point, the sex chromosomes cause the cortex of the primordial gonads to develop into ovaries or its medulla into testes.

Once the gonads are differentiated, the gender of the fetus is

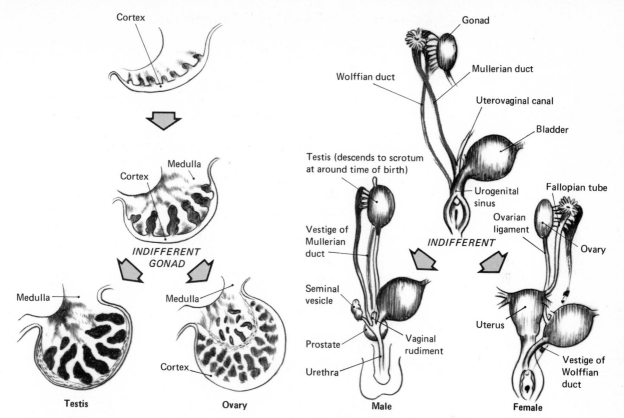

FIGURE 11.4 Development of internal sex organs. (Adapted from Burns, R. K. In *Analysis of Development*, edited by B. H. Willier, P. A. Weiss, and V. Hamberger. Philadelphia: Saunders, 1955; and from Corning, H. K., *Lehrbuch der Entwicklungsgeschichte des Menschen.* Munich: J. F. Bergman, 1921.)

determined. The presence of ovaries or testes controls the rest of the development of the internal and external genitalia. If, at this point, the gonads are removed (as you might expect, this is an exceedingly delicate operation), the organism will be born female in appearance, with uterus, vagina, labia, and clitoris—regardless of genetic sex. A male is produced only if the fetus contains functioning—that is, hormone-secreting—testes. A female is produced otherwise, even if the ovaries are removed. Speaking metaphorically, Nature's impulse is to produce a female, and she will fail to do so only if stimulated, with the appropriate hormones, to produce a male.

During the stage of development of the internal genitalia, the testes of the developing male fetus appear to produce a hormone, **Müllerian-inhibiting substance,** which causes the regression (disappearance) of the female internal sex organs. The testes also secrete **androgens** ("male producers"), which stimulate development of the male internal sex organs. Jost (1969) has shown that the effects of the Müllerian-inhibiting substance are local. If one testis is removed

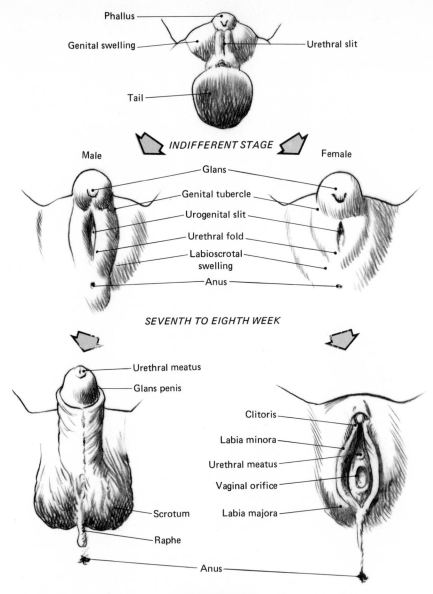

Phallus

Genital swelling

Urethral slit

Tail

INDIFFERENT STAGE

Male

Female

Glans

Genital tubercle

Urogenital slit

Urethral fold

Labioscrotal swelling

Anus

SEVENTH TO EIGHTH WEEK

Urethral meatus

Glans penis

Clitoris

Labia minora

Urethral meatus

Vaginal orifice

Scrotum

Labia majora

Raphe

Anus

TWELFTH WEEK

FIGURE 11.5   Development of the external genitalia. (Adapted from Spaulding, M. H. In *Contributions to Embryology*, Vol. 13. Washington, D. C.: Carnegie Institute of Washington, 1921.)

from a developing rabbit fetus, the Müllerian system (female), but not the Wolffian system (male), will develop on that side. On the other side, which contains an intact testis, the Wolffian system will develop into the internal male sex organs, while the Müllerian system will regress.

Whereas male and female internal sex organs develop from different sets of precursors, the external genitalia develop from indifferent, bipotential primordia, which are capable of assuming either male or female form. Figure 11.5 illustrates this process; note that the principal analogous structures are penis/clitoris, urethral tube of the penis/labia minora, and scrotum/labia majora. (See **FIGURE 11.5.**) Without hormonal stimulation, the external genitalia will become female, regardless of chromosomal sex. However, the presence of androgens will result in masculine development. Thus, even though one of Jost's unilaterally castrated rabbits had female internal sex organs on one side, the external genitalia were male. The androgens secreted by the single testis were sufficient to stimulate masculinization of the external genitalia, even though the effects of Müllerian-inhibiting substance on the internal sex organs were local.

I should note that the ovaries of a developing female do not secrete "Wolffian-inhibiting substance." In fact, the Wolffian system does not disappear—it can still be found in grown women. Without androgens, however, the Wolffian system fails to develop, and it remains about as small as it was in the fetus.

HORMONAL CONTROL OF SEXUAL MATURATION. Secondary sex characteristics do no appear until the onset of puberty. If the genitals are hidden from view, the sex of a prepubescent child is mostly determined from the child's haircut and the way he or she dresses; the bodies of young boys and girls are rather similar. However, at puberty the gonads are stimulated to produce their hormones. Cells in the hypothalamus secrete **gonadotrophic releasing hormones,** which stimulate the production and release of two **gonadotrophic hormones** by the anterior pituitary gland. In turn, the gonadotrophic hormones (**FSH,** or **follicle-stimulating hormone,** and **LH,** or **luteinizing hormone**) stimulate the gonads to produce *their* hormones. The gonadotrophic hormones are named for the effects they produce in the female (**follicle** production and its subsequent **luteinization,** to be described later). However, these same hormones are produced in the male, and they affect the production of sperms and testosterone in the testes. One can exchange male and female pituitary glands in rats, and observe that the ovaries and testes respond perfectly to the new glands (Harris and Jacobsohn, 1951–1952).

In response to the gonadotrophic hormones the ovaries chiefly produce **estrogens** (a generic term for a series of related hormones) and the testes chiefly produce **testosterone** (an androgen). Both glands also produce a small amount of the hormones of the opposite sex. Both estrogens and testosterone initiate closure of the epiphyses (growing portions of the bones) and thus halt skeletal growth. Estrogens also cause breast development, changes in the deposition of body fat, and maturation of the female genitalia. Testosterone stimulates growth of facial, axillary (underarm), and pubic hair, lowers the voice, alters the hairline on the head (sometimes culminating in

TABLE 11.1   Classification of sex steroid hormones

Class	Principal hormone in humans (where produced)	Examples of effects
Androgens	Testosterone (testes)	Maturation of male genitalia, production of sperm, growth of facial, pubic, and axillary hair, enlargement of larynx, inhibition of bone growth
	Androstenedione (adrenal glands)	In females, growth of pubic and axillary hair. Less important than testosterone in males
Estrogens	Estradiol (ovaries)	Maturation of female genitalia, growth of breasts, alterations in fat deposits, growth of uterine lining, inhibition of bone growth
Gestagens	Progesterone (ovaries)	Maintenance of uterine lining

baldness in later life), stimulates muscular development, and causes genital growth. This description leaves out two of the female secondary sex characteristics: axillary and pubic hair. These are not produced by estrogens, but rather by **androstenedione** (AD), a "male" sex hormone (androgen) that is produced by the adrenal glands. Even a prepubertally castrated male (eunuch) will grow axillary and pubic hair, in response to his own androstenedione. (See **TABLE 11.1.**)

The bipotentiality of many of the secondary sex characteristics remains throughout life. A man who is treated with an estrogen (to control an androgen-dependent tumor, for example) will grow breasts. Facial hair will become finer and softer. However, his voice will remain low, since the enlargement of the larynx is permanent. A woman who receives high levels of androgen (usually from a tumor that secretes AD) will grow a beard, and her voice will become lower (permanently, unfortunately).

A number of pathological conditions can prevent the normal development of an individual into male or female. Such people (or animals) are called **hermaphrodites** (from the mythical bisexual offspring of Hermes and Aphrodite). Originally, hermaphroditism referred to the ability to be reproductively competent as both a male and a female (seen in some animals, but not in mammals), but the term has been extended to individuals with ambiguous genital structure, internally and/or externally (Money and Ehrhardt, 1972). Nature's experiments in the production of such individuals have added much to our knowledge of the biological bases of differences in the behavior of human males and females.

As we have seen in this section, Nature would, but for the presence of the Y chromosome, produce a female body. However, if a Y chromosome is present, it causes the primordial gonads to become testes, which secrete androgens and Müllerian-inhibiting substance, resulting in the development of a male. As we shall see in the next section, there is at least some degree of sexual dimorphism in the brain. Without prenatal secretion of androgens from the testes, the brain will be born "female."

Organizational Effects of Sex Hormones on the Brain

We have seen that hormones have organizational and activational effects on internal sex organs, genitals, and secondary sex characteristics, all of which influence a person's behavior. Hormones also affect behavior by interacting with the nervous system. Again, there appear to be two effects, organizational and activational. In humans, prenatal sex hormones probably influence the development of the nervous system, but the evidence is indirect. (Evidence gathered from research with experimental animals is much more straightforward.) However, there is no doubt about the *activational* effects of sex hormones on the human nervous system; thus, our behavior is directly influenced by our hormones.

It is very difficult to study the interactions between sex hormones and development of the human brain. We must turn to three sources of information: experiments with animals; various developmental disorders in humans, which serve as Nature's own "experiments"; and the side effects of medical treatments of pregnant women. Let us first consider the evidence that has been gathered from research with laboratory animals.

Much of what we know about the neural and hormonal control of sexual behavior has been obtained from rodents: rats, mice, and hamsters. In part, this is due to their ready availability, low cost, and the fact that they are extensively used in other kinds of research. But another feature of these species makes them especially valuable for research on the organizing effects of hormones on neural development: most sexual differentiation of their nervous system takes place *after* birth. Needless to say, it is much more difficult to experiment with an unborn animal than one that has already emerged from its mother. Therefore, such procedures as castration, transplantation of gonads, or administration of hormones can be performed prior to sexual differentiation of the brain in a newborn laboratory rodent.

Rodents serve as very useful models for the study of human sexual development, but we must recognize the fact that there are substantial differences between the rat brain and the human brain. For example, great progress has been made in determining how the female rat brain participates in the cyclical release of hormones that occurs during the *estrous cycle* (the rough equivalent of the primate menstrual cycle). However, subsequent experiments with monkeys and observations of humans with hormone abnormalities have shown that primate brains work somewhat differently in this respect.

The most clear-cut evidence for sex differences in brain function is seen in the control of the pituitary gland by the hypothalamic releasing hormones in the rodent brain. The output of gonadotrophic hormones by the anterior pituitary gland is cyclic in females. In males it is relatively constant. (There are, in males, changes in response to stress or environmental stimuli, but there is no regular cyclicity in secretion.)

Before we look at the evidence concerning sexual dimorphism in brain function, it would be best to review the mechanisms that control the female reproductive cycle. Humans and other primates have menstrual cycles, which are characterized by periodic loss and regrowth of a richly-vascularized layer of epithelial tissue that lines the uterus. Other mammalian species have estrous cycles; these are hormonally similar to menstrual cycles, but the uterus does not shed its lining.

CONTROL OF THE MENSTRUAL CYCLE. A menstrual cycle (or estrous cycle) begins with the growth of *ovarian follicles,* small spheres of epithelial cells surrounding ova. The follicles grow in response to FSH (follicle-stimulating hormone) secretion by the anterior pituitary gland, and they begin to secrete estrogens. In turn, the estrogens stimulate the hypothalamus to cause the production and release of a surge of LH (luteinizing hormone) by the anterior pituitary gland. Evidence for the stimulating effects of estrogens on LH release comes from studies (Barraclough, 1966; Everett and Nichols, 1968) that show that estrogen injections will advance the time of ovulation (i.e., cause an earlier surge of LH), and that electrical stimulation of portions of the hypothalamus produces the same effect. More recent studies, utilizing sensitive assays of LH present in the blood, have confirmed that an injection of estrogen does indeed produce a surge of LH.

In response to the LH surge, one of the ovarian follicles will rupture, releasing the ovum. The ruptured follicle, under the continued influence of LH, becomes a *corpus luteum* ("yellow body"), which produces progesterone. This steroid hormone (a *gestagen*, or pregnancy-promoting hormone) prepares the lining of the uterus for implantation of the ovum, should it be fertilized by a sperm. Meanwhile, the ovum, which is released out into the abdominal cavity, enters one of the fallopian tubes and begins its progress down toward the uterus. The ovum is directed into the fallopian tube by the "rowing" action of the ciliated cells around the opening. This process works remarkably well; if a woman lacks a right ovary and a left fallopian tube, her ova will nevertheless find their way across the abdominal cavity and into the fallopian tube. If an ovum meets sperm cells during its travel down the fallopian tube and becomes fertilized, it will begin to divide, and several days later it will attach itself to the uterine wall.

If the ovum is not fertilized, or if it is fertilized too late for it to develop sufficiently by the time it gets to the uterus, or if implantation is prevented by the presence of an intrauterine device (IUD), the estrogen and progesterone levels will fall, and the lining of the walls of the uterus will slough off. Menstruation (in the case of pri-

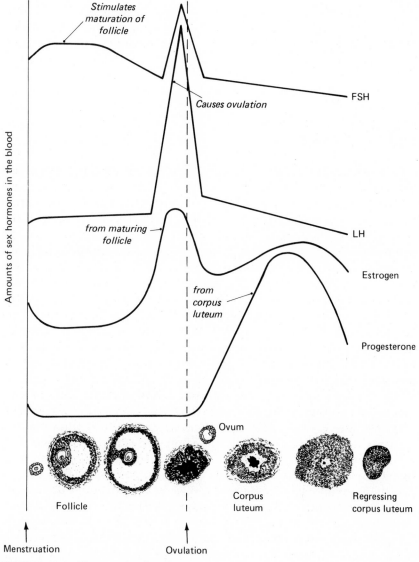

FIGURE 11.6 The menstrual cycle.

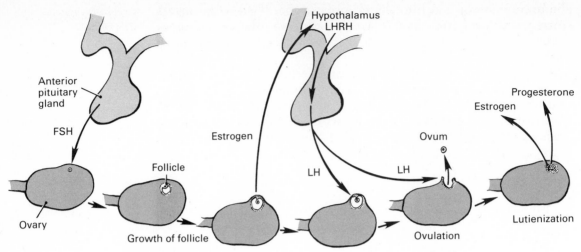

FIGURE 11.7   Neuroendocrine control of the menstrual cycle.

mates) will commence. (See **FIGURE 11.6** on the preceding page and **FIGURE 11.7** above.)

ORGANIZATIONAL EFFECTS OF ANDROGENS ON GONADOTRO-PHIN RELEASE.   The cyclicity of the female is the "normal" condition. If newborn male rats are castrated, they can be stimulated to produce the pituitary gonadotrophins (FSH and LH) in a cyclic manner, as a female does (Pfeiffer, 1936). The castration prevents the brain from becoming "masculinized," and so it responds in the female way. On the other hand, if a female rat is ovariectomized and given a testicular transplant at birth, her brain becomes masculinized, and she fails to show cycles in release of FSH and LH.

What accounts for the cyclicity in gonadotrophin release that characterizes the female pituitary gland? The gland itself is clearly not responsible; if a male pituitary gland is transplanted into a female, or a female pituitary gland into a male, the result will be gonadotrophin release that is characteristic of the sex of the animal receiving the transplant. The transplanted organ will cycle in a female body, but not in a male body (Harris and Jacobsohn, 1951–1952; Martinez and Bittner, 1956). Furthermore, a pituitary gland that is transplanted into a female will cycle only if it is placed close enough to the hypothalamus to allow revascularization (i.e., regrowth of blood vessels from hypothalamus to pituitary gland). Thus, the cyclicity of gonadotrophin secretion must be a result of cyclic release of the hypothalamic releasing hormones. The source of the cyclicity is in the brain.

A number of studies have shown that androgenization of the brain and of the external genitalia can occur only during a critical period (Gorski and Wagner, 1965; Plapinger and McEwen, 1978). As the brain matures, larger doses of androgens are needed to masculinize it, until approximately two weeks of age, when even massive doses fail to have an organizational effect. Furthermore, masculinization of the genitals and the brain take place at different times. Thus, it is possible to produce noncycling females with normal female genitalia, or cycling females with masculinized genitalia. Arai and Gorski (1968) have also shown that the masculinization of the brain takes only a few hours; a single injection of testosterone that was deactivated 6 hours later with a substance that inhibits the effects of androgens still caused the brain to convert to the male type.

SEX DIFFERENCES IN THE ANATOMY OF THE BRAIN. Since the anterior pituitary gland is controlled by the hypothalamus, the evidence that I just cited suggests that the hypothalamus may be the location of the sex differences in function. The basic difference between the cycling and noncycling rodent brain is that the cycling (nonandrogenized) brain responds to estrogen by releasing *luteinizing hormone releasing hormone (LHRH),* which causes the LH surge that produces ovulation. A noncycling (androgenized) brain fails to do so; estrogen has no effect on the release of LH (Mennin and Gorski, 1975). (Refer back to Figure 11.7 for an illustration of the triggering effect of estrogen on the LH surge.)

A few studies have found anatomical differences between androgenized and nonandrogenized brains that may be related to the differences in responsiveness to estrogen. Field and Raisman (1973) observed that more dendritic spines were present in some regions of the nonandrogenized preoptic area (a region of the brain just anterior to the hypothalamus). Similarly, Greenough, Carter, Steerman, and DeVoogd (1977) observed sex differences in the shapes of dendritic branches in the same region. Even in tissue cultures, Toran-Allerand (1976) has shown that growth of cells taken from the hypothalamus and preoptic areas is influenced by the presence of sex hormones. Cells taken from other parts of the brain were *not* affected. Whether these anatomical differences are responsible for the differences between cycling and noncycling brains, or for sex differences in behavior, is not known.

As I mentioned earlier, unlike rodent brains, male and female primate brains do not appear to differ in their response to estrogen. If a female rhesus monkey is androgenized before birth, she will show masculinized genitalia, and effects will be seen on her behavior. However, when sexual maturity is reached, she will begin to cycle; the androgenization does not prevent her brain from responding to a rise in estrogen with a surge of LHRH (Goy, 1968). Direct measurements of LH in the blood plasma have shown LH surges to injections of estrogen in both male and female rhesus monkeys (Karsch, Dierschke, and Knobil, 1973) and in castrated men (Stearns, Winter, and Faiman, 1973). Furthermore, Knobil (1974) has shown that the preoptic area does not play an essential role in this response;

surgical disconnection of the hypothalamus and preoptic area does not abolish the LH surge in response to estrogen in rhesus monkeys.

These experiments leave little doubt that androgens can alter characteristics of the brain, at least in rodents. As we shall see, the sexual behavior of female subprimate mammals is dependent on the estrous cycle, and hence the sexual behavior of a female rat with an androgenized hypothalamus is quite different from that of a normal animal. However, the same invariant relationship is not seen between the sexual behavior and menstrual cycle of primates. In the next section I shall describe the sexual behavior of mammals with estrous and menstrual cycles, and in later sections I shall relate these behaviors to neural and hormonal processes. We shall also see that there is good evidence that the behavior of humans can be affected by prenatal androgens. Presumably, these behavioral effects are produced by structural changes within the brain, but we lack direct evidence.

SEXUAL BEHAVIOR

Sexual Behavior in the Male

In order for fertilization to occur, a male mammal must emit sperm-containing semen into the female's vagina. Some male mammals are ready and willing to do so at nearly any season of the year, depending on the receptivity of the female. Seasonal breeders, like the deer, will be sexually active only at certain times of the year; during the off-season their testes regress and produce almost no testosterone.

Male sexual behavior is quite varied, although the essential features of intromission (entry of the penis into the female's vagina), pelvic thrusting (rhythmic movement of the hindquarters, causing genital friction), and ejaculation are characteristic of all male mammals. (Humans, of course, have invented all kinds of copulatory and noncopulatory sexual behavior. For example, the pelvic movements leading to ejaculation may be performed by the woman, and sex play can lead to orgasm without intromission.)

Of all the laboratory animals, the sexual behavior of rats has been studied the most. When a male rat encounters a receptive female, he will, after spending some time nuzzling her and sniffing and licking her genitals, mount her and engage in pelvic thrusting. He will mount her several times, achieving intromission on most of the mountings. After eight to fifteen intromissions approximately one minute apart (each lasting only about one quarter of a second) the male will ejaculate. At the time of ejaculation he shows a deep pelvic thrust and arches backward. The copulatory behavior of a mouse is similar, and even more dramatic. During the final intromission the male takes all four feet off the floor, climbing completely on top of the female. When he ejaculates he quivers and falls sideways to the ground. (Sometimes he takes the female with him.)

The male rat (along with many other male mammals) is most responsive to females who are "in heat." An ovariectomized female will be ignored, but an injection of estrogen will increase her sex appeal (and also change her behavior toward the male). As we shall see later, the stimuli that arouse his sexual interest are largely olfactory in nature, although visible changes, such as the swollen sex skin of a female monkey, also affect sex appeal.

Following ejaculation, the male refrains from sexual activity for a period of time (minutes, in a rat). Most mammals will return to copulate again, and again, showing a longer pause after each ejaculation. An interesting phenomenon can be demonstrated in some mammals. If the male, after finally becoming "exhausted" by repeated copulation with the same female, is presented with a new female, he begins to respond quickly—often as fast as he did with his initial contact with the first female. Successive introductions of new females can keep up his performance for prolonged periods of time. (The phenomenon, also seen in roosters, has been called the *Coolidge effect*. The following story is reputed to be true, but I cannot vouch for that fact. If it isn't true, it ought to be. Calvin Coolidge and his wife were touring a farm, when Mrs. Coolidge asked the farmer whether the continuous and vigorous sexual activity among the flock of hens was really the work of just one rooster. The reply was yes. "You might point that out to Mr. Coolidge," she said. Calvin Coolidge then asked the farmer whether a different hen was involved each time. The answer, again, was yes. "You might point that out to Mrs. Coolidge.")

The Coolidge effect is pronounced in the ram (male sheep). If a ram is given a new female after each ejaculation, he keeps up his performance (ejaculations in less than 2 minutes) with at least twelve different ewes. The experimenters, rather than the ram, got tired of shuffling sheep around (Bermant and Davidson, 1974). Figure 11.8 shows the striking difference in latency to ejaculate after reintroduction of the same female (upper curve) as opposed to introduction of new females (lower curve). (See **FIGURE 11.8** on the next page.) Beamer, Bermant, and Clegg (1969) tried to fool the rams by putting different clothing (coats and Halloween face masks) on the same female. The males were not fooled. They apparently smelled the same ewe, despite her varied disguise. (Yes, the ewes, dressed in coats and masks, looked just as ridiculous as you might imagine.)

These phenomena—a renewal of interest in sexual behavior on introduction of a new female and a good memory for females already copulated with—are undoubtedly useful for species in which a single male inseminates all the members of his harem. He thus gets around to all of his females. Other mammalian species with approximately equal numbers of reproductively active males and females are less likely to show these phenomena.

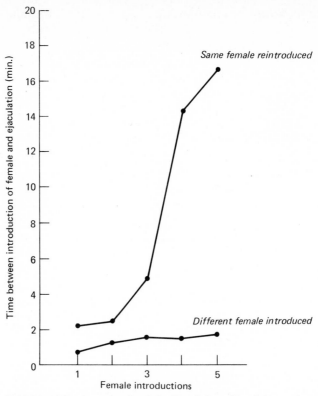

FIGURE 11.8   An example of the "Coolidge effect." (From Beamer, W., Bermant, G., and Clegg, M. *Animal Behaviour*, 1969, *17*, 706–711.)

Sexual Behavior in the Female

The mammalian female is generally described as being the passive participant in copulation. Although it is true that, in many species, her role during mounting and intromission itself is merely to assume a posture that exposes her genitals to the male (**lordosis** response), move the tail away (if she has one), and stand rigidly enough to support the weight of the male, her behavior in initiating copulation is often very active. Certainly if copulation with a nonestrous sub-primate is attempted, she will either actively flee or rebuff the male. When in a receptive state, she will often approach the male, nuzzle him, sniff his genitals and show behaviors characteristic of her species. For example, a female rat will exhibit quick, short, hopping movements and rapid ear-wiggling, which male rats find irresistible. And a human female, depending on her social history, might take very active measures to arouse a male's interest in sexual activity with her.

An ingenious set of experiments by Bermant (1961a, 1961b)

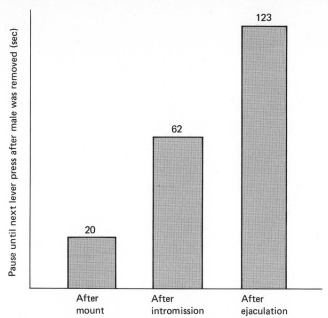

FIGURE 11.9   Time until lever press by the female rat after mounts, intromissions, and ejaculations. (Data from Bermant, G., *Science*, 1961, *133*, 1771–1773.)

tested the preferred rate of sexual contact of female rats. Males normally pause for a while after intromissions, and for a longer time after intromissions that culminate in ejaculations. What about the females? Would they also, given a choice, delay sexual contact in a similar fashion? Bermant provided female rats with a lever they could press to produce a male rat. After a mount (regardless of whether it resulted in intromission) the male was removed. As shown in Figure 11.9, the females quickly pressed the lever after the male was removed following a mount (without intromission), paused a bit more after an intromission (without ejaculation), and waited the longest time before summoning a male after an ejaculation. (See **FIGURE 11.9.**) Thus it appears that male and female rats "prefer" about the same frequency of sexual contact, and their behavior is similarly regulated by genital stimulation. What superficially looked like passivity in the female really reflected a congruence in sexual appetite with that of the male.

The sex steroids are of paramount importance to sexual behavior. We have already seen the necessity of androgens in the masculinization of a fetus, and of androgens and estrogens in sexual maturation and normal sexual functioning. In this section we shall see how these hormones regulate behavior with their organizational and activational effects.

First, let us examine the metabolic pathways for the sex steroids.

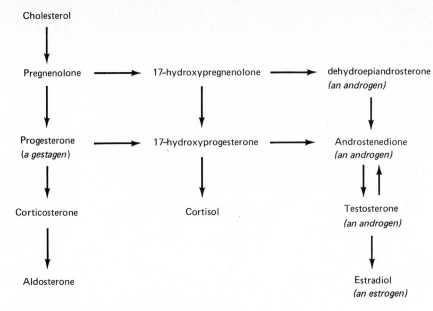

FIGURE 11.10   Biosynthesis of sex steroids.

As indicated in Figure 11.10, the precursor for biosynthesis of steroid hormones is cholesterol. The appropriate enzymes can produce from cholesterol the entire complement of sex steroids. (See **FIGURE 11.10.**) All of these hormones are produced in both males and females, but they are produced in different amounts. The adrenal glands mainly produce such nonsex steroids as corticosterone and aldosterone, but they also produce sex hormones; androstenedione, produced by the adrenal glands, is responsible for a woman's pubic and axillary hair. The testes produce androgens (mainly testosterone), but they also make a little **estradiol** (the principal human estrogen). Note that the ovaries are obliged to produce testosterone in making estradiol. (See **FIGURE 11.10.**) The blood testosterone level of women is approximately one-tenth that of men.

ORGANIZATIONAL EFFECTS OF SEX HORMONES.   A myth that should be dispelled immediately is that men and women would exchange their behavioral roles if their hormonal balances were reversed (subject, of course, to anatomical differences). Nothing of this sort would happen. A loss of testosterone, and its replacement with estrogen, would alter a man's body (as we have seen), but he would not become a woman. He would despise his enlarging breasts. He would also regret his loss of potency, but would not become interested in performing female sexual behavior. Homosexuality will *not* be produced in a heterosexual adult male by removal of testosterone and the administration of estrogen. Similarly, a heterosexual woman will not lose her sexual interest in men and want to engage in sexual activity with other women as a result of testosterone treatment. Nor will she lose her sex drive (even though men might be turned off by her beard and husky voice). In fact, she might become even more interested in sex than she was before.

The fact that sex hormones affect males and females in different ways can be attributed to the organizational effects of androgens during fetal development. You will remember that without androgens, Nature makes a female. The same is true for the behavior subsequently expressed in adulthood; prenatal androgens predispose the organisms to male behavior (regardless of genetic sex), whereas their lack predisposes toward female behavior. (I say "predispose," because we can only make probability statements about subsequent behavior. The effects of learning, as we shall see, can reverse prenatal organizational effects of androgens, especially in humans.)

Rats, mice, and hamsters are the animals of choice in studies of the organizational effects of sex steroids on later behavior, since, as I noted earlier, their "critical period" with respect to sensitivity to androgens does not end until some time after birth. The critical period ends prenatally in most other mammals, including humans. A normal, nonandrogenized female hamster will not exhibit male behavior (that is, attempts to mount an estrus female), even if her ovaries are removed and she is given injections of testosterone. However, if a female hamster is given an injection of testosterone shortly after birth, she *will* respond (with male sexual behavior) to administration of testosterone in adulthood. Similarly, male hamsters who are castrated immediately after birth (before normal androgenization takes place), will not mount females in adulthood, even if they are given replacement therapy with testosterone (Carter, Clemens, and Hoekema, 1972; Tiefer and Johnson, 1975). Thus, an adult hamster shows male sexual behavior only if it receives an androgen in the first few days of life.

Early androgenization is not only necessary for the expression of male sexual behavior; it also suppresses female sexual behavior. If a female hamster is androgenized just after birth, she fails to show lordosis responses to a normal male hamster later in life, even if she is given injections of progesterone and estrogen. Conversely, a male hamster who is castrated right after birth (preventing normal androgenization) *will* exhibit high levels of lordosis to a normal male later in life. Of course, he must be given injections of estrogen and progesterone in adulthood, to stimulate the female sexual behavior (Carter et al., 1972; Tiefer and Johnson, 1975).

The results of these studies (and many others—see Plapinger and McEwen, 1978, for a more complete list) are summarized in Table 11.2. (See **TABLE 11.2** on the following page.)

It seems most likely that the behavioral effects of early androgenization are produced in the brain, but we must not overlook the fact that early testosterone treatment also produces clitoral enlargement in females; the role of altered sensory feedback from this organ might very well play a role in the female's mounting behavior

TABLE 11.2 Organizational and activational effects of sex hormones on the sexual behavior of hamsters that were castrated or ovariectomized immediately after birth and treated with hormones at adulthood

Sex	Hormones at birth	Hormones at adulthood	Sexual behavior
Male	Estrogen or no hormone treatment	Progesterone + estrogen	Female
		Testosterone	None*
		None	None
	Testosterone	Progesterone + estrogen	None*
		Testosterone	Male
		None	None
Female	Estrogen or no hormone treatment	Progesterone + estrogen	Female
		Testosterone	None*
		None	None
	Testosterone	Progesterone + estrogen	None*
		Testosterone	Male
		None	None

*These treatments may produce weak effects but are ignored here for the sake of clarity.

(Beach, 1968; Swanson, 1971). However, some androgens can stimulate genital growth in castrated males without facilitating male sex behavior (Parrott, 1975). This observation supports the conclusion that testosterone affects male sexual behavior (at least in part) by acting directly on the brain.

Prenatal androgenization of human females. What about humans? Are we exempt from the hormonal rules that govern the sexual behavior of other mammals? As we shall see, there is good evidence for effects of prenatal androgenization on morphology and behavior in humans. There is also good evidence for partial reversal of the behavioral effects as a result of learning. A person's gender identity and gender role (i.e., self-perceived identity as a male or female, and its expression in behavior and outward appearance) can be contrary to genetic sex, or contrary to the hormonal exposure of the developing brain.

If a female human fetus is exposed to androgens, she will be born with varying degrees of clitoral hypertrophy (enlargement in size), and perhaps with some degree of labial fusion. (Remember, the scrotum and labia majora develop from the same primordia.) Such effects were seen in the cases of ten girls whose mothers received a (now-obsolete) synthetic gestagen in order to prevent miscarriage (Money and Ehrhardt, 1972). Unfortunately it was subsequently discovered that this gestagen sometimes had androgenizing effects. The

TABLE 11.3   Sex-role related behavior in fetally androgenized girls and girls with adrenogenital syndrome.* (Table adapted from Money, J., and Ehrhardt, A. *Man & Woman, Boy & Girl*. Baltimore: The Johns Hopkins University Press, 1972.)

Behavioral signs	Androgenized	Adrenogenital
Known to self and mother as tomboy	more	more
Satisfaction with female sex role	same	less
Athletic interests and skills	more	more
Preference for male versus female playmates	prefer male	prefer male
Childhood fights	same	same
Preference for slacks versus dresses	prefer slacks	prefer slacks
Interest in jewelry, perfume, and hair styling	same	same

*All comparisons are made with a matched group of control subjects.

clitoral enlargement and labial fusion were surgically corrected after birth.

The same syndrome is also seen in girls whose adrenal glands secrete abnormal amounts of androgens—the **adrenogenital syndrome.** The androgenized girls were raised and dressed as girls, and their gender identity was female. However, they described themselves as "tomboys." They preferred toy trucks and guns to dolls. They tended to choose boys, rather than other girls, as playmates. The girls did not dislike wearing dresses on special occasions, but they mostly preferred slacks and shorts. (If this study were conducted now, such a finding would cause no surprise. Almost *all* young girls— at least in North America—prefer slacks and shorts nowadays.) No special differences were seen in their childhood sex play, but since such behavior—in both males and females—is suppressed in our society, one could hardly expect otherwise. (See **TABLE 11.3.**) A follow-up study of the ten girls who were androgenized with the synthetic gestagen found no evidence of homosexuality (Mathews and Money, 1978). However, it appears that the adrenogenital syndrome may increase the incidence of bisexuality in females, although the data are not conclusive (Erhardt and Meyer-Bahlburg, 1979).

It should be noted that the behavior of none of the androgenized girls was "abnormal." Many girls prefer toy trucks and cars to dolls, and like to play baseball with boys. What would not be expected, however, is so much male-like interest from all members of a group of normal girls. One would expect to find a *range* of interests, from "masculine" to "feminine." The fact that all of the androgenized girls were from the masculine end of the scale suggests that in humans, prenatal androgens do indeed affect later behavior.

The case of the androgenized girls does not provide *conclusive* evidence that prenatal androgens have organizational effects on human behavior. The girls' parents knew that they had been affected by the drug, or by their own adrenal glands, since their genitals had been somewhat masculinized. It is possible that because of this knowledge they treated their daughters somewhat differently, and that this treatment affected their behavior. (I must confess that I would expect a parent to treat them *more*, rather than *less*, like girls, to compensate for their masculinization. However, we must still consider the issue as unproved.)

Goy and Goldfoot (1973) report about an attempt to assess the effects of prenatal androgenization on the social behavior of young female monkeys, where socialization by the parents is less likely. Testosterone was administered to pregnant monkeys, so that their female offspring would be androgenized.

The behavior of young male and female monkeys is very different—much more so than human boys and girls. Compared with females, male monkeys are much more likely to initiate periods of play, to engage in rough-and-tumble play, and to make playful threat gestures. The androgenized females acted like males. They even showed more mounting (male sexual behavior) than presentation of the hindquarters (female sexual behavior), just as males typically do. (Young monkeys engage in bisexual play; they mount and present to each other indiscriminately. However, normal females present more than they mount, and males mount more than they present.) We cannot be certain that the effects on sexual play are solely due to changes in the brain; the monkeys' clitorises were enlarged, and sensory feedback may have encouraged mounting behavior.

Other behaviors.   Early androgenization can affect behaviors other than copulatory activity. Maternal behavior (care of young pups) is easier to elicit from female or neonatally-castrated male rats than from normally androgenized males (Bridges, Zarrow, and Denenberg, 1973). Aggressive behavior in adult mice can be stimulated by testosterone administration only if the animals have been androgenized immediately after birth (Bronson and Desjardins, 1969, 1970; Barkley, 1974). Normal male rats, and females who have been androgenized soon after birth, show more timidity in leaving their home cage to explore a new environment than normal females do (Pfaff and Zigmond, 1971). The sexually dimorphic posture shown by an adult dog when urinating is also determined by exposure of the developing brain. A female dog that receives androgens at the time of birth will lift a leg and urinate like a male once she reaches adulthood (Beach, 1970). And as we saw earlier, juvenile social behavior of rhesus monkeys is affected by prenatal androgenization (Goy and Goldfoot, 1973). Thus, a variety of sex differences in behavior are determined by the organizational effects of androgens. The weight of this evidence means that we must seriously consider the possibility that human behaviors, too, might be somewhat affected by the action of androgens on the developing brain.

Prenatal estrogens and human development. Prenatal estrogens do not appear to be any more important in producing a female human than they are in producing a female in other mammalian species. Humans with only one X chromosome will develop what is called **Turner's syndrome.** Because of the missing X chromosome, these people will not develop ovaries, and hence no estrogen will be produced. Nevertheless, these women will become just as "feminine" as normal XX females; in fact, Money and Ehrhardt (1972) reported that young girls with Turner's syndrome participated in fewer childhood fights than normal girls did, and showed more interest in jewelry, perfume, and hair styling. If anything, they were more "feminine" in their behavior than XX girls.

Females with Turner's sydrome will respond well to estrogen treatment in adulthood, and will mature as normal (but sterile) women. Women with Turner's syndrome are typically short in stature, and some individuals show deficits in visual perception, so one cannot say that there are *no* effects other than the lack of ovaries.

Although fetal estrogens do not appear to be necessary for normal development of a human female (or other female mammals, for that matter), prenatal exposure to abnormally high levels of estrogens does seem to affect the subsequent behavior of human males. Yalon, Green, and Fisk (1973) investigated the behavior of 16-year-old boys whose diabetic mothers had received an estrogen during pregnancy in order to prevent spontaneous abortions. The boys were found to be somewhat less aggressive and assertive than normal boys, and tended to be less skillful in athletic activities. One case is particularly illustrative, although it is not conclusive, in itself. A diabetic mother had given birth to two boys. Her diabetes had not been recognized during the first pregnancy, and consequently she had not received estrogen therapy. However, she was treated with estrogen during the second pregnancy. The mother remarked to the experimenter that her younger son (who was exposed to exogenous estrogen) was regarded as a "sissy" by his peers, was poor at athletics, and would not fight back when other boys pushed him.

The results of the study suggest that excessively high levels of prenatal estrogen might decrease a male's degree of behavioral androgenization. It must be emphasized that the differences that were reported in this study were not very impressive statistically, and that an adequate control group of subjects could not be obtained, since almost all of the diabetic mothers in the area received estrogen treatment during pregnancy. Even if the apparent decrease in behavioral androgenization is a real effect, we do not know whether it is a direct effect of estrogen on the developing fetus, or whether the estrogen somehow competes with androgens.

Failure of androgenization in human males. Nature has performed the equivalent of the prenatal castration experiment in males (Money and Ehrhardt, 1972). Some males are insensitive to androgen **(androgenic insensitivity syndrome**—one of the more aptly named

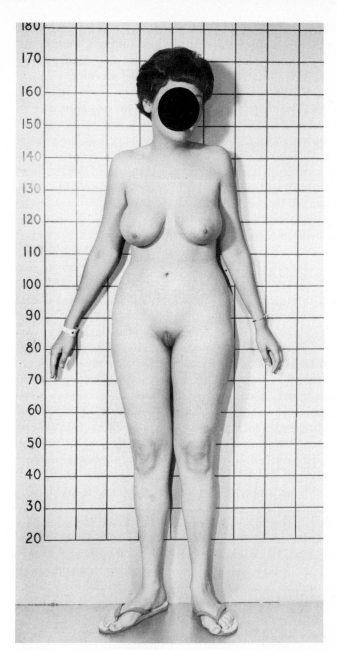

FIGURE 11.11 An XY female displaying the androgenic insensitivity syndrome. The absence of pubic hair can be explained by the person's insensitivity to androstenedione (AD). (From Money, J., and Ehrhardt, A. A., *Man & Woman, Boy & Girl.* Copyright 1973 by The Johns Hopkins University Press, Baltimore, Maryland. By permission.)

disorders). Their cells appear to lack functioning androgen receptors. Thus, even though the primordial sex organs become testes, the testosterone they produce fails to masculinize the body. There is an atrophic (undeveloped) uterus and a very shallow vagina. The lack of female internal sex organs may be explained by the effects of Müllerian-inhibiting substance produced by the testes; the Müllerian system is suppressed.

If an individual with this syndrome is raised as a girl, all is well. At puberty the body will become feminized by the small amounts of estrogen produced by the testes. (In normal males, this estrogen is counteracted by the far greater amounts of testosterone.) At adulthood the individual will function sexually as a woman although surgical enlargement of the vagina may be necessary. Women with this syndrome report average sex drives, including a normal frequency of orgasm in intercourse with a male. Homosexuality has not been reported. (We must define homosexuality carefully here. Chromosomally, copulation with a male would constitute "homosexuality," but since these people are morphologically and behaviorially female, homosexuality would entail sexual relations with another woman).

The success of testicular estrogen in producing a female body is attested to in Figure 11.11. (See **FIGURE 11.11.**) I think this photograph illustrates why it would be a tragedy to raise such an individual as a boy. Testosterone treatment at puberty would be ineffective; all that could be done would be to prevent development of the breasts. The voice would remain high, and no beard would grow. As a matter of fact, the woman in this figure lacks pubic hair because of an insensitivity to AD, the androgen that normally stimulates its growth.

The picture that emerges is this: prenatal androgens are necessary for production of a male body and a male brain, with its subsequent propensity for masculine behavior. However, as seen in the case of the prenatally androgenized females, this process is not absolute. If she is raised as a female, a partially androgenized girl will have more than the normal amount of masculine interests, but she will nevertheless perceive herself as female.

Reversal of the behavioral effects of prenatal androgenization by socialization. The power of a human's upbringing to contradict the effects of complete, natural, prenatal androgenization is seen in the following case (Money and Ehrhardt, 1972.) Identical twin boys were born to a couple and were raised normally until seven months of age, at which time one of the boys suffered accidental removal of his penis (a surgeon carelessly removed far too much tissue during an attempted circumcision). The cautery (a device that cuts tissue by means of electric current) was adjusted too high, and instead of removing the foreskin, the current burned off the entire penis. After a period of agonized indecision, the boy was (at seventeen months) subsequently raised as a girl. The first stage of plastic surgery, in creating a vagina, was performed. (The child will be given estrogen at puberty, and the final stages of surgery will be completed.) The child almost immediately responded to being treated as a girl, manifesting many behaviors typical of girls. She is neat and tidy (as opposed to her rather messy twin brother) and models her behavior on that of her mother. Since the two children are genetically identical, their differences in behavior must be attributed to the powerful effects of differential treatment of children who are perceived as boys and girls.

The effort that the parents made to encourage their daughter to become more "feminine" is attested to in the following quotes from her mother: (Money and Ehrhardt, 1972, pp. 119–120):

> I started dressing her not in dresses, but, you know, in little pink slacks and frilly blouses . . . and letting her hair grow . . . She likes for me to wipe her face. She doesn't like to be dirty, and yet my son is quite different. I can't wash his face for anything . . . She seems to be daintier. Maybe it's because I encourage it . . . One thing that really amazes me is that she is so feminine. I've never seen a little girl so neat and tidy as she can be when she wants to be . . . She is very proud of herself, when she puts on a new dress, or I set her hair. She just loves to have her hair set; she could sit under the dryer all day long to have her hair set. She just loves it.

ACTIVATIONAL EFFECTS OF SEX HORMONES IN FEMALES. So far, we have seen more similarities than differences in the general role of sex hormones on morphological and behavioral development of various mammalian species. In adulthood, however, the sex steroids have somewhat different effects in higher and lower mammals. The most obvious difference is between females who have menstrual cycles and those who have estrous cycles.

The menstrual and estrous cycles are similar, in that they are characterized by similar interactions between pituitary gonadotrophins and estrogen and progesterone from the ovarian follicle (which subsequently becomes a corpus luteum). The cycle of the female rat, like that of the woman, is characterized by follicular estrogen, which stimulates an LH surge, followed by a minor secondary estrogen rise and an increase in progesterone level. However, unlike the primate, the female rat will accept the male only at one stage of her cycle, near the time of ovulation. The hormones that are most important in controlling her sexual receptivity are estrogen and progesterone. Powers (1970) found that the duration of estrus depended principally upon the amount of estrogen, but that a high level of receptivity required that estrogen and progesterone both be present. In mammals who ovulate in response to mating (rabbits and cats, for example) an estrogen alone is necessary. In fact, in such animals, progesterone will have an inhibiting effect on mating behavior if it is given to an ovariectomized, estrogen-treated female just prior to mating tests (Young, 1961).

The sexual behavior of primates is not regulated in this way by the ovarian hormones. However, that is not to say that there are no hormonal effects. Female monkeys show more interest in sexual activity during midcycle (around the time of ovulation). In addition, many women report preferences for intercourse during different portions of the menstrual cycle, but no strong cyclicity in sexual desire can be generalized to all women. Early studies found a peak of sexual activity at midcycle (around the time of ovulation) and immediately before and after menstruation (Davis, 1929; Hart, 1960; Udry and Morris, 1968). These results suggested that estrogen (which peaks around the time of ovulation) might facilitate a woman's sex drive (or, conceivably, might make her appear more attractive to a man). The peak around the time of menstruation might be produced by AD, whose effects were previously suppressed by estrogen and progesterone. Consistent with this hypothesis was the claim that many women reported that their sexual desire was more of a passive need to be made love *to* at midcycle, whereas just before menstruation they felt more aggressive and were inclined to take the initiative in sexual activity (Money and Ehrhardt, 1972).

Later studies have cast doubt on these findings. James (1971) found no evidence for a midcycle peak, but *did* find one just after menstruation. Spitz, Gold, and Adams (1975) confirmed these results, and suggested that reluctance for some couples to engage in intercourse during menstruation could account for an increase in sexual activity once the menstrual flow was over. (Similarly, a couple could decide to make love just before the woman's period started, since they would not wish to do so for several days once the flow began.) Udry and Morris (1977) found that the differences in the length of women's menstrual cycles, and in the precise time of ovulation, made it difficult to analyze data with any assurance of being correct. They concluded that very different results could be obtained from the same data, depending on the methods of analysis that were used. Thus, we are a long way from solving this very interesting problem.

The fact that female primates do not show the cyclic changes in sexual receptivity that characterize mammals with estrous cycles suggests that their receptivity is not particularly dependent on estrogen. Indeed, this appears to be the case. Ovariectomy (carried out all too often in the course of hysterectomy) does not abolish a woman's libido. (Nor does menopause, Nature's own "ovariectomy.") However, vaginal dryness and other physiological changes may occur, which might produce indirect effects on a woman's interest in sexual activity. For some menopausal women, estrogen administration restores interest in sexual activity that had previously been lost because of these side effects (Bakke, 1965). The other ovarian sex hormone, progesterone, might have an inhibitory effect on sexual desire (Kane, 1968; McCauley and Ehrhardt, 1976). For this reason, some women report that a decrease in sexual desire accompanies the use of birth control pills (many of which contain some form of synthetic gestagen). However, for most women the release from the fear of pregnancy facilitates sexual desire much more than any direct suppressive effects that gestagens may have.

Another difference between women and animals with estrous cycles is indicated by their contrasting reactions to androgens. Testosterone produces a small increase in malelike mounting behavior in female rats, even if they were not androgenized early in life. However, androgens do not stimulate *male* sexual behavior in women. On the contrary, these hormones appear to increase heterosexual desire. Removal of the adrenal glands (which secrete an androgen—androstenedione—in both men and women) appears to decrease libido in women, even though ovariectomy does not (Waxenberg, Drellich, and Sutherland, 1959). However, the adrenalectomies were performed for treatment of cancer, so it is difficult to draw definitive conclusions.

More recent research (Bancroft and Skakkebaek, 1978) has supported the suggestion that women's sex drive responds to androgens, but the data are not compelling. In one study, the authors found that testosterone, but not a tranquilizer, increased sexual interest in women who had complained of a lack of responsiveness. A second study investigated women who reported a loss of sexual interest once they began taking oral contraceptives. These women were given either testosterone pills or a placebo to take with their oral contraceptives. No significant differences were found. It is possible that the contraceptive hormones may have blocked any effects that the testosterone might have had, so this study does not rule out a role for androgens in stimulating a woman's sexual interest. We must conclude that the role of sex hormones in human female sexual behavior is very much an open question.

A considerable amount of research has been performed with rhesus macaques, a common species of laboratory monkey. In general, results have confirmed a role for androgens in the sexual behavior of this species. Everitt, Herbert, and Hamer (1972) found that removal of the adrenal glands decreased the sexual interest of female rhesus macaques who had previously been ovariectomized. The effect was seen most strikingly in the animals' soliciting behavior. A smaller effect was seen on receptivity—the animal's willingness to engage in sexual activity with a male who initiated the behavior. Administration of testosterone reinstituted these behaviors to normal levels. A subsequent study (Everitt and Herbert, 1975) showed that the effective site of action of the testosterone was in the anterior hypothalamus. Intracerebral implants of the hormone were effective when they were placed in this region, but did not affect the behavior of adrenalectomized and ovariectomized monkeys when they were placed in the posterior hypothalamus, thalamus, or midbrain.

Unfortunately for those who prefer consistency in experimental findings, a subsequent study found that the results observed in rhesus macaques cannot even be generalized to another species of monkeys, stumptailed macaques. Baum, Slob, de Jong, and Westbroek

(1978) found that adrenalectomy had no significant effect on either sexual receptivity or soliciting behavior of these monkeys. Thus, it is not surprising that subsequent replacement therapy with testosterone had no effect. The authors suggest that some of the discrepancy might be accounted for in differences in the soliciting behavior of the two species of macaques, but it is clear that we cannot hope to settle the issue of whether hormones affect the sex drive of human females by studying nonhuman primates. If we cannot readily generalize between two species of macaques, we certainly cannot generalize from macaque to human.

ACTIVATIONAL EFFECTS OF SEX HORMONES IN MALES. Although women and female rodents are very different in their responsiveness to sex hormones, men and male rodents (and other mammals, for that matter) appear to resemble each other in their responsiveness to testosterone. With normal levels they are potent and fertile; without testosterone they become sterile (very quickly) and then impotent. Much emphasis has been placed on man's "emancipation from hormones." It is often said that humans, with their "high degree of corticalization" do not require hormones for sexual performance. These statements are not supported by the facts, however; they appear to result more from the typically human desire to see ourselves as not quite so dependent on our "biology" as other animals are.

The decline in copulatory ability after castration varies considerably among individuals, even of the same species. Most rats cease to copulate within a few weeks, but some retain this ability for up to five months (Davidson, 1966b). Since the average life-span of a rat is a little over two years, this performance compares favorably with that of castrated humans, taking the different life-spans into account. As reported by Money and Ehrhardt (1972), some men lose potency immediately, whereas others show a slow, gradual decline over the years. The fact that the changes are a result of the loss of testosterone is shown by the rapid return of libido and potency after the male hormone, and not a placebo, is administered.

Prior experience has an effect on the decline, at least in some mammals; Rosenblatt and Aronson (1958a,b) found that high levels of sexual activity before castration significantly prolonged subsequent potency in cats.

There are individual fluctuations in testosterone level, as was nicely demonstrated by a researcher who was stationed on a remote island (Anonymous, 1970). Just before he left for visits to London (and to female company) his beard began growing faster. Anticipation of sexual activity stimulated testosterone production. Similarly, stress (during wartime) can lower testosterone levels (Rose, Bourne, Poe, Mougey, Collins, and Mason, 1969). It is difficult, however, to obtain a relationship between testosterone level and sexual behavior. Male sexual behavior appears to be essentially an all-or-none phenomenon; a given male will exhibit a particular level of behavior if he has sufficient testosterone (usually considerably less than the amount normally present in his blood). Extra testosterone has little effect on his sexual behavior (Young, 1961; Bermant and Davidson, 1974).

Homosexuality

It is impossible to say for certain why some people become homosexual while others become heterosexual. Homosexual behavior (male and female) has been seen in almost all mammalian species that have been studied. Homosexual episodes are often seen in humans (especially males) who are essentially heterosexual. In some societies, homosexual behavior is the norm during adolescence, followed by marriage and a normal heterosexual relationship (Money and Ehrhardt, 1972). However, some people become exclusively homosexual in desire and practice (*obligative,* as opposed to *facultative,* homosexuality). It is this condition which has received considerable study.

Attempts have been made to relate hormone levels with male homosexuality, but the results of these studies are mixed (Bancroft, 1978). If homosexuality has a physiological basis, it is likely to be a more subtle difference in brain structure caused by deficiencies in prenatal androgenization. There is no doubt about the ability of the organizational effects of androgens to "bias" the later sexual proclivities of many species of animals. Female rats who are androgenized just after birth will be much more likely to mount other receptive females. Similarly, neonatally castrated male rats can be made to act like females, especially if they are given estrogen and progesterone treatment as adults.

More subtle effects can be seen, also. Clemens (1971) found that the probability of malelike mounting was highest in female rats that shared their mother's uterus with several brothers; fewer brothers resulted in less male sexual behavior. Presumably, the females were partially androgenized by their brothers' testosterone. Since most humans do not have any company in the uterus, this factor apparently is not of much importance in human female homosexuality.

As we saw earlier, there is evidence that suggests a slightly increased incidence of bisexuality in women who are affected with the adrenogenital syndrome. These women received an increased dose of androgens from their own adrenal glands before birth, which may have caused some androgenization of their brain. The bisexuality may be a manifestation of this androgenization. However, I must emphasize that the data are not conclusive.

There are also factors that might suppress the degree of androgenization of male fetuses. Ward (1972) put pregnant rats under stressful conditions (a bright light). The stress increased the mothers' secretion of nonsex steroid hormones, which presumably suppressed androgen production in the male fetuses. The male rats born to the stressed mothers had smaller external genitalia as adults and were deficient in their male sexual behavior. They were also more responsive to treatment with estrogen and progesterone, showing more lordosis than is seen in a normal male given these hormones.

It is impossible to say whether results such as those of Ward are relevant to human male homosexuality. The fact remains that genetic males and females, androgenized or not, can be successfully raised as members of either sex (with plastic surgery early in life and hormonal administration later, if necessary). These people will typically be attracted to members of the sex opposite to that of their upbringing; the most important factor in their development is that both parents unambiguously treat them as male or female and provide them with a suitable image upon which to model their own sexual identity (Money and Ehrhardt, 1972).

Neural Mechanisms Controlling Sexual Behavior

SPINAL MECHANISMS. Genital stimulation can elicit sexual movements and postures in female cats and rats even after transection of the spinal cord below the brain (Beach, 1967; Hart 1969). Thus, at least some elements of sexual behavior are organized at the level of the spinal cord.

The reflex mechanisms responsible for control of penile erection and ejaculation are located in the spinal cord of the male animal. Hart (1967a) severed the spinal cords of dogs and observed not only erection and ejaculaton, but also the characteristic pelvic thrusting, leg kicking, and arching back. A refractory period of 5 to 30 minutes followed ejaculations, during which time the animal was unresponsive. This contrasts with sexual behavior in the intact dog, which typically shows a pause of more than 30 minutes before sexual activity is resumed. Thus, the spinal cord is ready for more before the brain is.

Hart obtained evidence that the brain inhibits spinal sexual reflexes. He found that intense ejaculatory reactions, which can easily be obtained by means of mechanical stimulation of the penis of a dog with a spinal transection, cannot be produced in a normal intact dog unless a receptive bitch is present. The odor and sight of a receptive female apparently disinhibit the cerebral mechanisms that normally prevent the expression of the intense ejaculatory response. (See **FIGURE 11.12** on the following page.)

Emission of semen, as we have seen, is produced by a spinal reflex. Humans with spinal damage occasionally produce an erection and ejaculate. They have even become fathers, by artificially inseminating their wives with semen obtained by mechanical stimulation

Dog with severed spinal cord

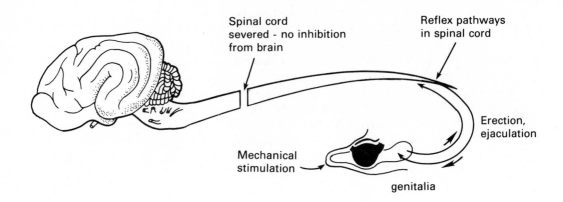

Spinal cord
severed - no inhibition
from brain

Reflex pathways
in spinal cord

Erection,
ejaculation

Mechanical
stimulation

genitalia

Normal dog

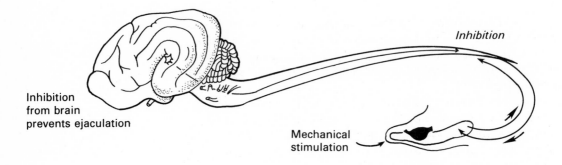

Inhibition

Inhibition
from brain
prevents ejaculation

Mechanical
stimulation

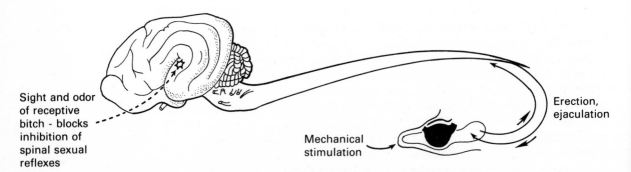

Sight and odor
of receptive
bitch - blocks
inhibition of
spinal sexual
reflexes

Mechanical
stimulation

Erection,
ejaculation

FIGURE 11.12   A schematic explanation of the experiment by Hart.

(Hart, 1978). They do not experience an orgasm as a result, and they will be unaware of the erection and ejaculation unless they see it happening. However, they do occasionally experience a "phantom erection" along with an orgasm, despite penile quiescence (Money, 1960; Comarr, 1970). Nothing is happening to their genitals or internal sex organs, but the spontaneous activity of various brain mechanisms gives rise to feelings of arousal and orgasm.

BRAIN MECHANISMS.   Penile erection and emission of semen appear to be controlled by different mechanisms. Herberg (1963) found that electrical stimulation of parts of the **medial forebrain bundle** (a fiber system running in a rostral-caudal direction through the lateral hypothalamus) resulted in slow emission of semen, without penile erection, in rats. The electrical stimulation was also reinforcing; the rats would press a lever to turn on the current. The nature of the reinforcing effect is unclear. One cannot assume that the rats pressed the lever because the emission was reinforcing (as an orgasm might be), since rats will press a lever for electrical stimulation of other hypothalamic areas even when such stimulation does not result in emission (see Chapter 17).

The neural circuitry involved in erection and ejaculation in the squirrel monkey was studied by MacLean and his coworkers. Locations were found all through the central nervous system, from the frontal lobe to the medulla, where electrical stimulation would produce erection and ejaculation. Many locations in the limbic system also give positive results; the investigators found locations where stimulation would produce signs of fear followed by an erection (as an apparent rebound phenomenon) when the stimulation was turned off. (In monkeys, penile erection is often used as a threat gesture that is directed toward other monkeys, so the occurrence of a penile erection does not necessarily indicate that the brain region being stimulated is involved in sexual arousal.) The major neural system involved in penile erection appears to travel from the dorsomedial nucleus of the thalamus (which communicates with prefrontal cortex) to the hypothalamus to the ventral tegmentum to the anterior medulla (MacLean, Dua, and Denniston, 1963; Dua and MacLean, 1964).

Increases in male sexual behavior can be produced by stimulation in the vicinity of the **preoptic area** (just anterior to the hypothalamus) in rats (Van Dis and Larsson, 1971). Conversely, large lesions of the preoptic area abolish sexual behavior in male rats (Heimer and Larsson, 1966/1967). Testosterone therapy does not restore this behavior. Female lordosis is not abolished by these lesions, but is instead facilitated. However, it is interesting to note that the occasional mounting behavior seen in normal females disappears after lesions of the preoptic region (Singer, 1968).

Female sexual behavior is disrupted by lesions posterior to the preoptic area, in the anterior hypothalamus (Kalra and Sawyer,

TABLE 11.4   Effects of some monoaminergic agonists and antagonists on sexual behavior of rats

Neurotransmitter	Action of drug	Sex of animal	Effect
Serotonin	agonist	male	inhibition
		female	inhibition
	antagonist	male	facilitation
		female	facilitation
Dopamine	agonist	male	facilitation
		female	inhibition
	antagonist	male	inhibition
		female	facilitation

1970), or in the ventromedial region (Law and Meagher, 1958). Male sexual behavior is not affected by anterior hypothalamic lesions (Paxinos and Bindra, 1972).

Attempts have been made to determine whether the various systems of monoaminergic neurons (NE, DA, and 5-HT) that have been revealed by the histofluorescence technique might play a role in sexual behavior. Their participation seems likely. Serotonin (5-HT) agonists tend to inhibit male sexual behavior in rats, whereas dopamine agonists facilitate it (Myerson and Malmnäs, 1978). In contrast, serotonin antagonists facilitate male sexual behavior and dopamine antagonists reduce it. In female rats, *both* serotonergic and dopaminergic neurons appear to inhibit sexual behavior; antagonists for these transmitter substances facilitate lordosis, while agonists inhibit it (Myerson and Malmnäs, 1978). The role of NE is uncertain. (See **TABLE 11.4.**)

Attempts have been made to destroy particular aminergic pathways to assess their specific roles in sexual behavior, but the results are still confusing and somewhat contradictory (see Everitt, 1978, for a general review).

Neocortex obviously plays an important role in sexual behavior, since learning and sensorimotor integration affect sexual activity to a very large extent. Large neocortical lesions seriously impair copulatory behavior of male rats (Larsson, 1964). However, less of an effect is seen when these are produced in females. Indeed, lordosis appears to be enhanced by neocortical removal (Beach, 1944).

The disparity in the role of cortex in male and female sexual activity probably can be accounted for by the different behaviors that are performed by males and females. And, in part, the difference seems to be a function of the behaviors that are measured by the experimenters. In order for a male rat to copulate successfully, he must find the female, determine that she is receptive, approach her, follow her around until she holds still, mount her, insert his penis, and so forth. These behaviors are complex, and require a considerable amount of sensory and integrative capabilities. Thus, neocortical lesions would be expected to impair male sexual behavior. In contrast, a female rat can remain relatively passive, merely exhibiting a spinal reflex (lordosis), and copulation can successfully occur. Since active participation is not required by the female, it is not surprising that neocortical lesions in female rats do not prevent copulation.

However, just because female rats do not need to engage in complex behaviors for copulation to occur, we cannot conclude that complex behaviors do not occur. As I mentioned earlier, estrous female rats normally entice males with hopping movements, ear-wiggling, and other seductive behaviors. Studies that have looked at the female rats as more than passive objects of the more interesting behavior of the male have found that female rodents engage in a surprisingly complex set of behaviors (McClintock, 1978). It is very likely that neocortical lesions would impair the performance of these behaviors, even though the brain damage does not prevent the occurrence of the lordosis response in an otherwise-incapacitated animal.

The temporal lobes of the brain appear to play a role in the modulation of sexual arousal, and in its direction toward an appropriate goal object, especially in higher mammals. We have already seen (in Chapter 9) that temporal cortex plays a role in the visual identification of the significance of objects. (Monkeys with temporal lobe damage can "see" objects; they can orient in space quite normally and can pick up small objects. However, they cannot visually discriminate nuts and bolts from raisins and other small pieces of food. They must put everything into the mouth first, rejecting inedible objects and chewing and swallowing the edible ones.) Temporal lobe damage appears to impair an animal's ability to choose an appropriate sex object; male cats with these brain lesions have been reported to attempt copulation with everything in sight—the experimenter, a teddy bear, furniture—everything remotely mountable (Schreiner and Kling, 1956; Green, Clemente, and DeGroot, 1957).

In humans, temporal lobe dysfunctions are often correlated with decreased sex drives. For example, focal epilepsy (brain seizures that originate from localized, irritative lesions) of the temporal lobes is sometimes associated with impotence (Hierons and Saunders, 1966). Surgical removal of the affected tissue sometimes leads to hypersexuality (Blumer, 1970). Temporal lobe dysfunction is also occasionally found in people with bizarre fetishes. A particularly striking example was that of a man with a safety pin fetish. He suffered from a compulsion to gaze at a safety pin, and the staring would then trigger a seizure. After the seizure he would often dress in his wife's clothing. Surgery corrected the epilepsy, and along with it went the sexual aberrations (Mitchell, Falconer, and Hill, 1954). A similar case was reported by Hunter, Logue, and McMenemy (1963). Surgical removal of a patient's diseased temporal lobe eliminated his prior epilepsy and transvestitism.

Kolářský, Freund, Machek, and Polák (1967) examined the cases of men with sexual disorders and found a strong correlation between the disorders and actual or presumptive temporal lobe damage, especially if the damage occurred early in life.

It is difficult to provide a clear and concise summary of the results of study of brain mechanisms involved in sexual behavior. It is safe to say that the spinal cord contains the circuitry necessary for the basic sexual reflexes: lordosis, erection, ejaculation, pelvic thrusting, etc. The brain (particularly, the preoptic area in males, and the anterior and ventromedial hypothalamus in females) regulates these reflexes, removing them from inhibitory control under the appropriate circumstances. The activity of dopaminergic neurons appears to facilitate male sexual behavior, and inhibit female sexual behavior. Serotonergic neurons depress the sexual behavior of both males and females. The temporal lobe appears to be particularly important in the analysis of what circumstances are appropriate for sexual activity.

The limbic system plays an important role in the regulation of sexual behavior, but it is impossible, without much more study, to assess just what the role is. For example, a lesion in a particular portion of the limbic system might increase sexual behavior not by affecting circuitry directly involved in copulatory behavior, but by decreasing the expression of some state (fear, for example) that normally competes with the expression of sexual behavior.

Neuroendocrine Control of Sexual Behavior

The final section of this chapter will consider the ways in which sex hormones produce their organizational and activational effects on the nervous system.

MECHANISM OF PRENATAL ANDROGENIZATION.   Steroid hormones exert their effects by entering cells, attaching to receptors within the cytoplasm, and then being transported to the nucleus, where they turn genes on or off, thus altering the production of proteins. As we have seen, androgens are responsible for the induction of masculine physical and behavioral traits. Without androgens, the developing organism becomes a female. It is a bit of Nature's irony that androgenization appears to be accomplished by *estrogens*.

Why would one ever suspect that estrogens might be responsible for androgenization? An early study (Wilson, Hamilton, and Young, 1941) found that if estrogens were administered to female rats immediately after birth, the physical and behavioral effects resembled those produced by androgens. Later studies have confirmed these findings (see Plapinger and McEwen, 1978, for a review). The most consistent effects are seen in production of anovulatory sterility; in adulthood, the females do not respond to increases in estrogen with a surge of LH.

By now, you are undoubtedly confused. Has it not just been

established that *androgens* produce masculinization? When we look at a little more biochemistry, the apparent contradictions will be resolved. Figure 11.13 shows that some androgens (testosterone and AD) can be converted to estrogens. The conversion process is called **aromatization**; aromatic compounds (such as estrogens) contain a benzene ring in their molecular structure. Estradiol and estrone are both estrogens. (See **FIGURE 11.13.**)

Aromatization of androgens to estrogens has been shown to take place in the brain (Naftolin, Ryan, Davies, Reddy, Flores, Petro, and Kuhn, 1975). Therefore, it is possible that the organizational effects of testosterone could actually be produced by estrogens. In support of this suggestion, administration of aromatizable androgens (such as testosterone and AD) early in life produces anovulatory sterility

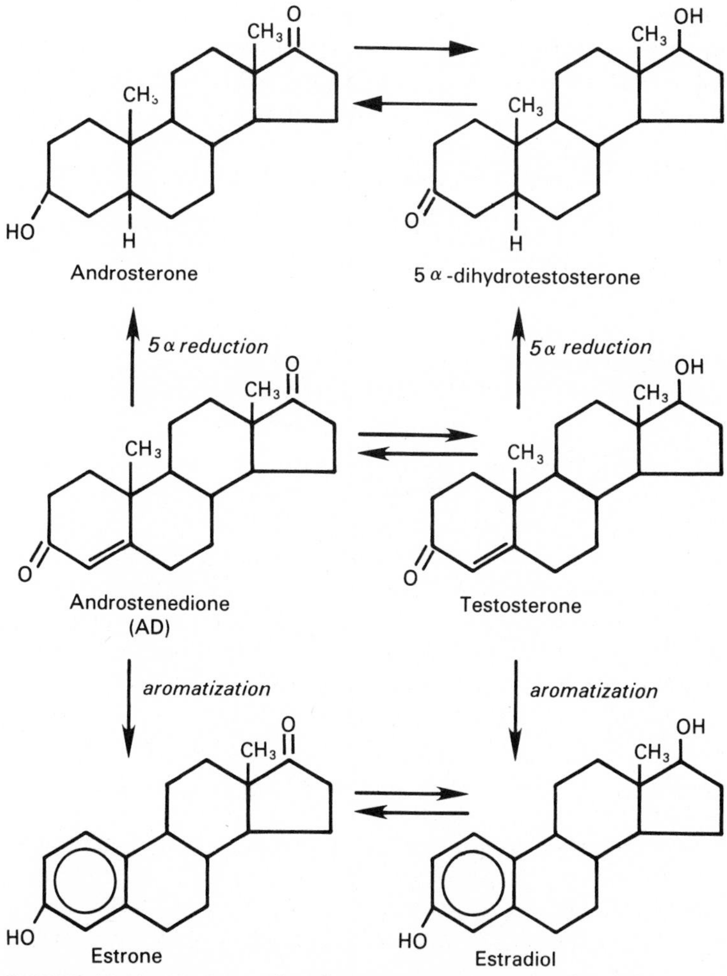

FIGURE 11.13 Biochemical pathways: aromatization and 5-alpha reduction of testosterone and androstenedione.

in females; the rats are defeminized. However, nonaromatizable androgens (such as *dihydrotestosterone* and *androsterone*) do not have this effect (McDonald and Doughty, 1974). Figure 11.13 shows why dihydrotestosterone and androsterone cannot be aromatized. These hormones are *5-α reduced* forms of testosterone and AD. Note the single-ended arrows, which indicate that 5-α reduction is not reversible. (See **FIGURE 11.13** on the preceding page.)

An even more convincing piece of evidence is the finding that masculinization of the male rat brain is prevented by estrogen antagonists or by drugs that block the aromatization of testosterone to estrogen (Lieberburg, Wallach, and McEwen, 1977). This indicates that testosterone by itself cannot masculinize the brain.

The results cited in the preceding two paragraphs refer to the defeminization of the brain with respect to anovulatory sterility; that is, failure of the brain to produce an LH surge in response to estrogen, as the female rat brain normally does. However, the active agent in organizing masculine sex behavior is less certain. McDonald and Doughty (1974) found that dihydrotestosterone (which is not aromatizable to estrogen) does not organize male-like behavior, whereas testosterone (which *is* aromatizable) does. Estrogen itself

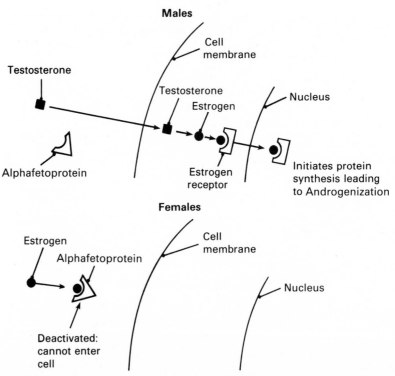

FIGURE 11.14 Intracellular aromatization of testosterone and extracellular deactivation of estradiol by alphafetoprotein.

can produce similar effects (Doughty, Booth, McDonald, and Parrott, 1975). However, androsterone, which is *not* aromatizable, also organized masculine sex behavior. These contradictory results indicate that the issue is not settled.

You are undoubtedly wondering, if estrogen is the effective agent in androgenization of at least some masculine characteristics, why all female fetuses are not also masculinized. After all, they possess ovaries, and their mothers are producing estrogens (which do, indeed, enter the fetal blood supply). The answer appears to be that they are protected by a special protein that is present only in the blood of fetuses. This substance, **alphafetoprotein,** has the ability to bind with estrogens, thus deactivating them (Raynaud, Mercier-Bodard, and Baulieu, 1971; Aussel, Uriel, and Mercier-Bodard, 1973). It was thus suggested that alphafetoprotein prevents a female fetus from being masculinized by the estrogen that is naturally present. Androgens, which are *not* bound by alphafetoprotein, enter the cells readily, where they can be aromatized. (Soloff, Morrison, and Swartz, 1972; Plapinger and McEwen, 1978). (See **FIGURE 11.14.**)

Presumably, a sufficient amount of alphafetoprotein is present in the fetal blood supply to prevent naturally-occurring amounts of estrogen from entering the cells. Experimental injection of large amounts of estrogens exceed the capacity of the alphafetoprotein, and estrogens enter the cells, where they produce their masculinizing effects. In support of this hypothesis, it was found that synthetic estrogens that are not bound by alphafetoprotein are able to produce masculinizing effects at much lower doses than natural estrogens, which *are* bound by alphafetoprotein (Raynaud, 1973). In fact, natural estrogens are less effective when levels of alphafetoprotein are high, and most effective when the levels decline. However, the effectiveness of the *synthetic* estrogens is not influenced by the amount of alphafetoprotein that is present in the fetus's blood.

The mechanism that has been outlined in this section may ex-

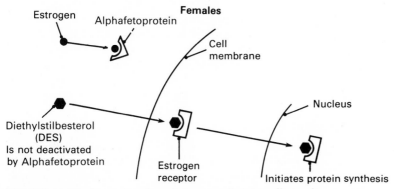

FIGURE 11.15   An explanation of the masculinizing effects of diethylstilbesterol.

plain at least some of the adverse effects of a synthetic estrogen that is not bound by alphafetoprotein: *diethylstilbesterol (DES)*, which was formerly used in an attempt to prevent spontaneous abortions in women who had previously miscarried. Unfortunately, the female offspring of these women had a rather high incidence of vaginal cancer later in life (Herbst, Kurman, Scully, and Poskanzer, 1972). More important for our purposes, some of them showed signs of masculinization, such as enlarged clitorises (Bongiovanni, DiGeorge, and Grumbach, 1959). Presumably, masculinization was produced because DES is not bound by alphafetoprotein. (See **FIGURE 11.15** on the preceding page.)

ACTIVATIONAL EFFECTS OF SEX HORMONES ON THE NERVOUS SYSTEM.   The spinal cord contains the first level of control of sexual behavior. As we saw earlier, male and female sexual behavior can be elicited by genital stimulation from animals whose spinal cords have been severed. Estrogen and testosterone both have direct effects on neurons of the spinal cord. The appropriate hormone facilitates sexual reflexes in males or females, whether administered systemically, or implanted directly into the spinal cord itself (Hart, 1978). Stumpf and Sar (1976) and Sar and Stumpf (1977) obtained direct evidence, from autoradiographic studies, that spinal cord neurons take up these hormones selectively, so their stimulating effect, when administered experimentally, appears to be a physiological one. Electrical recording of the activity of the pudendal nerve (which conveys somatosensory information from the genital area to the spinal cord) has shown that estrogen increases the sensitivity of this region of skin to tactile stimuli (Komisaruk, Adler, and Hutchison, 1972). Thus, estrogen appears to affect sensitivity to sexual stimuli at the most peripheral level. It is not known how this effect of estrogen is produced.

The effects of steroid hormones on the brain are undoubtedly more important to sex behavior than their effects on the spinal cord. In recent years, many studies have been devoted to locating the regions of the brain that contain neurons that possess receptors for the important steroid hormones: estrogens, androgens, and gestagens. Earlier studies have relied on autoradiographic methods to locate neurons that take up the sex hormones. (See Pfaff and Keiner, 1973; McEwen, 1978; Plapinger and McEwen, 1978, for reviews.) Radioactive hormones are injected into the animal, and a photographic emulsion is used to determine the whereabouts of radioactive cells in slices of the brain. These studies tell us the location of neurons that have an affinity for the hormones. Subsequent studies have confirmed that these regions contain neurons with cytoplasmic receptors that bind specifically to these hormones. (See McEwen, 1978, for a review.) The identification of progesterone receptors has been a problem until recently. Progesterone binds with cytoplasmic receptors, travels to the nucleus, initiates protein synthesis, and then rapidly disassociates from the receptors. The dissociation process is so rapid that by the time investigators extract receptors from the cells, the radioactive progesterone they have injected has split away from the receptors. Fortunately, a synthetic progesterone, R5020, binds with progesterone receptors more tightly than natural progesterone does. Previous suggestions that progesterone does not interact with specific receptors have been dismissed; there are receptors for this hormone, just as there are for androgens and estrogens (MacLusky and McEwen, 1978; Blaustein and Feder, 1979).

The distribution of cells that take up estrogens and testosterone is very similar. The most important regions (in the rat) are the basomedial hypothalamus (near the attachment of the stalk of the pituitary gland), the medial preoptic area, and amygdala. These hormones are also taken up to a much lesser extent by neurons in the midbrain region (*periaqueductal gray*), hippocampus, and septum. You will recall that testosterone can be aromatized to estrogen; in fact, much of the radioactivity that is found in the brain after an injection of labelled testosterone is actually coming from estrogen, bound with estrogen receptors (Lieberburg and McEwen, 1977).

Progesterone is taken up by neurons in many parts of the brain: hypothalamus, preoptic area, periaqueductal gray, neocortex, amygdala, hippocampus, basal ganglia, and cerebellum (MacLusky and McEwen, 1978).

The interesting question, of course, is what happens to an animal's behavior when the sex steroids are taken up by neurons in various brain regions. It appears that the primary effects of testosterone and estrogen on sexual behavior are produced by uptake in the ventromedial hypothalamus and preoptic regions. A very small quantity of estrogen placed directly in to these brain regions will stimulate sexual receptivity in an ovariectomized female rat (Malsbury and Pfaff, 1973; Davis, McEwen, and Pfaff, 1979). If estrogen is implanted into the anterior hypothalamus or preoptic area, it will stimulate a rat's activity in a running wheel, and if it is implanted into the ventromedial hypothalamus, it will decrease food intake (Wade and Zucker, 1970), just as systemic administrations do. Anterior hypothalamic implants of testosterone will restore male sexual behavior in rats who are castrated after puberty (Johnston and Davidson, 1972). Progesterone, implanted into the midbrain, will facilitate the effects of estrogen injections in ovariectomized female rats (Ross, Claybough, Clemens, and Gorski, 1971). No effects were observed when progesterone was implanted in various locations in the hypothalamus and preoptic area, although Powers (1972) found that basomedial hypothalamic implants had a facilitatory effect. A review of the literature by McEwen, Davis, Parsons, and Pfaff (1979) concluded that the locations of the effective sites of action of progesterone are still uncertain.

As we saw earlier, the sexual interest of ovariectomized, adrenalectomized female rhesus monkeys is facilitated by injections or brain implants of testosterone. The anterior hypothalamus was found to be an effective site, but the female rhesus macaque brain has not been extensively mapped for sensitivity to testosterone. The significance of these findings is obscured by the fact that females of another species of macaque monkey do not appear to be responsive to testosterone.

It must be noted that steroid hormones diffuse very quickly through brain tissue, so the precise location of neurons that mediate the effects of these hormones cannot be determined from implant studies alone. A combination of implant, lesion, and radioactive tracer studies will be needed to determine the sites of action of the sex steroids. For example, Davis, McEwen, and Pfaff (1979) implanted radioactive estrogen into the rat brain, and measured radioactivity of various regions of the brain after the animals' behavior had been assessed. They found that the hormone had not diffused very far from the sites of implantation, and were even able to estimate that only 4 percent of the estrogen receptors in the hypothalamus were occupied by the labelled estrogen. Thus, the behavioral effects that they observed had been triggered by stimulation of a very limited number of cells.

We know a phenomenal amount about the neural and hormonal control of sexual behavior, and we are learning more each year. The basis for sexual dimorphism—namely, early androgenization—has been determined. The role of aromatization in the mediation of many of the effects of testosterone has been discovered. Techniques are now available for locating specific receptors for the various sex hormones. The considerable attention that physiological psychologists and neuroendrocrinologists are giving to this area of inquiry promises to produce more interesting information in future years.

KEY TERMS

activational effect p. 316

adrenogenital syndrome (*a dree no GEN i tul*) p. 337

alphafetoprotein p. 355

5-α reduction p. 354

androgen (*AN dro jen*) p. 321

androgenic insensitivity syndrome p. 339

androstenedione (*an dro steen DY ohn*) p. 324

androsterone (*an DRAHSS tur ohn*) p. 354

aromatization (*air o mat i ZA shun*) p. 353

corpus luteum (*LEW tee um*) p. 326

diethylstilbesterol (DES) p. 356

dihydrotestosterone p. 354

estradiol (*ess tra DY all*) p. 334

estrogen p. 323

estrous cycle p. 326

facultative homosexuality (*fa KUL tuh tiv*) p. 346

follicle p. 323

follicle-stimulating hormone (FSH) p. 323

gamete (*GAM eet*) p. 317

gestagen p. 326

gonadotrophic hormone p. 323

gonadotrophic releasing hormone p. 323

hermaphrodite (*her MAFF row dite*) p. 324

karyotype (*KAIR ee o type*) p. 318

lordosis p. 332

SUGGESTED READINGS

BERMANT, G., AND DAVIDSON, J. M. *Biological Bases of Sexual Behavior.* New York: Harper & Row, 1974.

MONEY, J., AND EHRHARDT, A. A. *Man & Woman, Boy & Girl.* Baltimore: The Johns Hopkins University Press, 1972.

These two books are very well written. Readers of this book should have no trouble understanding them. Although some of the information that they contain is getting out of date, the books are still excellent for the basics.

HUTCHISON, J. B. (Ed.) *Biological Determinants of Sexual Behavior.* Chichester: Wiley & Sons, 1978.

The information in this volume, which contains chapters by various experts in the field, is much more current, but is heavier going. You will also find articles in most issues of *Annual Review of Neuroscience* that summarize the most recent data.

12

Regulation and the Control of Food Intake

"The constancy of the internal milieu is a necessary condition for a free life" is the way Claude Bernard put it, back in the late nineteenth century. This famous quote very succinctly states what organisms must do if they intend to roam about in environments hostile to the living cells that compose them (i.e., live a "free life")—they must regulate the internal fluid that bathes their cells (the so-called *interstitial fluid*).

Our cells require very precise control of their environment. The evolutionary process has not produced "hardier" cells; instead, it has produced the means by which the cells' environment can be regulated. The proper concentration of solutes is necessary, or the cells will lose water and shrink or gain water and swell as a result of osmosis. Nutrients and oxygen must be available, along with the proper hormones necessary for the entry of these substances into the cell, and waste products from the cell must not be allowed to accumulate. Temperature must be kept within a very small range, especially for proper functioning of nerve cells.

Single-celled organisms living in the ocean are, obviously, not able to regulate characteristics of their "extracellular fluid," the sea itself. It is possible for these organisms to change their buoyancy and thus ascend or descend to regions with a more favorable temperature or oxygen content, and some can swim a limited distance, but the existence of these organisms relies on the fact that their environment remains relatively constant. If it changes radically, they die; or in some cases they assume an inactive form until favorable conditions return.

As our single-celled ancestors evolved into more complex, multicellular organisms, they began to require regulatory mechanisms. Cells in the interior of their bodies could not rely on the process of diffusion to bring them nutrients and remove waste products. The animals evolved digestive, respiratory, circulatory, and excretory systems to facilitate the exchange of materials with the environment. Eventually, land-dwelling mammals evolved. Unlike sea-dwellers, who only have to seek sources of food, these animals must locate sources of water and periodically ingest it to prevent dehydration. And, unlike their poikilothermic ("varied-heat" or, more colloquially, "cold-blooded") counterparts, mammals have evolved mechanisms by which they may regulate their temperature by generating heat or losing it to the environment is required. These homoiotherms can therefore venture into regions far removed in temperature from their 37°–38° C interiors. Poikilotherms, on the other hand, must hibernate in cold weather, since their metabolic rates can fall so low that the animals cannot even move. In hot, sunny weather these animals must hide in the shade or immerse themselves in water.

The topic of this chapter and the next will be the means by which we mammals regulate our extracellular fluid by controlling our food and water intake. Judging from the number of articles in scientific journals, regulatory behavior (especially feeding) is one of the most intensely studied problems in physiological psychology. There are several reasons for such interest. First, regulation is of primary importance to life; other functions (sexual behavior, for example) cannot be carried out unless our interstitial fluid is kept constant. Understanding the physiological mechanisms that control regulatory behaviors might give us insight into other processes—sensory mechanisms, motor mechanisms, learning—that were probably evolved, in large part, to provide a regular supply of food and water as the organisms ventured into environments of increasing complexity and variability. There is a reasonably good relationship between an animal's sensory and motor capacities and learning ability and the way in which it obtains its food. This fact does not mean that a species evolved mechanisms *in order to* get at a new source of food; it means, instead, that a selective advantage, in terms of additional sources of food, accrued to an animal that was capable of seeing better, or remembering better, or becoming more agile. A better understanding of the control of regulatory behavior might, then, assist investigation of mechanisms that underlie other behaviors.

The Regulatory Process

A regulatory mechanism is necessary whenever a system requires constancy of some substance in the face of varying availability and/or utilization of that substance. A regulatory mechanism consists of four essentials: a substance (or characteristic such as temperature) to be regulated, usually referred to as a *system variable*; a *set point*, or "optimal value" around which that system variable is regulated; a *detector* that is sensitive to deviations of the system variable above or below the set point; and a *correctional mechanism* capable of restoring the system variable to the set point.

A good example of a regulatory system is a room whose temperature is controlled by a thermostatically regulated heater, as illustrated in Figure 12.1. The system variable is air temperature of the room, and the detector for this variable is a thermostat. This device can be adjusted so that contacts of a switch will be closed when the temperature falls below a preset value (the set point). Closure of the switch contacts engages the correctional mechanism (the heating coils). (See **FIGURE 12.1**.)

If the room cools below the set point of the thermostat, the heater is turned on, which warms the room. The rise in room temperature causes the thermostat to turn the heater off. The effects initiated by the thermostat thus feed back to the thermostat and cause it to stop the correctional mechanism it started. This process

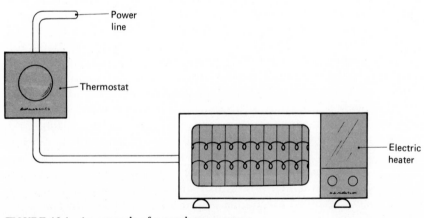

FIGURE 12.1   An example of a regulatory system.

is an example of **negative feedback,** and it is an essential characteristic of regulatory systems.

The term *feedback* should be self-explanatory; it refers to the consequences of an action affecting the factors that initiate that action. The rise in temperature produced by switch contact closure acts upon the thermostat to open the contacts (turn the switch off) again. The feedback is described as *negative* because the consequences (temperature rise) of the action (turning on the heater) oppose the action (the heater is turned off again).

Regulation of Intake: An Overview

O'Kelly, in a review of the problem of water regulation (O'Kelly, 1963), constructed a diagram that outlines the regulatory process in an organism's intake of food and water. (See **FIGURE 12.2.**) Living beings are **open systems;** they exchange energy and matter with their environment. Hence, there are input and output arrows shown above the box labelled "system variables." For example, water is gained by ingestion and lost by excretion by the kidneys and sweat glands, and through the lungs. System variables are monitored by the detectors, which initiate correction internally or by means of behavior. Internal regulation includes such activities as the secretion of hormones that regulate the balance of various minerals, such as potassium and calcium. Behavioral regulation requires the animal to act. The box labelled "discrimination and orientation" implies a mechanism whereby the animal acts differentially toward environmental stimuli previously associated with the substance that is needed. (See **FIGURE 12.2.**) Ultimately, the animal ingests the substance required and thus alters the system variables. O'Kelly has included another box, labelled "satiety," which interacts with the correctional mechanism. (See **FIGURE 12.2.**) The satiety mechanism serves to turn off corrective behavior even before the system variables, which are responsible for the initiation of ingestive behavior, are returned to the set point.

Why do we need a satiety mechanism? The reason lies in the physiology of our digestive system. Let us suppose (for the sake of argument) that we get hungry when some characteristics of our internal environment (for example, in the blood, interstitial fluid, cellular contents, or fat deposits) signal lowered energy stores. We begin to eat. But what makes us stop? Clearly, we do not work like the room-heating system I outlined. We do not continue eating until the energy stores of our cells are replenished, since it takes around four hours to digest a meal. We must possess a mechanism that says, "Stop—this meal, when digested and assimilated, will eventually provide enough nutrition to restore things." The fact that we stop eating before a significant amount of food is digested makes it necessary to postulate a satiety mechanism.

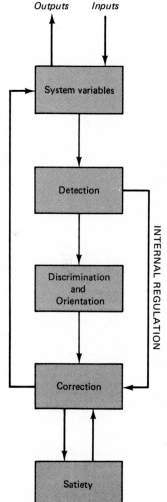

FIGURE 12.2 O'Kelly's model of regulation. (From O'Kelly, L. I., *Annual Review of Psychology,* 1963, *14*, 57–92. Reproduced with permission. Copyright © 1963 by Annual Reviews, Inc. All rights reserved.)

**EXPERIMENTAL ANALYSIS
OF THE REGULATION OF INTAKE**

How does one go about studying the physiological basis of regulation of intake? To put it most succinctly, one must find out what is regulated (and what the set point is), locate the detectors, and find out how the brain controls the initiation and suppression of ingestive behavior in response to signals from these detectors. Then one also has to find the way in which the satiety mechanism operates.

As you can imagine, the procedure is not nearly so simple as I have just outlined. Many sets of detectors monitor the system variables, and many different system variables appear to be implicated in ingestive behavior. Satiety has many sources, from several kinds of detectors. And the whole process may be overridden; even after a big dinner most people will eat a tasty dessert.

To find out whether a given variable (e.g., amount of glucose in the blood) is a factor in regulation of intake, one can experimentally alter this variable and observe the effects of such alteration on ingestive behavior. To search for detectors, one can change the value of the system variable in the vicinity of these hypothesized detectors and see whether ingestive behavior is altered. For example, if injections of water in a particular region of the brain caused a previously thirsty animal to cease drinking, the results would suggest that there were detectors sensitive to concentration of the interstitial fluid in that part of the brain. Finally, to locate brain mechanisms controlling ingestive behavior, one can perform brain lesion or brain stimulation studies, or locally block or stimulate particular types of synapses.

In this chapter and the next we shall see examples of a variety of ingenious experiments designed to isolate the critical variables and physiological mechanisms controlling intake. As we shall see, our ingestive behavior is affected by the sight, taste, odor, and texture of food, and the amount of food received by the stomach. The most important factor, however, is the quantity of nutrients stored in the cells of the body. For many years, physiological psychologists have assumed that detectors that measure the amount of stored nutrients are located within the brain. As we shall see, however, the most recent evidence suggests that it is the liver, not the brain, that monitors the store of nutrients. The detectors located in the brain do not appear to be responsible for control of feeding, but instead control hormonal secretion, which alters the availability of stored nutrients. The detectors in the brain are therefore part of the process O'Kelly calls "internal regulation."

In order to understand the newly found evidence concerning the control of food intake, we must first examine the processes of digestion, assimilation, and metabolism. Recent evidence, especially, has shown that it is essential that we understand how food is digested and absorbed, and how the body alternately stores and breaks down its energy supply.

The Digestive Process

First, look at an overall drawing of the digestive system. (See **FIGURE 12.3.**) Digestion begins in the mouth. We use our teeth to break down the food into pieces small enough to swallow safely and, in so doing, mix the food with saliva. The saliva serves several functions. It lubricates dry food and adds a digestive enzyme that begins the process of breaking down starches into sugars. Saliva also dissolves molecules of the food so that the taste buds can be stimulated. Besides stimulating taste receptors, of course, the food stimulates ol-

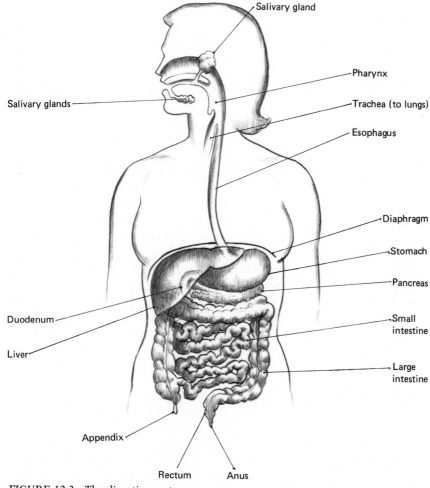

FIGURE 12.3 The digestive system.

factory receptors. Odor molecules enter the nose even before we begin to eat.

Swallowing involves a very complex set of reflexes that are voluntarily initiated but that are automatically controlled by the brain. Once the food is swallowed, it is propelled by waves of contractions of the circular smooth muscles of the esophagus into the stomach.

The stomach, in response to both the bulk and the chemical nature of the food it receives, begins to secrete hydrochloric acid and the enzyme **pepsin.** Hydrochloric acid breaks the food into small particles, and pepsin breaks **peptide bonds,** thus beginning the process of breaking proteins in the food into their constituent amino acids. The stomach also becomes very active, churning the food around so that it becomes well mixed with digestive juices.

Gastric secretion and motility are controlled by several factors. The stomach is innervated by branches of the autonomic nervous system, and efferent fibers in these nerves provide one source of control. By this means, gastric secretion is stimulated by the sight and odor of food. Once food is in the stomach, stretch receptors and **chemoreceptors** (i.e., receptors sensitive to various chemicals) in the wall of the stomach trigger reflex activation (via the brain) of further gastric secretion.

The stomach empties into the **duodenum,** the upper portion (approximately 8 to 11 inches long) of the small intestine. (*Duodeni* means "twelve"; the duodenum is approximately twelve finger-widths long.) The rate of gastric emptying is controlled by the composition of the foodstuffs received by the duodenum. The duodenum contains chemoreceptors that regulate gastric activity by means of neural reflexes, and by the release of various hormones. For example, if the duodenum receives large concentrations of fat from the stomach, gastric emptying is severely inhibited, since it takes much longer to digest this nutrient. Proteins and carbohydrates are much more readily digested, so the duodenum permits the stomach to empty more quickly when a meal low in fats is eaten. One duodenal hormone, **cholecystokinin (CCK),** has been nominated for a role in the control of hunger. This hormone received its name because it causes the gall bladder (also called the *cholecyst*) to contract, releasing bile into the duodenum. The secretion of CCK also retards gastric emptying.

The pancreas, which is located below the stomach, communicates with the duodenum by means of the pancreatic duct. (The pancreas, as we shall see, also produces two hormones, insulin and glucagon, and thus qualifies as an endocrine gland as well as an exocrine gland.) Pancreatic enzymes break down proteins, lipids, starch, and nucleic acids, thus continuing the digestive process. The pancreas also secretes bicarbonate, which neutralizes stomach acid. As the products of digestion begin to be absorbed into the bloodstream, and as the acid is neutralized by the bicarbonate, the stomach empties more foodstuff into the duodenum.

Digestive juices from the stomach and pancreas do a good job of breaking down proteins into amino acids, and starches and complex sugars into simple sugars. These water-soluble nutrients enter the capillaries of the intestinal *villi*, fingerlike structures that protrude into the intestine. (See **FIGURE 12.4**.) These capillaries drain into veins that travel directly to the liver. The veins branch into another set of capillaries within the liver; thus, the nutrients extracted from a meal reach the liver via the ***hepatic portal system*** before reaching any other portion of the body. (See **FIGURE 12.5** on the following page.)

Unlike the water-soluble nutrients, fats remain in fairly large droplets and cannot be absorbed to any extent without the presence of ***bile.*** The liver produces bile, which is subsequently stored in the gall bladder. When chemoreceptors in the duodenum detect the presence of fats, CCK is secreted, which causes the bile to be released

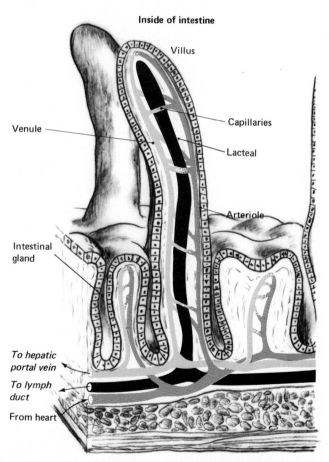

FIGURE 12.4 Intestinal villi. (Adapted from Vander, A. J., Sherman, J. H., and Luciano, D. S., *Human Physiology: The Mechanisms of Body Function*, 2nd edition. McGraw-Hill, 1975.)

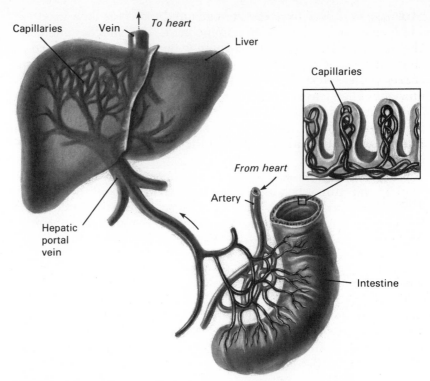

FIGURE 12.5 A schematic representation of the hepatic portal blood supply.

into the intestine. Bile breaks down lipid droplets into very small particles—a process called **emulsification.** For example, milk is an emulsion, containing very fine particles of butterfat suspended in water. (In fact, the word *emulsion* comes from the Latin word *mulgēre,* which means "to milk.") Emulsified fats can be absorbed into the *lacteals* located within the villi; the lacteals communicate with the lymphatic system, which then brings the fats to a duct that empties into veins in the neck.

The meal passes through the small intestine, where most of the available nutrients are absorbed. The residue enters the large intestine. Hardly any nutrients are absorbed in the large intestine, but a considerable amount of water and electrolytes is reabsorbed. (This reabsorption is extremely important. Cholera kills people by producing severe diarrhea, which results in the loss of fluids and electrolytes through the feces.) The intestinal bacteria live mainly on undigested cellulose. They produce some vitamins (especially vitamin K, important in the clotting of blood) that are absorbed into the body. Note that I said "into the body." The contents of the digestive tract are not, strictly speaking, within the body. If we are reduced, topologically, to our simplest form, we find that we are doughnut-shaped. The gastrointestinal tract is the hole in the doughnut.

Storage and Utilization of Nutrients

The metabolic processes of the body are beautifully interrelated. At different times, the processes cooperate to produce (1) the **absorptive phase** (which occurs while a meal is being absorbed from the intestine) and (2) the **fasting phase** (which occurs after the nutrients have been absorbed, and which usually leads to hunger and ingestion of the next meal). As we shall see, the factors that cause metabolism to enter the fasting phase are the ones that also induce hunger; hence, we must understand metabolic processes if we are to understand the physiological control of food intake.

ABSORPTIVE PHASE.   During the absorptive phase of metabolism, the body receives three types of nutrients from the intestines if a well-balanced meal is eaten: glucose (derived from carbohydrates), amino acids (derived from proteins), and fats. Glucose is the most easily metabolized nutrient, so this substance serves as the principal source of fuel. During the absorptive phase, more nutrients are received from the digestive system than the body needs right then, so excess amounts are converted to fats and glycogen, which can be broken down and metabolized later when they are needed.

Figure 12.6 illustrates the important metabolic pathways that characterize the absorptive phase. (For clarity, many intermediate steps have been omitted.) The numbers below correspond to the numbers in the figure. (See **FIGURE 12.6** on the next page.)

1.   The principal source of energy for almost all tissue is glucose.
2.   Excess glucose is converted to **glycogen** in the liver (2a) and muscles (2b). Glycogen is an insoluble carbohydrate that is often referred to as "animal starch." Glucose is also converted to fats in the liver (2c) or in **adipose** (fat) **tissue** (2d). The liver does not store fats; they are sent on to adipose tissue for storage.
3.   Amino acids are used in the construction of protein, mainly in muscles.
4.   Amino acids are used as a fuel by the liver.
5.   Excess amino acids are converted to fats by the liver, and are stored in adipose tissue.
6.   Fats that are received from the digestive system are stored in adipose tissue.

To summarize: Glucose is the primary fuel. Excess glucose is stored as glycogen or fat. Amino acids are used for protein synthesis as needed, and excess amounts are used as a fuel by the liver and/or stored as fat. Fats are not used as a fuel, but are put into storage in adipose tissue.

FASTING PHASE.   After a meal has been absorbed from the digestive system, glucose is no longer available in abundance, and other fuels must be used. The principal source of energy is received from adipose tissue in the form of **free fatty acids,** which are derived from stored fats. All tissues except for the brain can use free fatty acids.

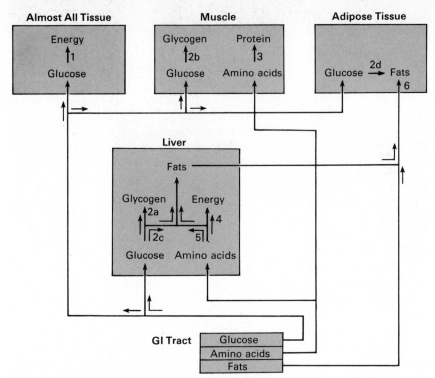

FIGURE 12.6   Metabolism during the absorptive phase.

The most important fuel for the brain is glucose, and in order to ensure that this substance is available for the brain, the other tissues of the body are prevented from using it.

Figure 12.7 illustrates the metabolic pathways that characterize the fasting phase. Again, many intermediate steps are omitted for the sake of clarity. As in the previous figure, numbers below correspond to numbers on the illustration. (See **FIGURE 12.7.**)

1.   The principal fuel for most tissues (including the liver but excluding the brain) is free fatty acid, released from adipose tissue.
2.   The principal fuel for the brain is glucose. The glucose comes from three primary sources: Glycerol (derived from fats) is released by adipose tissue and is converted to glucose in the liver (2a). Liver glycogen is converted into glucose (2b). In periods of prolonged fasting, muscle protein is broken down to amino acids, which are converted to glucose by the liver (2c).
3.   The liver converts some of the free fatty acids it receives into *ketones.* The liver cannot metabolize ketones, but the rest of the body can. Ketones are an important source of fuel for the brain if fasting is prolonged. This fact has only recently been appreciated; it is often said that the brain can use only glucose as a fuel, but this statement is not true.
4.   Muscle tissue can use glycogen as a source of energy. This met-

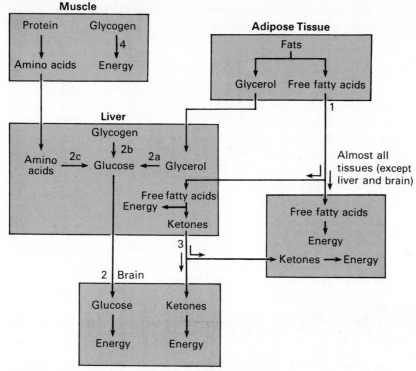

FIGURE 12.7   Metabolism during the fasting phase.

abolic pathway is more complicated than I have indicated, but the intricacies are not important for our purposes.

To summarize: During the fasting phase, most of the body uses fatty acids as fuel. Fats, amino acids, and glycogen can be converted to glucose by the liver. The sole user of the glucose is the brain. The liver can also convert free fatty acids to ketones, which it cannot metabolize, but which can be used by the rest of the body (including the brain).

Control of Metabolism

As we have just seen, the metabolic pathways followed during fasting and during absorption of nutrients are very different. There obviously must be mechanisms that control these metabolic pathways. As we shall see, this control is effected by the nervous system and by several hormones.

INSULIN AND THE ABSORPTIVE PHASE.   The absorptive phase of metabolism is produced primarily by the effects of insulin. Insulin is one of the two hormones secreted by the *islet cells* of the pancreas in its role as an endocrine gland. Insulin is a vital hormone. If the pancreas loses the ability to produce insulin, the disease ***diabetes mellitus*** results. Severe diabetes, if not treated with injections of insulin, can result in death. The lack of insulin results in a very high blood level of glucose, primarily because glucose cannot enter the cells of the body (except for cells of the nervous system, which do not depend on insulin).

Deprived of glucose, the other cells of a diabetic rely principally on fats for energy. The glucose is passed through the kidneys and, in the process, causes a loss of sodium. If severe enough, the mineral loss can lead to low blood pressure, reduced blood flow, and, ultimately, death. Diabetes mellitus literally means "passing through of honey," because the urine of an untreated diabetic tastes sweet. (The diagnosis of this disease, fortunately, can be accomplished by chemical procedures, and no longer relies on the sense of taste.) The presence of sugar in the urine led an observant physician to note, several hundred years ago, that diabetes could be diagnosed (at least in men) by noting whether a person's shoes attracted flies.

Insulin has many metabolic effects, as shown in **Figure 12.8.**

1. Insulin facilitates the entry of glucose into the cell. Without insulin, body tissue cannot readily utilize glucose in metabolic processes. The nervous system is an exception; it can use glucose independently of insulin.
2. Insulin increases the conversion of glucose to glycogen.
3. The conversion of glucose into fats is increased, and the buildup of fat storage is thus facilitated.
4. The breakdown of fats is inhibited.
5. The transport of amino acids into cells is facilitated, permitting protein synthesis to occur.

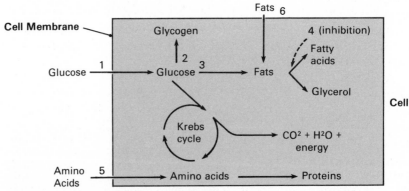

FIGURE 12.8 Effects of insulin on metabolism. (Adapted from Vander, A. J., Sherman, J. H., and Luciano, D. S., *Human Physiology: The Mechanisms of Body Function,* 2nd edition. McGraw-Hill, 1975.)

6. The transport of fats into adipose cells is facilitated, causing fat storage.

Insulin thus facilitates the metabolic pathways involved in the absorptive phase of metabolism: storage of nutrients and utilization of glucose.

Control of insulin secretion. A number of factors control insulin secretion: the level of glucose in the blood, the level of amino acids in the blood, and activity of the autonomic fibers that innervate the pancreas. During the absorptive phase of metabolism, glucose and amino acids are received from the intestines, and the blood level of these substances consequently rises. Cells within the pancreas respond directly to the presence of these substances, releasing insulin. During the fasting phase, the levels of these nutrients fall, and the secretion of insulin decreases.

The brain can also control the secretion of insulin, through the sympathetic innervation of the pancreas. One of the most important functions served by this innervation is the secretion of insulin in anticipation of the receipt of nutrients. When an animal begins to eat a meal, saliva and gastric juices are secreted, the stomach begins to churn, and insulin is secreted. This secretion of insulin starts to prepare the animal for the absorptive phase that is just commencing. This phenomenon probably provides at least part of the explanation for the fact that a small portion of a tasty food (like a potato chip or two) makes a hungry person feel even hungrier. The food causes the release of insulin. If the potato chip is not followed up by a meal, the blood glucose level will fall, which will increase hunger. I shall have more to say about this phenomenon later.

CONTROL OF THE FASTING PHASE. The most important control factor in the fasting phase of metabolism is a low level of insulin, for the following reasons: (1) Without insulin, glucose cannot enter cells (except those in the brain), so blood glucose is spared for use by the brain. (2) Without the inhibiting effects of insulin, fats are broken down to fatty acids and glycerol. Fatty acids are used by most tissues for fuel, and the glycerol is converted to glucose by the liver. Of course, the brain gets this glucose.

A second control factor in the fasting phase of metabolism is the hormone **glucagon** which, like insulin, is produced by the pancreas. When the blood glucose level falls (as it starts to do when all the food is absorbed from the digestive system) the pancreas not only ceases to secrete insulin; it also begins to secrete glucagon. This hormone causes liver glycogen to be converted to glucose (which the brain uses). It also promotes the breakdown of fats, so that fatty acids are available for the metabolic requirements of other tissues of the body.

Two other factors operate in parallel with glucagon: growth hormone, secreted by the anterior pituitary gland, and the sympathetic innervation of the liver and adipose tissue. However, these factors are less important than insulin and glucagon. In addition, under stressful conditions, or in times of strenuous exertion, epinephrine from the adrenal medulla and glucocorticoids from the adrenal cortex stimulate a rapid breakdown of the body's supplies of stored fuel.

Now that I have reviewed the important aspects of metabolism and its control, we can examine the various attempts that have been made to understand what makes an organism begin to eat, and what makes it stop. As you shall see, this background is a prerequisite for understanding fruitful recent research.

THE SEARCH FOR CENTRAL SYSTEM VARIABLES FOR HUNGER

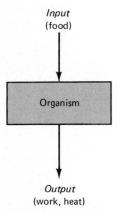

Input
(food)

Organism

Output
(work, heat)

FIGURE 12.9 Input must equal output if the organism is to maintain a constant weight.

If body weight is to be regulated, the system variables that are monitored must be related to the body's supply of nutrients. Body weight is ultimately determined by two factors: the amount of input, in the form of food, and the amount of output, in the form of work and heat production. If *calories in* exceed *calories out*, the organism gains weight. (See **FIGURE 12.9.**) Since output can vary, depending on environmental conditions and the amount of work the organism performs, it is not sufficient to monitor food intake. Let me suggest an analogy. If you have a houseplant in a room whose relative humidity can change, thus affecting the rate of water loss, you do not control the soil's moisture by carefully measuring the amount of water that you put on it. Instead, you feel the soil (i.e., measure a central system variable) and add water as needed. This example emphasizes that the most important factors for eliciting hunger must be centrally located—that is, within the body itself, rather than in the digestive tract. Perhaps hunger occurs when the level of the relevant system variables in the blood, and thus the amount that is available to the cells, falls below a given level. This fall would be detected by receptors whose activity causes neural feeding mechanisms to be aroused. (However, since a meal must stop before the food is digested, it is reasonable to propose that *satiety* could be produced by receptors in the digestive system. What stops a meal does not have to be what started it.)

Research on the nature of the system variables in hunger have concentrated on three of the principal sources of energy: glucose, lipids (fats), and amino acids. Unfortunately, there has been too much emphasis in the past on identifying which *single* factor is responsible for hunger. Some people have said that glucose is important, and have advocated the **glucostatic** hypothesis. (The suffix *-static* is used analogously to thermo*static*; a *glucostat* is a hypothetical detector of blood glucose, just as a thermostat is a detector of temperature.) Similarly, there have been advocates of **lipostatic** and **aminostatic** hypotheses, arguing for hunger produced by low levels of lipids or amino acids in the blood.

Things are not so simple. As we have just seen, metabolism is an exceedingly complex system, with many alternate pathways, whose characteristics depend on the nature of a meal, the amount of various stored nutrients already in the body, and the rate of breakdown caused by stress and exercise. Why should we expect hunger to be an on-or-off phenomenon triggered by a single metabolic fuel?

The Importance of Glucose

It has been suggested that glucose, a nutrient vital to the brain, might be an important system variable regulated by alterations in the intake of food. It has been known for a long time that a fall in blood glucose (produced by an injection of insulin) leads to hunger (Morgan and Morgan, 1940), whereas injections of epinephrine or glucagon, which raise blood sugar level, produce satiety (Mayer, 1955). The suggestion was made that low blood glucose levels meant hunger, whereas high blood glucose levels meant satiety.

This hypothesis was soon modified. Patients suffering from untreated diabetes have an extremely high blood sugar level, and yet these people are usually very hungry. If blood glucose is the system variable that is being measured and regulated, then the diabetics should not be hungry. However, despite the fact that the blood glucose level is high, the cells of the body are not able to use the glucose. Without insulin, glucose cannot enter the cells to be metabolized. Thus, the body is in the fasting phase despite high blood glucose levels.

This consideration led Jean Mayer (1955) to hypothesize that the relevant system variable is the degree to which glucose can be utilized, not the mere amount of glucose that is present in the blood. When one samples the arterial and venous blood of an untreated diabetic, one finds that the glucose levels in arterial and venous blood are almost identical. This "low A-V difference" means that the organs did not extract any significant amount of glucose from their arterial blood supply.

Eating is also produced by injections of **2-deoxyglucose**, or **2-DG** (Smith and Epstein, 1969). This substance is similar to glucose in its molecular shape, but it cannot be metabolized. (You will recall from Chapter 9 that radioactive 2-DG can be used to identify neurons with high rates of metabolism.) Injections of 2-DG block glucose metabolism and therefore reduce the availability of glucose to the body without lowering the blood sugar level. The fact that injections of 2-DG into the ventricles of the brain result in hunger (Miselis and Epstein, 1970) suggests that there may be glucoreceptors in the brain that control feeding.

SEARCH FOR BRAIN GLUCORECEPTORS. If glucose availability provides an important signal for the regulation of food intake, where are the receptors located? As I have noted, the nervous system is insensitive to the effects of insulin on glucose metabolism. The nervous system can utilize glucose even in the absence of insulin. Thus, although there is some electrophysiological evidence for hypothalamic neurons that respond to glucose (Desiraju, Banerjee, and Anand, 1968), one would not expect them to respond to glucose *availability*. Glucose availability depends on insulin, and if cells of the brain do not require insulin in order to utilize glucose then how can these cells monitor glucose availability? Mayer (1955) has suggested that there are neurons in the hypothalamus that *are* sensitive to the effects of insulin, metabolizing glucose only when insulin is present. Presumably, their firing rate depends on their level of metabolism, thus coding for glucose availability. However, good evidence for the existence of these insulin-dependent neurons is still lacking.

If the brain contained "glucostats" whose activity was important for hunger, one would expect that injections of glucose into the brain of a hungry animal should turn off the signal for hunger. However, experiments have not been able to find a brain region where glucose injections prevent eating in a hungry animal (Balagura, 1973). This failure represents a very serious gap in the hypothesis that glucoreceptors in the brain have a controlling role in hunger.

The fact that injections of 2-DG into the brain will cause a sated animal to eat remains the best evidence for the existence of brain glucoreceptors. But since 2-DG causes a *drastic* decrease in the utilization of glucose by the brain, we cannot be sure that brain glucoreceptors play an important role in normal day-to-day hunger.

An injection of insulin is a very potent stimulus for hunger. Insulin sharply lowers the level of glucose in the blood, because it causes glucose to enter cells and be converted into insoluble glycogen. Presumably, the eating is caused by the low blood glucose level. Early studies reported that this response is lost after lesions of the lateral hypothalamus (Epstein and Teitelbaum, 1967), perhaps because the lesions destroyed some glucoreceptors that were located there. However, subsequent studies showed that these lateral hypothalamic glucoreceptors (if they indeed exist) are not very important in normal regulation of food intake.

Blass and Kraly (1974) destroyed a fiber system of the lateral hypothalamus, which eliminated the feeding response to 2-DG. That is, depriving the cells of glucose did not increase the animals' food intake. If a fall in the blood glucose level is an important system variable in the regulation of food intake, then rats with this type of brain lesion should have difficulty with such regulation. However, they did not. The animals ate more when given a diluted liquid diet, and they ate less of a concentrated diet. They regulated their food intake in response to changes in room temperature that produced alterations in their expenditure of energy. When body weight was temporarily raised by forced feeding (through a stomach tube) or lowered by starvation, the animals altered their food intake appropriately when food was freely available *ad libitum,* and their weight returned to normal. By every measure, their regulation of food intake was normal. Why, then, did the rats fail to eat in response to injections of 2-DG?

A study by Stricker, Friedman, and Zigmond (1975) suggests an answer to this question. These investigators found that rats with lesions of the lateral hypothalamus *would* respond to injections of insulin with increases in eating. However, instead of giving the rats a single, large dose of insulin, they administered gradually increasing doses. Single, large doses of insulin apparently produce too much stress for these animals to cope with; in fact large doses of insulin are often lethal to rats with these brain lesions.

All of these results cast a strong doubt on the importance of *brain* glucoreceptors as the only detectors whose activity produces hunger. As we shall see, other important receptors (and these do not seem to be simply *glucose* receptors) appear to be located in the liver.

The Importance of Lipids

Calories that are ingested in excess of tissue need will be converted to fat (lipids), regardless of whether the calories are ingested in the form of fats, carbohydrates, or protein. And during fasting, the body's reserves of fat are broken down to glycerol and free fatty acids and used as a source of energy. Hence, it seems quite plausible that the amount of fat deposits might be a factor in long-term control of food intake. There is good evidence that this is the case; fat deposits do appear to be regulated. There is even some preliminary evidence that lipids (or other metabolities that are derived from them) may serve as system variables for control of food intake.

Lipids and Control of Food Intake

If a rat is made diabetic (by injecting it with *alloxan,* a substance that destroys the islet cells of the pancreas), the animal will become *hyperphagic;* that is, it will eat much more than a normal rat will. However, since glucose cannot be metabolized by most cells without the presence of insulin, large amounts of sugar are lost in the urine. Thus, the animal overeats because it needs nutrients to sustain itself.

The principal fuel for diabetic animals is lipids, since these are the only nutrients that can easily enter cells without the presence of insulin. Friedman (1978) made rats diabetic and fed them diets that contained various concentrations of nutrients. Normal rats will respond to changes in caloric value of their food; they will eat less of a concentrated diet, and more of a dilute diet. This is true whether the calories are available as fats or carbohydrates. However, Friedman found that diabetic rats did not respond to a reduction in the amount of carbohydrates in their diet, but greatly increased their intake if the amount of fats was reduced.

The rats were given a basal diet of 23 percent fat, 52 percent cornstarch, and 20 percent casein (a protein that is found in milk). This mixture yields approximately equal numbers of calories from fats and carbohydrates. Diabetics ate relatively normal amounts of this diet, presumably because the fat content was so high. When most of the fat content was replaced with nonnutrient petroleum jelly (Vaseline) or most of the carbohydrate content of the diet was replaced with powdered cellulose (which contains no significant caloric value), normal rats ate more food, thus keeping their daily caloric intake relatively constant. (See **FIGURE 12.10.**) However, diabetic rats ate only slightly more food when they were fed a diet that contained fewer calories of carbohydrate. Since their blood already contained excessive amounts of unusable glucose, a reduction in dietary carbohydrates was not of any consequence. But when the fat content was reduced, they ate much more food. (See **FIGURE 12.10.**) In fact, when the amount of fat in the diet was systematically varied, the diabetic rats altered their food intake in a regular fashion, keep-

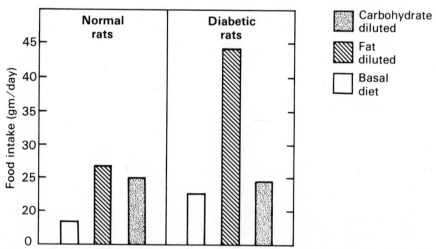

FIGURE 12.10   Effects of dilution of fats or carbohydrates in the diet on food intake of normal and diabetic rats. (From Friedman, M. I. Hyperphagia in rats with experimental diabetes mellitus: a response to a decreased supply of utilizable fuels. *Journals of Comparative and Physiological Psychology*, 1978, *92*, 109–117. Copyright © 1978 by the American Psychological Association. Reprinted by permission.)

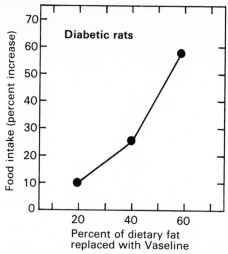

FIGURE 12.11 Food intake of diabetic rats as a function of dietary fat content. (From Friedman, M. I. Hyperphagia in rats with experimental diabetes mellitus: a response to a decreased supply of utilizable fuels. *Journal of Comparative and Physiological Psychology*, 1978, *92*, 109–117. Copyright © 1978 by the American Psychological Association. Reprinted by permission.)

ing total fat intake very close to a constant value. (See **FIGURE 12.11.**)

This experiment shows very nicely that even for short-term regulation, food intake is modulated by available calories, whether they are in the form of carbohydrates or lipids. One might suggest that the lipids could eventually be converted into glucose, which would then be monitored by "glucostats." However, the data from the diabetic rats argue against this suggestion. The diabetics had a high blood glucose level under all conditions, but nevertheless their eating was controlled by the lipid content of their diet. Either there are "lipostats" *and* "glucostats," or else the detectors of the central system variables are more general, responding to the presence of *any* available nutrient. As we shall see, the evidence favors the latter hypothesis.

EVIDENCE FOR THE REGULATION OF FAT DEPOSITS. First, let us examine the evidence that the amount of body fat is regulated. Liebelt, Bordelon, and Liebelt (1973) have described a number of studies they have performed concerning the regulation of adipose tissue. They found that the body's total amount of fat appears to be regulated. When a piece of adipose tissue was transplanted from one mouse into another, the transplanted tissue normally withered away. However, when the experimenters first removed some of the adipose tissue belonging to the recipient animal, the transplants "took." In other words, when the total amount of fat tissue in the recipient mouse was surgically reduced, an implant grew—presumably in response to mechanisms encouraging growth of such tissue. Furthermore, when mice were given a brain lesion that causes overeating and subsequent obesity (more about this phenomenon later), they accepted a graft of adipose tissue while they were gaining weight, as though the lesion had raised the "adipose tissue set point."

Another piece of evidence for regulation of body fat is that animals forced to eat—and to become fat—will reduce their food intake until their weight returns to normal levels (Hoebel and Teitelbaum, 1966; Steffens, 1975). Since excessive weight is carried mainly as body fat, perhaps adipose tissue gives rise to a satiety signal when the amount of tissue exceeds some set point. The evidence for mechanisms that regulate storage of fat seems, therefore, quite good.

The Importance of Amino Acids

We obtain amino acids from the breakdown of proteins that we eat. We use these amino acids as a source of energy and in the synthesis of our own protein. Many amino acids can be synthesized in our bodies, but some (the essential amino acids) cannot, and these must be obtained from our diet. Since proteins are constructed by our cells during the absorptive phase of metabolism, and since amino acids cannot be stored from one meal to the next (amino acids not used for protein synthesis are used for energy or are converted into fat), ingestion of a meal that is deficient in some essential amino acids will result in decreased protein synthesis. Protein is therefore an essential component of our diet—especially high-quality protein, containing all the essential amino acids.

Assimilation of a meal rich in amino acids is very satiating, despite the fact that it results in a rather low blood glucose level (Mellinkoff, Frankland, Boyle, and Greipel, 1956). The large amount of amino acids being absorbed stimulates insulin release, which causes amino acids (but also glucose) to enter the cells. In response, glucagon is also secreted, and this hormone keeps the blood glucose level high enough to nourish the brain, but there is still a certain amount of *hypoglycemia* (low level of blood glucose). The satiating effect of amino acids is capitalized on by high-protein reducing diets.

These observations led Mellinkoff to hypothesize that blood level of amino acids was the system variable that regulated food intake. There is no doubt about amino acids playing a role in feeding behavior. If an animal is deprived of one or more essential amino acids, its food intake will gradually decrease until the missing amino acids are put back into the diet. As we shall see in Chapter 13, in a section on specific hungers, animals learn to select foods that contain particular substances (such as vitamins, minerals, and essential amino acids) their bodies require. Regulation of these substances does not require a specific detector for each of them—only the ability to associate "feeling sick" or "feeling well" with ingestion of food with a particular flavor.

It would be interesting to determine whether normal rats and diabetic rats would respond to reduction of the protein content in a diet like the one used by Friedman (1978). One would predict that normal rats would increase their intake, but that diabetic rats would not, since diabetics utilize amino acids as poorly as glucose. If such evidence were obtained, it would provide strong support for the importance of available amino acids in the control of food intake.

WHERE ARE THE DETECTORS? A TOUR FROM MOUTH TO LARGE INTESTINE

So far we have seen that food intake cannot be regulated by signals arising from an empty stomach; there must be a monitoring of some system variables located within the body. On a long-term basis, it seems plausible that some factor related to deposits of fat in adipose tissue is regulated, but the evidence is still scanty. There appear to be brain glucoreceptors, but convincing evidence that they respond to insulin is lacking. Glucoreceptors that respond to glucoprivation may exist in the lateral hypothalamus, but feeding in response to a sudden and dramatic fall in blood glucose level (the glucoprivic response) is not necessary for normal regulation of body weight and appears to be part of an emergency system. Thus, our search for detection mechanisms in the brain has not been very successful. And we have no evidence for a satiety mechanism that terminates a meal before the food is digested and assimilated. Let us therefore turn our attention to the progress of a meal from mouth to large intestine and look at the evidence concerning the effects of the various ingestive, digestive, and assimilative mechanisms on subsequent food intake.

Head Factors

The head contains several sets of receptors that play a role in food intake. We respond to the sight, odor, taste, and texture of food, and we can monitor the amount of food that is swallowed. As we saw in Chapter 8, flavor is jointly determined by the odor and taste of a food. Only in rare cases (for example, ingestion of a pure sugar solution) are the taste buds alone stimulated.

For many animals, olfaction plays a large role in locating food. In rats, for example, more electrical activity is recorded from the olfactory bulbs (in response to an odor associated with food) when the animal is hungry. When it is satiated, less activity is seen. Furthermore, this differential response is *not* seen for olfactory stimuli not associated with food (Giachetti, MacLeod, and LeMagnen, 1970). And olfaction plays a very significant role in judging the suitability of a given substance as a source of food; we will avoid eating food that smells rotten or rancid.

As we saw earlier, regulation requires the existence of negative feedback. In early stages of the ingestion of a meal, however, *positive feedback* is seen. That is, the consequences of ingestion of a palatable food lead to *increased* intake. We are all aware of the fact that we feel the hungriest as we begin to eat, and that *hors d'oeuvres* served just before a meal (often appropriately called appetizers) increase our interest in the main course. An even clearer example of positive feedback is seen in sexual activity. A person's sex drive, which might lead to contact with a partner, is increased, not reduced, by such contact. Increased sexual activities with the partner lead to more arousal and more activity. Ultimately, negative feedback may be applied by an orgasm, or by some other factors (reluctance to actually engage in intercourse, for example), but the fact is quite clear that early stages are characterized by positive feedback.

This is not to say that head factors are only excitatory upon eating. Negative feedback is also aroused, but more slowly. When an animal is *esophagotomized,* head factors can be studied independently. In this preparation the esophagus is severed, and the cut ends are brought out through incisions in the skin. (See **FIGURE 12.12.**) When an esophagotomized animal eats, the food that is swallowed falls to the ground or is collected through a tube placed into the open end of the esophagus (we call this procedure *sham eating*). An esophagotomized animal will not eat indefinitely; a somewhat larger-than-normal meal is swallowed, then the animal stops eating (Janowitz and Grossman, 1949). This study thus demonstrates the exis-

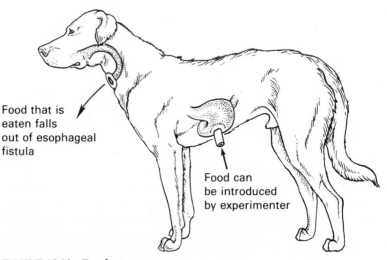

Food that is
eaten falls
out of esophageal
fistula

Food can
be introduced
by experimenter

FIGURE 12.12   Esophagotomy.

tence of head factors involved in satiety. However, the animal soon returns to eat again, which indicates that these satiety factors are short-lived and normally must be superseded by satiety factors (for example, gastric ones) farther down the digestive tract.

The head factors involved in satiety are even specific to taste. I suppose we all have had the experience of not being able to "touch another bite" of a particular food we just ate to excess, but of being able, nevertheless, to eat some dessert. Rats also demonstrate this phenomenon. If they are given four different flavors of food (actually the same food with different flavors added to it) in succession at a given meal, they will eat more than they normally do if given only one of them (Le Magnen, 1956). The animals will eat up to 270 percent of their normal intake, attesting to the power of the positive feedback, before it is counteracted by flavor-specific satiety (and by the increasing satiety produced by an enormously full stomach).

Gastric and Duodenal Factors

We have already seen that regulation of food intake cannot be accounted for solely by "hunger pangs" originating in the stomach. In fact, the stomach is not even necessary for feelings of hunger. Humans whose stomachs have been removed (because of cancer or the presence of large ulcers) still periodically get hungry (Ingelfinger, 1944). These people eat frequent, small meals, because of the absence of a stomach. In fact, a large meal causes nausea and discomfort, apparently because the duodenum quickly fills up. Although the stomach may not be especially important in producing hunger, it does appear to play a significant role in satiety.

It has been known for a long time that a hungry animal will eat less if nutrients are injected directly into its stomach just before it is given access to food (Berkun, Kessen, and Miller, 1952). Less of an effect is seen if a nonnutritive saline solution is loaded into the stomach, which suggests that there are receptors that respond to the presence of nutrients. There also appear to be stretch receptors whose activity will suppress food intake, but these are probably not important when the animal eats meals of normal size. Janowitz and Hollander (1953) were able to suppress food intake by inflating a balloon that had been placed in a dog's stomach, but the stomach had to be distended more than it would be if a normal meal had been ingested.

Other studies (Campbell and Davis, 1974; Novin, Sanderson, and VanderWeele, 1974) have shown that food intake is suppressed by injections of nutrients, but not nonnutritive control substances, into the duodenum. Since some food is received by the duodenum very soon after it reaches the stomach, it is even possible that the satiety mechanisms that were investigated in stomach-loading experiments might actually be triggered by receptors in the duodenum, and not in the stomach. In fact, a study by Liebling, Eisner, Gibbs, and Smith (1975) appears to argue against stomach factors in favor of duodenal ones. Rats were equipped with drainage plugs that permitted food to spill out of the stomach as it was eaten. These animals ate large volumes of food. When glucose was injected into the duodenum, the rats stopped eating. Thus, it would appear that gastric factors are not as potent as duodenal factors in producing satiety.

However, Deutsch, Young, and Kalogeris (1978) suggest an alternative explanation for these results. First, they note that since food spilled out of the stomach drainage plug very easily, not very much was allowed to accumulate, and since none of it stayed long, not much digestion could take place. If stomach factors are sensitive to the total amount of food, and/or if they respond to the breakdown products of a meal, and not to the raw food itself, this experimental procedure would not be expected to stimulate them adequately.

There is a second problem with both intragastric and intraduodenal injections of food. If an animal does not eat after food is placed directly into its digestive system, we cannot necessarily conclude that the eating is suppressed by *satiety*. Perhaps a direct injection of food simply makes the animal feel sick. Thus, Campbell and Davis, and Liebling and his colleagues may have suppressed the rats' food intake by making them feel ill.

To test this possibility, Deutsch, Molina, and Puerto (1976) implanted intragastric fistulas so that food could be injected directly into the stomach of unrestrained rats. (See **FIGURE 12.13**.) The rats were allowed to drink from two bottles, each of which contained water to which a flavoring had been added. One of the flavors (flavor A) was paired with an intragastric injection of sesame oil. As the rat

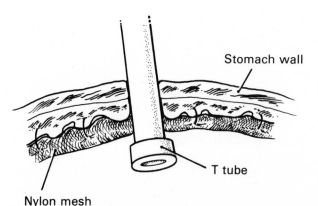

FIGURE 12.13   An intragastric fistula used by Deutsch and his colleagues.

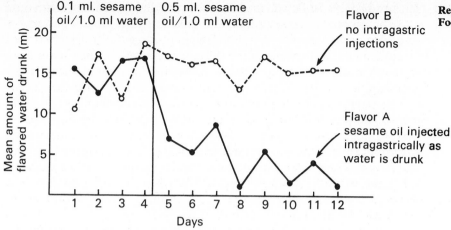

FIGURE 12.14   Effects of intragastric injections of sesame oil on flavor preference in rats. (From Deutsch, J. A., Molina, F., and Puerto, A. Conditioned taste aversion caused by palatable nontoxic nutrients. *Behavioral Biology*, 1976, *16*, 161–174.)

drank flavor A, sesame oil was simultaneously injected into the stomach, at a rate of 0.5 ml for every 1.0 ml of water drunk. No oil was injected when the water with flavor B was drunk. Sesame oil is a very palatable substance for rats; they will normally drink 4 ml a day, even if water and other food is continuously available. However, the rats in the present experiment quickly learned to *avoid* flavor A, which was paired with the intragastric injections of sesame oil. Figure 12.14 presents the data. (See **FIGURE 12.14.**)

The data indicate that the intragastric injection of sesame oil punished the rats; they apparently associated aversive effects of the injection with the drinking of flavor A. This means that the suppression of feeding caused by intragastric injections of nutrients might very well be caused by discomfort, and not by satiety. When food is ingested normally, head factors trigger activity in parasympathetic innervation to the stomach, which causes secretion of gastric acid (Pavlov, 1910; Powley, 1977). When food mixed with gastric acid is received by the duodenum, it causes the release of various hormones (including CCK). These hormones act on the stomach, slowing its rate of emptying. However, when food is administered intragastrically, it bypasses the head factors. Gastric acid secretion is *not* stimulated, and the duodenum consequently receives a sudden jolt of raw nutrients. The effect appears to be an aversive one.

Puerto, Deutsch, Molina, and Roll (1976a,b) found that intragastric injections of milk, like those of sesame oil, were aversive. However, another nutrient could be injected intragastrically without any evidence of aversion. Donor rats were allowed to drink milk, which was later pumped out of their stomachs. Unlike fresh milk, this predigested milk was found to be rewarding when it was injected into the stomachs of recipient animals. Animals would perform a response for intragastric injections of predigested milk, but they would not do so for fresh milk. Predigested milk either leaves the stomach more slowly, or else its entry into the duodenum is not aversive.

Deutsch, Puerto, and Wang (1977) performed a more definitive test of the possible aversive effects of raw food in the duodenum. They placed small sections of rubber tubing inside the *pyloric sphincter,* a ring of smooth muscle located between stomach and duodenum. The pyloric sphincter normally controls the rate of stomach emptying. The rubber tube kept the sphincter open, so that food received by the stomach was quickly dumped into the duodenum (See **FIGURE 12.15.**)

All rats in the experiment were given free access to dry food and water. In addition, fresh or predigested milk was available. After five days of this regimen, control animals (with no tube in the pyloric sphincter) drank approximately 12 ml of fresh milk each day. In contrast, operated animals drank less than 3 ml per day. However, operated animals who were given predigested milk drank as much as normals—in excess of 12 ml per day. Clearly, the entry of raw nutrients into the duodenum is aversive, but the entry of predigested milk is not. As Deutsch (1978) puts it, "We are only likely to observe normal satiation through stomach injection when the nutrient arriving in the upper gastrointestinal tract is in a physiologically appropriate form or when it evokes physiologically appropriate processes. It seems we cannot simply cut . . . one step out of the digestive sequence and still have the following steps occur normally" (p. 143).

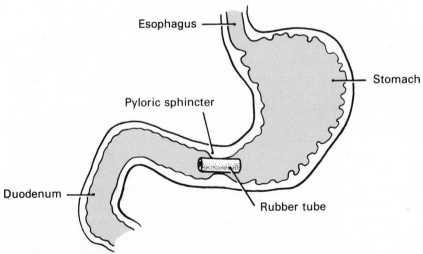

FIGURE 12.15 Deutsch and his colleagues placed a piece of rubber tubing in the pyloric sphincter so that the stomach contents could readily enter the duodenum.

EVIDENCE FOR GASTRIC FACTORS THAT PRODUCE SATIETY. Fortunately for our understanding of satiety mechanisms, Deutsch and his colleagues have done more than shed doubt on previous studies that used intragastric or intraduodenal injections of raw nutrients. They have also provided solid evidence for the existence of gastric receptors. The results of their research also suggest that duodenal factors, if they exist, are of lesser importance.

Davis and Campbell (1973) allowed rats to eat their fill, and shortly thereafter removed food from their stomachs through an implanted tube. When the rats were permitted to eat again, they ate almost exactly the same amount of food that had been taken out. This strongly suggests that the animals are able to monitor the amount of food that is held in their stomachs. The fact that rats ate in response to the removal of even small amounts of food suggested to the authors that stretch receptors, and not nutrient receptors, were important. After all, even after some food is removed from the stomach, nutrient receptors should still be stimulated by the food that remains. As we shall see, Deutsch and his colleagues suggest another possibility.

Deutsch and his colleagues (Deutsch, 1978; Deutsch, Young, and Kalogeris, 1978) attached an inflatable cuff around the pyloric region that worked on the principle of the cuff that is used to obtain a person's blood pressure. That is, when water was pumped through the tube, the cuff would inflate and compress the gut, thus preventing the stomach from emptying. Therefore, food could be confined to the stomach, eliminating duodenal factors. After observations were made, the water could be released so that the cuff would deflate and allow the stomach to empty normally.

The authors also attached a pressure-relief valve to the stomach tube, so that overfilling of the stomach would not be a factor in stopping the animals' eating. One might argue that when the pylorus is artificially blocked, pressure within the stomach could build up to abnormally high levels as the animal eats, thus shutting off further consumption even in the face of continued hunger. A normal rat has a sort of built-in safety valve; food can escape to the duodenum. The artificial pressure-relief valve was devised to allow food to escape through the stomach tube before pressure in the stomach would reach abnormally high levels.

Deutsch and his colleagues found that under these conditions hungry rats would eat less food when 8 ml of milk was injected into their stomachs 30 minutes before they were given the opportunity to eat. A preload of a similar amount of saline did not suppress food intake. Since the pyloric cuff prevented food from entering the duodenum, it is unlikely that the suppressive effect was simply an aversive one, since the aversion appears to be produced by entry of raw food into the duodenum. Thus, it appears that receptors in the stomach were responsible for suppressing subsequent food intake. In another experiment using the pyloric cuff, and thus ruling out duodenal factors, the investigators confirmed the results of Davis and Campbell; rats will replace food that is withdrawn from the stomach. The stomach appears to contain receptors that can monitor both volume and nutritive value of its contents.

We are left with the fact that the stomach signals satiety only when it is full of nutrients. An animal eats when its stomach is only partially full of nutrients (some having been removed), which indicates that volume must be measured. However, nonnutritive bulk does *not* produce satiety, so simple volume receptors cannot explain gastrically-mediated satiety. Deutsch (1978) suggests a possible mechanism to account for these results. The stomach walls contain nutrient receptors whose activity is proportional to the concentration of nutrients in the gastric contents. However, many receptors are buried deep in the walls of the mucous membrane that lines the stomach. As the stomach fills and expands, more and more of these receptors are exposed to the gastric contents. Thus, small volumes of rich nutrients stimulate high rates of firing in small numbers of receptors, producing little satiety. Similarly, large volumes of dilute nutrients stimulate low rates of firing in large numbers of receptors, producing little satiety. Only large volumes of rich nutrients stimulate high rates of firing in large numbers of receptors, and hence produce satiety. This hypothesis has yet to be proved, but it does fit the data very nicely. I am sure that it will be experimentally tested in the next few years.

A role for cholecystokinin?   As we saw in an earlier section, cholecystokinin (CCK), a hormone, is released by the duodenum when food is received from the stomach. CCK release is particularly stimulated by the presence of fats and stomach acid. This hormone causes the gallbladder to contract and inhibits gastric emptying. Since blood levels of CCK might be expected to be related to the amount of nutrients that the duodenum receives from the stomach, perhaps CCK is monitored by the brain as a satiety signal. In fact, many studies have indeed found that injections of CCK suppress eating (Gibbs, Young, and Smith, 1973a,b; Mueller and Hsiao, 1978).

The suppressive effect of CCK on eating is well established. However, Deutsch and Hardy (1977) found that the suppressive effect does not appear to be identical with natural satiety. They gave thirsty rats drinks of flavored water. Flavors A and B were received on alternate days. Experimental rats received injections of CCK after drinking flavor A. Control rats received injections of a saline solution. No injections were paired with flavor B for either group. On the test day that followed, the rats were allowed to choose between both flavors. Control rats drank similar amounts of both flavors: 8.6 ml of flavor A (previously paired with saline) and 8.9 ml of the unpaired flavor B. In contrast, experimental rats drank 3.3 ml of flavor A (previously paired with CCK) and 9.4 ml of flavor B. The results provide good evidence that CCK has an aversive effect.

Sturdevant and Goetz (1976) have also obtained evidence that CCK injections are unpleasant. Human volunteers were given doses of CCK that are rather small, in comparison with the amounts that are generally used to suppress eating in experimental animals. Eating was *increased* by 22%. Larger doses produced nausea and abdominal cramping. Therefore, large doses of CCK would undoubtedly suppress eating in humans, but one would could not refer to this suppression as *satiety*.

The available data do not rule out the possibility that CCK produces a satiety signal to the brain, but since it appears that amounts of CCK that reduce food intake also produce an aversive effect, it is possible that the experimentally-administered doses are much larger than the amounts of CCK that are released by the duodenum. This issue will have to be resolved by further research.

Liver Factors

The role of the liver in control of food intake has only recently been appreciated. Studies in the past few years have shown that besides being a biochemical factory of crucial importance in metabolism, the liver also appears to be the source of signals that are important to the control of food intake.

Mauricio Russek is the scientist who called people's attention to the role of the liver in regulation of food intake (Russek, 1971). He first noted that although intravenous (IV) injections of glucose were relatively ineffective in reducing food intake, **intraperitoneal** (IP) injections (that is, into the abdominal cavity) were quite effective in producing satiety. This observation certainly does not favor the hypothesis that brain glucoreceptors are important in food regulation. An IV injection of glucose raises the blood sugar level considerably, whereas an IP injection raises it very little. Most of the glucose injected into the abdominal cavity is taken up by the liver and stored as glycogen. The fact that the glucose that was injected IP probably got no farther than the liver, but nevertheless produced satiety, suggested to Russek that the liver might contain receptors that were sensitive to glucose.

Russek then attached two chronic cannulas in a dog, one in the hepatic portal vein (the system that carries blood from the intestines to the liver) and another in the jugular vein. An injection of glucose into the hepatic portal vein produced long-lasting satiety, whereas a similar injection into the jugular vein had no effect on food intake. Some other authors (Stephens and Baldwin, 1974; Bellinger, Trietley, and Bernardis, 1976) were unable to replicate Russek's findings, but the discrepancy was later shown to be caused by procedural differences (Russek, Lora-Vilchis, and Islas-Chaires, 1980).

Peñaloza-Rojas and Russek (1963) found that blocking neural conduction in the vagus nerve, which connects the liver to the brain, decreased eating in hungry cats. Thus, it seems likely that receptors in the liver monitor the level of nutrients that are being received, via the hepatic portal blood supply, from the intestines. If food is being received, then the liver passes this information to the brain via the vagus nerve, and hunger is suppressed.

NATURE OF THE SYSTEM VARIABLE.   What is the nature of the substance or substances that appear to be monitored by receptors in the liver? Most neuroscientists have not thought of the liver as an organ with any significant sensory functions, but a recent review (Sawchenko and Friedman, 1979) shows that nerve endings in the liver are sensitive to many different stimuli, including the osmotic pressure of the blood plasma, the concentration of the sodium ion, blood pressure, and various metabolic fuels. The sensitivity of the liver receptors to metabolites is of the most relevance for this chapter, but, unfortunately, the data are the least clear on this topic. Behavioral studies, some of which I shall describe later, suggest that the liver is sensitive to a variety of nutrients. That is, the receptors appear to inform the brain about the amount of available nutrients, regardless of their exact nature. However, electrophysiological studies, in which recordings are made from single axons in the nerves that serve the liver, have not yet provided decisive evidence.

Niijima (1969) found that intraportal infusions of glucose, but not other sugars, would alter the rate of firing of axons in the vagus nerve. Some of the sugars that failed to alter the rate of firing could nevertheless be metabolized by the liver. However, later experiments by the same author (Niijima, 1977) indicated that liver receptors were also sensitive to the presence of the substances into which glucose is converted during metabolism. Receptors in the liver even responded to injections of ATP (adenosine triphosphate), which is the energy-storing molecule that is produced by *all* metabolic fuels. This finding suggests that the rate of firing of at least some receptors is determined by the amount of energy that is available, regardless of its original source. Other investigators have failed to find evidence for receptors that are sensitive to glucose (Andrews and Orbach, 1974), but have found some that respond to another metabolic fuel, fatty acids (Orbach and Andrews, 1973). It is obvious that we need more electrophysiological studies to clarify the sensory functions of the liver.

Russek and Racotta (1980) have suggested a mechanism that could serve to unify much of the evidence concerning the nature of the system variables that are monitored by the liver. They suggest that there is a class of liver cells whose membrane potential is proportional to the presence of intermediate products of metabolism, such as pyruvate. In turn, the membrane potential of the liver cell would determine the rate of firing in nerve fibers with which they communicate. Pyruvate serves as an intermediate step in the metabolism of glucose, glycogen, and amino acids. It is not directly in-

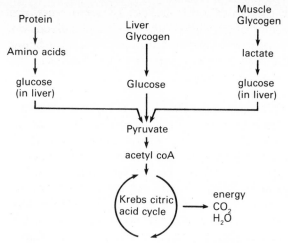

FIGURE 12.16   Metabolism of protein, liver glycogen, and muscle glycogen.

volved in lipid metabolism. (See **FIGURE 12.16.**)

Russek and Racotta hypothesize that the presence of lactate and/ or pyruvate would hyperpolarize the membrane potential of the receptors in the liver, thus decreasing the rate of firing of the "hunger fibers" connecting the liver with the brain. Feeding would consequently be suppressed. They note a number of pieces of evidence that support this hypothesis.

1. Intraperitoneal injections of lactate and pyruvate are even more satiating than injections of glucose.
2. Strenuous exercise is followed by a period of *anorexia* (loss of appetite). This might be accounted for by the fact that prolonged muscular activity results in the metabolism of glycogen in the muscles, which results in an increased level of lactate in the blood. The lactate might hyperpolarize the receptors in the liver, as pyruvate is hypothesized to do, and suppress hunger.
3. Injections of epinephrine suppress hunger if the liver contains sufficient stores of glycogen. The epinephrine, which causes the glycogen to be broken down and metabolized, increases the level of pyruvate.
4. Injections of 2-DG, which elicit eating, decrease the level of pyruvate in the liver.

Russek and Racotta's hypothesis is very interesting. It accounts for the fact that a variety of different nutrients can cause satiety, since most of them increase the level of pyruvate in the liver. However, the hypothesis does not account for the findings of the study by Friedman (1978). As we saw in an earlier section of this chapter, he found that diabetic rats, which cannot utilize carbohydrates, will regulate their intake of food in proportion to the amount of fat that the diet contains. Since lipids are not converted to pyruvate when they are metabolized, this study provides evidence for control of hunger that is independent of this substance. However, this finding does not necessarily contradict Russek and Racotta's hypothesis; after all, hunger might very well be controlled by more than one metabolic signal.

LIVER RECEPTORS AND THE CONTROL OF EATING.   The liver appears to contain receptors that monitor some aspect of metabolism, and thus measure the availability of a variety of nutrients. Since the liver is the first organ to receive water-soluble nutrients that are absorbed from the digestive system, it is in a good location to supply a satiety signal to follow up satiety signals that are produced by a stomach that is full of nutrients. Receptors in the liver might also play a role in the onset of hunger, since they are the first to "know" that the liver's supply of glycogen is being depleted, and that no more food is being absorbed from the intestines. Electrophysiological studies suggest that these receptors fire at a high rate when the amount of nutrients in the portal blood supply is low, and this neural activity, transmitted to the brain, might be one of the factors that stimulate food intake. Perhaps the fact that liver diseases such as hepatitis cause a severe decrease in appetite (Kassil, Ugolev, and Chernigovskii, 1970) is a result of temporary depression of the firing rate of the hepatic receptors, caused by inflammation of the liver.

There is good evidence that a treatment that decreases the availability of food to the liver, but leaves the brain unaffected, causes hunger. Novin, VanderWeele, and Rezak (1973) injected 2-DG directly into the hepatic portal vein. (As we saw, 2-DG suppresses glucose metabolism) The injections led to *immediate* large (over 200 percent) increases in food intake. The latency between injection and feeding was short indeed; the animal usually began to eat during the injection.

Another study, (Friedman, Rowland, Saller, and Stricker, 1976; Stricker, Rowland, Saller, and Friedman, 1976) demonstrated the fact that hunger is controlled by fuels that can be utilized by the liver, and not by the brain. The investigators injected rats with insulin, which made them hypoglycemic, and then administered injections of several different nutrients. Injections of insulin normally produce eating. The question was this: which metabolic fuels would satisfy the hunger produced by the hypoglycemia, and thus prevent eating? Injections of the sugars glucose, fructose, or mannose all produced satiety; they blocked the eating that normally follows injections of insulin. Since fructose cannot cross the blood-brain barrier, the brain should have been "hungry." And yet, the animals did not eat. Injections of a ketone that can be used by the brain but not by the liver did not abolish the feeding response. The animals' brains were "satiated" by the ketone; nevertheless, the rats ate. Apparently the "hungry" liver stimulated the feeding. (A large dose of the ketone

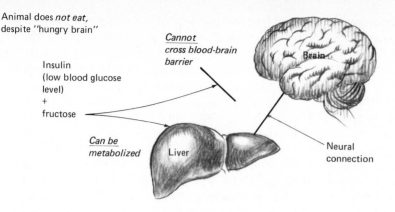

Animal does *not eat,*
despite "hungry brain"

Cannot
cross blood-brain barrier

Insulin
(low blood glucose level)
+
fructose

Brain

Can be metabolized

Liver

Neural
connection

Animal eats,
despite "sated brain"

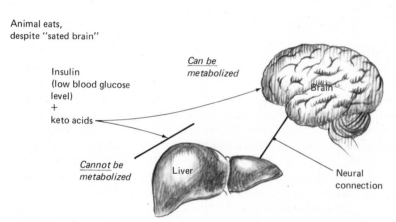

Can be metabolized

Insulin
(low blood glucose level)
+
keto acids

Brain

Cannot be metabolized

Liver

Neural
connection

FIGURE 12.17   Summary of the experiments by Friedman and his colleagues.

did inhibit eating, but so did a control dose of hypertonic sodium chloride. This suggests that the suppressive effect was produced by osmotic stress, and hence was not related to true satiety.) (See **FIGURE 12.17.**)

It is significant that fructose produced satiety in rats that had a low blood glucose level. Fructose can be metabolized without first being converted to glucose. This fact strongly suggests that the liver receptors are not simply glucose receptors. They might, as Russek hypothesizes, monitor the level of intermediate products of metabolism, such as pyruvate.

Friedman and his colleagues also found evidence for nutrient receptors in the brain, but these receptors did not appear to play a role in suppressing hunger. The investigators examined the effects of the metabolic fuels on the output of adrenal epinephrine in response to the low blood sugar level that had been produced by the insulin injections. Secretion of epinephrine is a compensatory response that occurs when an animal's blood glucose level is low; the hormone causes liver glycogen to be converted into glucose. The epinephrine response appeared to be controlled by receptors in the brain. Injections of ketone and glucose, which are fuels that the brain can use, abolished the epinephrine response. (The lower dose of ketone, which did not suppress eating, successfully blocked secretion of epinephrine.) However, fructose, which does not enter the brain, had no effect. Therefore, brain receptors apparently control the release of epinephrine by the adrenal glands in response to its own metabolic need. And like the liver receptors, the brain receptors appear to monitor metabolic rate, and not simply glucose level, since they respond to ketones as well as to glucose.

In a follow-up study, Granneman and Friedman (1978) operated on a group of rats, cutting the branch of the vagus nerve that supplies the liver. This operation prevented liver receptors from sending information to the brain. The investigators made the rats hungry with injections of insulin, and then attempted to suppress this hunger with fructose injections. Despite the injections of fructose, the vagotomized animals continued to eat. (Control rats, whose vagus nerves had not been cut, did not eat.) Thus, hunger persisted in animals that were not receiving afferent information from the liver. This observation might be explained as follows: The hypoglycemia that is produced by insulin made the brain and liver hungry. The fructose, which was given next, fed the liver. In normal, unoperated animals, the liver sent a satiety signal to the brain that suppressed eating. In rats whose vagus nerves had been cut, this satiety signal never reached the brain.

The results implicate brain receptors in hunger. If the *only* receptors for hunger were located in the liver, a vagotomized animal should never eat. But, as we saw, vagotomized animals do indeed eat in response to hypoglycemia. The results of this study, along with the previous one, suggest that the brain contains receptors that can stimulate hunger even in the absence of information from the liver, while the liver contains receptors that can produce both hunger and satiety.

Conclusions Concerning the Nature of the System Variables and Detectors

It really is not necessary for me to say that the control of food intake is extremely complex; I am sure that you are convinced of that fact. I will try to summarize what we have learned about the factors that produce hunger and satiety. Hunger seems to be stimulated by the relative lack of available fuels, as determined by receptors whose rate of activity depends on their metabolic rate. If fuel is readily present, they fire at a high rate. If it is absent, they fire at a low rate.

The low rate of firing is translated, by the brain, into hunger. The animal seeks out food.

Head factors (sight, smell, taste, texture, feedback from swallowing) determine the palatability of a food, and, if the food is readily accepted, engage a positive feedback mechanism that is later overridden by negative feedback, or satiety. Some satiety signals originate from head factors (as is shown in sham-feeding experiments), but gastric factors seem, by far, to be more important. Gastric distention causes neural inhibition of intake, but this factor appears to operate only at levels of distention higher than are encountered when an animal eats a normal meal. As Deutsch and his colleagues have shown, gastric nutrient receptors seem to supply the most important peripheral satiety signal. Duodenal factors appear to be less significant, although the issue is by no means settled. Finally, the liver, with its receptors, responds to the presence of nutrients in the blood received via the hepatic portal system, and it continues to suppress eating until the meal is digested and absorbed.

The work of Liebelt, who showed that total body fat is regulated, suggests that lipids appear to be important in long-term food regulation. How this factor interacts with the onset or cessation of a meal is unknown.

Finally, I should mention the importance of social factors and of learning. I said little about these factors because we do not know much at all about their physiological bases. Our ingestion of three meals a day, at regular times, probably defeats some of our regulatory mechanisms. And our habit of following a meal with another course of increased palatability (dessert) raises our caloric intake above what we need. (You will recall that a rat, successively given four flavors of the same diet, will eat much more than it would normally take of only one of them.) Finally, as we all know, social factors are important. Even a chicken, satiated by a large meal, will begin to eat again if it sees another chicken eating nearby.

BRAIN MECHANISMS INVOLVED IN FOOD INTAKE

I have summarized our current knowledge of the system variables involved in hunger and satiety, and the nature and location of the detectors that monitor these variables. Now it is time to turn to neural mechanisms that underlie the process of eating itself. Traditionally, research emphasis has focussed on a ventromedial hypothalamic "satiety center" and on a lateral hypothalamic "feeding center." As we shall see, these notions are being abandoned as a result of more recent evidence.

Anatomy of the Hypothalamus

The hypothalamus is of major importance in the regulation of food intake. Since this section will describe the results of physiological manipulations of hypothalamic nuclei and associated fiber systems, a brief introduction to the anatomy of the hypothalamus is in order. (See **FIGURE 12.18.**)

The hypothalamus, which lies at the base of the brain, can be divided into three zones along the lateral-medial axis. The periventricular region surrounds the third ventricle. The cells of this area

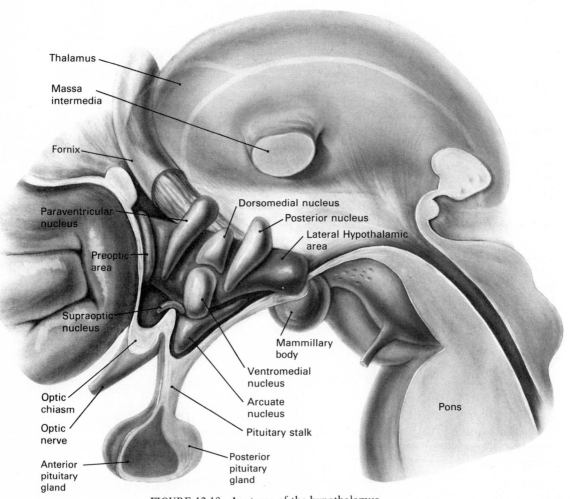

FIGURE 12.18 Anatomy of the hypothalamus.

are generally small. One part of this region (the arcuate nucleus) contains many neurosecretory cells involved in control of the pituitary gland. The second zone, the medial region, contains most of the hypothalamic nuclei, including the supraoptic and paraventricular nuclei (which, among other things, are involved in control of the posterior pituitary gland), the ventromedial nucleus (implicated, perhaps, in mechanisms of satiety), and the dorsomedial nucleus. The lateral region, through which pass the fibers of the medial forebrain bundle, is characterized by a diffuse organization of cell bodies and axons.

Rostrally, the hypothalamus is bounded by the preoptic region; there are pairs of medial and lateral preoptic nuclei. Some people consider the preoptic region to be a part of the hypothalamus, although these two brain regions have different embryological origins. The mammillary bodies are located at the caudal, ventral end of the hypothalamus. Within these bulges are located several sets of nuclei, the most prominent being the lateral and medial mammillary nuclei.

I shall not say too much, at this point, about the fiber connections of the hypothalamus, since an extensive description would not be of much use in discussing what we presently know about the role of the hypothalamus in food intake. For present purposes, we should realize that the various hypothalamic nuclei are intricately interconnected, and that the hypothalamus receives information from motor systems and from olfactory, gustatory, visual, and somatosensory systems. It also receives input from extensive areas of the limbic system, which is involved in motivation and emotional expression. The hypothalamus sends fibers to many parts of the brain, including thalamus, cortex, and motor systems. The most visible fiber tracts are the medial forebrain bundle (which connects the hypothalamus with midbrain structures), the fornix (which interconnects hippocampus, septum, and hypothalamus), and an efferent system that divides into the mammillothalamic and mammillotegmental tracts (connecting the mammillary bodies of the hypothalamus with the anterior thalamus and the motor systems of the tegmentum).

The Ventromedial Hypothalamic Syndrome

Tumors in the hypothalamic-pituitary stalk region were long ago shown to be associated with extreme obesity and, often, with genital atrophy. It was later shown that the genital atrophy resulted from loss of gonadotrophin secretion, but the obesity was independent of hormonal changes. The overeating and weight gain could be produced by lesions in the region of the **ventromedial nucleus** of the hypothalamus—the VMH (Hetherington and Ranson, 1939). (As we shall see, it is the *region* of the VMH, and not the nuclei themselves, that is involved in this syndrome.)

The VMH obesity syndrome seems to follow two stages (Brobeck, Tepperman, and Long, 1943). During the **_dynamic phase_** (a period of hyperphagia and weight gain lasting 4 to 12 weeks) the rats eat avidly, but as their weight increases, their motivation for food decreases (Miller, Bailey, and Stevenson, 1950). They become less willing to work for food or to tolerate aversive stimuli presented along with the food. Given a choice between food plus footshock and nothing, they will take nothing.

As the animals enter the **_static phase_** of the VMH obesity syndrome, they begin to eat less food, and their weight stabilizes. At this point the animals are very finicky; they will not eat food that is adulterated with quinine or diluted with cellulose. Only good-tasting food is consumed, and then only if it is easily available.

Some people have suggested that VMH lesions produce obesity by enhancing the rewarding properties of food rather than by damaging satiety mechanisms. Powley (1977) has marshalled a considerable amount of evidence that suggests that VMH lesions enhance the **_cephalic phase_** of metabolism. The cephalic phase refers to salivation, insulin secretion, gastric acid secretion, and other physiological changes that are triggered by head factors before and during eating. These changes provide positive feedback that increases hunger and thus reinforces an animal's intake of palatable foods. Although this mechanism is undoubtedly an important component of the VMH syndrome, it is probably not solely responsible for the overeating and obesity. For example, brain lesions in other areas cause similar enhancements in the appetitive value of food without producing obesity (Beatty and Schwartzbaum, 1968). Furthermore, rats with VMH lesions will overeat (by pressing a lever) even if the food is injected into the stomach and thus cannot be tasted (McGinty, Epstein, and Teitelbaum, 1965). And, as we shall see, VMH lesions appear to produce primary changes in an animal's metabolism.

The fact of VMH obesity, along with the observation that the region contains cells that might be directly sensitive to glucose levels and to stimulation of chemoreceptors and stretch receptors in the digestive tract, has led to suggestions that the VMH might be a "satiety center." Electrical stimulation of the VMH has been shown to produce a cessation of eating, further suggesting that the normal role of this region is to inhibit food intake (Wyrwicka and Dobrzecka, 1960). However, no strong conclusions can be drawn from the effects of VMH stimulation. Such stimulation has been shown to be aversive (Krasne, 1962); an animal will work to turn the electrical current off. The aversive effects might just be incompatible with eating. (Suppose, for example, that someone applied an electric shock to your big toe during a meal. The fact that you stopped eating would not implicate your toe in satiety mechanisms.)

Two findings have cast doubt on the role of the VMH as a satiety

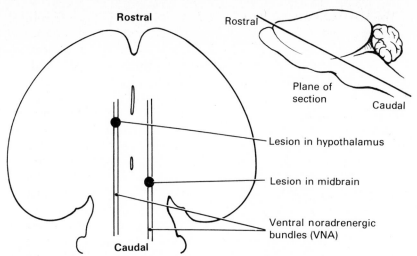

FIGURE 12.19   Schematic representation of the rationale for the asymmetrical lesion technique used by Kapatos and Gold (1973).

center. It now appears that the lesion destroys fibers that travel in a rostral-caudal direction along the dorsal border of the nucleus. Lesions restricted to the VMH were shown to be ineffective in producing hyperphagia and obesity (Gold, 1973). However, electrolytic lesions or knife cuts placed along the ***ventral noradrenergic bundle (VNA)*** *did* produce obesity (Kapatos and Gold, 1973). The lesions did not have to be bilaterally symmetrical to produce hyperphagia; that is, obesity occurred after a unilateral cut of the VNA at the level of the ventromedial nuclei along with a unilateral lesion, on the other side of the brain, in the VNA at the level of the midbrain. The asymmetrical lesions each cut half of the fibers, at different levels along the neuraxis. (See **FIGURE 12.19.**)

A more recent study (Gold, Jones, Sawchenko, and Kapatos, 1977) indicates that the fiber system whose destruction results in obesity travels from the midbrain through the ventral noradrenergic bundle, terminating in the region of the paraventricular nuclei (PVN). A knife cut was made just lateral to the PVN, and a contralateral knife cut was made in a wide variety of locations in the hypothalamus (in different animals, of course). Figure 12.20 illustrates the results. The black bars represent the knife cuts; the numbers beside them indicate the average daily intake of food. Note that knife cuts anterior to the paraventricular nuclei did not produce overeating. Also, note that parasagittal cuts (parallel to the midline—that is, up-and-down in the illustration) did not produce overeating except for one specific location: just even with the paraventricular nuclei. (See **FIGURE 12.20** on the following page.) This implies that the fibers ascend and then turn medially to terminate in the paraventricular nuclei.

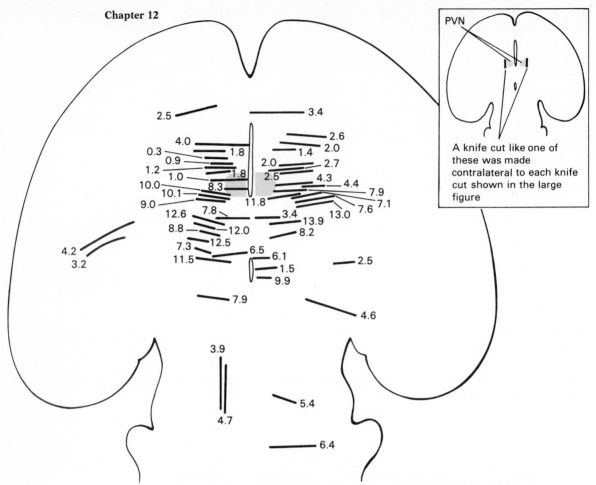

FIGURE 12.20 Results of the study by Gold, Jones, Sawchenko, and Kapatos (1977). A knife cut was placed parallel to the midline, just lateral to the paraventricular nucleus, on one side of the brain (insert). Another knife cut was made perpendicular to the midline, on the other side of the brain. Black lines represent the knife cuts; numbers adjacent to them represent average daily food intake.

A study by Powley and Opsahl (1974) appeared to indicate that the overeating produced by VMH lesions is mediated by the vagus nerve. They found that VMH obesity could be abolished by severing the vagus nerve. However, King, Carpenter, Stamoutsos, Frohman, and Grossman (1978) found that if vagotomies were performed 90 days before the VMH lesions were made, the rats overate and became obese. The authors pointed out that vagotomy interferes with swallowing and digestive functions of the stomach, so perhaps vagotomy inhibits food intake in a nonspecific way. If vagotomies were done 90 days before the VMH lesions were made, the animals would have time to adapt to these effects, and would be able to overeat immediately after the VMH destruction. Whatever the correct explanation may be, the study does prove that the VMH syndrome can occur even after the vagus nerve is severed.

Another possible explanation for VMH overeating and obesity is hyperinsulinemia. If insulin is administered every day, an animal will overeat and become obese (MacKay, Calloway, and Barnes, 1940). Since VMH lesions result in an increased secretion of insulin (Frohman and Bernardis, 1968), perhaps the increased food intake is a secondary response to the insulin. However, a number of studies have shown that diabetic rats who are given controlled doses of insulin begin to overeat when their VMH is destroyed (e.g., Vilberg and Beatty, 1975). The hyperinsulinemia that follows VMH lesions may be a contributing factor to the overeating and obesity, but it does not appear to be the sole cause.

According to Friedman and Stricker (1976), the best explanation for VMH obesity appears to be that the lesions alter metabolism so that the animal is constantly in the absorptive phase. These authors review evidence (e.g., Frohman, Goldman, and Bernardis, 1972) that shows that VMH lesions cause an increased conversion of glucose into fat and also reduce the breakdown and metabolism of fats. The growth of fat tissue occurs even when food is not available—when the animal should be in the fasting phase. The means by which VMH lesions cause the uptake of nutrients by adipose tissue are still unknown; control of the known factors (insulin and growth hormone) does *not* prevent the effect. An animal that is continuously in the absorptive phase, then, overeats because of *tissue need*, and not because of damage to a satiety center. VMH lesions appear to prevent adipose tissue from acting as "a reversible sponge," taking up excess nutrients during the absorptive phase and releasing them during the fasting phase. Instead of performing its normal function, adipose tissue continuously absorbs nutrients. Consequently, the animal with a VMH lesion must eat more often in order to maintain a proper energy balance.

Whatever the cause of the overeating and obesity that are produced by VMH lesions, it is not simply damage to a satiety mechanism. The many hundreds of studies devoted to the VMH syndrome have told us much about what VMH-lesioned animals are like, but unfortunately they have told us precious little about the neural mechanisms of satiety.

OBESITY IN HUMANS. Schachter (1971) has drawn a number of parallels between the etiology of the VMH obesity syndrome and obesity in humans. For instance, obese human patients drink much less of a milkshake adulterated with quinine and much more of a normal, good-tasting milkshake (i.e., obese humans are more finicky than normal subjects). Furthermore, they eat fewer nuts than do slim people if the nuts have shells on them, but they eat more nuts if the shells have already been removed (they are less willing to work for their food). However, the fact that other brain lesions produce finickiness and overeating in response to palatable food without producing hyperphagia and obesity suggests that Schachter's comparisons might be misleading.

A possible explanation for at least some cases of obesity in humans is given by Rowland and Antelman (1976). These authors report that a mild (and apparently *not* painful) pinch applied to rats' tails twice a day caused daily caloric intake to go up by 129 percent, as compared with unpinched control subjects that had access to the same food (sweetened condensed milk, which rats appear to find delicious). This tail-pinch phenomenon appears to be related to release of brain dopamine in response to mild stress; I shall discuss it in more detail in the next section. Rowland and Antelman suggest that stress might similarly cause overeating in humans. Perhaps obesity in itself (and the negative social factors that accompany it) provides enough stress to cause overeating and, hence, to produce a vicious circle from which it is difficult to escape.

The Lateral Hypothalamic Syndrome

Whereas destruction of the ventromedial region of the hypothalamus (actually, the ventral noradrenergic bundle) causes hyperphagia, destruction of the lateral area causes **adipsia** (absence of drinking), **aphagia** (absence of eating), and weight loss (Anand and Brobeck, 1951). Electrical stimulation of the same region was found to elicit eating. This area very quickly came to be called the "feeding center," in contrast to the VMH "satiety center." It was found, however, that rats with lateral hypothalamic (LH) lesions, if they were carefully nursed with intragastric injections of food, would gradually recover (Teitelbaum and Stellar, 1954). The recovery was characterized by several stages. First the animals began to accept wet, very palatable food (water-soaked Sunshine chocolate-chip cookies seemed to work best); then they would accept their normal diet, although they still would not drink. Consequently, the food had to be moist. Finally, the animals began to drink and to eat dry food. However, all the water was consumed with meals, suggesting that they only drank in order to eat (Teitelbaum and Epstein, 1962).

"Recovered" animals with lateral hypothalamic lesions are also abnormal in other ways. They do not eat in response to 2-DG (as we saw earlier). However, we also saw earlier that rats with LH lesions, who do not eat in response to a sudden, drastic fall in the blood glucose level, can nevertheless regulate their food intake very nicely so long as they are not subjected to undue stress.

What accounts for the temporary aphagia and for the recovery of eating? Teitelbaum (1971) has suggested that the recovery depends on the "encephalization" of hunger; that is, cortical areas take over the function of the damaged tissue. In support of this notion, Teitelbaum and Cytawa (1965) found that temporary depression of cortical activity (by applications of potassium chloride to the surface of the cortex) produced long-lasting suppression of food intake in rats that had "recovered" from the effects of LH lesions. This experiment suggests that the cortex had taken over some of the functions of damaged tissue. Teitelbaum has noted that the recovery from LH lesions resembles the development of many systems in infant animals (animals with these lesions also show profound sensory and motor disturbances, as well as the aphagia). He has suggested that investigation of the similarities might provide important insights into both the developmental process and the nature of the recovery process.

Powley and Keesey (1970) made an observation that suggests that the slow recovery of ingestive behavior might not depend on time, but on a gradual reduction in body weight to a new set point. They noted that animals that had recovered from LH lesions were able to regulate their weight, albeit at a lower level than normal. They produced LH lesions in two groups of animals, a normally fed group and a group that had been starved down to a weight lower than the weight that recovered LH animals usually attained. Their results are shown in Figure 12.21. Note that the prestarved animals immediately began to eat and in fact gained a little weight. (See **FIGURE 12.21.**) The lesioned animals previously fed *ad libitum* were aphagic until their weight fell to the level at which they eventually stabilized (**FIGURE 12.21**).

These results suggest that somehow the set point for body weight

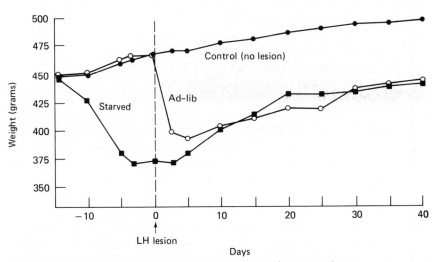

FIGURE 12.21   Effects of preoperative starvation on the LH syndrome. (From Powley, T. L., and Keesey, R. E. *Journal of Comparative and Physiological Psychology*, 1970, *70*, 25–36. Copyright © 1970 by the American Psychological Association. Reprinted by permission.)

is altered by LH lesions. At least, the threshold of hunger mechanisms appears to be raised; a stronger signal is needed to elicit eating. The stages of recovery might, therefore, only reflect the time it takes to lose weight. However, as we shall see, there is a simpler explanation for the entire LH syndrome.

I noted earlier that animals with LH lesions exhibit sensory and motor deficits. In fact, some studies indicate that aphagia and adipsia can be produced by damage to neural systems that have generally been characterized as sensory or motor in function. It appears that the ***nigrostriatal bundle***, a collection of dopaminergic fibers from the substantia nigra to caudate nucleus (a part of the extrapyramidal motor system), plays an activating role in feeding behavior (and also in other behaviors). This fiber bundle is usually damaged by LH lesions that produce aphagia. Electrolytic lesions of the substantia nigra and chemical destruction of the nigrostriatal fibers with 6-hydroxydopamine (6-HD) result in aphagia, adipsia, and decreases in movement (Ungerstedt, 1971). Other studies have shown that these lesions produce long-lasting finickiness, drinking only during meals, and other symptoms seen in animals with LH lesions (Fibiger, Zis, and McGeer, 1973). Furthermore, other lesions (such as those of the globus pallidus, another part of the extrapyramidal motor system) that produce adipsia and aphagia (Morgane, 1961) lower the concentration of dopamine in the caudate nucleus (Anden, Fuxe, Hamberger, and Hökfelt, 1966). The nigrostriatal dopaminergic system thus appears to be very important in feeding behavior and in other motor deficits seen after LH lesions.

In the previous section I mentioned the rather incredible fact that mild tail pinch will cause a satiated rat to begin eating. Antelman, Rowland, and Fisher (1976) have shown that such tail pinch can reverse the effects of lesions of the lateral hypothalamus or the nigrostriatal dopamine system on eating. When their tails were gently pinched, the aphagic animals would begin to eat. If they were stimulated often enough, a large proportion of these animals would eat sufficient amounts of food to "nurse themselves" to recovery.

In a series of experiments, Antelman, Szechtman, Chin, and Fisher (1975) demonstrated that eating induced by tail pinch results from stimulation of dopaminergic fibers. Drug studies showed that the effect could be blocked by dopamine antagonists but not by norepinephrine antagonists. The authors suggested that the nigrostriatal dopamine system should not be considered a "feeding system," since, in the presence of appropriate goal objects, tail pinch (or electrical shock delivered to the tail) can elicit a variety of other behaviors, such as copulation or aggression (Caggiula, 1972). Perhaps the nigrostriatal dopaminergic system plays a role in the attention to stimuli related to these behaviors or in the arousal of motor systems necessary for their performance.

Yet another piece of evidence against the concept of an "LH feeding center" comes from a study by Zeigler and Karten (1974), who noted that a large number of ascending sensory fibers of the trigeminal system (which mediates somatosensory information from the head and neck region) pass through the brain in the vicinity of the lateral hypothalamus. The authors made lesions in various locations that interrupted the projections of trigeminal information to the ventral posterior thalamus. The lesions (which did not damage the hypothalamus) produced a period of adipsia and aphagia, followed by recovery. The animals subsequently regulated their weight at a new (lower) level. Perhaps the activity of the trigeminal system provides tonic stimulation similar to that produced by pinching the tail. Thus, loss of this tonic input would decrease the activity of the dopaminergic system.

This hypothesis is supported by an experiment by Stricker, Cooper, Marshall, and Zigmond (1979). These authors made lesions in the lateral hypothalamus, which produced a period of aphagia and adipsia, followed by eventual recovery. As I noted earlier, recovered animals with LH lesions do not eat in response to sudden metabolic challenges, such as glucoprivation produced by injections of 2-DG or hypoglycemia produced by injections of insulin, although they will regulate their food intake if the metabolic changes occur gradually. Stricker and his colleagues injected their recovered rats with 2-DG and performed a series of neurological examinations. They found that the animals showed gross sensorimotor impairments similar to those that occur immediately after the lesion is made. Thus, the metabolic challenge reinstates the original sensorimotor deficit, perhaps by interfering with the activity of neurons that play a role in nonspecific arousal. Furthermore, these deficits could also be reinstated by injections of a drug that blocks postsynaptic dopamine receptors (spiroperidol). The evidence supports the hypothesis that LH lesions impair feeding by interfering with brain mechanisms that are not specific to food intake.

So—does the lateral hypothalamus play a special role in hunger, or does it just happen to be located near fibers of some important sensory and motor systems that are involved in a variety of behaviors? We do not yet have a definitive answer to that question. The answer, when it comes, will undoubtedly show that the process is extremely complex, involving many different mechanisms. The hypothalamus itself surely plays some role in hunger, but we certainly must abandon our concepts of feeding centers and satiety centers that, in reciprocal fashion, determine when an animal eats.

The unraveling of the causes of the LH syndrome is proving to be a very tedious process. An excellent review by Grossman (1979) shows just how complicated the story is, and tells us that it is far from complete. We should bear in mind that an animal can be prevented from eating in many ways, including damage to detectors for the relevant system variables that elicit hunger, difficulty in swallowing, inability to interpret food-related environmental stimuli, disturbances in metabolism, as well as "loss of appetite" (whatever that means). When the dust finally settles, we will undoubtedly find out that lateral hypothalamic lesions suppress eating for many reasons, not all of which are of direct relevance to our understanding of the neural control of food intake.

CONCLUSIONS

Physiological psychologists began their study of the control of eating by concentrating on mechanisms above the neck. In particular, they looked for receptors in the brain, and devoted much effort to investigation of the VMH and LH syndromes. Many of the people who studied the neural control of hunger and satiety knew very little about metabolism or the physiology of the digestive system. However, this state of affairs is rapidly changing. In recent years investigators have begun to study the physiology of the entire organism, and not just the brain. These efforts have been very fruitful, as we saw in the case of gastrically-mediated satiety and the important role of the liver in the control of hunger. The sophisticated techniques we now have to study the brain will be complemented by new techniques for studying peripheral metabolism and the neural and hormonal control of the digestive system, which will undoubtedly bring us many more answers to the problems of hunger and satiety.

KEY TERMS

absorptive phase of metabolism p. 369

adipose tissue p. 369

adipsia (*ay DIP see uh*) p. 402

ad libitum p. 377

alloxan p. 377

aminostatic hypothesis (*a MEE no*) p. 375

aphagia (*a FAY ja*) p. 402

bile p. 367

cephalic phase (*seff AL ik*) p. 398

chemoreceptor (*KEE mo*) p. 366

cholecystokinin (CCK) (*koh li sis toh KY nin*) p. 366

correctional mechanism p. 362

detector p. 362

2-deoxyglucose (2-DG) p. 375

diabetes mellitus (*mell EYE tis*) p. 372

duodenum (*doo oh DEE num*) p. 366

dynamic phase p. 398

emulsification p. 368

esophagotomy (*ee soff a GOT a mee*) p. 382

fasting phase of metabolism p. 369

free fatty acids p. 369

glucagon (*GLOO ka gon*) p. 373

glucostatic hypothesis p. 375

glycogen (*GLY ko jen*) p. 369

hepatic portal system p. 367

hyperphagia (*hy per FAY ja*) p. 377

hypoglycemia p. 380

interstitial fluid (*in ter STI shul*) p. 360

intraperitoneal (*in tra pair a tow NEE ul*) p. 389

islet cell (*EYE let*) p. 372

ketones (*KEY tones*) p. 370

lipostatic hypothesis p. 375

negative feedback p. 363

nigrostriatal bundle (*NY grow stry AY tul*) p. 404

SUGGESTED READINGS

BALAGURA, S. *Hunger: A Biopsychological Analysis*. New York: Basic Books, 1973.

FRIEDMAN, M. I., AND STRICKER, E. M. The physiological psychology of hunger: A physiological perspective. *Psychological Review*, 1976, *83*, 409–431.

NOVIN, D., WYRWICKA, W., AND BRAY, G. A. (editors) *Hunger: Basic Mechanisms and Clinical Implications*. New York: Raven Press, 1976.

Annual Review of Neuroscience and *Annual Review of Psychology* often publish up-to-date reviews of research on the physiology of hunger.

13

Thirst and
the Control of
Mineral Intake

As we saw in the introduction to the previous chapter, mammals have evolved regulatory mechanisms to maintain the constancy of their extracellular fluid. The regulation of the intake of nutrients is an extremely complex process, involving several different mechanisms. However, as we shall see in this chapter, much more is known about regulation of the water and mineral content of the extracellular fluid.

In Chapter 12 it was necessary to describe the ingestion, absorption, and assimilation of food, as well as the general metabolic processes by which nutrients are stored and utilized. Similarly, in this chapter I shall describe the physiological processes that control water and sodium balance in the body. Fortunately, these processes are much simpler than the ones involved in regulation of nutrients. I shall restrict my discussion to water and sodium balance, since these substances are most important in water intake, the behavioral process we are interested in. I shall not discuss the specific regulation of minerals other than sodium. However, in the final section I shall describe the process by which an animal can compensate for various deficiencies (of minerals, vitamins, specific amino acids, etc.) by altering its intake of various foodstuffs.

PHYSIOLOGICAL REGULATION OF WATER AND SODIUM BALANCE

Fluid Compartments of the Body

The fluid surrounding the cells of the human body (the interstitial fluid) is approximately iso-osmotic with a 0.87 percent solution of sodium chloride. In other words, if extracellular fluid were placed on one side of a semipermeable membrane (which allowed water, but not other molecules, to pass through) and 0.87 percent NaCl solution were placed on the other, there would be no osmotic gradient and no net migration of water. Interstitial fluid is a protein-free filtrate of blood plasma; it is "squeezed" out of the porous capillaries by the action of blood pressure.

Two forces are involved in the formation (and, as we shall see, circulation) of interstitial fluid: hydrostatic pressure (blood pressure) and osmotic pressure. Hydrostatic pressure, in a typical capillary, is equal to approximately 35 mm Hg at the arterial end, and gradually decreases to approximately 15 mm Hg at the venous end. (The unit of measurement for hydrostatic pressure is the mm Hg—the amount of pressure that will lift a column of mercury one millimeter.) There is extra room in the interstitial space for more fluid, so a "back pressure" to the flow of fluid out of the capillaries is negligible (in normal circumstances). If this process were to go unchecked, of course, all of the fluid would leave the vascular system and enter the interstitial space. This does not happen because the hydrostatic pressure is opposed by osmotic pressure.

As water is squeezed out of the capillaries, so are most of the substances that are dissolved in it. The exception is protein; the walls of the capillaries are impermeable to these large molecules, so they remain in the plasma. The presence of proteins in the plasma, but not in the interstitial fluid, means that there is a concentration difference, and hence a difference in osmotic pressure, between the two fluid compartments. Thus, there is an osmotic pressure gradient that tends to force the interstitial fluid back into the capillaries. This force has been calculated to be equal to a hydrostatic pressure of 25 mm Hg.

Figure 13.1 illustrates the process of plasma filtration. At the arterial end of the capillary an outward hydrostatic pressure of 35

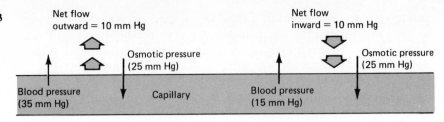

FIGURE 13.1 Filtration of plasma and the ensuing circulation of interstitial fluid.

mm Hg is opposed by an inward osmotic pressure of only 25 mm Hg, so plasma flows out. At the venous end the hydrostatic pressure has decreased to 15 mm Hg, so there is now a net *inward* pressure of 10 mm Hg. The outward flow (at the start of the capillary) and the inward flow (at the end) perfectly balance. (See **FIGURE 13.1.**)

Since this process is characterized by a balance of opposing forces, changes in one of the forces will result in a change in the balance of fluids. For example, if the barrier to protein provided by the walls of the capillaries were eliminated, protein would be filtered out with the plasma, and the differences in the osmotic pressure of the two fluid compartments would decrease. There would then be a net flow of fluid from the blood plasma into the interstitial fluid, since the hydrostatic pressure would be unopposed. If you would like, you can demonstrate this process. Place your thumb on a hard surface, and hit it sharply with a hammer. You will note a rapid increase in the size of your thumb, resulting from leakage of protein through the damaged capillaries and a subsequent accumulation of interstitial fluid. (My publisher advises me to inform you that this suggested "demonstration" should not really be tried.)

There is a very close balance between the osmotic pressure of the extracellular fluid and that of the intracellular fluid. Increases in extracellular water will cause cells to swell as they absorb water, which is travelling down its concentration gradient. Removal of water from the extracellular fluid, on the other hand, causes cells to lose water.

The relative sizes of the fluid compartments of the body are shown in Figure 13.2. Note that 67 percent of the body water is intracellular and that approximately 80 percent of the extracellular fluid is interstitial. The balance of the extracellular fluid, blood plasma, constitutes the smallest fluid compartment, only approximately 7 percent of total body water. (See **FIGURE 13.2.**)

As we have seen, the fluid compartments of the body are closely interrelated. Changes in the characteristics of one will cause changes in the others. And the total volume of two of these compartments,

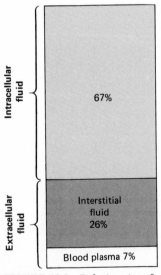

FIGURE 13.2 Relative size of the fluid compartments of the body.

the blood plasma and intracellular fluid, must be closely regulated. A fair amount of interstitial fluid—so long as it is *isotonic* and does not alter the other fluid compartments—can accumulate without immediate harm. However, too much or too little fluid in the blood plasma can lead to heart failure or a disastrous fall in blood pressure, while too much or too little fluid in the cells can damage them irreparably. As we shall see, the body's mechanisms of fluid regulation are organized to maintain the volume and electrolyte concentration of the blood plasma and intracellular fluid at their proper levels.

Regulatory Mechanisms

Fluid regulation is achieved principally by controlling the intake and excretion of two substances: water and the sodium ion. Under most conditions we drink more water than our body needs, and the excess is excreted by the kidneys. Similarly, we ingest more sodium than we need, and the kidneys get rid of the surplus. Let us examine the way the kidneys handle these regulatory processes.

The anatomy of the kidney is shown in Figure 13.3. This organ consists of a large number (approximately one million in the human) of individual functional units called **nephrons.** Each of these nephrons extracts urine from the blood and carries the urine, via collecting ducts, to the ureter. The ureters, in turn, connect the kidneys to the urinary bladder. (See **FIGURE 13.3** on following page.) During urination, the *urethra* passes the urine to the outside of the body. For our purposes, urine is outside the body once it reaches the bladder. Our society has developed customs pertaining to the release of urine from the bladder, but these customs have nothing to do with water regulation.

Production of urine begins in the **glomerulus.** Protein-free plasma is filtered from the capillaries and enters **Bowman's capsule.** (See **FIGURE 13.3.**) If this fluid were then passed out, unaltered, to the bladder, we would urinate ourselves to death in short order. Each day approximately 45 gallons of water (180 liters) is filtered into Bowman's capsule, so it should be quite obvious that most of the fluid reenters the capillaries somewhere.

As a matter of fact, approximately 99 percent of the water filtered by the glomeruli is subsequently reabsorbed. Other substances are also conserved; over 99 percent of the sodium and 100 percent of the glucose are reabsorbed. Reabsorption occurs in the **renal tubules** (*renal* means "of the kidney") and **collecting ducts**, and it may occur by means of active or passive processes. For example, sodium is reabsorbed by means of an active process. A pump in the renal tubules is capable of reabsorbing sodium against its concentration gradient. The negatively charged chloride ion passively follows the positively charged sodium ion, so the net result is that salt is retained by the body.

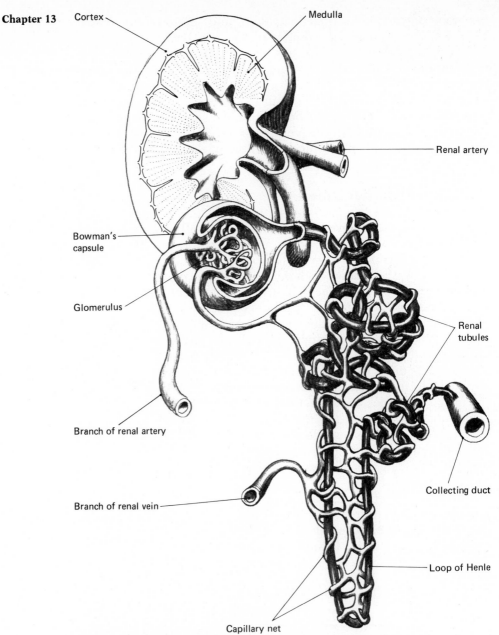

FIGURE 13.3 Anatomy of the kidney and nephron. (From Orians, G. H., *The Study of Life*. Copyright 1973 by Allyn and Bacon, Inc., Boston, Massachusetts.)

The active reabsorption of sodium is the event that is responsible for the passive reabsorption of water from the renal tubules and collecting ducts. As NaCl is pumped out of the urine, it accumulates in the interstitial fluid that surrounds the renal tubules and collect-

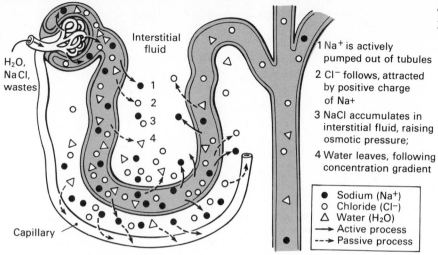

Interstitial
fluid

H_2O,
NaCl,
wastes

1 Na^+ is actively
pumped out of tubules

2 Cl^- follows, attracted
by positive charge
of Na+

3 NaCl accumulates in
interstitial fluid, raising
osmotic pressure;

4 Water leaves, following
concentration gradient

● Sodium (Na^+)
○ Chloride (Cl^-)
△ Water (H_2O)
→ Active process
--→ Passive process

Capillary

FIGURE 13.4   A schematic representation of the functions of a nephron.

ing ducts. This causes an osmotic gradient to develop. If it is permitted to, water will travel down its concentration gradient and thus
re-enter the interstitial space. (The reabsorbed sodium and water
will then enter the capillaries again. See **FIGURE 13.4.**)

Other substances are excreted, of course. Urea, the waste product from metabolism of amino acids, is excreted by passive means.
This substance is not regulated in the blood, but is carried away by
the excess water that is excreted. Many foreign substances, such as
penicillin, are actively *secreted* by the kidney. That is, independent
of glomerular filtration, some substances are transported across the
walls of the renal tubules and are deposited into the urine.

The two renal processes that we are interested in are (1) the
active reabsorption of sodium and (2) the passive reabsorption of
water, which depends on this transport of sodium. Control of water
and sodium balance is achieved by (1) altering the rate of sodium
transport and (2) varying the permeability of the walls of the distal
tubules and collecting ducts. If these parts of the nephrons are
permeable to water, it is reabsorbed. If their permeability to water
decreases, a large amount of water is excreted in the urine. The
dependency of water reabsorption tells us something else; unless
sodium is actively reabsorbed, water cannot be retained by the body.
Sodium deficits, then, will result in loss of water by the body. This
fact explains why water balance cannot be discussed independently
of sodium balance (and it explains why one must ingest salt in a hot
climate when working—and sweating away salt—in order to prevent
dehydration).

Now that we know how the kidney regulates sodium and water
excretion (and I use the word "know" very loosely—I have not done
justice to this marvelous organ in my brief description of its func-

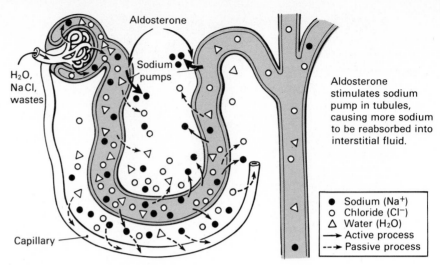

FIGURE 13.5 A schematic representation of the effects of aldosterone on the reabsorption of sodium into the interstitial fluid.

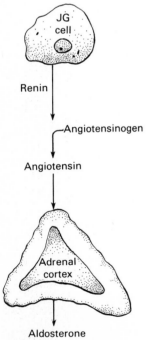

FIGURE 13.6 Control of aldosterone secretion by the kidney.

tions), we must turn our attention to the factors that control sodium reabsorption and the permeability of the distal tubules and collecting ducts.

The rate of sodium reabsorption is controlled by **aldosterone**, a hormone that is secreted by the adrenal cortex. Aldosterone stimulates reabsorption of sodium; therefore, its absence will result in a loss of this ion. (See **FIGURE 13.5.**) A person with damaged adrenal glands will ingest large quantities of salt in an attempt to maintain a normal sodium balance. This **sodium appetite** seems to be innate. A young boy with faulty adrenal glands ate great amounts of salt each day. His parents were concerned, and his physician had him admitted to a hospital. Unfortunately, he was denied access to salt while in the hospital, and, very shortly, he died.

Now that we know that aldosterone levels control sodium reabsorption, we must ask what controls aldosterone. The answer is somewhat complicated. Two factors—increased activity of the sympathetic fibers to the kidney and a fall in renal blood flow (we shall see shortly why these factors are significant)—result in secretion of the hormone **renin** by the **juxtaglomerular cells** (JG cells) of the kidney. Renin enters the blood supply and converts a substance called **angiotensinogen** into **angiotensin**. (There are two forms of the latter, angiotensin I and angiotensin II, but I shall ignore this fact for the sake of simplicity.) Angiotensin has several effects, one of which is to stimulate the adrenal cortex to produce aldosterone. Therefore, a reduction in renal blood flow or increased activity of the sympathetic efferents to the kidney causes sodium to be retained by the body. (See **FIGURE 13.6.**) The significance of this system will be explained shortly.

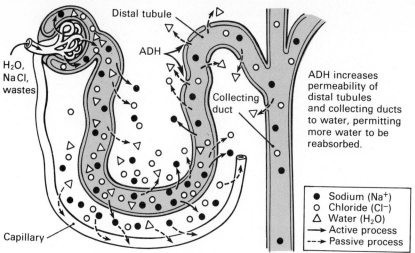

ADH increases
permeability of
distal tubules
and collecting ducts
to water, permitting
more water to be
reabsorbed.

● Sodium (Na⁺)
○ Chloride (Cl⁻)
△ Water (H₂O)
→ Active process
--→ Passive process

FIGURE 13.7   A schematic representation of the effects of antidiuretic
hormone (ADH) on reabsorption of water from the urine into the
interstitial fluid.

As we have seen, water reabsorption requires (1) that sodium be
reabsorbed (so that the interstitial fluid around the renal tubules
and collecting ducts contains a hypertonic salt solution) and (2) that
the walls of the renal tubules and collecting ducts be permeable to
water. The permeability of these structures is controlled by a secretion of the posterior pituitary gland: *antidiuretic hormone,* or **ADH.**
(See **FIGURE 13.7.**)

Antidiuretic hormone is produced by neurons in the supraoptic
nucleus of the hypothalamus and transported, in vesicles, down
through the axoplasm of these neurons. ADH collects in the posterior
pituitary gland and is released by neural stimulation. (See **FIGURE
13.8** on the following page.)

The importance of ADH in the reabsorption of water is demonstrated by the disease produced by the lack of this hormone—*diabetes insipidus* (or "a not-tasty passing through"—the urine of a person with diabetes insipidus has very little taste, since it is so dilute).
You will recall that 180 liters of fluid is filtered through the glomeruli
each day. Approximately 155 liters is reabsorbed by the proximal
tubules (which are always permeable to water, regardless of the presence or absence of ADH). This leaves 25 liters. Lacking ADH, a person
with diabetes insipidus will excrete these 25 liters. Having to pass
over 6.5 gallons of urine each day (and, of course, having to drink an
equal volume of water) makes it necessary to stay pretty close to a
bathroom and a source of water. Injections of ADH increase the
permeability of the distal tubules and collecting ducts to water and
allow it to be reabsorbed. A normal excretion rate (approximately
1.5 liters, or under 2 quarts per day) is restored.

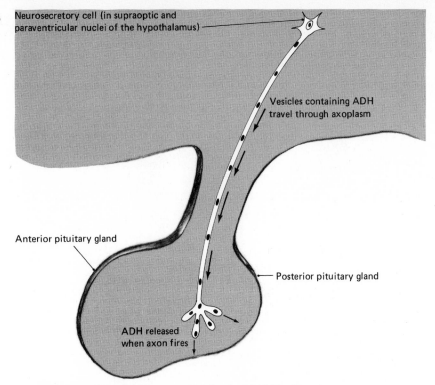

Neurosecretory cell (in supraoptic and paraventricular nuclei of the hypothalamus)

Vesicles containing ADH travel through axoplasm

Anterior pituitary gland

Posterior pituitary gland

ADH released when axon fires

FIGURE 13.8   The posterior pituitary gland and ADH release.

To summarize, sodium retention is regulated by aldosterone, which is released in response to angiotensin. (Aldosterone release is also controlled by plasma potassium, but we shall ignore this factor.) Angiotensin is produced by secretion of renin by the kidneys, in response to neural stimulation or to decreased renal blood flow. Control of water retention is simpler; the level of activity of the neurosecretory cells in the supraoptic nucleus of the hypothalamus determines how much ADH is released, which in turn determines how much water is resorbed.

Detectors

Let us consider events that would produce disturbances in the body's water and salt balance, and then see what detection mechanisms might respond to these disturbances and what the appropriate correctional mechanisms should be. Let us suppose a person with diarrhea loses a considerable amount of body fluid. The fluid loss would be approximately isotonic, so the body's store of water and salt would be equally depleted. From which compartment(s) would the fluid loss come? During digestion, water and electrolytes are with-

further loss of water and sodium through the kidneys. As we shall see, this hypovolemia also produces thirst. That is, mechanisms that monitor the volume of the blood plasma can produce thirst: **volumetric thirst**.

The easiest way to produce hypovolemia in experimental animals is to inject *colloids* into the peritoneal (abdominal) cavity (Fitzsimons, 1961). Colloids are gluelike substances made up of large molecules that cannot traverse cell membranes. Thus, they stay in the intraperitoneal cavity, and, because of osmotic pressure, extracellular fluid is drawn into the abdomen. Within an hour, the urine volume drops as a result of the mechanisms described in the previous section (increased secretion of ADH and renin). At the same time, the animal begins to drink and continues to do so until the volume of fluid stolen from the extracellular fluid is replaced.

Fitzsimons injected a colloid (**polyethylene glycol**) in rats and then drained the fluid from the peritoneal cavity, getting rid of the colloid, along with the water and sodium it drew from the extracellular fluid. He allowed the rats to recover for a day or two, with food and water available *ad libitum*. During this time the animals drank and excreted copious amounts of water. He then presented the animals with two liquids, water and a hypertonic 1.8 percent saline solution (which the rats did not drink earlier). The rats avidly consumed the saline solution. This sodium ingestion was useful; at the start of the experiment, isotonic fluid (containing sodium) was drawn into the peritoneal cavity and was subsequently drained out of the body. The rats thus were suffering from a sodium deficiency.

Thus, a treatment that reduces the volume of the extracellular fluid causes a retention of sodium and water by the kidneys and stimulates thirst. When sodium is permanently lost (by removing the fluid collected in the peritoneal cavity), the ingested water cannot be retained; remember, water can be reabsorbed from the distal tubules and collecting ducts only if sodium is first transported out of the tubules and into the surrounding extracellular fluid. If the animal lacks sodium, water cannot be reabsorbed. Thus, the volume of the extracellular fluid cannot be maintained, and the animal is chronically thirsty. This thirst can be satisfied only if sodium is ingested; the fact that the animals will drink a concentrated salt solution that is normally rejected suggests that a "sodium hunger" is elicited as well as thirst. In the last section of this chapter I shall discuss the way intake of sodium and other minerals is adjusted to meet the body's requirements.

We have seen that the detectors that initiate internal corrective mechanisms (retention of sodium and water) in response to hypovolemia reside in the left atrium of the heart and in the kidney itself. It would seem plausible that one or both of these detectors might also be involved in extracellular thirst.

In an article on the physiology of thirst, Fitzsimons (1972) reviewed a series of experiments he performed to isolate mechanisms

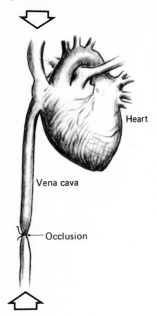

Venous blood pressure
falls, rats drink

Heart

Vena cava

Occlusion

From lower part
of body

FIGURE 13.10 A schematic
representation of the first
experiment by Fitzsimons.

of extracellular thirst. First, he occluded the vena cava (the vein that returns blood from the lower part of the body) below the liver. In a short time the rats began to drink. (See **FIGURE 13.10.**) These animals began drinking in response to a fall in venous blood pressure (a result of the obstruction of a large vein). They were not dehydrated, nor were they hypovolemic. This experiment thus provides strong evidence that the baroreceptors and/or receptors in the kidney produce thirst.

Next, he restricted the blood flow to the kidneys by partially constricting the abdominal aorta (a major artery) above the renal arteries, or by partially constricting the renal arteries themselves. (Neither of these procedures lowered venous blood pressure.) The animals drank. However, when the constriction of the abdominal aorta was made below the renal arteries, thus not affecting renal blood flow, drinking did not occur. Similarly, arterial constriction had no effect at all in rats whose kidneys had been removed. (See **FIGURE 13.11.**)

Thus, it appears that the secretion of renin, which occurs in response to a decreased renal blood flow, can produce thirst. However, the rats with restricted kidney blood flow did not drink as much water as the rats with lowered venous blood pressure, which suggests that the baroreceptors might also contribute to osmometric thirst. To test this hypothesis, Stricker (1973) injected **nephrectomized** rats (those whose kidneys had been removed) with intraperitoneal polyethylene glycol (a colloid that removes fluid from the extracellular compartments, but not from the cells). The reduced volume of the extracellular fluid and the corresponding fall in venous blood pressure could thus stimulate only the baroreceptors, since the kidneys

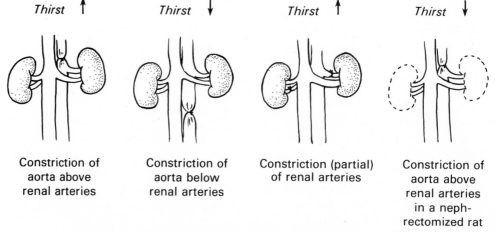

Thirst ↑	*Thirst* ↓	*Thirst* ↑	*Thirst* ↓
Constriction of aorta above renal arteries	Constriction of aorta below renal arteries	Constriction (partial) of renal arteries	Constriction of aorta above renal arteries in a nephrectomized rat

FIGURE 13.11 A schematic representation of the second experiment by Fitzsimons.

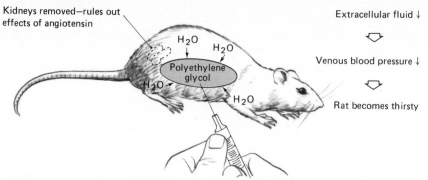

Kidneys removed—rules out
effects of angiotensin

H_2O

H_2O

Polyethylene glycol

H_2O

H_2O

Extracellular fluid ↓

Venous blood pressure ↓

Rat becomes thirsty

FIGURE 13.12 A schematic representation of the experiment by Stricker (1973).

were gone. These rats became thirsty. (See **FIGURE 13.12.**) The experiments thus confirm that the baroreceptors and renal blood flow detectors are independently responsible for drinking that is produced by reduction of extracellular fluid. The neural mechanisms of volumetric thirst will be discussed later.

Osmometric Thirst

As we saw earlier, thirst can be produced by an injection of hypertonic saline, which draws fluid out of the cells but does not reduce the volume of the extracellular fluid. Therefore, some mechanism must detect either the increased osmotic pressure of the extracellular fluid (as a result of the injection) or the ensuing loss of cellular fluid. The craving for water that ensues is called *osmometric thirst.*

In 1937, Gilman injected dogs with either hypertonic saline or hypertonic urea. The hypertonic saline produced twice as much drinking as did the urea, although both treatments increased the osmotic pressure of the extracellular fluid. The results suggest that cell shrinkage, and not an increased osmotic pressure, produces thirst.

Let us see why this is so. Sodium is excluded from cells and thus remains in the extracellular fluid. Its presence there produces an osmotic gradient, which causes cellular dehydration—and thirst. On the other hand, urea easily crosses the cell membrane. Although the injected urea increases the *total* osmotic pressure of the body fluid, there is no net movement of water out of the cells. If this analysis is correct, urea should produce no thirst at all. As we shall see, the fact that it produces *some* thirst sheds light on the location of the detectors for thirst produced by cellular dehydration.

Fitzsimons nephrectomized a group of rats (to prevent the kidneys from contributing to fluid regulation) and injected the animals with hypertonic solutions of substances that can enter cells (such as glucose and methyl glucose) and substances that cannot (sodium

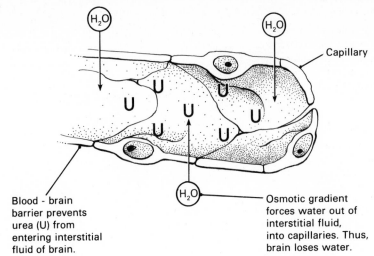

Blood - brain
barrier prevents
urea (U) from
entering interstitial
fluid of brain.

Osmotic gradient
forces water out of
interstitial fluid,
into capillaries. Thus,
brain loses water.

FIGURE 13.13   A schematic representation of the way in which injections of urea remove water from the interstitial fluid of the brain.

chloride, sodium sulfate, and sucrose—table sugar). He also injected rats with urea, which, as we shall see, presents a special case.

The results confirmed that cellular dehydration produces thirst; the only animals that drank excessively were those receiving substances that could not enter the cells. Glucose and methyl glucose (which can enter the cells) did not produce thirst. As Gilman had previously found, urea injections produced thirst, but not as much as was produced by the substances excluded from the cells.

Although urea can freely enter the cells of the body, it does not easily cross the blood-brain barrier (Reed and Woodbury, 1962). All other substances Fitzsimons tested can. Therefore, urea present in the rest of the body slowly withdraws water from the cells of the brain. This fact suggests that the osmoreceptive cells are within the brain—on the other side of the blood-brain barrier. (See **FIGURE 13.13.**)

Satiety

Anticipatory mechanisms also are involved in regulation of water intake, but these mechanisms are not so important as the ones needed for the regulation of food intake. The time interval between ingestion of water and its absorption into the fluid compartments of the body is much shorter than the 4 hours required to digest an average meal. In the case of water regulation, peripheral and central factors are not so far removed. Oropharyngeal satiety factors do not appear to be very important in water regulation; in 1856, Claude

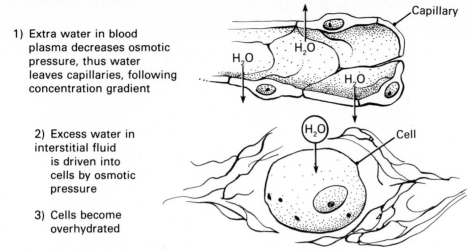

1) Extra water in blood plasma decreases osmotic pressure, thus water leaves capillaries, following concentration gradient

2) Excess water in interstitial fluid is driven into cells by osmotic pressure

3) Cells become overhydrated

FIGURE 13.14   Excessive drinking causes overhydration of cells.

Bernard found that an esophagotamized horse (whose ingested water spilled to the ground) drank to exhaustion. Intubation of water into the stomach, however, led to immediate satiety.

A hypovolemic animal will drink water, but as the volume of the extracellular fluid increases, the fluid becomes more dilute and thus produces a migration of water into the cells (cellular overhydration). (See **FIGURE 13.14**.) This overhydration is the reverse of the signal for osmometric thirst; will it inhibit drinking produced by hypovolemia?

Stricker (1969) found that overhydration did inhibit drinking in rats that were made hypovolemic by intraperitoneal injections of polyethylene glycol. The rats stopped drinking when their body fluid was diluted by 8 to 10 percent, even though plasma volume was still lower than normal. There seemed to be a trade-off in the excitatory effect produced by hypovolemia and the inhibitory effect produced by cellular overhydration. When the rats were given some sodium chloride (which raised the osmotic pressure in the extracellular fluid and took some of the extra water out of the cells), they again started to drink water.

In the normal animal, with normal kidneys, volumetric and osmometric dehydration generally go hand-in-hand. When the animal is deprived of water, it loses water through the lungs and skin. Despite the secretion of ADH, it loses an irreducible amount of water through the kidneys. As the extracellular fluid begins to lose volume and become more concentrated, it draws water from the cells. The ensuing cellular dehydration and slight reduction in blood plasma volume both produce thirst, and all fluid compartments are returned to normal by a drink of water. If the dehydration also causes a loss of sodium (through sweat), a sodium appetite will also occur.

Anticipation of Future Needs

As we all know, a meal without some beverage is thirst-provoking. In acknowledgment of this fact we almost always drink water (in one form or another) with each meal. So does a rat. Fitzsimons and Le Magnen (1969) showed that rats appear to ingest as much water as they will *subsequently* need to counteract the dehydrating effects of their meal.

All meals produce a certain degree of hypovolemia. Secretion of digestive juices into the digestive tract entails a temporary loss of water (remember, the inside of the gut is actually outside the body). The hypovolemia leads to thirst. A meal rich in protein is particularly dehydrating. The presence of amino acids in the digestive tract produces a considerable rise in the osmotic pressure, thus withdrawing fluid from the extracellular space. If you ate a large steak (even without any added salt) without taking any beverage, you would later become thirsty. (You might prefer this experiment to hitting your thumb.)

Rats ingest an amount of water that is closely related to the osmotic demand placed upon them by their meal (Le Magnen and Tallon, 1966). It appears, however, that they drink the water *before* the body fluid enters the gut (Oakley and Toates, 1969). Fitzsimons and Le Magnen (1969) showed that the matching of water intake to water need appears to be learned. They maintained rats on a carbohydrate diet and noted that the rats ingested almost equal amounts of water and food. The rats were then shifted to a protein diet, and the animals now took 1.47 times as much water as food. At first, they drank most of the water *after* the meal, when the osmotic demands were being felt. However, after a few days the animals began to drink more water with meals. They apparently learned to associate the new diet with subsequent thirst, and they drank to prevent that thirst.

It is interesting to note that when animals were shifted from a protein diet to a carbohydrate diet, they decreased their water intake very slowly. Since the kidneys quickly get rid of any excess water, there is no great pressure on the animal to drink less water; a slight degree of overhydration does not appear to be aversive. The animals probably began to drink less water because of the effort involved in its ingestion.

There also appears to be a mechanism that produces more immediate anticipation of future water need. Nicholaidis (1968) found that the mouth contains osmoreceptors that convey information to an osmo-sensitive region of the hypothalamus. Possibly, ingestion of hypertonic foods could stimulate drinking by means of this pathway.

Volumetric Thirst

THE ROLE OF ANGIOTENSIN.   Hypovolemia is detected by baro-receptors on the venous side of the blood supply, and by a renal mechanism that monitors blood flow through the kidneys. Both of these mechanisms cause the kidney to release renin; the barorecep-tors do so via a reflex circuit (which has not been traced) that in-creases sympathetic activity in the kidneys, and the decreased renal blood flow does so directly. It therefore seems plausible to suggest that renin (or angiotensin, whose synthesis is initiated by renin) acts on some cells in the brain and initiates thirst.

Early studies (e.g., Asscher and Anson, 1963) showed that injec-tions of kidney extracts produced increases in drinking. Later, Fitz-simons and Simons (1969) showed that angiotensin could produce drinking in normally hydrated rats. Drinking resulted whether the angiotensin was administered in a single dose or given very slowly for up to 5 hours. Angiotensin produces increased blood pressure (in fact, that is how this hormone got its name; *angeion* means "vessel" and *tensio*, "tension.") But if anything, increased blood pressure should result in *less* drinking, since it removes a stimulus for volu-metric thirst.

All investigators agree that injections of angiotensin can elicit drinking. However, there is some debate over the significance of this observation. Stricker (1977, 1978) suggests that the amount of renin that is released by the kidneys of a water-deprived animal does not stimulate enough angiotensin production to cause a significant amount of drinking. The real role of the angiotensin, he says, is to constrict the low-pressure (venous) portion of the vascular system, keeping the blood pressure near normal levels, despite the loss of volume of the blood plasma. Others dispute his conclusions. Fitz-simons, Kucharczyk, and Richards (1978) infused angiotensin into the veins of dogs at a rate that kept the blood level of this hormone within the range that is typically observed in thirsty dogs. (The ac-tual level of angiotensin in the blood was confirmed by an assay procedure.) The dogs drank in response to the infusions. Another study (Malvin, Mouw, and Vander, 1977) has shown that the admin-istration of **Saralasin**, a drug that blocks angiotensin receptors, de-creases the amount of water that is drunk by rats who have been deprived of water for 30 hours. (The Saralasin was injected into the ventricular system of the brain; presumably, it blocked the receptors in the brain that normally respond to angiotensin.) Injections of Sar-alasin did not affect the food intake of hungry animals, which sug-gests that the effect is specific to drinking, and does not merely make the animals sick.

The location of angiotensin receptors. Angiotensin is a polypeptide chain of eight amino acids, and, so far as we know, all polypeptides that directly affect behavior do so by interacting with specific receptors in the membrane of neurons. Therefore, it seems likely that some region or regions of the brain contain neurons that are capable of initiating thirst when they detect the presence of angiotensin.

Epstein, Fitzsimons, and Rolls (1970) injected angiotensin into various parts of the brain and successfully elicited drinking. However, it was later shown that angiotensin is especially effective when it is infused into the ventricular system. Since many of the cannulas implanted by Epstein and his colleagues passed through the walls of the ventricular system on their way to their target, the angiotensin could have flowed back along the puncture wound and entered the cerebrospinal fluid (Johnson, 1972). It now seems clear that this was the case.

Studies have shown that angiotensin does not readily cross the blood-brain barrier (Volicer and Loew, 1971). Similarly, it does not pass through the choroid plexus into the cerebrospinal fluid (Shelling et al. 1976). So how can it excite neurons in the brain? The blood-brain to angiotensin is low in several places, including the **subfor-**

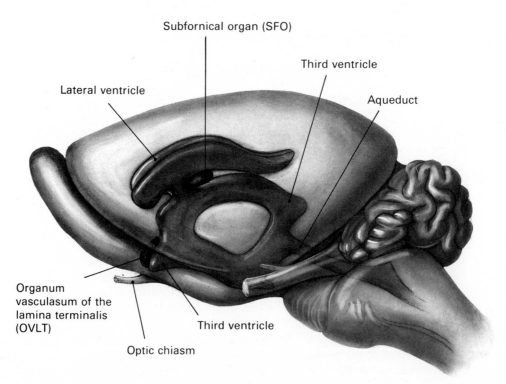

FIGURE 13.15 Location of the subfornical organ (SFO) and organum vasculosum of the lamina terminalis (OVLT).

nical organ (SFO) and the ***organum vasculosum of the lamina terminalis (OVLT).*** Both of these organs are found on the border of the ventricular system; the SFO is located near the junction of the lateral ventricles with the third ventricle, and the OVLT is found in the third ventricle, just rostral to the optic chiasm. (See **FIGURE 13.15.**)

Injection of very low doses of angiotensin into the vicinity of either the SFO or the OVLT will elicit drinking (Simpson, Epstein, and Camardo, 1978; Phillips, 1978). Furthermore, there is direct evidence, from iontophoretic injections of angiotensin along with simultaneous recording of unit electrical activity, that both the SFO and OVLT contain neurons that are sensitive to this hormone (Phillips and Felix, 1976; Felix and Phillips, 1978). However, various studies suggest that the region of the OVLT may be more important in the elicitation of drinking.

First, let us consider the site of action of intraventricularly-administered angiotensin. If a plug of cold cream is injected into the anteroventral third ventricle (the region of the OVLT), injections of angiotensin into the ventricles do not produce drinking. However, angiotensin *does* produce drinking when cold cream plugs are placed around the SFO (Johnson and Buggy, 1977). (See **FIGURE 13.16.**) Injections of angiotensin into the CSF appear to produce drinking by stimulating neurons in the vicinity of the OVLT, but not the SFO.

Although the SFO does not appear to play a role in drinking produced by antiotensin that is present in the CSF, this structure might be responsible for the drinking produced by angiotensin that is present in the blood. However, a study by Swanson, Kucharczyk, and Mogenson (1978) provides evidence that suggests that this is not

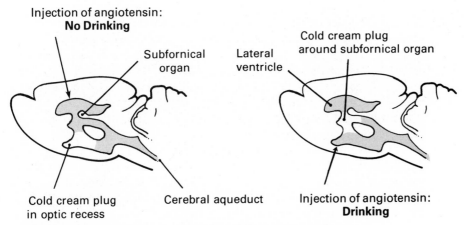

FIGURE 13.16   A schematic representation of the experiment that suggested that the subfornical organ is not the receptive organ for angiotensin-produced thirst.

the case. Swanson and his colleagues injected angiotensin into the SFO. In some animals, the injections produced drinking. The angiotensin was mixed with a radioactive amino acid so that diffusion of the injected substance could be traced by autoradiography. The investigators found that the angiotensin injections never caused drinking when the injected fluid remained in the vicinity of the SFO. Drinking occurred only when radioactivity was found in the medial preoptic ares—a result of diffusion into the ventricular system and uptake through the walls of the anteroventral third ventricle into the brain. Thus, even direct injections of angiotensin into the SFO appear to exert their effects in the region of the brain that surrounds the anteroventral third ventricle.

Swanson and his colleagues also injected angiotensin (plus a labeled amino acid) into various sites in the basal forebrain region. Drinking occurred only when radioactivity was detected in the posterior region of the medial preoptic area, which borders the anteroventral third ventricle. Radioactivity did not have to be found in the OVLT in order for drinking to occur.

Brody and Johnson (1980) attempted to determine the locations of receptive tissue by producing small lesions around the walls of the anteroventral third ventricle. The immediate effects of such lesions were a complete adipsia (cessation of drinking), with little or no effect on eating, grooming, and response to handling. (Of course, adipsia causes a secondary decrease in the intake of dry food.) The rats would drink water that had been sweetened with saccharine, so the lack of drinking cannot be a result of motor deficits. After several days, some rats began to drink spontaneously. Depending on the size of the lesion, various deficits would still remain. In some cases, the rats would drink in response to an injection of polyethylene glycol (which, you will recall, produces hypovolemia) but not to an injection of angiotensin or hypertonic saline. Presumably, the hypovolemia elicited drinking by an intact mechanism stimulated by baroreceptors in the venous blood supply. In other cases, the rats would respond to hypertonic saline (osmometric thirst) but not to injections of angiotensin, ligation of the vena cava, or injections of isoproterenol (a beta-adrenergic agonist that has been shown to cause the release of renin by the kidneys). The results support the suggestion that angiotensin-sensitive neurons can be found in the vicinity of the anteroventral third ventricle.

Many questions are still not answered. Stricker's objections point out that we still cannot assess the relative importance of angiotensin-induced drinking, as compared with drinking stimulated by the activity of baroreceptors in the venous blood supply. Although the evidence of Johnson and his colleagues, and Swanson and his colleagues, suggests that the angiotensin-sensitive neurons are in the vicinity of the anteroventral third ventricle, the SFO still has its adherents (Simpson, Mangiapane, and Dellman, 1978). Finally, the route by which angiotensin enters the brain is not yet known. Although the OVLT has a low blood-brain barrier, and contains neurons that respond to angiotensin, Swanson et al. found that angiotensin injections did not have to reach this organ in order to elicit drinking. Perhaps, as Brody and Johnson suggest, cells in the OVLT actively transport angiotensin to neurons in the adjacent brain tissue. The participants in the controversy are busy in their laboratories trying to determine the truth of the matter.

Osmometric Thirst

As we have seen, injections of hypertonic solutions produce drinking, so long as the solute administered cannot penetrate cells. The ensuing cellular fluid loss leads to thirst.

THE NATURE OF RECEPTORS FOR OSMOMETRIC THIRST. The investigation of physiological mechanisms of drinking seems to be full of controversy, and here comes another one. Most investigators believe that osmotically-produced thirst is produced by receptors that respond to loss of intracellular water. That is, the extracellular fluid becomes more concentrated (either by losing water or by gaining solutes like sodium that cannot cross cell membranes), which draws water out of the cells. Some set of neurons alter their rate of firing in response to this loss of volume. But a prominent neuroscientist, Bengt Andersson, suggests that the "osmoreceptors" are actually sodium receptors, altering their rate of firing in response to the extracellular concentration of this ion. This suggestion makes sense, because sodium is the most important ion in the extracellular fluid that is normally excluded from cells. Naturally-occurring dehydration does indeed increase the concentration of the sodium ion in the extracellular fluid, so, theoretically at least, this could be the signal for osmometric thirst.

The argument for sodium as a stimulus for osmometric thirst is complex, but one of the most important pieces of evidence is that the injection of hypertonic saline into the ventricular system of goats produces drinking, whereas injection of hypertonic sucrose does not. Since sucrose cannot cross the cell membrane, it dehydrates cells and should, therefore, elicit osmometric thirst (Andersson, 1971; 1978).

When the solutions are injected into the blood supply rather than into the cerebral ventricles, hypertonic saline and hypertonic sucrose (equal in osmotic pressure) produce equivalent amounts of drinking in sheep (McKinley, Denton, and Weisinger, 1978). These results suggest that the stimulus for osmometric thirst is cell dehydration. However, when the solutions are injected into the third ventricle of sheep (McKinley, Denton, and Weisinger, 1978) or rats (Buggy, Hoffman, Phillips, Fisher, and Johnson, 1979), hypertonic saline produces much more drinking than hypertonic sucrose does.

Therefore, we still do not know what the precise stimulus is for osmotic thirst: extracellular sodium, loss of intracellular water, or both.

The location of receptors for osmometric thirst. In 1947, Verney found that injections of hypertonic saline solution into the blood supply of the diencephalon resulted in water retention, triggered by increased ADH secretion. The results suggested the presence of osmoreceptors, probably in the hypothalamus. Andersson (1953) found that injections of hypertonic saline solution into the rostrolateral hypothalamus produced drinking (but not ADH release), whereas injections into caudal hypothalamic regions stimulated ADH secretion, but not thirst.

Further studies (Peck and Novin, 1971; Blass and Epstein, 1971) suggested that the critical sites for osmometric thirst were in the lateral preoptic area, but it was later found that the medial preoptic area was more important. In an extensive mapping study, Peck and Blass (1975) found that most of the effective sites for drinking produced by intracerebral injections were close to the midline, and not in the lateral preoptic area. Similarly, midline lesions in the basal forebrain that extended only as far as the medial border of the medial preoptic area were shown to produce long-term deficits in drinking to osmometric stimuli in goats and rats (Andersson, Leksell, and Lishajko, 1975; Buggy and Johnson, 1977). Lesions of the lateral preoptic area alone produce a deficit in short-term tests, but long-term drinking in response to injections of hypertonic saline is normal (Coburn and Striker, 1978).

Dehydration produces drinking, which replenishes lost water, but it also causes the release of antidiuretic hormone by the posterior pituitary gland. The ADH secretion, by increasing the permeability of the distal tubules and collecting ducts in the kidneys, causes less water to be excreted in the urine. There is evidence for the existence of a pair of specific sensory organs in the hypothalamus, very near the walls of the third ventricle, that regulate the secretion of ADH (Hatton, 1976). The organ is called the **nucleus circularis.** This nucleus is very small, consisting of approximately 275 cells. None of the standard sterotaxic atlases make mention of it; most people (myself included, before I learned of Hatton's work) tended, if they noticed this structure in a Nissl-stained section, to dismiss it as a staining artefact.

The nucleus circularis appears ringlike in cross section, with a hollow interior. It is actually tubeshaped, its long axis running in a rostral-caudal direction. The neurons in this nucleus lie within a capillary bed and are surrounded by heavily myelinated fibers. These fibers appear to enclose the cells, along with the capillary bed and surrounding interstitial fluid, in a water-tight compartment. (See **FIGURE 13.17.**) According to Hatton, the nucleus lies in the area that, when destroyed, produces deficits in osmometric control of

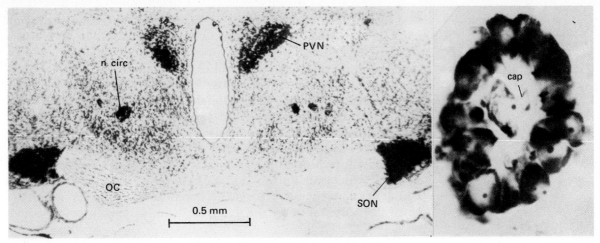

FIGURE 13.17   The location and appearance of nucleus circularis. (From Hatton, G. I., *Brain Research Bulletin*, 1976, *1*, 123–131.)

ADH release. The nucleus sends fibers to the supraoptic nucleus (wherein reside the neurons that produce ADH) and also toward the posterior pituitary gland itself.

Electrical stimulation of the nucleus circularis, but not of adjacent regions, produced retention of water, which constitutes evidence for ADH secretion. Furthermore, water deprivation led to an increase in the number of nucleoli in the cells of the nucleus. You will recall from Chapter 2 that the ribosomes (sites of protein synthesis) are produced by the nucleolus; hence, increases in the number of nucleoli suggest increased protein synthesis, presumably a result of stimulation produced by increased osmotic pressure.

Neural Control of Drinking

As we saw in Chapter 12, lesions of the lateral hypothalamus produce temporary aphagia and adipsia, followed by gradual recovery (or rapid recovery, if the animals are first starved to a weight below their new "set point"). When drinking returns, it occurs only during meals. Thirst is no longer produced by injections of hypertonic saline solution (Epstein and Teitelbaum, 1964) or by hypovolemia induced by IP injections of polyethylene glycol (Stricker and Wolf, 1967). Neither of the normal signals, then, produces drinking.

It appears likely that the lateral hypothalamic lesion interrupts a system, directed toward brainstem motor mechanisms, that controls drinking behavior. Earlier evidence suggested that volumetric and osmometric thirst were both mediated by the same mechanisms at the level of the hypothalamus (Peck and Novin, 1971; Blass and Epstein, 1971; Peck, 1973). However, more recent studies found that lesions of the midlateral hypothalamus disrupt drinking that is elicited by intracerebral injections of angiotensin, or by peripheral injections of renin or isoproterenol (which causes renin to be released by the kidneys), but do not disrupt osmometric thirst. In contrast, more lateral lesions do the opposite; they disrupt osmometric drinking, but not drinking that is elicited by angiotensin (Kucharczyk and Mogenson, 1975; 1976). More caudally, lesions that include the **ventral tegmental area** disrupt drinking produced by angiotensin, but do not affect drinking that follows injection of hypertonic saline (Kucharczyk and Mogenson, 1977).

You will recall that Swanson, Kucharczyk, and Mogenson (1978) used a radioactive amino acid as a tracer to indicate the sites of action of angiotensin. They also used amino acid autoradiography to trace the route taken by the axons of the angiotensin-sensitive neurons.

Two efferent pathways were found: a lateral pathway that passed through the lateral hypothalamus, terminating in the ventral tegmental area; and a medial pathway that terminated in the **periaqueductal gray** of the midbrain. (See **FIGURE 13.18.**) These results,

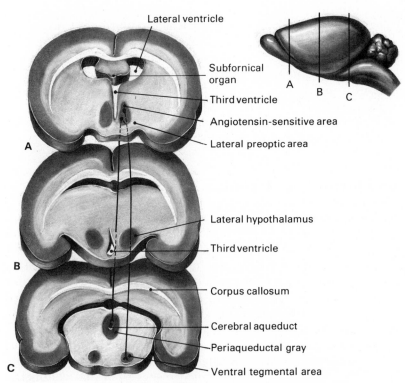

FIGURE 13.18 Efferent pathways from the angiotensin-sensitive area near the anteroventral third ventricle to the ventral tegmental area and periaqueductal gray. (From Swanson, L. W., Kucharczyk, J., and Mogenson, G. J. Autoradiographic evidence for pathways from the medial preoptic area to the midbrain involved in the drinking response to angiotensin II. *Journal of Comparative Neurology*, 1978, *178*, 645–660.)

along with the results of the lesion study described two paragraphs ago, suggest that the lateral pathway may be related to drinking in response to angiotensin. Brody and Johnson (1980) suggest that the medial pathway may be involved in the effects of angiotensin on brain mechanisms that control blood pressure.

We should also remember that not all drinking is produced by depletion of extracellular or intracellular fluid; much drinking is anticipatory, and accompanies the ingestion of food. The fact that lateral hypothalamic lesions do not disrupt drinking that is necessary for the ingestion of dry food suggests that there are other brain mechanisms that can mediate drinking.

One other brain region has been shown to play a role in the control of drinking—the septum. Damage to this structure, a part of the limbic system, generally results in polydipsia. The increased drinking appears to be primary, rather than a result of changes in kidney function (Lubar, Boyce, and Schaefer, 1968). The septum thus appears to play an inhibitory role on water intake. Mogenson (1973) has suggested that this control may be important in mediating the effects of learning, taste, and smell on drinking. It is interesting to note that the septum appears to exert very specific effects; septal lesions enhance water intake only when drinking is stimulated by angiotensin administration or by experimental procedures that cause its release (Blass, Nussbaum, and Hanson, 1974).

CONTROL OF MINERAL AND VITAMIN INTAKE

As we saw in the first part of this chapter, hypovolemia entails a loss of isotonic fluid from the extracellular fluid compartments. A drink of water restores only one of the missing substances. Unless the animal can also obtain some sodium, its extracellular fluid will remain somewhat dilute. Since cells will be damaged by extracellular fluid that is too dilute, the animal will stop drinking before all the water deficit is made up. Therefore, the animal must ingest some sodium before its fluid need can be completely restored. The animal will develop a specific appetite for sodium; it will show a strong preference for salty food. In contrast, an animal that has been deprived of other substances that it needs to function properly (such as calcium) must learn to find this substance through a trial-and-error process.

Innate Recognition of Sodium

If an adrenalectomized rat is deprived of sodium, it will develop a strong sodium deficiency because the lack of aldosterone prevents sodium retention by the kidneys. Nachman (1962) found that adrenalectomized rats showed an immediate preference for sodium over other minerals, and he later showed (Nachman, 1963) that sodium-deficient rats also preferred lithium chloride, which has a salty taste, Krieckhaus and Wolf (1968) trained thirsty rats to press a lever for water that contained either sodium or other minerals. A sodium deficiency was then induced in these animals (by giving them a subcutaneous injection of formalin, which damages capillaries and withdraws sodium from the extracellular fluid). The sodium-deficient rats were allowed to press the lever, but now the fluid delivery mechanism was disconnected. Those animals that had previously received a mineral other than sodium when they pressed the lever soon stopped pressing it, now that no fluid was delivered. However, the rats that had previously obtained sodium continued to press the now-defunct lever for prolonged periods of time. They appeared to be able to associate present sodium need with the prior availability of sodium, even though it was not needed earlier, when it was available.

These studies suggest that sodium recognition is innate. They cannot *prove* this is so, since one would have to demonstrate that the rats had never previously suffered from sodium imbalance that was subsequently relieved by ingestion of sodium (and thus provide the basis for a learned sodium preference in response to sodium need). However, the fact that animals respond so differently to other mineral and vitamin deficiencies suggests that sodium appetite is indeed a special case. And we should consider the fact that the tongues of most animals have four kinds of specific taste receptors, for salty, bitter, sweet, and sour tastes. The receptor for saltiness seems to be the peripheral mechanism by which an animal detects the presence of this vital ion.

Location of Receptors for Sodium Appetite

A series of experiments by Stricker (Stricker 1980a,b) provides strong evidence that the receptors that induce a sodium appetite are different from the ones that mediate volumetric thirst, which suggests that they are located within the brain. As we saw, volumetric thirst is mediated by receptors in the low-pressure side of the blood supply (probably in the right atrium of the heart) and by the kidneys, which release angiotensin in response to a decreased blood flow. (Recall that Stricker doubts that this angiotensin directly stimulates thirst.)

Stricker gave rats subcutaneous injections of polyethylene glycol, which caused a reduction in the blood plasma. The animals were permitted to drink water and a solution of sodium chloride. Thirst began almost immediately, as soon as the plasma volume decreased. However, the rats did not drink saline until a few more hours had elapsed. For a while, the rats alternately drank water and saline.

They would drink water until their plasma became dilute, whereupon they would drink some salt water, apparently in response to this dilution. Finally, they began to drink large amounts of saline solution.

Stricker explained the results in the following way. The polyethylene glycol produced a rapid depletion of fluid from the plasma, which stimulated thirst through peripheral receptors. However, since the rats did not consume a saline solution, these receptors apparently do not stimulate sodium appetite. As they drank water, their blood plasma became diluted. Thus sodium was drawn from the interstitial fluid into the capillaries.

Since the volume of interstitial fluid is almost four times the volume of the blood plasma, there is a large reservoir of sodium upon which to draw. The delay in the sodium appetite, Stricker says, is due to this reservoir. Not until it is sufficiently depleted, so that the brain is not receiving normal amounts of sodium, will a sodium appetite develop. In support of this conclusion, Stricker found that rats that had been fed a sodium-deficient diet for several days prior to testing drank a saline solution much sooner than rats who had been maintained on the sodium-rich rat food that the subjects normally receive. Presumably, the rats who had eaten the sodium-deficient diet had a smaller reserve of sodium in their interstitial fluid.

These findings do not tell us the location of the receptors that are responsible for sodium appetite, but by ruling out the peripheral receptors that lead to volumetric thirst, they suggest that the brain is the most probable location.

Selection of Other Minerals and Vitamins

It has been known for a long time that it is difficult to kill rats with poison (i.e., poison left out for them in their natural environment). Rats typically exhibit a dietary **neophobia**—fear of a substance with an unfamiliar flavor. More precisely, it should be said that they avoid a novel substance, at first tasting only a small amount of it. A day or two later, assuming nothing bad happens to them, they will eat more of the new diet and add it to their "list" of acceptable foods.

I said they would subsequently eat the new food if "nothing bad happens to them." This is an important qualification. If the rats become ill after eating the new diet, during a period lasting for several hours, they will thereafter permanently avoid the food. (I am assuming, of course, that the animal survives the illness. A dead animal cannot be referred to as "avoiding" anything.) Rats that show this avoidance are referred to as "bait shy."

We humans are probably also subject to this phenomenon (which is generally called **conditioned aversion**). Probably all of us have one or another aversion to specific foods, and some of these aversions may be a result of fortuitous association of the taste with later illness. Also, we have probably all heard someone (perhaps ourselves) say something like this: "When I was around 13 years old I drank a bottle of cheap sweet wine, and I puked all night. I can't even stand the sight of the stuff now." At first glance, this possible example of a conditioned aversion might seem to be contradicted by the countless people who get sick after a bout of drinking, only to do it again. This can probably be explained by the fact that *novel* substances are best associated with ensuing illness. Once a substance has been consumed several times without illness, it is difficult to produce a conditioned aversion to it.

Garcia, Kimmeldorf, and Koelling (1955) showed that rats were indeed capable of associating a novel taste stimulus with sickness that develops several hours later. This caused consternation among some learning theorists; it had been known for many years that a stimulus must precede reinforcement by a very short interval (on the order of seconds) in order for an association to be made. It was suggested that the association could be explained by the presence of postingestional cues around the time of the sickness. For example, the animals might regurgitate when they were sick and thus taste the ingested substance in their vomitus. This explanation does not seem to work, however. Garcia, Green, and McGowan (1969) used a 0.05 percent solution of hydrochloric acid as a taste stimulus, followed one hour later by an injection of lithium chloride (which causes nausea and illness). Hydrochloric acid does not put any novel ions into the stomach (gastric juices contain a high concentration of hydrochloric acid), so nothing is added to the taste of vomitus. (Anyway, rats do not have an effective vomiting mechanism.) Nevertheless, the hydrochloric acid was later avoided only by rats that had been made sick.

Garcia and Koelling (1966) found that taste cues, but not visual or auditory cues, could be associated with illness after a long delay. When visual, auditory, and gustatory stimuli were simultaneously presented prior to illness, only the gustatory stimulus was later avoided. However, a painful electric shock was best associated with visual or auditory stimuli, but only after a short delay interval. Foot-shock delivered long after the presentation of any of the stimuli had no effect upon the animal's subsequent approach to any of them.

Paul Rozin has used the phenomenon of conditioned aversion to explain how rats learn selectively to consume diets that provide substances (vitamins or minerals) that their bodies require (Rozin and Kalat, 1971). He fed rats a thiamine-deficient diet. (Thiamine, also called vitamin B_1, is an essential vitamin). The rats began to eat less food as they began to suffer the ill effects of the vitamin deficiency, and they acted as though the food, formerly quite palatable, had become aversive. They spilled it out of the dish and often ran away from the food, chewing on portions of the cage at the end opposite the food. (Rats given a diet made unpalatable with quinine show much the same behavior.) When the rats were given a new (thiamine-supplemented) food, they ate it avidly, ignoring the old diet.

Why do the rats choose the new food? The best explanation is that the developing illness produces a conditioned aversion to the deficient diet. This aversion overrides their natural neophobia (fear of novelty), and they show a preference for a new diet. Rozin (1968) raised rats on a thiamine-supplemented diet A. He then gave them thiamine-free diet B and allowed a deficiency to develop. When they were later tested with diets A, B, and C (a novel, thiamine-supplemented diet), they preferred A the most, followed by C. B was not eaten at all. The fact that they chose A over C shows that there was still some neophobia (good, old, safe diet A was best), but the rats would take C in preference to the now-aversive B.

Besides the aversion toward a diet associated with a deficiency, rats seem to form a positive association with a diet that makes them feel better. Garcia, Ervin, Yorke, and Koelling (1967) placed rats on a thiamine-deficient diet. After several days the animals were given a drink of saccharine, followed by an injection of thiamine, which produced a quick recovery from their deficiency. These rats subsequently drank more saccharine than did control animals, providing evidence for conditioned preference.

Other investigators have shown that rats respond to mineral deficiencies in a similar way. Rodgers (1967) fed rats a calcium-deficient diet. When given a choice between a novel diet and their old one, they chose the novel one—even if calcium had now been added to the old diet. Rats cannot taste calcium and will choose a new flavor, even if it does not contain calcium.

Nature has evolved an appealing, elegant (and I use the word in its original sense—simple) mechanism for regulation of intake of vital substances. If the regular diet lacks an essential ingredient, an aversion to this diet develops. In the wild, the animal would probably seek other sources of food. If ingestion of a new food is associated with recovery, that food is subsequently preferred. This mechanism appears to operate for calcium intake (Rodgers, 1967) and for specific amino acids (Harper, 1967), as well as for thiamine. I think that it is safe to assume that this mechanism operates for *any* substance we require in our diets.

KEY TERMS

aldosterone (*al DOSS ter own*) p. 414

angiotensin (*an jee o TEN sin*) p. 414

angiotensinogen p. 414

antidiuretic hormone (*dy ur ET ik*) p. 415

baroreceptor p. 417

Bowman's capsule p. 411

collecting duct p. 411

conditioned aversion p. 435

diabetes insipidus p. 415

glomerulus (*glow MARE you luss*) p. 411

hypovolemia (*hy po vo LEEM ee a*) p. 417

juxtaglomerular cell p. 414

neophobia p. 435

nephrectomy p. 420

nephron p. 411
nucleus circularis p. 430
organum vasculosum of the lamina terminalis
    (OVLT) p. 427
osmometric thirst p. 421
periaqueductal gray p. 432
polyethylene glycol p. 419

renal tubules (*REE nul*) p. 411
renin (*REE nin*) p. 414
Saralasin p. 425
sodium appetite p. 414
subfornical organ (SFO) p. 427
ventral tegmental area p. 432
volumetric thirst p. 419

SUGGESTED READINGS

EPSTEIN, A. N., KISSILEFF, H. R., AND STELLAR, E. *The Neuropsychology of Thirst: New Findings and Advances in Concepts.* Washington: Winston & Sons, 1973.

FITZSIMONS, J. T. *Proceedings of the Society for Experimental Biology and Medicine, 37,* 13 (November, 1978). This volume contains the proceedings of a symposium on angiotensin-produced thirst. The controversies are presented very well.

ROZIN, P., AND KALAT, J. W. Specific hungers and poison avoidance as adaptive specializations of learning. *Psychological Review,* 1971, *78,* 459–486.
    This article discusses the topics of conditioned aversion and dietary selection.

14

The Nature and Functions of Sleep

Why do we sleep? Why do we spend at least one-third of our lives in a state that provides most of us with only a few, fleeting memories? I shall attempt to answer this question in several ways. In this chapter, I shall describe what is known about the phenomenon of sleep: How much do we sleep? What do we do while asleep? What happens if we do not get enough sleep? What factors affect the duration and quality of sleep? How effective are sleeping medications? Does sleep perform a restorative function? What about sleepwalking and other sleep-related disorders? When we talk in our sleep, does it mean we are dreaming? What makes us get sleepy at the end of a day—are there chemicals in our bloodstream that make us sleep?

In Chapter 15, I shall attempt to answer a different question: Given that we do sleep, what brain mechanisms are involved in sleep and in its counterpart, arousal? This question is by no means completely answered, but the issue is actually much more straightforward and amenable to experimentation than the question of *why* we sleep.

SLEEP: A PHYSIOLOGICAL AND BEHAVIORAL DESCRIPTION

The best research concerning sleep in humans comes from a "sleep laboratory." Usually in a university setting, a sleep lab consists of one or several small, comfortable, homey bedrooms adjacent to an observation room, where the experimenter spends the night (trying to stay awake). The volunteer (or patient, in the cases where observations are made in an attempt to diagnose sleep disorders) is first prepared for electrophysiological measurements. Scalp electrodes are glued on, and electrodes to monitor muscle activity are taped to the chin. Eye movements are monitored from electrodes taped to the face around the eyes. Other autonomic measures (heart rate, respiration, and skin conductance) are occasionally monitored. Wires from the electrodes are bundled together in a "ponytail," which is then plugged into a junction box at the head of the bed.

As you might imagine, a subject in a sleep experiment often has difficulty in sleeping during the first night. If dreams are monitored then, they typically involve the laboratory situation. Careful investigators always take account of the "first-night phenomenon" and avoid making conclusions based on observations made during this time.

During wakefulness, the electroencephalogram (EEG) of a normal person shows two basic patterns of activity: **alpha** and **beta.** Alpha activity is observed when the person is resting quietly, not particularly aroused or excited and not engaged in strenuous mental activity (such as problem solving). Although alpha waves may be recorded from a person whose eyes are open, they are seen far more often when the eyes are closed. The other type of EEG pattern seen during waking, beta activity, is seen while the person is alert and aroused. Alpha activity will usually be disrupted and be replaced by beta activity when the person is asked to solve a problem (such as the mental addition of two three-digit numbers), or when a sudden loud noise is presented.

Figure 14.1 illustrates these two forms of the waking EEG. Note that beta consists of low-voltage, irregular activity, consisting mostly of high frequencies (13–30 Hz). Alpha activity, on the other hand, is much more regular. A lower frequency (8–12 Hz) of higher voltage predominates. (See **FIGURE 14.1.**)

beta

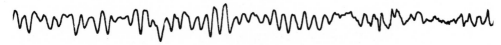

alpha

FIGURE 14.1   Alpha activity (relaxed) and beta activity (aroused, alert) from the EEG.

Stages of Sleep

EEG STAGES 1–4.   As a person progresses into sleep, the EEG contains an increasing amount of low-frequency, high-voltage activity. *Alpha* During the transition from waking to sleep, the EEG record contains mostly irregular, low-voltage waves, but as sleep becomes deeper, more and more slow, high-voltage **delta activity** (1–4 Hz) is seen. Sleep has been divided into stages 1 through 4, according to the type of EEG activity that is present, but since there are no clear-cut boundaries between adjacent stages, the precise definitions of each stage are not of importance to us here. What is important is to note that deeper stages of sleep (stages 3 and 4) are accompanied by increasing amounts of delta activity (indicated by the horizontal lines beneath the records). (See **FIGURE 14.2** on the next page.)

THE SIGNIFICANCE OF SYNCHRONY.   Since the EEG is produced by the summed postsynaptic activity of neurons in the brain, low frequency, high-voltage activity (such as delta waves) is usually *delta* referred to as reflecting neural **synchrony.** There is, presumably, a great deal of similarity in the temporal pattern of activity in a large number of neurons. The activity of the individual neurons is similar *beta* to a large number of people chanting the same words (speaking synchronously). Beta activity is, for the same reason, referred to as **desynchrony;** it is more similar to a large number of people broken into many small groups, each carrying on an individual conversation. The analogy helps explain why desynchrony is generally taken to represent activation, whereas synchrony reflects a resting or depressed state. The group of people who are all chanting the same thing will process very little information; only one message is being produced. The desynchronized group, on the other hand, will process

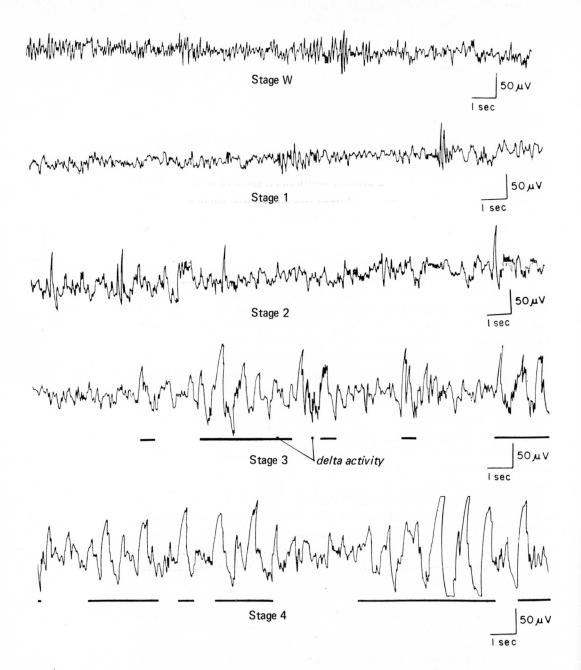

FIGURE 14.2 Stages of sleep. Stage W indicates drowsy wakefulness, with a large percentage of alpha activity, immediately preceding the onset of stage 1 of S sleep. (These records were obtained from *A Manual of Standardized Terminology, Techniques and Scoring System for Sleep Stages of Human Subjects*, edited by A. Rechtschaffen and A. Kales. Washington, D.C.: U. S. Government Printing Office, 1968.)

and transmit many different messages. The waking state of the brain is more like the desynchronized group of people, with much information processing going on. During delta sleep the neurons of the resting brain (especially the cortex) quietly murmur the same message in unison (following the lead of a group of pacemaker cells, as we shall see in the next chapter).

Description of a Night's Sleep

Let us follow the progress of a volunteer (a male college student) in the sleep lab. Our subject has already slept there, so the sleep patterns we observe will not be unduly influenced by a new, unfamiliar environment. The wires are attached, the lights are turned off, and the door is closed. Our relaxed subject shows mostly alpha activity, which is soon replaced by evidence of stage 1 sleep. Around 10 minutes later he enters stage 2, followed, 15 minutes later, by the occurrence of a few delta waves, signalling entry into stage 3. Gradually, over a 15-minute period, delta waves predominate (stage 4). The subject is sleeping soundly now, but if awakened, he might report that he had not been asleep. (This phenomenon is often reported by nurses who awaken loudly snoring patients early in the night—probably to administer a sleeping pill—and find that the patients insist that they were lying there awake all the time.)

We do not gradually slip from waking into sleeping, by the way. The onset of sleep appears to be sudden; it is seen as gradual only in retrospect. Of course, we gradually become sleepier, and gradually become less restless as sleep approaches, but the border between wakefulness and sleep appears to be distinct. Dement (1974) reports a study in which a subject's eyes were taped open (which, Dement assures us, is not uncomfortable). The subject was asked to press a switch every time he saw a bright flash from a strobe light placed 6 inches from his face. The subject pressed repeatedly and then suddenly stopped. The curtain of sleep had suddenly dropped. If the subject were awakened, he would not realize that he had stopped pressing the switch. The cessation of button pressing corresponds with the onset of slow, rolling eye movements (not the rapid eye movements I shall shortly describe) and an EEG record characteristic of stage 1 sleep.

About 90 minutes after the onset of sleep, an abrupt change is suddenly seen in a number of physiological measures. The EEG suddenly becomes desynchronized; if we did not see our subject lying there, asleep, we would assume he was awake. We also note that his eyes are rapidly darting back and forth beneath his closed eyelids. (We can see this in the *electro-oculogram,* or *EOG,* recorded from electrodes taped to the skin around his eyes, or we can observe the eye movements directly. The cornea produces a bulge in the closed eyelids that can be seen to move about.) We also see that the EMG (a measure of muscular activity) becomes silent; there is a profound loss of muscle tonus. However, we occasionally see brief twitching movements of the hands and feet, and our subject probably has an erection.

This peculiar stage of sleep is quite distinct from the quiet, slow wave sleep we saw earlier. It is usually referred to as **REM sleep** (for the *rapid eye movements* that characterize it). It has also been called paradoxical sleep, because of the presence of a "waking" EEG during sleep. The term "paradoxical" merely reflects surprise at observing an unexpected phenomenon, but the years since its first discovery (reported by Aserinsky and Kleitman in 1953) have blunted the surprise value. In accordance with a suggestion by Hartmann (1973), I shall use the terms **S sleep** and **D sleep** to refer to the two states. S sleep is slow-wave, synchronized, spindling sleep. D sleep, on the other hand, is characterized by desynchrony. Some would call it *deep* sleep. And, as we shall see, D also refers to *dreaming*.

If we arouse our volunteer during D sleep and ask him what was going on, he will almost certainly report that he had been dreaming. The dreams of D sleep tend to be narrative in form; there is a storylike progression of events. If we wake him during S sleep and ask, "Were you dreaming?" he will most likely say no. However, if we question him more carefully, he might report the presence of a thought, or an image, or some emotion. I shall return to this issue later.

During the rest of the night our subject will alternate between

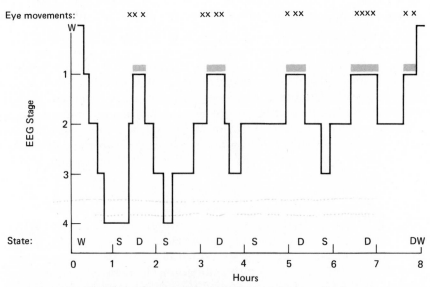

FIGURE 14.3 A typical pattern of the stages of sleep during a single night. (From Hartmann, E., *The Biology of Dreaming*, 1967. Courtesy of Charles C Thomas, Publisher, Springfield, Illinois.)

periods of S sleep and D sleep, a cycle being approximately 90 minutes long, and containing a 20- to 30-minute bout of D sleep. That means an 8-hour sleep will contain four or five periods of D sleep. Hartmann (1967) has drawn a graph of a typical night's sleep, as shown in Figure 14.3. Note that most stage 4 sleep is accomplished early in the night. D sleep is represented by the heavy horizontal bars. (See **FIGURE 14.3**.)

The regular cyclicity of D sleep suggests that there is some intrinsic brain mechanism that alternately produces the D and S states. Dement and his colleagues (reported by Dement, 1972) tabulated the intervals between successive periods of 1000 cases of D sleep. They found that once a period of D sleep was over for at least 5 minutes (indicating that it was truly over, and not just interrupted temporarily), there was a 95 to 98 percent probability that D sleep would not occur again for at least 30 minutes. By 80 minutes there is almost a 100 percent certainty that another period of D sleep will occur (unless the subject awakens). D sleep, then, cannot randomly occur at any time. There seems to be a refractory period after each occurrence, during which time D sleep cannot again take place.

D Sleep

More attention has been paid to D sleep than to S sleep, partly because it is more interesting. During D sleep we become paralyzed; there is massive inhibition of alpha motor neurons. Tendon reflexes cannot be elicited. At the same time, the brain is very active. Cerebral blood flow and oxygen consumption are accelerated; in fact, a physician long ago reported

> the case of a man who, some time after receiving a severe injury of the head by which a considerable portion of the skull was lost, came under my professional care. Standing by his bedside one evening just after he had gone to sleep, I observed the scalp rise slightly from the chasm in which it was deeply depressed. I was sure he was going to wake, but he did not, and very soon he became restless and agitated while continuing to sleep. Presently he began to talk, and it was evident that he was dreaming. In a few minutes the scalp sank down to its ordinary level when he was asleep, and he became quiet. I called his wife's attention to the circumstance and desired her to observe this condition thereafter when he slept. She subsequently informed me that she could always tell when he was dreaming from the appearance of the scalp. (Hammond, 1883, p. 145, quoted by Freemon, 1972)

EYE MOVEMENTS DURING DREAMS.   Eye movements appear to be related to the visual content of the dream. Roffwarg, Dement, Muzio, and Fisher (1962) awakened subjects during D sleep. They carefully elicited a report of the dream, and they determined from the narration what eye movements would have occurred if the dream had been reality. The polygraph record was examined (by a person who did *not* hear the story) and a description of the eye movements was prepared. The investigators found a high degree of concordance between predicted and observed eye movement. A later study by Pivik, Bussel, and Dement (described by Dement, 1974) found that the relationship between EOG and predicted eye movements during D sleep was just as good as the relationship between EOG and actual eye movements during waking. Other investigators have failed to obtain such good agreement, but it is not clear that they questioned their subjects about their dreams so carefully as Dement's group did.

PASSAGE OF TIME DURING DREAMS.   It is sometimes said that time is compressed in a dream—that what seems to take hours really occurs in a few seconds. This does not appear to be true, however. Dreams take the same amount of time as they seem to. Dement and Wolpert (1958) sprayed cold water on sleeping subjects during D sleep. They awakened the subjects at various times after the stimulus and questioned them about the dream. When the spray of water was incorporated into the dream, it was found that the perception of time between the stimulus and the subsequent awakening was quite accurate. Time seems to go by at the same rate during dreams as it does during waking. The authors give the following example:

> The S [subject] was sleeping on his stomach. His back was uncovered. An eye movement period started and after it had persisted for 10 minutes, cold water was sprayed on his back. Exactly 30 seconds later he was awakened. The first part of the dream involved a rather complex description of acting in a play. Then, "I was walking behind the leading lady when she suddenly collapsed and water was dripping on her face. I ran over to her and felt water dripping on my back and head. The roof was leaking. I was very puzzled why she fell down and decided some plaster must have fallen on her. I looked up and there was a hole in the roof. I dragged her over to the side of the stage and began pulling the curtains. Just then I woke up." (Dement and Wolpert, 1958, p. 550)

OTHER D SLEEP PHENOMENA.   What about the penile erections that occur in male dreamers? Are they related to dreams with sexual content? This does not appear to be the case; erections occur during almost all periods of D sleep, even when the accompanying dreams are devoid of sexual content (Fisher, Gross, and Zuch, 1965). Of course, there are also dreams with frank sexual content that culminate in ejaculation–the so-called nocturnal emissions, or "wet dreams." A women's sex organs are also affected during D sleep; vaginal blood flow, and ensuing fluid secretion, increase at this time.

The fact that penile erections occur during D sleep, independent of sexual arousal, has been used clinically to assess the causes of impotence (Karacan, Salis, and Williams, 1978). A subject sleeps in the laboratory with a device attached to him that measures the circumference of his penis. If penile enlargement occurs during D sleep, then his failure to obtain an erection during attempts at intercourse is not caused by neurological or circulatory disorders. Often, the knowledge that he is physically able to obtain an erection is therapeutic in itself.

I have been describing dreaming as a universal phenomenon of D sleep, but there are many people who insist that they never dream. They are wrong—everyone dreams. What does happen, however, is that most dreams are subsequently forgotten. Unless a person awakens during or immediately after a dream, memory for the dream is most likely to be lost. Many people who thought that they had not had a dream for years have been startled by the vivid narrations they are able to supply when roused during D sleep in the laboratory. It is a very interesting fact that such vivid experiences can be so completely erased from consciousness. I am sure that many of you have had the experience of waking during a particularly vivid (and perhaps amusingly bizarre) dream. You decide to tell your friends about it—and all of a sudden, it slips away. You can't even remember the general gist of the dream, which was so vivid and real just a few seconds ago. You have the feeling that if you could remember just one thing about it, everything would come back. I think that an understanding of this phenomenon would tell us a lot about the way in which memories get stored or lost.

What about dreams themselves? Primitive people thought (and many modern people still think) that dreams provide magical insights. Prophetic dreams, for example, have altered military campaigns and changed the course of history. However, Freemon (1972) classifies prophetic dreams into three categories: (1) after-the-fact dreams, which can be culled out of a large store of dreams to suit the occasion (if the facts had been otherwise, a different prophetic dream would have been selected); (2) statistical dreams, which are remembered if the predicted event comes true, and conveniently forgotten (or at least not advertised) if it does not; and (3) inner knowledge dreams, which help us to recognize formally something we really know (and are ignoring, perhaps actively). The latter category often includes dreams that are useful to a person or to his or her therapist, since they sometimes reveal thoughts or feelings that are suppressed during waking hours. Dement (1979) gives an example of another useful function of dreaming—a glimpse into a possible future. Here are his own words:

> . . . [A] dream . . . may have saved my life. Some years ago I was smoking three packs of cigarettes a day. I coughed up blood. I had an X-ray, my lungs were full of cancer, I felt poignantly the intense reality of the premature termination

of my life, and then *I woke up*. Given the opportunity to experience this alternative, the choice was obvious. I have not smoked a cigarette in the intervening 13 years. (p. 289)

S Sleep

I have been speaking of dreaming as being synonymous with D sleep. This does indeed appear to be the case, but mental activity also occurs during S sleep. The probability of obtaining a report of a dream during S sleep depends greatly upon one's definition of what constitutes a dream. Freemon (1972) reviewed 12 studies that compared reports of dreaming during awakenings from D and S sleep. The percentages varied considerably, but all studies reported considerably more dreaming during D sleep.

FIGURE 14.4 *The Nightmare*, 1781, by Henry Fuseli, Swiss, 1741–1825. (Oil on canvas; H 40″, W 50″. Gift of Mr. and Mrs. Bert L. Smokler and Mr. and Mrs. Lawrence A. Fleischman, Acc. No. 55.5. Courtesy of The Detroit Institute of Arts.)

Some of the most terrifying nightmares occur during S sleep, especially in stage 4 (Fisher, Byrne, Edwards, and Kahn, 1970). As is typical of mental activity during S sleep, the people do not report a storylike dream as they generally do after D sleep dreaming, but rather a situation, such as being crushed or suffocated. This common sensation is reflected in the terms that some languages use for describing what we call a nightmare. For example, the word is *cauchmar*, or "pressing devil," in French. Figure 14.4 shows a victim of a nightmare (undoubtedly, in the throes of stage 4 S sleep) being squashed by an *incubus* (Latin *incubare* = "to lie upon"). (See **FIGURE 14.4.**) Often, when people wake from a stage 4 nightmare, they cannot even remember what frightened them.

Ontogeny and Phylogeny of Sleep Stages

All mammals and birds sleep, and all of them, except for the most primitive mammals, have periods of D sleep. Reptiles also sleep, and fish and amphibians enter periods of quiescence that probably is a form of sleep. However, these animals do not appear to engage in D sleep. (See Rojas-Ramírez and Drucker-Colín, 1977.) Thus, D sleep appears to be of more recent phylogenetic origin, since it is seen only in higher animals (mammals and birds).

The ontogeny of sleep contradicts Haeckel's hypothesis that "ontogeny recapitulates phylogeny" (that is, a recently-evolved animal resembles primitive species in early prenatal development, and, in stages, resembles more and more advanced species as it grows). D sleep appears to be of recent evolutionary origin, but it is more prevalent in young organisms. Premature human infants appear to spend all their time in a stage that resembles D sleep (Dreyfus-Brisac, 1968). Newborn full-term infants (humans and other mammals) have a very high proportion of D sleep. The proportion decreases with maturation of the central nervous system. In contrast, infant guinea pigs, which are born with an almost fully mature nervous system, do not show such a high percentage of D sleep (Jouvet-Mounier, Astic, and Lacote, 1970). Perhaps, as was suggested by Roffwarg, Muzio, and Dement (1966), D sleep facilitates neural growth in young animals. As we shall see, the rate of synthesis of brain proteins is highest during D sleep.

WHY DO WE SLEEP?

Sleepiness: An Insistent Drive

We all know how insistent the urge to sleep can be, and how uncomfortable we feel when we have to resist it and stay awake. With the exception of the effects of severe pain and the need to breathe, sleepiness is probably the most insistent drive. A person can commit suicide by refusing to eat or drink, but even the most stoic person cannot indefinitely defy the urge to sleep. Sleep will come, sooner or later, no matter how hard the person tries to stay awake. The urgent and insistent nature of sleepiness suggests that sleep is a necessity of life. If this is true, then it should be possible to deprive people of sleep and (following the same logic that underlies brain lesions studies) see what capacities are disrupted. We should then be able to infer the role that sleep plays.

We all know that there is a difference between sleepiness and tiredness. We might want to rest after tennis or a vigorous swim, but the feeling is quite different from the sleepiness we feel at the end of a day—a sleepiness that occurs even if we have been relatively inactive. What we must do in order to study the role of sleep as opposed to the restorative function of rest, is to rest without sleeping. Unfortunately, that is not possible. When Kleitman first began studying sleep in the early 1920s, he hoped to have subjects undress and lie quietly in bed. They would remain awake, however, so that it would be possible to observe the effects of "pure" sleep deprivation. It did not work. People cannot stay awake without engaging in physical activity, no matter how hard they try. So Kleitman had to contend with the fact that his subjects were rest-deprived as well as sleep-deprived.

Kleitman observed that his subjects did not show a steady progression of sleepiness throughout the deprivation period (Kleitman, 1963). They were sleepiest at night, but they recovered considerably during the day. After two sleepless days there did not appear to be any significant increase in sleepiness; there continued to be a cycle of sleepiness, with the worst period occurring at night, but Kleitman noted that after 62–65 hours a person "was as sleepy as he was likely to be."

The amazing thing about these sleep-deprived subjects was how well they could perform, so long as the tasks were short. Prolonged, boring tasks were difficult to do, but if the subjects were properly motivated, their performance on short tasks was as good as that of rested subjects.

Sleep Deprivation and Personality Disorders

Personality changes are sometimes seen during sleep deprivation, and during the Korean War many American servicemen were induced to sign false confessions by means of techniques that included sleep deprivation. A famous case that occurred in 1959 lent further support to the belief that sleep deprivation would produce psychotic reactions. Peter Tripp, a New York disc jockey, stayed awake for 200 hours as a publicity stunt to raise money for charity. He made his

broadcasts from a glass booth in view of the public, and he was
attended at all times, to prevent surreptitious sleep. He developed
a severe paranoid psychosis, and believed that he was being poisoned. His suspicion became so marked that he could not even be
tested during the later stages of sleep deprivation (Dement, 1974).

However, a subsequent case (reported by Gulevich, Dement, and
Johnson, 1966) showed that personality changes did not necessarily
occur as a result of sleep deprivation. Randy Gardner, a 17-year-old
boy, stayed awake for 264 hours (so that he could find a place in the
Guinness Book of World Records). He found it difficult to stay awake
and had to engage in physical activity. He beat Dr. Dement at 100
straight games on a baseball machine in a penny arcade during the
final night of sleeplessness, which suggests that his coordination was
not severely impaired. (I have no idea how well Dr. Dement plays
this game.) After this ordeal, Randy Gardner slept for a little under
15 hours and awoke feeling fine. He slept a normal 8 hours the following night. At no time did Randy exhibit any psychotic symptoms.

The results of this observation suggest that the psychotic symptoms sometimes seen during sleep deprivation result from the stress
produced by the deprivation and not from the lack of some basic
function carried out by sleep. Presumably the general mental and
physical health of the subject (and the conditions under which the
sleep deprivation occurs) determines whether the stress will produce
psychotic reactions. The servicemen in prisoner-of-war camps in
Korea were certainly in an environment different from the one
Randy Gardner was in.

Sleep-Deprivation Studies with Animals

Sleep-deprivation studies with animals have provided us with no
more insight into the role of sleep than have the human studies. The
fact that animals cannot be "persuaded" to stay awake makes it even
more difficult to separate the effects of sleep deprivation from the
effects produced by the method by which the animal is kept awake.
We can ask a human volunteer to try to stay awake, and expect some
cooperation. He or she will say, "I'm getting sleepy—help me to stay
awake." The animal, however, is interested only in getting to sleep
and must be constantly stimulated—and hence, stressed. Webb
(1971) summarized the sleep-deprivation literature with the statement that "the effect of sleep deprivation is to make the subject fall
asleep."

The insistent, unyielding nature of sleepiness seems to be a convincing argument that we *need* to sleep. This may be true, but the
effects of sleep deprivation on human performance have not pointed
to any obvious function that is performed during sleep. The changes
in performance after sleep deprivation do not prove that some neural
mechanisms are "fatigued." It is just as likely that the decrements

are a result of the distraction that is produced by having to fight to stay awake, and by the intrusion of periods of *microsleep.* This latter phenomenon is characterized by EEG signs of sleep and loss of attention in an apparently awake person. Therefore, the decrements might simply be the result of sleepiness itself, and not the result of a lack of whatever it is sleep does for a person.

There are many hypotheses about what the functions of sleep are, but most of them are variations on two themes. (1) Sleep is a period of restoration, during which certain anabolic physiological processes occur. Sleep is seen as being analogous to eating, another activity that is necessary for survival. (2) Sleep is an adaptive response—a behavior that serves a useful purpose. For example, energy is not wasted during a time of day that food is not available. In the next two sections, I shall consider these two general hypotheses: sleep as construction or repair, and sleep as an adaptive response.

Sleep as Construction or Repair

A person who misses a meal will probably get hungry, even though his or her body has enough reserves to sustain many days of fasting. Similarly, sleep deprivation might produce sleepiness long before any serious physical symptoms can be detected. Thus, the lack of serious effects of sleep deprivation does not prove that sleep is not a physiologically necessary state. Just as hunger and satiety are related to the availability of various nutrients, perhaps sleepiness and its opposite are related to the presence of some substances in the blood, or perhaps in the fluid compartments of the brain. It may be that some substance that is necessary for proper functioning of the brain can be produced only during sleep. Waking uses it up, so sleepiness is produced by its depletion. Or perhaps it goes the other way around; wakefulness produces a substance that interferes with normal brain functions. This substance produces sleep, during which time it is deactivated. The possibility that one (or both) of these hypotheses might be true suggests that a search should be made for substances in the body whose concentrations vary with cycles of sleep and waking.

BLOOD-BORNE FACTORS.   Some studies have shown that, under special conditions, the brain might release some substances into the blood that can affect the brain activity of another animal. Low-frequency electrical stimulation of the midline nuclei of the thalamus produces cortical synchrony, which is one of the manifestations of sleep. A number of studies (e.g., Monnier, Koller, and Graber, 1963) have shown that thalamic stimulation of one member of a pair of animals with surgically interconnected circulatory systems will increase the occurrence of cortical synchrony in the other member. The converse effect can also be seen; stimulation of the midbrain reticular formation (which, as we shall see, produces arousal) de-creases cortical synchrony in the nonstimulated member. Monnier and Hösli (1964, 1965) were able to concentrate a substance from the cerebral blood of a thalamically stimulated rabbit that would pro-duce this synchronizing effect in a recipient animal. Later studies (Monnier, Dudler, Gächter, Maier, Tobler, and Schoenenberger, 1977) identified the sleep-promoting substance as a simple 9-amino acid polypeptide, which was given the name of **_delta sleep-inducing_** **_peptide._** Synthetic DSIP was found to produce EEG signs of slow wave sleep in recipient rabbits.

DSIP

It does not appear that the substance discovered by Monnier and his colleagues plays a crucial role in the regulation of sleep cycles. Ringle and Herndon (1968) were unable to find this substance in the blood of rabbits that had been deprived of sleep; thus, the substance appears to be secreted (at least, in significant amounts) only during thalamic stimulation. Even more crucial is the observation that Si-amese twins who share a common cerebral blood supply have in-dependent sleep cycles (Lenard and Schulte, 1972). If sleep and wakefulness were controlled by factors in the blood, one would ex-pect that the sleep cycles of Siamese twins would be synchronized.

FACTORS EXTRACTED FROM CEREBROSPINAL FLUID.   It is possible that sleep is regulated by factors that are produced within the brain, and kept out of the blood supply by the blood-brain bar-rier. This hypothesis was tested a long time ago (Piéron, 1913; Le-grende and Piéron, 1913). These investigators kept dogs awake for several days. Cerebrospinal fluid was removed from these animals and was injected into the ventricular system of recipient dogs, who subsequently went to sleep for 2 to 6 hours. In 1939, Schnedorf and Ivy replicated the phenomenon, but noted that hyperthermia usually followed injections of CSF (as a result of increased intracerebral pres-sure), so the notion of a central sleep-producing substance fell into disrepute. More recently, however, Fencl, Koski, and Pappenheimer (1971) have shown that a sleep-promoting factor (with a molecular weight of less than 500) can be concentrated by means of selective filtration from the CSF of sleep-deprived (but not control) goats. This factor (which does not appear to be one of the neurotransmitters thought to be involved in neural sleep mechanisms) increases the duration of sleep and decreases locomotor activity in recipient sub-jects. The investigators also discovered a wakefulness-promoting fac-tor (with a molecular weight somewhere between 500 and 10,000) that produces hyperactivity in the recipient that lasts for several days. They found this factor in both control and sleep-deprived goats. However, until we find out where the substances are produced and where they exert their effects, we shall not know whether they play a role in an animal's sleep-waking cycle. At normal physiological concentrations these factors might or might not affect the brain of the goat from which the fluid is taken. The brain makes many substances, and many of them will have abnormal effects if they are administered in unusually high concentrations to another animal.

FACTORS EXTRACTED DIRECTLY FROM THE BRAIN.   A special technique developed by Myers (1970) permits the extraction of substances from the brain of a freely-moving animal. A ***push-pull cannula*** is implanted into the brain of a donor animal. This device consists of a small metal tube placed inside a slightly larger one. ***Ringer's solution*** (a solution of salts that closely duplicates those found in normal interstitial fluid) is slowly pumped through the smaller tube, and is recovered from the larger one. As the Ringer's solution flows from one tube to the other, it picks up substances that are present in the interstitial fluid of the brain. (See **FIGURE 14.5**.)

Drucker-Colín and his colleagues implanted push-pull cannulas in cats (Drucker-Colín and Spanis, 1976). Ringer's solution was perfused through the midbrain reticular formation of donor cats, and was then injected into the brains of recipient cats. This procedure would presumably introduce the substances that were present in the brains of donor animals into the brains of the recipients. The perfusions were done while the donors and recipients were asleep or awake.

The results suggested that there are, indeed, sleep-promoting and wakefulness-promoting substances that are present at different parts of the sleep-waking cycle. Ringer's solution that had been perfused through a sleeping donor would cause the recipient to go to sleep faster than control injections of plain Ringer's solution. Conversely, Ringer's solution perfused through an awake donor would cause a sleeping recipient to wake up. (See **FIGURE 14.6**.)

These results are interesting and provocative. But the sleep-promoting factor is not necessarily the kind of sleep-producing substance that I described earlier—a chemical that is produced by waking activity that must be deactivated during sleep. Similarly, the wakefulness-promoting substance is not necessarily the product that

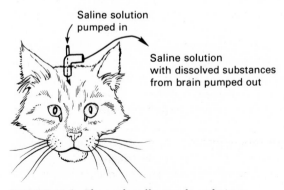

Saline solution pumped in

Saline solution with dissolved substances from brain pumped out

FIGURE 14.5   The push-pull cannula technique.

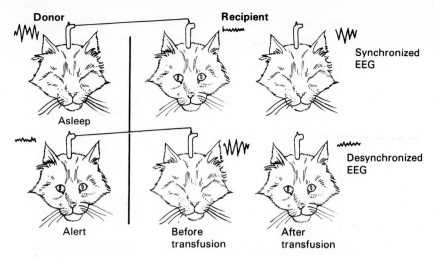

Donor   Recipient

Asleep

Synchronized EEG

Alert   Before transfusion   After transfusion

Desynchronized EEG

FIGURE 14.6 Effects of transfusion of perfusate obtained by use of the push-pull cannula technique. (From Drucker-Colín, R. R., and Spanis, C. W. Is there a sleep transmitter? *Progress in Neurobiology*, 1976, *6*, 1–12.)

sleep is designed to produce. Both of these substances might be necessary components of the states of sleep and wakefulness, but not the factors that *cause* them. Let me use the feeding analogy again. Suppose that we did not know that hunger was related to nutrients. We might perfuse fluid through various parts of the brain and find that some substances were present during eating that would initiate eating when they were injected into the same brain regions of recipients. We might be tempted to conclude that we had found the hunger-promoting substance, which eating serves to bring back down to normal levels. But in fact we might have merely extracted neurotransmitters or neuromodulators that are secreted when feeding mechanisms have already been activated by other means. We would simply be stimulating motor systems that function in eating. Similarly, the substances that Drucker-Colín and his colleagues extracted might not be the initiators of sleep and waking cycles, but rather the neurotransmitters and/or neuromodulators that sleep and waking mechanisms use to perform their functions.

PRODUCTS SYNTHESIZED DURING SLEEP. As we have seen, experimental investigations have not yet identified any substances in the body that are directly responsible for sleep-waking cycles. The search will undoubtedly be continued, but there is another approach to the investigation of the functions of sleep that can be carried out independently. It is possible to study what happens during sleep, physiologically. If certain biochemical events occur only then, perhaps the role of sleep is to perform these reactions. The products need not have sleep-promoting or wakefulness-promoting functions of their own.

There are several substances that are secreted during sleep. An important one, growth hormones, is secreted shortly after the first occurrence of delta activity in S sleep. Figure 14.7 shows the secretion of growth hormone (GH), along with stages of sleep and waking, over a 24-hour period in two different human subjects (Takahashi, 1979). Subject T.F., whose data are shown in the lower part of the figure, slept continuously, with only a few brief awakenings, and showed a single phase of GH secretion. (See **FIGURE 14.7.**) The sleep of subject B.B. (above) was interrupted for several hours; his GH secretion was biphasic, occurring principally during stages 3 and 4 of S sleep. (See **FIGURE 14.7.**)

Others studies have shown that GH secretion is related to sleep, and not just to time of day. If subjects stay awake during the night and sleep during daylight hours, their GH secretion follows their new sleep pattern (Sassin, Parker, Mace, Gotlin, Johnson, and Rossman, 1969). GH is even secreted during daytime naps of people with normal nighttime schedules of sleep (Karacan, Rosenbloom, Londono, Williams, and Salis, 1975). More GH is secreted during after-

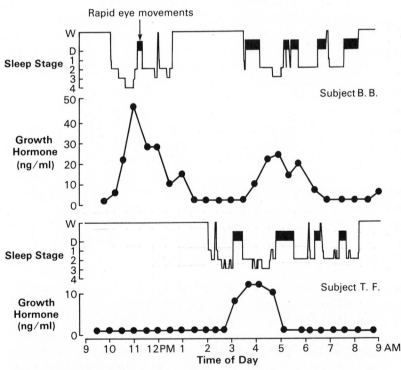

FIGURE 14.7   Plasma levels of growth hormone as a function of stages of sleep and waking in two subjects. Note that the hormone is secreted during the deepest stages of S sleep. (From Takahashi, Y. Growth hormone secretion related to the sleep waking rhythm. In *The Functions of Sleep*, edited by R. Drucker-Colín, M. Shkurovich, and M. B. Sterman. New York: Academic Press, 1979.)

noon naps than morning naps. Since afternoon naps contain more S sleep than morning naps do, the results further support GH secretion as a phenomenon related to S sleep.

It would be gratifying to be able to conclude that the function of sleep is to provide a state during which GH (and, possibly, other substances) can be secreted. But things do not appear to be so simple. Children younger than 4 years of age, and many older people, do not secrete GH only during sleep (Finkelstein, Roffwarg, Boyar, Kream, and Hellman, 1972; Carlson, Gillin, Gorden, and Snyder, 1972). Furthermore, the correlation between sleep and GH secretion breaks down during many pathological conditions, including various pituitary disorders, dwarfism, narcolepsy (a sleep disorder that will be discussed later), schizophrenia, and depression (Takahashi, 1979).

The push-pull cannula technique has been used to measure the rate of protein synthesis in the brain during stages of sleep and waking (Drucker-Colín and Spanis, 1976). These studies (using cats as subjects) have shown that much more protein synthesis occurs during sleep than during wakefulness. Specifically, the highest rate occurs during D sleep. (See **FIGURE 14.8**.)

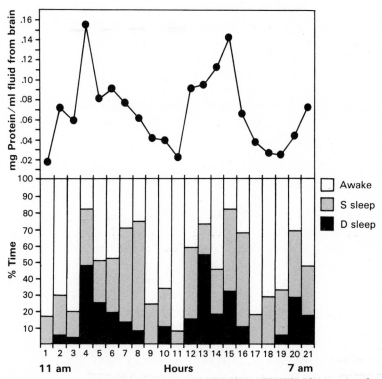

FIGURE 14.8 Amounts of proteins obtained from the brain by means of the push-pull technique during stages of sleep and waking. (From Drucker-Colín, R. R., and Spanis, C. W. Is there a sleep transmitter? *Progress in Neurobiology*, 1976, 6, 1–12.)

The proteins that are produced during D sleep do not appear to be unique to this state; Drucker-Colín, Spanis, Cotman, and Mc-Gaugh (1975) performed chemical analyses that indicated that the same classes of proteins are synthesized during other parts of the sleep-waking cycle, but are produced in smaller amounts.

The results of these studies favor the hypothesis that sleep serves as repair. In particular, D sleep might facilitate the synthesis of proteins in the brain. This suggests that sleep (especially D sleep) might be a response to the depletion of brain proteins. However, an experimental finding suggests that cause-and-effect might be reversed: D sleep might be a *result* of increased protein synthesis rather than its cause. Drucker-Colín et al. (1975) administered a drug that inhibits protein synthesis (anisomycin). The result was a decrease in D sleep. Similarly, starvation (which reduces protein synthesis) also reduces D sleep (McFayden, Oswald, and Lewis, 1973).

So far, no one has found any substances in the blood of normal animals that induce sleepiness. Thus, there is no evidence that sleep is a response to the products of general metabolism. Substances have been found in the CSF and interstitial fluid of the brain, but we do not know whether these cause the onset of sleep, or whether they are produced by neural sleep mechanisms that have already been turned on by some other means. Growth hormone is normally secreted during stages 3 and 4 of S sleep, but since this is not always the case, it would appear that the sole function of sleep cannot be the secretion of this hormone. (And since GH is secreted during the early stages of sleep, some other function would have to explain the fact that we sleep so long.) Finally, D sleep seems to be associated with protein synthesis. There is evidence from other sources that D sleep has special functions (which might require protein synthesis), so let us examine these studies next.

Selective Deprivation of D Sleep

It may seem strange to say that more is known about the effects of depriving an organism of a particular kind of sleep—D sleep—than of the effects of total sleep deprivation. This is because an organism can be selectively awakened during D sleep but be permitted to obtain S sleep. Control subjects can be awakened the same number of times, at random intervals. Theoretically, at least, effects produced by this differential treatment will result from the presence or absence of D sleep. (As we shall see later, some evidence suggests that D sleep deprivation is not a pure phenomenon—there are at least some complications.)

THE REBOUND PHENOMENON. An early report on the effects of D sleep deprivation (Dement, 1960) noted that as the deprivation progressed, **the subject had to be awakened more frequently;** the "pressure" to enter D sleep built up. Furthermore, after several days of D deprivation the subject would show a "rebound" phenomenon when permitted to sleep normally; a much greater percentage of the recovery night was spent in D sleep than would normally occur. These phenomena were subsequently replicated many times. This rebound suggests that there is a need for a certain amount of D sleep. A deficiency that is produced by selective deprivation is made up for later, when uninterrupted sleep is permitted.

Dement also reported anxiety and irritability, and an impaired ability to concentrate. However, later studies (including investigations by Dement and his colleagues) found no such changes, in either humans or animals. Cats have been deprived of D sleep for 70 days, and humans for 16 days, without any major impairment in psychological functions (Dement, 1974).

Some of the phenomena that normally occur only during D sleep will "escape" during D sleep deprivation and occur during S sleep, or even during waking. For example, penile erections begin to occur during S sleep (Fisher, 1966). The D sleep of many animals (particularly cats) contains brief, phasic bursts of electrical activity found in the pons, lateral geniculate nucleus, and visual (occipital) cortex (Brooks and Bizzi, 1963). These **PGO waves** are often the first manifestations of D sleep in cats. Whether or not humans also have PGO waves is not known, since their measurement would require the introduction of depth electrodes into the brain—and curiosity is not a good enough reason for such a procedure. Dement, Ferguson, Cohen, and Barchas (1969) found that in cats, PGO waves, normally restricted to D sleep, occur during most S sleep and waking after D sleep deprivation. It is quite possible that some of the functions normally accomplished during D sleep are transferred to other times when D sleep deprivation is attempted. Thus, the role played by D sleep may be greater than what we might infer from the missing functions that result from its (only partial) absence.

EMOTIONAL CHANGES. There do, however, appear to be some subtle impairments after D sleep deprivation. Performance on simple verbal tasks with no emotional content is not impaired (see McGrath and Cohen, 1978, for a review). However, Greiser, Greenberg, and Harrison (1972) found that D sleep deprivation impaired the recall of emotionally toned words. Cartwright, Gore, and Weiner (1973) found that subjects who sorted a group of adjectives according to how accurately the words described themselves had a lower recall for these words if the subjects were deprived of D sleep. Another study examined the effects of D sleep deprivation on the assimilation of anxiety-producing events. The subjects viewed a film that generally produces anxiety in the observers (a particularly gruesome circumcision rite performed by a primitive tribe). Normally, less anxiety is shown during the second viewing. Greenberg, Pillard, and Pearlman (1972) found that subjects permitted to engage in D sleep between the first and second viewings of the film showed more reduction in anxiety than subjects who were deprived of D sleep. Breger, Hunter, and Lane (1971) found that the dream content of subjects viewing the film was affected by the anxiety-producing material. Taken together, the studies suggest that D sleep somehow assists the integration of emotional material with memories of other experiences, and allows habituation of the anxiety-producing value of the material. This is a very fuzzy statement—especially for a textbook of physiological psychology. Fortunately, I can discuss a similar (but more objective) conclusion based on studies with animals.

LEARNING AND MEMORY.   D sleep appears to have functions that are related to learning and memory. D sleep deprivation retards memory formation, and learning a new task results in increases in D sleep.

A special procedure has been developed to deprive laboratory animals of D sleep. The subjects are placed in water-filled cages, on small platforms that are just large enough to hold them. (Often, inverted flower pots are used—hence the term *flower-pot technique.*) During S sleep, the animal snoozes quietly, its head held above the pool of water. But as you will recall, one of the components of D sleep is a loss of muscle tonus. When the animal enters D sleep, its head slumps downward, splashing into the water. The animal wakes up, and soon resumes S sleep. (See **FIGURE 14.9.**) Once the subject is adapted to the apparatus, it learns to awaken as soon as its head begins to nod; it no longer gets its face wet.

Many studies have used this technique, with varied results. The most consistent findings have been obtained with mice. These animals appear to adapt to life on a small island much better than rats

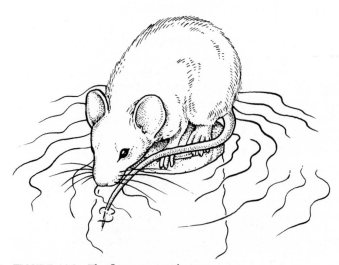

FIGURE 14.9   The flower-pot technique.

do. As Fishbein and Gutwein (1977) note, mice habitually jump up to the wire mesh cage top just above them, and obtain exercise by crawling upside down. (Mice are amazing acrobats.) Rats do not do this; a D sleep–deprived rat is also an exercise-deprived rat (at least, when the flower pot technique is used). This means that mice are better subjects for these experiments.

A study by Rideout (1979) shows the effects that D sleep deprivation can have on memory formation. He trained mice to run through a complex maze to find food. Each day's training session (one trial) was followed by 3 hours of D sleep deprivation, using the flower-pot technique. Figure 14.10 shows the performance of D sleep–deprived (gray curve) and nondeprived control (black curve) subjects. The graph depicts the time to traverse the maze as a function of days; hence, lower values represent superior performance. Note that D sleep deprivation retarded the rate of acquisition. (See **FIGURE 14.10.**)

Rideout also noted that some animals failed to fall asleep on the platform. These animals received *total* sleep deprivation. Surprisingly, they performed as well as control subjects did. (See dashed curve, **FIGURE 14.10.**) The results suggest that D sleep deprivation by itself was not responsible for the retarded performance. Rather, S sleep *without D sleep* seems to have deleterious effects. Perhaps one

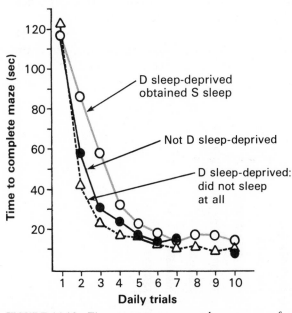

FIGURE 14.10   Time to traverse a complex maze as a function of daily training sessions in nonsleep-deprived mice and mice who were deprived of D sleep, or of both D and S sleep. (From Rideout, B. Non-REM sleep as a source of learning deficits induced by REM sleep deprivation. *Physiology & Behavior*, 1979, *22*, 1043–1047. Copyright 1979, Pergamon Press, Ltd.)

of the functions of D sleep is to reverse harmful effects that S sleep has on memory formation. (If this is so, then why do we engage in S sleep?)

Studies have also shown that new learning experiences produce an increase in subsequent D sleep. Lucero (1970) found that a 1½–2-hour maze learning session increased D sleep during the 3 hours immediately afterwards. S sleep was unaffected. Bloch, Hennevin, and Leconte (1977) gave rats daily training trials in a complex maze. They found that the experience enhanced subsequent D sleep. Moreover, there was a correlation between daily performance and D sleep. Figure 14.11 shows the results. The lower curve represents D sleep as a percentage of total sleep. (Sleep stages were determined by EEG recording during a 2-hour recording session immediately after the training trials.) The upper curve illustrates the animals' performance in the maze. Unlike Rideout's graph (previous figure), this one shows *speed* rather than time to traverse the maze. Thus, higher values represent superior performance. Note that once the task was well learned (after day 6) D sleep declined back to baseline levels. (See **FIGURE 14.11.**)

The role of D sleep in memory formation is certainly not all-or-none. In animals, D sleep deprivation retards memory formation, but does not prevent its occurrence. In humans, D sleep seems to be related to learning experiences that involve emotionally-relevant material. Perhaps the relationship between protein synthesis and D sleep is significant in this regard. As we shall see in Chapter 18, memory formation almost certainly involves protein synthesis. D

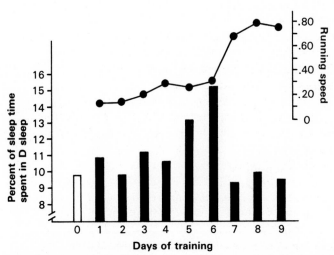

FIGURE 14.11 Percent of sleep time spent in D sleep (black vertical bars) as a function of maze learning performance (curve). (From Lucero, M. A. Lengthening REM sleep duration consecutive to learning in the rat. *Brain Research*, 1970, *20*, 319–322.)

sleep might provide a favorable condition for memory formation, or perhaps a more subtle integration of new memories with old ones.

Selective Deprivation of S Sleep

Since S sleep normally precedes D sleep, it is not possible selectively to deprive an organism of S sleep alone. What can be done is to deprive an organism of stage 4 sleep. A buzzer (loud enough to lighten sleep, but not loud enough to awaken) can be sounded whenever the EEG record becomes dominated by synchronous delta waves. The procedure seems effective; people deprived of stage 4 sleep show a rebound phenomenon—they engage in more of this type of sleep when they are permitted to sleep normally (Agnew, Webb, and Williams, 1964). This rebound suggests that there is a need for a certain amount of S sleep. No severe deficits appear to follow deprivation of stage 4 (the deepest stage of S sleep); at the most, some physical lethargy and depression is seen (Agnew, Webb, and Williams, 1967). Moldofsky and Scarisbrick (1976) found that subjects who had been S sleep deprived complained of muscle and joint pains, and reported a general increase in sensitivity to pain. The results suggest that S sleep might indeed serve a restorative effect on the wear and tear that occurs during waking, but the effects are rather modest, and do not seem to explain the irresistible pressure of the urge to sleep.

Sleep as an Adaptive Response

An excellent case can be made for sleep being an adaptive response. One of the most active proponents of this hypothesis is Wilse Webb. (See Webb, 1974; 1975.) Sleep might not have any special restorative properties, but might simply function to keep an animal out of harm's way when there is nothing important to do. For example, our primitive ancestors were probably well served by the presence of a state that kept them from stumbling around in the dark, where predators were harder to see, and injuries were more likely to occur. For other animals, food could be obtained only during part of the day-night cycle. These animals would profit from a period of inactivity, during which less energy would be expended. Animals who have safe hiding places (e.g., rabbits) sleep a lot, unless they are very small, and need to eat much of the time (e.g., shrews). Large predators (e.g., lions) can safely sleep wherever and whenever they choose, and indeed, sleep many hours of the day. In contrast, large animals who are preyed upon and have no place to hide, and who consequently must remain on guard (e.g., cattle), sleep very little. In fact, Allison and Chichetti (1976) found that body weight and danger of attack accounted for 58 percent of species variations in length of sleep.

The hypothesis that sleep serves simply as an adaptive response has led to the suggestion (Meddis, 1977) that we could (theoretically, at least) dispense with sleep altogether, now that we have artificial lighting, and are not endangered by predators (nonhuman ones, anyway). However, most sleep researchers disagree, and doubt that we will ever be able to eliminate sleep by chemical or surgical means. After all, even if sleep originally evolved for one reason, it is still likely that it came to serve other functions once the mechanisms that produce it became established. The evidence for specific functions that are performed by sleep is still rather unimpressive, but the insistence of the urge to sleep remains a convincing argument for most people that sleep must somehow be very important.

Although I have presented the two major hypotheses (sleep as construction or repair and sleep as an adaptive response) separately, it must be stressed that they need not be mutually exclusive. They could both be correct. Even if a biochemical need for sleep is eventually established, this will not rule out the fact that this state keeps many animals out of harm's way during part of the day. Since we cannot go back to the early times during which mechanisms of sleep began to evolve, we will never know for certain what forces shaped their development. All we will ever be able to determine is what significance sleep has now.

SLEEP PATHOLOGY

Insomnia

Insomnia is a problem that affects at least 20 percent of the population at some time (Raybin and Detre, 1969), but unfortunately, little is known about its causes. At the outset, it must be emphasized that there is no single definition of insomnia that can apply to all people. The amount of sleep that individuals require is quite variable. A short sleeper may feel fine with 5 hours, while a long sleeper may still feel unrefreshed after 10 hours of sleep. Insomnia must be defined relative to a person's particular sleep needs. Some short sleepers have been led to seek medical assistance because they thought that they should be getting more sleep, even though they felt fine. They should be reassured that whatever amount of sleep seems to be enough *is* enough. Meddis, Pearson, and Langford (1973) reported the case of a 70-year-old woman who slept approximately one hour each day (documented by sessions in a sleep laboratory). She felt fine, and was of the opinion that most people "wasted much time" in bed.

Probably the most important cause of insomnia is sleeping medication. Insomnia is not a disease that can be corrected with a medicine the way that diabetes can be treated with insulin. Insomnia is a symptom. If it is caused by pain or discomfort, the physical ailment that leads to the sleeplessness should be treated. If insomnia is secondary to personal problems or psychological disorders like depression, these should be dealt with directly. William Dement says that one of the things he hopes his students will remember, if they remember nothing else from his course, is that "sleeping pills cause insomnia" (Dement, 1974). This fact, unfortunately, is not appreciated by a very large number of people (or by their physicians, either). According to Freemon (1972), "the promiscuous prescribing of sleep medications is the most common error in medicine." Solomon (1956) classes it as "perhaps the commonest of iatrogenic disorders" (*iatrogenic* means "physician-produced").

I cite these authorities to emphasize the fact that sleeping medications do not induce normal sleep. They suppress D sleep at first, but this effect later habituates—D sleep reemerges (probably because of increased activity of some compensatory mechanism). The D sleep that does reemerge is apparently different from normal D sleep; Carroll, Lewis, and Oswald (1969) found a decrease in the vividness of dreams while subjects were under the effects of barbiturates. When sleeping medications are discontinued, a large rebound of D sleep is seen; in fact, the increased D activity leads to sleeplessness and vivid nightmares (Oswald, 1968). The patient then becomes frightened and demands more pills in order to get to sleep again.

There are two special forms of insomnia (one of them might be called pseudoinsomnia) that present interesting problems. Unfortunately, some people dream that they are awake. They do not dream that they are running around in some Alice-in-Wonderland fantasy, but that they are lying in bed, trying unsuccessfully to fall asleep. In the morning, their memories are of a night of insomnia, and they feel as unrefreshed as if they had really been awake. Others might not dream of being awake, but they nevertheless grossly underestimate the amount of time actually spent asleep. As Hartmann (1973) notes, it is absolutely necessary actually to observe someone's sleep before it is possible to decide whether they get more or less sleep than normal. Fortunately, the knowledge (obtained in the sleep lab) that they are really getting enough sleep is sufficient to make some of these people feel much better.

dreaming about lying awake in bed

The second type of special insomnia is rather pathetic. Some people are unable to sleep and breathe at the same time. They fall asleep and cease to breathe—they exhibit what is called ***sleep apnea.*** (Nearly all people have occasional episodes of sleep apnea—especially people who snore—but not to the extent that it interferes with sleep.) The level of carbon dioxide in the blood stimulates chemoreceptors, and the person wakes up, gasping for air. The oxygen level of the blood returns to normal, the person falls asleep, and the whole cycle begins again. Some of these people are aware of a sleep prob-

inability to sleep and breathe at the same time.

lem and complain of insomnia. Some others, who rapidly forget such awakening, complain of sleeping too much at night. Their total sleep period is prolonged by the large number of awakenings.

Although the issue has not been settled yet, many investigators believe that one of the principal causes for the *sudden infant death syndrome* (SIDS) is sleep apnea; in these cases, however, a high level of carbon dioxide in the blood fails to awaken. Occasionally, infants are found dead in their cribs, without any apparent signs of illness. Baker and McGinty (1977) maintained kittens for 8 hours per day in a chamber that contained an atmosphere that was deficient in oxygen. Most kittens did well, increasing their rate of respiration to compensate for the lowered oxygen level. Their episodes of sleep apnea (which are present in kittens just as they are in humans) decreased. However, some kittens did *not* compensate during sleep. Their periods of sleep apnea continued, and in a few cases, led to death. Perhaps a low-grade illness, producing some depression of respiratory mechanisms and an increased tissue need for oxygen, may cause susceptible human infants to succumb, just as some kittens did. The development of monitoring devices that sound an alarm when a susceptible infant stops breathing during sleep has saved a number of lives.

Narcolepsy

This disorder is characterized by several symptoms; not all of the symptoms of *narcolepsy* (narke, "numbness"; lepsis, "seizure") need occur in one individual. Although the physiological causes for this disorder have not yet been explained, the symptoms can be described in terms of what we know about the phenomena of sleep. The primary symptom is the **narcoleptic sleep attack.** This is an overwhelming urge to sleep that can happen at any time, but occurs most often under monotonous, boring conditions, especially when the patient is in a situation of passivity, such as riding a bus. Sleep (which appears to be entirely normal) usually lasts for 2–5 minutes. The patient wakes up feeling refreshed.

There is another disorder, **hypersomnia,** which should not be confused with narcolepsy. Hypersomnic patients wake up with great difficulty (they usually require assistance in getting up and out of bed), and feel tired all day. They take frequent naps, but waken unrefreshed. They do not experience the other symptoms of narcolepsy (which I shall describe now).

Another symptom of narcolepsy is **cataplexy** (kata, "down"; plēxis, "stroke"). During a cataplectic attack an apparently normal, waking person will suddenly wilt and fall like a sack of flour. The person will lie there, conscious, for several minutes. The patient is capable of moving his or her eyes, and can do so in response to requests from another person. What apparently happens to these people is that one of the phenomena of D sleep—muscular paralysis—intrudes into wakefulness. You will recall that the EMG taken during D sleep indicates a loss of muscle tonus, as a result of massive inhibition of motor neurons. (This phenomenon is taken advantage of in the flower-pot technique of D sleep deprivation.) When muscular paralysis occurs during the day, the victim of a cataplectic attack falls as suddenly as if a switch has been thrown.

Cataplexy is often precipitated by strong emotion, or by sudden physical effort, especially if the patient is caught unaware. Laughter, anger, or trying to catch a suddenly-thrown object can trigger a cataplectic attack. Common situations that bring on cataplexy are attempting to discipline one's children or making love (which, you must admit, is an awkward time to become paralyzed). Completely spontaneous attacks are rare.

D sleep paralysis sometimes intrudes into waking, but at a time that does not present any physical danger—just before or after normal sleep. This symptom is referred to as *sleep paralysis,* an inability to move just before the onset of sleep, or upon waking in the morning. A person can be snapped out of sleep paralysis by being touched, or hearing someone call his or her name. A good description of sleep paralysis was written by Edward Binns, a 19th-century physician who suffered from this disorder:

> During the intensely hot summer of 1825, I experienced an attack of this affection. Immediately after dining, I threw myself on my back upon a sofa, and, before I was aware, was seized with difficult respiration, extreme dread, and utter incapability of motion or speech. I could neither move nor cry, while the breath came from my chest in broken and suffocating paroxysms. During all this time I was perfectly awake; I saw the light glaring in at the windows in broad sultry streams; I felt the intense heat of the day pervading my frame; and heard distinctly the different noises in the street, and even the ticking of my own watch, which I had placed on the cushion beside me; I had, at the same time, the consciousness of flies buzzing around, and settling with annoying pertinacity on my face. During the whole fit, judgment was never for a moment suspended. I felt assured that I labored under incubus. I even endeavored to reason myself out of the feeling of dread which filled my mind, and longed, with insufferable ardour, for some one to open the door, and dissolve the spell which bound me in its fetters. The fit did not continue above five minutes: by degrees I recovered the use of sense and motion; and, as soon as they were so far restored as to enable me to call out and move my limbs, it wore insensibly away. (Binns, 1852, p. 156)

Sometimes the mental components of D sleep intrude into sleep attacks; the patient dreams while lying awake, paralyzed. These *hypnogogic hallucinations* are often alarming or even terrifying—an unpleasant way to fall asleep, I should think. This explanation is supported by the observations of Rechtschaffen, Wolpert, Dement, Mitchell, and Fisher (1963), who found that narcoleptic patients generally skip the S sleep that normally begins a night's sleep; instead, they go right into D sleep from waking.

The symptoms of narcolepsy can be successfully treated with drugs. Narcoleptic sleep attacks respond well to stimulants like amphetamine, a catecholamine agonist. The D sleep phenomena (cataplexy, sleep paralysis, and hypnogogic hallucinations) can be alleviated by imipramine, which facilitates serotonergic, as well as catecholaminergic, activity. Most often, the drugs are given together. I shall discuss the implications of the effects of these drugs on neural mechanisms of sleep in Chapter 15.

Problems Associated with S Sleep

Although you might think that sleep talking should occur during D sleep (the acting out of a dream, for example), most of it occurs during S sleep—during the deepest stage, in fact. (I suppose sleep talking should not be classified as a problem, unless you tend to give away secrets, but since it usually occurs during stage 4 sleep, this seemed to be the best place to mention it.) Bed-wetting (**nocturnal enuresis**) and sleepwalking (**somnambulism**) usually occur during stage 4 sleep, also. So do **night terrors,** a condition that is seen most often in children. Night terrors are characterized by anguished screams, trembling, and a rapid pulse (and often no idea at all about what caused the terror). Bed-wetting often can be cured by training methods (having a bell ring when the first few drops of urine are detected in the bedsheet by a special electronic circuit—a few drops usually precede the ensuing flood). Night terrors and somnambulism usually cure themselves as the child gets older. Neither of these phenomena is related to D sleep; a sleepwalking person is *not* acting out a dream. Dement firmly advises that the best treatment for these two disorders is no treatment at all. There is no evidence that they are associated with mental disorders or personality variables.

Broughton (1968) has suggested that the disorders of S sleep are related not to the state of sleep itself, but to sudden *arousal* from deep slow wave sleep. Polygraphic records from his laboratory indicated that enuresis, sleepwalking, and night terrors were usually triggered by a sudden sign of EEG arousal in the midst of deep slow wave sleep. He notes that people who are suddenly aroused from D sleep are generally lucid and coherent. In contrast, people who are suddenly aroused from S sleep act confused, speak indistinctly, and exhibit poorly coordinated behavior. In addition, they generally do not remember the awakening later, whereas people awakened from D sleep usually do.

To test his hypothesis, Broughton gave large drinks of water to a number of normal children just before they went to bed. They were subsequently awakened during sleep by a full bladder. When the awakening came during S sleep, the children got out of bed, walked to the toilet, urinated, and returned to bed. They looked and acted like sleepwalkers, and, like spontaneous sleepwalkers, did not remember the episode the next day. Broughton hypothesized that some people enter such deep stages of S sleep that they spontaneously awaken, presumably as a defensive measure. The awakening triggers the disorders that are associated with S sleep.

CONCLUSIONS

I have attempted to describe sleep (which is a rather straightforward task) and to summarize the evidence concerning its utility (which is considerably more difficult). I suspect you have come to conclusions similar to my own; yes, sleep seems to perform some function—perhaps to assist in the integration of new material into the existing storehouse of memories—but if sleepiness is such an overwhelming drive, and if it can even take precedence over hunger and sex, then why aren't its functions more obvious?

In the next chapter I shall ignore the mysterious *why* of sleep and concentrate on studies attempting to elucidate the *how* of it.

KEY TERMS

alpha activity p. 440

beta activity p. 440

cataplexy p. 466

D sleep p. 444

delta activity p. 441

delta sleep-inducing peptide p. 453

desynchrony p. 441

electro-oculogram p. 443

flower-pot technique p. 460

hypersomnia p. 466

hypnogogic hallucinations p. 467

microsleep p. 452

narcoleptic sleep attack p. 466

night terrors p. 468

nocturnal enuresis p. 468

PGO wave p. 459

push-pull cannula p. 454

REM sleep p. 444

Ringer's solution p. 454

S sleep p. 444

sleep apnea p. 465

sleep paralysis p. 467

somnambulism p. 468

synchrony p. 441

SUGGESTED READINGS

DEMENT, W. C. *Some Must Watch While Some Must Sleep.* San Francisco: W. H. Freeman, 1974.

WEBB, W. B. *Sleep: The Gentle Tyrant.* Englewood Cliffs, N. J.: Prentice-Hall, 1975.

These books provide a general introduction to the topic of sleep.

DRUCKER-COLÍN, R., SHKUROVICH, M., AND STERMAN, N. B. (Eds.) *The Functions of Sleep.* New York: Academic Press, 1979.

MENDELSON, W. B., GILLIN, J. C., AND WYATT, R. J. *Human Sleep and Its Disorders*. New York: Plenum Press, 1977.

These books are more specialized. The one edited by Drucker-Colín and his colleagues contains chapters by researchers who are concerned with the functions performed by sleep. The one by Mendelson and his colleagues discusses disorders of human sleep (as you may have guessed from its title).

15

Neural Mechanisms of the Sleep-Waking Cycle

BIOLOGICAL CLOCKS

Circadian Rhythms and Zeitgebers

As we saw in Chapter 14, the search for chemical factors that are responsible for cycles of sleep and waking has so far been unsuccessful. The possibility exists, therefore, that sleep cycles are regulated neurally. (Of course, neural control must involve biochemistry, in the form of intracellular processes, ionic flow, and chemical activity at synapses.) A considerable amount of research effort has been devoted to the investigation of biological clocks that provide the basis for **circadian rhythms.** *Circa* = "about," *dies* = "day;" therefore, a circadian rhythm is one that varies on a 24-hour cycle. Perhaps patterns of sleep and waking are controlled by some kind of neural clock.

Circadian rhythms are found throughout the plant and animal world. Some of them (such as the rate of plant growth) are a direct consequence of variations in the level of illumination, and are of no interest in the study of sleep. Other rhythms are intrinsic to the organism, and are controlled by an internal clock. Figure 15.1 shows the activity of a rat during various conditions of illumination. Each horizontal line represents 24 hours. Vertical tick marks represent activity (in a running wheel). The upper portion of the figure shows the activity of the rat during a normal day-night cycle with alternating 12-hour periods of light and dark. Note that the animal is active during the night. (See **FIGURE 15.1.**) Next, the day was moved forward by six hours; the animal's activity cycle quickly followed the change. (See **FIGURE 15.1.**) Finally, dim lights were turned on continuously. The cyclicity in the rat's activity remains even when no changes in the level of light provide clues as to the time of day. However, you can see that the rat's clock is not set precisely to 24

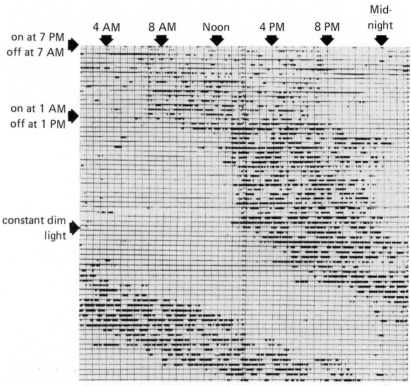

FIGURE 15.1   Wheel-running activity of a rat. Note that the animal's activity occurs at night, and that the active period is reset when the light period is changed. When the animal is maintained in constant dim illumination it displays a free-running activity rhythm of approximately 25 hours. (From Groblewski, T. A. Nuñez, A., and Gold, R. M. Circadian rhythms of activity and food intake in vasopressin deficient Brattleboro rats. Paper presented at the meeting of the Eastern Psychological Association, April, 1980. Photograph courtesy of A. Nuñez.)

hours; when the illumination is held constant, it runs a bit slow. The animal begins its bout of activity, about one hour later each day. (See **FIGURE 15.1**.)

The phenomenon that is illustrated in Figure 15.1 is typical of circadian rhythms in a great number of species. A free-running clock, with a cycle of approximately 24 hours, controls some biologic functions (in this case, motor activity). Changes in the level of illumination (that is, the onset of day and night) keep the clock adjusted to 24 hours. In the parlance of scientists who study circadian rhythms, light serves as a *Zeitgeber* (German for "time giver"); it synchronizes the endogenous rhythm. Studies have shown that if many different kinds of animals are maintained in constant darkness, a brief flash of light will reset their biological clock, advancing or retarding it, depending upon the biological time that the light flash occurs (Aschoff, 1979). In the absence of light, animals may use other environmental stimuli (such as fluctuations in temperature) to synchronize their rhythms.

Humans, like other animals, exhibit circadian rhythms. Our normal period of inactivity begins several hours after the start of the dark portion of the day/night cycle, and persists for a variable amount of time into the light portion. Under conditions of constant illumination, our biological clocks will run free, gaining or losing time like a not-too-accurate watch. Different people have different

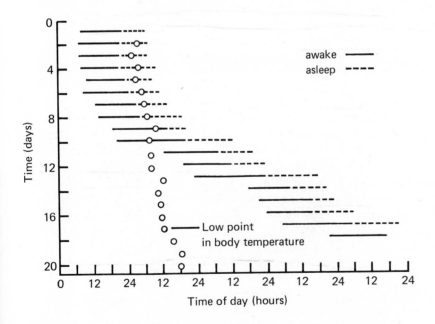

FIGURE 15.2 Light-entrained and free-running rhythms in a human. Note that cycles in body temperature and sleep/activity follow different rhythms. (From Aschoff, J. Desynchronization and resynchronization of human circadian rhythms. *Aerospace Medicine*, 1969, *40*, 844–849.)

cycle lengths, but most people will begin to live a "day" that is approximately 25 hours long. Figure 15.2 shows the sleep/waking cycles and body temperature cycles of a subject who was kept for 17 days in an isolated room with constant illumination. Note that the clock that controlled sleeping ran a little slow; it lost about seven hours over the 17-day period. (See **FIGURE 15.2** on the preceding page.) However, temperature followed an independent rhythm; the clock controlling body temperature gained about 36 hours during the same time. (See **FIGURE 15.2.**) Thus, although body temperature and sleep cycles are normally correlated when a *Zeitgeber* is present to synchronize them, these functions are apparently controlled by different clocks.

The Neural Basis of Biological Clocks

It has been suspected for some time that the biological clocks of mammals are neural mechanisms, and that light is the primary *Zeitgeber*. If rats are blinded, their activity cycles become free-running (Browman, 1937; Richter, 1965). Since destruction of all the (then) known projection regions of the visual system did not abolish the relationship of activity cycles to the light/dark cycle even though the animals were blind, it appeared that the biological clock must have its own private projection from the retina (Stephan and Zucker, 1972a). The finding that medial hypothalamic lesions abolished a rat's cycles in activity (Richter, 1965, 1967) suggested that the clock may be located there.

It was subsequently discovered by researchers working independently in two laboratories (Moore and Eichler, 1972; Stephan and Zucker, 1972b) that the primary biological clock of the rat is located in the *suprachiasmatic nucleus* of the hypothalamus (**SCN**).

The suprachiasmatic nuclei of the rat consist of approximately 10,000 small, tightly-packed neurons. The dendrites of these neurons synapse with each other—a phenomenon that is found only in this part of the hypothalamus, and which undoubtedly relates to the special function of these nuclei (Güldner and Wolff, 1974; Güldner, 1976). Autoradiographic techniques have revealed a direct projection of fibers from the retina to the SCN (Hendrickson, Wagoner, and Cowan, 1972). Figure 15.3 shows the appearance of the suprachiasmatic nuclei in a transverse section through the hypothalamus of a mouse; they appear as two clusters of dark-staining neurons at the base of the brain, just above the optic chiasm. Also note the small dark spots within the optic chiasm, just ventral and medial to the base of the suprachiasmatic nuclei; these are cell bodies of oligodendroglia that serve axons that enter the SCN and provide information from the retina. (See **FIGURE 15.3.**)

The SCN also receives input from the ventrolateral geniculate nucleus and the raphe nuclei (Aghajanian, Bloom, and Sheard,

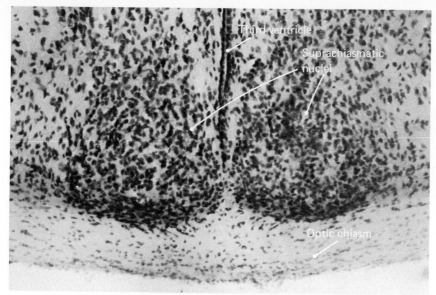

FIGURE 15.3 A transverse section through a mouse brain, showing the location and appearance of the suprachiasmatic nuclei. Cresyl violet stain.

1969; Ribak and Peters, 1975). Neither of these inputs appears to be a necessary component of the biological clock; circadian rhythms are still seen after destruction of the lateral geniculate nucleus or lesions of the raphe nuclei (Coindet, Chouvet, and Mouret, 1975; Block and Zucker, 1976).

Neurons of the SCN project caudally to the ventromedial and posterior hypothalamus and interpeduncular nucleus, dorsally to the dorsomedial thalamus and lateral habenula, and rostrally to the lateral septum and nucleus of the diagonal band (Swanson and Cowan, 1975b; Sofroniew and Weindl, 1978). Knife cuts that interrupt the caudal efferents of the SCN abolish hormonal rhythms, including those which control estrous cycles in female rats (Moore and Eichler, 1972), but do not affect cycles of feeding, drinking, and activity (Nuñez and Casati, 1979). Therefore, not all functions that the SCN regulates are controlled by the same set of efferent fibers. The roles that are played by the various output pathways of the SCN have yet to be determined. Furthermore, the SCN contains neurosecretory cells (Sofroniew and Weindl, 1978), so it is possible that some of its control is exerted chemically.

A study of Schwartz and Gainer (1977) nicely demonstrates day/night fluctuations in the activity of the SCN. These investigators injected rats with radioactive 2-DG. As we saw in Chapter 12, this chemical is structurally similar to ordinary glucose, and is thus taken up by cells that are metabolically active. However, it cannot be utilized, nor can it leave the cell. Therefore, metabolically active cells will accumulate radioactivity. (This technique was also used in

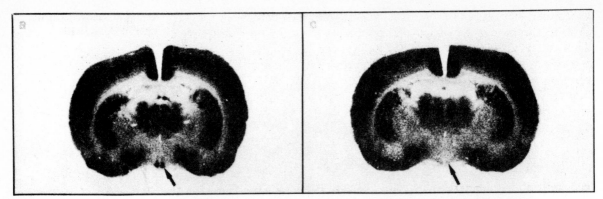

FIGURE 15.4 Autoradiographs of transverse sections through the brains of rats that had been injected with ^{14}C-2-deoxyglucose during the day (left) or the night (right). The dark region at the base of the brain indicates increased metabolic activity of the suprachiasmatic nuclei. (From Schwartz, W. J., and Gainer, H. Suprachiasmatic nucleus: use of ^{14}C-labelled deoxyglucose uptake as a functional marker. *Science*, 1977, *197*, 1089–1091.)

a study on the visual mechanisms of cortex, reported in Chapter 9.)

The investigators injected some rats with radioactive 2-DG during the day, and injected others at night. The animals were then killed, and autoradiographs of sections through the brain were prepared. Figure 15.4 shows photographs of two of these sections. Note the evidence of radioactivity (and hence, a high metabolic rate) in the SCN of the brain that was injected during the day (left). (See **FIGURE 15.4.**)

THE PHYSIOLOGICAL BASIS OF CIRCADIAN RHYTHMS.   We do not know what physiological processes provide the time base for biological clocks. It is not even known whether the rhythmicity is a property of individual cells that constitute a biological clock, or whether circadian oscillation can occur only in an interconnected circuit of cells (Jacklet, 1978). One hypothesis that has received some preliminary support is that a time base may be provided by the synthesis of proteins. It is possible that cells could begin to synthesize protein, and as the level of this product rises, negative feedback would shut down the synthetic process. The proteins would be degraded or dispersed over time, and then the low level of proteins would permit the process to start the cycle over again. The presence of the protein could affect some characteristic of the neurons (for example, their membrane permeability) that would alter their rate of firing.

This hypothesis remains speculative, but Jacklet (1977) has shown that anisomycin (an inhibitor of protein synthesis that we encountered in Chapter 14) stops the biological clock that is contained in the isolated eye of *Aplysia* (a marine invertebrate). Furthermore, even brief exposure to anisomycin, which causes a brief drop in the rate of protein synthesis, resets the clock to a new time. This observation suggests that the intrinsic activity of the *Aplysia* eye might use the process of protein synthesis for its time base. The physiological basis of circadian activity in SCN is completely unknown, but I am sure that many laboratories are studying this problem right now.

Circadian Rhythms of Human Sleep

Does the human SCN control sleep cycles? It is possible—perhaps even likely—but it will be some time until this question can be answered. It had been thought for some time that the human brain did not contain a SCN, but fortunately (for those who hope that the rat

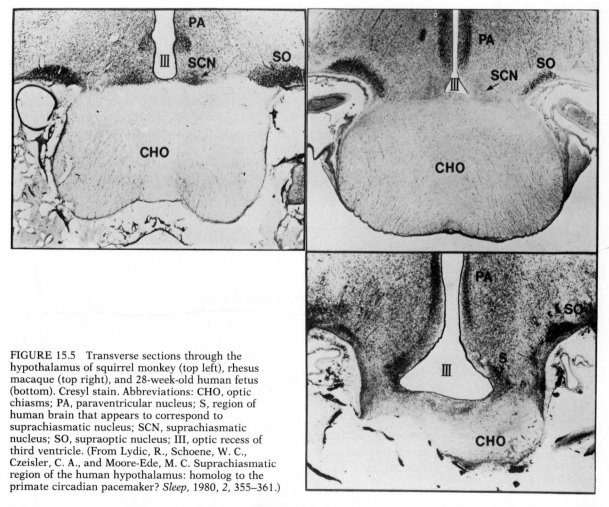

FIGURE 15.5 Transverse sections through the hypothalamus of squirrel monkey (top left), rhesus macaque (top right), and 28-week-old human fetus (bottom). Cresyl stain. Abbreviations: CHO, optic chiasms; PA, paraventricular nucleus; S, region of human brain that appears to correspond to suprachiasmatic nucleus; SCN, suprachiasmatic nucleus; SO, supraoptic nucleus; III, optic recess of third ventricle. (From Lydic, R., Schoene, W. C., Czeisler, C. A., and Moore-Ede, M. C. Suprachiasmatic region of the human hypothalamus: homolog to the primate circadian pacemaker? *Sleep*, 1980, *2*, 355–361.)

research will someday have relevance to the human brain) careful investigation has established that we humans have what appears to be a suprachiasmatic nucleus (Lydic, Schoene, Czeisler, and Moore-Ede, 1980). Figure 15.5 illustrates the SCN of a squirrel monkey (top left), rhesus macaque (top right), and 28-week-old human fetus (bottom right). (See **FIGURE 15.5** on the preceding page.)

At least in rats, the SCN appears to provide the primary control over the timing of sleep and waking cycles. Rats are nocturnal animals; most of their sleep occurs during the day, and they forage and feed during the night. Lesions of the SCN abolish this pattern; sleep occurs in bouts randomly dispersed throughout both day and night (Ibuka and Kawamura, 1975; Stephan and Nuñez, 1977). However, these rats still obtain the same amount of sleep that normal animals do. The temporal pattern, but not the total amount of sleep, is disrupted.

This is a very important finding, both for the immediate results, and for the promise that it holds for future research on the mechanisms of sleep. The fact that the total amount of sleep per day is not altered when the biological clock that controls it is destroyed means that an animal's quantity of sleep is *regulated*. This means that there must be system variables, set points, detectors, and correctional mechanisms for sleep, just as there are for feeding and drinking. Clearly, the SCN does not perform any of these functions, but exerts an overriding control that either sets the period during which sleep is possible, or keeps the animal awake during its active period, thus indirectly causing the sleep regulating mechanism to put the animal to sleep during its inactive period. The fact that such a discrete region of the brain is able to control sleep and waking means that we should be able to learn much about mechanisms of sleep by studying the efferent pathways of the SCN. If this structure controls the daily cycles of sleep, then it may be that its efferent fibers synapse with neurons that are directly involved with the production of sleep. Thus, neuroanatomical investigations, accompanied by selective knife cuts, would be worth pursuing. On the other hand, if the SCN affects sleep cycles by chemical means, then it would be profitable to study the neuromodulators it secretes, and look for receptors for these substances elsewhere in the brain.

So far, the SCN has not received much attention from scientists whose primary interest is in mechanisms of sleep. One of the reasons for this neglect is that most sleep researchers use cats as experimental subjects. There are many reasons for using cats: they have brains that are large enough for practical electrophysiological investigations, they are inexpensive and relatively easy to obtain, and they have excellent toilet habits, keeping themselves clean and well-groomed, and confining their waste products to a pan of kitty litter. Unfortunately, cats do not have circadian rhythms of sleep; they doze on and off throughout all portions of the day/night cycle. (The term "catnap" has a basis in reality.) Hence, the SCN probably does not have much to do with sleep mechanisms in these animals. Rats have not been used much for sleep studies because they are too small for some kinds of electrophysiological studies, and monkeys are expensive to procure and maintain in an animal colony.

The fact that body temperature and sleep can cycle independently in humans (Figure 15.2) suggests that when all the answers are in, we shall not find that there is a single biological clock that is responsible for all circadian rhythms. In fact, Moore-Ede, Lydic, Czeisler, Fuller, and Albers (1980) found that destruction of the SCN in squirrel monkeys abolished circadian rhythms in sleep-activity, but did not affect daily rhythms in body temperature. Thus, the biological clock that controls body temperature in at least one species of primates appears to be located outside the SCN. Perhaps the SCN of the human also controls sleep-waking cycles, but not cycles in body temperature. The next few years promise to provide some interesting experimental results.

SLEEP AS AN ACTIVE PROCESS

We do not know what sets the basic rhythm of the sleep/waking cycle, although the SCN seems to be the best candidate. Thus, there is another question that we can ask: what neural mechanisms are directly responsible for waking and for the two major stages of sleep?

We can very quickly dispense with the notion that the brain sleeps because it "runs down." There do not appear to be any neurons that get tired (e.g., run out of energy stores for the sodium-potassium pump or deplete the supply of the neurotransmitter) and consequently need a period of sleep in which to rest. Moruzzi (1972) reviewed electrophysiological data that show that neurons in the neocortex, lateral geniculate nucleus, reticular formation, and hypothalamus generally decrease their rate of firing during S sleep, but during D sleep fire as fast as (or faster than) they do during wakefulness. For example, neurons in motor cortex fire more often during sleep than during wakefulness, except when a movement is in progress (Evarts, 1965). Other units (such as those of the hippocampus) show a low rate of firing during D sleep. The point to be made is that there is not a *universal* decline in firing rate during sleep.

Bremer (1937), in an investigation of neural sleep mechanisms, showed that brain mechanisms important for arousal were located in the brainstem. He severed the brainstem of a cat between the superior and inferior colliculi. The cat survived for only a few days and showed a permanently synchronized EEG and pupillary constriction. The effects of this **midcollicular transection** (which Bremer

Cat is
comatose

FIGURE 15.6   The cerveau isolé. (After Bremer, F., *Bulletin de l'Academie Royale de Beligique*, 1937, 4, 68–86.)

called a *cerveau isolé*) are shown in **FIGURE 15.6.**

A transection made at the caudal end of the medulla, just above the spinal cord (the *encéphale isolé* preparation), produced a cat that demonstrated normal sleep and waking cycles. The animal was paralyzed, of course, since the brain and spinal cord were disconnected, so waking could not be evaluated in the normal fashion. However, the EEG showed periodic episodes of desynchronized and synchronized activity. During desynchrony, the cat's pupils were dilated, and the eyes followed a moving object. During synchrony, the pupils were constricted, and the eyes showed no reaction to visual stimuli. Therefore, the animal apparently slept and woke in alternate fashion. Figure 15.7 shows the location of the brainstem section and the arousing effects of the experimenter handling the cat's head. (Somatosensory information is received from the head region by means of the trigeminal nerve.) (See **FIGURE 15.7.**)

At the time, Bremer believed that the difference in the arousability of these two preparations could be accounted for by the dif-

Asleep   Awake

FIGURE 15.7   The éncephale isolé. (After Bremer, F., *Bulletin de l'Academie Royale de Beligique*, 1937, 4, 68–86.)

ferent amounts of sensory input that could be received by the cerebrum. He believed that the brain was normally inactive, and that it took the tonically arousing effects of sensory stimulation to keep the brain awake. The *encéphale isolé* had all its cranial nerve inputs intact. However, the midcollicular section (*cerveau isolé*) isolated the cerebrum from all sensory input except for olfaction and vision since the other cranial nerves enter the brain caudal to the plane of transection.

Subsequent studies showed that sleep is not a passive process; it is an active one. Batini, Moruzzi, Palestini, Rossi, and Zanchetti (1958, 1959) cut through the midbrain a few millimeters caudal to the midcollicular transection of Bremer. This operation, which the authors called a **midpontine pretrigeminal transection,** produced a cat with insomnia; the EEG and other signs indicated wakefulness 70 to 90 percent of the time. (In contrast, a normal cat is awake approximately 30 to 40 percent of the time.) These animals had exactly the same sensory input as did Bremer's cats with the midcollicular section (*cerveau isolé*). However, they showed definite wakefulness. Furthermore, if the remaining sensory inputs were removed (Batini, Palestini, Rossi, and Zanchetti, 1959), the animal continued to exhibit signs of wakefulness (after exhibiting a transient period of EEG synchrony). Therefore, sleep appears to be produced by the activity of some neural system, and not by the loss of incoming sensory information. Sleep is an active process, and not a passive one.

Midpontine cats (as these animals are called), like *encéphale isolé* (spinal-sectioned) cats will, during periods of cortical desynchrony, follow an object with their eyes, show pupillary dilation when a visual stimulus is presented, and show accommodation (adjustment of the lens of the eye) to a near object. However, during periods of cortical synchrony, visual stimuli have no effect. Ocular responses also show signs of habituation; after a while, a repetitive stimulus fails to elicit any ocular responses (perhaps even a midpontine cat can get bored.) Thus, it appears to be safe to conclude that midpontine cats demonstrate sleep and waking cycles.

Three conclusions can be based on these studies:

1. There is a brain region between the midcollicular section of Bremer and the midpontine section of Batini et al. that is important in producing wakefulness. The rostral section disconnects the cerebrum from this region, whereas the midpontine section does not.
2. There is a sleep-producing region that lies somewhere between the midpontine section and the caudal part of the medulla—the *encéphale isolé* section of Bremer. Animals with the midpontine cut show very little sleep, whereas those with the caudal cut show normal sleep/waking cycles.
3. The brain does not need sensory input in order to show signs of wakefulness. The midpontine section and the midcollicular section (*cerveau isolé*) produce the same effects on sensory input,

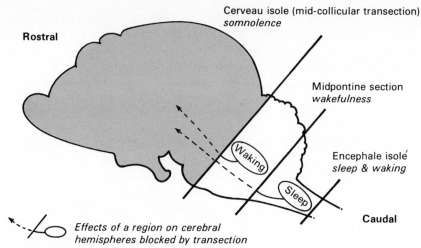

FIGURE 15.8   A schematic summary of the results of the three brainstem transections.

and deafferentation of the midpontine animal does not produce permanent sleep.

These results are summarized in **FIGURE 15.8.**

It would appear, then, that some rostral pontine region is necessary for waking, and some caudal pontine (and/or medullary) region is necessary for sleep. However, some evidence suggests that there is more to the story than that. Some investigators (notably Giuseppe Moruzzi, of Pisa, Italy) believe that there are diencephalic structures that are also involved in sleep and wakefulness, while others (notably Michel Jouvet, of Lyons, France) believe that there are not.

BRAINSTEM MECHANISMS

First, let us examine further evidence that the brainstem contains mechanisms that are important for producing sleep and arousal. Magni, Moruzzi, Rossi, and Zanchetti (1959) tied off some cerebral blood vessels so that the arterial blood supply to the medulla and lower pons was separated from the blood supply to the upper pons and cerebrum. (See **FIGURE 15.9.**) Injections of a general anesthetic (thiopental) into the blood supply of the rostral pons and cerebrum anesthetized the cat. (Obviously, this is not a surprising fact.) However, when only the caudal brainstem was anesthetized, a sleeping cat would wake up. A sleeping (that is, synchronous) EEG would be replaced by an aroused (that is, desynchronized) one. How can an-

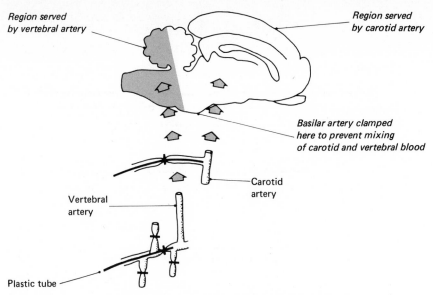

Region served by vertebral artery

Region served by carotid artery

Basilar artery clamped here to prevent mixing of carotid and vertebral blood

Carotid artery

Vertebral artery

Plastic tube

FIGURE 15.9   The procedure of Magni et al., which allowed the injection of an anesthetic into different regions of the brain. (Adapted from Magni, F., Moruzzi, G., Rossi, G. N., and Zanchetti, A., *Archives Italiennes de Biologie*. 1959, 97, 33–46.)

esthetization of part of the brain awaken a sleeping animal?

The answer seems to be this: There must be some region in the caudal brainstem whose activity is necessary to put an animal to sleep. When the activity of this region is temporarily suppressed by an injection of thiopental, the sleeping brain awakens. Like the mid-pontine pretrigeminal preparation, this observation provides further evidence that sleep is an active process.

The brainstem has been implicated in arousal mechanisms for a long time. In 1949, Moruzzi and Magoun found that electrical stimulation of the brainstem reticular formation would produce arousal. The reticular formation, in the core of the brainstem, is known to receive collaterals from ascending sensory pathways. Lindsley, Schreiner, Knowles, and Magoun (1950) produced lesions that disrupted the lateral sensory pathways (medial and lateral lemniscus, and trigeminal lemniscus). Tactile stimulation of these subjects produced long-lasting arousal. Since the lateral lesions destroyed the direct sensory pathways (which go to thalamus, and thence to sensory cortex), the arousal was obviously mediated by the reticular formation. Although the animals presumably could not "feel" the stimulation, they nevertheless were aroused by it. If medial lesions were produced (destroying the reticular formation), only a very transient arousal was produced by sensory stimulation. (See **FIGURE 15.10** on the following page.) It appears, then, that the reticular formation of the brainstem plays a role in arousal.

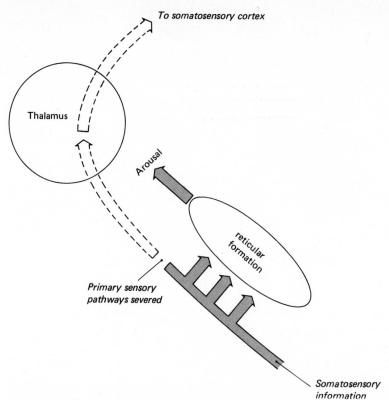

To somatosensory cortex

Thalamus

Arousal

reticular formation

Primary sensory pathways severed

Somatosensory information

FIGURE 15.10   A schematic representation of the experiment by Lindsley et al. (1950).

Brainstem Arousal Mechanisms

THE DOPAMINERGIC SYSTEM OF SUBSTANTIA NIGRA.   There is evidence for two distinct arousal mechanisms situated in the brainstem: a dopaminergic system and a noradrenergic system. According to Jouvet (1972), there is a system of dopaminergic neurons (whose cell bodies lie in the substantia nigra) that contributes to behavioral arousal. Lesions that destroy the substantia nigra (and also damage the ventral noradrenergic bundle) result in a temporarily or permanently comatose state; the cats remain stuporous and behaviorally unresponsive to sensory stimuli (Jones, 1969; Jones, Bobillier, and Jouvet, 1969). The location of these structures is shown in **FIGURE 15.11.** In the animals that remained permanently comatose, the level of dopamine in the rostral brain was reduced by 90 percent, attesting to the effectiveness of the lesion in disrupting this system. Sensory stimulation produced long-lasting cortical desynchrony without any signs of behavioral arousal. However, lesions of the substantia nigra in rats (as opposed to cats) do not severely affect behavioral arousal; Simpson and Iversen (1971) found that the ani-

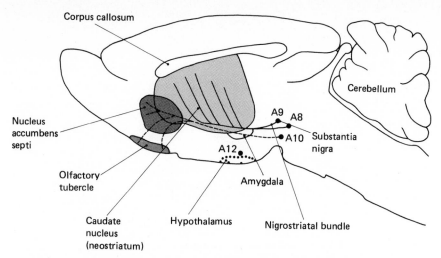

FIGURE 15.11   The pathways followed by dopaminergic neurons in the rat brain. (Redrawn from Livett, B., *British Medical Bulletin*, 1973, *29*, 93–99.)

mals actually increased their activity. Thus, the dopaminergic system that originates in substantia nigra is not necessary for behavioral arousal.

The effects of lesions of substantia nigra in rats and cats has yet to be reconciled. However, although this dopaminergic system may not be necessary for an animal to remain awake and active, there is plenty of evidence that dopaminergic neurons play a role in mechanisms of arousal. As we saw in Chapter 12, dopaminergic neurons appear to mediate the arousing effects of mild stress (pinching of the tail). And, as we shall see in Chapter 17, dopamine-secreting neurons play an important role in mediating the effects of reinforcement. They are also necessary for some of the activating effects of catecholamine agonists like amphetamine and cocaine. Many investigators believe that dopaminergic neurons play a role in mechanisms of attention, and that their hyperactivity can lead to schizophrenia. (This hypothesis will be discussed in Chapter 20.) However, we do not know what role these neurons play in controlling cycles of sleep and arousal.

THE NORADRENERGIC SYSTEM OF LOCUS COERULEUS.   It has also been suggested that a noradrenergic system plays a role in arousal. This system arises mainly from the rostral portion of *locus coeruleus,* a structure in the dorsal pons. (See **FIGURE 15.12** on the following page.) If a lesion is made in the dorsal noradrenergic pathway (interrupting the ascending axons from the neurons of the rostral locus coeruleus) the animals show hypersomnia; D sleep and S sleep increase dramatically (Jones, Bobillier, and Jouvet, 1969). Thus, there is good evidence that the dorsal noradrenergic pathway is involved in arousal (or, at least, in the suppression of sleep).

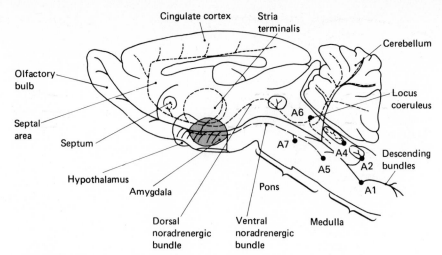

FIGURE 15.12   The pathways followed by noradrenergic neurons in the rat brain. (Redrawn from Livett, B., *British Medical Bulletin*, 1973, *29*, 93–99.)

Stimulation of this region produces arousal; indeed, when Moruzzi and Magoun (1949) stimulated the reticular formation, their electrodes were very near the dorsal noradrenergic pathway. However, the effects of electrical stimulation are not very localized, so it is impossible to say whether this system mediates all of the arousing effects that are produced by stimulation of the reticular formation. It is very possible that there are other, noncatecholaminergic, arousal systems that project from the brainstem to the cerebral hemispheres. The development of the histofluorescence technique has concentrated research efforts on the monoaminergic neurotransmitters; we must not lose sight of the fact that there are many brain functions that are *not* mediated by these chemicals.

Brainstem Sleep Mechanisms

As we saw earlier (from the insomnia that is produced by the midpontine transection and from the sleep-opposing effects of caudal brainstem anesthesia), there appears to be a brainstem mechanism, located behind the middle of the pons, that is important for the occurrence of sleep. The precise location of that mechanism is not known for certain, but there are two candidates: the ***raphe*** and the ***nucleus of the solitary tract.*** The raphe (say "ruh-FAY") is a complex of nuclei running through the core of the brainstem, from the medulla to the back of the midbrain. The nucleus of the solitary tract is a structure located in the medulla, which receives taste information and visceral sensation. A role in D sleep has been suggested for other structures in the brainstem, but the evidence is not at all clear.

Sagittal section

Transverse section

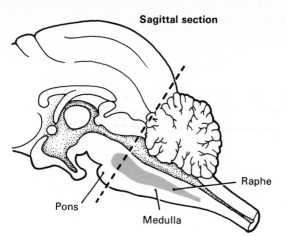

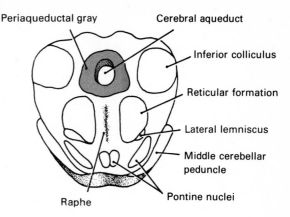

Periaqueductal gray

Cerebral aqueduct

Inferior colliculus

Reticular formation

Lateral lemniscus

Middle cerebellar peduncle

Pontine nuclei

Raphe

Raphe

Pons

Medulla

FIGURE 15.13   The raphe.

THE RAPHE.   Figure 15.13 shows the location of the raphe in a midsagittal section through the brainstem; you can see how the mid-pontine section separates most of the raphe from the cerebrum. (See **FIGURE 15.13.**) The transverse section in the same figure shows how this structure got its name—raphe means crease, or seam. (See **FIGURE 15.13.**) Jouvet and Renault (1966) produced large lesions that destroyed 80 to 90 percent of the raphe in cats. They observed complete insomnia for 3 to 4 days. S sleep (but not D sleep) gradually returned, but never exceeded approximately 2.5 hours per day. Smaller lesions resulted in more recovery, but D sleep did not reappear until S sleep totalled approximately 3.5 hours per day. Attempts have been made to produce sleep by electrically stimulating the raphe, but results have been mixed; some investigators (e.g., Gumulka, Samanin, Valzelli, and Consolo, 1971) have met with success, while others (e.g., Polc and Monnier, 1966) produced *arousal* with raphe stimulation.

The nuclei of the raphe are rich in cells that contain the neurotransmitter 5-HT (serotonin). Jouvet (1968) found that raphe lesions depressed cerebral levels of 5-HT, and that the amount of time the animals spent sleeping correlated with the amount of 5-HT that was present in the brain. (Decreases in 5-HT were associated with decreases in sleep.) This fact suggested that sleep might be produced by the activity of serotonergic neurons whose cell bodies were located in the raphe. In support of this hypothesis, administration of a single dose of para-chloro-phenalanine (PCPA), which suppresses the biosynthesis of 5-HT, was found to suppress sleep. Figure 15.14 shows the effects of PCPA on the brain levels of 5-HT and on sleep; note that the percentage of sleep (top curve) and the percentage of 5-HT (bottom curve) both drop at approximately 16 hours after the

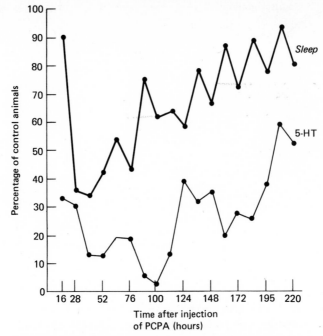

FIGURE 15.14   Effects of a single injection of PCPA on sleep and the amount of 5-HT in the brain. (From Mouret, J. R., Bobillier, P, and Jouvet, M., *European Journal of Pharmacology*, 1968, *5*, 17–22.)

injection and show a parallel recovery course. (See **FIGURE 15.14.**)

Brain levels of 5-HT can be restored after treatment with PCPA by injecting 5-HTP, the immediate precursor of 5-HT. Figure 15.15 illustrates the biosynthetic pathway for 5-HT and shows how PCPA blocks, but 5-HTP restores, production of 5-HT. (See **FIGURE 15.15.**) It would be much more straightforward, of course, to inject 5-HT itself, but this substance does not cross the blood-brain barrier,

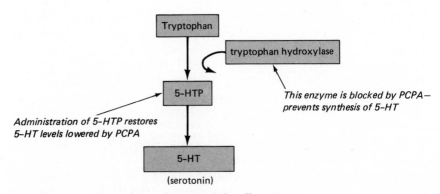

FIGURE 15.15   Biosynthesis of 5-HT and the effects PCPA.

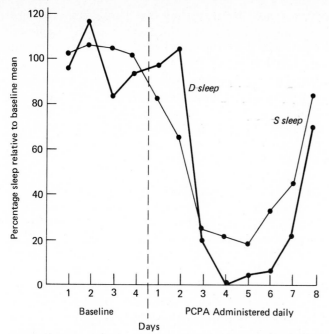

FIGURE 15.16 Effects of daily administration of PCPA on sleep. (From
Dement, W., Mitler, M., and Henriksen, S., *Revue Canadienne de Biologie,*
1972, *31*, 239–246.)

whereas 5-HTP will. Insomnia produced by an injection of PCPA can
be quickly reversed (that is, sleep is restored) by an injection of 5-
HTP (Pujol, Buguet, Froment, Jones, and Jouvet, 1971).

Unfortunately, the story becomes complicated when PCPA is administered chronically. When a cat is given daily injections of PCPA,
the insomnia eventually disappears; both D sleep and S sleep return
to approximately 70 percent of their normal levels (Dement, Mitler,
and Henriksen, 1972). This occurs despite a chronically depressed
(by 90 percent) level of 5-HT. (See **FIGURE 15.16.**) These results
indicate that the activity of serotonergic neurons of the raphe cannot
be essential for sleep.

Why does treatment with PCPA cause temporary insomnia? The
apparent answer seems to be that serotonergic neurons play a role
in controlling some components of D sleep, and that inhibition of
serotonin synthesis disrupts this control, thus interfering with sleep.
Henriksen, Dement, and Barchas (1974) found that although sleep
returns to a cat that receives injections of PCPA every day, the phasic
effects of D sleep (PGO waves, muscular twitches, rapid eye movements) "escape" and begin to intrude during S sleep and even during
waking. The cats have sudden attacks of "hallucinatory episodes";
they jump and exhibit emotional behavior, such as snarling and hissing. We cannot determine whether the cats are really having hallucinations—all we can say is that if they were to have them, the an-

Before medullary
cooling (cat is asleep)

During medullary
cooling (cat is aroused)

FIGURE 15.17   Effects of
medullary cooling on the
arousal level of the éncephale
isolé cat. (Drawings based on
photographs from Berlucchi,
G., Maffei, L., Moruzzi, G., and
Strata, P. *Archives Italiennes de
Biologie*, 1964, *102*, 372–392.)

imals would probably act the way they do. These episodes are probably very disturbing to the cat, and thus keep the 5-HT-depleted animal awake. After a few days the cats become habituated to the attacks and manage to sleep.

The best evidence that we have at this time suggests that neurons of the rostral raphe are involved in restricting phasic D sleep phenomena to the appropriate time for their occurrence—during D sleep. We shall examine some of this evidence in a later section. Evidence for a more direct role in the initiation of sleep is still lacking.

THE NUCLEUS OF THE SOLITARY TRACT.   There is evidence for another brainstem sleep mechanism located in the nucleus of the solitary tract. Berlucchi, Maffei, Moruzzi, and Strata (1964) found that cooling of the medulla (which produces a temporary lesion) at the floor of the fourth ventricle produced behavioral and EEG arousal. (The cat's spinal cord was severed just below the medulla to prevent peripheral effects that are controlled by the medulla from indirectly producing arousal by means of sensory feedback.) Figure 15.17 shows the appearance of one of these cats during sleep and during arousal produced by cooling of the medulla. (See **FIGURE 15.17.**)

Magnes, Moruzzi, and Pompeiano (1961) used low-frequency electrical current to stimulate the region of the nucleus of the solitary tract (which is located within the medulla, and which was cooled in the study by Berlucchi et al.). These investigators observed a synchronizing effect of the stimulation on the EEG, which is usually synonymous with deactivation. The synchrony often outlasted the stimulation. Figure 15.18 shows the effects of such stimulation on the EEG. The tick marks on the lower line represent the stimulation; note that the cortical synchrony outlasts the stimulation. (See **FIGURE 15.18.**)

Bonvallet and Allen (1963) placed small transverse sections rostral to this region and found that the lesions enchanced and prolonged the phasic arousal that is produced by electrical stimulation of the reticular formation. These results suggest that the region of the nucleus of the solitary tract normally exerts an inhibitory effect upon the reticular activating system.

This portion of the medulla receives sensory information from the tongue and from various internal organs. Stimulation of afferent fibers of the vagus nerve (which sends fibers to this region) produces EEG synchrony (Bonvallet and Sigg, 1958). Pompeiano and Swett (1962) stimulated cutaneous nerves in unanesthetized, freely moving cats and found that repetitive, low-frequency (3 to 8 pulses per second) stimulation produced EEG synchrony when the cat was in a relatively quiet state. They subsequently found that medullary neurons responded to the cutaneous nerve stimulation that produced synchrony (Pompeiano and Swett, 1963). However, high-intensity

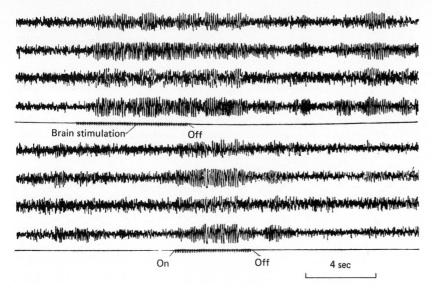

Brain stimulation Off

On Off 4 sec

FIGURE 15.18   The effects of stimulation of the region of the nucleus of
the solitary tract. (From Magnes, J., Moruzzi, G., and Pompeiano, O.,
Archives Italiennes de Biologie, 1961, 99, 33–67.)

nerve stimulation, which produces arousal, activated neurons in the
pontine and midbrain reticular formation. It would seem to be very
likely that the calming effects of gentle rocking (which usually
soothes a baby) are mediated by the mechanism located in the vicinity of the nucleus of the solitary tract. Perhaps we can also thus
account for the fact that a large meal makes us sleepy; we receive
increased activity from afferents of the digestive tract, which stimulates the nucleus of the solitary tract.

Koella (1974) reported that evoked potentials could be recorded
in the nucleus of the solitary tract after the application of brief pulses
of stimulation in the midbrain reticular formation, and that the information was subsequently relayed back to the reticular formation.
This evidence (along with that previously presented) suggests that
the nucleus of the solitary tract is capable of detecting the activity
of the reticular activating system and of modulating reticular activity. The nature of this modulation depends on other factors such as
the nature of the sensory input to this region. Whether this region
contains the control mechanism for sleep is not yet known.

Brainstem Mechanisms in D Sleep

LESION STUDIES.   It appears quite certain that D sleep is produced by a mechanism that resides within the brainstem, somewhere between the upper border of the spinal cord and the anterior
pons. Most likely, it lies within the pons. Jouvet (1962) made a brain-

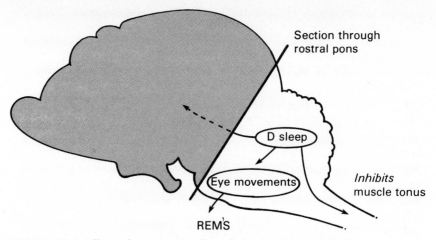

FIGURE 15.19   Effects of a transection through the rostral pons on D sleep.

stem transection through the rostral pons and observed periods of D sleep, characterized by loss of neck muscle tonus and rapid eye movements. Since the nerves that control the eye and neck muscles leave the CNS caudal to the location of the transection, the results mean that the neural mechanisms that trigger D sleep must also be located caudal to the transection. If the D sleep mechanisms were located rostral to the transection, they would no longer have control over these muscles. (See **FIGURE 15.19.**)

Although we can be fairly certain that the neural mechanisms that trigger D sleep are located caudal to the upper pons, we cannot be sure how far caudally they extend. Jouvet (1962) made sections just caudal to the pons. These animals did not exhibit loss of muscle tonus, so perhaps the transection isolated a D sleep mechanism in the pons from the inhibitory motor system, which is known to be located in the medulla. The cats also showed periodic signs of cortical desynchrony, which suggests that the transection did not isolate the D sleep mechanism from the cerebral cortex. This argues that the mechanism is located rostral to the location of the transection— that is, in the pons. However, the animals did not show unequivocal signs of D sleep, so the results of the caudal transection are not conclusive.

Jouvet also reported that large lesions of the pontine reticular formation abolished both behavioral and electrophysiological signs of D sleep. Smaller lesions caudal to this region abolished the profound muscular inhibition that normally accompanies D sleep; as a matter of fact, the cats "acted out their dreams."

> ... to a naive observer, the cat, which is standing, looks awake since it may attack unknown enemies, play with an absent mouse, or display flight behavior. There are orient-

ing movements of the head or eyes toward imaginary stimuli, although the animal does not respond to visual or auditory stimuli. These extraordinary episodes . . . are a good
argument that "dreaming" occurs during [D sleep] in the
cat . . ." (Jouvet, 1972, pp. 236–237)

On the basis of later evidence (Jouvet, 1972), Jouvet concluded
that the locus coeruleus was responsible for D sleep; the large lesions
of his earlier study included this structure. However, a more recent
study (Jones, Harper, and Halaris, 1977) found that lesions of locus
coeruleus prevented the muscular inhibition of D sleep, but did not
abolish D sleep itself. Thus, the data from lesion studies strongly
suggest that descending efferents of locus coeruleus are important in
mediating the muscular inhibition that is normally observed during
D sleep, but that it does not appear to be part of the mechanism that
initiates D sleep. In fact, evidence discussed in the next section suggests that noradrenergic mechanisms (presumably, ascending efferents of locus coeruleus) have an *inhibitory* effect on D sleep.

PHARMACOLOGICAL STUDIES.   Pharmacological data suggest
that noradrenergic and cholinergic neurons may be involved in D
sleep, but the evidence does not permit us to conclude that they play
an essential role. And, as we saw in the previous section, there is
evidence that serotonergic neurons are important in preventing D
sleep phenomena from occurring at inappropriate times.

Norepinephrine. ⟨In general, noradrenergic antagonists increase
D sleep and noradrenergic agonists decrease it, which suggests that
D sleep is inhibited by noradrenergic neurons.⟩ AMPT, which blocks
the synthesis of the catecholamines, causes a temporary increase in
D sleep in humans (Vaughan, Wyatt, and Green, 1972). There is a
rebound after the drug is withdrawn. Reserpine, which prevents the
storage of the catecholamines (and also serotonin) also increases D
sleep in humans (Coulter, Lester, and Williams, 1971). The effect
appears to be produced by alpha-adrenergic synapses, since D sleep
is facilitated by drugs that block alpha-adrenergic synapses (Oswald,
Adam, Allen, Burack, Spence, and Thacore, 1974), but is not affected
by propranolol, a drug that blocks beta-adrenergic synapses (Mendelson, Gillin, and Wyatt, 1977). A drug that blocks dopaminergic
receptors (pimozide), does not affect sleep (Sagales and Erill, 1975).
This fact suggests that the drugs that affect both catecholamines
alter D sleep by their effect on norepinephrine, and not on dopamine.

In contrast to the effects of noradrenergic antagonists, noradrenergic agonists, such as amphetamine and tricyclic antedepres-
sant drugs, cause temporary decreases in D sleep (Baekeland, 1967;
Dunleavy, Brezinova, Oswald, MacLean, and Tinker, 1972). Thus, the
evidence for an inhibitory role for noradrenergic neurons in D sleep
appears to be substantial.

It is very interesting that reserpine, which, as we saw, increases
D sleep, can also produce depression in humans. Reserpine is often

used to treat high blood pressure, and one of its side effects is clinical depression. Chronic, naturally-occurring depression (that is, not produced by drugs) is also characterized by earlier and more frequent episodes of D sleep (Vogel, Augustine, McAbee, and Thurmond, 1977). It is even more remarkable that D sleep deprivation can actually alleviate the symptoms of chronic depression (Vogel, 1979). Perhaps the D sleep deprivation affects the activity of neurons that perform functions that are related to this disorder. It will be very interesting to see whether these promising results will be repeated by other investigators.

Serotonin. As we saw earlier, PCPA, which blocks the synthesis of serotonin, also permits D sleep phenomena to occur at inappropriate times. This appears to be true in humans as well as cats. Cancer patients who had tumors (outside the nervous system) that secreted high levels of serotonin were given PCPA. The drug caused the development of hallucinations, which might be interpreted as D sleep phenomena (that is, dreaming) that were intruding into a waking stage (Engelman, Lovenberg, and Sjoerdsma, 1967). In addition, various hallucinogens, such as LSD and mescaline, also appear to be potent serotonin antagonists. Jacobs and Trulson (1979) have suggested that hallucinations and dreams may share some common mechanisms.

On the other hand, MAO inhibitors that serve as serotonin agonists (but also, to a somewhat lesser extent, as catecholamine agonists) cause decreases in D sleep (Wyatt, 1972). And as we saw in Chapter 14, imipramine (a monoamine agonist) has been successfully used to treat cataplexy, sleep paralysis, and hypnagogic hallucinations, which have been interpreted as manifestations of D sleep. Amphetamine, which is a potent catecholamine agonist, with little effect on serotonergic neurons, is not very effective in treating these symptoms, although it does cause a temporary decrease in D sleep in people without a sleep disorder. Thus, the data strongly suggest that the serotonergic neurons exert an inhibitory effect on brain mechanisms that are active during D sleep.

Acetylcholine. Drugs that excite cholinergic synapses appear to facilitate D sleep. People who have been exposed to organophosphate insecticides, which inhibit AChE, and therefore increase postsynaptic activity that is mediated by acetylcholine, spend an increased time in D sleep (Stoyva and Metcalf, 1968.) In a controlled experiment, Sitaram, Wyatt, Dawson, and Gillin (1976) administered intravenous injections of an AChE inhibitor (physostigmine) to human subjects during sleep. If the subjects were in S sleep, the injections induced D sleep. If the subjects were in D sleep, the injections awoke them.

The location of the cholinergic synapses appears to be in the pons; local stimulation of this region by means of a drug that provides longlasting stimulation of cholinergic synapses (carbachol) produces muscular flaccidity, cortical desynchrony, and rapid eye movements in cats (McKenna, McCarley, Amatruda, Black, and Hobson, 1974). These results do not prove that D sleep mechanisms contain cholinergic neurons, but they do suggest that cholinergic neurons have an excitatory effect on these mechanisms.

As we saw in an earlier section, people with clinical depression exhibit increased amounts of D sleep, along with earlier awakening in the morning. Sitaram, Nurnberger, Gershon, and Gillin (1980) gave injections of a drug that stimulates muscarinic cholinergic receptors (arecoline) to normal people and people who had a history of depression, but were presently in a state of remission. The injections were given through an intravenous catheter while the subjects were sleeping, immediately after the first period of D sleep was over. The drug reduced the interval until the next bout of D sleep in both groups, but had much more of an effect on the subjects with a history of depression. The study is very interesting, because it provides evidence for increased sensitivity of a brain mechanism that is related to a clinically-significant symptom even at a time when the patient is not suffering from depression, and is showing normal patterns of sleep. The authors suggest that a hypersensitive cholinergic mechanism might be the underlying factor that predisposes the people to bouts of clinical depression. The hypothesis is certainly worth pursuing.

ELECTROPHYSIOLOGICAL STUDIES.   It has been observed that the firing rate of some neurons in the dorsal raphe goes down drastically during D sleep, which suggests that they might inhibit this state during waking and S sleep (McGinty, Harper, and Fairbanks, 1974). Recordings made in Hobson's laboratory revealed the existence of neurons that apparently fired at their highest rate during D sleep (Hobson, McCarley, Pivik, and Freedman, 1974). The authors suggested that these neurons were responsible for the initiation of D sleep. These neurons were located in the **gigantocellular tegmental field**, or **FTG**. The term "gigantocellular" is well-earned, as figure 15.20 shows. One of these cells is shown in a sagittal section through a rat brain; the cell body (which has many dendrites with prominent spines) is located near the letter R (the location of FTG). Note how the axon divides and sends processes throughout the telencephalon, diencephalon, midbrain, and hindbrain. (See **FIGURE 15.20** on the following page.) As Scheibel and Scheibel (1961) note, 300 FTG cells can potentially affect 9 million cells of the brainstem reticular formation. If there were ever a system whose anatomy suggested some sort of controlling role in sleep/waking mechanisms, this is it.

However, more recent evidence has shown that the FTG neurons do not appear to be responsible for the initiation of D sleep. It has been shown in the cat (Siegel and McGinty, 1977) and in the rat

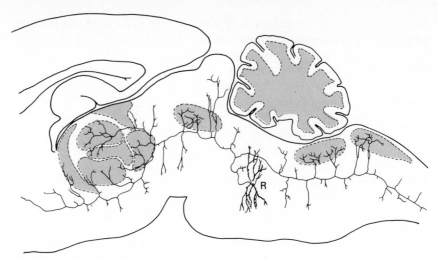

FIGURE 15.20   The extensive arborizations of an FTG neuron. (From Brazier, M. A. B., *The Electrical Activity of the Nervous System*, 4th edition. London: Pitman Medical Publishing Co., 1976.)

(Vertes, 1977) that FTG neurons actually fire at their highest rate when the animal is awake and moving. The rate of firing was low when the animal was motionless, whether awake or asleep. During D sleep, the rate went up again.

Figure 15.21 shows the activity of an FTG neuron during waking, S sleep, and D sleep. Note the close association between movement (shown by the EMG record) and unit activity during wakefulness. During S sleep the unit was silent. However, it became very active during D sleep, especially when the EMG showed some slight signs of muscular movement. (See **FIGURE 15.21.**)

The authors concluded that FTG neurons were primarily associated with motor mechanisms, and not with D sleep. But if this is so, why do these neurons respond at such a high rate during D sleep, when the animal is certainly moving less than it does during waking? You will recall that some slight twitching movements are made during D sleep, but that most muscles are inhibited—in fact, paralyzed—by a mechanism that appears to involve locus coeruleus. (As we saw, lesions of locus coeruleus remove this inhibition, and the animals' muscles move during D sleep; the cats act out their dreams, so to speak.) This means that motor systems (which appear to include FTG) are very active during D sleep, but that their *output*, in the form of muscular movement, is inhibited by neurons in locus coeruleus.

Why do the data from Hobson's laboratory show a low rate of activity in FTG neurons during waking? The answer is simple—the cats were restrained and were unable to move. A cat that is held in a restraining device will struggle at first, but it soon learns to stop.

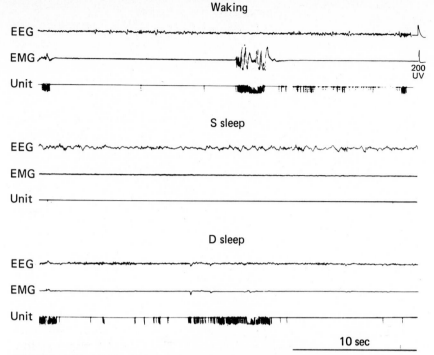

FIGURE 15.21 EMG activity and activity of an FTG neuron during
waking, S sleep, and D sleep. (From Siegel, J. M., McGinty, D. J., and
Breedlove, S. M., *Experimental Neurology*, 1977, *56*, 553–573.)

Siegel and McGinty (1977) found that the activity of FTG neurons
was very high in newly-restrained cats, but as the cats adapted to
restraint and stopped struggling, the activity of these neurons
ceased. Figure 15.22 shows the activity of an FTG neuron during
adaptation to restraint. Note the decrease in struggling (EMG activity) and the firing rate of the FTG neuron. (See **FIGURE 15.22** on the
following page.)

For a while it appeared that we knew what the primary D sleep
mechanism was. Now we are back to the beginning again. This situation makes for a less satisfying discussion in a textbook, but it
goes to show us how difficult research on the physiology of sleep has
proven to be. We can conclude that (a) serotonergic neurons in the
dorsal raphe nuclei may restrict the phasic components of D sleep
(for example, rapid eye movements) to the proper time, (b) descending noradrenergic efferents of locus coeruleus may inhibit muscular
movements during D sleep, (c) ascending efferents of locus coeruleus
may play an inhibitory role on the initiation of D sleep, (d) cholinergic neurons in the central pons may play a role in initiating or
facilitating D sleep, and (e) the executive mechanism for D sleep lies
caudal to the rostral pons, probably within the pons itself.

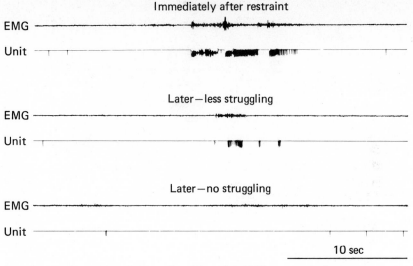

FIGURE 15.22 EMG activity and activity of an FTG neuron during adaptation to restraint. (From Siegel, J. M., McGinty, D. J., and Breedlove, S. M., *Experimental Neurology*, 1977, *56*, 553–573.)

FOREBRAIN MECHANISMS

Forebrain Waking Mechanisms

EVENTUAL RECOVERY OF THE CERVEAU ISOLÉ.   There is evidence that the isolated cerebrum is capable of periodic wakefulness. Genovesi, Moruzzi, Palestini, Rossi, and Zanchetti (1956) found that multistage lesions (i.e., a little bit at a time) that eventually transected the brainstem between the superior and inferior colliculi did not produce totally comatose animals; the animals showed periodic ocular and electroencephalographic signs of arousal.

Batsel (1960) produced single-stage brainstem transections in dogs, which he was able to keep alive for up to 73 days by very careful nursing. As time passed, the animals showed increasing amounts of cortical desynchrony. The level of transection was too far rostral to allow observation of ocular behavior, so only the EEG could be used to determine whether the animal was asleep or awake. When a more caudal transection was made (Batsel, 1964), the EEG again showed a gradual recovery of desynchronized activity. This time some ocular signs of arousal could be seen. These results suggest that there is a forebrain waking mechanism that is able to produce arousal independently of the brainstem.

POSTERIOR HYPOTHALAMUS. Where might these mechanisms be? In 1936, Ingram, Barris, and Ranson reported that lesions in the vicinity of the posterior hypothalamus resulted in a state of somnolence, with hypoactivity. The animals could be aroused by sensory stimulation, but the arousal was minor and short-lived. Nauta (1946) reported that posterior hypothalamic lesions produced a state of lethargy; the animals could be aroused by pinching the tail, but soon afterward "they would yawn and stretch and settle down in a comfortable position to go to sleep again" (Nauta, 1946, p. 292). These results were further corroborated by Naquet, Denavit, and Albe-Fessard (1966), who found that lesions in the vicinity of the posterior hypothalamus of the cat produced almost total somnolence for 8 to 10 days. The cats could be aroused briefly by sensory stimulation. D sleep was not disrupted.

However, the issue is not at all clear. To produce hypersomnia effectively, the lesions must include both the medial and lateral portions of the posterior hypothalamus (Sweet and Hobson, 1968). Ascending catecholaminergic fibers and descending fibers of the extrapyramidal motor system are also destroyed by these lesions. And one often sees "somnolent" behavior even when the EEG is desynchronized.

Robinson and Whishaw (1974) found that rats with posterior hypothalamic lesions showed deficits in voluntary movements, but performed reflexive or species-typical movements very well. For example, the animals would not walk, but they would groom themselves vigorously when they were sprayed with water. The posterior hypothalamus appears to play a role in the initiation of movement rather than act as a "waking center." The hypoactivity seen in animals with posterior hypothalamic lesions cannot really be called sleep. These lesions sometimes also enhance true sleep, but we cannot rule out the possibility that they do so by interrupting fibers that convey activating influences from brainstem structures.

Forebrain Sleep Mechanisms

An area of the brain just rostral to the hypothalamus—the preoptic-basal forebrain region—appears to play a role in sleep. Destruction of this area resulted in total insomnia in rats (Nauta, 1946). The animals subsequently fell into a coma and died; the average survival time was only 3 days. McGinty and Sterman (1968) found that cats reacted somewhat differently; the animals did not show any insomnia until several days after the lesion. Figure 15.23 illustrates this effect. (See **FIGURE 15.23** on the next page.)

The effects of basal forebrain stimulation are consistent with the lesion experiments. Sterman and Clemente (1962a,b) found that electrical stimulation of this region in an unanesthetized, freely moving cat produced cortical synchrony, which is usually indicative of deactivation or sleep. The average latency between stimulation and EEG synchrony was 30 seconds; often, the effect was immediate. Drow-

see page 120 diagram.

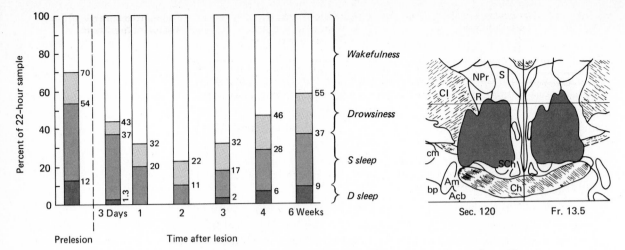

FIGURE 15.23   Effects of lesions of the preoptic area on sleep. (From McGinty, D. J., and Sterman, M. B., *Science*, 1968, *160*, 1253–1255. Copyright 1968 by the American Association for the Advancement of Science.)

siness and behavioral sleep often followed. (See **FIGURE 15.24**.)

It is difficult to prove that electrical stimulation has put an animal to sleep; after all, perhaps it was ready to doze off anyway. However, there are several pieces of evidence that suggest that the stimulation actually produces sleep. Clemente, Sterman, and Wyrwicka (1963) presented a tone for a total of 30 seconds. During the last 20 seconds they also electrically stimulated a basal forebrain region that produced cortical synchrony. After several pairings of tone and stimulation, the tone presented by itself became capable of producing EEG synchrony and behavior associated with the prepa-

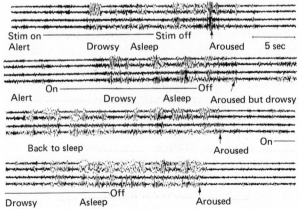

FIGURE 15.24   Sleep produced by prolonged electrical stimulation of the basal forebrain region. (From Sterman, M. B., and Clemente, C. D., *Experimental Neurology*, 1962, *6*, 103–117.)

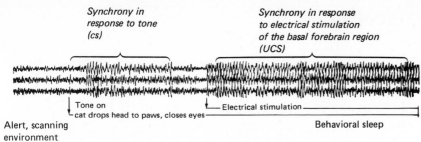

FIGURE 15.25 Conditioning of the effects of stimulation of the basal
forebrain region to a neutral stimulus. (From Clemente, C. D., Sterman, M.
B., and Wyrwicka, W., *Experimental Neurology*, 1963, 7, 404–417.)

ration for sleep; the cat would lie down, drop its head, and close its
eyes. This constitutes classical conditioning of sleep as a *response*,
and thus provides very strong evidence that the basal forebrain region contains sleep-producing mechanisms. (See **FIGURE 15.25**.)

Another study provides further support. Roberts and Robinson
(1969) warmed the preoptic area, which stimulates thermoreceptors
that are located in this region, and observed drowsiness and EEG
synchrony. Thus, a more "natural" stimulation mimics the effects of
electrical stimulation. Perhaps the excessive sleep that accompanies
a fever is produced by this mechanism. Thermoreceptors in the skin
also relay information to the preoptic area; perhaps preoptic stimulation accounts for the drowsiness and lassitude we feel on a hot
day. Pavlov (1923) even noted that when thermal stimulation of the
skin was used as a conditioned stimulus, the dogs being trained often
fell asleep.

Bremer (1970) provided further support for the synchronizing
effects of preoptic stimulation. He obtained electrophysiological evidence for an inhibitory influence of the preoptic region on activating
effects of the reticular formation. When the preoptic stimulation was
presented simultaneously with sensory stimulation or electrical
stimulation of the reticular formation, the arousing effects of the
latter two procedures were diminished. The opposite effect was also
seen; stimulation of the reticular formation was shown to produce
electrophysiological signs of inhibition in the preoptic sleep-producing area (Bremer, 1975). Moreover, transection of the brainstem just
rostral to the midbrain reticular formation caused an immediate
increase in the activity of neurons in the preoptic area, which suggests that the transection removed a tonic inhibitory influence on
the preoptic sleep-producing area, probably coming from the reticular activating system. Bremer also found that activity of preoptic
neurons was related to the animal's sleep cycles, being highest during S sleep. This relationship was disrupted by brainstem transections placed just rostral to the midbrain reticular formation. As you
can see, the evidence for a forebrain sleep mechanism is very strong.

CONCLUSIONS

There is good evidence that the suprachiasmatic nucleus controls the distribution of sleep and waking in animals who display circadian rhythms of these states. There is also good evidence for brainstem sleep and waking mechanisms, and for a forebrain sleep mechanism. The evidence for a forebrain waking mechanism (in the posterior hypothalamus) is not very compelling. Of course, this leaves open the issue of what causes the eventual recovery of periods of arousal in the chronic *cerveau isolé* (midcollicular transection). And, although it appears that the total amount of sleep that an animal obtains is regulated, we have no idea what mechanisms monitor total sleep time. We also have no idea what produces the cyclicity of S and D sleep. The past few years have provided us with much information about neural mechanisms of the sleep/waking cycle, but the basic process still remains a mystery.

KEY TERMS

cerveau isolé (*sair voh ee soh lay*) p. 480
circadian rhythm p. 471
encéphale isolé (*ah seh fahl ee soh lay*) p. 480
gigantocellular tegmental field p. 495
locus coeruleus (*sur OO lee us*) p. 485
midcollicular transection p. 479

midpontine pretrigeminal transection p. 481
nucleus of the solitary tract p. 486
raphe (*ruh FAY*) p. 486
suprachiasmatic nucleus (*ky az MAT ik*) p. 474
Zeitgeber (*TSYTE gay ber*) p. 473

SUGGESTED READINGS

DEMENT, W. C. *Some Must Watch While Some Must Sleep.* San Francisco: W. H. Freeman, 1974.

WEBB, W. B. *Sleep: The Gentle Tyrant.* Englewood Cliffs, N. J.: Prentice-Hall, 1975.

DRUCKER-COLIN, R., SHKUROVICH, M., AND STERMAN, N. B. (eds.) *The Functions of Sleep.* New York: Academic Press, 1979.

MENDELSON, W. B., GILLIN, J. C., AND WYATT, R. J. *Human Sleep and its Disorders.* New York: Plenum Press, 1977.

In addition to the suggested readings that I listed for Chapter 14, you might want to consult recent volumes of *Annual Review of Neuroscience.* This publication often carries chapters that review recent research in biological rhythms.

16

Emotion, Aggression, and Species-Typical Behavior

Emotions constitute a very important part of our lives. Basic motives, such as those produced by lack of food or suitable shelter from the cold, are not important issues for most people in industrially developed countries. We tend to evaluate our lives by our jobs, our incomes, our friends, and our material possessions. But ultimately, we ask ourselves whether we are happy or satisfied with our lives. We all acknowledge a relationship between wealth and happiness, but most of us would agree that a rich, lonely, unhappy person is worse off than a materially poor but happy and contented member of a "less civilized" society.

I do not intend to present here a brief for the life of the "innocent savage." I merely wish to make the point that a surfeit of the goal objects sought by people to satisfy their biological drives—for example, plenty of food, warm clothing, and shelter—does not guarantee "happiness." Although countless writings and lectures have been devoted to telling us how to be "happy," we still cannot specify the conditions that lead to this state. And we do not know how to eliminate hate, jealousy, and other negative emotions that exist even in people who are able to satisfy their material needs.

We know that lack of food causes hunger; consequently, we have an idea of how to investigate its physiological basis. As we saw in Chapter 12, much progress has been made. In fact, the relationship between food deprivation and hunger is so clear that we do not refer to hunger as an emotion. However, we cannot analyze the causes for other feelings so easily. What makes someone happy might bore someone else. A situation that terrifies one person might merely amuse another. We have no knowledge of the physiological mechanisms that determine why a person reacts in a given way to emotion-producing stimuli.

Where we *have* made some progress is in the investigation of the physiological mechanisms controlling behaviors that accompany various emotions. For example, we know something about the physiological basis of aggressive behavior, although we do not know why a given situation makes a person angry. Similarly, we know quite a bit about neural and hormonal mechanisms of sexual behavior, but not of love.

In this chapter I shall discuss what is known about the physiological bases of aggressive and defensive behavior, and of parental behavior. (Sexual behavior has already been covered in Chapter 11.) The stress is on behavior, not on the emotions that presumably accompany the behavior. There has been some study of the effects of physiological manipulations (brain damage, alteration of hormone levels, administration of drugs) on emotions expressed by humans, but most of the objective work has been performed with animals, and this work is necessarily restricted to observable behavior. The problem of disturbances in mood (depression and mania) will be covered in Chapter 20.

This chapter will also contain a discussion of attempts to control undesirable human behavior—specifically, violent aggressive behavior—by surgical and chemical means. As we shall see, serious questions have been raised about the effectiveness of these procedures and about the moral and ethical issues involved in producing irreversible brain damage.

FEELINGS OF EMOTION

The James-Lange Theory

An early theory attempted to explain feelings of emotion as feedback from peripheral effects of the autonomic nervous system, and from skeletal muscles. This theory, formulated independently by William James and Carl Lange in the late nineteenth century, argued that environmental stimuli produced reflex autonomic and somatic effects, and that afferent feedback from these effects constituted the experience of emotion. James put it this way:

> ... *the bodily changes follow directly the perception of the exciting fact, and ... our feelings of the same changes as they occur is the emotion.* Common sense says we lose our fortune, are sorry, and weep; we meet a bear, are frightened, and run; we are insulted by a rival, are angry, and strike. The hypothesis here to be defended says that this order of sequence is incorrect, that the one mental state is not immediately induced by the other and that the bodily manifestations must first be interposed between. The more rational statement is that we feel sorry because we cry, angry because we strike, afraid because we tremble, and not that we cry, strike, or tremble, because we are sorry, angry or fearful, as the case may be.
>
> ... If we fancy some strong emotion, and then try to abstract from our consciousness of it all the feelings of its bodily symptoms, we find we have nothing left behind, no "mind stuff" out of which the emotion can be constituted and that a cold and neutral state of intellectual perception is all that remains. (James, 1890, pp. 449–451)

William Cannon offered a critique of the James-Lange theory that persuaded many people that it could not be correct (Cannon, 1927). He noted that the viscera are not very sensitive, that identical visceral changes can be seen in a variety of emotional states, that artificial changes (by means of drugs) in the state of the internal organs will not produce an emotion, and that visceral changes take place slower than changes in emotional states. In addition, if feedback from internal organs were surgically disconnected from the brain, an animal would still exhibit emotional responses to threatening or aversive stimuli.

interior organs

These criticisms are not conclusive. More recent evidence shows that Cannon underestimated the sensitivity and complexity of afferent feedback from the visceral organs. Furthermore, James did not say that the internal organs were the sole source of feedback; information from the activity of skeletal muscles was also important. More importantly, the fact that animals with deafferented organs still *display* emotions does not address the issue of whether they *feel* them. It is a common assumption that the expression of emotional behavior is synonymous with feelings of emotion. However, this is not always the case.

Hohman (1966) collected data from people whose spinal cords had been severed. He reasoned that if feedback from the body is important for emotions, then people with high (rostral) transections should experience less intense emotional states than people with low (caudal) transections. (See **FIGURE 16.1** on the next page.)

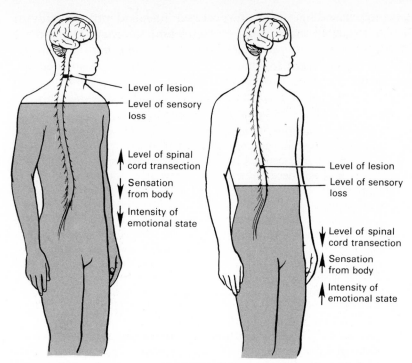

FIGURE 16.1   Schematic representation of the rationale for the study by Hohman.

His prediction was confirmed; the higher the transection, the lower the reported intensity of emotion. Here are some examples of what his patients reported: "I sit around and build things up in my mind, and I worry a lot, but it's not much but the power of thought. I was at home alone in bed one day and dropped a cigarette where I couldn't reach it. I finally managed to scrounge around and put it out. I could have burned up right there, but the funny thing is, I didn't get all shook up about it. I just didn't feel afraid at all, like you would suppose." Another man was caught in a sinking fishing boat during a storm. "I knew I was sinking, and I was afraid all right, but somehow I didn't have that feeling of trapped panic that I know I would have had before." Feelings of anger were also diminished. "Now, I don't get a feeling of physical animation, it's sort of cold anger. Sometimes I act angry when I see some injustice. I yell and cuss and raise hell, because if you don't do it sometimes, I've learned people will take advantage of you, but it doesn't have the heat to it that it used to. It's a mental kind of anger" (Hohman, 1966, pp. 150–151).

Although the feelings of emotion were diminished by high spinal cord transection, it is interesting to note that the *expression* of emotion was not necessarily impaired (subject to the limits of the person's physical disability, of course). The patient who yelled and cussed and raised hell did so not because he felt angry, but because a display of anger was the appropriate response. Perhaps Cannon's animals with deafferented visceral organs also did not feel angry when they hissed and snarled.

A report by Sweet (1966) describes the case of a person whose sympathetic nervous system was severed on one side of the body, for therapeutic reasons. The patient subsequently found that the shivering sensation he previously felt while listening to music now occurred only on the unoperated side. Thus, autonomic feedback appears to play a role in positive emotions. It would be nonsense to say that his shivering *caused* his emotion; a shiver that was produced by a cold draft is certainly not to be confused with the feeling a music lover gets from a stirring passage of music. Similarly, a music-hater would not shiver when he or she heard the same music. However, it would also be foolish to dismiss the shivering sensation as irrelevant to emotion.

It appears that feelings of emotion do, indeed, depend at least partially on feedback from body to brain. It also appears that emotional behavior can occur without this feedback. In any event, physiological psychologists can only observe the *behavior* of their animals; thus, investigations of emotion are almost exclusively devoted to emotional behavior, with little attention being paid to what the animal might be experiencing.

AGGRESSION AND FLIGHT

Aggression is difficult to define precisely. Some events—a fight between two dogs, for example—are clear. Others are not so clear. What about a tennis player who smashes his racket against the ground when he misses an important point? No one is hurt, but the feelings that accompany this display are probably very similar to those that would accompany an attack against another person. People can express aggression verbally. A person's silence could even be interpreted as aggression, if that person knows that the silence will frustrate and annoy the other person.

So we need not restrict a definition of aggression to the production (or threat) of bodily harm. Nor should all behaviors that intentionally result in harm to another be interpreted as aggressive. Certainly a surgeon is not expressing aggression by cutting a patient's abdomen open. Nor is an employee in a slaughterhouse displaying hostility by killing dozens of animals each day.

Types of Aggression

Moyer (1976) has identified seven kinds of aggression, classified partly by the object of the attack, partly by the situation that elicits the attack, and partly by the way in which the animal carries out the attack.

1. *Predatory attack* is different from all the others, and a good case could be made for its not really being a form of aggression. When a lion attacks a zebra, or a fox attacks a rabbit, the predator does not appear to be "angry" at its prey.

2. *Inter-male aggression* is usually restricted to fights between strangers. A male mouse raised in isolation will attack other males, but it will not fight with other males if it is raised with them.

3. *Fear-induced aggression* can be seen in a cornered animal; even a rat will turn and fight, as the saying goes.

4. *Irritable aggression* can be elicited by frustration or (more frequently, at least in experimental settings) by pain. If a pair of normally docile animals are given a painful foot shock, they will often attack each other viciously. (We are all familiar with the warning not to approach a wounded animal too closely.) This category also includes attack upon an inanimate object (frustrated tennis players breaking their rackets, for example).

5. *Sex-related aggression*, when it occurs, is almost always displayed by males. The copulatory behavior of minks and ferrets looks especially vicious; after a prolonged battle, the male attacks the female, subduing her with a bite to the neck (Etkin, 1964). And, as we all know, there have been many cases of men (heterosexual and homosexual) brutally killing the victims of their sexual assaults.

6. *Maternal aggression* is displayed by a lactating mammal when disturbed near her nesting site, or near her young.

7. *Instrumental aggression* is an attack that can be trained by conditioning procedures. For example, a dog can be trained to be vicious by rewarding it for attacking, and punishing it for being affectionate toward, a (well-padded) human.

As we shall see, very little is known about the physiological bases of some of these types of aggression; predatory attack, inter-male aggression, and irritable aggression have been the subjects of most of the investigations. And sometimes the type of aggression seen in a study is not specifiable. For example, if electrical stimulation of some brain region makes a cat hiss and snarl (at no particular object), we would conclude that we were stimulating neural circuits involved in aggression, but not necessarily any particular type.

Attack Elicited by Electrical Brain Stimulation

It is possible to stimulate various parts of the brain electrically and elicit at least three types of aggressive behavior: irritable aggression, predatory attack, and fear-induced aggression. Although electrical stimulation occasionally produces only elements of attack, or a halfhearted display of aggression, full-blown attack, when it occurs, appears identical to that which is elicited naturally. Flynn, Vanegas, Foote, and Edwards (1970) have reviewed a series of investigations on the neural bases of rat killing by cats. They studied two general types of attack that can be elicited by brain stimulation: **affective attack** and **quiet-biting attack,** which appear to correspond to Moyer's irritable aggression and predatory attack, respectively. It may seem surprising that a cat should need any special treatment to induce it to attack a rat, but most laboratory cats do *not* spontaneously attack rats.

Affective attack (irritable aggression) is by far the more dramatic. The animal adopts a "Halloween-cat" posture, with arched back, erect fur on the back and neck, dilated pupils, and bared teeth. A nearby rat will be viciously attacked with the claws, sometimes to the accompaniment of screams. If the stimulation continues, the cat will often begin biting the rat. We do not know how the cat feels, of course, but it acts as if it were brimming over with rage. The form of irritable aggression that is elicited by electrical stimulation of the brain appears very similar to attack elicited by pain or removal of an anticipated reward. Furthermore, both forms of attack are accentuated by amphetamine and are diminished by chlordiazepoxide, a tranquilizer (Horovitz, Piala, High, Burke, and Leaf, 1966; Panksepp, 1971a).

release of ACh accentuates affective & quiet biting attack!

Quiet-biting attack (predatory attack) is quite different. This type of attack is not accompanied by a strong display of emotion. The cat begins searching, and it suddenly pounces on the rat, directing powerful bites to the head and neck region. The cat does not growl or scream, and it stops attacking once the rat ceases to move. This type of attack appears more cold-blooded and ruthless than affective attack (it is more efficient, too, in terms of the probability of the rat being killed). Almost certainly, this behavior is identical with the attack normally seen when a cat is hunting.

Fear-induced aggression is seen when an animal is frightened, and is prevented from escaping. Electrical brain stimulation can elicit vigorous attempts to run and hide, and an experimenter who is so foolish as to get between the animal and its means of escape will get clawed or bitten.

Many investigators who have studied this phenomenon refer to it as *escape* or *flight* rather than fear-induced aggression. If the experimenter does not put a hand into the cage, and if no other animal is present for the subject to attack, the only behavior that will be seen is attempts to hide or escape. It is presumed that attack would occur in these studies if an appropriate target were provided, but that is a supposition.

SPECIES-TYPICAL NATURE OF ATTACK. The targets of an animal's attack can certainly be modified by experience, and the method an animal uses to kill its prey or subdue another member of its own species is improved through learning, but the basic patterns of behavior do not appear to have to be learned. That is to say, the behavior patterns are **species-typical**—characteristic of the species. A cat kills a rat with an efficient bite to the back of the neck, but attacks another cat with its claws. A male deer attacks a rival with his horns, but defends himself against a predator with his hooves. A mouse chases a cockroach and kills it by biting its head off.

Moyer and his colleagues attempted to determine whether patterns of attack behavior could be learned. They chose a group of rats that did not spontaneously kill mice, and rewarded them for attacking a mouse. As Moyer (1976) put it, "It is relatively easy to induce the rat to chase, harass, and nip at the mouse. However, although we have tried repeatedly we have been unable to induce mouse killing by this procedure" (p. 15). These results and evidence that will soon be presented suggest that the neural circuitry that controls at least some forms of attack are present in the brain of certain species of animals. Presumably, some individual animals do not attack because their neural mechanisms are not as easily aroused. Many studies have shown that an animal that does not spontaneously attack other animals will do so if the appropriate region of its brain is electrically stimulated. These observations support the suggestion that the neural mechanisms for attack are present even in animals who do not display these behaviors under normal conditions.

Another piece of evidence that supports the contention that some forms of attack are inborn, species-typical behavior, is the observation that frog killing is not observed in rats until they reach the age of 50 days, and that prior experience with frogs does not alter the time at which the behavior emerges. Johnson, DeSisto, and Koenig (1972) raised rats in close proximity with frogs. In one condition the rats lived on a raft floating in a bathtub full of water. Before the rats were 50 days old, the frogs were not molested when they climbed onto the platform. But once the rats reached this age, they began killing the frogs. As the authors put it: "Rats began to patrol the edge of the platform in an effort to snare passing frogs, and they became so skillful that replacement frogs were quickly captured and devoured" (Johnson et al., 1972, p. 239). The age at which the attacks began was the same if the rats were not raised with frogs. Thus, it would appear that the emergence of this behavior is controlled by maturational factors, such as development of the nervous system, or the activational effects of a hormone secreted in adulthood.

Predatory attack and hunger. Predatory attack is related to feeding behavior, but these behaviors are clearly mediated by different neural circuitry. Food deprivation will increase the probability of spontaneous mouse killing in rats (Paul, Miley, and Baenninger, 1971), but some rats never become killers and will starve to death in the presence of live mice—even if they have learned to eat dead ones (Karli, 1956).

Hutchinson and Renfrew (1966) found that all brain stimulating electrodes that elicited eating in cats would produce predatory attack on rats if the intensity of the stimulating current were increased. However, the animals did *not* eat the rats they killed. Roberts and Kiess (1964) implanted electrodes in the brain of cats and obtained quiet-biting attacks on rats when electrical stimulation was delivered through the electrodes. The cats learned to traverse a maze in order to obtain a rat to attack, but they would do so only while the brain stimulation was turned on. When it was off, they would not seek out the rat. When stimulation was turned on, a hungry cat would even leave a dish of food to run through the maze and attack the rat, which it would not subsequently eat. Therefore, predatory attack is not synonymous with hunger. It provides a means for predatory carnivores to obtain food, but the neural mechanisms for attack and eating are different.

The distinction between predatory attack and irritable aggression. Panksepp (1971b) found that both predatory (quiet-biting) attack and irritable aggression (affective attack) could be obtained from rats by electrical brain stimulation. Most rats would quickly learn to turn off the stimulation that produced irritable aggression (suggesting that this stimulation was aversive) but they would press a lever that turned *on* stimulation that elicited predatory attack. Furthermore, administration of amphetamine accentuated irritable aggression but diminished predatory attack (Panksepp, 1971a). These physiological data only reinforce the behavioral differences between these two forms of electrically elicited attack.

Neural Pathways Mediating Electrically Elicited Attack

HYPOTHALAMUS AND MIDBRAIN. Attack can most readily be elicited from hypothalamic stimulation; in general, irritable aggression is produced by medial hypothalamic stimulation, whereas predatory attack occurs after stimulation of the lateral hypothalamus. This appears to be true for cats (Flynn et al., 1970), opossums (Roberts, Steinberg, and Means, 1967), monkeys (Delgado, 1969), and rats (Panksepp, 1971b). Furthermore, electrodes in the dorsal portion of the hypothalamus often produce flight when current is passed through them; the animal breathes rapidly, its pupils dilate, it might urinate or defecate, and it exhibits frantic attempts to escape the chamber in which it is confined (Clemente and Chase, 1973). If restrained, the animal will frequently attack in an attempt to flee. (This fear-induced aggression is thus not directly elicited by the brain stimulation, but resembles the natural attack of a cornered animal.)

The effects of hypothalamic stimulation appear to be mediated caudally, through the midbrain. Both irritable and predatory attack

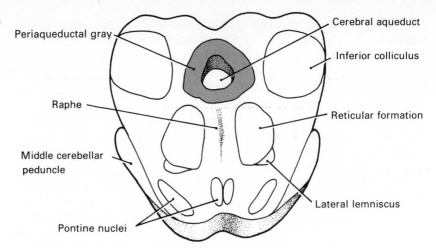

FIGURE 16.2   Location of the periaqueductal gray of the midbrain.

can be elicited by stimulation in various midbrain locations (Clemente and Chase, 1973; Hutchinson and Renfrew, 1978), especially in the *periaqueductal gray* of the tegmentum, shown in **FIGURE 16.2.** This region of the midbrain receives fibers from preoptic and hypothalamic regions. It is extensively connected with midbrain sensory and motor systems. When lesions are placed in the periaqueductal gray, affective attack, previously produced by hypothalamic stimulation, can no longer be elicited. Nor will such a cat show an emotional reaction to the presence of a dog.

In order to ascertain whether the hypothalamus is uniquely involved in organizing aggressive behavior, Ellison and Flynn (1968) produced a hypothalamic "island" with a specially designed surgical instrument. The device contained two small knives that could be rotated around the hypothalamus, thus severing all connections between the hypothalamus and the rest of the brain. (This complicated technique is used because large hypothalamic lesions will kill the animal because of severe disruption of the endocrine system. The hypothalamic "island" technique leaves the pituitary gland attached to the isolated hypothalamus.)

Even after hypothalamic isolation, midbrain stimulation could still effectively elicit both affective and quiet-biting attack, although higher levels of electrical current had to be used than were necessary before the isolation. The cats also reacted aggressively to a tail pinch, and natural mouse-killers continued to attack mice. The behavior was not quite so intense as before the surgery, but it is safe to conclude from these results that the hypothalamus is not necessary for the organization of motor control of aggressive behavior. The fact that the cats were inactive when not stimulated (naturally or electrically) suggests that the hypothalamus plays a role in the initiation or facilitation of behaviors that are organized and integrated elsewhere.

These studies show that lesions of the periaqueductal gray block the effects of hypothalamic stimulation but that isolation of the hypothalamus from the rest of the brain does not prevent the occurrence of normal attack or attack elicited by electrical stimulation of the periaqueductal gray. Therefore, the data suggest that the sequence of movements that constitute attack behaviors are principally organized in the midbrain, and that hypothalamic mechanisms play some kind of role in turning these circuits on and off, perhaps in response to environmental stimuli. Stimulation of the lateral hypothalamus elicits predatory attack, stimulation of the medial hypothalamus elicits irritable aggression, and stimulation of the dorsal hypothalamus elicits flight.

INTERACTION BETWEEN BRAIN STIMULATION AND SENSORY AND MOTOR MECHANISMS. Flynn and his colleagues have investigated interactions between electrical stimulation of the brain and the sensory and motor systems that mediate attack. They report that stimulation of various locations in the midbrain leads to attack with the forepaw contralateral to the stimulating electrode. The cats never used the ipsilateral forepaw. This result strongly argues against the possibility that the stimulation merely produced some emotional effects (making cat "angry" at the rat), which then triggered the normal attack sequence. Instead, the motor systems involved in the attack were directly facilitated; otherwise, the cat would always have used its preferred paw (cats and other animals usually show some degree of "handedness," just as humans do). Even when the pyramidal tract was cut on one side, resulting in the cat's reluctance to use the contralateral limb, electrical stimulation of the midbrain would elicit attack with the normally disused paw.

Most extensive region from which head–orienting response could be elicited during brain stimulation

A clear picture of the facilitatory effects of brain stimulation was shown by MacDonnell and Flynn (1966). They noted that a cat normally turns its head away when a stick is touched to the side of its cheek. If a cat is stimulated through an electrode in the hypothalamus that normally elicits a quiet-biting attack, the animal will instead turn toward the stick so that the object meets its lips; when this contact occurs, the cat's mouth opens. (See **FIGURE 16.3**.) At low levels of stimulation a rather small region of the cat's face will produce this set of responses when touched, but as the intensity increases, the sequence can be elicited by touching the cat farther and farther from the front of its mouth.

Bandler and Flynn (1972) observed a similar sensory-motor interaction in a cat's attack with its paw. When its hypothalamus is electrically stimulated, a cat shows a reflexive striking movement when a specific region of its front leg is touched. If the intensity of hypothalamic stimulation is increased, the reflex can be elicited by

Region from which jaw-opening response could be elicited during brain stimulation

FIGURE 16.3 Tactile stimuli applied to these regions of the cat's face caused head turning or mouth opening during electrical stimulation of the hypothalamus. (From MacDonnell, M. F., and Flynn, J. P., *Science*, 1966, *152*, 1406–1408. Copyright 1966 by the American Association for the Advancement of Science.)

touching more widespread areas of the cat's leg. The reflex could be triggered by touching either leg, but the leg that was contralateral to the hypothalamic stimulation was more sensitive.

The experiments cited so far show that electrical stimulation of the hypothalamus or midbrain will facilitate sensory and motor mechanisms involved in attack and defensive behavior. Furthermore, hypothalamic stimulation appears also to facilitate some sort of drive mechanism; the animal will learn a maze to find a rat it can then attack. The periaqueductal gray of the midbrain appears to be the most important region for the organization of the behavior patterns that constitute attack. If the brainstem is transected beneath the midbrain, fragments of emotional behavior can be seen, but they are not coordinated. It is impossible to say at this time precisely what role is played by the hypothalamus in the organization of the response patterns.

Modulation of Emotional Behavior by the Limbic System

Various portions of the limbic system appear to be involved in the modulation of emotional behavior. Concerning aggressive and defensive behavior, the most important regions appear to be the septum and the amygdala. The septum appears to be involved in suppression of these behaviors; the amygdala appears to be involved in both suppression and facilitation, but it appears, on the balance, principally to play a facilitatory role.

THE AMYGDALA.   The amygdala is located in the rostromedial temporal lobe in humans, and in analogous locations in other mammals. There are several nuclei in the amygdaloid complex, but they are divided into two principal groups: the *corticomedial group* (phylogenetically older) and the *basolateral group* (evolved more recently). The separation between the nuclear groups is not complete, but the *stria terminalis* conveys efferent information principally from the corticomedial group to the hypothalamus and other forebrain structures, while the outflow from the basolateral group is chiefly carried by the *ventral amygdalofugal pathway.* The ventral pathway reaches the hypothalamus and preoptic region and also sends fibers to the midbrain tegmentum and central gray. The amygdalar connections are shown schematically in **FIGURE 16.4.** The amygdala receives fibers from the olfactory system, from surrounding cortex, and from the thalamus and hypothalamus. Electrical recordings have shown that the amygdala is responsive to a variety of sensory stimuli. The anatomical evidence thus provides a basis for a role for the amygdala in the modulation of hypothalamic-midbrain mechanisms in aggressive and defensive behavior.

[handwritten margin note: Septum + amygdala are associated with aggressive + defensive behavior]

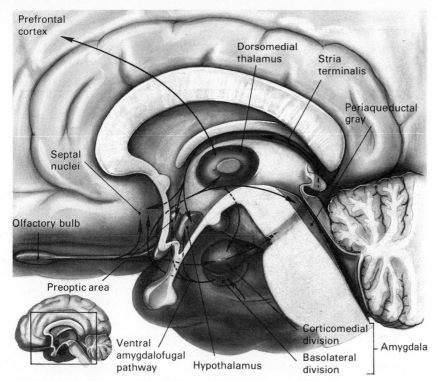

Prefrontal cortex

Dorsomedial thalamus

Stria terminalis

Periaqueductal gray

Septal nuclei

Olfactory bulb

Preoptic area

Corticomedial division

Amygdala

Ventral amygdalofugal pathway

Hypothalamus

Basolateral division

FIGURE 16.4   The two efferent fiber systems of the amygdala.

Effects of electrical stimulation of the amygdala.   Electrical stimulation of the amygdala has been found to produce either affective attack (usually characterized as "defensive reactions" by the authors of these studies) or escape behavior (usually called "fear reactions"). Predatory attack does not appear to occur as a result of amygdaloid stimulation, but the amygdala nevertheless appears to exert some influence on this behavior. Egger and Flynn (1963) found that stimulation of different portions of the lateral nucleus (a part of the basolateral complex) either facilitated or inhibited predatory attack elicited by hypothalamic stimulation. Furthermore, predatory attack was shown to be facilitated by lesions of the amygdalar area in which stimulation produced an inhibitory effect (Egger and Flynn, 1967).

A wide variety of locations within the basolateral amygdala will produce alerting and increased attentiveness when stimulated (Ursin and Kaada, 1960). At higher current intensities about half of the electrode locations that produce alerting will elicit an affective reaction: fear or attack. The regions that elicit these behaviors do not correspond precisely to anatomical subdivisions, but the basolateral electrodes that produce attack tend to be located caudally and laterally to those that produce escape. Lesions in either of these regions tend to suppress (but do not eliminate) the corresponding emotional behaviors.

Neural pathways mediating the effects of amygdala stimulation. <u>The amygdala appears to exert most of its effects on emotional behavior via the hypothalamus;</u> hypothalamic lesions abolish the effects of amygdaloid stimulation, and the behaviors elicited by stimulation of the amygdala tend to begin gradually and continue for a while after the stimulus is turned off, whereas the behaviors associated with hypothalamic stimulation generally start and stop more abruptly (Clemente and Chase, 1973). Hilton and Zbrozyna (1963) found that affective attack could be elicited by amygdaloid stimulation even after the stria terminalis was cut, so they suggested that the relevant connections between the amygdala and hypothalamus were made by means of the ventral amygdalofugal pathway.

THE SEPTUM. Although discrete regions of the amygdala exert either excitatory or inhibitory effects on emotional behavior, large lesions of the amygdala tend to suppress attack and flight. The septum, on the other hand, has been assigned a role as a mediator of inhibitory influence on the same behaviors, but as we shall see, this statement must be qualified. Electrical stimulation of the septum has been shown to inhibit attack elicited by hypothalamic stimulation in cats, which suggests that it plays an inhibitory role (Siegal and Skog, 1970). Brady and Nauta (1953) found that septal lesions produce a profound lowering of a rat's "rage threshold." These animals will show signs of extreme emotional arousal if someone approaches their cage, and they will usually scream and jump wildly around the cage if poked at with some object or even if disturbed with a puff of air. If someone is so foolhardy as to put a hand into the cage, the rat will launch a vicious, bloody attack upon it. (One of my most memorable experiences as an undergraduate was to watch an escaped "septal" rat chase two laboratory instructors around the animal room.) This increased emotionality subsides within a couple of weeks, however, and the speed with which the syndrome disappears seems to be a function of environmental variables such as handling (Gotsick and Marshall, 1972).

Other rodents such as mice (Slotnick, McMullen, and Fleischer, 1974), show increases in flight behavior, rather than increases in what appears to be irritable aggression, when the septum is destroyed. A "septal" mouse, for example, will jump wildly around in its cage if it is disturbed, and it is extremely difficult to catch if (perhaps I should say when, because these animals are faster than most experimenters) it escapes, in contrast with the more placid unoperated laboratory mouse. The animals will not attack a human, although they will attempt to bite if they are held. Also in contrast with rats, these animals remain hyperemotional indefinitely; in fact, they get better at escaping if they are handled repeatedly. If a battle is staged between a normal and a "septal" mouse, the brain-damaged animal will invariably lose (Slotnick and McMullen, 1972), so these animals cannot be called "aggressive."

The hyperemotionality (increased aggressiveness or fearfulness) seen in rodents is not seen in most other animals; some investigators report slightly increased rage in cats with septal lesions (Moore, 1964), whereas others find that some of these cats act more affectionate (Glendenning, 1972). Monkeys apparently show no change in emotionality after septal lesions (Buddington, King, and Roberts, 1967). The effects of these lesions appear to depend upon the species used; therefore, no general statement can be made about the role of the septum in emotional behavior.

Hormonal Influences on Aggression

TESTOSTERONE. In most species the male is more aggressive than the female. Furthermore, castration generally reduces this aggressiveness. We saw in Chapter 11 that early androgenization modifies the developing brain, making neural circuits that underlie male sexual behavior more responsive to testosterone. Similarly, early androgenization appears to have an organizational effect that stimulates the development of testosterone-sensitive neural circuits underlying aggressive behavior. Connor and Levine (1969) found that neonatal castration reduced irritable (pain-elicited) aggression when the rats were tested as adults, as compared with rats that were castrated after puberty. The late-castrated rats could be made as aggressive as normal animals if they were given injections of testosterone, but this replacement therapy did not increase the aggressiveness of the early-castrated rats. Edwards (1968) found that the androgen-sensitive neural circuits underlying aggression could also be "masculinized" in females; administration of testosterone to newborn female mice resulted in animals that reacted like males (i.e., with increased aggression) when given injections of testosterone during adulthood. Female mice given control injections immediately after birth did *not* respond this way when they were given testosterone as adults.

Inter-male (isolation-induced) aggression appears to be very similar to the irritable aggression (elicited by pain) just described. This form of aggression does not appear until puberty (Fredericson, 1950) and does not occur after castration unless testosterone is administered (Beeman, 1947).

Sex-related aggression is also related to testosterone levels. Some countries prescribe castration (or suppression of the effects of testosterone by the administration of androgen antagonists) for men who have been convicted of repeated sex crimes. Both heterosexual and homosexual aggressive attack have been reported to disappear, along with the offender's sex drive (Hawke, 1950; Sturup, 1961;

Laschet, 1973). In one study, injections of testosterone were given to castrated men. The treatment appeared to restore the patients's violent behavior (Hawke, 1950).

ADRENOCORTICOTROPHIC HORMONE. Other endocrine systems appear to be involved in inter-male aggression. Adrenalectomy decreases aggressiveness in male mice even after prolonged periods of isolation (Harding and Leshner, 1972). Administration of *corticosterone* (one of the principal hormones of the adrenal cortex) to the adrenalectomized mice will restore the aggressiveness (Walker and Leshner, 1972). One cannot conclude that the decreased corticosterone levels are responsible for the decreased aggression seen in the adrenalectomized mouse, since at least two other hormonal changes accompany a fall in corticosterone. (1) A fall in adrenal glucocorticoids (corticosterone is one of them) leads to a compensatory increase in **ACTH (adrenocorticotrophic hormone)**, the pituitary hormone that stimulates the adrenal gland to produce more corticosteroids. Normally the adrenal gland responds, and the subsequent rise in glucocorticoid level in the blood causes a fall in the pituitary output of ACTH. Adrenalectomy prevents the production of glucocorticoids, and the levels of ACTH remains high. (2) It appears that high levels of ACTH suppress production of testosterone (Bullock and New, 1971); since testosterone is involved in inter-male aggression, a fall in this hormone could be responsible for the inhib-

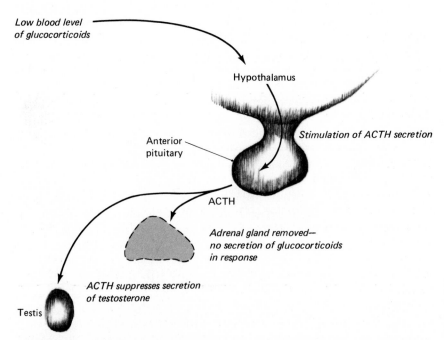

FIGURE 16.5 A schematic representation of the effects of adrenalectomy on production of ACTH and testosterone.

itory effects of adrenalectomy on aggressiveness. These hormonal interactions are schematized in **FIGURE 16.5**.

Leshner, Walker, Johnson, Kelling, Kreisler, and Svare (1973) performed a series of experiments to determine the hormone(s) responsible for the decreased inter-male aggressiveness seen after adrenalectomy. They adrenalectomized and castrated a group of mice and gave these animals controlled amounts of corticosterone and testosterone as replacement therapy. When these animals were also given ACTH, their aggressiveness declined. Since the increase in ACTH did not alter the blood levels of either testosterone or corticosterone (since these hormones were administered daily by injection), we must conclude that ACTH itself has an antiaggressiveness effect. They also showed that replacement injections of a synthetic glucocorticoid restored aggressiveness in adrenalectomized mice, apparently because the hormone, by feeding back on the hypothalamus-pituitary system, inhibited further secretion of ACTH. However, when mice were adrenalectomized *and* castrated, the glucocorticoid did *not* restore aggressiveness. Without testosterone, the mice did not fight.

The study by Leshner and his colleagues allows us to conclude that inter-male aggression is suppressed by ACTH, and that testosterone is necessary for inter-male aggression—aggression is not seen without this hormone, no matter what the levels of ACTH and glucocorticoids are.

PSYCHOSURGERY AND THE SUPPRESSION OF HUMAN VIOLENCE

No reasonable person would deny that there is too much violence in the world. Wars and individual acts of aggression produce a great deal of human misery. It is questionable, however, whether the neuroscientist can contribute very much, at least at the present time, to solving this problem. First of all, even the environmental variables that produce violence have not been identified by social scientists. Although studies of mice and rats have found that overcrowding causes increased aggression and even cannibalism, it is not at all clear that crowding produces violence in humans. In fact, careful studies that control for socioeconomic level and ethnic origin have failed to find any effects on violence and crime that can be attributed solely to increased population density (Lawrence, 1974). A crowded, squalid slum with poorly maintained buildings and a high level of unemployment will surely tend to produce frustration and violence in many of its inhabitants, but luxury high-rise apartment buildings, with even higher population densities, will probably be very peaceful. Whatever the causes of violence are, we are certainly a long way from isolating the variables and identifying their physiological bases.

Even more people are killed by wars than by individual acts of violence, but the factors that cause wars are probably very different from those that are responsible for the attack of one organism upon another. Fear-induced aggression is certainly prominent in battle, but even a complete understanding of this phenomenon will not explain why politicians and rulers send the soldiers off to engage in these battles.

Aside from wars (which probably are not started primarily by aggressive emotions) and crimes of violence endemic to many slums, there are individual acts of irrational violence that might very well have a basis in some kind of pathology. For example, Charles Whitman, who killed fourteen people (and then himself, in addition) from a tower at the University of Texas, was found to have a malignant tumor in the vicinity of the amygdala. The obvious inference is that the tumor caused the violence by stimulating neural circuits involved in aggressive behavior, and that prompt diagnosis and surgery could have prevented the tragic results of this violence. It is to issues like these that the following discussion is addressed.

Control of Violence by Brain Surgery

Most brain surgery is performed for reasons that are not at all controversial—removal of brain tumors, repair of *aneurysms* of blood vessels (balloon-shaped widenings due to local weakness), removal of scar tissue triggering epileptic attacks, reduction of intractable pain by destruction of brain circuits conveying this sensation, and alleviation of motor disturbances by localized brain lesions. **Psychosurgery**—intentional damage to brain tissue in an attempt to alter behavior, but in the absence of any observable neuropathology—has, however, been criticized a great deal in recent years.

Elliot Valenstein has written an excellent review and critique of attempts to correct human aggressive behavior by means of brain stimulation and ablation procedures (Valenstein, 1973). His conclusions are rather pessimistic; a good relationship between malfunctioning limbic system mechanisms and violent behavior has not been established, and most of the clinical reports that purport to show improvement after psychosurgical correction of this behavior do not contain adequate, impartial descriptions of preoperative and postoperative behavior.

One of the first neurosurgeons to attempt to control violent behavior by means of ablation of portions of the limbic system was Dr. Hirataro Narabayashi, of Japan. At first he performed amygdalectomies on people who showed signs of temporal lobe epilepsy along with hyperexcitability and aggressive tendencies. Later he began to try to treat people with behavior disorders, but without accompanying signs of neuropathology (such as epilepsy or severe EEG abnormalities). His earlier technique was to position the tip of a hypodermic needle in the amygdala by means of a stereotaxic apparatus. An oil-wax mixture (which causes localized brain destruction) was then injected into the brain. He currently uses a radiofrequency lesion maker.

The following quotation is a case history cited by Dr. Narabayashi and his colleagues.

> This 7-year-old boy had been diagnosed as having symptomatic epilepsy with right spastic hemiplegia and imbecility. The child maintained a severe behavior disturbance with ... explosiveness and uncontrollable hyperactivity. The EEG revealed a general dysrhythmia with spike discharges bilaterally. . . .
>
> When he reached the age of 5, he began to have convulsions once or twice a week, such attacks lasting approximately one hour. . . .
>
> [Following surgery] the change in behavior was so complete, the patient had become so obedient and cooperative, that it was almost impossible to imagine that it was the same child who had been so wild and uncontrollable preoperatively. The electroencephalogram no longer revealed spike discharges on either side. (Narabayashi, Nagao, Saito, Yoshida, and Nagahata, 1963, pp. 6–7)

Narabayashi reported that a "marked and relatively satisfactory improvement consisting in calming and taming effects" was seen in twenty-seven out of forty cases reported on in a conference on psychosurgery (Narabayashi, 1972). Since his patients were classified as idiots or imbeciles to begin with, it is difficult to assess effects of the lesions on intellectual capacities.

PROBLEMS WITH EVALUATION.   In general, attempts to evaluate the effects of psychosurgical procedures on intellectual capacities and on behaviors other than aggressiveness have not been very serious or concerted. For example, Balasubramaniam, Kanaka, and Ramamurthi (1970) use the following rating form in assessing the effects of surgery on the hyperactive or violent behavior of their patients:

Grade	Criteria
A	There is no need of any drug. Patient is able to mingle with others.
B	Very much docile and given to occasional outburst only.
C	Manageable when given drugs through not leading a useful life.
D	Transient improvement.
E	No change.
F	Died.

Note that there is no category for patients whose condition was made worse by the surgical procedure, and that the manageability of the patient is the most prominent characteristic that is rated. For example, the following case would be rated as "Grade A" by the classification system.

> Case 34 was admitted and kept in Mental Hospital. . . . He was a young man of 25 years . . . admitted into the Mental Hospital because he was always violent. He was constantly aggressive and was destructive. He could not be kept in the general wards and had to be nursed in an isolated cell. It was difficult to establish any sort of communication with him. Bilateral stereotaxic amygdalectomy was performed. Following the operation he was very quiet and could be safely left in the general wards. He started answering questions in slow syllables. (Balasubramaniam et al., 1969, p. 381)

This person can hardly be classified as cured, although he is certainly easier to manage (and I do not want to suggest that control of violent patients is a trivial matter). Dr. Balasubramaniam not only performs amygdalectomies, but also destroys the posterior hypothalamus and even the periaqueductal gray of the midbrain if the patient is not sufficiently calmed down by destruction of the amygdala. Many of his patients are children; some of them were operated on when they were under the age of 5. He does not provide documentation to show that drug therapy was tried and found to be inadequate. This is a crucial issue, since tranquilizers, anticonvulsants, and amphetamines have been found to reduce various forms of violence and hyperactivity, especially in children. Certainly it is better to try to find a drug treatment that will control a child's behavior than to produce irreversible damage to the brain. *Amen*

ALTERNATIVES TO PSYCHOSURGERY. An example of the beneficial effects that can often be achieved with drugs is given in the following case, in which an anticonvulsant drug was used to treat a child who exhibited violent behavior, apparently associated with a seizure-producing focus in the temporal lobe.

> A good example is Jimmy, whom we saw at age 9 because of serious rage reactions and aggressiveness leading to threatened expulsion from school. He was one of nine children, and the only one who had any behavior problem. In fact, the other children were considered outstanding in the community and at school. Jimmy had an identical twin, Johnny, who was considered a "model child." Jimmy's EEG showed left temporal spikes, Johnny's was negative. On methsuximide, Jimmy became an entirely different boy, a "model child" like his twin. When medication was omitted for a short time, he reverted to his old self by the third day. . . . (Gross, 1971, p. 89)

Not many cases are so clear-cut as this; we have evidence that the home environment of the nine children was such that they did not all become violent. In fact, the contrary was seen. The violent boy even had a normal identical twin, so we cannot blame any genetic factors. Furthermore, his EEG was abnormal, indicating that there very probably had been some degree of brain damage. I think that everyone would agree that this child is much better off with the drug treatment, and that this means of therapy should be explored thoroughly before resorting to psychosurgery.

THE "DYSCONTROL SYNDROME." Doctors Vernon Mark and Frank Ervin are among the most prominent supporters of the use of psychosurgery to treat violent and aggressive behavior. They believe that a substantial number of people suffering from the "dyscontrol syndrome" have localized brain abnormalities that trigger neural circuits responsible for aggressive behavior (Mark and Ervin, 1970).

The apparent helplessness of some people to control their bouts of rage and violence is attested to in the following case history:

> We had a patient whom we had in fact operated on. We had done the diagnostic procedure and had wires in his brain. A guy who had a very dramatic "flip-flop" in his personality state, he was either aggressive, paranoid, litigious, difficult to deal with or he was a very sweet, reasonable, passive, dependent kind of neurotic. These were his two modes of existence. Long before he had also happened to have epilepsy which is why he had come to us. In fact, there were two patients. I had a choice as to which one to deal with. These two patients were a great stress on the wife of the single body in whom they were contained. On this occasion she broke down and wept and said, "Who are you, honey? I don't know who you are." He had gotten extremely upset on the ward. We could not hold him against his will in the hospital since our hospital is a voluntary hospital and he could only be there by his own choice. He threatened to leave and was, in fact, in the process of leaving, ostensibly to kill his wife. At least that is what he said he was going to do, and I rather believed him. In the course of the day, we managed to get him into the laboratory and stimulate [his amygdala]. . . . In about a minute he visibly relaxed. He took a deep breath and said, "You know you nearly let me get out of here?" . . . I said, "Yes, I couldn't have held you." He said, "You've got to do something to keep me from getting out of here. I think if you had the nurse hide my pants, I wouldn't have left." I thought that was a good suggestion and followed it. We had a very reasonable discussion and he was very grateful for my having stopped him. I said, "Well, I guess we won't have to go through this very much longer because tomorrow morning we have planned to make the definitive lesion and I wanted

to talk to you about that. What we are going to do is burn out this little part of the brain that causes all the trouble." He said, "Yeah, that's great." Well, the next morning about 9:00 he was brought down and he said, "You're going to burn what out of my brain? Not on your bloody life you're not!" He would easily at the earlier point have signed anything I asked for. He was guilty; he was sweet; perhaps he was reasonable. But which of those two states I should deal with posed a real problem for me. So informed consent isn't all it's cracked up to be. Voluntary understanding has its problems. (Ervin, 1973, pp. 177–178)

Mark and Ervin appear to be careful and conservative in their choice of candidates for surgery, but even their results do not appear to be uniformly good. Mark, Sweet, and Ervin (1972) report on patients exhibiting violent or fearful behavior who were treated by lesions of the amygdala. Most of the patients showed some improvement, but side effects (hyperphagia, impotence) were also seen in some cases. Some people were not helped, or showed only a temporary reduction of violent behavior. The most successfully treated case was that of a woman who fell and injured her head. She subsequently had seizures and assaulted people violently. Amygdalectomy reduced the number of seizures and eliminated her attacks of rage. In this case, however, there was clear evidence of brain injury.

Even some clear cases of brain pathology might not necessarily be related to accompanying violent behavior. Charles Whitman did indeed have a rapidly growing temporal lobe tumor, but his diary shows that he had been carefully planning the shooting episode. His diary also shows evidence of severely disordered thought processes. Valenstein points out that we cannot ever determine what triggered Whitman's behavior, but it certainly does not appear that he was "in the throes of a sudden, episodic attack of violence." The history of careful planning stands in contrast with the spasms of rage we might expect to see if the developing brain tumor were stimulating neural circuits underlying aggressive behavior.

PROBLEMS WITH INTERPRETATION OF ANIMAL STUDIES. Valenstein notes that neurosurgeons generally base their psychosurgical procedures on data from experiments performed with animals. He also points out that many of them have "tunnel vision" when they look at the data. They tend to see only the potentially beneficial aspects of the operation and to ignore its harmful aspects. For example, some neurosurgeons have produced lesions in the region of the anterior cingulate gyrus (a portion of limbic cortex) based on studies with monkeys that reported taming effects from these lesions. However, Ward (1948) notes that "tameness" is a poor word to use in describing the postoperative behavior of the monkeys.

... the most marked change was in social behavior. The monkey's mimetic activity decreased and it lost its preop-

erative shyness and fear of man. It would approach me and
curiously examine my finger instead of cowering in the far
corner of the cage. It was more inquisitive than the normal
monkey of the same age. In a large cage with other monkeys
of the same size it showed no grooming or acts of affection
toward its companions. In fact, it behaved as though they
were inanimate. It would walk over them, walk on them if
they happened to be in the way, and would even sit on
them. It would openly take food from its companions and
appeared surprised when they retaliated. . . . (p. 15)

Another example of the selective use of animal data is provided
by the use of ventromedial hypothalamic (VMH) lesions by Roeder,
Orthner, and Müller (1972). These surgeons saw a film of cats and
monkeys that became hypersexual after lesions of the amygdala.
Subsequent VMH lesions reversed the hypersexuality. As we saw in
Chapters 11 and 12, the region of the VMH is implicated in endocrine
control and perhaps in food regulation. It certainly cannot be characterized as a region responsible for pedophilic homosexuality, the
condition Roeder and his colleagues wished to treat. They describe
the operation as restoring a balance between "male and female sex-
behavior centers." First of all, it has certainly not been established
that pedophilic homosexuality (the desire of a man to engage in sexual activity with young boys) is a result of overactive neural circuits
mediating female sexual behavior. Secondly, although lesions of the
ventromedial hypothalamus do disrupt sexual behavior in female
rats, it is simplistic to conclude that there is a distinct "female sex-
behavior center" in the VMH. Finally, even the results of the surgery
refute the notion that a balance has been achieved; as Valenstein
notes, "The major effect of the surgery seems to be a general reduction of sexual drive to the point where it is possible for the patient
to control his deviant behavior." If a lowering of sexual drive is the
result of the operation, it would appear that it would be preferable
to administer antiandrogenic drugs instead.

The rationale for amygdalectomy in humans is, of course, that
it suppresses violent behavior. However, Kling, Lancaster, and Benitone (1970) reported that amygdalectomized monkeys do appear
less aggressive and more friendly toward humans, but that these
animals did not get along very well with their peers in the wild. They
acted confused and exhibited *more*, rather than less, fear. They sometimes responded to dominance gestures from higher-ranked monkeys
in an inappropriate way and consequently got thrashed. In general,
they appeared to have trouble interpreting the signs by which monkeys communicate with each other, and they eventually became out-
casts. The amygdalectomized monkeys all subsequently died. Although humans with amygdala lesions do not show these severe
defects, it would be absurd to attend solely to the interaction of the
brain-damaged monkeys with humans in the laboratory, ignoring
their fate in the wild, and thus conclude that "amygdalectomy has
a taming affect on monkeys."

Further evidence obtained from monkeys suggests that the "taming" seen might be a result of impaired visual recognition. (You will recall from Chapter 9 that the temporal lobe is involved in visual recognition.) Downer (1962) cut the corpus callosum and other forebrain commissural fibers, including the optic chiasm. Thus, visual information received by one eye was relayed solely to the ipsilateral hemisphere. A unilateral amygdalectomy was also performed. The rather wild monkey continued to react emotionally when visual stimuli were presented to the intact side of the brain. The same stimuli were treated with aplomb when they were presented to the amygdalectomized side; the monkey appeared quite tame. However, the animal reacted aggressively to tactile stimuli, suggesting that the *mechanism* for violence was not impaired, but rather the means by which visual stimuli can trigger this mechanism.

It should be noted that amygdaloid stimulation in humans does *not* appear to result in aggressive behavior unless the patient normally shows episodes of violent behavior (Kim and Umbach, 1972). These results suggest that the stimulation, instead of directly triggering aggressive circuits, might be producing some aversive effects, which, in turn, trigger aggression in a person who has a low threshold for violent behavior. Alternatively, there could be pathology at some other brain region, and the amygdaloid stimulation might summate with the effects of this pathology (just as subthreshold brain stimulation in the hypothalamus and amygdala can summate).

In conclusion, it would appear that although psychosurgery might indeed ameliorate some instances of violent behavior, preoperative and postoperative observations are generally inadequate to provide conclusive evidence. It would appear that surgery is most effective when there are definite signs of some form of neuropathology, above and beyond the behavioral manifestation of violence. Animal research has not yet provided us with a clear enough picture of the neural mechanisms underlying aggressive behavior to justify any particular types of brain ablation. It would be a wonderful thing if we could indeed locate and isolate neural circuits that mediate violence, and nothing else, and then destroy these circuits in people who wish to be free of uncontrollable fits of rage (and there is no doubt about the existence of such people, who would gladly be rid of these irresistible bouts of violence). However, until such a happy occurrence comes about (and I doubt that it will), we should be very cautious in our use of neurosurgical procedures and concentrate on controlling these forms of violence by means of chemical treatments.

MATERNAL BEHAVIOR

We can now turn our attention to a much more pleasant topic. Maternal behavior is the antithesis of aggressive behavior. It involves nurturance and protection rather than attack and injury (although the mother may very well direct some violent behavior toward animals that appear to endanger her brood). This section will examine the role of hormones in the initiation and maintenance of maternal behavior, and the role of the underlying neural circuits that are responsible for their expression. The research is principally restricted to rodents; very little is known about the neural and endocrine bases of maternal behavior in primates. Harlow has shown the importance of experience (especially of interaction between a young monkey and its peers) in the later development of proper sexual and maternal behavior (Harlow, 1973). Maternal behavior, especially, is much more complex in primates than the relatively stereotyped behavior sequences seen in rodents, and the underlying neural and hormonal substrates will undoubtedly be found to be different in these species.

In concentrating on maternal behavior I do not deny the existence of paternal behavior, but male parental behavior is seen most prominently in higher primates like humans, where the least is known neurologically. Male rodents do not show parental behavior except under special circumstances. There are, of course, other animals (e.g., some species of mouth-breeding fish) in which the male takes care of the young, and in many species of birds the task of caring for the offspring is shared equally. However, neural mechanisms of parental behavior have not received much study in these species.

Maternal Behavior in Rodents

The final test of the "superiority" of a given animal's genes is the number of offspring that survive to a reproductive age. Just as reproductively competent animals are selected for, so are animals that will care adequately for their young (if their young in fact require any care). Rat and mouse pups certainly do require care; they cannot survive without a mother who will attend to their needs.

At birth, rats and mice resemble fetuses. The infants are blind (their eyes are still shut) and they can only helplessly wriggle. They are poikilothermic; their brain is not yet developed enough to regulate body temperature. They even lack the ability to spontaneously release their own urine and feces; they must be helped to do so by their mother. As we shall see shortly, this phenomenon does not appear to be merely one of the consequences of bearing young that are quite immature; it serves a useful function.

Why, in fact, are rodent young so immature at birth? They do not even look like mice or rats at all. We might speculate as follows: There will obviously be genetic selection for an organism that produces a large number of young. This means large litters, spaced closely together. A mouse can carry a litter of twelve (or even more) pups. A pregnant mouse looks like she swallowed a golf ball; one cannot imagine the mouse carrying any more weight than she does. Therefore, if mice (or rats) gave birth to young that were larger and

FIGURE 16.6 A mouse's brood nest. Beside it is a length of the kind of rope from which it was constructed.

more mature, they would have to carry a smaller number of them. Also, the sooner the uterus is cleared out, the sooner the mouse can start cooking up another batch. As a matter of fact, some mice have a rather clever trick: they can become pregnant on top of a current pregnancy. This phenomenon, called **superfetation,** is supposedly rare, but I have observed it commonly in a breeding colony I ran in my laboratory when I left a male with the females for around 15 days. One litter of pups would be born, and just around the time the first litter was weaned (around 15 days of age), a second litter was born.

During gestation, female rats and mice build a nest. The form this structure takes depends, naturally, on the material available for its construction. In the laboratory the animals are usually given strips of paper or lengths of rope or twine. A good *brood nest*, as they are called, is shown in Figure 16.6. This nest is made of a piece of hemp rope, shown below. The mouse laboriously shredded the rope and then wove an enclosed nest, with a small hole for access to the interior. (See **FIGURE 16.6.**)

At the time of parturition the female begins to groom and lick the area around the vagina. As a pup begins to emerge, she assists the uterine contractions by pulling the pup out with her teeth. She then eats the placenta and umbilical cord, and cleans off the fetal membranes—a quite delicate operation. (A newborn pup looks like it is sealed in very thin Saran Wrap.) After all the pups are born and cleaned up, the mother will probably nurse them. Milk is usually present very near the time of birth.

Periodically, the mother will lick the anogenital region of the pups which stimulates reflexive urination and defecation. Friedman and Bruno (1976) have shown the utility of this mechanism. They noted that a lactating female rat produces approximately 48 gm of milk (containing approximately 35 ml of water) on the tenth day of lactation. They injected some of the pups with radioactive tritiated water and later found radioactivity in the mother and in the litter mates. Friedman and Bruno calculated that a lactating rat normally consumes 21 ml of water in the urine of her young, thus recycling approximately two-thirds of the water she gives to the pups in the form of milk. The water, traded back and fourth between mother and young, serves essentially as a vehicle for the nutrients contained in milk. Since the milk production of a lactating rat each day is approximately 14 percent of her body weight (for a human weighing 120 pounds that would be around 2 gallons), the recycling is extremely useful, especially where the availability of water might be a problem.

Besides cleaning, nursing, and purging her offspring, a female rodent will retrieve pups if they leave or are removed from the nest. The mother will even construct another nest in a new location and move her litter there, should the conditions at the old site become unfavorable (e.g., when an inconsiderate experimenter puts a heat lamp over it). The way in which a female rodent picks up her pup is quite consistent; she gingerly grasps the animal by the back, managing not to injure it with her very sharp teeth (I can personally attest to the ease with which these teeth can penetrate skin) and carries the pup with a characteristic prancing walk, her head held high. (See **FIGURE 16.7** on the following page.) The pup is brought back to the nest and is left there. The female then leaves the nest again to search for another pup. She will continue to retrieve pups until she finds no more; she does not count her pups and stop retrieving when they are all back. A mouse or rat does not appear to discriminate between her own young and that of another female, but will accept all the pups she is offered. I once observed two lactating female mice with nests in corners of the same cage, diagonally opposite each other. I disturbed their nests, which triggered a long bout of retrieving, during which each mother stole youngsters from the other's nest. The mothers kept up their exchange for a long time, passing each other in the middle of the cage.

Maternal behavior begins to wane as the pups become more active and begin to look more like adult mice. At around 16 to 18 days of age they are able to get about easily by themselves, and they begin to obtain their own food. The mother ceases to retrieve them when they leave the nest (although they still return to the nest to

FIGURE 16.7   A female mouse carrying one of her pups.

nurse), and she will eventually run away from them if they attempt to nurse.

Stimuli that Elicit and Maintain Maternal Behavior

Most virgin female rats will begin to retrieve and care for young pups after having infants placed with them for several days, a process called sensitization or **concaveation** (Wiesner and Sheard, 1933). The same phenomenon can be observed in mice, but a higher percentage of these animals are spontaneously "maternal" anyway. Olfaction and audition appear to be the primary senses involved in sensitization, but they act in different ways. Noirot (1970) exposed virgin female mice to the sound of the distress calls of an isolated pup; these mice engaged in more nest-building behavior than did controls. On the other hand, exposure to the odor, but not the sound, of pups (the presence of a nest that previously held a litter of mice) enhanced subsequent handling and licking of pups, but not nest-building.

Mouse, rat, and hamster pups emit at least two different kinds of ultrasonic calls (see Noirot, 1972). These sounds cannot be heard by humans, but have to be translated into lower frequencies by a special device (a "bat detector") in order to be perceived by the experimenter. The mother can, of course, hear these calls. When a pup gets cold (as it would if removed from the nest), it emits a characteristic call that brings the mother (or any other sensitized female) out of her nest. The sound is so effective that female mice have been observed to chew the cover off a loudspeaker that is transmitting a recording of this call. Once out of the nest, the female uses olfactory cues as well as auditory ones to find the pups, since she can even find a buried, anesthetized baby mouse that is unable to make any noise. The second call is made in response to rough handling. When a mother hears this sound, she stops what she is doing. Typically, it is she who is administering the rough handling, and the distress call makes her stop. This mechanism undoubtedly plays an important role in the training of mother mice in the proper handling of pups.

Olfaction and the olfactory bulbs play an important role in controlling maternal behavior, but the means by which this occurs is uncertain. Removal of the olfactory bulbs causes pup killing and eating in female rats (Fleming and Rosenblatt, 1974a) and mice (Gandelman, Zarrow, Denenberg, and Myers, 1971). However, elimination of olfactory sensitivity by application of zinc sulfate to the olfactory mucosa was found to facilitate sensitization of virgin female mice by pups (Fleming and Rosenblatt, 1974b). These rats showed less of the ambivalence normally shown by naive rats toward pups; instead of approaching them gingerly, sniffing them, and then suddenly jumping back, the animals approached them more boldly. Apparently some aspect of the odor of the pups had an inhibitory effect. The fact that removal of the olfactory bulbs induces killing and cannibalism attests to the fact that this structure does more than simply transmit olfactory information to the brain.

Another aspect of rodent pups, besides their odor, appears to identify them as juveniles. A lactating mouse will often exhibit maternal aggression, attacking other mice that venture near her nest. However, she will almost never attack a baby mouse, even if it belongs to a litter other than her own. Svare and Gandelman (1973) showed that the most important distinguishing characteristic that controlled a mouse's attack was the presence or absence of hair; a strange 14-day-old mouse will be attacked unless its hair has been shaved off. The absence of hair appears to identify a mouse as an infant, thus protecting it from attack.

Hormonal Control of Maternal Behavior

Nest building behavior appears to be primarily dependent on progesterone, the principal hormone of pregnancy. Lisk, Pretlow, and Friedman (1969) found that nonpregnant female mice built brood nests after a pellet of progesterone was implanted under the skin. The pellet slowly dissolved, maintaining a continuously high level of progesterone. The enhanced nest building was suppressed by the administration of estrogen. After parturition, mothers continue to maintain their nests, and they construct new nests if necessary. Their blood level of progesterone is very low then, but Voci and Carlson (1973) found that hypothalamic implants of prolactin (the principal hormone of the lactation period) as well as progesterone facilitated nest building in mice. Presumably, nest building can be facilitated by either hormone: progesterone during pregnancy and prolactin after parturition.

A pregnant female rat might retrieve and take care of a young pup if one is offered to her (she cannot provide milk, of course); the probability that she will do so increases as the gestational period proceeds (Rosenblatt, 1969). She will not attend to the pups immediately, but only after they have been around for a while. Figure 16.8 shows the percentage of females that began to retrieve pups as a function of days of pregnancy. (A fresh batch of pups was presented each day.) Note that the females given pups on day 17 of the gestational period (which takes 21 to 23 days) show an interest in them sooner than do females that receive the pups on day 11. (See **FIGURE 16.8.**)

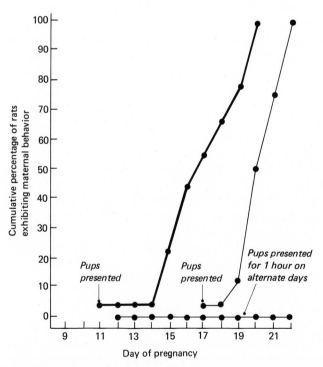

FIGURE 16.8   Cumulative percentage of pregnant female rats exhibiting maternal behavior as a function of days of exposure to pups. (From Rosenblatt, J. S., *American Journal of Orthopsychiatry*, 1969, *39*, 36–56. Copyright 1969 by the American Orthopsychiatric Association.)

At the time of parturition, the mother is immediately ready to care for her pups. She will do so even if her pups are taken by caesarian section and presented to her later (Moltz, Robbins, and Parks, 1966); the experience of parturition is not necessary for the initiation of maternal behavior.

It would appear likely that some hormone is responsible for the initiation of maternal behavior, and the assumption is commonly made that this hormone is prolactin, which is present in high levels around the time of birth. Riddle, Lahr, and Bates (1942) found that injections of prolactin induced maternal behavior in virgin female rats, but later investigators criticized their experimental procedure and found that prolactin alone did *not* initiate maternal behavior (Lott and Fuchs, 1962; Beach and Wilson, 1963). In a series of hormonal administrations that attempted to simulate the estrogen-progesterone-prolactin sequence of pregnancy and parturition, Moltz, Lubin, Leon, and Numan (1970) were able to facilitate maternal behavior in virgin female rats. Almost all rats will eventually exhibit maternal behavior, but they require around 7 days of sensitization, as we saw earlier. Moltz and his colleagues were able to shorten this sensitization time to between 1 and 2 days. Rats that received only estrogen and progesterone in the sequence also showed facilitated maternal behavior, but this sequence itself has been shown to stimulate prolactin secretion (Amenomori, Chen, and Meites, 1970).

The entire story of the role of the endocrine system in the initiation of maternal behavior has not yet been told, however. Terkel and Rosenblatt (1968) found that blood plasma taken from a lactating rat induced maternal behavior (within an average of 2 days) when injected into a virgin female. Only one injection was necessary, as opposed to the long series of injections used by Moltz and his colleagues. And a special technique that allows the continuous exchange of blood between a virgin female and one giving birth induced maternal behavior in about half a day (Terkel, 1970; 1972). No studies using pure hormones have been able to accomplish such rapid sensitization, so it remains distinctly possible that some unidentified humoral factors (besides the known female hormones) are responsible for the induction of maternal behavior.

Neural Control of Maternal Behavior

The most critical brain region responsible for maternal behavior appears to be the medial preoptic area. Numan (1974) found that medial preoptic lesions, or knife cuts that isolated this region from the medial forebrain bundle, disrupted both nest building and pup care. The mothers simply ignored their offspring. However, female sexual behavior was unaffected by these lesions. Numan noted that similar lesions have been shown to disrupt male sexual behavior (Hitt, Hen-

Emotion, Aggression, and Species-Typical Behavior

533

dricks, Ginsberg, and Lewis, 1970; Paxinos and Bindra, 1973). More caudal hypothalamic damage was shown to disrupt female sexual behavior (Kalra and Sawyer, 1970) but not maternal behavior or male sexual behavior (Yokoyama, Halasz, and Sawyer, 1967; Paxinos and Bindra, 1972).

Various lesions of the limbic system will disrupt the sequence, but not the elements, of maternal behavior. Slotnick (1967) found that lesions of the cingulate cortex in rats resulted in a scrambling of the normal sequence of pup retrieval. The mother would pick up a pup, enter the nest, walk out again still carrying the pup, drop it, try to nurse one pup outside the nest, remove pups from the nest, and, in general, act extremely confused. The component behavioral acts that make up pup care (e.g., picking up the pup and licking its anogenital region) were still present, but the behaviors appeared to occur at random. Carlson and Thomas (1968) observed an even more severe deficit after lesions of the septum in mice. Besides exhibiting a disordered sequence of pup-retrieval, the mice failed to build nests. The lack of nest building does not occur because of damage to progesterone receptors: I have found that even when mice with septal lesions are placed in a refrigerator, they fail to build nests. Normal mice (males and females) will build nests that conserve their body heat. (I should note that the refrigerated mice did not appear to suffer any ill effects from their ordeal.) In contrast with the effects of septal lesions, cingulate lesions produced only a slight deficit. The animals with septal lesions did not improve in their pup-retrieving performance, as do both rats and mice (Slotnick and Nigrosh, 1975) with cingulate lesions. Portnoy and Carlson found that even experienced mothers showed this confused behavior after they were given septal lesions.

Blaustein and Carlson (unpublished data) found that another species-typical behavior that requires spatial integration (hoarding of food) was disrupted by septal lesions. When normal mice were permitted to enter a small food-filled chamber that was attached to their home cage for one hour each day, they would carry blocks of food to their home. Mice with septal lesions repeatedly entered the small chamber, and moved blocks of food around, but did not succeed in getting many pieces back to their home cage. In contrast, another species-typical behavior, cockroach killing, was not impaired by septal lesions. This predatory behavior consists of an extremely excited chase and a stereotyped bite at the head, very similar to rat killing by cats. The mice did not have to integrate their behavior spatially; they only had to follow the cockroach (an enormous variety of *Blattis americanis*) as it ran around the cage. These studies suggest that septal lesions do not impair all sequences of species-typical behavior—only those that require spatial orientation.

It must be concluded that the analysis of the hormonal and neural mechanisms involved in maternal behavior is still in its infancy (please notice my pun) and that much more work must be done before a clearer picture of its physiological control will emerge.

KEY TERMS

adrenocorticotrophic hormone (ACTH) p. 518
affective attack p. 509
basolateral group (*BAY zo LAT er ul*) p. 514
concaveation p. 530
corticomedial group p. 514
psychosurgery p. 520

quiet-biting attack p. 509
species-typical behavior p. 510
stria terminalis (*STREE uh*) p. 514
superfetation p. 528
ventral amygdalofugal pathway p. 514

SUGGESTED READINGS

CLEMENTE, C. D., AND CHASE, M. H. Neurological substrates of aggressive behavior. *Annual Review of Physiology*, 1973, *35*, 329–356.

MOYER, K. E. *The Psychobiology of Aggression*. New York: Harper & Row, 1976.

SLOTNICK, B. M. Neural and hormonal basis of maternal behavior in the rat. In *Hormonal Correlates of Behavior, Vol. 2*, edited by B. E. Eleftheriou and R. L. Sprott, New York: Plenum Press, 1975.

VALENSTEIN, E. S. *Brain Control*. New York: Wiley, 1973.

The chapter by Clemente and Chase and the book by Moyer review the physiology of aggressive behavior. The chapter by Slotnick reviews the physiology of maternal behavior in rats. The book by Valenstein is a must for people who are interested in attempts to control human behavior by means of brain stimulation and ablation. If you are interested in the use of psychosurgery for problems other than aggression (which were not covered in this chapter) then you should read the *Report and Recommendations* and *Appendix: Psychosurgery* by the National Commission for the Protection of Human Subjects of Biomedical and Behavioral Research, DHEW Publication No. (OS) 77-0001 and (OS) 77-0002, U. S. Government Printing Office.

17

Reward and Punishment

The phenomenon of reinforcement is very important in the lives of higher organisms. It is the mechanism through which an animal's behavior is shaped by the requirements of its environment. It is as important for the life of an individual animal as the process of natural selection was for the evolution of that animal's species.

The discovery of the principle of natural selection provided biologists with a key to understanding the way in which so many diverse species, fitting such an enormous variety of ecological niches, could have evolved. The process of reproduction is not perfect; mutations can occur. Unless the mutations produce fatal flaws, these alterations in the structure of the chromosomes cause the development of offspring that differ from their parents in some way. If these offspring are better suited for their environment, and are able to reproduce more successfully because of this superiority, then their genes, with the new changes, will be carried by more and more members of the species.

The principle of natural selection explains how species are able to adapt to environmental changes. When conditions change, the only animals who survive are those whose characteristics are suited for the new environment. In a very similar fashion, the process of reinforcement explains how the behavior of an individual organism adapts itself to the environment. Although an animal is capable of performing a variety of behaviors, it learns which behaviors to perform, and which ones not to perform in a particular environment. This selection process is controlled by the consequences that the behaviors have for the animal. Some consequences **reinforce** the behavior, making it more likely that under the present conditions, it will occur again. Other consequences **punish** the behavior, making it less likely that it will occur. Thus the behaviors of an individual animal are shaped by its history, just as the selection of its genes was shaped by the history of its ancestors.

It is easy to perceive the importance of the phenomenon of reinforcement. It is much more difficult to understand how it works. In a very general way, we can say that the consequences of a behavior (that is, the stimuli that are produced as a result of what the animal has just done) somehow facilitate or inhibit the activity of the brain mechanisms that produce that behavior. The problem that this chapter addresses is the following: What is the nature of reinforcing stimuli, and what neural mechanisms initiate the physical changes that alter the probability that the reinforced response will occur?

Psychologists have been concerned with the first part of this problem for many years, but we cannot say that it has really been solved yet. There is general agreement that there is no such thing as a "reinforcing stimulus." For example, if a rat that has not eaten for two days enters a small chamber in an experimental apparatus and finds food there, the rat will be very likely to enter that chamber on a later occasion. The food reinforces the "chamber-entering behavior." However, although a piece of food will reinforce the behavior of an animal that has not eaten for a long time, it will have no effect on the behavior of an animal that has just finished a large meal. Similarly, the presence of an estrous female rat can reinforce the behavior of a normal adult male rat, but not one that has been castrated. And the opportunity to look out of a window can reinforce the behavior of a monkey that has been confined to a bleak cage for several days, but not one that has just come in from a long romp outside. Clearly, the state of the animal determines whether a particular class of stimuli (such as food, potential sex partners, or views of the outside world) can serve as reinforcers.

Thus, any attempt to understand the process of reinforcement must consider the state of the organism. Everyone, not just psychologists, understands this fact. This understanding is shown by the names that we have for many of these states. For example, we would say that organism would have to be hungry, sexually aroused, or bored for the previously-mentioned stimuli to serve as reinforcers. Do these states have something in common, or are they independent?

Even after many years of investigation, psychologists still do not agree on an answer to this question.

The general term that has been applied to a state that makes it possible for a particular stimulus to serve as a reinforcer is called *drive.* For a variety of reasons, some of which I shall discuss, this term has fallen into disrepute among many psychologists today. But before I discuss the problems with the concept of drive, let us look at the historical reasons for its development. Although it is possible that the process of reinforcement works differently for different classes of reinforcing stimuli, such as food, water, or sex partners, it seems much more reasonable that some common mechanisms are responsible for the effects of all classes of reinforcing stimuli. Thus, early in this century, psychologists suggested that reinforcing stimuli were those that reduced an organism's homeostatic needs. For example, food clearly reduced a need in an animal whose supply of stored nutrients was depleted. The suggestion became known as the ***need-reduction*** hypothesis.

It soon became clear that not all classes of reinforcing stimuli satisfied homeostatic needs. For example, a sexually aroused animal does not correct a homeostatic imbalance when it engages in sexual activity. The behavior obviously has biological utility in ensuring the continued existence of the species, but it certainly cannot be said that an animal "needs" to copulate in the same way that it needs to eat. And other classes of reinforcing stimuli are even farther removed from homeostasis; a monkey who presses a lever in order to gain a glimpse of the world outside its cage does not do so because of tissue need. Thus, the concept of "drive" was developed.

An animal who has not eaten for several hours, or who has been given an injection of insulin (which lowers the level of glucose in the blood) is described as being hungry, or as having a "hunger drive." Similarly, a rat who eagerly copulates with a willing partner is said to possess a "sex drive." A monkey who works at some task in order to look out the window of its cage is described as being bored, or as possessing a "curiosity drive." The reinforcing stimuli in each of these cases are said to satisfy (reduce) the drive. Thus, reinforcement can be explained as a process of ***drive reduction.*** Drives are seen as aversive states which motivate the animal to seek stimuli that can reduce them.

The drive-reduction hypothesis of reinforcement was much more useful than the need-reduction hypothesis. It suggested a common mechanism for the effects of a wide variety of reinforcing stimuli. However, it soon ran into problems. Hypotheses must be tested by empirical observations. In order to validate the drive-reduction hypothesis it is necessary to identify the presence of a "drive" in an animal, and then determine whether all stimuli that reduce that drive are capable of reinforcing the animal's behavior.

Here is where the difficulty lies. How can "drive" be measured? The drive-reduction hypothesis postulates an internal state that cannot be independently measured by any techniques that we presently possess. The only way that we can be sure that an animal has an elevated "sex drive" is to see whether it attempts to copulate with a receptive partner. The only way we can be sure that an animal's "curiosity drive" is aroused is to see whether it looks out the window. The logical dilemma is this: We have hypothesized that state A produces phenomenon B. If the only way we can determine whether state A exists is to look for phenomenon B, then we can prove nothing. In order for an experimental test to be valid, the test must be capable of rejecting the hypothesis if it is not true.

Let me give a concrete example of the logical difficulty. Suppose we put a monkey into an empty cage for several days (a treatment that would be expected to increase the animal's "exploratory drive"), and find that the monkey cannot be taught to press a lever that causes a window to be opened briefly. What can we conclude? Do we decide that the drive-reduction hypothesis is incorrect? No, we do not, because we do not know that the animal's "curiosity drive" was really aroused. There is no way that the experiment could turn out that would permit us to conclude that the hypothesis is false. Therefore, the hypothesis is not testable. Until drive, and its reduction, can be measured by procedures that are independent of the effectiveness of the reinforcer, we have no way of verifying or disproving the drive-reduction hypothesis.

As we shall see later in this chapter, physiological psychologists have obtained evidence that argues against the drive-reduction hypothesis. However, the concept of "drive" itself has not disappeared. Drive is a state that cannot be measured independently of the behaviors that it is hypothesized to motivate, which has led many psychologists to conclude that the term is useless and should be abandoned. The problem is, many phenomena are difficult to explain without some reference to internal states of the organism. For example, if a hungry rat is given one small piece of food to eat, and no more, it becomes very excited, and actively explores the place in which the food was delivered. It is hard to watch such an occurrence and not conclude that the food increased the animal's "drive." Certainly, the delivery of the piece of food caused an increase in exploratory behavior (that is an objective fact), and therefore, it must have caused a corresponding increase in the activity of neural mechanisms that control this behavior. What is the harm in calling this neural activity "drive"?

The potential harm is this: "Drive" is a very poorly defined concept, embracing the arousing effects of many different events, such as the delivery of a small piece of food to a hungry animal, the presence of a receptive sex partner, administration of a concentrated salt solution, and confinement in a small, barren cage. Therefore, use of the term in discussing the reinforcement process may have the effect of equating phenomena that are, in fact, very different, and thus make the identification of the actual brain mechanisms much more difficult. In addition, the drive-reduction hypothesis is held in very low esteem these days, and thus the term "drive" itself is viewed with suspicion.

The fact that the term "drive" reminds some people of the discredited drive-reduction hypothesis is not a sufficient cause for abandoning the term, if it proves useful for other reaons. The danger of confusing distinctly different phenomena by giving them all the same label is a very real one. However, I see no way to avoid using the term. My job is to present important experiments, and, as much as possible, to tie the results together so that they make sense. As you shall see, the concept of drive provides me with the necessary string.

The first part of this chapter will be concerned with the phenomenon of reinforcing brain stimulation, and with a discussion of what this phenomenon tells us about drive and reward. I shall also consider the implications of such brain stimulation for human society. In the second part of the chapter I shall deal with aversive stimulation: pain and brain mechanisms of punishment.

REWARDING BRAIN STIMULATION

Its Discovery

The discovery of the rewarding properties of brain stimulation was made by accident. Dr. James Olds was investigating whether electrical stimulation of the reticular formation might increase arousal and thus facilitate the learning process. He was assisted in this project by Peter Milner, who was a graduate student at the time. Olds had heard a paper, presented by Neal Miller, that described the aversive effects of electrical stimulation of the brain. Therefore, he decided to make sure that stimulation of the reticular formation was not aversive—if it were, the effects of this stimulation on the speed of learning would be difficult to assess. Fortunately for the investigators, one of the electrodes missed its target; the tip wound up some millimeters away, probably in the hypothalamus (the brain of this animal was lost).

Here is Olds's description of what happened when he tested this animal to see if the brain stimulation was aversive:

> I applied a brief train of 60-cycle sine-wave electrical current whenever the animal entered one corner of the enclosure. The animal did not stay away from that corner, but rather came back quickly after a brief sortie which followed the first stimulation and came back even more quickly after a briefer sortie which followed the second stimulation. By the time the third electrical stimulus had been applied the

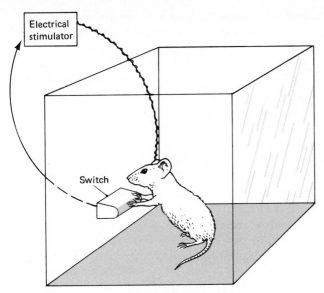

FIGURE 17.1   A rat in a self-stimulation apparatus.

animal seemed indubitably to be "coming back for more."
(Olds, 1973, p. 81)

Olds and Milner then implanted electrodes in the brains of a group of rats and allowed the animals to administer their own brain stimulation by pressing a lever-operated switch in an operant chamber. (See **FIGURE 17.1.**) The animals readily pressed the lever; in the initial study (Olds and Milner, 1954) they reported response rates of over 700 per hour. In subsequent studies, rates of many thousands of responses per hour have been obtained.

The Potency of Rewarding Brain Stimulation

As we have seen, ESB is very reinforcing; behaviors rewarded by ESB will persist at a very high rate for many hours, until the animal becomes exhausted. The only natural reward that comes close to sustaining long bouts of responding is a mixture of saccharin and dilute glucose (Valenstein, Cox, and Kakolewski, 1967).

The reinforcing effect of brain stimulation is so potent that a hungry rat will often ignore food when it is able to press a lever for shocks to the brain. Routtenberg and Lindy (1965) trained rats to press two levers—one delivered rewarding electrical brain stimulation (ESB), whereas presses on the other one produced food pellets. The animals were allowed to press the food lever for only one hour each day, and some of them spent so much time at the lever that administered ESB that they starved to death. However, when food and ESB are continuously available, rats will alternately eat, press the lever for ESB, and sleep; they will not remain at the lever to the exclusion of other activities (Valenstein and Beer, 1964). ESB is a powerful reinforcer, but subjects will not starve in preference to leaving the lever, except under very special circumstances (i.e., when the opportunity to obtain food is sharply restricted).

How Does ESB Reinforce Behavior?

There has been much conflict and controversy about the way in which electrical brain stimulation can reinforce behavior. Does it cause pleasure? Does it reduce drive? Does it increase drive? Is it possible that each stimulation produces a drive that the next one reduces? Are the effects specifically related to a particular kind of natural reward, or does reinforcement produced by brain stimulation activate a reward system that is common to all kinds of reinforcement? There is evidence that favors each of these hypotheses, but there is also evidence that *contradicts* each of them, as well. How is one to choose among them?

The answer, I think, is that one *cannot* choose any one of these hypotheses. It is likely that there is some value in each of them. In this section, I shall not support a single, unifying explanation for the rewarding effect of ESB. Every explanation that I have read can be challenged by experimental evidence, and I have not been able to devise a comprehensive one of my own. The truth of the matter is almost certainly that the way in which ESB reinforces behavior depends upon several factors, including the particular neural systems that are being stimulated and the state of the animal. For example, stimulation of different parts of the brain probably reinforces behavior for a variety of different reasons. Moreover, electrical stimulation can be reinforcing when the animal is in one physiological state, and fail to be reinforcing, or even be aversive, when the animal is in another state. Therefore, I shall organize the discussion around the general kinds of phenomena that have been investigated with regard to rewarding brain stimulation. Unfortunately, most of the studies have investigated the effects of stimulating one area of the brain—the medial forebrain bundle. This fiber system contains axons of many different ascending and descending systems, several of which can independently support self-stimulation. (The MFB was previously described in Chapter 12, in the context of feeding.) Thus, it is very difficult to correlate the behavioral effects of a particular study with the neural systems that are mediating the effect.

Recent neuroanatomical and biochemical evidence (which will be reviewed in a later section) indicates that there are several different neural systems whose stimulation can be reinforcing. Therefore, the particular way in which ESB reinforces behavior may depend upon the system that is being stimulated. And if stimulation in a particular region activates more than one system, the behavioral effects could be complex, indeed. Unfortunately, neuroanatomical and biochemical studies have not yet been able to correlate different behavioral effects of reinforcing brain stimulation with particular neural systems. This integration will undoubtedly occur in the future, but in the meanwhile, my discussion of the behavioral effects of ESB cannot yet be tied to the discussion of the anatomy and biochemistry of reinforcement mechanisms that follows in the next section.

REINFORCEMENT THAT INTERACTS WITH BIOLOGICAL DRIVES.

Reinforcing brain stimulation that also produces drive. Many studies have shown that the rewarding effect of ESB can be related to the arousal of natural drives. For example, brain stimulation that elicits such behaviors as eating and drinking will usually be found to be reinforcing, if the animal is permitted to press a lever for its delivery. But why should an animal press a lever in order to make itself hungry or thirsty? This fact is certainly not consistent with the drive-reduction hypothesis of reinforcement. This hypothesis would predict that electrical stimulation that makes an animal become more hungry or more thirsty should be aversive, not reinforcing.

In fact, the immediate effect of most natural reinforcers is not to reduce drive, but to increase it. Homeostatic mechanisms (like those which control feeding or drinking) are, of course, characterized by negative feedback loops: an imbalance elicits a correctional mechanism, which restores the balance and thus shuts itself off. Such a system is schematized in **FIGURE 17.2.** But as we saw in Chapter 12, there also appear to be positive feedback mechanisms; the effect of the sight, odor, and taste of food, and of the act of ingestion itself, is to *increase* hunger, not to decrease it. A particularly dramatic example of this positive feedback is the "feeding frenzy" sometimes

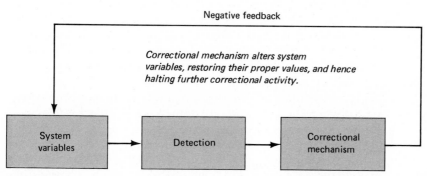

FIGURE 17.2   A negative feedback loop

seen in groups of sharks or other fish. Ultimately, of course, the satiety signals that are produced by the ingestion of food suppress further eating; the positive feedback loop is overwhelmed by a more powerful negative feedback loop.

As I noted in Chapter 12, a clearer example of positive feedback is provided by sexual behavior. One can only conclude that increases in sexual arousal are positively reinforcing—otherwise, why would people engage in foreplay? Why not get that unpleasant drive over with right away? Similarly, most people enjoy momentary frights, or "thrills," such as roller coasters or horror films. Children generally like to be thrown into the air, or to be startled by a loud "Boo!" They may run away, but they will usually come back to say, "Do it again!" So, perhaps, at least some increases in drive are reinforcing.

There is good, objective evidence for the drive-inducing effects of reinforcement. Hunsiker and Reid (1974) found that thirsty rats would run faster for water reinforcement when the individual trials were separated by 7 seconds than when they were separated by 95 seconds. Presumably, positive feedback (arousal) from each reinforcement lasted long enough to facilitate performance on the next trial. That is, one reinforcement "primed" the performance for the next trial.

Of course, not *all* increases in drive are reinforcing. To take an obvious example, painful stimuli increase drive, but not very many people seek out pain. (Yes, some people do, and it would be interesting to know why.) Even drive increases that are clearly reinforcing can become aversive if they become too intense or too prolonged. Too much of even a good thing can make peace and quiet become a very reinforcing state of affairs. This fact only confirms my earlier statement: an attempt to find a unitary explanation for reinforcement is probably doomed to failure.

Effects of natural drives on responding for ESB. Treatments that alter the level of a natural drive also affect responding for ESB, at least if the electrical stimulation affects neural systems that are related to the natural drive. For example, when a rat is pressing a lever for electrical stimulation of the MFB, its rate of responding will covary with alterations in hunger, no matter how these alterations are produced. Increases in both hunger and response rate were produced by food deprivation and injections of insulin. Decreases were produced by normal feeding, gastric intubation of food, and glucagon injections (Balagura and Hoebel, 1967; Mount and Hoebel, 1967; Hoebel, 1968). Furthermore, Deutsch and DiCara (1967) found that it took longer to extinguish behavior that had previously been reinforced by ESB when the animals were deprived of food prior to starting the extinction procedure. ESB thus appears to be even more reinforcing when the animal's drive state is increased.

Caggiula (1970) showed that increases or decreases in specific drives have effects on specific kinds of ESB; that is, the changes in response rate are not produced merely by changes in general arousal.

Caggiula trained male rats to respond for lateral hypothalamic or posterior hypothalamic stimulation. (When the experimenter turned on the current, the lateral stimulation produced drinking, and the posterior stimulation produced copulation.) Castration reduced the rate of responding for self-administered posterior (but not lateral) hypothalamic stimulation. Replacement therapy with testosterone reinstated the responding. As you might predict, the contrary is also true; food deprivation facilitates responding for lateral (but not posterior) hypothalamic stimulation (Herberg, 1963).

There are other ways to increase an animal's drive, besides depriving it of food or altering its hormonal levels. In many cases, the presence of an ***appetitive stimulus*** (that is, one that serves as the target of motivated behavior) can increase drive. If an animal is even moderately hungry, the sight or odor of food can cause drastic increases in drive. And the presence of a potential sex partner obviously plays a significant role in an animal's sexual arousal.

The presence of a particular goal has been shown to increase the reinforcing effects of ESB that elicit behaviors directed toward that object. DeSisto and Zweig (1974) tested lever pressing of rats that would either (a) eat food or (b) kill a frog when brain stimulation was administered. The "eaters" pressed the lever for a longer time when food (but not a frog) was present, whereas the "killers" pressed the lever for a longer time when a frog (but not food) was present. (See **FIGURE 17.3**.)

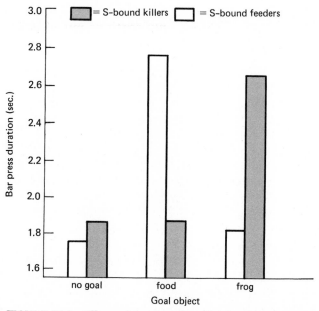

FIGURE 17.3   Effects of the presence of food or frogs on lever pressing of rats that eat or kill frogs in response to electrical stimulation of the brain. (From DeSisto, M. J. and Zweig, M. *Physiological Psychology*, 1974, *2*, 67–70.)

Most studies that have investigated the effects of drive on responding for ESB have investigated appetitive drives—that is, drives that involve approach toward a goal object that is itself reinforcing. Hunger, thirst, and sexual arousal would all be considered examples of appetitive drives. An interesting study has shown that aversive stimuli can also facilitate responding for ESB, if the stimuli are related to the effects that the brain stimulation has on the animal's behavior.

Maxim (1972) investigated the effects of stimulating the anterior lateral hypothalamus of rhesus monkeys by means of a radio-controlled stimulating device that was chronically attached to their heads. He found that electrical stimulation appeared to reduce the animals' fear; they showed fewer signs of arousal and distress when they were confronted by a more dominant member of the colony.

In a later study (Maxim, 1977), he implanted electrodes in the same region of the hypothalamus. He attached a lever to the wall of the colony cage that could be programmed to transmit radio signals that turned on ESB in the hypothalamus or the locus coeruleus. The animals quickly found the lever and learned to press it. The interesting thing is that they pressed it much more during interactions with a dominant monkey. In addition, the presentation (by the experimenter) of a live snake, which is universally feared by monkeys, also increased the animals' rate of responding for hypothalamic stimulation. However, the monkeys did not increase their rate of responding when the stimulation was delivered through electrodes that had been implanted in locus coeruleus. ESB delivered to locus coeruleus is reinforcing, but it does not appear to reduce fear. Thus, the study shows that when an animal is confronted by a fear-inducing situation, ESB that reduces fear becomes more reinforcing.

EFFECTS OF ESB ON SUBSEQUENT DELIVERY OF ESB.   As we have seen, reinforcing ESB often elicits goal-directed behavior, which suggests that it increases drive. We have also seen that increases in drive can facilitate the effects of reinforcing ESB. Therefore, one might expect that the drive-inducing effects of a shot of reinforcing ESB might increase the potency of the *next* shot of ESB. Indeed, this is often the case, as we shall see.

Right from the very beginning, it was noticed that despite the fact that ESB seems to be very reinforcing, an animal that has responded at high rates on one day is often observed to ignore the lever the next day when put into the testing chamber. If the animal is then given one or two "free" *priming* shots of ESB, it will run to the lever and begin responding again. (The term comes from the fact that a pump that has not been used for a while often needs to be "primed" with a bit of water to get it working again.) These phenomena, overnight decrements and priming, have been observed in a large number of studies (e.g., Olds and Milner, 1954; Wetzel, 1963). However,

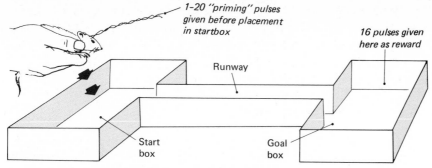

1–20 "priming" pulses
given before placement
in startbox

16 pulses given
here as reward

Runway

Start
box

Goal
box

FIGURE 17.4   The procedure used by Gallistel (1969).

it should be noted that not all animals show these phenomena. The performance of some rats does not decline overnight, and hence these animals do not require priming (Kent and Grossman, 1969).

An explanation for the phenomena of overnight decrements and priming was presented by Deutsch and Howarth (1963) and by Gallistel (1973). These investigators suggested that an animal would work for ESB only if it were motivated by a drive. When a previously-trained animal was placed into the testing chamber, it would not approach the lever because its drive state was low. However, a few free priming shots of ESB would activate a drive, and thus motivate the animal to begin to respond. Thereafter, drive was maintained by delivery of the reinforcing brain stimulation. Thus, ESB was seen to have two effects—reinforcement and drive-induction.

Gallistel (1969) investigated the decay of the hypothesized drive-inducing effects of ESB. He reinforced the animals with "trains" of brief pulses of current to the brain. The rats were taught to traverse a runway for ESB that was delivered when the animal reached the far end. First, the rats were given a variable number (one to twenty) of priming pulse trains of ESB (sixty-four pulses in each train) just before they were placed in the starting box. Once they reached the goalbox at the end of the alley they were then given a fixed number of pulses (sixteen) as a reward. The procedure is outlined in **FIGURE 17.4.**

Gallistel found that the speed of running depended on two factors: the number of priming shocks the subject received and the amount of time that had elapsed since the priming shocks were administered. Figure 17.5 shows data obtained from one rat; note how one priming pulse train facilitated running only slightly (the effect lasted less than 30 seconds), whereas twenty priming pulse trains produced a much more striking increase in running speed. (See **FIGURE 17.5** on the following page.) The data appear to show that ESB has an aftereffect that can influence subsequent responding that is reinforced by ESB.

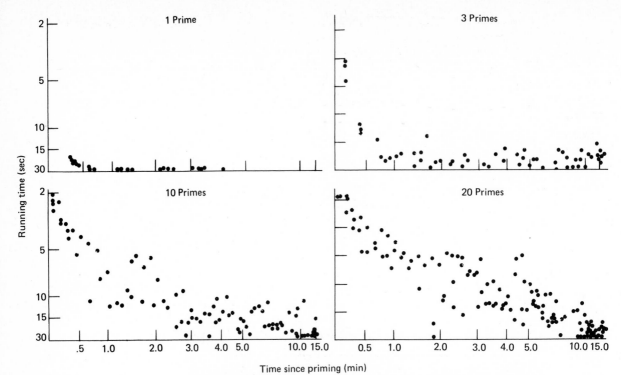

FIGURE 17.5 Data from the procedure shown in Figure 17.4: Running time as a function of delay after varying amounts of priming stimulation. (From Gallistel, C. R., *Journal of Comparative and Physiological Psychology,* 1969, *69,* 713–721. Copyright © 1969 by the American Psychological Association. Reprinted by permission.)

EFFECTS OF NONSPECIFIC AROUSAL ON REWARD. For many years, psychologists have debated whether drives are specific to the particular physiological conditions that produce them, or whether drive is simply a generalized state of neural activation. For example, it is possible that what we call "thirst" is the activity of a particular system of neurons in the brain. These neurons would be activated only when the animal had a need for water, and thus they would participate in only one drive. Alternatively, it is possible that the physiological conditions that give rise to thirst activate a system of neurons that is involved in many different kinds of drive, including thirst, hunger, and sexual arousal. This arousal system would provide the motivation for the animal to engage in behavior, while stimuli arising from the tissue need for water would provide the direction for this energy—that is, toward sources of water. (See Bolles, 1975, for a review of the behavioral literature.) Figure 17.6 presents the nonspecific drive hypothesis in schematic form. (See **FIGURE 17.6.**)

A phenomenon that I mentioned in Chapter 12 (feeding) appears to be related to this issue: stress produced by tail-pinch. As we saw earlier, a rat can be induced to eat or drink (depending on whether food or water is present) merely by pinching its tail. Other behaviors

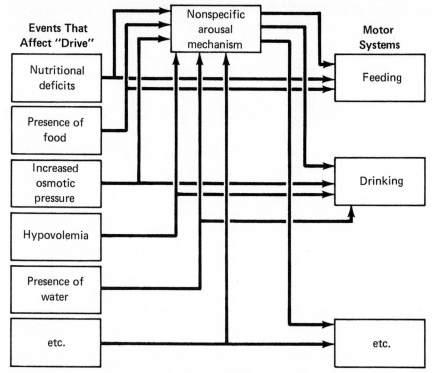

FIGURE 17.6   A schematic representation of the concept of nonspecific arousal.

can also be elicited, including the gnawing of blocks of wood, sexual activity, or maternal behavior (Antelman, Rowland, and Fisher, 1976). Again, the behavior that is observed depends upon the goal objects that are present at the time of testing.

The arousal that is produced by this form of mild stress appears to be mediated by a system of dopamine-secreting neurons. Antelman and his colleagues found that dopamine antagonists, but not norepinephrine antagonists, blocked the behavior-eliciting effects of tail-pinch. (As we shall see, dopaminergic mechanisms have been implicated in the reinforcing effects of electrical brain stimulation, which provides independent evidence that reinforcement and arousal might be related.)

Koob, Fray, and Iversen (1976) attached paper clips to the tails of rats and found that this source of stimulation would induce them to run through a maze in order to find a block of wood, which they would then proceed to gnaw on. This study did not contain a control group of rats whose tails were not pinched, but fortunately, the authors later reported that they had run the proper control group, and that the unstimulated rats did not learn the maze, nor did they engage in any significant amount of gnawing when they came across the block of wood (Fray, Koob, and Iversen, 1978). (I am sure that you can see the necessity of this control group.)

As the authors noted, reinforcement, at its simplest, appears only to require nonspecific arousal. (Of course, it must also require the presence of a stimulus object that will support a species-typical behavior—in this case, a block of wood.) Previously, Valenstein, Cox, and Kakolewski (1970) showed that a similar case can be made for the nonspecificity of behaviors that are elicited by electrical stimulation of the brain. These investigators found that a rat that ate food in response to electrical stimulation of the MFB could be induced to drink if the food was removed and water was presented instead. At least under some conditions, ESB appears to facilitate a *range* of behaviors, and does not specifically elicit only one. The particular behavior that the animal performs depends upon the environment. Perhaps such electrical stimulation activates the same dopaminergic mechanism that is aroused by tail-pinch.

It must be pointed out that there are many instances of the elicitation of specific behaviors by electrical brain stimulation that cannot be accounted for by a general arousal mechanism. As we saw in Chapter 16, predatory attack, affective attack, and flight are clearly different behaviors that are made in response to stimulation of different brain regions. And male rats have been observed to attack or attempt to mate with a female, depending upon which brain region is stimulated. In this case the electrode, and not the particular goal object, determined the behavior, since the goal object (female rat) remained the same. Thus, it does not appear that electrical stimulation of the brain produces *only* nonspecific arousal.

MIXED EFFECTS OF ESB.   It is an interesting fact that the same animal that will respond for a burst of ESB will also respond to *turn off* the same stimulation. Roberts (1958a) found that cats would escape from hypothalamic stimulation but, curiously, did not learn to *avoid* it. That is, the cats waited until it came on, and then they responded to turn it off. In a later study, Roberts (1958b) found that the animals would, in fact, work to turn on the same stimulation from which they would subsequently escape. Similarly, Bower and Miller (1958) were able to train rats to press one lever to turn ESB on, and to press another to turn it off. Valenstein and Valenstein (1964) found that a wide variety of placements gave these ambivalent effects. Thus, it does not appear that the electrode must be positioned between rewarding and punishing circuits, since it is not likely that so many electrodes would just happen to be placed in that way. Rather, prolonged ESB appears to be aversive.

Some investigators (e.g., Kent and Grossman, 1969) have suggested that the drive-inducing effect hypothesized by Deutsch and Gallistel might not be drive at all; their results might merely be a consequence of the mixed rewarding and punishing effects of ESB. Kent and Grossman noted that some of their rats required priming shots of ESB after they had been away from the lever for a while.

They suggested that ESB was purely rewarding for the rats that did not show overnight decrements, and hence did not require priming, but that it was both rewarding and punishing for rats that did show overnight decrements, and hence required priming.

The investigators reasoned as follows: If ESB had both aversive and reinforcing effects, the reinforcing effect might overcome the aversive one during the session in the experimental chamber. When the animal was placed in the chamber the next day, its behavior would be controlled by the punishing effects of the ESB rather than by the reinforcing effects. However, once a few priming shots of ESB were received, the reinforcing effect would become dominant, and the animal would begin responding.

To test this hypothesis, the investigators attached a pair of shock electrodes to the tails of the "nonprimers." The animals were then given painful tail shocks along with ESB when they pressed the levers. The rats continued to respond, but they now acted like primers. They avoided the lever after they had been away from it for a while, but would return and commence pressing when a "free" shot of ESB plus tail shock was administered. It is interesting to note that a tail shock alone, or even a loud noise, would send the rats back to the lever. At the time Kent and Grossman performed their experiment, the tail-pinch phenomenon was not known. Now, we can explain the priming effects of the tail shock and loud noise as increases in nonspecific arousal, probably mediated by dopaminergic neurons.

Kent and Grossman's hypothesis provides an alternative to Deutsch and Gallistel's explanation for overnight decrements. It is possible that animals are reluctant to approach the lever after being away from the testing chamber for several hours because the ESB produces an aversive, as well as a reinforcing, effect. However, their data do not contradict the conclusion that ESB increases drive; on the contrary, they support it. Whatever the reason for the nonresponding after a time out may be, the animals can be made to approach the lever by giving them a free shot of ESB, shocking their tails, or presenting a loud noise. The most plausible explanation for this reinstatement of responding is that of all these treatments increase the animals' drive.

Drive-inducing brain stimulation that is not reinforcing. Although ESB that elicits a behavior such as eating or drinking is usually also reinforcing, at least one study suggests that a link between drive and reward is not inevitable. Olds, Allan, and Briese (1971) placed extremely small electrodes (62.5 μm., or approximately 0.002 inches in diameter) in various locations in rats' brains. They hoped that use of such small electrodes would make it likely that stimulation would not affect a wide variety of neural systems. They observed the effect of stimulation on eating and drinking, and they noted whether the stimulation was reinforcing—that is, whether the rats would press a lever for ESB.

Olds and his colleagues found several regions where stimulation had distinctly different effects: (1) eating, but not drinking or self-stimulation; (2) drinking, but not eating or self-stimulation; (3) self-stimulation alone; (4) eating, drinking, and self-stimulation (all from the same electrode); and (5) *suppression* of eating and drinking.

These results suggest that hunger and thirst mechanisms can be separated from reward mechanisms; it is possible to elicit eating and drinking with brain stimulation that is not also reinforcing.

Although Olds and his colleagues found brain locations where reinforcement was associated with induction of eating or drinking, they did not find any places where reinforcement was associated with the *inhibition* of ingestive behaviors. In fact, other studies have shown that stimulation that inhibits feeding and drinking is also aversive. However, these results are not conclusive; we have to be careful not to make unwarranted assumptions about cause-and-effect. It is possible that aversive effects of ESB might simply interfere with ingestive behavior in a nonspecific way, just as a painful foot shock would do. A suppression of eating or drinking is not necessarily synonymous with a reduction in hunger or thirst. Nevertheless, we can still conclude that the investigators found no evidence that drive-reducing stimulation is also reinforcing.

There is one final note of caution to be made: The fact that stimulation delivered at some electrode locations produced a rewarding effect without making the rat hungry or thirsty does not prove that ESB can produce a reinforcing effect without also eliciting drive. The animals in this study were not given an opportunity to kill mice or frogs, to copulate, or to perform any one of a variety of behaviors that might have resulted from the brain stimulation. It is possible that drive was actually increased in all cases, but without the presence of the appropriate goal object, some of the behaviors associated with the elicited drives could not be seen. Olds and his colleagues appear to have demonstrated drive without apparent reward, but we cannot conclude that they demonstrated reward without drive.

REINFORCEMENT WITHOUT APPARENT DRIVE. Almost everyone has a sweet tooth, and many people enjoy the odor of roses, or lilacs, or a particular brand of cologne. Similarly, it feels good to stroke the fur of a cat, or feel a mink pelt against one's cheek. People will work hard to grow beautiful flowers, or will pay a lot of money for a beautiful work of art. There are a number of sensory stimuli that are pleasant, even in the absence of any internal state that could plausibly be called a drive. In fact, most of these stimuli give pleasure (and hence can serve as reinforcers) only in the relative *absence* of drive. An extremely hungry person is not likely to be interested in the smell of flowers or in tasting a small after-dinner mint. However, a thick, juicy steak (or a nice plate of soybean curds, depending on one's food preference) will certainly produce a drive-dependent reinforcing effect.

Some kinds of electrical brain stimulation appear to be related to pleasant sensory stimuli. Needless to say, the experience such stimulation produces cannot resemble complex stimuli like beautiful paintings, but even some simple stimuli can give pleasure. For example, electrical stimulation of the olfactory bulb can reinforce lever-pressing behavior of rats (Phillips, 1970). And Sem-Jacobsen (1968) reported that humans will repeatedly press a button for the delivery of electrical brain stimulation that elicits the sensation of a pleasant odor.

Panksepp and Trowill (1967b) found that electrical stimulation can have effects that are similar to those produced by tasty sweets delivered to an animal that is not hungry. Responding for electrical brain stimulation often extinguishes very rapidly; that is, a rat who has been pressing a lever at a very high rate for ESB will quickly cease responding if the stimulator is suddenly disconnected from the circuit. This contrasts with the performance of a hungry rat that has been responding for food, who will persist much longer after the food dispenser has been disconnected.

Panksepp and Trowill trained nonhungry rats to press a lever that caused chocolate milk to be squirted into their mouths through previously-implanted fistulas. Rats, like humans, have a sweet tooth, and will drink some chocolate milk even if they have just finished a meal. Panksepp and Trowill compared the rates of extinction of hungry rats and sated rats that were pressing a lever for injections of chocolate milk into the mouth. When the lever was turned off, the hungry rats continued to press it almost three times more than the sated rats did. Furthermore, the responding of sated rats extinguished even when the animals were prevented from pressing the lever for a period of time. Thus, the animals showed "overnight decrements." (The period of time was not actually overnight.)

Decrements in responding are not normally seen in hungry animals that are responding for food; if the lever is removed for a while, the hungry animal will begin responding again as soon as the lever is replaced. In Panksepp and Trowill's experiment, the animals whose responses had been extinguished could even be "primed" to begin responding again. A "free" squirt of chocolate milk sent them back to the lever. It appears that the behavior shown by sated animals receiving a very palatable food is thus similar at that of animals that are receiving reinforcing brain stimulation. Thus, ESB appears to be able to reinforce behavior in the same way that a palatable food (and presumably, other pleasant stimuli as well) can be reinforcing in the absence of any obvious drive.

ESB and Drive Reduction

The last issue that I want to address in this section is whether the phenomenon of reinforcing brain stimulation can help us evaluate

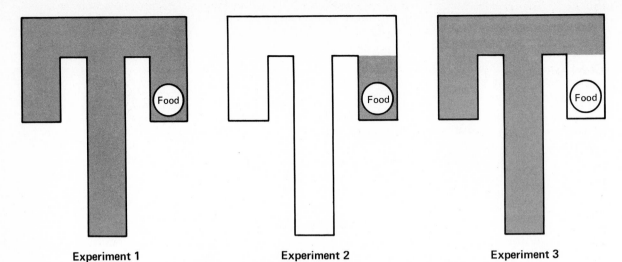

Experiment 1   Experiment 2   Experiment 3

FIGURE 17.7   A schematic representation of the experiment by
Mendelson (1966). When the animal was in the shaded portions of the
maze, electrical brain stimulation (which elicited eating) was turned on.

the drive-reduction hypothesis of reward. The fact that reinforcing
brain stimulation often appears to *increase* drive certainly argues
against this hypothesis. Further evidence was provided in a study by
Mendelson (1966).

Mendelson's subjects were rats that could be made to eat by
stimulating their brains through chronically-implanted electrodes.
The rats (who were not food deprived) were trained to run through
a maze under three different conditions, as shown in Figure 17.7.
(The shaded areas indicate where the rats received brain stimulation.) (1) The stimulation was turned on as soon as the rats entered
the maze. Food was available in the goalbox. (See **FIGURE 17.7
LEFT.**) Under this condition, the rats readily learned the task. (2)
The stimulation was not turned on until the rats reached the goalbox,
which, again, contained food. (See **FIGURE 17.7 MIDDLE.**) These
animals also learned the task. We can conclude that drive (at least,
drive produced by food deprivation or by the brain stimulation) was
not necessary for performance. Despite the fact that the rats trained
under condition (2) entered the maze "cold," they quickly ran to the
goalbox. Furthermore, the results do not support the drive-reduction
hypothesis. If drive is an aversive state, why not just stay in the start
box?

The last condition bears even more directly on the drive-reduction hypothesis. (3) The stimulation was turned on as soon as the rat
entered the maze, and was turned off as soon as the animal reached
the goalbox. Thus, the drive that was induced by brain stimulation
was reduced by entry into the goalbox. (See **FIGURE 17.7 RIGHT.**)
Despite the reduction in drive, the animals did *not* learn the maze
under this condition. Drive reduction (at least, as it was produced in

this experiment) does not appear to be a sufficient condition for re-inforcement.

Conclusions

As I noted in the introduction, it is difficult to make sense of the many experiments that have been done on the neural mechanisms of reinforcement without using the concept of drive. I think that the context in which I have used this term has explained what I mean by it, but I must confess that I have not adequately defined it. I have used the term drive to refer to a variety of states, such as those that are produced by depriving an animal of food or water, presenting it with a receptive sex partner, permitting a hungry animal to consume a small amount of food, delivering brain stimulation, and pinching an animal's tail. My justification for doing so is not that we can independently measure drive and show it to be present during all of these treatments, but that all of these treatments appear to interact with each other, and with mechanisms of reinforcement.

Increases in drive that are associated with appetitive states are reinforcing, and they facilitate responding for ESB that also elicits these drives. There appears to be some degree of nonspecificity in drive, because even electrical shock or pinching of the tail enhances the effects of reinforcing brain stimulation. On the other hand, drive-*reducing* brain stimulation can be reinforcing, if it reduces a clearly aversive state that is produced by environmental situations (i.e., the experiment with the dominant monkey). Reduction of an appetitive drive (hunger) does not, by itself, appear to be reinforcing. Finally, there are instances of reinforcement that do not appear to be associated with drive (such as the smell of a rose, or the taste of chocolate).

I am optimistic about the future of research on the problem of reinforcement. We have come a long way since the need-reduction and drive-reduction hypotheses. Things are not so simple as they once seemed, but that is the way things are; Nature did not design a simple mechanism to control the ways in which environmental stimuli come to direct an animal's behavior. I believe that the discovery of the tail-pinch phenomenon, which appears to be mediated by a system of dopaminergic neurons, will be one of the keys that will ultimately unlock the puzzle of reinforcement and drive.

Anatomy of Rewarding Brain Stimulation

In recent years, much progress has been made in the investigation of the neural systems that mediate the reinforcing effects of ESB. With the advent of the histofluorescence techniques, it was soon discovered that the distribution of rewarding electrode sites nicely coincided with the distribution of catecholaminergic neurons. Furthermore, the administration of amphetamine, a potent catecholamine agonist, greatly increased the rate at which animals would respond for reinforcing brain stimulation (Stein, 1964).

If an increase in rate of responding can be taken as a measure of strength of reinforcement, a reasonable explanation for the effect of amphetamine is that reward is mediated, or at least facilitated, by the activity of catecholaminergic neurons. Since stimulation of regions that contain dopaminergic or noradrenergic neurons is very likely to be reinforcing, and since amphetamine enhances the effectiveness of these neurons and also increases the reinforcing effects of ESB, perhaps these neurons are themselves responsible for reward. We shall see in this section that there is good evidence for involvement of dopaminergic neurons in mechanisms of reinforcement. However, evidence for noradrenergic involvement is less satisfactory.

In the last section we saw that ESB can apparently reinforce behaviors by more than one means. As we shall see in this section, more than one neuroanatomical system of neurons is capable of mediating reinforcement. Unfortunately, these statements cannot be connected. That is, we do not yet know which neural system is associated with which aspect of reinforcement. This correlation awaits future research.

NEUROANATOMY OF CATECHOLAMINERGIC SYSTEMS. There are several systems of neurons that secrete dopamine or norepinephrine at their terminal buttons. First, let us consider the noradrenergic pathways (Clavier and Routtenberg, 1980; Moore and Bloom, 1979). There are two principal systems, the **central tegmental tract** and the **dorsal tegmental bundle.** The central tegmental tract (previously called the ventral noradrenergic bundle) arises in the medulla, from groups of cell bodies labeled by Ungerstedt (1971) as A1 and A2, and from cells in the pons (group A5 and the **subcoeruleus cell group,** located just ventral to **locus coeruleus**). The axons of almost all of these neurons terminate in the hypothalamus. (See **FIGURE 17.8.**) The cell bodies of the second major noradrenergic system, the dorsal tegmental bundle, are contained in locus coeruleus. The neurons project to neocortex, hippocampus, thalamus, cerebellar cortex, and medulla. (See **FIGURE 17.8.**) You should note that the major catecholaminergic pathways pass through the medial forebrain bundle on their way forward. Thus, stimulation of the MFB activates axons of all of these systems.

The two major pathways of dopaminergic neurons are the **nigrostriatal system** and the **mesolimbic system** (Moore and Bloom, 1978; Clavier and Routtenberg, 1980). The nigrostriatal system starts in the **pars compacta** of the substantia nigra (usually abbreviated as *SNC*, for "substantia nigra compacta"), and projects to the **neostriatum**—the caudate nucleus and putamen. As we saw earlier, this

The central tegmental tract arises in the medulla and ends in the hypothalamus

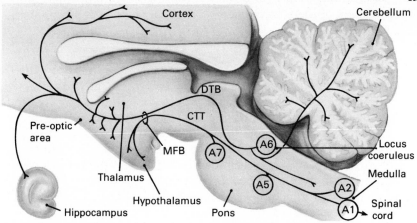

FIGURE 17.8   A semi-schematic diagram of the principal noradrenergic
pathways.(CTT: central tegmental tract; DTB: dorsal tegmental bundle;
MFB: medial forebrain bundle.)

system is important in the control of movement; its degeneration
results in Parkinson's disease. (See **FIGURE 17.9.**) The cell bodies of
the second major dopaminergic pathway, the mesolimbic system,
reside in the ***ventral tegmental area,*** or VTA. They project to a variety
of forebrain structures, including hypothalamus, nucleus accumbens, olfactory tubercle, septum, limbic cortex, and neocortex. Many
investigators believe that this system is involved in arousal and attention, and that excessive activity of the mesolimbic dopamine system in humans can result in schizophrenia. (More about this hypothesis in Chapter 20). (See **FIGURE 17.9.**) There are three other
dopamine pathways, one in the retina, one from hypothalamus to

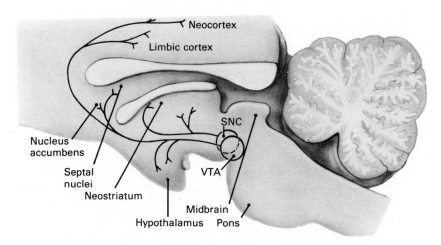

FIGURE 17.9   A semi-schematic diagram of the principal dopaminergic
pathways. SNC is lateral to (behind) VTA.

the pituitary stalk, and one entirely within the tegmentum, but so far none of them have been implicated in reward mechanisms.

NORADRENERGIC SYSTEMS.

Anatomical studies. The noradrenergic system of locus coeruleus was the first one to be implicated specifically in reinforcement (Stein, 1964; 1968). Animals will work for stimulation delivered there (Crow, Spear, and Arbuthnott, 1972). Furthermore, after one hour of electrical stimulation delivered to locus coeruleus, increased metabolism of norepinephrine is observed in the cortex, which suggests that the reinforcing effect may be related to the release of NE (Anlezark, Arbuthnott, Crow, Eccleston, and Walter, 1973). In addition, responding for locus coeruleus stimulation is facilitated by injections of amphetamine (Ritter and Stein, 1973).

Although electrical stimulation of locus coeruleus is reinforcing, a number of studies suggest that the effect is not mediated by noradrenergic neurons. Clavier, Fibiger, and Phillips (1976) destroyed noradrenergic axons of the dorsal tegmental bundle by means of bilateral injections of 6-hydroxydopamine (6-HD). The efficacy of the lesions was confirmed by the fact that the content of norepinephrine in neocortex and hippocampus fell by an average of 96.7 percent (100 percent depletion, in some animals). Despite the fact that the noradrenergic efferents of locus coeruleus were completely destroyed, the rats continued to respond for electrical stimulation of the locus coeruleus. (See **FIGURE 17.10.**) The investigators also found that even after the 6-HD lesions were produced, injections of amphetamine increased the animals' rate of responding. Thus, we can conclude that the ascending noradrenergic efferents of locus coeruleus are not necessary for the reinforcing effects of electrical stimulation

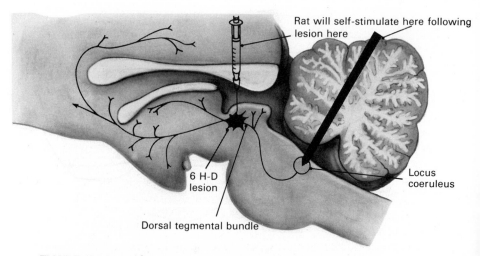

FIGURE 17.10 A schematic representation of the experiment by Clavier, Fibiger, and Phillips (1976).

of that region, or for the enhancing effects of amphetamine.

It is possible that the reinforcing effects of electrical stimulation of locus coeruleus are mediated by *descending* noradrenergic efferents, which were not damaged by the localized injections of 6-HD. However, Cooper, Konkol, and Breese (1978) found that rats continued to press a lever for electrical stimulation of locus coeruleus even after the norepinephrine content of the entire brain had been reduced by 92 percent (by injections of 6-HD into the ventricular system along with injections of a dopamine-β-hydroxylase inhibitor, which blocks the biosynthesis of norepinephrine). This finding makes it unlikely that noradrenergic neurons, ascending or descending, mediate the reinforcing effect of electrical stimulation of locus coeruleus. Presumably, the effect is mediated by nonnoradrenergic neurons (an awkward word, I must admit) located in the vicinity of locus coeruleus.

Rats will respond for electrical stimulation of the hippocampus, a structure that receives noradrenergic input, but has no dopaminergic input. Injections of amphetamine enhance responding for hippocampal stimulation. If this effect is mediated by the noradrenergic input to the hippocampus, it should be abolished if this input were disconnected. Phillips, Van der Kooy, and Fibiger (1977) destroyed the dorsal tegmental bundle with local injections of 6-HD, which produced a 97 percent depletion of norepinephrine in the hippocampus. Nevertheless, injections of amphetamine still enhanced responding for hippocampal stimulation. (See **FIGURE 17.11**.) Thus, noradrenergic input to the hippocampus does not appear to modulate the reinforcing value of hippocampal self-stimulation.

Pharmacological studies. Pharmacological studies implicate the catecholamines in rewarding ESB. Agonists such as MAO inhibitors, amphetamine, and cocaine increase response rates whereas antago-

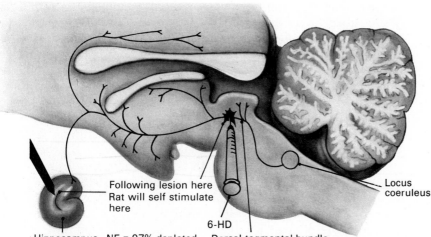

FIGURE 17.11 A schematic representation of the experiment by Phillips, Van der Kooy, and Fibiger (1977).

nists such as α-methyl-paratyrosine, chlorpromazine, or reserpine decrease it. (See Fibiger, 1978, for references, and see Chapter 5 to review the ways in which these drugs work, in case you have forgotten.) Wise and Stein (1968) suggested that the important neurotransmitter is norepinephrine; they found that an inhibitor of dopamine-β-hydrophylase, disulfiram, decreased the rate of responding for ESB. Dopamine-β-hydroxylase catalyzes the final step in the biosynthesis of norepinephrine—the conversion of dopamine to norepinephrine. Therefore, disulfiram decreases the levels of norepinephrine that are produced in the brain.

However, Roll (1970) observed that disulfiram appeared to put rats to sleep. If she woke them, they would temporarily resume their bar-pressing. A newer, more specific inhibitor of dopamine-β-hydroxylase, FLA 63, does not put animals to sleep. Lippa, Antelman, Fisher, and Canfield (1973) administered FLA-63 to rats and observed that their rate of responding for ESB was unchanged, despite the fact that the norepinephrine level was reduced by 70 percent in the brain. And, as we saw in the previous section, intraventricular 6-HD plus another dopamine-β-hydroxylase inhibitor without sedative side effects did not affect the rate of self-stimulation (Cooper and Breese, 1976).

These pharmacological studies do not support the hypothesis that noradrenergic neurons are involved in a significant way in the mediation of reward, or its enhancement by amphetamine. As we shall see in the next section, there is better evidence for the involvement of dopamine.

DOPAMINERGIC SYSTEMS.

Anatomical studies. A number of studies have implicated dopaminergic systems in brain mechanisms of reinforcement. Electrical stimulation of areas of the brain that contain the cell bodies of the principal dopaminergic systems, SNC and VTA, is reinforcing (Routtenberg and Malsbury, 1969; Crow, 1972). Furthermore, stimulation of many of the regions in which these neurons terminate (e.g., caudate nucleus, nucleus accumbens, septum, prefrontal cortex) is also reinforcing (Olds, 1977). And the medial forebrain bundle, which consistently and reliably supports self-stimulation, is rich in dopaminergic axons (along with many other kinds of axons, of course).

Several studies have suggested that the mesolimbic dopaminergic system may play a role in reward mechanisms, whereas the nigrostriatal dopaminergic system is more important in motor performance. Mora, Sanguinetti, Rolls, and Shaw (1975) placed stimulating electrodes in the lateral hypothalamus. They observed that local injections of spiroperidol (a potent blocker of dopamine receptors) into the nucleus accumbens abolished responding for lateral hypothalamic stimulation. (See **FIGURE 17.12**.) The results suggest that the rewarding effect was transmitted by means of dopaminergic

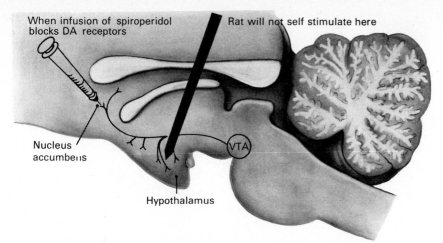

When infusion of spiroperidol blocks DA receptors

Rat will not self stimulate here

VTA

Nucleus accumbens

Hypothalamus

FIGURE 17.12   A schematic representation of the experiment by Mora, Sanguinetti, Rolls, and Shaw (1975).

fibers that pass through the medial forebrain bundle and synapse in the nucleus accumbens. In a related study, Roberts, Zis, and Fibiger (1977) found that lesions of the nucleus accumbens (produced by 6-HD, and hence at least somewhat specific to dopaminergic terminals) abolished the rewarding effects of cocaine; rats with these lesions would no longer respond in order to inject themselves with this drug. (Cocaine, like amphetamine, is a potent dopaminergic agonist.)

Self-stimulation in another area appears to depend upon dopaminergic innervation from VTA. In a neuroanatomical study, Clavier and Gerfen (1979a) showed that a particular region of prefrontal cortex in the rat, *sulcal cortex*, receives projections from VTA. In a later behavioral study (1979b) they found that *ipsilateral* damage to VTA (by means of 6-HD) abolished responding for stimulation of sulcal cortex. However, *contralateral* damage to VTA did not. (See **FIGURE 17.13** on the following page.) These results are very interesting, since they cannot be explained by general motor deficits that might be produced by unilateral damage to VTA.

Animals will also respond for electrical stimulation of SNC or the neostriatum (caudate nucleus and putamen), the two ends of the nigrostriatal pathway. Clavier and Fibiger (1977) found that either ipsilateral *or* contralateral 6-HD lesions of the fibers of the nigrostriatal pathway would abolish responding for electrical stimulation of the SNC. The effect was temporary; responding recovered in 8 to 10 days. However, the ipsilateral lesions *permanently* blocked the enhancing effects of amphetamine on self-stimulation of the SNC. (See **FIGURE 17.14** on p. 563.) The authors concluded that: (1) The temporary suppression was probably a result of interference with motor mechanisms, and not reward mechanisms, since ipsilateral and contralateral lesions had similar effects. (2) The reinforcing effect of SNC stimulation does not depend upon dopaminergic mechanisms, since responding recovered even though dopamine levels in

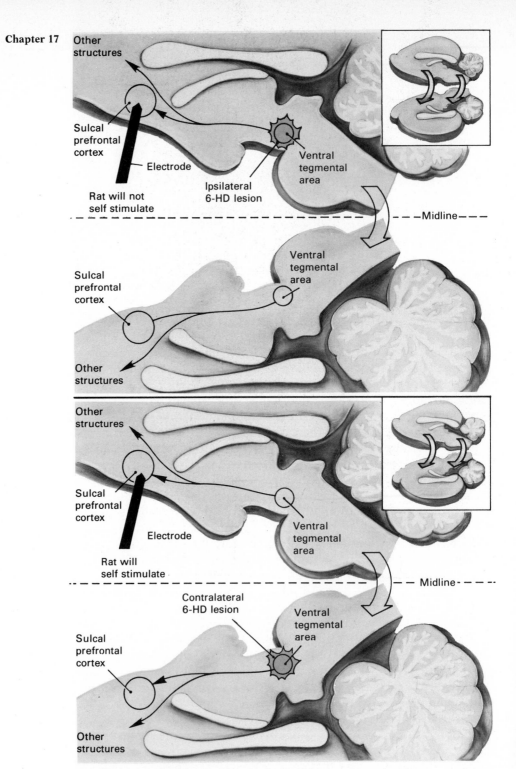

FIGURE 17.13 A schematic representation of the experiment by Clavier and Gerfen (1978b).

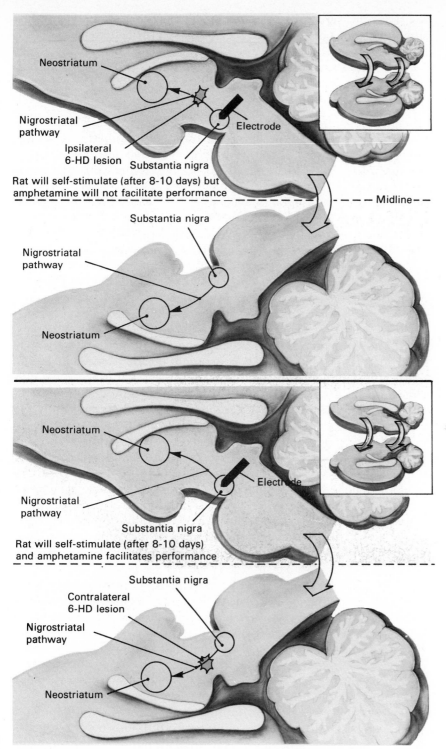

Neostriatum

Nigrostriatal
pathway

Ipsilateral
6-HD lesion   Substantia nigra

Electrode

Rat will self-stimulate (after 8-10 days) but
amphetamine will not facilitate performance

— — — Midline — —

Substantia nigra

Nigrostriatal
pathway

Neostriatum

Neostriatum

Nigrostriatal
pathway

Substantia nigra

Electrode

Rat will self-stimulate (after 8-10 days)
and amphetamine facilitates performance

Substantia nigra

Contralateral
6-HD lesion

Nigrostriatal
pathway

Neostriatum

FIGURE 17.14   A schematic representation of the experiment by Clavier
and Fibiger (1977).

the striatum were still depressed by an average of 98.5 percent. Presumably, some nondopaminergic neurons in the neighborhood of the SNC mediate the effect. (3) The excitatory effect of amphetamine on SNC self-stimulation is mediated by dopaminergic neurons that pass through the nigrostriatal pathway, since destruction of these neurons with 6-HD abolishes this effect.

Pharmacological studies. A variety of dopaminergic receptor blocking agents, including pimozide, spiroperidol, and haloperidol, suppress responding for ESB (Fibiger, 1978). The question is this: Is the suppression due to an inhibition of reward mechanisms, or simply of motor systems? As we shall see, the results favor the reward hypothesis, but not enough evidence has been gathered to provide a definite answer.

Rolls, Rolls, Kelly, Shaw, Wood, and Dale (1974) evaluated the effects of spiroperidol on responding for food, water, and electrical stimulation of VTA, hippocampus, hypothalamus, septum, and nucleus accumbens. (No wonder thay had so many authors!) The drug (a blocker of dopamine receptors) suppressed responding for *all* kinds of reward. The question is, did the animals stop responding because the food, water, or electrical stimulation was no longer reinforcing, or did they stop because the drug interfered with motor systems that controlled their behavior? We just cannot decide from their data.

A study by Fouriezos and Wise (1976) was somewhat more conclusive. If an animal is trained to respond for a reward, and if the reward is then terminated (for example, by disconnecting the food-dispensing mechanism), the animal will respond for a while at the normal rate, and then will gradually cease to respond. In other words, extinction is not immediate. Fouriezos and Wise reasoned that if a drug that blocks dopamine receptors suppresses neural systems that mediate reinforcement, then a drugged animal should act like one that has been placed on an extinction schedule—it should respond normally for a while, and then gradually stop. On the other hand, if the drug merely interfered with motor performance, then the animal's rate of responding should be low right from the time it is placed in the chamber.

Fouriezos and Wise trained rats to press a lever for ESB. Next, they gave the animals injections of pimozide (a blocker of dopamine receptors). When the rats were placed in the operant chamber, they began pressing at a normal rate, but soon slowed down. (See **FIGURE 17.15.**) Their behavior resembled that of rats placed on an extinction schedule. Therefore, the results favor the hypothesis that dopamine antagonists depress neural mechanisms of reinforcement, and do not simply interfere with motor systems.

As we saw earlier, amphetamine, a dopamine (and norepinephrine) agonist, facilitates responding for ESB. Is this facilitation specific to reward mechanisms, or is it simply a motor effect? Stein and Ray (1960) tested rats in a special apparatus that was equipped with

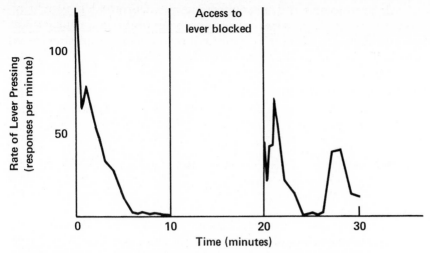

FIGURE 17.15 Lever-press responses made by a rat that has received an injected of pimozide, a blocker of dopaminergic receptors. Note that the rate of responding is high in the early part of the session. (From Fouriezos, G., and Wise, R. A. *Brain Research*, 1976, *103*, 377–380.)

two levers. The first lever delivered shots of ESB, but each time it was pressed, the current level got weaker and weaker. If the second lever was pressed, the current was reset to the highest level. Presumably, rats would press the first lever until the ESB was too weak to be rewarding, and would then press the second one to increase it again. When the rats were given amphetamine, they would make more presses on the first lever before they would go to the second one to reset the current level. It appears that the amphetamine enhanced the effectiveness of the rewarding brain stimulation, and hence made lower levels of current more reinforcing. The results suggest that amphetamine acts on neural mechanisms of reward, and not simply on motor systems.

These results were confirmed by Poschel and Ninteman (1966), who trained rats in a two-compartment apparatus. When the rats were in one compartment, they received reinforcing ESB. (No response was necessary.) No ESB was received when they were in the other compartment. When the current level was low, rats spent only slightly more time in the compartment where reinforcement was received. Administration of amphetamine greatly increased the time they spent in this compartment. Since the animals were not required to make a repetitive response for ESB, it does not seem likely that the result could simply be due to motor facilitation. Instead, amphetamine appears to facilitate reward mechanisms.

Since amphetamine is an agonist of norepinephrine as well as dopamine, we cannot conclude from this study that dopamine is the relevant neurotransmitter. However, the evidence in the previous section suggests that the effects of amphetamine on ESB depend upon dopaminergic, and not noradrenergic, neurons. <u>The weight of</u> <u>the available evidence thus suggests that dopaminergic neurons are</u> <u>involved in reward,</u> and that their facilitation increases reward, and not simply motor mechanisms that are involved in performance of the appropriate responses.

Neuroanatomical and pharmacological studies on catecholaminergic mechanisms of reinforcement have made real progress. However, work has just begun. Study has concentrated on the catecholamine pathways partly because histofluorescence techniques have made it possible to identify them. There are many neurotransmitters used by the brain, and when techniques to map them become available (as we all hope they will) we will undoubtedly find that other systems of neurons play a role in the process of reinforcement.

IMPLICATIONS OF ESB FOR THE CONTROL OF HUMAN BEHAVIOR

The fact that ESB can be so reinforcing—that it is possible to arrange conditions so that a self-stimulating rat will ignore food (Routtenberg and Lindy, 1965) or even neglect its newborn pups (Sonderegger, 1970)—has raised fears that ESB could possibly become, in the hands of a tyrant, a means by which human behavior could be absolutely controlled. Fortunately, there are two sets of arguments against this fear.

Practical Arguments against the Use of ESB

The first argument is practical. If the population can be controlled well enough so that people will submit to surgical implantation of stimulating electrodes, the tyrant's objective has already been achieved. There is no need for inserting these wires. Furthermore, even if electrodes were somehow placed in the brains of unwilling subjects, who would control the buttons? It is conceivable that a factory worker could automatically be administered a shot of ESB whenever a part was assembled, but there would have to be some process that detected that the part was assembled correctly. If the part were simple and could be easily evaluated by machinery, then it is probably the case that a human was not needed to construct the part—it could have been constructed more cheaply by a machine. If the work is more complicated, then it would probably be evaluated only by an inspector, and this means that one would need people present to adminster the ESB at the appropriate times. But why bother with all this rigamarole? Hitler had no trouble getting a considerable amount of work out of slave laborers, without recourse to

ESB. Fear and pain can be used very effectively to get people to work.

A more fundamental fear might be that our thoughts and attitudes could be controlled by use of ESB. Again, it is difficult to imagine how this could be accomplished. One can only reinforce observable responses, and thoughts are the most private phenomena there are. To control a person's political or social behavior effectively, that person would have to be followed around and given ESB whenever a "correct" behavior occurred. This means that half the population would have to monitor the behavior of the other half. Remember, behaviors reinforced by ESB are subject to rapid extinction; it would not be sufficient merely to monitor and reinforce behaviors periodically. The techniques of social control that are available now seem much more effective than ESB could ever be. If, in a totalitarian society, a person's livelihood, physical amenities, and possibility of advancement depend upon making the "correct" social and political responses, that person's behavior is very likely to come into line.

Rewarding Effects of ESB in Humans

The second argument against the possible use of ESB as an agent for control has been raised by Valenstein (1973). He notes, in a review of the effects of psychosurgery and brain stimulation in humans, that studies have not yet shown ESB to be all that effective in humans. There are problems with this conclusion, since most of the people who have received brain-stimulating electrodes have received them because of some pathology, so the effects of ESB in normal people might be different. Furthermore, the brain of humans has not been mapped the way the rat brain has; perhaps the really effective sites have yet to be discovered. With these reservations in mind, the data argue that whereas ESB can elicit reactions of pleasure and can sustain self-stimulation (button pushing), people can "take it or leave it." For example, Sem-Jacobsen (1968) noted: "In man, curiosity is probably the most dominant causative factor in initiating self-stimulation. If a patient feels 'something,' he might wonder 'Precisely what is the nature of this sensation? What am I feeling? Let me try it once more. Once more! Is it tickling? Is it real pleasure?' " Heath (1964) noted that one of his patients repeatedly pressed the switch because each shot of ESB evoked a vague memory—he kept responding in order to bring the memory more clearly in focus. Another reported: "I have a glowing feeling. I feel good."

Thus, there is not yet any evidence that ESB produces an overwhelming reinforcing effect that could be used to control human behavior to any significant degree. It is interesting that no particular fears have been raised over a manipulation that can, demonstrably, produce a very powerful reinforcing effect—injection of drugs such as heroin. Certainly heroin is a much-feared and much-legislated-against substance. But one does not read articles about the dangers of a dictator first making us become addicted to heroin and then attaching infusion devices to us that could administer small shots of heroin when we perform the "correct" behavior. And yet this is much more practical than ESB ever could be. A shot of heroin is extremely reinforcing, especially if some time has elapsed since the last shot. People could be induced to work very hard toward some goal in the hopes of getting their injection of heroin, whereas no such over-whelming drive accompanies the absence of ESB. (In fact, to the extent that a "drive" for ESB occurs, it decays quickly after the previous stimulation, as Deutsch and Gallistel have shown.) Perhaps the reason for the fear of ESB (and Valenstein quotes a number of popular articles that express this fear) and the lack of fear of drug addiction as a means of social control can be attributed to relative ignorance about the surgical and technical procedures necessary for the administration of ESB, and a general feeling that electrical stim-ulation could be used to "control the brain."

PAIN

Pain is a curious phenomenon. It is more than a mere sensation; it can be defined only by some sort of withdrawal reaction or, in hu-mans, by verbal report. Pain can be modified by opiates, by hypnosis, by the administration of pharmacologically-inert sugar pills, by emotions, and even by other forms of stimulation such as acupunc-ture. Recent research efforts have made remarkable progress in dis-covering the physiological bases of these phenomena.

The importance of emotional and other "psychological" factors in the perception of pain (documented very well by Sternbach, 1968) suggests that there must be neural mechanisms that modify either the transmission of pain or the translation of central pain messages into negative affect. As we shall see, there is excellent evidence for both types of interactions.

We might reasonably ask *why* we experience pain. A person suf-fering from the terminal stages of cancer presents a particularly dis-tressing case. Pain, for this person, serves no useful purpose. Damage has already occurred, and no action can be taken on the part of the patient to avoid further damage. The pain only turns the patient's last days into misery.

In most cases, however, pain serves a more constructive role. The best example of the importance of pain come from cases of peo-ple who have congenital insensitivity to pain. These people suffer an abnormally large number of injuries, such as cuts and burns. One woman eventually died because she did not make the normal shifts in posture that we normally do when our joints start to ache. As a consequence, she suffered damage to the spine that ultimately resulted in death. Other people have died from ruptured appendixes and ensuing peritonitis (infection within the abdomen) that they did not feel (Sternbach, 1968). I am sure that a person who is passing a kidney stone would not find much comfort in the fact that pain does more good than ill, but it is, nevertheless, very important to our existence.

Neural Pathways for Pain

As we saw in Chapter 8, there is general agreement that the free nerve endings generate the messages that are ultimately interpreted as pain. These sensory endings are found in the skin, in the sheath surrounding muscles, in the internal organs, and in the membrane surrounding bones. They are also located in the cornea of the eye, and (as many of you know only too well from a visit to the dentist) in the pulp of the teeth. As we saw in Chapter 8, just about any kind of manipulation that causes tissue damage will cause pain, and thus most investigators believe that pain receptors are chemically stimulated by substances that are liberated by the damaged tissue.

Pain messages arise from two types of peripheral pain fibers, the

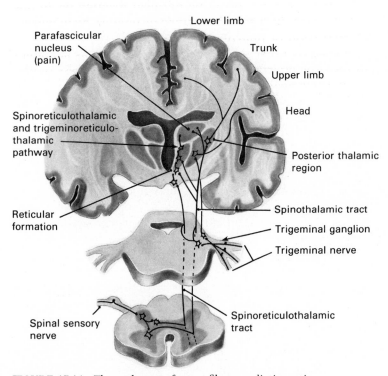

FIGURE 17.16 The pathways of nerve fibers mediating pain.

C fibers (thin, unmyelinated, slow-conducting) and the *A-delta fibers* (thicker, faster-conducting, myelinated). Pain fibers (from the regions below the head) enter the dorsal roots of the spinal cord and synapse in the dorsal horn. The second-order neurons located there send axons to the other side of the spinal cord and ascend via the contralateral *spinothalamic tract.* As this tract ascends, it sends collaterals into the reticular formation. A secondary route to the thalamus thus consists of a polysynaptic pathway: the *spinoreticulothalamic tract.* The thalamus appears to be the "end station" for pain, in that projections to the cortex do not seem necessary for its perception. The pathways are shown in Figure 17.16; note that there is a considerable amount of divergent branching (up and down the spinal cord) by the primary sensory neurons. (See **FIGURE 17.16** on the previous page.)

Pain fibers originating in the trigeminal nerves (i.e., those which serve the face and head) follow a similar set of pathways. There are multisynaptic projections through the reticular formation (the *trigeminoreticulothalamic tract*) and a straight-through pathway that lies alongside the spinothalamic tract (the *anterior trigeminothalamic tract*). (See **FIGURE 17.16.**)

Perception and Tolerance of Pain

There are manipulations that diminish the perception of pain. For example, Beecher (1959) noted that wounded American soldiers back from the battle at Anzio reported that they felt no pain from their wounds—they did not even want medication. It would appear that their perception of pain was diminished by the relief felt from surviving such an ordeal. There are other instances where people still report the perception of pain but are not bothered by it. Some tranquilizers have this effect.

There is clear-cut physiological evidence for such a distinction between the perception and tolerance of pain. Mark, Ervin, and Yakovlev (1962) made stereotaxically placed lesions in the thalamus in an attempt to relieve the pain of patients suffering from the advanced stages of cancer. Damage to the sensory relay nuclei (VPM and VPL) produced a loss of cutaneous senses: touch, temperature, and cutaneous pain (the ability to detect pinpricks). However, patients obtained no relief from deep, chronic pain. Lesions in the parafascicular nucleus and in the intralaminar nucleus were successful; pain, but not cutaneous sensitivity, was gone. Finally, destruction of the dorsomedial and anterior thalamic nuclei left cutaneous sensitivity and the perception of pain intact. However, the patients did not pay much attention to the pain. The lesions appeared to reduce or remove its emotional component. It is noteworthy that these nuclei are intimately involved with the limbic system, and that the dorsomedial nuclei project to prefrontal cortex. Removal of this region

(*prefrontal lobotomy* or *prefrontal leucotomy*) also reduces the emotional aspects of pain perception.

It seems clear that pain perception and pain tolerance are separate phenomena. It would appear that the intralaminar and parafascicular nuclei are the thalamic nuclei necessary for the perception
of pain, and that the limbic system and prefrontal cortex mediate its
emotional component. Mark and his colleagues also noted a very
interesting fact: electrical stimulation of the thalamus never resulted
in reports of pain. Neither did the patients report any specific sense,
such as temperature or touch. Instead, they reported tingling sensations like "pins and needles." This fact suggests that the *pattern*,
temporal and/or spatial, of incoming activity is of paramount importance in somatic sensation. It is not enough to produce excitation
of thalamic neurons—they must be stimulated in a particular way
that has not been duplicated by electrical stimulation. This complex
coding of pain in the thalamus contrasts with the simple and direct
way that noxious sensory information is coded in the spinal cord
and ascending pain pathways. Electrical stimulation of the dorsal
horn (in humans) elicits pain, whereas lesions made at the same site
through the stimulating electrodes cause permanent analgesia for
parts of the body that are served by spinal nerves caudal to that
location (Mayer, Price, and Becker, 1975; Mayer, Price, Becker, and
Young, 1975; Price and Dubner, 1977). (Obviously, these procedures
were carried out in the course of surgical treatment of chronic pain,
and not simply as experimental exercises.)

The study by Mark and his colleagues confirms the long-standing
supposition that there are two types of pain: a rapidly felt "sharp"
pain and a more gradual, but more aversive, "dull" pain. Stub your
toe and you will see what I mean. The first flash of pain subsides
fairly quickly, to be replaced by another that is longer-lived and
more poorly localized. The fact that lesions of VPM and VPL abolished pain felt from pinpricks, but not deep-seated pain, suggests
that the "bright pain" but not the "dull pain" component is mediated by these nuclei.

Treatments That Relieve Pain

As I mentioned earlier, there are several treatments that can produce
analgesia—that is, reduce or eliminate the perception of pain. (*An*
= "not"; *algos* = "pain.") The various methods of producing analgesia appear to work through two different brain mechanisms, one
of which involves a special class of neuromodulators called *endogenous opiates,* or *endorphins.*

THE ACTION OF OPIATES. As you know, opiates such as morphine
produce analgesia. The neural basis of this effect takes place in spe-

cific locations within the brain, particularly in the periaqueductal gray. (As we saw in Chapter 16, electrical stimulation of this region produces a variety of species-typical behaviors, including attack.) Microinjections of morphine into the periaqueductal gray produce analgesia, whereas injections into many other regions are ineffective (Tsou and Jang, 1964; Herz, Albus, Metys, Schubert, and Teschermacher, 1970). It was soon discovered that the opiates produced analgesia by activating specialized opiate receptors on neurons in the brain. Pert, Snowman, and Snyder (1974) homogenized the brains of rats and extracted synaptosomes by means of differential centrifugation (described in Chapter 4). They further removed the terminal buttons from the postsynaptic membrane to which they were attached, and found that the membrane would selectively take up radioactive **naloxone** and **dihydromorphine**. Naloxone is a drug that reverses the effects of opiates. The fact that it, and dihydromorphine (a potent opiate) both bind with molecules in fragments of postsynaptic neural membrane is strong evidence for the existence of opiate receptors. Subsequent studies found that opiate receptors are found on neurons in the periaqueductal gray of the pons, neostriatum, and various parts of the limbic system (see Fields and Basbaum, 1978, for a review). Not all of these neurons are involved in the modulation of pain.

Opiates have been shown to produce analgesia in the following way: The drugs stimulate neurons in the periaqueductal gray. These neurons send axons to the medulla, where they synapse on serotonergic neurons in the **nucleus raphe magnus** (one of the nuclei of the raphe). These neurons send axons down the spinal cord through the **dorsolateral columns.** The terminal buttons of these serotonergic axons inhibit activity of neurons that transmit pain messages up to the brain. (As we shall see later, the inhibition is probably mediated indirectly, through interneurons.) Thus, when morphine activates neurons in the periaqueductal gray, it causes messages to be sent to the spinal cord that block the transmission of impulses that convey pain information to the brain. (See **FIGURE 17.17.**)

Now for the evidence: As we have seen, analgesia is produced by microinjections of morphine into the periaqueductal gray. Neural activity in this region is increased when animals are given a systemic injection of morphine (Criswell and Rogers, 1978; Urca and Nahin, 1978). Other treatments that activate neurons in the periaqueductal gray, such as electrical stimulation (Mayer and Liebeskind, 1974) or the administration of excitatory drugs (Behbehani and Fields, 1979), also produce analgesia. These treatments increase the electrical activity of neurons in the nucleus raphe magnus (Behbehani and Fields, 1979).

The pain-reducing effect of morphine is blocked by lesions of the nucleus raphe magnus (Proudfit and Anderson, 1975), so this structure appears to be an essential way station in morphine-produced analgesia. Electrical stimulation of the nucleus raphe inhibits the

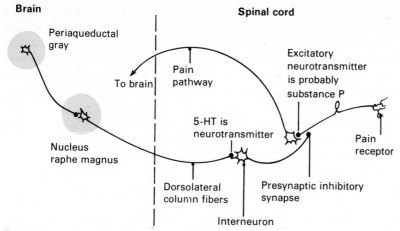

Brain   **Spinal cord**

FIGURE 17.17   A schematic representation of the pathway that mediates analgesia produced by opiates.

response of spinal cord neurons to painful stimuli, and *this* effect is blocked by lesions of the dorsolateral columns. Similarly, morphine analgesia is blocked by the administration of PCPA, which prevents the biosynthesis of serotonin (Tenen, 1968; Akil and Liebeskind, 1975). Lesions of the dorsolateral columns of the spinal cord abolish analgesia produced by morphine (Basbaum, Marley, O'Keefe, and Clanton, 1977). Finally, microinjections of serotonin into the spinal cord block the response of neurons there to painful stimuli (Randic and Yu, 1976).

ENDOGENOUS OPIATES.   Of course, Nature did not put opiate receptors in the brain for the amusement of neuroscientists. If there are receptors in the brain, then the brain must produce its own chemicals to occupy these receptors. And in fact, it does. Terenius and Wahlström (1975) reported the existence of a substance in human cerebrospinal fluid that had a specific affinity for opiate receptors that had been extracted from rat brain. They called this chemical "morphinelike factor." They found that less of this substance was found in the CSF of patients with trigeminal neuralgia, a disease that produces severe facial pain; perhaps the chronic pain was related to lower-than-normal levels of the endogenous opiates.

Hughes, Smith, Kosterlitz, Fothergill, Morgan, and Morris (1975) found that there were actually two morphinelike factors and identified them as two very small peptide chains, each containing five amino acids. They synthesized these substances and found that the artificial **enkephalins** (the authors' name for the substances) acted as potent opiates. The two enkephalins (labeled as *leu*-enkephalin and *met*-enkephalin) were found to bind with opiate receptors even more effectively than morphine. Since these two original reports, other endogenous opiates have been identified. (See Snyder and Childers, 1979, for a review). The most potent of these is called **β-endorphin.**

What is the purpose of the endogenous opiates? As you might expect, they appear to play a role in the reduction of pain. Many of the nonpharmacological treatments that produce analgesia do so by triggering the release of endogenous opiates. (For convenience, I shall use the term *endorphin* to refer to all of the naturally-occurring opiates.) Thus, the brain can administer its own drugs. But what circumstances cause endorphins to be secreted?

The standard procedure that has been used to assess the role of endorphins in analgesia is to perform the treatment that causes analgesia and then administer naloxone, which blocks opiate receptors. If the pain returns, then the analgesia must have been produced by the secretion of endorphin.

Pain can be reduced by stimulating regions other than those that hurt. For example, people often rub or scratch the area near a wound, in an apparent attempt to diminish the severity of the pain. And as you know, acupuncturists insert needles into various parts of the body in order to produce analgesia. The needle is usually then rotated, thus stimulating axons and nerve endings in the vicinity. Often, the region that is stimulated is far removed from the region that becomes less sensitive to pain.

Acupuncture can be effective, as a number of experimental studies have shown (Gaw, Chang, and Shaw, 1975; Mann, Bowsher, Mumford, Lipton, and Miles, 1973). Mayer, Price, Rafii, and Barber (1976) reported that the analgesic effects of acupuncture could be blocked by naloxone. However, when pain was reduced by hypnotic suggestion, naloxone had no effect. Thus, acupuncture, but not hypnosis, appears to cause the release of endorphins.

Pain can successfully be reduced, in some patients, by administering a **placebo,** or pharmacologically-inert substance. (The term "placebo" comes from *placere,* which means "to please." The physician pleases an anxious patient by giving him or her an innocuous substance.) The pain reduction seems to be mediated by endorphins, because it is blocked by naloxone (Levine, Gordon, and Fields, 1979). Somehow, when some patients take a medication that they think will reduce pain, it does so *pharmacologically,* by triggering the release of endorphin. This pharmacological effect is eliminated by the opiate receptor-blocker, naloxone. (For some people, it is incorrect to refer to a placebo as "inert.")

Experiments have shown that analgesia can be produced by the application of painful stimuli, or even by the presence of nonpainful stimuli that have been paired with painful ones. A remarkable observation has even shed some light on a phenomenon that has been of interest to psychologists—*learned helplessness,* Briefly, learned helplessness works like this. An animal is given a number of unavoidable, inescapable footshocks. Later, the animal is placed in a situation in which it can easily avoid or escape the shock. Unlike naive animals, who have *not* experienced unavoidable footshock, the previously-shocked subject learns the avoidance task very slowly (or sometimes, not at all). It is as if the experience with unavoidable pain taught the animal that there was nothing it could do to avoid the shock—struggling is futile. The animal learned to be helpless.

Maier and Jackson (1979) found that at least part of the phenomenon of learned helplessness is mediated by natural mechanisms of analgesia. They subjected groups of rats to (a) escapable or (b) inescapable shock. The next day the rats were given a "priming" footshock, and then their sensitivity to pain was tested. The animals that had received inescapable shock on the previous day were less sensitive to pain than those that had received shock that they could learn to escape. Thus, inescapable shock produces analgesia, but if the pain is escapable, it continues unabated, serving to motivate the animal to perform the escape or avoidance behavior. (That makes good sense, doesn't it?) The analgesia continued even after the rats received injections of naloxone. The results suggest that at least some forms of naturally-occurring analgesia do not require the secretion of endorphins.

In a related study, Fanselow (1979) found that another well-known phenomenon is related to natural analgesia. Normally, animals prefer signaled footshock to nonsignaled footshock. That is, a shock that is preceded by the sound of a buzzer is preferred to a sudden footshock that comes right out of the blue. In this case, the preference for signaled footshock is eliminated by naloxone injections; the rats do not care whether the shocks come with a warning. Presumably, the warning stimulus causes endorphin to be secreted, which causes a certain amount of analgesia. Thus, signaled shock is, in fact, less painful than nonsignaled shock. This physiological explanation is a lot simpler than the ones that had previously been advanced to account for this phenomenon.

THE BIOLOGICAL SIGNIFICANCE OF ENDOGENOUS MECHANISMS OF PAIN REDUCTION.   As we just saw, inescapable pain causes the release of endorphin. There are other analgesic mechanisms, as well. In the thick of battle, soldiers often suffer serious wounds, but do not discover them until later. The same is true for animals (including humans) who fight, or who are being attacked. Even sexual arousal, or, particularly, actual participation in sexual activity, causes analgesia. And although childbirth is not painless, most women report that the pain is not so unpleasant as one might expect. We now know something about the physiological mechanisms that produce these instances of analgesia.

Analgesia produced by brain stimulation.   It has been known for some years that electrical stimulation of the brain can attentuate *reduce in force* pain. In fact, Reynolds (1969) was able to use electrical brain stimulation as an anesthetic for surgery in rats. Since this study, analgesia has been produced at a wide variety of electrode locations in a wide variety of animals, including humans. It seems likely that stimulation at some loci produces true analgesia and increased pain tolerance at others. Mayer and Liebeskind (1974) reported that electrical stimulation of the periaqueductal gray of rats resulted in analgesia that was equivalent to that produced by at least 10 milligrams of morphine per kilogram of body weight (a large dose). The rats did not react to pain of any kind; the authors pinched the tails and paws, applied electric shock to the feet, and applied heat to the tail. Stimulation of the septum, dorsomedial thalamus, and ventral tegmentum produced a less striking analgesia. Stimulation of the lateral hypothalamus or ventrobasal thalamus (the somatosensory area) had no effect.

The investigators found no causal relationship between rewarding and analgesia-producing brain stimulation. Rats failed to press a lever for analgesic stimulation at some locations, and some rewarding stimulation was found not to produce analgesia. Furthermore, they noted that analgesia and self-stimulation elicited by stimulation of the central gray are differently affected by pharmacological means; depletion of serotonin diminishes analgesia without affecting self-stimulation (Margules, 1969; Akil and Mayer, 1972).

Analgesic brain stimulation apparently triggers the neural mechanisms that mediate analgesia that is normally produced through more natural means. For example, electrical stimulation of the periaqueductal gray (the best site for analgesic brain stimulation) elicits various species-typical behaviors, including attack or sexual activity. When an animal normally engages in these behaviors, a certain degree of analgesia is produced. Komisaruk (1974) demonstrated a nice relationship between genital stimulation and analgesia. He gently probed the cervix of female rats with a glass rod, and observed a diminished sensitivity to pain, along with increased activity of neurons in the periventricular gray of the hypothalamus and periaqueductal gray of the midbrain. Presumably, copulation (and/or the birth process) stimulates the cervical region, activates neural mechanisms in the central gray, and produces analgesia.

Analgesia that is produced by electrical stimulation of the brain is partly, but not completely, mediated by endorphins. Akil, Mayer, and Liebeskind (1976) found that naloxone reduced the analgesic effect of stimulation of the periaqueductal gray, but did not completely eliminate it. Presumably, there are at least two brain mechanisms for analgesia: one that involves the release of endorphins and one that does not. In contrast, when analgesia is produced by stimulation of the nucleus raphe magnus, the analgesia is completely eliminated by naloxone. But you will recall that neurons of the nucleus raphe magnus send axons down the spinal cord that terminate in the dorsal horn, and that these neurons are *serotonergic*. It is easy to understand why PCPA and other serotonin antagonists block the

yet endorphins may be involved in reward mechanisms (p 578)

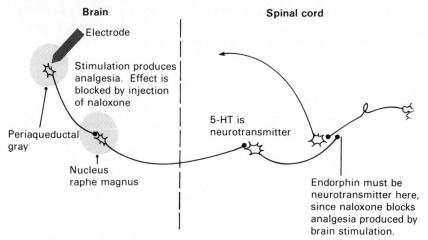

Brain
Electrode

Stimulation produces analgesia. Effect is blocked by injection of naloxone

Periaqueductal gray

Nucleus raphe magnus

Spinal cord

5-HT is neurotransmitter

Endorphin must be neurotransmitter here, since naloxone blocks analgesia produced by brain stimulation.

FIGURE 17.18   A schematic representation of the evidence for analgesia being mediated presynaptically by endorphinergic interneurons in the spinal cord.

analgesia that occurs when these neurons are stimulated. But why should naloxone do so, too?

The likeliest explanation is that the serotonergic terminal buttons synapse on interneurons in the dorsal horn, which, in turn, synapse on the neurons that transmit pain information to the brain. Apparently, these interneurons use endorphin as a neurotransmitter. Naloxone thus blocks the postsynaptic opiate receptors on the neurons in the dorsal horn. (See **FIGURE 17.18.**)

This hypothesis is not yet proved, but it has some additional support. Opiate receptors have been found in the dorsal horn (Atweh and Kuhar, 1977), and endorphins are produced there (Hökfelt, Ljungdahl, Terenius, Elde, and Nilson, 1977). If the dorsal roots are severed, fewer opiate receptors are found in the dorsal gray matter (Lamotte, Pert, and Snyder, 1976). This finding suggests that the endorphin-secreting terminals synapse presynaptically on afferent pain fibers, which degenerate after the dorsal roots are cut. In addition, analgesia can be produced by microinjections of enkephalin and morphine directly into the spinal cord (Duggan, 1979), and spinal injections of naloxone can block the analgesic effect of systemically administered morphine (Yaksh, 1979). The details have yet to be worked out, but it appears that endorphins are specific neurotransmitters as well as analgesia-producing neuromodulators.

Other Effects of Endogenous Opiates

Endorphins are found in several regions of the brain that do not appear to play a primary role in the perception of pain. Undoubtedly, they take part in many other functions. The investigation of the endogenous opiates is still very new, but already some hints of their additional roles are emerging. Endorphins may play a role in thermoregulation, since local injection into the hypothalamus produces changes in body temperature (Martin and Bacino, 1978), and the administration of naloxone disrupts thermoregulatory responses (Holaday, Wei, Loh, and Li, 1978). <u>The endorphins may even be involved in reward mechanisms</u>. This should not come as a complete surprise, since many people have found injections of opiates like morphine or heroin to be extremely pleasurable. Belluzzi and Stein (1977) found that rats will perform a response in order to receive injections of met- or leu-enkephalin into the lateral ventricles. The authors note that many of the brain regions that have been implicated in self-stimulation also contain endorphins (for example, the periaqueductal gray, SNC, and nucleus accumbens). In fact, they found that rats would cease responding for electrical stimulation of the periaqueductal gray when they were given injections of naloxone. As we saw in an earlier part of this chapter, dopamine appears to play a central role in the mediation of reward. It appears quite possible that endorphins do so, too.

It is interesting that the discussion has turned full circle, from mechanisms of reinforcement to mechanisms of pain to mechanisms that modulate pain and also produce reinforcement in their own right. The past few years have seen tremendous growth in our understanding of the physiology of reinforcement and pain, and undoubtedly, the next few years will see even more.

KEY TERMS

A-delta fiber p. 570

analgesia *(an ul JEEZ ee uh)* p. 571

anterior trigeminothalamic tract p. 570

appetitive stimulus p. 545

C fiber p. 570

central tegmental tract p. 556

dihydromorphine p. 572

dorsal tegmental bundle p. 556

dorsolateral columns p. 572

drive p. 538

drive reduction hypothesis p. 538

endogenous opiate p. 571

endorphin p. 571

β-endorphin p. 574

enkephalin p. 573

locus coeruleus *(sur OO lee us)* p. 556

mesolimbic system *(MEE zo)* p. 556

naloxone *(na LOX ohn)* p. 572

need-reduction hypothesis p. 538

neostriatum p. 556

nigrostriatal system *(NY grow stry AY tul)* p. 556

nucleus raphe magnus *(ruh FAY)* p. 572

pars compacta of the substantia nigra (SNC) *(NY gra)* p. 556

placebo *(pla SEE bo)* p. 574

prefrontal leucotomy p. 571

prefrontal lobotomy p. 571

priming p. 546

punishment p. 537

reinforcement p. 537

spinoreticulothalamic tract p. 570

spinothalamic tract p. 570

subcoeruleus cell group *(sub suh ROO lee us)* p. 556

sulcal cortex *(SUL kul)* p. 561

trigeminoreticulothalamic tract p. 570

ventral tegmental area p. 557

SUGGESTED READINGS

OLDS, J. *Drives and Reinforcements: Behavioral Studies of Hypothalamic Functions.* New York: Raven Press, 1977.

WAUQUIER, A., AND ROLLS, E. T. (eds.) *Brain Stimulation Reward.* Amsterdam: North Holland, 1976.

Recent volumes of *Annual Review of Pharmacology and Toxicology, Annual Review of Psychology*, and *Annual Review of Neuroscience* have contained chapters covering the physiology of reinforcement, pain, and endogenous opiates.

18

Physiology of Learning and Memory

The ability to learn and remember is one of the brain's most astonishing capabilities. It is not possible to measure the number of pieces of information that can be retained by the human brain, but it certainly must be in the millions. Think of how many items you can remember: words of songs and poems, events, places, faces. You can remember events from early childhood. You can remember the last meal you ate. And you can recall most memories rapidly and accurately; you know where to find them, among the millions of others. If you are trying to remember a friend's telephone number you do not come up with a recollection of a childhood birthday party. Your millions of memories are nicely sorted and classified, so that when you look for a member of a particular category, you retrieve only memories that are related. The human memory system is an incredibly complex and efficient system of information storage, rivaled only by the chromosomes (which, after all, contain the information that is needed to construct an organism, including its brain).

This chapter and the next one summarize what is known about the physiology of learning and memory. The emphasis is on human learning and memory, which means that a lot of excellent work is not cited here. For example, many investigators are studying simple forms of learning in marine molluscs, which have relatively simple nervous systems. These experiments will almost certainly provide us with basic information about the physical basis of memory that will help in the study of mammalian brains. But since the potential for direct applications of these findings to mammalian brains has yet to be realized, I will not describe the research on invertebrate memory here.

There are three essential steps in learning and memory: perception, storage, and retrieval. Let us look at a simple example. Suppose that you pass by the window of a department store and happen to notice an object that you have never seen before—a child's toy golf club made out of blue plastic. A day later you see a child playing with a blue plastic golf club, and you recognize that it is the same as the one that you saw the day before. Something like the following must have happened. Sensory information was transmitted to your brain, and a neural representation of a blue object of a particular size and shape was established. Since you were already familiar with the shape of a golf club, you perceived the object for what it was. Your brain did not merely register an abstract shape. A particular three-dimensional object was perceived. Therefore, your perception was based upon prior knowledge—it was based upon memories that were already stored in your brain. So, with the possible exception of the most limited kinds of abstract, unrecognizable stimuli, the first step in memory formation depends upon memories that already exist.

The second step in the process of learning and memory is that of storage. Somehow, your perception of the toy golf club must have produced some changes in your brain, because when you saw a child playing with a plastic golf club the next day (that is, when you again perceived a blue plastic golf club), you realized that it was just like the one you had seen before. The first perception left something behind, so that the second perception was that of a familiar object, not a new one. We can say that the second perception caused the retrieval of a memory.

Of course, memories can be retrieved by other means. Suppose that you stayed overnight with some relatives who had small children. During the night you left your room to go to the bathroom, and, in the dark, you tripped over an object on the floor. You reached down to feel what it was. Turning it over in your hands, you identified it as a plastic golf club, just like the one you saw the previous day. (Of course, since it was dark, you were not able to tell whether it was of the same color.) Consider what this anecdote means. You can perceive a stimulus through one sense modality, store the information in memory, and retrieve it when you perceive another stimulus through an entirely different sense modality. The operation is so automatic that we take it for granted. But it is very difficult to conceive of the way that a particular pattern of neural activity in one sensory system can be recognized as being equivalent to a different pattern of neural activity in a different sensory system. As we shall see in Chapter 19, the phenomenon of cross-modal memory retrieval ("modal" referring to sensory modality, like vision, audition, touch, etc.), and its failure in certain instances of brain damage, allow us to make some conclusions about the location of various kinds of memories in the human brain.

This chapter is devoted to a discussion of physiology of the memory process: Why do we remember some, but not all memories? Why are brand new memories more distinct and easier to recall than older memories? What is the physiological basis for memory? Is there physiological evidence for the linkage of memories that are related to each other? The next chapter is concerned with the anatomy of the memory process: Where are new and old memories stored? Are all memories located in one place in the brain, are specific kinds of memories located in separate regions, or are all kinds of memories scattered throughout the brain? Do some parts of the brain play special roles during perception, storage, or retrieval processes?

STAGES OF MEMORY

Most learning theorists believe that there are two stages of memory: **short-term memory** and **long-term memory.** (A sensory stage, lasting for a fraction of a second, precedes short-term memory, but this complication is not important for the present discussion.) Short-term memory can be loosely defined as our present awareness of what has just happened. William James put it this way: an object in short-term memory "never was lost; its date was never cut off in consciousness from that of the immediately present moment. In fact it comes to us as belonging to the rearward portion of the present space of time, and not to the genuine past" (James, 1890). In contrast, long-term memories are those that were acquired in the past and dropped from consciousness for some time, "the knowledge of an event, or fact, of which meantime we have not been thinking, with the additional consciousness that we have thought or experienced it before" (James, 1890).

Let me give a concrete example. Suppose that you look up a number in the telephone directory. You repeat the number to yourself and dial it. The number does not ring, so you dial it again;

obviously, you still have the number in short-term memory. This time you hear the number ringing and you relax a little, because you do not need to keep rehearsing the number. Then, you hear a voice saying, "The number you have reached is not in service. What number did you dial?" And you don't remember. Once you relaxed and stopped rehearsing, the number left your short-term memory.

We can, of course, memorize telephone numbers so that they will remain in memory without having to be constantly rehearsed. The process of rehearsal, carried out for a long enough period of time, seems to produce a stable, long-term memory for the information. We can do our rehearsing all in one bout, or we can learn a telephone number by looking it up on repeated occasions. The more time a given piece of information spends in short-term memory, the more likely it becomes that it can be retrieved later from a long-term memory. The transition from short-term memory into long-term memory has the appearance of a gradual process.

Effects of Head Injury in Humans

A severe enough blow to a person's head produces disturbances in memory. These disturbances are most likely to occur when there is diffuse, rather than localized, head injury, such as that produced by a blow to the head with a blunt object (Russell and Nathan, 1946). The following example (which I have adapted from a talk delivered by Hans-Lukas Teuber to a meeting of the New England Psychological Society in 1969) shows how temporary disruption of normal brain functions produces effects that can best be understood in terms of a two-process model of memory.

Bill and John are walking down the street, engaged in conversation. They pass the drugstore, the hardware store, and the dress shop on the corner, and start to cross the street. Suddenly John looks up and shouts, "Look out!" jumping back to the curb as he does so. Bill reacts too late and is struck by an oncoming car. He is thrown several feet, and his head strikes the pavement.

When he awakens in the hospital, Bill has enough presence of mind to avoid the obvious "where am I" question. Instead, he says to a nurse entering the room, "What happened to me? My head hurts."

"You were hit by a car."

"I was? Am I badly hurt?"

"No, we don't think so. Some X-rays have been taken to be sure."

"Will you see that my wife is called? I don't want her to worry. We were supposed to meet for lunch." And he gives the nurse the telephone number she should call. The nurse leaves the room to place the call.

Several minutes later she returns. Before she can say anything, Bill says, "What happened to me?"

"Why, you were hit by a car."

"Will you call my wife for me? I was supposed to meet her for lunch, and I don't want her to worry."

Obviously, Bill has a memory problem.

There are still other symptoms. Some days later, after recovering from his injury, Bill discusses his accident with John. He says, "For the life of me, I can't remember what happened. The last thing I can recall was passing the hardware store."

"But don't you remember watching that deliveryman rolling the rack of clothes into the dress shop? You were looking at them and said you'd have to remember to mention them to your wife. As a matter of fact, I think that's why you didn't notice the car—you were looking back at the dress shop."

"I don't remember that at all."

So Bill has more than one memory problem. Let us review the evidence. (1) He has forgotten events immediately before the time of his head injury, but he remembers more remote events. He has ***retrograde amnesia***—an amnesia back in time, prior to this head injury. He remembers passing the drugstore and the hardware store, but he does not remember anything after that. Conveniently, we have a witness who can attest to the fact that there were events that Bill would have been expected to remember later. (2) For a time after regaining consciousness, Bill was able to converse and could remember past events (his telephone number, the fact that he was supposed to meet his wife). However, he did not later remember things that occurred during this period; he repeated his conversations with the nurse. He had ***anterograde,*** or ***posttraumatic, amnesia***—a temporary

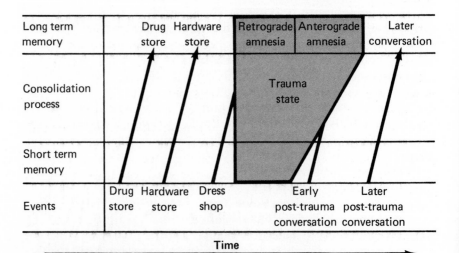

FIGURE 18.1 A schematic representation of retrograde and posttraumatic amnesia.

inability, after the accident, to convert short-term memories into long-term memories. (See **FIGURE 18.1.**)

IMPLICATIONS FOR THE MEMORY PROCESS.   The best way to explain these symptoms is to hypothesize that (1) immediate memory is retained in short-term memory, which is different from long-term memory; (2) short-term memory can be converted into long-term memory, but the process takes time; and (3) events that are stored in short-term memory can be disrupted by diffuse brain damage, but events that are stored in long-term memory are much more durable. The reasoning for these statements is as follows: While walking down the street, Bill had experiences that entered short-term memory. Some of these temporarily retained memories were stored in long-term memory (remember, we do not remember *everything* that we perceive), but the transition takes time. The head injury interrupted this transition (from short-term memory to long-term memory) in one of two ways: either (a) events represented in short-term memory were lost (i.e., the "slate was cleared") or (b) short-term memory decayed normally, but the brain mechanism that transfers the memories into long-term memory was temporarily disrupted by the trauma. Let us refer to the transition from short-term memory to long-term memory as the process of _consolidation_. Some evidence that provides a bit of support for alternative (b) comes from the fact that Bill forgot his first conversation with the nurse. Clearly his short-term memory was working, or he could not carry on a conversation. In order to respond to something said by someone else, it is necessary to retain information about what was said long enough to phrase an answer. If we hypothesize that some brain mechanism must function normally in order for consolidation to occur, and that this mechanism is disrupted by the injury for a period of time, we can account for the forgetting of the conversation with the nurse. The short-term memory gradually decays, but no transfer to long-term memory occurs.

I must emphasize that the evidence concerning the consolidation process is not conclusive. During the state of posttraumatic amnesia, patients are usually confused to varying degrees, and one could just as well attribute the lack of consolidation to a more rapid decay of memories in short-term memory; perhaps they do not stay around long enough for consolidation to take place. However, whether consolidation requires operation of a special mechanism, or whether it occurs automatically when memories remain long enough in short-term storage, we can still conclude that short-term memory and long-term memory are separate processes.

Experimental Amnesia

The evidence obtained from the effects of head injury in humans tells us something about the memory process, but further study with animals obviously requires a better method for producing amnesia. One *could*, I suppose, train some rats and then hit them on the head, but fortunately there are better techniques available.

ELECTROCONVULSIVE SHOCK: HISTORY. For many years, attempts have been made to treat schizophrenia and other mental disorders with various sorts of shock treatments, such as dunking the patient in cold water, exposing him to snakes, producing a fever—doing something to shake the patient up and initiate some change, hopefully for the better (Valenstein, 1973). Particularly popular types of shock treatment were the induction of comas by injections of insulin or of seizures by injections of *metrazol* (the active ingredient in camphor). The rationale for the therapeutic application derived partly from the fact that schizophrenia and epilepsy appeared to occur infrequently in the same person, and from the observation that a seizure appeared to produce a remission from the psychotic symptoms (von Meduna, 1938).

The production of electrically elicited seizures was first performed by Ugo Cerletti, an Italian psychiatrist (Cerletti and Bini, 1938). He noted that pigs in the local slaughterhouse were first made unconscious by electric shock across the temples and were then killed with a knife. He tried the same treatment (the electric shock, that is, not the stabbing) on experimental dogs and observed that seizures could be produced by application of current to the head for a few tenths of a second. The dogs did not appear to suffer any long-term ill effects.

Cerletti then went on to try the procedure on a schizophrenic patient, who apparently experienced hallucinations, and whose speech was full of meaningless babbling. He applied a low-current shock to the head, which was insufficient to produce unconsciousness. When the patient heard Cerletti say that he would try it again tomorrow with higher current, the patient said, "Not another one! It's deadly!" Encouraged by this sudden display of rational speech, Cerleti immediately tried a more intense shock. Here are his observations:

> We observed the same instantaneous, brief, generalized spasm, and soon after, the onset of the classic epileptic convulsion. We were all breathless during the tonic phase of the attack, and really overwhelmed during the apnea as we watched the cadaverous cyanosis of the patient's face; the apnea of the spontaneous epileptic convulsion is always impressive, but at that moment it seemed to all of us painfully endless. Finally, with the first stertorous breathing and the first clonic spasm, the blood flowed better not only in the patient's vessels but also in our own. Thereupon we observed with the most intensely gratifying sensation the characteristic gradual awakening of the patient "by steps." He rose to sitting position and looked at us, calm and smil-

ing, as though to inquire what we wanted of him. We asked: "What happened to you?" He answered: "I don't know. Maybe I was asleep." Thus occurred the first electrically produced convulsion in man, which I at once named "electroshock." (Cerletti, 1956)

Electroconvulsive shock, or *ECS,* soon caught on and largely supplanted the use of insulin or metrazol. Some patients suffered broken limbs or backs as a result of the convulsions, but physicians soon learned to use paralyzing drugs in order to prevent these injuries. The interesting thing about ECS is that it was soon discovered that patients forgot events that occurred shortly before production of the seizure. That is, they showed a retrograde amnesia. Apparently, the ECS did something that disrupted the process of consolidation.

A number of studies have documented the amnesic effect of ECS and have demonstrated a *temporal gradient* of retrograde amnesia. The shorter the learning–ECS interval, the greater the likelihood that the material will be forgotten. ECS also produces posttraumatic amnesia; patients later do not remember events that occurred during a period of time after the seizure (Cronholm, 1969). The effects of ECS, then, are quite similar to those of blunt head injury. I should note that the use of ECS is usually restricted to cases of severe depression. Most clinicians agree that ECS has very little therapeutic value for other disorders. Moreover, repeated ECS treatments cause a general deterioration in the ability to recall *all* memories, recent or old (Squire, 1974). For these reasons, ECS should be used sparingly and judiciously. (I shall say more about this treatment in Chapter 20.)

ELECTROCONVULSIVE SHOCK: USE IN RESEARCH ON THE MEMORY PROCESS.   As you might expect, ECS has been used by many investigators to study the consolidation process in animals. The procedure that is most commonly used today uses a *passive avoidance task* that was devised by Chorover and Schiller (1965). These investigators used a special task that could be learned in a single trial. The apparatus consisted of a chamber with a grid floor, in the middle of which was a small wooden platform. (See **FIGURE 18.2.**) The rat was placed on the platform and was given a brief foot shock shortly after it stepped off the platform onto the floor (rats do not have to be trained to step off the platform—they do so normally). The animal was removed from the chamber after receiving the foot shock, and was put back into its home cage. The next day the rat was again placed on the platform. This time the animal stayed put for a relatively long time; it showed evidence of remembering the fact that it previously received a painful shock. Animals that had *not* been previously shocked stepped off the platform almost immediately.

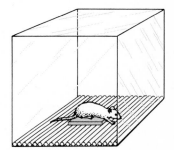

FIGURE 18.2   The step-down apparatus for testing retention of passive avoidance.

587

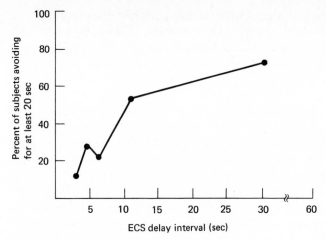

FIGURE 18.3   Data obtained from an experiment using an apparatus similar to that pictured in Figure 18.2. (From Chorover, S. L., and Schiller, P. H., *Journal of Comparative and Physiological Psychology,* 1965, *59,* 73–78.)

Figure 18.3 shows the data from the experiment by Chorover and Schiller. Animals that received ECS within a few seconds of getting shocked through the feet stepped off the platform quickly the next day; they did not appear to remember the shock. (The ECS was delivered through metal snaps that had been fastened to the rats' ears before the start of the experiment.) As the foot shock–ECS interval lengthened, more and more subjects remained on the platform when tested the next day. (See **FIGURE 18.3.**)

DOES ECS IMPAIR CONSOLIDATION OR CATALOGING?   Although thousands of rats and mice have received ECS treatment, not all investigators agree that the observed impairment in subsequent performance is a result of disrupted consolidation. We can say that ECS produces amnesia, but we do not know whether information was never stored (a failure in consolidation) or whether it was stored in such a way that it cannot be retrieved. In fact, it is logically impossible to prove that a particular memory never got from short-term memory to long-term memory. If something is not found after a search, we cannot know for sure that it does not exist; perhaps we were looking in the wrong places.

Some investigators (e.g., Miller and Springer, 1973) believe that short-term memories are consolidated in a fraction of a second and that some sort of cataloging function, which is essential for subsequent retrieval of the information, is performed subsequent to this consolidation. The cataloging process, and not long-term memory storage, is disrupted by ECS. The authors do not explicitly define what is meant by cataloging, but they present an analogy that explains what they mean. Consolidation is similar to the placement of a new book on the shelf of a library. It is there, but unless you know of its existence and how to find it, it will not be available. The cataloging process of memory presumably makes the consolidated information somehow *locatable*.

Let me give a more concrete example. Short-term memory is shaped and modified by existing long-term memories, which means that the ongoing neural activity represents retrieved information (from long-term memory) as well as information that was just perceived. (For this reason some psychologists prefer the term "working memory" to short-term memory.) Suppose that you are browsing through a store, and see some attractive and interesting house plants on sale. You remember that a friend of yours loves house plants, and would probably appreciate having one of these. You decide to buy one of the plants for her. Right at the time, your immediate, working memory is filled with images of the plant, of your friend, and of yourself giving the plant to her. You purchase the plant and bring it home. Later, you meet your friend on the street and suddenly think of the plant. The sight of your friend serves as a retrieval cue for your memory of it.

When you bought the plant, your short-term memory became consolidated into long-term memory, and, in the process, formed some links with existing long-term memories that concerned your friend. Thus, when you recognized your friend (that is, activated the neural circuits that encode long-term information about her) the recently-formed linkage between these circuits and your long-term memories of the plant caused the memory of the plant to be retrieved.

It is possible, as Miller and Springer suggest, that short-term memories consolidate very quickly, but that the cataloging process takes more time. Thus, ECS prevents retrieval by interfering with the process by which newly formed long-term memories are linked with old ones. However, although we still do not have enough evidence to decide the issue, I think that it is not necessary to conceive of consolidation and cataloging as separate processes. After all, short-term memory is not a simple representation of a stimulus; the perception of an object or event is guided by long-term memories that already exist. Thus, our immediate memory contains new *and* old information. During the consolidation process, all of this information is stored. The relevance of the new information to existing long-term memories determines how and where the information is stored—new memories build on old ones. Thus, it is probably an artificial distinction to speak of consolidation *versus* cataloging.

There is some good evidence that consolidation does indeed entail cataloging with respect to old memories, and that ECS can disrupt this process. Robbins and Meyer (1970) trained several groups of rats in a series of three different discrimination tasks. The tasks

Training	Test	Retention?
S_1 F_2 S_3 + ECS	S_1	NO
	F_2	YES
S_1 S_2 F_3 + ECS	S_1	YES
	S_2	YES
F_1 S_2 S_3 + ECS	F_1	YES
	S_2	NO
F_1 S_2 F_3 + ECS	F_1	NO
	S_2	YES
F_1 F_2 S_3 + ECS	F_1	YES
	F_2	YES
S_1 F_2 F_3 + ECS	S_1	YES
	F_2	NO

FIGURE 18.4 A summary of the data from the experiment by Robbins and Meyer (1970).

could be appetitively or aversively motivated; the animals could be taught to choose the proper stimulus in order to get a food pellet or to avoid mild foot shock. The animals learned each of the three discriminations, but for different reinforcers. For example, some rats first learned a shock-motivated task (S_1), then a second, food-motivated, task (F_2), and finally a third, shock-motivated, task (S_3). The sequence for these rats would be $S_1F_2S_3$. There were six groups of animals learning the three tasks under different orders of motivational conditions.

After learning the third task, the rats received an ECS treatment. They were then tested on problems 1 and 2. The interesting result was that an impairment was seen in the animals' performance on the tasks learned earlier. One would not expect the ECS to affect performance on a task learned several days ago. Furthermore, the impairment was selective; if task 3 (the one followed by ECS treatment) had been food-motivated, performance on the previous food-motivated task (but *not* the shock-motivated task) was impaired. Similarly, if task 3 had been shock-motivated, performance on the previous shock-motivated task was impaired. (See **FIGURE 18.4.**)

ECS treatment was found by Robbins and Meyer to affect not just the most recent memories, but also memories for previously learned habits that were acquired under similar motivational conditions. The investigators suggested that what was disrupted was not the long-term memories themselves, but their accessibility. Perhaps some sort of cataloging mechanism was active at the time, and its disruption somehow altered the "entries" relating to tasks learned under a particular motivational condition. Consider the $S_1F_2S_3$ condition. When the rats learned the last task (S_3), information related to an earlier shock-motivated task (S_1) was probably activated. That is, the animals' immediate, working memory contained new information about the present task and old information about the earlier one. All of this information began to be consolidated. The process was disrupted by ECS before it was complete. Perhaps nonsense information, in the form of meaningless electrical activity produced by the seizure, was linked to the old information about S_1 that was being activated at the time. This would explain why the rats were unable to retrieve the appropriate information when they were tested on S_1 again. The information was probably still there, but it was modified in such a way that it could not be retrieved.

The "library" analogy of Miller and Springer might make this point clearer. Perhaps the entries for books of a particular category were being altered in the card catalog because a new book had been added to the collection. If we disrupt the person who is doing the cataloging, we lose the entry not only to the new book, but also to the other books in that category. In Chapter 19 I shall describe a human amnesia syndrome that appears to involve deficits in the same sort of cataloging process.

THE PHYSICAL CHARACTERISTICS OF SHORT-TERM MEMORY

The Reverberatory Hypothesis

As we have seen, short-term memory contains information about events that were just perceived. Since this phase of memory is an active, ongoing process, and since it can be prolonged indefinitely by rehearsing or "thinking about" the information, it is probably safe to say that short-term memory requires neural activity. The particular pattern of activity represents the particular information. In contrast, long-term memories do not appear to be maintained by continuous neural activity. Instead, some stable changes must occur whereby information can be stored away until such time as it is activated. If the neural activity that represents information in short-term memory persists long enough, it causes long-term physical changes in neurons—the ones that are representing the short-term memory, or perhaps other neurons that receive inputs from them. These long-term changes alter the "circuitry" of the brain and hence change the way it responds to a subsequent presentation of that stimulus, or to related stimuli.

Short-term memories appear to be disrupted by blunt head injury or ECS. This fact suggests that short-term memory involves coherent neural activity that can be disrupted by treatments that temporarily suppress neural activity or that induce incoherent, meaningless firing of neurons (as ECS would do). Head injury and ECS are not the only treatments that disrupt short-term memory; anesthesia, cooling of the brain, anoxia, and treatment with various drugs will also do so (Jarvik, 1972). However, long-term memories are not susceptible to damage from these treatments unless, as we have seen, they are in the process of being updated with new information.

What kind of neural activity can represent short-term memories? The mechanism that has most often been proposed is ***reverberation.*** If activity is initiated in a complex, interconnected network of neurons, it is very likely that loops of interconnected neurons will recurrently circulate bursts of excitation. Figure 18.5 presents that concept more concretely. *Theoretically,* the initiation of an action potential in neuron A could cause incessant firing of the circuit; A excites B, which fires and then excites C . . . and so on, back to A, where the process begins again. (See **FIGURE 18.5.**)

Of course, we know that a circuit of four neurons would not continue to fire in this manner. It takes more than one EPSP from a single terminal button to trigger an action potential in a neuron. But consider Figure 18.6, which is merely a more redundant version

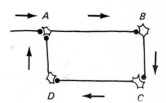

FIGURE 18.5 A schematic representation of a reverberatory circuit.

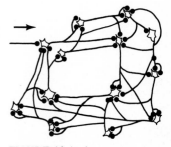

FIGURE 18.6 A representation of a reverberatory circuit, slightly more realistic than that shown in Figure 18.5.

of the simpler circuit. Neural activity might continue in this circuit for a much longer time. (See **FIGURE 18.6** on the previous page.)

DO REVERBERATORY CIRCUITS EXIST?   First of all, is there any evidence that reverberation does, in fact, take place? Burns (1958) studied the properties of the ***isolated cortical slab*** and obtained data that argue very strongly for the existence of reverberatory activity. An isolated cortical slab is produced by undercutting a section of cortex so that it is not connected neurally to any other region of the brain. Care is taken to preserve the blood supply that runs along the top of the cortex. (See **FIGURE 18.7.**) The tissue can be recorded from after it has recovered from the immediate effects of surgery. Neurons in these slabs are normally silent, but if the tissue is stimulated with a train of electrical pulses, bursts of activity can be recorded. If the stimulus is intense enough, firing will continue for up to 30 minutes.

This prolonged neural activity suggests that reverberation can indeed occur in networks of neural tissue. Burns noted that if the activity is a result of recirculation of excitation in loops of neurons, then one should be able to halt this activity by stimulating the neurons to discharge all at the same time. That way, the cells would simultaneously be in the refractory period, during which an action potential cannot be elicited. The neurons would all quickly recover, but none of them would be firing. (Picture what would happen if all four neurons in Figure 18.5 fired simultaneously. Reverberation could not continue.) (See **FIGURE 18.5.**) Burns' prediction was borne out; a single shock of sufficient intensity, delivered to the center of the slab, halted the neural activity of the slab.

CAN REVERBERATORY CIRCUITS ENCODE INFORMATION?   It would appear from the experiment by Burns that reverberation *can*

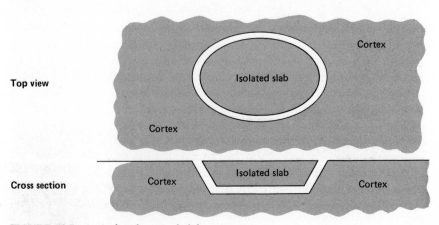

FIGURE 18.7   An isolated cortical slab.

occur in the brain, even in small isolated regions. But can a reverberatory circuit represent information? First, we should make sure that information, coded by the sensory system, can (potentially, at least) produce reverberatory activity unique to the perceived stimulus. In other words, can the hypothesized mechanism represent, in different and distinct ways, the great variety of stimuli that an organism can perceive? Secondly, is there any experimental evidence that recurrent activity in fact bears any relationship to perceived stimuli?

There are some data, such as that obtained by Verzeano and his colleagues, that suggest that specific reverberatory activity can occur in response to sensory stimulation (Verzeano and Negishi, 1960; Verzeano, Laufer, Spear, and McDonald, 1970). These investigators believe that short-term memories are represented by circulating activity between cortex and thalamus. They have recorded unit activity from closely spaced electrodes (30–200 μm apart) arranged in a row, in order to determine the firing patterns of neurons in relation to their neighbors. There appeared to be consistent patterns of discharge; it was possible to follow recurrent waves of excitation from one location to the next, suggesting some sort of recirculating activity. Moreover, the observed patterns appeared to vary as a function of the stimulus that was being presented. There is, then, some evidence that reverberatory activity can indeed take place in the brain, and, furthermore, that this activity is related to the stimuli being perceived by the animal. The evidence is, of course, not conclusive; we are far from being able to specify the way in which reverberatory activity encodes short-term memory.

So far we have seen evidence that reverberatory circuits exist and that they appear to be able to encode information. The next question is how neural stimulation can produce changes in the structure of neurons. It is time to turn our attention to the physiological basis of long-term memory.

THE PHYSICAL CHARACTERISTICS OF LONG-TERM MEMORY

Very few neuroscientists question the assumption that long-term memory storage entails some physical changes in neurons. The most elemental unit of long-term memory storage would seem to be the following. After participating in the storage of a memory, a neuron will respond differently to a particular input. Thus, when changes have taken place in a circuit of interconnected neurons, a cue for retrieval (such as the perception of the originally-learned stimulus) will evoke a specific pattern of activity in these neurons, which constitutes recognition. In many instances, this activity will elicit a particular response. And where should we look for these changes? The most obvious location would be the synapse. Neurons communicate with each other through synapses, so structural changes in synapses will affect the ways in which neurons respond to the activity of cells that communicate with them. As Sherrington put it nearly a century ago, "Shut off from all opportunity of reproducing itself and adding to its number by mitosis or otherwise, the nerve cell directs its pent-up energy towards amplifying its connection with its fellows, in response to the events which stir it up. Hence, it is capable of an education unknown to other tissue" (Sherrington, 1887, quoted by Rosenzweig, 1976).

Although the notion of a neuron's "pent-up energy" for reproducing itself would not be taken seriously today, it certainly seems likely that "education" requires synaptic changes. For example, synapses could change in size, presynaptically, postsynaptically, or both; there could be alterations in the number of receptor proteins in the postsynaptic membrane; terminal buttons could sprout off of axons and establish entirely new synapses with other neurons. Any of these alterations would affect the way circuits of neurons would interact with each other, and could serve as the basis of long-term memory.

How does one approach the problem of isolating the physical changes that constitute long-term memory? The problem is exceptionally complex. Despite decades of research, no one has been able to point to particular structural aspects of neurons and say, "There—that is what memory is." Here are some of the ways the problem has been attacked.

1. Train animals, and look for changes in the morphology of neurons. The problem here, of course, is detecting a few changes in synaptic connections among the several billions of neurons in the brain. Where does one look? No two brains are exactly alike, which means that it is not possible to compare the brain of an animal that has learned a particular task with one that has not. They will look different even if both animals have received the *same* training. Attempts have been made to get around this problem by teaching animals a specific task very early in life, before the brain has attained its adult state. As we shall see, early experience appears to have a much more profound effect than experiences that occur later in life. Whether the early changes are simply larger-scale versions of the changes that occur later in life is an open question.

2. Train animals and look for biochemical changes rather than structural ones. As we shall see, structural changes must, of necessity, entail biochemical changes. Perhaps chemical analyses that are performed before and after learning experiences can give us a clue to the physical basis of memory. The success of this approach has been modest, partly for the reasons that I outlined in the previous paragraph—which of the thousands of

chemicals in the brain should be analyzed? And even if biochemical changes are found, how can we know that they represent long-term memory storage, and not other physiological states that accompany the learning process, such as activity of sensory systems, attentional changes, or arousal? As Agranoff, Burrell, Dokas, and Springer (1978) have put it, there is a "Catch-22" in research in this field: if a biochemical change is large enough to be detected, then it is almost certainly *not* specific to memory storage. This means that positive results are suspect, but, of course, negative ones tell us nothing. (The term "Catch-22" comes from a fictitious army regulation in Joseph Heller's novel; if you apply to get out of the army on the grounds of insanity, the fact that you want to leave proves that you are sane. Only crazy people want to stay in the army.)

3. Inhibit biochemical systems that are necessary links in the chain of events that ultimately produce certain physical changes. If a failure in learning is observed, then perhaps the physical changes that were blocked are those that mediate the memory storage. Unfortunately, the biochemical inhibitors that are presently available are not very specific in their effects; they block a wide range of physical changes and do not permit us to conclude much about the nature of the changes that take place during the storage of memory.

4. Facilitate certain biochemical systems and thereby accomplish effects opposite to the ones proposed in (3); the acquisition of a memory should thus take place more rapidly. Unfortunately, it is not at all easy to speed up biosynthetic processes in a living organism, and attempts such as this have not met with success. *General* facilitators, such as strychnine (which blocks many inhibitory synapses) or amphetamine, sometime increase the rate of learning, but no one would argue that the effects of these drugs are specific to memory formation.

Biochemistry of Physical Alterations

THE ROLE OF PROTEINS. We cannot evaluate any of the above approaches without first being acquainted with the way in which cells may undertake physical changes. Since long-term memories can last for many decades, the changes must be permanently maintained. One type of substance is particularly important in the production of physical alterations in cells, and in their continued maintenance—protein.

TYPES OF PROTEINS. Proteins serve two general functions, both of which are essential for physical alterations in cells. First, almost all chemical reactions that take place in cells are controlled by enzymes, which are proteins. Any change in the structure of neurons would require the presence of the appropriate enzymes. Secondly, many of the constituents of cells are made of protein. Thus, the growth of new synaptic connections, for example, would require both classes of protein: enzymes and structural proteins. If we are to understand how neurons change in order to store long-term memories, we must understand how proteins are produced.

Amino Acids: The Elements of Proteins

Proteins are constructed of long chains of ***amino acids.*** A few representative amino acids (most proteins are constructed from a group of twenty of these) are shown in Figure 18.8. Note that each amino acid contains a ***carboxyl group*** (COOH), shown on the right, and an ***amino group*** (NH_2), shown on the left. (Small gray dots represent hydrogen atoms; large black dots represent carbon atoms; O and N represent oxygen and nitrogen.) (See **FIGURE 18.8.**)

The carboxyl group of one amino acid can couple with the amino group of another. This junction, the ***peptide bond,*** is the link that connects a string of amino acids together to make proteins. (A molecule that consists of a few amino acids is referred to as a ***polypeptide*** rather than a protein; a protein consists of a larger number of them— somewhere over fifty.) Figure 18.9 shows how two amino acids (serine and glycine) can be linked together with a peptide bond. (See **FIGURE 18.9.**)

THE THREE-DIMENSIONAL STRUCTURE OF PROTEINS.   A protein can consist of many hundreds of amino acids; the particular sequence of the amino acids specifies the nature of the protein. How-

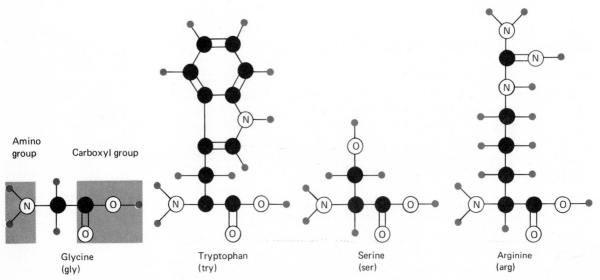

Amino group   Carboxyl group

Glycine (gly)

Tryptophan (try)

Serine (ser)

Arginine (arg)

FIGURE 18.8   Structures of some amino acids.

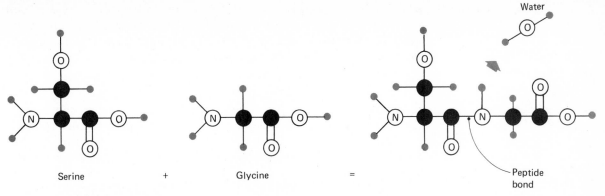

Serine   +   Glycine   =   Peptide bond

FIGURE 18.9   A peptide bond between two amino acids.

ever, a protein molecule is not shaped like a long rod; it has a complex three-dimensional structure.

The three-dimensional structure of a protein is determined by the interactions among its constituent amino acids. Some amino acids join with each other, forming stable bonds. These bonds serve as points of attachment for adjacent loops of the polypeptide chain. (See **FIGURE 18.10.**) The location of the amino acids that are capable of bonding with each other determines which parts of the polypeptide chain stick together, and thus determines the protein's three-dimensional structure.

FIGURE 18.10   The three-dimensional structure of a protein.

The three-dimensional structure of a protein is of extreme importance; an enzyme that has been straightened out by chemical means no longer acts as an enzyme. (The term for the straightening-out process, in fact, describes this loss of activity; the protein is said to be **denatured**.) An enzyme attaches only to substrates that are able to nestle into its three-dimensional conformation. Thus, enzymes act only on certain substrates. When the substrate and enzyme join, the enzyme apparently bends; in some cases this also causes the substrate to bend, exposing some part of it so that another molecule can now be attached (synthesis). In other cases, some portion of the substrate is exposed to the destructive effects of other molecules, as in the case of enzymes that facilitate hydrolysis (dissolution by water) of their substrates. The important aspect of the three-dimensional structure of enzymes is that specificity for a given substrate is achieved, and the substrate is subsequently altered in a particular way.

We have seen that proteins are essential for initiating long-term alterations in the properties of cells; they serve both enzymatic and structural roles. We have also seen that the nature of a protein is determined by its sequence of amino acids. Therefore, we must understand the process by which a particular protein is produced and also the mechanisms that initiate and terminate its production.

proteins serve both enzymatic & structural roles

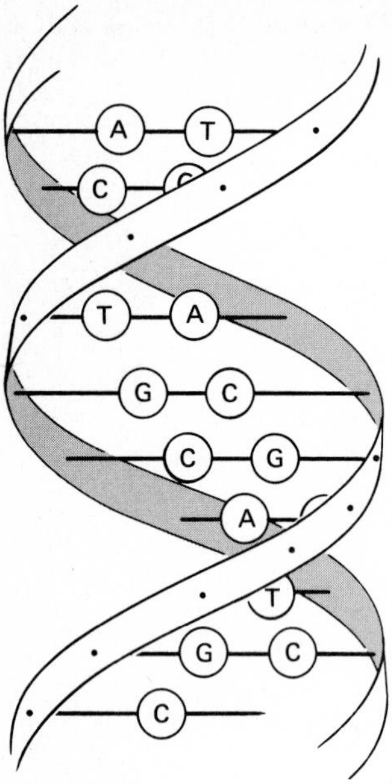

FIGURE 18.11   A molecule of deoxyribonucleic acid (DNA).

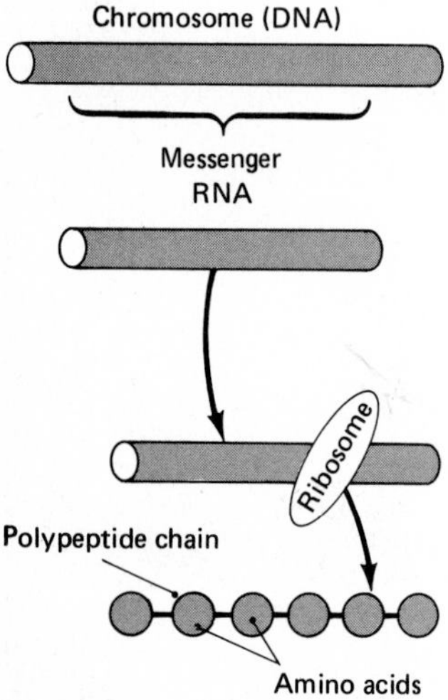

FIGURE 18.12   A schematic overview of the process of protein synthesis.

Protein Synthesis

THE STRUCTURE OF DNA.   The recipe for a particular protein consists of a list of amino acids. Therefore, there must be a set of lists contained somewhere in the cell that directs the construction of each protein the cell can produce. These lists are found in the chromosomes. Each of our twenty-three pairs of chromosomes consists of a double-stranded helix (coil) of **deoxyribonucleic acid (DNA).** The important thing about DNA is not its helical structure, or the sugar molecule (2-deoxy-D-ribose) that serves as its backbone. What is important is the fact that DNA contains a set of four nucleotide bases, **adenine** (*A*), **guanine** (*G*), **cytosine** (*C*), and **thymine** (*T*). These nucleotide bases are arranged in pairs down the center of the DNA strand, linked together by an easily broken hydrogen bond. Note that nucleotide bases are complementary; A and T bond together, as do G and C. (See **FIGURE 18.11.**)

An overview of protein synthesis. The synthesis of protein is accomplished by means of two processes. A portion of the DNA strand (a gene) contains a coded list of the amino acids that specify a particular protein. The information contained in this stretch of DNA is copied onto a strand of RNA (**ribonucleic acid,** which uses the sugar *ribose* instead of 2-deoxy-D-ribose). During this process, the two strands of DNA temporarily break apart. The RNA just assembled (called **messenger RNA,** or *mRNA*, since it conveys coded information from the DNA strand) then travels to the ribosome, where the proper amino acids are assembled into protein. (See **FIGURE 18.12.**)

THE GENETIC CODE.   What is the nature of the code? A set of painstaking experiments have shown that information on the DNA strands (and on RNA, also) is represented in three-letter words, the letters being represented by nucleotide bases. Thus, since there are four different letters (A, G, C, and T) sixty-four unique three-letter words can be formed. These three-letter words are called **codons.** Since only twenty amino acids need to be specified, that means there are more than enough words in this language to encode a protein's sequence of amino acids. The codes (in terms of the base sequences of RNA, and not of DNA) are shown in Figure 18.13. Note that there is redundancy; a particular amino acid can be specified in more than one way. Also, there are special codes for *start* and *stop*, which serve as punctuation. The *start* and *stop* codes indicate the beginning and end of the protein chain. (See **FIGURE 18.13.**)

TRANSCRIPTION.   When protein synthesis is initiated, the strands of DNA pull apart, and a strand of RNA is produced. RNA is very similar to DNA, except that it consists of only one chain, and, as I already noted, its sugar is ribose. There is one other exception; one of the nucleotide bases of RNA is different. Thymine is replaced by **uracil.** So the letters used by RNA are A, G, C, and *U*.

Ala — GCU, GCC, GCA, GCG

His — CAU, CAC

Ile — AUU, AUC, AUA

Thr — ACU, ACC, ACA, ACG

Arg — AGA, AGG, CGU, CGC, CGA, CGG

Leu — CUU, CUC, CUA, CUG, UUA, UUG

Tyr — UAU, UAC

Try — UGG

Asn — AAU, AAC

Lys — AAA, AAG

Val — GUU, GUC, GUA

ASP — GAU, GAC

Met START — AUG

Val START — GUG

Cys — UGU, UGC

Phe — UUU, UUC

STOP — UAA, UAG, UGA

Glu — GAA, GAG

Pro — CCU, CCC, CCA, CCG

Gln — CAA, CAG

Gly — GGU, GGC, GGA, GGG

Ser — AGU, AGC, UCU, UCC, UCA, UCG

FIGURE 18.13   The genetic code. These sequences of bases on the RNA molecule specify corresponding amino acids.

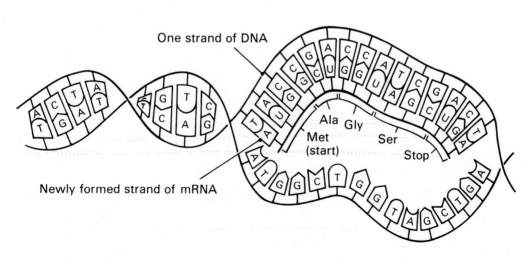

FIGURE 18.14   Transcription: The synthesis of messenger RNA from a portion of the DNA strand.

Figure 18.14 gives an example of this process of RNA synthesis, which is called ***transcription*** because the code is copied onto the messenger RNA. Five codons are shown, specifying the following sequence of amino acids: methionine (*start*)—alanine—glycine—serine—(*stop*). (See **FIGURE 18.14** on the preceding page.)

TRANSLATION.   The newly synthesized messenger RNA now travels to a ribosome, to which it attaches. The ribosome, consisting chiefly of protein, somehow facilitates the process that follows. Since the information that specifies the sequence of amino acids is contained in a series of three-letter words, there must be some mechanism that reads these words. The reading (or ***translation,*** as this process is called) is accomplished by another form of RNA called ***transfer RNA.*** This substance does not resemble a long coil, as messenger RNA does, but is folded and bent back upon itself. Transfer RNA has two important working sites: the ***anticodon loop,*** which "fits" the codon on the messenger RNA that specifies a particular amino acid, and an ***amino acid acceptor end,*** to which is attached that amino acid. (See **FIGURE 18.15.**)

The sequence of translation works this way: The ribosome attaches to the end of the messenger RNA, which signifies a *start* code. This attachment somehow exposes the first codon, to which a molecule of transfer RNA with the appropriate anticodon gets attached. For example, transfer RNA with the anticodon UAC attaches to the codon, AUG, which specifies methionine and *start*. (See **FIGURE 18.16.**)

The ribosome then moves down the strand of messenger RNA, exposing the next codon (GCU), which specifies alanine. The anticodon should be CGA; however, transfer RNA uses a variety of nucleotide bases besides A, G, C, and U. In this case, the codon contains *inosine*, abbreviated as *I*. The transfer RNA with the anticodon CGI

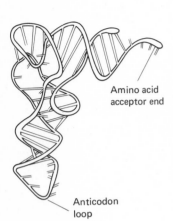

FIGURE 18.15   A molecule of transfer RNA.

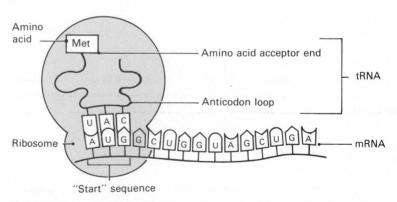

FIGURE 18.16   The beginning of translation: Production of a polypeptide from messenger RNA.

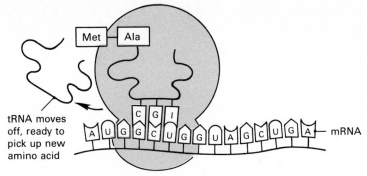

FIGURE 18.17   Translation: Attachment of the next amino acid.

tRNA moves off, ready to pick up new amino acid

mRNA

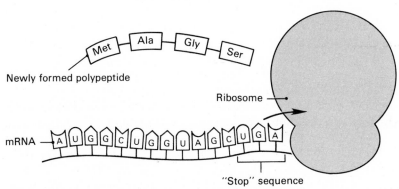

Newly formed polypeptide

Ribosome

mRNA

"Stop" sequence

FIGURE 18.18   Completion of the translation process.

has a molecule of alanine at its amino acid acceptor end. This molecule attaches to the second codon, its alanine attaches to the adjacent methionine, and the first molecule of transfer RNA drops off. This molecule, now amino acid–less, picks up another free amino acid and is ready to be used again. (See **FIGURE 18.17.**)

This process continues, the ribosome moving down the strand of messenger RNA and causing the assembly of the sequential amino acids until it gets to the *stop* codon, UGA, at which point it detaches. A new protein molecule has been produced. (See **FIGURE 18.18.**)

Control of Protein Synthesis

We have seen how proteins are synthesized; the message along a particular region of DNA (a gene) is transcribed into messenger RNA, which goes to a ribosome. The ribosome moves along the strand of messenger RNA, permitting the assembly of the coded-for amino acids, with transfer RNA serving as an intermediary that brings the amino acids into position. Various amino acids join together, and the protein adopts its active shape. The protein then goes out to do its stuff: to serve as a structural element or as an enzyme.

So far no control mechanism has been described. If all genes in a cell continuously produced messenger RNA, and thus synthesis of a protein, the situation would be chaotic. Too many different kinds of protein would be produced. In fact, only a small proportion of the genes are active at any one time. All cells of the body (except for special ones like sperm, ova, and blood cells) have identical genetic material. Therefore, cells are not differentiated on the basis of DNA. For example, a liver cell contains the system of genes that controls the synthesis of acetylcholine. Similarly, a neuron contains all it needs to know about how to produce bile. And yet cells are specialized. They are all constructed from the same set of blueprints, and they all use these same blueprints to maintain their structure and to respond to changes in their environment. How can such an enormous variety of cells be constructed from a single set of plans, and how can these same plans be used to respond appropriately to a variety of different environmental stimuli?

Chromosomes consist of more than naked strands of DNA; they also contain two varieties of protein—***histone*** and ***nonhistone*** protein. There are only a few kinds of histone proteins, and some of these are almost exactly alike in the cells of most living things, from pea plants to humans. The histone proteins attach to the DNA strands at fairly regular intervals. They form bonds with each other, which serves to make the DNA strands fold and coil upon themselves, which gathers them into discrete clusters. Although DNA strands are too small to be seen if they are spread out, the clusters (called ***nucleosomes***) are large enough to be photographed under an electron microscope. (See **FIGURE 18.19.**)

It appears that an important function of the histone proteins that are attached to the chromosomes is to "protect" some portions of DNA; that is, to make them unavailable for transcription. The strands of the "protected" regions of DNA cannot separate, and therefore the proteins that are coded for in these regions cannot be produced.

Nonhistone proteins are also associated with the chromosomes. There are many different kinds of nonhistone proteins. The cell nuclei of different species of animals, and even different types of cells in a single animal, contain different populations of nonhistone proteins. Nonhistone proteins appear to play a role in releasing portions of the DNA strand from the repressing effects of histone proteins. The evidence for this conclusion is as follows: If chromosomes are broken apart by special enzymes, two types of fragments are found. One type contains nonhistone proteins and newly-formed RNA.

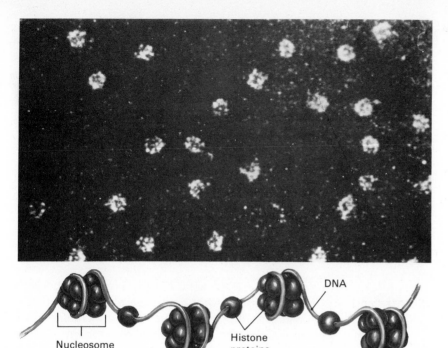

FIGURE 18.19 Dark-field electron micrograph of nucleosomes extracted from nuclei of chicken red blood cells, and a schematic representation of what appears to be shown in the micrograph. (From Olins, D. E., and Olins, A. L. *American Scientist*, 1978, *66*, 704–711. Reprinted by permission of *American Scientist*, journal of Sigma Xi, the Scientific Research Society.)

Thus, the fragments of this type contain active genes. Fragments of the second type contain neither nonhistone proteins nor newly-formed RNA; they are inactive (Olins and Olins, 1978).

It appears that the transformation of a region of DNA (containing one or more genes) from the inactive state to the active state entails the addition of a particular nonhistone protein. The precise way in which this occurs is uncertain, but it has been suggested that the process of phosphorylation plays an important role (Stein, Stein, and Kleinsmith, 1975). The addition of phosphate groups (PO_4) to proteins makes the proteins change their shape, and alters their electrical charge. Therefore, it is possible that the phosphorylation of nonhistone proteins in the nucleus is the first step in turning on a gene. The phosphorylation of a particular nonhistone protein could cause it to attach to a region of a chromosome, where it interacts with the histone proteins in such a way as to expose some DNA for the transcription process. The newly-formed RNA then goes to the ribosomes, where protein synthesis takes place.

Now we have to ask what might cause some proteins to become phosphorylated. This question takes us back to Chapter 5. As you will recall, some neurotransmitters produce their postsynaptic ef-

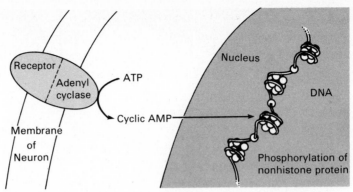

FIGURE 18.20 A schematic representation of the way in which extracellular chemicals can cause alterations in protein synthesis.

fects by converting ATP or GTP to cyclic AMP or cyclic GMP, which, in turn, activates protein kinases that cause the phosphorylation of proteins in the postsynaptic membrane. The resulting change in the shape of these proteins alters the permeability of the membrane to ions such as NA^+, K^+, and Cl^-, and produces postsynaptic potentials.

But recall that I mentioned in Chapter 5 that cyclic AMP or GMP might travel to the nucleus and cause some changes there. It is this phenomenon that is of special interest to the study of the memory process: could cyclic nucleotides that are produced as a result of synaptic activity cause the phosphorylation of nonhistone proteins in the nucleus, and thus initiate the synthesis of new proteins? If so, we have a possible link between neural activity and the production of long-term physical change. (See **FIGURE 18.20.**)

The sequence could go like this:

> Neural activity encoding short-term memory → increases in extracellular transmitter substance (or other external chemical events, such as ionic changes) → activated nucleotide cyclase → cyclic nucleotide → phosphorylated non-histone protein → synthesis of messenger RNA → protein synthesis → change in structure of cell → new properties of cell, contributing to the encoding of long-term memory.

Entingh and his colleagues (Entingh, Dunn, Glassman, Wilson, Hogan, and Damstra, 1975) have presented a diagram that very nicely puts together these various steps and provides some details that I have not discussed, such as the possibility that phosphorylation of synaptic proteins may provide the basis for short-term memory; perhaps reverberation is "kept in one track," for example, by temporarily facilitating particular synaptic connections. The diagram also refers to experiments that provide evidence for each of these hypothesized links. (See **FIGURE 18.21.**)

I. Resting state
   (no input)

   But cell is sensitive
   to electrolytes, hor-
   mones and metabo-
   lites.

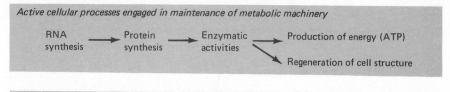

Active cellular processes engaged in maintenance of metabolic machinery

RNA synthesis → Protein synthesis → Enzymatic activities → Production of energy (ATP)

Regeneration of cell structure

I. Active state

   Neurotransmitter
   from presynaptic
   cell attaches to
   post–synaptic
   membrane.

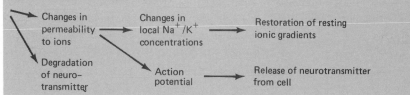

Responses to general synaptic input

Changes in permeability to ions → Changes in local Na⁺/K⁺ concentrations → Restoration of resting ionic gradients

Degradation of neuro-transmitter

Action potential → Release of neurotransmitter from cell

I. Reorganization state

   Characteristics
   unknown.

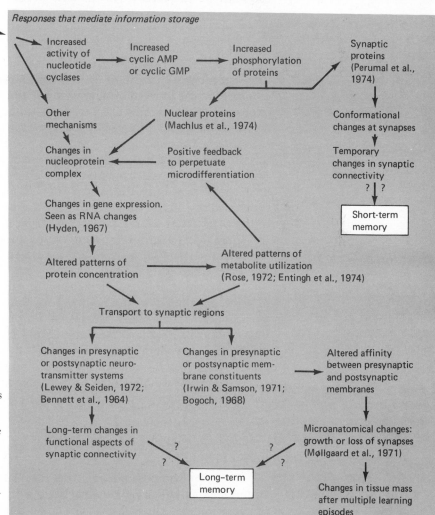

Responses that mediate information storage

Increased activity of nucleotide cyclases → Increased cyclic AMP or cyclic GMP → Increased phosphorylation of proteins → Synaptic proteins (Perumal et al., 1974)

Other mechanisms

Nuclear proteins (Machlus et al., 1974)

Conformational changes at synapses

Changes in nucleoprotein complex ← Positive feedback to perpetuate microdifferentiation

Temporary changes in synaptic connectivity

? ?

Changes in gene expression. Seen as RNA changes (Hyden, 1967)

Short-term memory

Altered patterns of protein concentration → Altered patterns of metabolite utilization (Rose, 1972; Entingh et al., 1974)

Transport to synaptic regions

Changes in presynaptic or postsynaptic neuro-transmitter systems (Lewey & Seiden, 1972; Bennett et al., 1964)

Changes in presynaptic or postsynaptic membrane constituents (Irwin & Samson, 1971; Bogoch, 1968)

Altered affinity between presynaptic and postsynaptic membranes

Long-term changes in functional aspects of synaptic connectivity

? ? ? ?

Long-term memory

Microanatomical changes: growth or loss of synapses (Møllgaard et al., 1971)

Changes in tissue mass after multiple learning episodes (Bennett et al., 1964)

FIGURE 18.21 Possible ways in which neurons might mediate memory. The references can be found in the bibliography. (From Entingh, D., Dunn, A., Glassman, E., Wilson, J. E., Hogan, E., and Damstra, T. In *Handbook of Psychobiology*, edited by M. S. Gazzaniga and C. Blakemore. Copyright 1975 by Academic Press, New York.)

605

NEURAL STIMULATION AND PHYSICAL CHANGE

Now it is time to see what we do know about the physical nature of memory. I shall consider each of the links in the chain from STM to morphological changes, and I shall review some representative studies that are relevant to these links. In particular, I shall review the effects of experience on each of the steps in the process of neural change, starting at the end of the chain (the changes themselves) and working backward.

Effects of Experiences on the Structure of Neurons

First of all, it should be made clear that neurons are indeed capable of altering their relationships with their neighbors. It is difficult to understand how induction of RNA synthesis can cause a change as complex as the growth of a new axonic process that finds its way to another cell and grows a new synapse there, but this does happen. (It must be noted that this process requires cooperation from the second cell, for no synapse is functional without postsynaptic receptor sites and the appropriate membrane mechanisms for producing PSPs). This process occurs during development, when a budding axon finds its way from one part of the nervous system to another. Synapses are made on the appropriate neurons or muscles, even though they may be far away. The genetic mechanisms that direct this growth during development could quite conceivably direct the establishment of similar connections later. The genes are there; they just have to be turned on.

Perhaps one of the ways they can be turned on is by the secretion of some substance by the neuron to which the new process is sent: we might call it a "come and get me" substance. For example, Raisman (1969) has shown that individual cells in the septum receive terminals from the hippocampus (via the fimbria) and from the hypothalamus (via the medial forebrain bundle). If the fimbria is destroyed, the terminals from the MFB axons will sprout and occupy the vacated synaptic sites. Similarly, MFB lesions will result in the sprouting of the axons from the fimbria and the establishment of new synapses on the soma, which does not normally receive terminals from these axons. These results are shown schematically in **FIGURE 18.22.**

This experiment indicates that neurons in the mature nervous system *can* establish new connections; whether the memory process involves similar kinds of changes has not yet been shown. However, two experiments by Spinelli and his colleagues have identified changes in the functional properties and physical characteristics of neurons that appear to be related to learning. Spinelli and Jensen

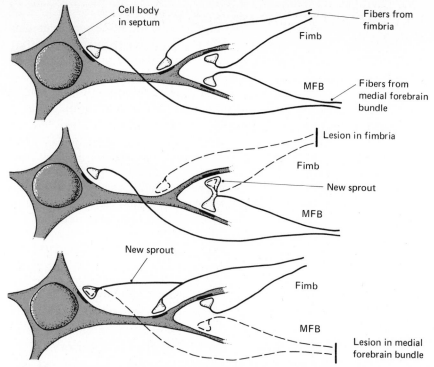

FIGURE 18.22   Sprouting of nerve terminals in the septum and attachment to new postsynaptic sites caused by degeneration of the fimbria or medial forebrain bundle. (From Raisman, G., *Brain Research*, 1969, *14*, 25–48.)

(1979) trained a group of kittens, four weeks of age, to flex a foreleg in order to avoid a mild shock. The kittens were suspended in a cloth sling, their legs protruding downward through four holes. A rubber band held a shock electrode on a foreleg. The kittens wore goggles, through which visual stimuli could be presented. When a kitten held its foreleg down it received electric shocks, and saw a pattern of horizontal lines with one eye. When the foreleg was flexed (pulled up) no shocks were received, and a pattern of vertical lines was seen with the other eye. (For some kittens, the patterns were reversed.) Thus, one eye saw a "danger" pattern, while the other eye saw a "safety" pattern. The kittens learned the task very quickly, keeping their foreleg up over 95 percent of the time. The training sessions were very brief: 8 minutes per day.

After 10 weeks of training the experimenters made single unit recordings of neurons in somatosensory cortex, primary visual cortex, and visual association cortex.

Changes were seen in neural responses to somatosensory and visual stimuli. Somatosensory representation was determined by recording from a cortical neuron, and then touching the animal's skin, to see which areas of the body surface produced a response in the cell. It was found that tactile stimuli that were applied to the shocked region of skin produced a response in a much larger area of somatosensory cortex than stimuli applied to the same region of skin on the unshocked forearm. In addition, 75 percent of the cells that responded to touch from the shocked areas also responded to visual stimuli, whereas only 30 percent of the cells on the other side of the brain did so. And most of the cells on the trained side responded selectively to the "danger" stimulus. As the authors noted, this is very significant, since the "danger" stimulus was seen, in total, for approximately 10 minutes, as opposed to 250 minutes for the "safety" stimulus. In fact, the kittens who learned the fastest (and thus saw the "danger" stimulus the least) had the largest percentage of cells in somatosensory cortex that responded to the "danger" stimulus. Similar results were found in primary visual cortex and visual association cortex; cells tended to respond to lines oriented like the two stimuli, especially the "danger" stimulus.

In a follow-up study, Spinelli, Jensen, and DiPrisco (1980) trained kittens in a similar manner, and examined the architecture of neurons in the animals' somatosensory cortex. The brains were stained with the Golgi-Cox technique, which makes the entire membrane of the neuron visible, including the dendritic branches. The investigators found significant increases in the number and complexity of dendritic branches in the region of somatosensory cortex that received information from the shocked forearm. The effect was striking, as Figure 18.23 shows. The photomicrograph on the left is from "untrained" somatosensory cortex; the right is from "trained" cortex. (See **FIGURE 18.23.**)

The results of these studies are very interesting, and suggest that the method might profitably be used to investigate the physical basis of memory. However, we cannot be certain that the observed changes constitute "memory." The authors used immature animals in order to increase the likelihood of obtaining measurable changes. Indeed, they also trained an older kitten, beginning at eleven weeks of age, but failed to detect any changes in the response characteristics of cortical neurons. Obviously, since the cat learned the task, some kinds of changes had to be there, but they were apparently too subtle to be detected. (Catch-22 again?) One hopes that the changes that occur in the kitten brain are simply "magnifications" of the changes that occur in the brains of older animals, but we cannot be sure that this is the case. However, until we have methods that permit the detection of subtle changes in a few of the billions of neurons in the brain, the use of young animals will undoubtedly play an important role in the investigation of the physical basis of memory.

Effects of Experience on Protein Synthesis

A number of studies have demonstrated that neural stimulation increases protein synthesis. A study by Wegener (1970) shows just how

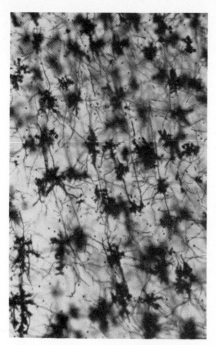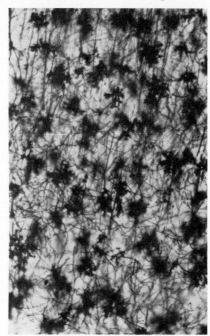

FIGURE 18.23   Cortical dendrites (Golgi-Cox stain) from "untrained" somatosensory cortex (left) and "trained" somatosensory cortex (right). (From Spinelli, D. N., Jensen, F. E., and DiPrisco, G. V. *Experimental Neurology*, 1980, *62*, 1–11.)

localized this effect can be. He removed the left eyes of a group of frogs and kept them in the dark for 10 days. The animals were then paralyzed with curare and given injections of radioactive histidine (an amino acid, and thus a constituent of protein). The frogs were divided into four groups. One group was exposed to diffuse light, the second saw a narrow vertical band of light, and the third saw two bands of light. The fourth group was kept in the dark. The stimuli were presented for 75 minutes. After time was allowed for protein synthesis to take place (and thus for the radioactive amino acid to be incorporated), the frogs were killed and fine-grain measurement was made of the radioactivity of various parts of the optic tectum.

The brains of frogs that were exposed to diffuse light contained radioactivity all over the optic tectum contralateral to the intact eye. Frogs that were kept in the dark showed the same amount of radioactivity in each hemisphere of the brain. The animals that saw the patterned light showed similar patterns of radioactivity on the optic tectum: a single wide band or two narrower bands, depending on the particular stimulus. Thus, increased activity of neurons produced by sensory input can stimulate protein synthesis in these cells.

LEARNING AND PROTEIN SYNTHESIS.   It is not surprising that neural activity increases protein synthesis; after all, neural activity requires the expenditure of energy. In addition, a variety of substances get used up. For example, some neurotransmitter is inevitably lost, and not all of the presynaptic membrane is recycled. The replacement of these substances and the replenishment of the cell's energy reserves requires protein synthesis. Thus, an increase in protein synthesis does not indicate that the cell has undergone any physical change that encoded memory.

When an animal is taught a task in order to determine whether learning might produce biochemical changes in the brain, the experience is usually a very significant one to the animal. That is, the subject receives painful footshock, or is made very hungry or thirsty, and is taught ways to obtain food or water. The reason for making the experience very significant is obvious: one wants to produce a change that will be large enough to be detectible. But making the experience very important to the animal also guarantees that nonspecific effects will be produced. That is, the experience will produce effects that are not directly related to the neural coding of the information that is to be incorporated into memory.

A series of studies on the effects of avoidance training illustrate this fact very nicely. Rees, Brogan, Entingh, Dunn, Shinkman, Damstra-Entingh, Wilson, and Glassman (1974) injected radioactive lysine (an amino acid) into mice, and trained them to avoid a painful footshock. They observed increased radioactivity (relative to control animals) in the protein that was extracted from the brain, which suggests that the training increased the rate of protein synthesis. However, they also discovered an increased rate of protein synthesis in liver cells, which no one believes to be involved in the formation of memory. The effect seemed to be produced by stress-related hormones. Removal of the adrenal glands did not prevent this effect (Rees and Dunn, 1977), but removal of the pituitary gland did (Dunn, 1980). Injections of ACTH (the pituitary hormone that stimulates the secretion of stress-related adrenal hormones) mimicked the effects of the avoidance training on protein synthesis (Dunn, Rees, and Iuvone, 1978). Thus, the stress that is inherent in avoidance training caused the release of ACTH, which increased the rate of protein synthesis in cells of the brain and liver (and, no doubt, many other cells of the body). Again, we are reminded of Agranoff's "Catch-22;" the change here is certainly too general to be associated specifically with memory. On the other hand, experiences that are trivial to the subjects will produce changes that are too small to be detected.

Effects of Experience on RNA Synthesis

Horn, Rose, and Bateson (1973) have described a series of experiments that provide excellent biochemical evidence that experience can produce changes in RNA synthesis. Their training method used the phenomenon of imprinting.

A young chick (or other precocial bird that must follow its mother around shortly after hatching) will become imprinted upon the first significant stimulus it sees. It will continue to follow that stimulus around—as if it were the chick's mother—whether that stimulus is a box or a toy train or even a human being. Usually the first thing the chick sees is its mother, so there are few problems of inappropriate imprinting in a natural setting. The advantage of using imprinting as the learning task is that the animal does not have to be given a painful shock or be made hungry or thirsty to learn an association. (Of course, it is possible that the imprinting causes many nonspecific effects, including the release of hormones, even though it is a "natural" experience.)

Newly-hatched chicks were injected with radioactive uracil (one of the nucleotide bases that make up RNA) and were exposed for varying amounts of time to a rotating, flashing light. The chicks were killed later, and estimates were made of the rate of RNA synthesis by measuring the radioactivity of RNA extracted from the animals' brains.

The results are seen in Figure 18.24. The curves represent the estimated amount of RNA produced in the roof of the brain (which includes the optic tectum) at varying times after the onset of the stimulus. Experimental animals (upper line) saw a flashing light; control subjects (lower line) saw a steady light. (Chicks will not become imprinted to a steady light.) Note that RNA synthesis appeared to be increased selectively after 76 minutes of exposure to the flashing light. (See **FIGURE 18.24.**)

Some of Horn's colleagues (Haywood, Rose, and Bateson, 1975) also extracted an enzyme called ***RNA polymerase*** from the brain of chicks. Enzymes direct all kinds of biosynthetic events, and RNA synthesis is no exception. As we saw, nuclear nonhistone proteins

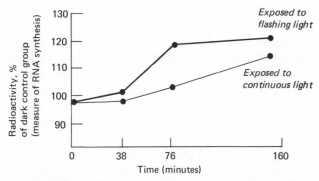

FIGURE 18.24 Alterations in RNA synthesis in the roof of the brain (tectum) after exposure of a chick to a flashing or steady light. (From Horn, G., Rose, S. P. R., and Bateson, P. P. G., *Science*, 10 August 1973, *181*, 506–514. Copyright 1973 by the American Association for the Advancement of Science.)

activate a region of DNA. However, before RNA synthesis can begin, the enzyme RNA polymerase must be present to facilitate the process. Thus, if RNA synthesis in the brain is increased, one should see an increase in RNA polymerase activity.

Indeed, this was found to be the case. An extract of proteins from the brains of the trained chicks was found to contain more active RNA polymerase than extracts from the brains of control chicks. The results confirm that imprinting increases the rate of RNA synthesis.

Subsequent studies have shown that RNA synthesis is most increased in regions of the brain that are most important for successful imprinting. Horn, McCabe, and Bateson (1979) injected chicks with radioactive uracil before the imprinting experience, and used autoradiographic techniques to determine which specific regions of the brain underwent the greatest increase in RNA synthesis. They found that the *medial hyperstriatum ventrale,* a part of the roof of the forebrain, had the greatest amount of radioactivity. The significance of this finding is enhanced by the fact that destruction of this region impairs imprinting (Bateson, Horn, and McCabe, 1978). The use of converging methods is especially interesting, and is a rarity in research on the biochemistry of memory.

Effects of Experience on Phosphorylation of Nonhistone Proteins

As we saw earlier, RNA synthesis is apparently initiated by the phosphorylation of nuclear nonhistone proteins. Machlus, Wilson, and Glassman (1974) trained rats to avoid a shock by jumping from the grid floor onto a platform located at one end of the chamber (the rats, and not the authors, did the jumping, of course). The authors measured the amount of radioactive phosphorus that became incorporated into nonhistone protein extracted from the rats' brains; thus, they obtained an estimate of the degree of protein phosphorylation. The results are shown in Figure 18.25. Note that the change is quite large (100 percent above the level obtained in control subjects) and is also quite brief, suggesting that a brief period of RNA synthesis (and subsequent protein synthesis) is initiated. These results are consistent with the hypothesis that a set of genes was turned on by the experience. (See **FIGURE 18.25.**)

Follow-up studies with mice (Gispen, Perumal, Wilson, and Glassman, 1977) showed that avoidance training produced phosphorylation of proteins that were associated with synaptosomal fractions of homogenized brain tissue. (The process by which synaptosomes are extracted was described in Chapter 4.) Control mice that received shocks, but not the avoidance training, did not show an increase in phosphorylation of synaptosomal proteins. Unfortunately, the observed changes are unlikely to be directly responsible for long-term memory storage for two reasons. (1) *Nuclear* nonhis-

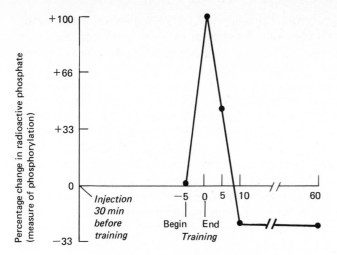

FIGURE 18.25   The amount of phosphorylated protein observed in the
brain before and after training, in the study by Machlus et al. (1974).
(From Entingh, D., Dunn, A., Glassman, E., Wilson, J. E., Hogan, E., and
Damstra, T. In *Handbook of Psychobiology*, edited by M. S. Gazzaniga and
C. Blakemore. Copyright 1975 by Academic Press, New York.)

tone proteins, and not synaptosomal proteins, must be phosphorylated in order for protein synthesis to be initiated. As we saw in
Chapter 5, the phosphorylation of membrane proteins causes
changes in membrane permeability to various ions, and the effect
seen in the study by Gispen et al. is likely to be related to this process.
(2) ACTH, which as we saw increases RNA synthesis, also has effects
on the phosphorylation of synaptosomal proteins (Zwiers, Veldhuis,
Schotman, and Gispen, 1976). Thus, it is possible that this hormone
produced the changes seen in the avoidance-training study.

   Since phosphorylation of nuclear nonhistone proteins is a necessary step in protein synthesis, future studies on the biochemistry
of memory formation will have to investigate this process. However,
studies that have been performed so far have failed to identify any
protein phosphorylation as being specific to memory.

Inhibition of RNA or Protein Synthesis

If long-term memory depends upon structural changes, then biochemical treatments that affect the mechanisms producing these
changes should have an effect on the formation of memory. Many
experiments have been performed to determine whether synthesis of
RNA and protein is necessary for the consolidation of memory. However, there are two problems inherent with all of these studies. First,
no inhibitor has been found that affects *only* RNA or protein synthesis; other biochemical systems are also affected. However, even if

suppression of protein synthesis were the sole effect of a particular drug, we could not conclude that protein synthesis was necessary for the consolidation of memory if this drug did block memory. The biochemistry of neurons could be so upset by cessation of protein synthesis that the brain could not function normally; lack of consolidation could be a secondary, rather than a primary, effect of the halted protein synthesis. Thus, we cannot reach any firm conclusions about the necessity of protein synthesis for memory consolidation, even if we find this function impaired by these drugs.

The second problem is just the opposite. No drug used so far succeeds in *completely* inhibiting RNA or protein synthesis; the suppression is generally 80 to 95 percent complete. As a matter of fact, total suppression of RNA or protein synthesis would probably be lethal. If we fail to obtain amnesia after drug treatment, it might be because enough protein synthesis took place to produce the necessary physical changes.

The results of the experiments carried out so far are just what might be expected, given the problems with the available techniques: Some studies demonstrate amnesia, others do not, and still others show that the degree of amnesia depends upon the procedure that was used to train the subjects.

Protein synthesis has been experimentally disrupted with a number of drugs, including **puromycin, cyclohexamide (CXM), acetoxycyclohexamide (AXM),** and **anisomycin.** Studies with all of these drugs have been plagued with inconsistencies and the results are impossible to interpret with any certainty. For example, puromycin causes abnormal electrical activity in the brain. Injections of Dilantin (an anticonvulsant drug) can prevent the abnormal activity from occurring, and also prevent puromycin from producing amnesia (Cohen, Ervin, and Barondes, 1966). Thus, the amnesic effects of the drug might not be a result of suppression of protein synthesis.

Dunn (1980) reviewed a number of studies that have demonstrated the reversal of the amnesic effects of inhibitors of protein synthesis. That is, the inhibitor of protein synthesis administered alone produces amnesia for a learned task, but when another substance is administered along with the protein synthesis inhibitor, no amnesia is seen. And it is especially significant that the substances that reverse the amnesic effect do not prevent the suppression of protein synthesis. That is, the animals manage to form long-term memories even though protein synthesis is inhibited (by 80 to 90 percent, in most cases) in their brain.

The nature of the chemicals that reverse the amnesic effects of inhibitors of protein synthesis is very diverse, including general stimulants like strychnine, nicotine, and caffeine; catecholamine agonists like amphetamine, L-DOPA, and imipramine; and hormones such as ACTH, antidiuretic hormone, and adrenocortical steroids. Even enkephalin has been shown to reverse the amnesic effect of puromycin. (See Dunn, 1980, for specific studies.)

Some studies have shown that inhibitors of protein synthesis produce amnesia only when they are injected into certain parts of the brain. For example, Eichenbaum, Quenon, Heacock, and Agranoff (1976) found that CXM produced amnesia when it was infused into the basal ganglia (either corpus striatum or amygdala) or posterior lateral thalamus. Infusions into the anterior lateral thalamus, medial cortex, or midbrain reticular formation were ineffective. The specificity suggests that these structures may vary in their importance to retention of the task that the animals were trained on. However, the fact that the amnesic effects of CXM can be reversed by other compounds means that we cannot conclude that suppression of protein synthesis was what caused the amnesia.

I personally believe that protein synthesis is essential to the formation of memory, because there appears to be no other way for long-term physical changes to be initiated and maintained. However, I must admit that there is no compelling experimental evidence that confirms this belief. It is quite possible that we will never find a drug that effectively prevents consolidation by general suppression of protein synthesis. Perhaps we will have to find out *which* proteins are involved in consolidation, and how they work, before we can prove that their *selective* inhibition can prevent short-term memories from becoming long-term ones. (We may be in for a long wait.)

CONCLUSIONS

In this chapter we saw that the memory process appears to consist of two stages: short-term memory and long-term memory. Short-term memory consists of immediately perceived information, which is encoded in terms of existing long-term memories. In addition, activated (that is, retrieved) long-term memories join the new information. Thus, some investigators refer to short-term memory as "working memory." Short-term memory, which appears to exist in the form of reverberating neural activity, is subsequently converted into long-term memory. This process includes modification of existing information as well as the entry of new information into long-term memory.

Long-term memory appears to consist of alterations in neuronal structure, such as growth of new synaptic connections or reinforcement (or perhaps deactivation) of existing ones. Since proteins play an essential role in structural changes, serving as enzymes and structural elements, the control of protein synthesis was described. Although molecular biologists have made much progress in their study of protein synthesis, neuroscientists have been less successful in applying this knowledge to the study of the memory process. We shall probably have to identify the particular structural changes that take place in neurons before the biochemical approach will yield important information. And that will not be easy.

KEY TERMS

acetoxycyclohexamide (AXM)
 (*ass et ox ee sy klo HEX a myde*) p. 614
adenine (*AD a neen*) p. 598
amino acid (*a MEE no*) p. 596
amino acid acceptor end (of transfer RNA) p. 600
amino group p. 596
anisomycin p. 614
anterograde amnesia p. 584
anticodon loop (of transfer RNA) p. 600
carboxyl group p. 596
codon p. 598
cyclohexamide (CXM) p. 614
cytosine (*SY to seen*) p. 598
denaturation p. 597
deoxyribonucleic acid (DNA) p. 598
electroconvulsive shock (ECS) p. 587
guanine (*GWA neen*) p. 598
histone protein p. 602
isolated cortical slab p. 592
long-term memory p. 582

messenger RNA p. 598
nonhistone protein p. 602
nucleosome (*NEW klee o sohm*) p. 602
passive avoidance task p. 587
peptide bond p. 596
polypeptide p. 596
posttraumatic amnesia p. 584
puromycin p. 614
retrograde amnesia p. 584
reverberation p. 591
ribonucleic acid (RNA) p. 598
RNA polymerase p. 611
short-term memory p. 582
temporal gradient (of retrograde amnesia) p. 587
thymine p. 598
transcription p. 600
transfer RNA p. 600
translation p. 600
uracil (*YOOR a sil*) p. 598

SUGGESTED READINGS

LEWIS, D. J. A cognitive approach to experimental amnesia. *American Journal of Psychology*, 1976, *89*, 51–80.

McGAUGH, J. L. and HERZ, M. J. *Memory Consolidation.* San Francisco: Albion, 1972.

These references present the conflicting views concerning the nature of the consolidation process and the reasons for experimentally produced amnesia.

ROSENZWEIG, M. R., and BENNETT, E. L. (eds.) *Neural Mechanisms of Learning and Memory.* Cambridge, Mass.: MIT Press, 1976.

This book contains a number of excellent chapters by experts in the field. You should also consult recent volumes of *Annual Review of Psychology* and *Annual Review of Neuroscience*.

19

Anatomy of Language and Memory

Where are memories located? Are short-term memories and long-term memories located in the same or different regions of the brain? I am happy to be able to report that we have some real answers to these questions. The search for the location of long-term memory has been a tedious one, but we are finally getting some interesting results, especially in the study of the human brain.

IN SEARCH OF THE ENGRAM

I have appropriated the title of this section from a provocative paper by a famous and very influential physiological psychologist, Karl Lashley (Lashley, 1950). By **engram,** Lashley meant the physical basis of long-term memory. In a series of experiments ranging over several decades, Lashley concluded, "It is not possible to demonstrate the isolated localization of a memory trace anywhere within the nervous system. Limited regions may be essential for learning or retention of a particular activity, but within such regions the parts are functionally equivalent. The engram is represented throughout the region" (p. 478).

Lashley called this principle *equipotentiality.* His conclusion was based on the results of many studies that showed that subtotal damage to any part of a particular region of cortex would not destroy individual memories. Let me be more specific. When a rat was trained to perform a visual discrimination and some, but not all, visual cortex was then removed, the animal would give evidence of some retention of the information it needed to perform the task. That is, it might not perform correctly after the surgery, but it would relearn the task in a few trials, indicating that some information was left in long-term memory. The point is that it did not matter which part of visual cortex was left intact; one part served as well as any other. Hence the term equipotentiality.

As we shall see, Lashley's conclusion still appears to be valid. Since the cortical areas of a rat brain are not so specialized as the brain of a primate, the "limited regions" that Lashley spoke of are very broad, indeed. For example, one must remove approximately one third of a rat's cortex to abolish performance on a simple brightness discrimination. In contrast, the brain of a human contains regions that are much more specialized in their functions. Damage to very restricted portions of neocortex can lead to very specific memory deficits. If one studies the localization of memories in the rat brain, one gets the impression that memories are very diffusely represented. However, if one studies the effects of human brain damage, the impression is quite the opposite; specific classes of memories appear to reside in specific regions of cortex.

Learning without Cortex

So far I have referred only to neocortex as a possible repository of long-term memory. Several experiments have shown that animals can learn and retain information even in the total absence of cortex. In an early study, Girden, Mettler, Finch, and Culler (1936) found that a decorticated dog could learn to make a leg withdrawal in anticipation of an electrical shock that was preceded by an auditory, thermal, or tactile signal. The animal could even learn to respond to a bell that was followed by a shock, but to withhold its response when a tone was presented. (The tone was not followed by a shock.) Thus, it could not be argued that the dog merely became sensitized to respond to all kinds of stimuli. (See **FIGURE 19.1.**)

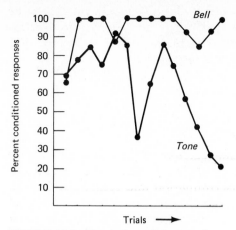

FIGURE 19.1 Classical conditioning of a dog whose cortex was removed. (From Girden, E., Mettler, F. A., Finch, G., and Culler, E., *Journal of Comparative Psychology*, 1936, *21*, 367–385.)

A more recent study (Norman, Buchwald, and Villablanca, 1977) demonstrated the acquisition of a classically-conditioned eyeblink response in cats with complete brainstem transection at the level of the pons. Since the sensory input to the brain and the motor output to eye muscles is through cranial nerves that are located caudal to the transection, the association had to be made within the lower brainstem. Clearly, cortex is not needed for learning.

Despite the fact that learning can occur in subcortical brain regions, most of the discussion in this chapter will concern the neocortex. Since sensory information must pass through subcortical regions, and since subcortical regions play such an important role in the control of movement, we could say that the entire brain participates in the memory process, in one way or other. But most of the available evidence suggests that memories themselves are most likely to be stored in neocortex. In humans, especially, cortical lesions produce the most profound kinds of memory loss. There is even some evidence that the neocortex normally suppresses the ability of subcortical regions to make associations, and that these areas do so only when the neocortex is removed.

Lashley (1950) found that rats could relearn a brightness discrimination after their visual cortex was removed. Apparently, the associations were made in subcortical regions (including the superior colliculi) that receive visual information. But since the animal had to learn the task again after the surgery, it appears that the subcortical associations were made only after the cortex was removed. In contrast, lesions of the superior colliculi and surrounding tissue had no effect on the performance of a rat with an intact neocortex. (See **FIGURE 19.2** on the following page.)

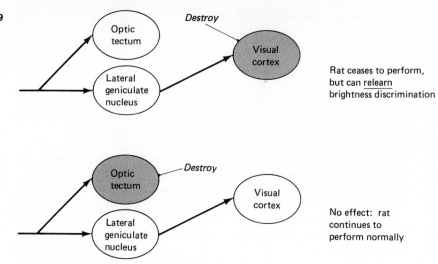

FIGURE 19.2   A schematic summary of the experiments by Lashley.

LONG-TERM MEMORY STORAGE

Human Brain Damage

The study of cases of damage to the human brain has provided some fascinating insight into the localization of various memory systems. First, here are some of the conclusions that I shall make.

1. Short-term memories and long-term memories are located in the same general regions.
2. Memories are stored on a sensory-specific basis, with visual memories being stored in visual association areas of cortex, auditory memories in auditory association areas of cortex, etc.
3. In humans (and in some other animals, as well), specific memories are not limited to particular sense modalities; a memory of a rose is an image, a smell, and perhaps the prick of a thorn. These separate sensory constituents of a single experience are linked together by means of connections between the various regions of association cortex.

   Some very specific deficits can be seen after brain lesions in humans. Most of these occur as a result of an occlusion of a cerebral artery by a blood clot, which destroys a region of brain by depriving it of oxygen and nutrients. Brain lesions can also be produced by tumors or, as we shall see, by the surgeon's knife. The behavioral deficits that follow damage to the human brain are very instructive. In some cases, specific kinds of memories appear to be lost. In other cases, memories seem to be intact, but specific kinds of memories can no longer control certain classes of motor responses. Finally, the interchange of information that is stored in various regions of the brain can be disrupted by lesions that destroy their interconnections.

Evidence for Sensory-Specific Memory Storage: The Aphasias

The area of human auditory association cortex that is most important for the analysis of speech is located in the posterior part of the **superior temporal gyrus** of the speech dominant hemisphere (in most cases, this is the left hemisphere). This auditory association area is usually referred to as **Wernicke's area,** after its discoverer. (See **FIGURE 19.3.**)

Damage to most or all of Wernicke's area does not impair a patient's ability to articulate speech sounds, but it does seem to remove the meaning of what he or she says. It also impairs recognition of the speech of other people. Here is an example of the speech of a person with severe Wernicke's aphasia. Note that the speech is filled with wordlike sounds that are devoid of any meaning at all.

Examiner: What kind of work did you do before you came into the hospital?

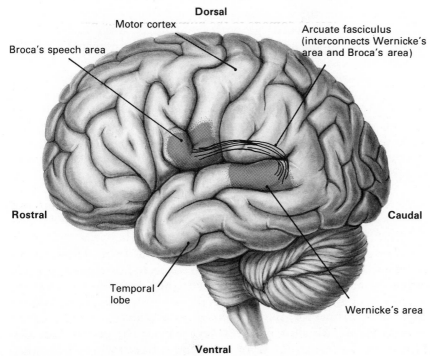

FIGURE 19.3   Speech areas within the left hemisphere.

Patient: Never, now mista oyge I wanna tell you this happened when happened when he rent. His—his kell come
down here and is—he got ren something. It happened. In
thesse ropiers were with him for hi—is friend—like was.
And it just happened so I don't know, he did not bring
around anything. And he did not pay it. And he roden all
o these arranjen from the pedis on from iss pescid. In these
floors now and so. He hadn't had em round here. (Kertesz,
1980)

The picture is this: fluent speech with very little content.

In contrast, damage to Broca's area, a region of frontal cortex
just anterior to the inferior portion of precentral gyrus (motor cortex)
leads to difficulties in speech *production*. (See **FIGURE 19.3.**) The
patient has great difficulty in producing words. The deficit is the
opposite of Wernicke's aphasia; the speech of a patient with Broca's
aphasia is anything but fluent, but the words usually convey meaning. For example, a patient said the following when he was asked
about a dental appointment: "Yes . . . Monday . . . Dad and Dick
. . . Wednesday nine o'clock . . . 10 o'clock . . . doctors . . . and . . .
teeth" (Geschwind, 1979, p. 186).

The simplest explanation for these disorders is the following:
Words are recognized by mechanisms in Wernicke's area; that is,
memories of the sounds of words are stored there. The "idea" of a
word is translated into reality by mechanisms in Broca's area; that
is, memories of the patterns of muscle movements of particular
words are stored there. Lesions in Wernicke's area impair a person's
memory for what words go with what ideas, whereas lesions in
Broca's area impair a person's memory for what movements of the
speech apparatus (tongue, lips, larynx, etc.) go with what word. I
must emphasize that the term "simplest explanation" really means
"simplistic explanation" in this case; these disorders are more complex than I have indicated here. For example, a patient with Broca's
aphasia usually suffers from some deficits in speech comprehension
as well as speech production. But, as they say, we must walk before
we run; complicating details can be appreciated only after the broad
outlines are understood.

WERNICKE'S APHASIA AND MEMORY FOR SPELLING.   Subtotal
damage to Wernicke's area (producing a very mild Wernicke's
aphasia) results in difficulty in discriminating between closely related sounds, such as *b* and *p* or *d* and *t* (Luria, 1970). These results
are not difficult to interpret; it would be expected that a considerable
degree of analysis is necessary to distinguish between such similar
sounds, and Wernicke's area, after all, is a part of *auditory* association cortex. However, minor damage to Wernicke's area produces
another effect, which is not quite so obvious. The patient will also
have difficulty in writing; words will tend to be misspelled. However,
the errors will not be random. The patient will tend to interchange

letters that represent similar sounds. For example, the patient might write *tip* instead of *dip*. In contrast, the letters *o* and *c*, which are certainly more similar in shape than *d* and *t* are, are less likely to be interchanged, since they represent different sounds. Why should the spelling errors that are made by people with damage to auditory association cortex be related to the sounds that are represented by the written letters?

Let us see how these results can be explained. Take the case of concrete nouns—words that represent physical, tangible objects. It seems plausible that the following assumptions are true: (a) These words are learned by associating auditory representations of the word in the brain (in Wernicke's area) with the corresponding sensory representations (visual, tactual, etc.) that are located elsewhere in the brain. (b) Writing, which is acquired later in life than the ability to understand spoken words, depends upon associations between writing movements made by the hand and acoustically learned representations of the words. If you observe children learning to write, you will notice that they spell out the sounds each letter makes as they write a word. Even when an adult spells a long, relatively unfamiliar word, the sounds of the letters are pronounced, or at least imagined in an acoustic manner.

Here are a few little experiments for you to try. Try to write a rather long word that you know how to pronounce, but do not write very often—for example, "antidisestablishmentarianism." You will probably find that you conjure up an "auditory image" of the way the individual syllables of the word sound, and then write out the corresponding letters. Now try writing the word again (on a fresh sheet of paper) but this time sing "Happy Birthday" (or any other song you know) as you attempt to write the word. Unless you are much more talented than I am, you will find that you have to write during pauses in the words of the song. However, if you try to draw a picture instead of writing a long word, there should be no interference. Draw a picture of a simple house, or anything else, while you sing your song. No problem, right?

I urge you to try these little experiments. They prove that talking and singing do not interfere significantly with complex hand and finger movements themselves—these activities only interfere with hand and finger movements that are producing *words* that are not the same as the ones that are being sung or spoken. Thus, writing and speaking (or singing) appear to draw upon the resources of the same brain mechanisms, whereas picture-drawing and speaking (or singing) do not.

Now try a final experiment. *Copy* the word "antidisestablishmentarianism" (or whatever else you chose), writing it beneath your first successful attempt. At the same time, sing your song. It was much easier, wasn't it? This experiment suggests that visual information (the sight of the written letters) has some fairly direct access to mechanisms that control writing movements, independent of the acoustic representation of the way the word sounds. These conclusions are summarized schematically in **FIGURE 19.4.**

If our ability to write depends upon a transfer of information between acoustically-acquired representations of sounds and motor movements that represent those sounds, then damage to auditory association cortex should not impair written symbols that represent visually acquired memories. (Remember, you can draw a picture or copy a word while singing; therefore, drawing or copying is at least somewhat independent of acoustic memories.) This does indeed seem to be true. Minor damage to Wernicke's area in the left hemisphere does not impair a person's ability to draw pictures that represent visual memories.

There is even more impressive evidence concerning the location of the memory of how to write words. Chinese patients with lesions in Wernicke's area, like Westerners, suffer from impairments in auditory recognition, mistaking words that sound similar; nevertheless, they are able to write very accurately, unlike Westerners with the same lesions (Luria, 1970). Chinese is written in the form of pictographs, which are not related to the sounds of the words they represent. When Chinese people write a word, they do not need to gain access to its acoustic representation in memory in order to spell it; a character that represents a concrete noun can be based upon associations between a visual representation of that word and the movements that are necessary to produce it. Similarly, the writing ability of deaf people is not affected by lesions in Wernicke's area. Obviously, lesions of Wernicke's area do not produce impairments in the understanding of spoken words, since deaf people cannot hear. However, the fact that the lesions do not affect their ability to spell words indicates that their memory for the spelling is not stored in Wernicke's area. Therefore, we can conclude that the location of a specific class of memories (how to write words) depends upon the sensory modality by which the information is received. A person who can hear uses Wernicke's area to store the spelling of words, whereas a deaf person does not. Presumably, they use visual association cortex. (Here is a problem for you to work out. Helen Keller, who was deaf and blind, learned to talk and write by means of touch. Her teacher spelled out words in her hand, and eventually she learned braille. The objects denoted by the words had to be felt, tasted, or smelled. Where were Helen Keller's memories for the spelling of words stored?)

A final example, which wraps things up very nicely, is provided by Japanese patients. The Japanese language makes use of two kinds of written symbols. *Kanji* symbols are pictographs, like Chinese characters. *Kana* symbols are phonetic representations of syllables; thus, they encode acoustic information. Left temporal lobe lesions in Japanese patients interfere with the writing of *kana* symbols (acoustic), but not *kanji* symbols (visual). In contrast, other brain damage (probably to visual association areas) impairs a patient's ability to write

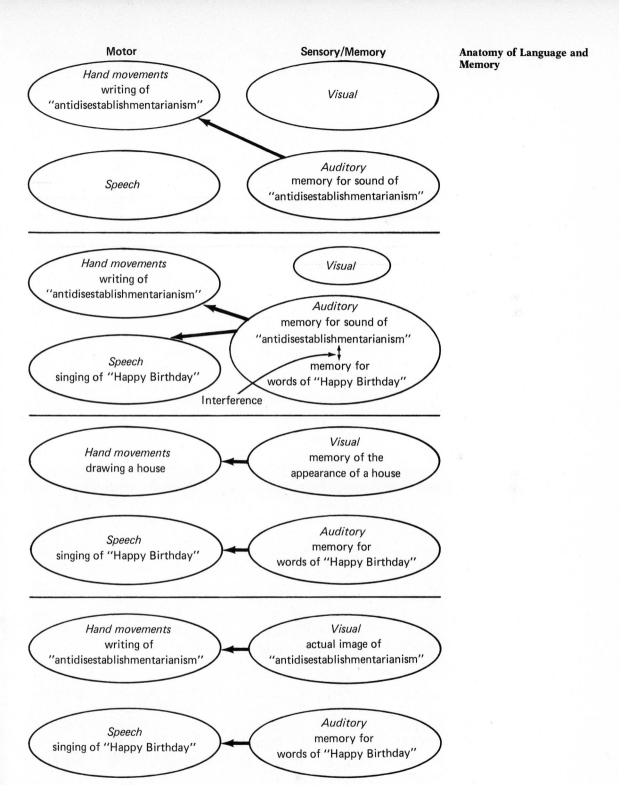

FIGURE 19.4   A schematic representation of a possible explanation for
the results of the do-it-yourself "experiment."

kanji symbols, but not *kana* symbols (Sasanuma, 1975). The results certainly suggest that different classes of memories are stored in different locations in the brain.

CONDUCTION APHASIA AND TRANSCORTICAL TRANSFER OF INFORMATION.   Norman Geschwind (1965; 1972; 1975) has described a number of specific disorders in humans that can best be explained in terms of interrupted connections between various cortical association areas. (See **Figure 19.5.**) One of these disorders, ***conduction aphasia,*** appears to result from damage to fibers of the ***arcuate fasciculus.*** This fiber bundle interconnects Wernicke's and Broca's areas. (Refer back to **FIGURE 19.3.**) Patients with conduction aphasia generally exhibit spontaneous speech; they can express their thoughts in words. They can usually carry out verbal commands by means of movements, so they obviously comprehend the meaning of the words they hear. However, they are very poor at *repeating* words. Geschwind cites the example of a patient who was asked to repeat the word "president." He replied, "I know who that is—Kennedy." (The patient was tested in the early 1960s.) So information can get from Wernicke's area to Broca's area, but only in an indirect fashion. Presumably the word "president" evoked an association between the word and some sort of representation (probably visual) of John Kennedy. Connections between visual association cor-

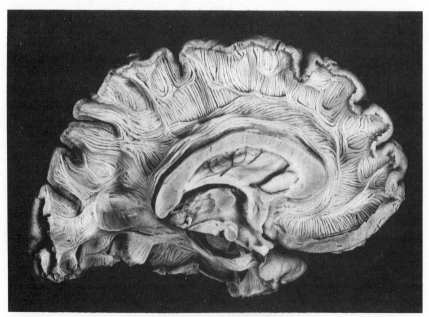

FIGURE 19.5   Bundles of cortical association fibers of the left hemisphere. (From Gluhbegovic, N., and Williams, T. H. *The Human Brain: A Photographic Atlas.* Hagerstown, Md.: Harper & Row, 1980.)

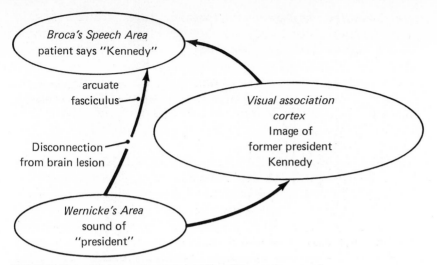

FIGURE 19.6 A schematic representation of the explanation for conduction aphasia.

tex and Broca's area conveyed this information, and the patient verbalized the image. (See **FIGURE 19.6**.)

A surprising fact is that patients who suffer from conduction aphasia are usually able to identify numbers. They can read them, and can repeat them back. In fact, they show obvious signs of relaxation when they are asked to repeat numbers, as opposed to nonnumerical words. In these patients, auditory association cortex is disconnected from Broca's area but remains connected to visual association and somatosensory association cortex, so the route from temporal lobe to speech center must be mediated indirectly, by arousing associations elsewhere. Geschwind points out that numbers are usually associated with specific visual and somatosensory representations. We teach a child to count on his or her fingers, and to associate numbers with images—the number "two" with a visual image of two objects, for example. Thus, there are previously learned mediators that represent numerical information. There are, of course, mediators for nonnumerical information, but they are generally more ambiguous. The image evoked by the word *chair* might be verbally described by a patient suffering from conduction aphasia as "sit," or "furniture."

The likelihood that this explanation is correct is enhanced by close examination of the errors patients make when asked to repeat numbers. "Fifty-five percent" becomes "fifty-five progum," and "eleven plus eight" becomes "eleven, eight . . . nineteen." "Three-quarters" becomes "three, four," even though the patient could say "penny, nickel, dime, *quarter*." Presumably, there are good visual associations available to mediate the transmission of the representation of the coins. A patient could read "28" or "twenty-eight" equally well, but could not, for instance, read "train." He said "travel" instead. Presumably, an image went from visual association cortex to auditory cortex, where the "meaning" of the image was evoked. The "meaning" was communicated back to visual cortex, and then Broca's area translated the concept into words: "travel" for "train." The numerical information survived the successive transformations much better. In more severe cases, patients lose the ability to say "sixty-eight" but will say "six, eight" instead.

ISOLATED SPEECH MECHANISMS: A SUFFICIENT CONDITION FOR LEARNING AND RETRIEVAL.   So far, a good case has been made for the importance of Wernicke's area for the comprehension of speech and for memories of particular words, whereas Broca's area is necessary for translating words into movements of the vocal apparatus. However, it is entirely possible that other cortical areas are necessary, as well. It is possible that the memories (i.e., for the sounds of words and motor patterns of speech) are stored elsewhere, in some general "filing cabinet" of memories, but that mechanisms in Wernicke's area and Broca's area "know" how to retrieve these memories. This possibility is entirely consistent with the data that have been presented so far, and is schematized in **FIGURE 19.7.**

There is evidence against this explanation. A fascinating case of brain damage suggests that Wernicke's area and Broca's area (and not much else) are the only cortical areas that are needed to identify spoken words, pronounce them, and learn new sequences of words. A woman suffered a brain lesion that did not damage Wernicke's area, or Broca's area, or their interconnections via the arcuate fasciculus, but did isolate both of these areas from visual cortex and much of somatosensory cortex (Geschwind, Quadfasel, and Segarra,

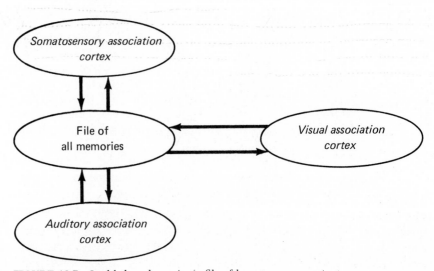

FIGURE 19.7   Could there be a single file of long-term memories?

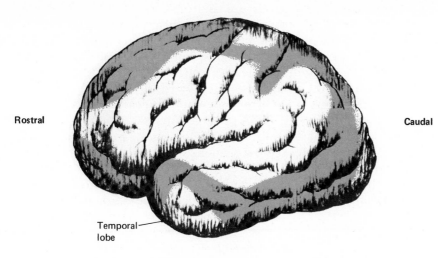

Dorsal

Rostral

Caudal

Temporal
lobe

Ventral

FIGURE 19.8   Brain damage produced by inhalation of a poison gas in
the patient reported by Geschwind, N., Quadfasel, F. A., and Segarra,
J. M., *Neuropsychologia*, 1968, *13*, 229–235.

1968). (See **FIGURE 19.8.**) The lesion, which was produced by the
inhalation of gas from a faulty water heater, was almost the reverse
of the lesion that causes conduction aphasia. Instead of damaging
the connections between Wernicke's area and Broca's area, leaving
everything else intact, the lesion destroyed the other connections,
leaving the Wernicke-Broca circuit intact.

The symptoms were also the reverse of conduction aphasia; what
this woman could do, the conduction aphasic could not do, and vice-
versa. She could repeat everything that was said to her, but she never
spoke spontaneously during the nine years she survived after the
lesion occurred. In addition, she gave no sign of understanding what
was said to her. If the first part of a poem were recited she would
continue with the last part. She could sing and was even capable of
learning new songs. In addition, she would correct errors of grammar; if someone said a sentence that was grammatically incorrect,
she would say it properly, rather than repeat it back word-for-word.
Thus, the area of cortex that was spared by the lesion contains the
rules of grammar, which is an exceedingly complex set of long-term
memories.

PHONEMIC APHASIA: EVIDENCE FOR TWO DISTINCT READING
MECHANISMS.   When we read, most familiar words are identified
quickly, "by sight." Unfamiliar words have to be "sounded out"
when they are first read; that is, we translate individual letters or
small groups of letters into sounds (linguists call these sounds ***phonemes***) and then put the sounds together to make a word. There is

evidence from human brain damage that the neural mechanisms that control these processes are anatomically distinct. Unfortunately, the sites of the crucial brain regions are not known.

A person with ***phonemic aphasia***, caused by damage to the left hemisphere, can read concrete nouns, but has difficulty in reading abstract words. In addition, when a concrete word is read, it often gets translated into a closely-related word; *dream* might become "sleep." Sometimes, visual errors are made; *origin* is read as "organ" (Patterson and Marcel, 1977). A possible interpretation is that these patients can recognize familiar words "by sight," but cannot translate a sequence of letters into a sequence of sounds (phonemes).

Patterson and Marcel (1977) obtained evidence from two aphasic patients that supports this hypothesis. First, the patients correctly read words that could be imagined, like *lecture* or *marriage*. They had

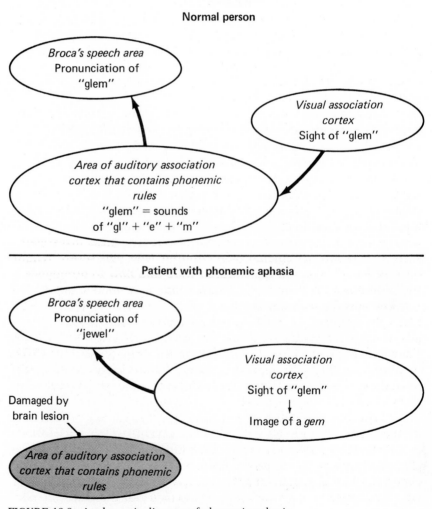

FIGURE 19.9   A schematic diagram of phonemic aphasia.

difficulty with words that were hard to imagine, like *event* or *scene*. Errors were sometimes related to the image that the word would evoke (*disaster*—"like danger, airplane, crashed;" *event*—"athletics the same;" or *fact*—"truth"). Sometimes, the errors were visually related to the word that was being read (*patent*—"patient;" *gain*—"grain;" or *agony*—"anger"). The patients did *not* have conduction aphasia; they were perfectly capable of repeating a word that was spoken to them. They were also good at repeating nonword sequences of sounds, like "dube," "plosh," or "widge." Only *reading* was impaired.

The crucial observation was that the patients failed hopelessly at reading nonword sequences of letters. When they were confronted with sequences like "dube," "plosh," or "widge," they just stared at them. One patient got 3 out of 20 sequences correct, and the other made no correct pronunciations. When attempts were made to pronounce the letter sequences, the errors indicated that the patients were trying to read "by sight," and were unable to sound the letters out. For example, *cit*—"city;" *frute*—"flute;" *rud*—"naughty." An especially instructive example is the following: *glem* became "jewel." The nonword *glem* looks like *gem*, and presumably the patient conjured up an image of a gem. This was transmitted from visual association cortex to Broca's area, in a manner that represented the *idea* (or perhaps the image) of a gem, but not a specific word. An appropriate word for this idea—"jewel"—was selected and spoken. Figure 19.9 presents this hypothesis in schematic form. (See **FIGURE 19.9.**)

It is possible that some congenital forms of **dyslexia** (reading difficulty) may be related to this disorder. Perhaps some people have difficulty in translating letters into phonemes, but can nevertheless associate visual images with words. Support for this hypothesis was provided by Rozin, Poritsky, and Sotsky (1971). These investigators studied a group of second-grade children who had severe reading disabilities. Even when the children were told how to pronounce *at*, they could not read the words *cat*, *fat*, *mat*, and *sat*.

The authors assembled a list of Chinese characters, and tried to teach the children to read them. The children were taught the English words that went with the characters, not the Chinese words. They did very well. After 2½ to 5½ hours of individual tutoring they were able to learn 30 characters, and could read sentences made from these characters. Figure 19.10 shows the final test that the children were given to read. (The English version is written for your information only; it was not included in the test.) (See **FIGURE 19.10** on the following page.)

This accomplishment is remarkable. The reading deficit that the children displayed was not simply due to poor teaching, because the experimenters also tutored them in reading English words, but met with no progress. It appears that (1) the children could readily learn words (after all, they were able to speak), (2) they could memorize

父　買　黑　車

Father　buys　black　car.

這　人　不　見　黑　家　跟　二刀

This　man　(does)not　see　black　house　and　two knives.

哥哥　說　母　用　白　書

Brother　says　mother　uses　white　book.

你　要　一　大　魚　跟　黑　家

You　want　one　big　fish　and　black　house.

他　說　"哥哥　有　小　口"

He　says　"brother　has　small　mouth."

好　哥哥　不　給　人　紅　車

Good　brother　(does)not　give　man　red　car.

FIGURE 19.10　The final examination used in the experiment by Rozin and his colleagues. The English words were not included in the test.

complex visual stimuli (the Chinese characters), and (3) they could learn associations between the two and use them to read sentences that they could comprehend. Their primary deficit appears to be the coding of letters and letter sequences into phonemes. They performed very much like the patients who were studied by Patterson and Marcel.

Is there any evidence that brain damage is associated with congenital dyslexia? The answer seems to be yes. Galaburda and Kemper (1979) obtained the brain of a patient with a congenital dyslexia (reading disability) who died in an accident. They found that the *planum temporale*, a part of Wernicke's area, was the same size on both sides of the brain. In contrast, the left planum temporale is usually much larger in normal patients with speech-dominant hemispheres (Galaburda, Le May, Kemper, and Geschwind, 1978). Per-

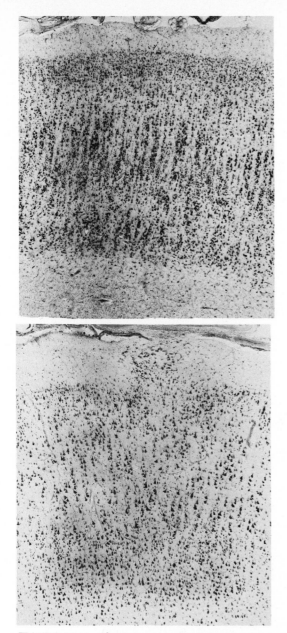

FIGURE 19.11 Photomicrographs of the left planum temporale (a portion of Wernicke's area) of a normal person (top) and a person who had dyslexia (bottom). Nissl stain. (Photographs courtesy of A. Galaburda.)

haps more significantly, the microscopic appearance of this region was abnormal. Figure 19.11 shows a section through the left planum temporale of a normal person (top) and the dyslexic accident victim (bottom). Notice that there is a regular columnar arrangement of cells in the normal brain, but not in the brain of the dyslexic accident victim. (See **FIGURE 19.11.**)

FIGURE 19.12 A few well-chosen lines can evoke the image of a well-known face. (Courtesy of United Press International Photos.)

PROSOPAGNOSIA: AN EXAMPLE OF A NONVERBAL MEMORY DEFICIT. Although human memory is most conveniently studied by means of language, there are many memories that are nonverbal in character, and that are usually retrieved by nonverbal means. So far, my examples have been related to verbal memories. It is conceivable that nonverbal memories might not be stored in specific regions of the cerebral cortex.

An incredibly specific disorder shows that this is not the case; at least one class of nonverbal memories requires a specific cortical area. The disorder is called *prosopagnosia,* from *prosopon* ("face"), *a-* ("not"), and *gnomon* ("one who knows"). Thus, a person with prosopagnosia does not know (i.e., cannot recognize) faces (Geschwind, 1979).

The recognition of faces is a very complex process. A skillful caricaturist can abstract the essential features of an individual face in a few strokes of a pen. (Most people will quickly recognize the face in **FIGURE 19.12.**) And consider the fact that each of us can identify hundreds—perhaps thousands—of faces. Any teacher learns the identity of a very large number of students over the course of his or her career. (This is not to say that we retain the names that go with the faces.)

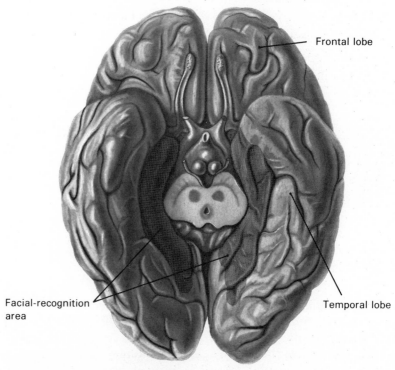

FIGURE 19.13 The "facial recognition area," destruction of which causes prosopagnosia. (From Geschwind, N. Specializations of the human brain. *Scientific American,* 1979, *241,* 180–201.)

Prosopagnosia is produced by lesions of the underside of portions of the temporal and occipital lobes, as illustrated in **FIGURE 19.13.** A person with a severe deficit cannot even recognize a member of his or her immediate family by sight. A husband might not recognize his wife when she walks into the room. He has not forgotten who she is, because he identifies her as soon as she speaks—he recognizes the sound of her voice. The disorder is not simply a general deficit in complex visual perception, since most patients can read and name objects. He or she can often correctly match a full-face photograph with a profile view of the same person, which is surely a very complex perceptual task. What is gone is the *recognition*—the identification of a particular face as belonging to a particular person. That certainly seems like a very specific memory deficit.

Geschwind reports that another investigator is now studying the anatomy of visual recognition of faces in monkeys. These animals can recognize the faces of other monkeys from photographs. It will be interesting to learn whether a specific region of cortex of the monkey brain can be identified with this function. If so, it will be possible to perform experimental analyses of its connections with other brain mechanisms that cannot be performed with humans.

HUMAN BRAIN DAMAGE AND LONG-TERM MEMORY: CONCLUSIONS. The foregoing data in this section support the following conclusions.

1. Different classes of memories appear to be located within different regions of the brain. The classification appears to be by sensory modality; visual memories are stored in visual association cortex, auditory memories are stored in auditory association cortex, and so on. In addition, "motor memories" (such as the patterns of movement that constitute speech) are stored in areas of frontal cortex.
2. Complex, multisensory memories require connections among the various regions of association cortex, and if these are disrupted, predictable memory deficits occur.
3. In order for retrieval of long-term memories to produce specific motor acts (like speech), there must be connections between the appropriate area of association cortex and regions of cortex that mediate the motor acts. Disruption of the connection prevents direct expression of the retrieved memory in a motor act. In some cases (such as conduction aphasia) the memory can follow an indirect route, passing first to another sensory association area, from which it is relayed to a motor area.

Lesion Studies with Monkeys

Obviously, we must study humans if we are to understand how and where the human brain stores memories. As you have seen, much has been learned by studying people whose brains have been damaged. However, this approach has some serious limitations. For one thing, the location and extent of a lesion cannot be determined with any degree of precision until the patient's brain is sliced and stained. It goes without saying that the patient must first die before this can happen, and some patients stubbornly outlive the investigators who are studying them. In addition, the family of a deceased patient does not always permit the brain to be removed. The technique of computer-assisted tomography (CAT scan), which was described in Chapter 7, has helped this situation somewhat, but the images of the brain that are obtained by this method do not permit precise localization of lesions. Perhaps techniques that are now under development will give investigators an accurate picture of lesions in living brains, but at present there is no substitute for postmortem histological examination.

A second problem will persist even if perfect localization of the lesion can be achieved. Although we know very much about the detailed anatomy of the brains of many species of animals, we know very little about the interconnections of various regions of the human brain. If you recall the methods that are used to investigate the detailed circuitry of the brain, you will realize the reason for our ignorance. Histological techniques such as Nissl stains and myelin stains permit the anatomist to locate nuclei and fiber bundles and identify the more obvious interconnections. However, in order to determine what is connected to what it is necessary to use degenerating-axon stains or the more recently developed techniques that use horseradish peroxidase and amino acid autoradiography. And these techniques require a lesion to be made, or a chemical to be injected into the *living brain*. The animal is then killed, and the pathways are traced. Obviously, we cannot use these methods on humans. Human brains are available only when a person dies, when it is too late to take advantage of axoplasmic flow or neural degeneration. What we know about connections between various parts of the human brain is based on rather crude myelin stains. The rest is conjecture, based on similarities between the human brain and the brains of nonhuman animals. This means that most of the detailed connections that you see listed in a neuroanatomy text are inferred from studies in animals other than humans; no one really knows that they are the same in human brains.

These limitations notwithstanding, studies of the organization of memory in the brains of monkeys have contributed to the understanding of human memory. Lesions can be produced much more precisely in experimental animals, whereas lesions in human brains depend upon accidents, and are seldom limited to a particular anatomical subdivision of the brain. And, of course, ethical issues preclude many kinds of investigations in humans.

VISUAL MEMORY.

Anatomy of the visual system. In monkeys, primary visual cortex contains only the first stage of cortical analysis of visual infor-

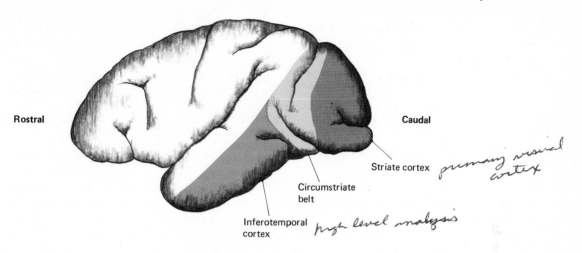

Rostral

Caudal

Striate cortex *primary visual cortex*

Circumstriate belt

Inferotemporal cortex *high level analysis*

Ventral

FIGURE 19.14   Primary visual cortex (striate cortex), inferotemporal cortex, and the circumstriate belt. (Redrawn from Gross, C. G. In Jung, R., editor, *Handbook of Physiology*. Vol. 7/38. Berlin: Springer–Verlag, 1973.)

mation. There is a second band of cortex around *striate cortex* (primary visual cortex) called the *circumstriate belt.* The most crucial region of the circumstriate belt has been called the *foveal prestriate area;* this area receives and processes information from the central (foveal) portion of the visual field, where a monkey's acuity is greatest. Indeed, when we "look at" an object, what we are doing is moving our eyes (and perhaps also our head) so that the image of the object falls within the foveal field of vision. Finally, there is a third cortical area important to the processing of visual information: *inferotemporal cortex.* These three regions are shown in **FIGURE 19.14.**

Visual information proceeds from striate cortex to the ipsilateral circumstriate belt by means of corticocortical connections. Circumstriate cortex is then connected to both ipsilateral and contralateral inferotemporal cortex; the contralateral connections are made by means of fibers of the corpus callosum. Furthermore, striate cortex and superior colliculus both project to the *pulvinar* (a thalamic nucleus), which in turn projects to and receives fibers from both circumstriate and inferotemporal cortex. (See **FIGURE 19.15** on the following page.)

Role of inferotemporal cortex in visual processing.   You will recall from Chapter 9 that monkeys with lesions of temporal cortex show "psychic blindness"; they can get around very nicely without bumping into things, but they appear to have trouble recognizing what these objects are. They will pick up items from a tray containing

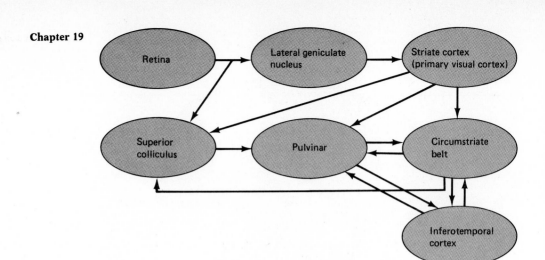

FIGURE 19.15 A schematic and simplified representation of the interconnections of various regions of the visual system.

small edible and inedible objects, eating the pieces of food and dropping the pieces of hardware. They show no fear of objects that monkeys normally avoid; Klüver and Bucy (1939) reported that these monkeys will even try to investigate the tongue of a living snake. Mishkin (1966) made brain lesions in a series of three stages and tested the monkeys' ability to perform a visual discrimination after each lesion. First he removed striate cortex on one side: no deficit. Then he removed the contralateral inferotemporal cortex: no deficit. Finally, he cut the corpus callosum. The last operation isolated the remaining inferotemporal cortex from the remaining striate cortex, and the animals now suffered from a deficit in visual pattern discrimination. Striate cortex alone could not mediate memory for this task. Neither could inferotemporal cortex that was deprived of its visual input. (See **FIGURE 19.16.**) However, lesions of the pulvinar (the only subcortical area that projects to inferotemporal cortex) do not produce such deficits (Chow, 1961). Thus, we must conclude that the transcortical connections from striate cortex to circumstriate cortex to inferotemporal cortex are the ones that convey the essential visual information.

Inferotemporal cortex appears to participate in a very high level of analysis of visual information; it is here, perhaps, that meaning is assigned to visual stimuli. Gross (1973) summarized the results of a series of studies performed in his laboratory. He reported that it was very difficult to characterize features of the stimuli that most effectively altered the firing rate of single units located there. For example, one unit responded most vigorously to the sight of a bottle brush, another to the shadow of a pair of hemostatic forceps (which are shaped somewhat like scissors), and another to a monkey hand. The shapes shown in Figure 19.17 are ordered according to the mag-

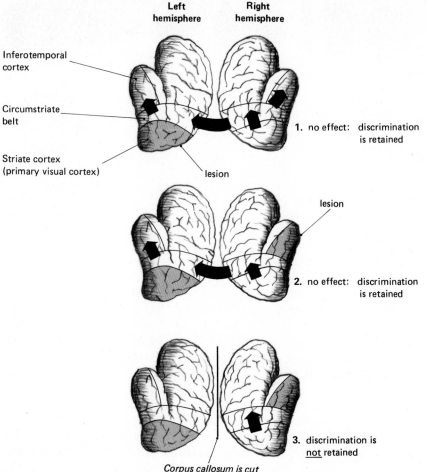

Left
hemisphere

Right
hemisphere

Inferotemporal
cortex

Circumstriate
belt

Striate cortex
(primary visual cortex)

lesion

1. no effect: discrimination
is retained

lesion

2. no effect: discrimination
is retained

3. discrimination is
not retained

Corpus callosum is cut

FIGURE 19.16   A schematic representation of the procedure used by
Mishkin et al. (1966). Not all of the control groups used in this experiment
are represented here. (Adapted from Mishkin, M., Visual mechanisms
beyond the striate cortex. In *Frontiers in Physiological Psychology,* edited
by R. W. Russell, New York: Academic Press, 1966.)

nitude of the response they produced in this last neuron. (See **FIG-
URE 19.17** on the following page.) Gross does not say that this neuron detects monkey hands, because this claim could be made only
after showing that no other stimulus produced a better response. But
it is the case, nevertheless, that neurons in inferotemporal cortex are
most idiosyncratic in the features they respond to. Many of them, as
Gross reports, respond better to a particular three-dimensional stimulus than to any silhouette made from it, suggesting a high level of
analysis of the visual information. It would be interesting to know
whether small lesions of inferotemporal cortex (they would have to
be made bilaterally, or in one hemisphere of a split-brain monkey)
would result in losses in the ability to recognize particular classes

Ineffective   Most effective

Relative effectiveness of various stimuli
in changing activity of unit

FIGURE 19.17 A rank-ordering of the effectiveness of various stimuli in producing a response in a single neuron located in inferotemporal cortex. (From Gross, C. P., Rocha-Miranda, C. E., and Bender, D. B., *Journal of Neurophysiology*, 1972, *35*, 96–111.)

of stimuli in the way that lesions of striate cortex produce blindness in particular parts of the visual field.

SOMATOSENSORY MEMORY.   There is good evidence that another area of sensory association cortex, the **second somatosensory projection area (SII)**, plays a special role in memory. SII lies just below primary somatosensory cortex (postcentral gyrus), in the depths of the lateral fissure, which divides the temporal lobe from the frontal and parietal lobes.

Bilateral destruction of SII does not produce significant deficits in simultaneous, side-by-side judgments that involve somatosensory stimuli; the monkeys can still detect differences in coarseness of various grades of sandpaper, they can discriminate the size of different objects by touch, and they can detect differences in the weight of objects of the same size (Ridley and Ettlinger, 1976; 1978). However, these animals showed a severe memory impairment; they were unable to learn which particular somatosensory stimulus was paired with reinforcement. They could not even continue to perform a task that they had learned before the operation. The deficit is not one that affects memory ability in general; the monkeys could perform a *visually*-guided discrimination task that required long-term memory. The results suggest that the second somatosensory projection area is necessary for long-term memories related to somatosensory stimuli.

A DISCONNECTION SYNDROME IN MONKEYS.   The effects of lesions of inferotemporal cortex on visual memory, and of area SII on somatosensory memory suggest rather strongly that memories are organized in monkey cortex on the basis of sensory modality, as they appear to be in humans. Of course, since monkeys do not have the verbal abilities that humans have, there is no such thing as Broca's area or Wernicke's area, so generalizations between monkeys and humans will be limited. However, despite this limitation, there is evidence that brain lesions can produce a disconnection syndrome in monkeys, as they can in humans.

Pairs of stimuli

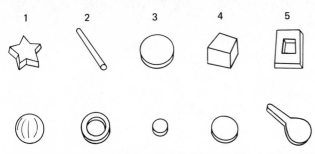

FIGURE 19.18   Pairs of stimulus cookies used by Cowey and Weiskrantz (1975). (Redrawn from Cowey, A., and Weiskrantz, L. *Neuropsychologia*, 1975, *13*, 117–120.)

Petrides and Iverson (1976) used a special procedure (developed by Cowey and Weiskrantz, 1975) to test the formation and retrieval of cross-modal memories in monkeys. They presented pairs of cookies to the animals, who were kept in absolute darkness; the shapes of these cookies are shown in **FIGURE 19.18.** One member of each pair (the positive stimulus) was molded of commercial monkey food. The other member (the negative stimulus) was made of equal parts of monkey food and sand, with added quinine for bitterness. (It should be obvious why this stimulus was referred to as negative.) The monkeys readily ate the positive stimulus and, not unexpectedly, left the negative stimulus alone after a few nibbles. When permitted to select the shapes visually (but not by touch) in a testing apparatus, the monkeys showed a distinct preference for the positive ones, which they had felt, but had never seen before. Thus, a memory that had been learned by means of tactual information was retrieved by means of a visual stimulus; hence the term cross-modal memory.

Petrides and Iverson trained two groups of monkeys, normal monkeys, and monkeys with lesions of **periarcuate cortex.** This region of cortex surrounds the **arcuate sulcus,** which is located in the frontal lobe of the monkey brain. (Do not confuse it with the arcuate fasciculus, the fiber bundle that connects Wernicke's area with Broca's area.) (See **FIGURE 19.19** on the following page.) Periarcuate cortex receives fibers from visual, auditory, and somatosensory cortex (Pandya and Kuypers, 1969). Destruction of this area prevented the monkeys from learning the cross-modal task. Perhaps they failed to do so because communication among the various sensory association areas was disrupted.

The data on the neural basis of cross-modality transfer of information in nonhuman primates is still limited, but the evidence reported so far is consistent with the human disconnection syndromes, insofar as nonhuman primates are capable of performing tasks that resemble human capabilities.

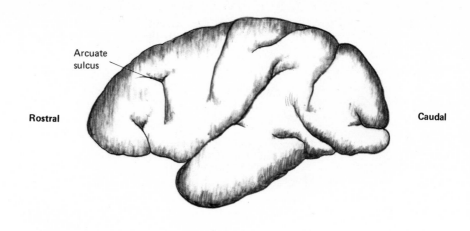

Dorsal

Arcuate
sulcus

Rostral

Caudal

Ventral

FIGURE 19.19   Location of the arcuate sulcus in the brain of the rhesus
monkey.

SHORT-TERM MEMORY STORAGE

The conclusions that were made about the locations of long-term
memories also appear to apply to short-term memories; that is, they
seem to reside in specific areas of association cortex. Although evidence about the location of short-term memories is difficult to obtain, what we do know supports the conclusion that the various
classes of short-term memories are located in the same regions as
the long-term memories that they produce.

To get you thinking about short-term memory, recall the experiments I had you do earlier. First, I asked you to write "antidisestablishmentarianism." Then I asked you to attempt to write it while
you were singing "Happy Birthday." Then I asked you to draw a
picture of a house while singing "Happy Birthday." Finally, I asked
you to *copy* "antidisestablishmentarianism" while singing "Happy
Birthday." The results of these experiments were used to support
some conclusions about long-term memory.

The reason I have brought up these experiments again is that
very similar ones can be used to demonstrate that two messages
compete with each other for storage in short-term memory only if
both messages are received by means of the same sensory modality,
or if they both must be encoded by the same neural mechanism. For
example, imagine what an artist does while sketching a picture. He
or she periodically looks up at the scene that is being drawn, and

then looks down to the sketch pad and continues work. Since it is not possible to examine the scene and the sketch pad at exactly the same time, the sketch is actually being constructed from short-term memory. Suppose that the artist is sitting next to a friend, and that the two of them are chatting. Carrying on a conversation demands a considerable amount of activity in short-term memory. The participant in a conversation must remember what the other person said, decide what to say, and then speak. Just uttering a sentence presupposes that by the time you get to the end of it, you still remember how it started. Obviously, the memory is short-term in nature. Since the artist is using two different forms of short-term memory, there is little competition between short-term memory of the scene and short-term memory of the conversation. (I should imagine, however, that if the artist's friend asked a question about the details of some famous painting, the artist would have to stop drawing while describing it.)

Now imagine a different situation. You are a stock-broker working on the floor of the exchange. Quotations are being briefly flashed on a screen in front of you. You see quotations and then write them on a pad of paper (from short-term memory, since each quotation is up on the screen very briefly). A friend comes up to you and tries to engage you in a conversation. Assuming that it was important for you to obtain the quotations, would you take up the conversation with your friend, or would you say something like "I can't talk now—I have to concentrate on this."? I imagine that you would not try to carry on a conversation while retaining verbal information in short-term memory, since the two tasks would compete. Unlike the artist, who can talk and maintain strictly visual information in short-term memory, you would have trouble remembering words and numbers that had been presented on the screen while carrying on a conversation.

We need not rely solely on my imaginary situations. Some data from people with brain damage support the conclusion that short-term memories are sensory specific, and anatomically distinct. Warrington and Shallice (1972) reported on a patient (K. F.) who had a lesion of the left parietal area. K. F. could repeat a single letter or a series of two letters if they were presented acoustically, but he had a considerable amount of trouble repeating a series of three letters. However, when he was presented with one-, two-, or three-letter sequences visually, he had no trouble repeating them. Warrington and Shallice presented patient K. F. with a stimulus and then made him perform a distraction task (reading numbers aloud) for 5, 10, 30, or 60 seconds. As soon as the distraction interval was over, the patient was asked to repeat the original stimulus. By plotting the percentage of correct responses that were given after various delay intervals, the authors could obtain an estimate of the rapidity of the decay of short-term memory. (The distraction task was used to prevent rehearsal, which can prolong short-term memory indefinitely.)

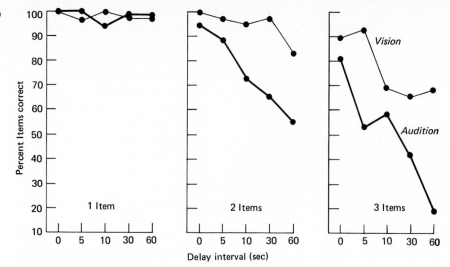

FIGURE 19.20 Short-term retention of one-, two-, and three-letter sequences presented acoustically and visually to patient K. F. (From Warrington, E. K., and Shallice, T., *Quarterly Journal of Experimental Psychology*, 1972, *24*, 30–40.)

The results are shown in Figure 19.20. K. F. had no trouble remembering a single letter presented acoustically or visually. However, short-term memory for two- and three-letter sequences showed signs of decay, the rate of forgetting being much more pronounced for acoustically-presented material. After a 60-second delay he remembered, on the average, less than one letter out of three that had been presented acoustically. (See **FIGURE 19.20.**)

Warrington and Shallice also found that K. F. did not show the kinds of acoustic confusion that a normal person would exhibit when attempting to recall visually presented material. For instance, the letters B and D, and Q and U sound very similar, whereas E and F, and O and Q are similar in appearance, but sound very different. When trying to recall letters that were seen earlier, normal people tend to make acoustic errors; that is, their incorrect responses are likely to be letters that sound like the correct responses. In contrast, patient K. F. made acoustic errors when the letters were presented acoustically, but he did not tend to do so when the letters were presented visually.

When a normal person sees a letter, he or she can easily "hear" it in the "mind's ear" as well as see it. (In fact, try looking at the letter A without thinking of what it sounds like.) Therefore, when someone sees a series of three letters, he or she can easily rehearse them acoustically *and* visually.

How can K. F.'s deficit be explained? First, let us examine some more information. Shallice and Warrington (1975) reported that patient K. F. was able to read concrete words aloud, but failed to read abstract ones. They diagnosed his condition as a form of conduction

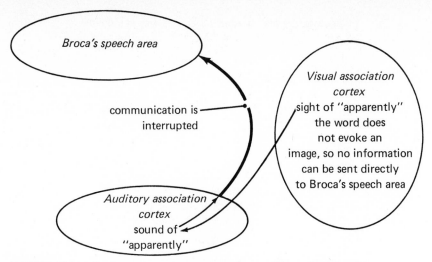

FIGURE 19.21   A schematic representation of a possible explanation for the inability of patient K. F. to read abstract words like "apparently."

aphasia. Anatomical evidence from a post-mortem was consistent with this diagnosis. Thus, when patient K. F. read a concrete noun, the visual information passed to auditory association cortex, where it was identified. Since the lesion interrupted the direct connection between auditory association cortex and Broca's area, the representation of the word could not be transmitted directly to neural mechanisms that control speech. Instead, the information was passed back to visual association cortex, where it was translated into an image. The image was conveyed to Broca's speech area and translated into a spoken word. However, K. F. was not able to read abstract words like "apparently" because they could not be translated into an image in visual association cortex. (See **FIGURE 19.21.**)

What about the earlier experiment? There are at least two explanations, and both of them support the hypothesis that visual and auditory short-term memory are anatomically distinct. One possibility, suggested by Warrington and Shallice (1972), is that the lesion interfered with auditory short-term memory, so that the "auditory image" of the three acoustically-presented letters faded more rapidly than usual. The other explanation does not make any assumptions about damage to auditory short-term memory. Instead, it hypothesizes that acoustically-presented information, which must be relayed from auditory short-term memory to visual short-term memory to Broca's area, travels a relatively inefficient route. In contrast, visually-presented information can travel directly from visual short-term memory to Broca's area. (See **FIGURE 19.22** on the next page.)

The same important conclusion can be made from this case, whichever alternative you choose: visual and auditory short-term memory, like visual and auditory long-term memory, seem to be located in different parts of the brain.

Acoustically-presented material: pathway is indirect

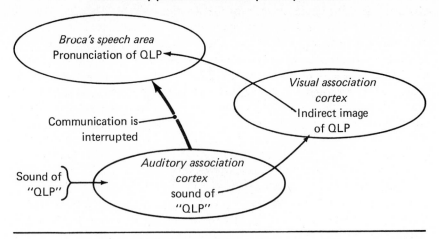

Visually-presented material: pathway is more direct

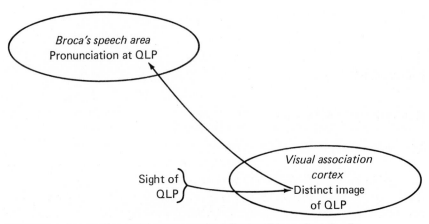

FIGURE 19.22   A schematic representation of a possible explanation for the results of the study by Warrington and Shallice (1972).

ANATOMY OF SHORT-TERM MEMORY IN ANIMALS.   Further evidence for anatomical specificity in short-term memory storage was provided by Kovner and Stamm (1972). These investigators trained monkeys to perform a delayed matching-to-sample task. The animals were shown a visual pattern (the sample), which was then turned off for several seconds. At the end of the delay interval, they were shown two different patterns, one of which was the same as the one just shown to them. If they responded to the appropriate pattern (i.e., if they chose the pattern that matched the sample), they were rewarded with a piece of food.

Electrical stimulation of inferotemporal cortex, delivered through chronically implanted electrodes just before or during the time the animal was shown the matching stimuli, produced severe decrements in performance. However, stimulation in foveal prestriate cortex (which projects to inferotemporal cortex) had no effect, nor did stimulation of prefrontal association cortex. The deficit produced by the interfering stimulation would not appear to be perceptual, since stimulation of inferotemporal cortex did not impair performance on a *simultaneous* matching-to-sample task. In this task, all three stimuli were present at the same time; the monkeys had to select which of two lower stimuli matched the upper stimulus. It was only when the animals had to compare visual stimuli with their short-term memory for a previously presented stimulus that the electrical stimulation had a deleterious effect. To put it crudely, the stimulation "erased" the short-term memory of the stimulus.

It would appear that inferotemporal cortex is important for visual short-term memory in monkeys. Since only visual stimuli were used in this study, we cannot conclude that only *visual* short-term memory is disrupted by inferotemporal stimulation, but this fact does seem likely, since lesions of inferotemporal cortex impair visual, but not auditory, tactile, or gustatory, discriminations (Gross, 1973).

ANATOMY OF THE CONSOLIDATION PROCESS

Korsakoff's Syndrome

A very dramatic memory deficit occurs when a person sustains certain forms of brain damage. Korsakoff, a Russian physician, first described the syndrome in 1889, and the disorder was given his name. The most profound symptom of ***Korsakoff's syndrome*** is a severe anterograde amnesia; the patients are unable to form new memories, although old ones still remain. They act like our head-injured patient did after waking in the hospital; they can converse normally and can remember past events but they cannot later recall events that occur after the brain damage. Unlike the head-injured patient, people with Korsakoff's syndrome suffer this impairment *permanently*.

Korsakoff's syndrome occurs most often in chronic alcoholics. The disorder is thought to result from a thiamine deficiency caused by the alcoholism (Adams, 1969). Since alcoholics receive a substantial number of calories from the alcohol they ingest, they usually eat a very poor diet, so their vitamin intake is consequently low. Furthermore, alcohol appears to interfere with intestinal absorption of thiamine (vitamin B_1), and the ensuing deficiency produces brain damage. There is a lively controversy about the critical brain regions whose destruction leads to Korsakoff's syndrome. I will discuss the controversy in some detail, after I have described a related syndrome that is produced by bilateral removal of the medial temporal lobes.

Scoville, in 1954, reported that bilateral removal of the medial temporal lobe produced a memory impairment that was apparently identical to that seen in Korsakoff's syndrome. Thirty operations had been performed on psychotic patients in an attempt to alleviate their mental disorder, but it was not until this operation was performed on patient H. M. that the anterograde amnesia was discovered.

Patient H. M. had very severe epilepsy before his operation. Even though he received medication at what is described as near-toxic levels, he suffered major convulsions approximately once a week and had dozens of minor attacks each day. After bilateral removal of the medial temporal lobes, he showed a considerable amount of recovery. He receives moderate doses of dilantin and has only one or two minor seizures each day. Major convulsions are seen very rarely— one every two or three years. In terms of treatment of his epilepsy, the surgery was very successful.

However, it quickly became apparent after surgery that the patient suffers a severe memory impairment. The deficit had not previously been seen in the psychotic patients because of the severity of their mental disorders. H. M., however, is quite normal, except for his epilepsy, and the memory impairment was only too easily detected. Subsequently, Scoville and Milner (1957) tested eight of the psychotic patients who were able to cooperate with them. They found that some of these patients also had anterograde amnesia; the deficit appeared to occur when the hippocampus was removed, but not when the amygdala, uncus, and overlying temporal cortex were removed, sparing the hippocampus. As we shall see, this conclusion is a point of controversy.

These results were compatible with findings in the case of patient P. B., reported by Penfield and Milner (1958), and Penfield and Mathieson (1974). Parts of the left temporal lobe of this patient were removed in two stages, on separate occasions. No memory disorder was seen after the first operation. The amygdala, uncus, and hippocampus were removed during the second operation and a severe anterograde amnesia resulted. Unilateral temporal damage does not normally result in memory impairment, so this effect was unexpected. Patient P. B. died 12 years later (of unrelated causes) and his brain was examined. The right hippocampus was found to have been damaged, probably as a result of trauma at birth. The left temporal lobectomy removed the only functioning hippocampus. The authors concluded that the hippocampal removal produced the amnesia.

THE CASE OF H. M.: HIS MEMORY DEFICITS.   Now let us examine H. M.'s case in more detail (Milner, Corkin, and Teuber, 1968; Milner, 1970). His intellectual ability and his short-term memory appear to be normal. He can carry on a conversation, rephrase sentences, and perform mental arithmetic. His immediate memory for a series of numbers *(digit span)* is low-normal; he can repeat seven numbers forward and five numbers backward. He has a partial retrograde amnesia for events of the 2 years preceding the operation, but he can retrieve older memories very well. He showed no personality change after the operation, and he appears to be a polite and well-mannered person.

However, since the operation H. M. has, with rare exceptions, been unable to learn anything new. He cannot identify people he met since the operation (performed in 1953, when he was 27 years old) nor can he find his way back home if he leaves his house (his family moved to a new house after his operation and he has been unable to learn how to get around the new neighborhood). He is aware of his disorder and often says something like this:

> Every day is alone in itself, whatever enjoyment I've had, and whatever sorrow I've had. . . . Right now, I'm wondering. Have I done or said anything amiss? You see, at this moment everything looks clear to me, but what happened just before? That's what worries me. It's like waking from a dream; I just don't remember. (Milner, 1970, p. 37)

H. M. is capable of remembering a piece of information if he is not distracted; constant rehearsal can keep an item in his short-term memory for a very long time. However, he does not show any long-term effects of this continuous rehearsal. If he is distracted for a moment, he will completely forget the item he had been rehearsing so long. He works very well at repetitive tasks; since he so quickly forgets what previously happened, he does not become bored easily. He can endlessly reread the same magazine or laugh at the same jokes, finding them fresh and new each time. H. M. also shows a few symptoms that do not appear to be related to his memory impairment. He has no interest in sexual behavior; he does not express feelings of hunger, although he will eat normally when food is in front of him; and he shows no reaction to pain. As a matter of fact, an attempt was made to ascertain whether H. M. could be classically conditioned to give an autonomic response to a stimulus paired with a painful shock. The attempt was abandoned when it was found that H. M. did not react to shock levels that normal people would find quite painful. He could feel the shock, but it did not appear to bother him. (Recall, from Chapter 17, that damage to the dorsomedial thalamus similarly disrupts reactivity to painful stimuli. As we shall see later, this fact may be significant.)

THE CASE OF H. M.: EVIDENCE OF CONSOLIDATION.   There *are* a few tasks on which H. M. shows some evidence of learning. Milner (1965) presented him with a mirror-drawing task. This procedure requires the subject to trace the outline of a figure (in this case, a

FIGURE 19.23   The mirror-drawing task.

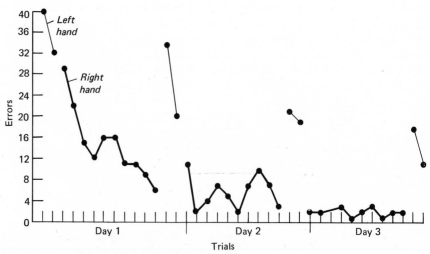

FIGURE 19.24   Data obtained from patient H. M. on the mirror-drawing task. (From Milner, B. In *Cognitive Processes in the Brain*, edited by P. M. Milner and S. Glickman. Princeton: Van Nostrand, 1965.)

star) with a pencil, being able to see the procedure in a mirror, but not directly. (See **FIGURE 19.23**.) The task may appear to be simple, but it is actually rather difficult. Left and right are maintained normally in the mirror, but movements toward or away from the body are reversed. That makes it somewhat complicated to follow a diagonal line. I have observed college students trying this task; when it was explained to them they began, confident that they would be able to quickly trace around the star. Instead, they found themselves repeatedly leaving the confines of the double lines. Some of the students got so rattled that they broke the points of their pencils while trying to figure out which way to move next. However, with practice, people can eventually become quite proficient at this task.

The interesting thing is that H. M., also, has become better at mirror-drawing. Figure 19.24 illustrates his improvement; his errors were reduced considerably during the first session, and his improvement was retained on subsequent days of testing. (See **FIGURE 19.24**.) However, H. M. did not remember having performed the task previously. He reported no sense of familiarity with it.

Another task on which H. M. has shown long-term improvement is the recognition of incomplete pictures. Figure 19.25 shows two sample items from this test; note how the drawings are successively more complete. (See **FIGURE 19.25**.) The subjects are first shown the least-complete version (set I) of each of twenty different drawings. They are asked to identify as many items as possible. They are then shown more complete versions until each of the items is identified. One hour later the subjects are tested again for retention, starting with set I.

H. M. was given this test and, when retested an hour later,

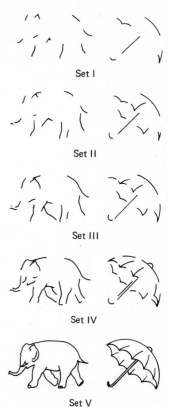

Set I

Set II

Set III

Set IV

Set V

FIGURE 19.25   Examples of broken drawings. (Reprinted by permission of author and publisher from Gollin, Eugene S., Developmental studies of visual recognition of incomplete objects. *Perceptual and Motor Skills*, 1960, *11*, 289–298.)

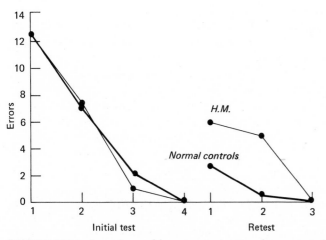

FIGURE 19.26   Learning and long-term retention demonstrated by patient H. M. on the broken-drawing task. (From Milner, B. In Pribram, K. H., and Broadbent, D. E., editors, *Biology of Memory*. Copyright 1970 by Academic Press, New York.)

showed considerable improvement (Milner, 1970). When he was re-tested 4 months later, he still showed this improvement. His performance was not so good as that of normal control subjects, but unmistakable evidence of long-term retention was obtained. (See **FIGURE 19.26** on the previous page.)

MILNER'S CONCLUSIONS. Milner, who has studied H. M. and other cases like his, concludes that the hippocampus plays a vital role in the consolidation process. It seems obvious, she notes, that neither short-term nor long-term memories are stored there. H. M. has a normal STM, as is shown by his digit span and by the fact that he can carry on a normal conversation. His long-term memories are not lost, since memories for events that occurred prior to the operation can be retrieved. Because these memories can be retrieved, we also know that the retrieval mechanism is still functioning. The hippocampus, Milner says, must exert some function that permits the transition of STM to LTM to occur; apparently, the transition itself occurs somewhere else.

Temporal Lobectomy in Animals

As you might have predicted, investigators have studied the effects of these lesions in animal subjects. Many thousands of monkeys, cats, and rats have been tested on a variety of tasks after receiving bilateral hippocampal lesions in an attempt to find out what role this structure plays (see Isaacson, 1974, for a general review). A number of deficits accompany these lesions, but there does not appear to be an overwhelming amnesia such as is seen in humans. Animals can learn many different kinds of tasks and show long-term retention later. For example, they have little or no difficulty with brightness or pattern discriminations. When the task involves spatial elements, more severe deficits are seen; hippocampal lesions severely impair the learning or retention of a maze. Conditional discriminations (if the alley is dark, turn right; if it is light, turn left) are also difficult for animals with these lesions. Nevertheless, the expectation that temporal lobe lesions would produce a permanent and global anterograde amnesia in animals was clearly not fulfilled.

There are at least three possible explanations for this discrepancy.

1. *The hippocampus of the human brain has different functions from the hippocampus of other animals.* I think that this suggestion should be seriously considered only if other attempts to explain the discrepancy fail. It is certainly true that our brains can do more than those of other animals, and thus the hippocampus could be called upon to do things in humans that it does not do in other species. However, other animals can consolidate memories. Why should this function have been transferred over to the

hippocampus so abruptly, so that hippocampal removals produce amnesia in humans but not in monkeys?

2. *Animals and humans have not been tested in comparable ways.*
Perhaps the tests that are passed by animals with hippocampal
lesions are not as challenging as the ones that are given to humans with Korsakoff's syndrome or bilateral temporal lobe damage.

3. *A structure other than the hippocampus is involved in the memory
deficit.* Perhaps destruction of the hippocampus is not what
causes the amnesia in humans; perhaps some nearby structure
is damaged when the hippocampus is removed.

In recent years alternatives (2) and (3) have received some support. On the one hand, some investigators have found that humans
with temporal lobe lesions do well when the behavioral tests are
made more similar to those used in research with animals. On the
other hand, a good case can be made that the hippocampus is not
the critical structure for temporal lobe amnesia. Either one of these
findings would resolve the discrepancy. Let us look at some of the
evidence.

The Human Amnesia Syndrome:
A Failure in Encoding?

It has been suggested that the human amnesia syndrome produced
by temporal lobectomy or chronic alcoholism is not a failure to consolidate short-term memories. Instead, the brain damage might disrupt the process whereby newly-consolidated memories are encoded
or cross-indexed with old ones. Such a deficit would be expected to
impair the performance of humans more than other animals, since
their memories are usually tested differently. Before I elaborate on
this suggestion I shall describe some experimental evidence.

DEFICITS IN ENCODING INFORMATION IN SHORT-TERM MEM-
ORY. Humans with the amnesic syndrome have difficulty in encoding newly-perceived information so that it can be stored effi-
ciently in long-term memory. This was nicely shown by Sidman,
Stoddard, and Mohr (1968), who tested H. M. on a delayed matching-
to-sample problem with verbal and nonverbal material. A stimulus
(the sample) would appear on a square of frosted Plexiglas in the
middle of an array of nine squares. H. M. would press the square,
and the stimulus would disappear. After a delay interval, a number
of stimuli would appear in the surrounding squares. One of these
stimuli would be the same as the sample stimulus, and depression
of the proper square would cause a penny to be delivered into a dish.
The apparatus is illustrated in Figure 19.27. Also shown are the two
types of items used: verbal stimuli (nonsense words composed of
three consonants) and nonverbal stimuli (ellipses of various shapes).
(See **FIGURE 19.27** on the following page.)

Training apparatus

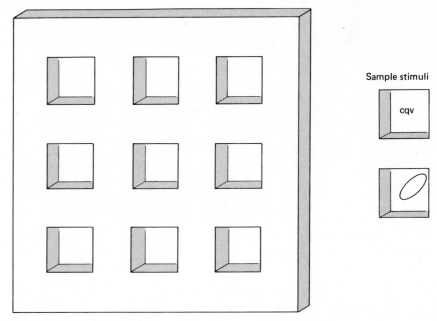

FIGURE 19.27 The apparatus used by Sidman, Stoddard, and Mohr (1968).

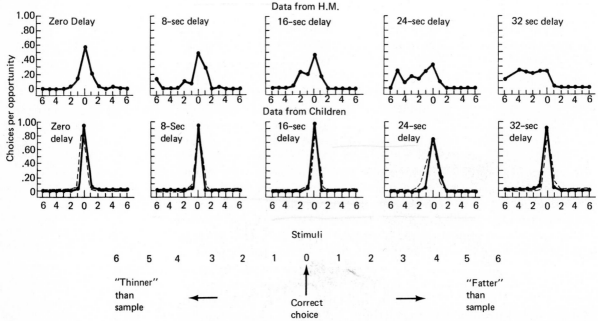

FIGURE 19.28 Data from the ellipse matching-to-sample test, (From Sidman, M., Stoddard, L. T., and Mohr, J. P., *Neuropsycholgia*, 1968, 6, 245–254)

H. M.'s performance on the matching-to-sample problem for the nonverbal stimuli (ellipses) is shown in Figure 19.28 along with data from two normal preadolescent children (shown below), for comparison. The numbers on the horizontal axis to the right and left of the 0 (correct choice) refer to responses made to ellipses "fatter" or "thinner" than the sample. Note that H. M. could do fairly well when there was no delay, but his performance quickly deteriorated as a delay was introduced. On the other hand, the children had no trouble remembering the sample stimulus. (See **FIGURE 19.28.**)

When H. M. was presented with a three-letter nonsense word, he had no trouble at all selecting the matching stimulus, even after a 40-second delay. Why should there be such a difference in his ability to retain verbal as opposed to nonverbal material in STM? As Sidman and his colleagues note, H. M. can easily rehearse verbal information. In fact, he can be observed to form the letters of the stimulus with his lips. If the sample is *cqv*, for example, he repeatedly says "c-q-v" to himself, and thus retains the information. There is no such verbal code presented by an ellipse of a particular shape. However, normal subjects manage to invent a code of their own. Sidman and his colleagues reported that normal subjects soon learn to identify the shapes and remember them as "the largest one, the next-to-largest, the smallest," etc., or give them numbers from 1 to 8. Then, they have a verbal code that can easily be retained. H. M., on the other hand, does *not* construct these verbal codes. He does not write notes to himself or even ask others to help him get around. His deficit in this short-term task is in the ability (or motivation) to construct an easily retained verbal code. He must retain an image of the ellipse, which is difficult to do.

An experiment by Stewart (1977) showed that when amnesic patients were supplied with strategies for encoding information, their performance on a test of long-term memory was facilitated. She showed a series of random figures to normal subjects and subjects with Korsakoff's syndrome. (See **FIGURE 19.29.**) Under the

FIGURE 19.29   An example of the random shapes that were used in the experiment by Stewart (1977).

"experimenter-encoded" condition, she pointed out to the subjects how various stimuli could be seen to resemble familiar objects. For example, she might say, "See, this one looks like a hammer. Here's the handle, here's the head, and here's the part that pulls out the nails." After a delay the subjects were shown pairs of stimuli (one old, one new), and were asked to point to the one they had seen before.

The amnesic patients remembered more of the stimuli for which an encoding cue had been supplied. In fact, under this condition their performance approached that of normals. However, the performance of the normals was made *worse* by the cues. Apparently, the normals formed their own images anyway, and only found the experimenter's suggestions to be confusing. The results clearly support the hypothesis that the amnesic syndrome entails a deficit in the encoding of information in short-term memory which, in turn, impairs storage and subsequent retrieval.

trouble encoding information in STM impairs storage & subsequent retrieval

DEFICITS IN VERBAL ACCESS TO STORED INFORMATION. As we already saw, amnesic patients are capable of storing and retrieving some kinds of information. For example, H. M. learned a mirror drawing task and an incomplete picture recognition task. In both of these tasks, H. M. was tested with the same stimuli that were used during training. In the incomplete picture task he was not simply asked whether he could recall the pictures (in fact, he said that he could not); rather, he was shown a few elements of a particular picture, and was asked what the stimulus looked like. The incomplete picture served as a precise retrieval cue for the originally-learned information.

Most animals are tested similarly. They are taught to do something in the presence of certain stimuli, and are tested later by presenting the same stimuli. Suppose that an experimenter showed a series of visual stimuli to a monkey or a rat, and then later asked the subject to tell her what it has previously seen. If the experimenter reported that the animal had no memory of the stimuli, since it was unable to say what it had seen, we would question her sanity. Obviously, since animals other than humans cannot talk, we must test their memory in a different way. But since language is such an easy and familiar way to test humans, we tend to use tests of memory with them that are different from the ones that are used with animals. Let us see what happens if an amnesic patient is trained and tested as an animal might be.

Sidman, Stoddard, and Mohr (1968) trained H. M. to press the square that contained the image of a circle (using the apparatus pictured in Figure 19.27), no matter where the circle appeared. No verbal instructions were offered, except for those necessary to seat the patient and get him started on the task. A penny was dispensed each time a correct response was made. H. M. quickly learned to press the circle, and he would select this stimulus from a display containing one circle and seven ellipses of various shapes. He was

then interrupted and asked to count his pennies (distraction task). **Anatomy of Language and**
Then he was asked how he earned them (Sidman, Stoddard, and **Memory**
Mohr, 1968).

> H. M.: Well, let's see. Something would flash up there and the idea was to pick out one of those squares and to point it toward dark. To tip it—to hit it with my finger tip and to match up. Each time the two matched a penny would drop in.
> E: Each time the two matched?
> H. M.: The two matched.
> E: Uh-huh. What was on them?
> H. M.: X.
> E: X was on them?
> H. M.: Yeah.

A few minutes later:

> E: Can you tell me once more what you did to earn all those pennies?
> H. M.: Well, one of those would flash up and actually I made a decision to point or to hit one of them with my finger tip . . .
> E: What were you pointing to—what were you pressing over there?
> H. M.: Well, one of these would light up and get one of them matched and every time one would match, of course, a penny would drop in.
> E: What did the one that matched look like?
> H. M: Cross.
> E: A cross. Uh-huh. A plus sign?
> H. M: Uh-huh.
> E: Or a multiplication sign?
> H. M.: Well, you'd say, uh, it wouldn't be multiplication—addition.

Judging by what H. M. said, he had already forgotten the task. However, when another series of stimuli was presented, he continued to select the circle. After several trials he was again asked what he had been doing.

> H. M.: Get the circles that were round. . . . Some were oval-shaped and definitely the roundest one.
> After this accurate description he counted his pennies, and two minutes later he was asked again to tell what he had done to earn the pennies:
> H.M.: Well, I pressed matching up to that would be exactly alike of, uh, well crosses. . . . There would be several of them on there, but two of them would be exactly alike. . . . Pointing to one of them would naturally mean that there was another one just like it.

He was tested once more, and again he selected the circle. H. M. showed amnesia for the task when tested verbally, but he showed perfect retention when tested nonverbally, as an animal would be.

El-Wakil (1975) found that the ability of amnesic patients to consolidate specific information is not limited to simple stimuli like circles. He prepared twenty-nine pairs of color slides: two different trees, two different automobiles, two different houses, etc. The patients were shown one member of each pair for 15 seconds. Twenty-four hours later they were shown each pair of slides, side by side, and were asked to point to the one they remembered. Some patients said that they hadn't seen the slides before, but nevertheless all of them pointed to the correct member of each pair. They were tested again 18 days later and this time none of them remembered seeing the slides; nevertheless, they again showed almost perfect retention.

The results in this section suggest that the memory deficit of patients with Korsakoff's syndrome and temporal lobe amnesia is not primarily a failure of consolidation. A patient who is trained with nonverbal stimuli, and is then tested with the identical stimuli, shows evidence of very good retention. Perhaps the important thing is that the test stimulus be the same as the training stimulus. Perhaps memories are normally consolidated in amnesic patients to the extent that they are encoded while they are in short-term memory. Since they are poorly encoded, and since they are not effectively cross-indexed with information that already exists in long-term memory, poor retention is shown later. When H. M. sees the circle, the stimulus serves to elicit the appropriate response that he has learned; obviously, he can form a long-term association between the presence of a circle and the performance of a response. However, when he is asked to describe what he did, the verbal input (the question) cannot initiate retrieval of the stored information.

An Alternative to the Hippocampal Hypothesis

Most investigators believe that Korsakoff's syndrome and temporal lobe amnesia result from bilateral damage to ***Papez's circuit,*** a set of interconnected structures that constitute an important part of the limbic system. Papez, a neuroanatomist, proposed that the hippocampus, mammillary bodies, anterior thalamic nuclei, cingulate cortex, and entorhinal cortex formed a loop that plays a critical role in the expression of emotion. (See **FIGURE 19.30.**) When it was discovered that the brains of patients with Korsakoff's syndrome often include damage to the mammillary bodies, and that bilateral temporal lobectomy causes a similar amnesic syndrome, it was proposed that bilateral damage to Papez's circuit was responsible for the memory deficit. Thus, these structures, and their interconnections, were implicated in the process of memory consolidation (or, as seems more likely now, some functions related to encoding and cross-indexing of information).

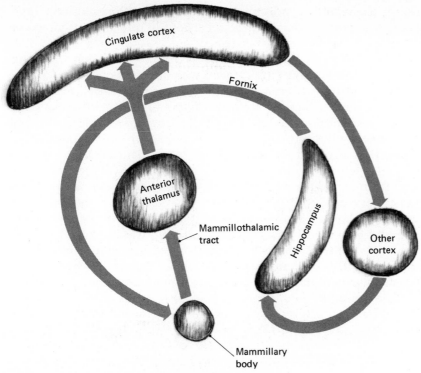

FIGURE 19.30   A schematic diagram of Papez's circuit.

There are two problems with this conclusion. First, newer neuroanatomical techniques have shown that Papez's circuit does not appear to exist, at least as originally outlined. Swanson and Cowan (1975a) injected radioactive amino acids into various portions of the hippocampus. The amino acids were taken up by cells, incorporated into proteins, transported, via axoplasmic flow, to the terminal buttons, which were revealed by means of autoradiography. Absolutely no radioactivity was found in the mammillary bodies; all of the efferent axons of the hippocampus appeared to synapse in the septum. However, when radioactive amino acids were injected into the **subiculum** (a region of limbic cortex adjacent to the hippocampus), radioactivity was found in the mammillary bodies and anterior thalamus. Thus the "wiring diagram" has to be revised. Papez's "circuit" is no longer a continuous loop of interconnected structures.

It is too early to assess the significance of these neuroanatomical findings. But there is a second problem with the hippocampal hypothesis. Horel (1978) has amassed an impressive amount of data that suggest that the **temporal stem,** and not the hippocampus, is the critical brain region that is damaged by temporal lobectomy. Furthermore, he notes that damage to the dorsomedial thalamic nuclei (to which the temporal stem projects) is correlated with memory loss in Korsakoff's syndrome much better than mammillary body damage is. Finally, the data are consistent in humans and monkeys.

HUMAN DATA: TEMPORAL LOBECTOMY. Horel notes that the approach that Scoville (the surgeon who operated on patient H. M.) used in performing his temporal lobectomies necessarily invaded the temporal stem, a bundle of efferent fibers of neurons that reside in anterior temporal neocortex. Therefore, Scoville's patients with temporal lobectomies must have suffered damage to the temporal stem. In addition, you will recall that Penfield's patient P. B. was given a unilateral left temporal lobectomy, but it was discovered later that birth trauma had previously destroyed the right hippocampus; thus, the surgery produced the amnesic syndrome. Horel notes that the published photographs of stained sections through the patient's right temporal lobe reveal that the temporal stem was completely degenerated. (In contrast, the posterior portion of the right hippocampus was intact.) Thus, the amnesia could just as reasonably be attributed to temporal stem damage as to hippocampal damage.

The temporal stem connects temporal cortex with a variety of diencephalic structures, including the dorsomedial thalamic nuclei (specifically, the *magnocellular* portion). In turn, these nuclei project

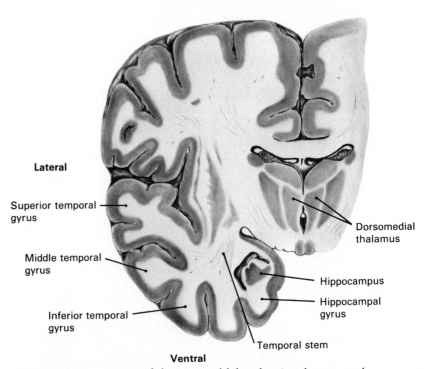

FIGURE 19.31 Anatomy of the temporal lobe, showing the temporal stem. (From Horel, J. *Brain*, 1978 *101*, 403–445.)

to *orbital* and *insular prefrontal cortex.* (See **FIGURE 19.31**.) Bilateral lesions of temporal neocortex have profound effects on memory. Sabourand, Gagnepain, and Merault (1976) described the case of a 15-year-old boy who suffered brain damage during a motorcycle accident. Although he could copy pictures and repeat words that were spoken to him, he could not associate written labels with the appropriate pictures; he could not respond to spoken commands; and he was unable to produce intelligible speech on his own. He showed some evidence of memory; when he was given a picture to copy that he had already drawn, he would look for the drawing in his sketch book.

The patient died five years after the accident. Autopsy revealed that the injury had almost totally destroyed the middle and inferior temporal gyri of the left hemisphere, and parts of the superior, middle, and inferior temporal gyri of the right hemisphere. (The fact that there was less damage to the right temporal lobe is in accord with the fact that the patient showed evidence of memory for pictures that he had already drawn; in most people, the right hemisphere is specialized for perception of spatial relationships, and probably pictorial memory, as well.) It should be noted that the patient's hippocampus was *not* damaged.

Prefrontal cortex has roles that are related to the motor functions of the frontal lobes. It can be thought of as "motor association cortex." Some neurologists and neuropsychologists have suggested that planning and problem-solving are accomplished by symbolic motor processes that involve prefrontal cortex; the consequences of various actions can be "tested" there without involving overt motor movements. Similarly, memories for some sequences of movements, and for plans and strategies that have been devised, may be located there. In addition, prefrontal cortex may be involved in cross-indexing memories in terms of contextual cues, especially their time of occurrence; damage to this area impairs a patient's ability to remember which stimulus was most recently seen or which response was most recently made (Rozin, 1976).

It thus seems plausible that the connection between anterior temporal cortex (which is important for sensory memory) and frontal cortex (which is important for motor memory, planning, and cross-indexing of information according to contextual cues) might result in the kinds of impairments that are shown by patient H. M.

HUMAN DATA: KORSAKOFF'S SYNDROME. Even before Horel drew attention to the temporal stem, the anatomy of Korsakoff's syndrome was in dispute. Because of the fact that temporal lobe amnesia seemed to be produced by hippocampal damage, some investigators tended to overlook the fact that bilateral damage to Papez's circuit sometimes did not produce amnesia, and that amnesia sometimes occurred after brain damage that left Papez's circuit intact.

First let us look at some cases of bilateral damage to Papez's circuit that did not produce amnesia. Victor, Adams, and Collins (1971) studied 82 patients with brain damage from chronic alcoholism. Five of them had severe bilateral degeneration in the mammillary bodies, but did not exhibit anterograde amnesia. Also, some neurosurgical procedures involve bilateral sectioning of the fornix. One would predict that these operations should produce amnesia, but they do not. For example, cysts (fluid-filled sacs) occasionally develop in the region of the third ventricle. Often, the neurosurgeon must sever both the right and left fornix in order to remove the cyst. Dott (1938) and Cairns and Mosberg (1951) described cases of bilateral fornix sectioning without any memory deficits. Garcia-Bengochea, Delattore, Esquivel, Vieta, and Claudio (1954) performed bilateral transections of the fornix in fourteen epileptic patients. They noted that "So far, in none of the twelve surviving cases has there been any neurological or psychological sequela."

On the other hand, bilateral damage to the magnocellular region of the dorsomedial thalamus is reliably associated with Korsakoff's syndrome (Victor, Adams, and Collins, 1971). Similarly, memory impairments have been seen after tumors of the medial thalamus (Smythe and Stern, 1938). In some cases, dorsomedial thalamic lesions, produced in an attempt to relieve chronic, unremitting pain, have also resulted in Korsakoff-like memory deficits (Spiegel, Wycis, Orchinik, and Freed, 1955). These reports strongly support the hypothesis that damage to the temporal stem or the dorsomedial thalamus, to which the temporal stem projects, produces the amnesic syndrome.

One of patient H. M.'s symptoms might be relevant here. You will recall that he shows no emotional reaction to painful stimuli. In this regard he resembles patients with damage to the dorsomedial thalamus. Perhaps his loss of reactivity to pain is a result of the destruction of the temporal stem, which contains fibers that project to the dorsomedial thalamus. It is unlikely that it is due to the hippocampal lesion, since animals with hippocampal lesions show no signs of diminished reactivity to pain.

I think that it is safe to conclude the following about human temporal lobe amnesia and Korsakoff's syndrome. The disorder is not simply a failure of consolidation of short-term memories. Instead, there are problems with efficient coding of information in short-term memory, which necessarily will impair its storage. Also, there appears to be a difficulty with cross-indexing of new information with old. Whether these are separate deficits or merely different sides of the same coin remains to be seen.

Secondly, although most investigators still refer to the hippocampus in particular, and Papez's circuit in general, as playing a role in the memory process, there is very good evidence that this might not be the case. The critical circuit may be (1) anterior medial temporal cortex to (2) dorsomedial thalamus (magnocellular part) to (3) orbital and insular prefrontal cortex.

Anatomy of Language and Memory

KEY TERMS

arcuate fasciculus *(AR kew et)* p. 626
arcuate sulcus p. 641
circumstriate belt p. 637
conduction aphasia p. 626
dyslexia p. 631
engram p. 617
equipotentiality p. 618
foveal prestriate area p. 637
inferotemporal cortex p. 637
insular prefrontal cortex p. 661
Korsakoff's syndrome p. 647
orbital prefrontal cortex p. 661

Papez's circuit p. 658
periarcuate cortex p. 641
phoneme *(FOE neem)* p. 629
phonemic aphasia p. 630
prosopagnosia *(pross o pag NO zha)* p. 634
pulvinar *(PULL vi nar)* p. 637
second somatosensory projection area (SII) p. 640
striate cortex p. 637
subiculum p. 659
superior temporal gyrus p. 621
temporal stem p. 659
Wernicke's area *(VAIR ni kee)* p. 621

SUGGESTED READINGS

BEACH, F. A., HEBB, D. O., MORGAN, C. T., and NISSEN, H. W. (eds.) *The Neuropsychology of Lashley*. New York: McGraw-Hill, 1960.

GESCHWIND, N. Disconnexion syndromes in animals and man. *Brain*, 1965, *88*, 237–294; 585–644.

HOREL, J. A. The neuroanatomy of amnesia. A critique of the hippocampal memory hypothesis. *Brain*, 1978, *101*, 403–445.

PRIBRAM, K. H., and BROADBENT, D. E. (eds.) *Biology of Memory*. New York: Academic Press, 1970.

ROSENZWEIG, M. R., and BENNETT, E. L. (eds.) *Neural Mechanisms of Learning and Memory*. Cambridge, MA.: MIT Press, 1976.

The volume edited by Beach et al. contains a selection of papers by Karl Lashley that should be read by every serious student of the physiology of memory, even though the experiments were done many years ago. I especially recommend *In Search of the Engram* and *The Problem of Serial Order in Behavior*. Geschwind's articles discuss the anatomy of association cortex in humans and other animals and describe the disconnection syndromes. The title of Horel's paper speaks for itself. The book edited by Pribram and Broadbent contains individual chapters concerning cognitive and physical aspects of the two phases of memory. Brenda Milner's chapter describes the case of patient H. M. in detail. The book edited by Rosenzweig and Bennett was recommended for Chapter 18, also.

20

Physiology of Mental Disorders

Most of the discussion in this book has concentrated on the physiology of normal, adaptive behavior. But there are many people whose behavior is often maladaptive. These people become incapable of engaging in coherent, logical thought, exhibit inappropriate emotional responses, or suffer from delusions and hallucinations. These abnormal conditions are referred to as mental disorders. Since normal thoughts, emotions, and perceptions have a physiological basis, perhaps mental disorders are actually physiological disorders—malfunctionings of physiological processes that must perform properly for normal thinking, feeling, and perception.

The most serious mental disorders are called *psychoses.* The two major psychoses, *schizophrenia* and the *affective psychoses,* can disrupt a person's behavior so severely that he or she cannot function in society, but must be kept in an institution. In contrast, *neuroses* are generally less severe. People with neuroses are often unhappy, but they can generally reason normally and do not have hallucinations or delusions. They often have good insight into their problems, and can articulate them very well. We can all understand, and perhaps sympathize with, neurotics, for their problems seem like exaggerations of the ones we all encounter from time to time. However, the thoughts and behaviors of a person with a psychosis are very different from our own. They sometimes seem to follow a different logic, which is probably why past generations attributed these disorders to the possession of inhuman devils.

SCHIZOPHRENIA

Description

Schizophrenia is the most common psychosis, afflicting approximately 1 percent of the population. The term "schizophrenia," invented by Paul Bleuler, a Swiss psychiatrist and neurologist, is a rather unfortunate one. Literally, it means "split mind," which suggests that the word refers to a case of multiple personality. In popular use, a desire for two different, incompatible goals is often referred to as "schizophrenic." ("She felt herself pulled, schizophrenically, by the desire to be taken care of by someone stronger than she was, and the need to establish and assert her own personality.") But this popular use is incorrect. Schizophrenia refers to a *break with reality*—an essential illogic, as characterized by delusions (beliefs that are contrary to normally-accepted evidence) and hallucinations (the perception of stimuli that cannot be detected by anyone else).

There are many different kinds of schizophrenia. **Paranoid schizophrenia** is characterized by delusions of grandeur, persecution, or control. A paranoid schizophrenic might believe that he or she is God, or is being spied upon by secret agents that have landed in flying saucers, or is being controlled by a miniature radio receiver that was secretly installed in his or her brain. **Hebephrenic schizophrenia** is nonsense and silliness raised to its ultimate, where it completely breaks with reality. The speech of a person with this form of schizophrenia deteriorates into meaningless jumbles of words, many of which are themselves meaningless; the term "word salad" has been coined to describe this phenomenon. **Catatonic schizophrenia** refers to unusual motor symptoms: rigidity or (more rarely) wild excitement and hyperactivity. A catatonic schizophrenic often displays **waxy flexibility;** his or her limbs show some resistance to being moved, but remain in whatever position they are placed. This physical inactivity does not necessarily indicate a cessation of all functions; these patients often reveal later that their thoughts were proceeding at a furious rate during their motionless period.

Not all schizophrenics can be placed into one of these discrete categories; many exhibit different symptoms at different times, and many others show no distinct pattern of symptoms other than thought disturbances, perhaps including delusions and hallucinations. Most clinicians agree that the present subcategories of schizophrenia do not reveal important distinctions among different types of patients. A much more important distinction is made by classifying the disorder according to its onset. If an episode of schizophrenia comes on suddenly, with serious, full-blown symptoms, it is called **acute schizophrenia.** This disorder has a good prognosis; the patient will probably get better soon, and may or may not suffer from another attack later. In contrast, **process schizophrenia** has a slow onset. At first, symptoms are sparse and not very severe. However, they gradually get worse, until the person needs to be put into an institution. Process schizophrenia has a bad prognosis; the patient may suffer from this disorder, with periodic remissions and relapses, for the rest of his or her life.

Heritability of Schizophrenia

Is schizophrenia a physiological disorder? It is certainly not a contagious disease, caused by a bacterium or virus. But perhaps it is an inherited disorder, like Huntington's chorea. If schizophrenia were shown to be heritable, we would have strong evidence that the disorder is physiological, perhaps caused by some sort of defect in the nervous system.

After years of investigation, the data clearly show that a tendency to develop schizophrenia can be inherited. However, the relationship between genetics and schizophrenia is not simple. There is no "schizophrenia gene," and no single biochemical deficit has been found to produce this disorder. The genetic basis of schizophrenia appears to be very complex, and probably involves many genes. In addition, hereditary predispositions to develop schizophrenia undoubtedly interact with environmental factors.

One of the best-known studies of the heritability of schizophrenia was performed by Kety, Rosenthal, Wender, and Schulsinger (1968). In Denmark, a record is kept of all citizens, which makes it possible to trace the family history of people with various disorders. Kety and his colleagues identified a number of schizophrenic people who had been adopted when they were children. Because of the Danish *folkeregister*, the investigators were able to identify the biological families, as well as the adopted families, of the patients. They found that the incidence of schizophrenia in the adopted families of the patients was exactly what would be expected in the general population. Thus, it did not appear that the patients became schizophrenic because they were raised in a family of other schizophrenics.

However, the investigators did find an unusually high incidence of schizophrenia in the patients' *biological* relatives, even though they were not raised by them—and probably, in most cases, did not even know them. The results clearly favor the conclusion that schizophrenia can be transmitted by means of heredity.

This study also found that chronic schizophrenia and acute schizophrenia appear to have different causes. People with chronic schizophrenia tended to have schizophrenic biological relatives, but people with acute schizophrenia did not. These results suggest that acute schizophrenia might be triggered by environmental causes, whereas the tendency to develop chronic schizophrenia may be inherited.

The study by Kety and his colleagues started with schizophrenic patients and looked back to their families. Another study (Heston, 1966) took the opposite approach. Heston identified schizophrenic mothers and looked at the incidence of schizophrenia in their children. In particular, he examined the offspring of schizophrenic women whose children had been given up for adoption within two weeks of birth. Thus, it would be unlikely that the mothers' behavior could have had a chance to affect their children. For comparison, he selected a group of nonschizophrenic mothers whose children had also been given up for adoption.

The results support the hypothesis that schizophrenia is heritable. Seventeen percent of the children of schizophrenic mothers later became schizophrenic, whereas *none* of the children of the nonschizophrenic mothers did. In addition, the children of the schizophrenic mothers tended to have an unusually high incidence of emotional disorders, even if they did not become schizophrenic. Many were arrested for antisocial crimes, and many became alcoholic or addicted to drugs. Surprisingly, many had exceptional musical ability.

So many studies have confirmed the heritability of a tendency toward schizophrenia that the matter is no longer in dispute among researchers in this area. However, there is disagreement about whether a high incidence of musical ability, or of criminality or other disorders, is found in the offspring of schizophrenic parents (Kringlen, 1978). And it must be pointed out that even when both parents are schizophrenic, the probability of developing schizophrenia is less than 50 percent. Most authorities would place the figure in the range of 25 to 35 percent.

If schizophrenia were a simple trait, produced by a single gene, we would expect to see schizophrenia in at least 50 percent of the children of two schizophrenic parents if the gene were dominant. If it were recessive, all children of two schizophrenic parents should become schizophrenic. The fact that the actual incidence is less than 50 percent has led investigators to suggest that the genetics of schizophrenia is complex, and that environmental factors probably play a role in its development. Perhaps people inherit some sort of structural or biochemical defect in their brain that may, under certain environmental circumstances, develop into schizophrenia. In fact, although we have some hints about what the biochemical defect may be, investigators have not yet identified the environmental events that can trigger schizophrenia. Severe stress (such as occurs during combat) may cause an acute episode of schizophrenic symptoms, but chronic, lifelong schizophrenia has not yet been shown to be related to any particular types of environmental factors.

In any event, the topic of this book is the *physiology* of behavior. Even if future investigations are successful in identifying some environmental factors in schizophrenia, we are interested in the biological factors here. The fact that a vulnerability to develop schizophrenia is heritable *must* mean that the disorder has a physiological basis; all we inherit are nucleic acids (with associated chromosomal proteins), and all these nucleic acids can do is direct the synthesis of proteins. Therefore, schizophrenia is at least partially a result of the production of too much of a particular protein, or too little of it, or the production of the wrong kind. (Or all of the above.)

DRUGS THAT ALLEVIATE SCHIZOPHRENIA.   The treatments for most physiological disorders are developed after we understand their causes. For example, once it was discovered that diabetes was caused by lack of a hormone produced by the pancreas, it was possible to find an extract of pancreatic tissue that would alleviate the symptoms of this disease. However, in some cases the treatments are discovered before the causes of the disease. It was discovered that a tea made from the bark of the cinchona tree could prevent death from malaria many years before it was discovered that this disease is caused by microscopic parasites that are transmitted in the saliva of a certain species of mosquito. (The bark of the cinchona tree contains quinine.)

In the case of schizophrenia, a treatment was discovered before its causes were understood. (In fact, its causes are *still* not understood.) The discovery was accidental (Snyder, 1974). The antihistamine drugs were discovered in the early 1940s, and were found to be extremely useful in the treatment of allergic reactions. Since one of the effects of histamine release is a lowering of blood pressure, a French surgeon named Henri Laborit began to study the effects of antihistamine drugs on the sometimes-fatal low blood pressure that can be produced by surgical shock. He found that one of the drugs, *promethazine,* had an unusual effect; it reduced anxiety in his presurgical patients without causing the sort of mental confusion and nausea that was often produced by morphine.

Paul Charpentier, a chemist with a French drug company, examined a number of chemicals that had been developed during research on the antihistaminic drugs, but had been discarded because they acted as sedatives. One of them, **chlorpromazine,** had profound calming effects but did not decrease the patient's alertness. This drug produced "not any loss in consciousness, nor any change in the patients' mentality but a slight tendency to sleep and above all 'disinterest' for all that goes on around him" (Laborit, 1950, quoted by Snyder, 1974). This drug is still used to prepare patients for the administration of a surgical anesthetic.

Chlorpromazine was tried on patients with a variety of mental disorders: mania, depression, anxiety neuroses, and schizophrenia (Delay and Deniker, 1952a; 1952b). The drug was not very effective in treating neuroses or affective psychoses, but it had dramatic effects on schizophrenia.

The discovery of the antischizophrenic effects of chlorpromazine had profound effects on the treatment of schizophrenia. Figure 20.1 illustrates the patient population of mental hospitals in the United States before and after the widespread adoption of antischizophrenic drugs around 1955. Note the drastic change in the growth rate of the curve after 1955. (See **FIGURE 20.1**.)

The efficacy of antischizophrenic drugs has been established in many double-blind studies (Baldesserini, 1977). It is a common misconception that chlorpromazine (and the many other antischizophrenic drugs that have been developed since the 1950s) merely tranquilize patients who have schizophrenia. (In fact, these drugs are often referred to as **major tranquilizers.**) However, this is not the

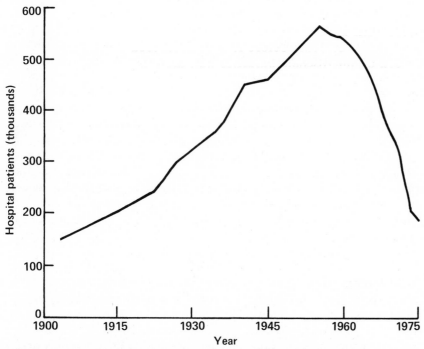

FIGURE 20.1   Number of patients in public mental hospitals from 1900 to 1975. (Redrawn from Bassuk, E. L., and Gerson, S. Deinstitutionalization and mental health services. *Scientific American*, 1978, *238*, 46–53.)

case. While the antischizophrenic drugs do often produce sedation, this effect is not related to the drugs' effects on the primary disorders of schizophrenia—delusions and hallucinations. The drugs that produce the most sedation are not necessarily the most effective antischizophrenic agents. **Benzodiazepines** are very effective tranquilizers that are used to treat anxiety, but they are ineffective in treating schizophrenia. Conversely, antischizophrenic drugs are not very good at alleviating neurotic anxiety. Moreover, antischizophrenic drugs can have either activating or calming effects, depending upon the patient's symptoms. An immobile patient with catatonic schizophrenia becomes more active, whereas a furiously active patient who is suffering from frightening hallucinations becomes more calm and placid. And the results are not just a change in the patient's attitudes; the hallucinations and delusions go away, or at least become less severe.

THE PHARMACOLOGICAL EFFECTS OF ANTISCHIZOPHRENIC DRUGS.   If we can find out what antischizophrenic drugs do, pharmacologically, perhaps we can infer what sort of defect in brain structure or metabolism might be responsible for schizophrenia. The one pharmacological effect that these drugs appear to have in common is a blocking of dopamine receptors. Carlsson and Lindqvist (1963) showed that antischizophrenic drugs increased the rate of dopamine synthesis in the brain, presumably because the blocking of postsynaptic dopamine receptors caused a compensatory increase by activating feedback mechanisms. In addition, a common side effect of the antischizophrenic drugs is the production of symptoms of Parkinson's disease. Parkinson's disease is known to be caused by degeneration of dopaminergic neurons of the nigrostriatal bundle; therefore, the induction of parkinsonism by antischizophrenic drugs could plausibly be attributed to blockage of dopamine receptors on neurons in the caudate nucleus. (As you will recall, the nigrostriatal bundle connects the substantia nigra with the caudate nucleus.)

Another piece of evidence is provided by the fact that the effectiveness of antipsychotic drugs is increased if patients are also treated with alpha-methyl paratyrosine (AMPT), a drug that blocks the activity of tyrosine hydroxylase, and thus interferes with the synthesis of dopamine and norepinephrine (Walinder, Skott, Carlsson, and Roos, 1974).

A particularly interesting study by Johnstone, Crow, Frith, Carney, and Price (1978) examined the effects of an antipsychotic drug called **flupenthixol.** This drug comes in two **isomeric** forms; that is, the molecules contain the same atoms, but parts are rotated in different ways. The *cis*-isomer of flupenthixol is an effective blocker of dopamine receptors, whereas the *trans*-isomer is not. (*Cis* and *trans* mean "on this side of" and "across," respectively.) A group of 45 patients with acute schizophrenia were treated (in a double-blind study) with *cis*-flupenthixol, *trans*-flupenthixol, or a placebo. Figure

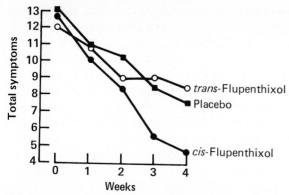

FIGURE 20.2   The effects of *cis*-flupenthixol, *trans*-flupenthixol, and a placebo on ratings of psychotic symptoms of patients with acute schizophrenia. (From Crow, T. J., Johnstone, E. C., Longden, A. J., and Owen, F. *Life Sciences*, 1978, *23*, 563–568. Copyright 1978, Pergamon Press, Ltd.)

20.2 illustrates the ratings of the patients' symptoms during the four-week treatment period. As would be expected for acute schizophrenia, all patients improved. However, the patients who were treated with *cis*-flupenthixol (which blocks dopamine receptors) improved faster than those who received either *trans*-flupenthixol or the placebo. (See **FIGURE 20.2.**)

More direct evidence for an antidopamine effect of the antischizophrenic drugs comes from studies on dopamine receptors. Kebabian, Petzold, and Greengard (1972) extracted an adenyl cyclase from the caudate nucleus of rat brain that served to convert ATP to cyclic AMP when it was exposed to dopamine. Presumably, the enzyme was bound to dopamine receptors whose effects are mediated by cyclic AMP. (See figure 5.10, page 95, if your memory needs to be refreshed.) When antischizophrenic drugs were also administered, dopamine was less effective in stimulating the production of cyclic AMP; the drugs appeared to block the dopamine receptors. However, one class of antischizophrenic drugs (the **butyrophenones**), which are clinically many times more potent than equal amounts of drugs like chlorpromazine, were *less* effective than chlorpromazine in inhibiting the production of cyclic AMP. What was the reason for this discrepancy?

Snyder, Burt, and Creese (1976) obtained evidence that there are two different binding sites on dopamine receptors, one of which is most effectively bound by dopamine or agonists like apomorphine, and the other of which is most effectively bound by an antagonist like **haloperidol,** one of the butyrophenones.

Presumably, once the antagonistic receptor site is occupied, agonists have difficulty in activating the agonistic receptor site. Their technique was as follows: Cell membranes from the caudate nucleus of the calf brain were exposed to dopamine or haloperidol that had been made radioactive by the addition of a tritium (^{3}H) ion. The membrane was then washed very quickly, to remove the labeled substance from every place but the receptor sites. Next, the radioactivity of the sample was measured; the level of radioactivity provided an estimate of the amount of binding of the labeled substance with receptor sites.

In order to determine how effectively various antischizophrenic drugs bound with postsynaptic dopamine receptors, ^{3}H-dopamine or ^{3}H-haloperidol was mixed with various concentrations of these drugs. The better a particular drug would bind with the dopamine receptors, the more effectively it would prevent the labeled substance from occupying these receptors. Thus, the affinity of a particular antischizophrenic drug for dopamine receptors could be determined by measuring the radioactivity of the sample of cell

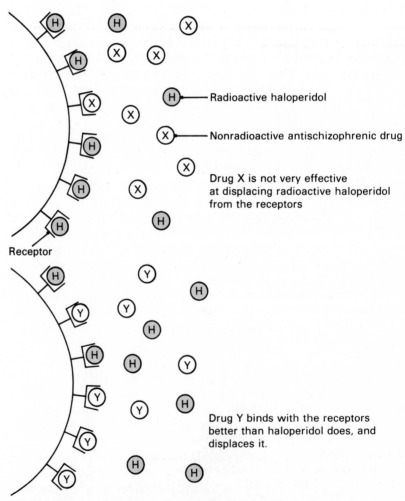

FIGURE 20.3 A schematic explanation for the study by Snyder, et al. (1976).

membrane that had been exposed to the ^{3}H-dopamine or ^{3}H-haloperidol along with the antischizophrenic drug—the *lower* the radioactivity, the *better* the drug bound with the receptor. (See **FIGURE 20.3.**)

The potency of a particular drug in inhibiting the binding of ^{3}H-haloperidol was found to be closely related to its clinical effectiveness. That is, drugs that alleviated the symptoms of schizophrenia were very effective in inhibiting the binding of ^{3}H-haloperidol. Drugs that had to be administered in higher doses displaced ^{3}H-haloperidol less. (See **FIGURE 20.4.**)

The experiment by Snyder and his colleagues cannot be said to constitute absolute proof that antischizophrenic drugs exert their therapeutic effects by blocking dopamine receptors. For one thing, it is unlikely (as we shall see) that the caudate nucleus is involved with schizophrenia in any important way, and it is possible that the pharmacological effects of these drugs in the caudate nucleus are different from the pharmacological effects that reduce the symptoms of schizophrenia. Furthermore, it has not been proven that the ^{3}H-haloperidol binding sites are dopamine receptors. The data suggest that they are, but it is possible that haloperidol and the other antischizophrenic drugs bind with some component of the membrane

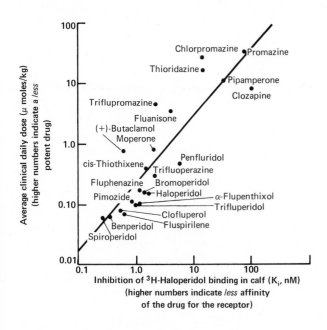

FIGURE 20.4 The relationship between the ability of an antischizophrenic drug to inhibit binding of ^{3}H-haloperidol with cell membranes extracted from calf caudate nucleus and the dose of the drug that is needed to achieve a therapeutic result. (From Snyder, S. H. *Journal of Continuing Education in Psychiatry*, 1978, *39*, 21–31.)

that is not directly related to the postsynaptic activity of dopamine. Perhaps the ^{3}H-haloperidol binding sites are on different neurons altogether. A definitive answer awaits further research.

Drugs that Produce Schizophrenia

It is a remarkable fact that we do not know how to produce schizophrenia by behavioral means. Stressful situations can, indeed, produce psychotic episodes in some people, but once the stress-producing stimuli are removed, the symptoms usually disappear. Long-lasting effects may occur, but they are not likely to resemble schizophrenia. In contrast, it is possible to administer drugs that will greatly intensify the symptoms of a person with schizophrenia, or even *induce* them in an apparently normal volunteer.

The drugs that can produce these symptoms have one known pharmacological effect in common: they act as dopamine agonists. Amphetamine, cocaine, L-DOPA and methylphenidate (Ritalin) are all capable of producing psychotic symptoms, and all are dopamine agonists. However, these drugs do not all work in the same way. L-DOPA, a precursor of the catecholamines, increases the synthesis of dopamine. It has little effect on the synthesis of norepinephrine, whose rate of production is limited by the quantity of dopamine-β-hydroxylase that is present. The other drugs affect the release and re-uptake of both norepinephrine and dopamine. Moreover, the symptoms that these drugs produce can be alleviated with anti-schizophrenic drugs, which further strengthens the argument that these drugs exert their therapeutic effects by blocking dopamine receptors.

Davis (1974) provided a particularly dramatic illustration of the way that symptoms of schizophrenia can be exacerbated by dopaminergic stimulation. He and his colleagues injected small doses of methylphenidate (which causes dopamine to be released by the terminals and which also inhibits re-uptake) into the veins of schizophrenic patients who were in a fairly quiet state. Within a minute after the injection, each patient was transformed "from a mild schizophrenic into a wild and very florid schizophrenic." One of the patients began to make a clacking noise after the injection. He then took a pad of paper which he pounded repeatedly, and ultimately shredded, with a pencil. He had been "sending and receiving messages from the ancient Egyptians." Catatonic patients became more catatonic, displaying their characteristic waxy flexibility. Thus, it cannot be said that the drug merely made the patients more active—whatever their symptoms were, they became worse.

If large doses of amphetamine or cocaine are given to normal, nonpsychotic people, the people will eventually develop symptoms that are difficult, if not impossible, to distinguish from those of schizophrenia. Griffith, Cavanaugh, Held, and Oates (1972) recruited a group of people who had a history of amphetamine use, and gave them large doses (10 mg) of dextroamphetamine every hour for up to five days. (Experimentally, it would have been better to study nonusers. Ethically, it was better not to introduce this drug to people who did not normally use it.) None of the subjects had prior histories of any psychotic behavior.

All seven volunteers became psychotic within two to five days. The first symptoms were sleeplessness and loss of appetite. However, the drug-induced psychoses could not be attributed to these effects, because some subjects became psychotic in less than 48 hours; neither a two-day fast nor a sleepless night will normally produce psychotic symptoms.

It is interesting to note that the immediate effects of amphetamine and cocaine are pleasurable—the drugs produce a sensation of euphoria. Snyder (1974) notes that a patient who is in the early stages of an acute schizophrenic episode often reports feelings of elation and euphoria. (Of course, this feeling may be masked by a dread of what is yet to come.) Perhaps these phenomena are related.

Abnormalities in the Brains of Schizophrenic Patients

What, physiologically, might be wrong with schizophrenic patients? The evidence that we just examined suggests that the disorder might be a result of hyperactivity in dopaminergic synapses. Perhaps too much dopamine is synthesized, or too much is released, or there are too many dopamine receptors on the postsynaptic neurons, or the receptors are too sensitive. Perhaps there is nothing wrong with dopamine itself, or with dopamine receptors; perhaps some enzyme abnormality in the brains of schizophrenics causes the production of a substance that acts like dopamine, stimulating postsynaptic receptors. Or, perhaps, dopamine and its receptors are not involved at all; there might be too little activity in some system elsewhere in the brain that can be compensated for by a reduction in the activity of dopamine neurons.

There is no evidence in support of the simplest hypothesis—that more dopamine is secreted in the brains of patients with schizophrenia. Subjects were first treated with probenecid, a drug that prevents the breakdown products of dopamine metabolism from being transported from the cerebrospinal fluid to the blood, where they will quickly be destroyed. Amphetamine treatments, which produce symptoms of schizophrenia, were found to increase the concentration of **homovanillic acid (HVA)** in the cerebrospinal fluid of normal subjects (Angrist, Sathananthan, Wilk, and Gershon, 1974). Homovanillic acid is a metabolite of dopamine; thus its increased concentration was evidence of an increased secretion of dopamine. However, when schizophrenic patients (who were not receiving an-

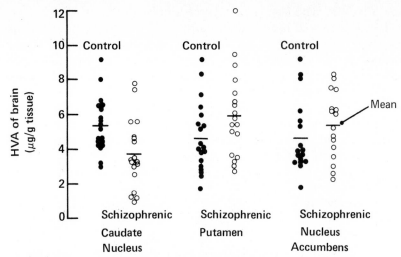

FIGURE 20.5   Amount of homovanillic acid found in the caudate nucleus, putamen, and nucleus accumbens of brains of deceased schizophrenics and nonschizophrenic controls. (From Crow, T. J., Johnstone, E. C., Longden, A. J., and Owen, F. *Life Sciences*, 1978, *23*, 563–568. Copyright 1978, Pergamon Press, Ltd.)

tischizophrenic medication at the time) were treated with probenecid to keep the dopamine metabolites in their cerebrospinal fluid, no increase was found in HVA levels. In fact, a *decreased* amount was found (Bowers, 1974; Post, Fink, Carpenter, and Goodwin, 1975). Similarly, a study of the brains of recently-deceased schizophrenic patients found no evidence of increased levels of HVA or **DOPAC (dihydroxyphenylacetic acid,** another metabolite of dopamine) in the neostriatum (caudate nucleus and putamen) or nucleus accumbens, which are two areas rich in dopaminergic terminals (Owen, Cross, Crow, Longden, Poulter, and Riley, 1978). (See **FIGURE 20.5.**)

In contrast to these negative results on the release of dopamine, Owen and his colleagues have reported an increase in the amount of binding of radioactive ^{3}H-spiroperidol in the brains of deceased schizophrenics, as compared with nonschizophrenic controls. (Spiroperidol is closely related to haloperidol, the drug used by Snyder and his colleagues in their study of the two sites on the dopamine receptor.) Five of the patients had not received antischizophrenic drugs for a year before they died, and two of them apparently had never received such medication, yet the rate of ^{3}H-spiroperidol binding was significantly higher in extracts from their brains. Therefore, the results do not seem to be secondary to the effects of treatment with antischizophrenic drugs. (See **FIGURE 20.6.**)

This study suggests that there may be more dopamine receptors in the brains of patients with schizophrenia, which suggests that schizophrenia occurs as a result of increased postsynaptic effects of dopamine. Of course, these results will have to be confirmed by other techniques before we can be certain that there is an increase in do-

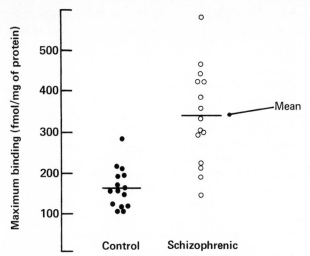

FIGURE 20.6   Amount of binding of ^{3}H-spiroperidol in the caudate nucleus of brains of deceased schizophrenics and nonschizophrenic controls. (From Crow, T. J., Johnstone, E. C., Longden, A. J., and Owen, F. *Life Sciences*, 1978, *23*, 563–568. Copyright 1978, Pergamon Press, Ltd.)

pamine receptors. Remember, we are not sure precisely what it is that ^{3}H-spiroperidol and ^{3}H-haloperidol bind with. And even if further studies confirm an increased number of dopamine receptors in the brains of schizophrenic patients, this will not prove that the primary cause of schizophrenia has been identified. Something else could be responsible for schizophrenia, and an increase in the number of dopamine receptors might merely be a secondary effect. It is even possible that ^{3}H-spiroperidol and ^{3}H-haloperidol bind with a class of receptors that are normally stimulated by a naturally-occurring chemical that has yet to be identified. Perhaps the brain produces its own "antischizophrenic" chemical in response to abnormal activity in some neural systems. The antischizophrenic drugs, then, would act as agonists at these sites.

Most recent investigations of the physiology of schizophrenia have been centered on dopamine, although all investigators acknowledge that the biochemical defect, if there is one, may be in a variety of places besides the dopaminergic synapse. However, at the present, the dopaminergic synapse is the most logical place to look for abnormalities. The scope of the search will undoubtedly be widened in the future, as investigators learn more about the ways in which dopaminergic neurons interact with their neighbors.

The Attentional Hypothesis

SCHIZOPHRENIA AS A DISTURBANCE IN SELECTIVE ATTENTION.   Schizophrenia has a variety of symptoms. Is there any single hypothesis that can unite them? Many investigators have suggested that the primary effect is a disturbance in the patient's control of his or her attentional processes.

> If there is any creature who can be accused of not seeing the forest for the trees, it is the schizophrenic. If he is of the paranoid persuasion, he sticks even more closely than the normal person to the path through the forest, examining each tree along the path, sometimes even each tree's leaves with meticulous care. . . . If at the other extreme he follows the hebephrenic pattern, then he acts as if there were no paths, for he strays off the obvious path entirely; he is attracted not only visually but even [by smell and taste], by any and all trees and even the undergrowth and floor of the forest, in a superficial flitting, apparently forgetting in the meantime about the place he wants to get to. (Shakow, 1962, p. 30)

An articulate woman who suffered from schizophrenic episodes appears to agree with this description.

> Each of us is capable of coping with a large number of stimuli, invading our being through any one of the senses. We could hear every sound within earshot and see every object, line, and color within the field of vision, and so on. It is obvious that we would be incapable of carrying on any of our daily activities if even one-hundredth of all these available stimuli invaded us at once. So the mind must have a filter which functions without our conscious thought, sorting stimuli, and allowing only those which are relevant to the situation in hand to disturb consciousness. And this filter must be working at maximum efficiency at all times, particularly when we require a high degree of concentration. What had happened to me . . . was a breakdown in the filter, and a hodge-podge of unrelated stimuli were distracting me from things which should have had my undivided attention. (McDonald, 1960, p. 218)

The literature on schizophrenia and attentional processes is too great to review here; if you are interested, you may want to consult one of the books listed in the suggested readings at the end of this chapter.

ATTENTION AND DOPAMINE.   If symptoms of schizophrenia can be diminished by dopamine antagonists, made worse by dopamine agonists, and are somehow related to defects in attention, then perhaps dopaminergic neurons themselves play some role in attention. Indeed, there is evidence that this is the case. Experimentally-produced damage to dopaminergic pathways results in "sensory neglect." For example, if a lesion is made by injecting 6-HD unilaterally into a rat's medial forebrain bundle, thus damaging axons of ascending catecholaminergic neurons, the animals will fail to react to visual, tactual, or auditory stimuli that are presented to the contra-lateral side (Ljungberg and Ungerstedt, 1976). This occurs despite the fact that the lesions do not involve any primary sensory path-ways.

The dopaminergic pathway from the substantia nigra to the cau-date nucleus is known to be important in motor control, since dam-age to this system is the apparent cause of Parkinson's disease. But several investigators have suggested that the dopaminergic pathway from the ventral tegmental area to the nucleus accumbens, olfactory tubercle, and frontal cortex is more likely to be related to attentional processes, and to schizophrenia. You will recall from Chapter 17 that this system has been shown to play a role in reinforcement, and that the reinforcing effects of amphetamine are abolished by its destruc-tion. One might expect that reinforcement and attention are closely related processes. It is very likely that the presentation of a rein-forcing stimulus causes an increase in attention. It would therefore seem plausible that abnormal activity in a system that plays a role in reinforcement might disrupt the processes of attention.

In fact, one of the two recognized therapeutic uses of drugs like amphetamine and methylphenidate is in cases of extreme distract-ability—inability to focus attention. The disorder is called the **hy-perkinetic syndrome,** and is seen in young children. Hyperkinetic children are always up and about, having great difficulty in remain-ing seated, and not being able to concentrate on a single task. They are continually distracted, dashing from one activity to the next. As a result, they come into conflict with teachers who try to restrain them, and who attempt to make them settle down and work. Am-phetamine and methylphenidate have remarkable effects on hyper-kinetic children; their ability to concentrate increases dramatically, and they cease their restless activity. Thus, hyperkinesis, in contrast to schizophrenia, can be seen as an attentional deficiency. Also in contrast to schizophrenia, it is successfully treated with dopamine *agonists.* (By the way, the only other recognized therapeutic use of these drugs is the treatment of narcoleptic sleep attacks.)

Some Remaining Problems

Since this is a textbook of physiological psychology, I have ignored the possible environmental causes of schizophrenia. (Anyway, my reading of the psychodynamic and behavioral literature leads me to believe that not too much progress has been made by these ap-proaches.) However, it is possible that the biological approach misses some of the important causes of this disorder.

There are also problems with the dopamine hypothesis. As I al-ready noted, there is no guarantee that the primary defect is in the dopaminergic systems—the real cause may lie elsewhere. In addi-tion, the time course of the therapeutic response to antischizophrenic drugs and the absence of a tolerance effect remain to be explained.

Let us examine these last two problems. When a schizophrenic patient is given an antischizophrenic drug, a therapeutic response is not seen immediately. In general, improvements in the symptoms of schizophrenia are not seen for a week or more. However, the effects of the drug on dopamine receptors appear to be very rapid. A study by Cotes, Crow, Johnstone, Bartlett, and Bourne (1978) illustrates this discrepancy very nicely. These investigators contrasted the time course of the antischizophrenic drug *cis*-flupenthixol with its effects on prolactin release.

Prolactin is a hormone that is produced and released by the anterior pituitary gland. This hormone causes the production of milk by the mammary glands. Like the other hormones of the anterior pituitary gland, prolactin release is controlled by hypothalamic hormones. In this case, the hypothalamus produces an *inhibiting hormone*. In the presence of this hormone, the anterior pituitary gland does not secrete prolactin. In its absence, the gland produces and secretes prolactin. The hypothalamic hormone has been identified as none other than dopamine (Moore and Bloom, 1978). Thus, as you might expect, a drug that blocks dopamine receptors antagonizes the effects of dopamine on the prolactin-producing cells of the anterior pituitary gland, and thus causes secretion of prolactin. (In fact, lactation is an occasional side effect of antischizophrenic medication in both male and female patients.) Figure 20.7 illustrates the time course of the effects of drug treatment on prolactin secretion and on the improvement in schizophrenic symptoms, as compared with control patients who received a placebo. Note that the effect of the drug

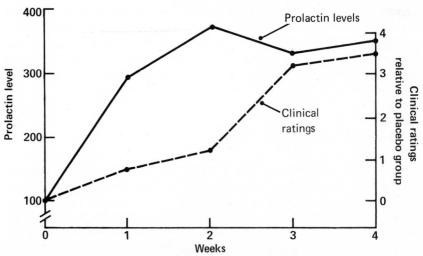

FIGURE 20.7 Prolactin levels and clinical ratings (relative to the ratings of patients who received a placebo) of patients who received an antischizophrenic drug (*cis*-flupenthixol). (From Cotes, P. M., Crow, T. J., Johnstone, E., Bartlett, W., and Bourne, R. C. *Psychological Medicine* (*London*), 1978, *8*, 657–665.)

on prolactin secretion is rapid, whereas the antischizophrenic effect is slow. The discrepancy has yet to be explained. (See **FIGURE 20.7.**)

The second problem with the dopamine hypothesis is the nature of the motor disturbances that are seen in patients who receive antischizophrenic medication. I have already mentioned the fact that a common side effect of these drugs is parkinsonian symptoms, including loss of facial expression and muscular rigidity and tremor. Approximately 10 percent of the patients who receive these drugs suffer from a motor disorder called *tardive dyskinesia. Tardus* = "slow" and *dyskinesia* = "faulty movement"; thus, tardive dyskinesia is a late-developing movement disorder. This syndrome includes peculiar facial tics and gestures, including tongue protrusion, cheek puffing, and pursing of the lips. In some cases, speech is affected. In addition, writhing movements of the hands and trunk are often seen. The symptoms generally occur only after many months of drug treatment. They are made *worse* by withdrawal of the antischizophrenic drug but are improved by increasing the dose. The symptoms are also made worse by administering dopamine agonists like L-DOPA or amphetamine. Thus, the disorder is the *opposite* of Parkinson's disease. Similar disturbances are seen in nonschizophrenic patients (such as hyperkinetic children or patients with Parkinson's disease) who receive an overdose of amphetamine or L-DOPA (Baldessarini, 1979). Therefore, the disorder appears to be produced by an overstimulation of dopamine receptors. It is made worse by dopamine agonists and better by dopamine antagonists. But, if this is so, why should it be originally caused by antischizophrenic drugs, which are dopamine antagonists?

The answer seems to be provided by a phenomenon called *denervation supersensitivity.* When the nerve to a muscle is cut, the muscle develops, within a few days, an increased sensitivity to acetylcholine. That is, a given amount of acetylcholine produces more of an effect than it previously did. This phenomenon seems to be related to an increase in the number of postsynaptic cholinergic receptors on the muscle (Miledi, 1959). Presumably, decreased incoming traffic causes a regulatory mechanism to increase the muscle's sensitivity to the transmitter substance to which it normally responds.

It has been suggested that neurons can also develop denervation supersensitivity. If a particular input ceases to be active, regulatory mechanisms increase the neuron's sensitivity to that input. Thus, in the case of tardive dyskinesia, blockage of the dopamine receptors in the caudate nucleus may result in the development of a compensatory receptor supersensitivity to this transmitter substance. This supersensitivity results in dyskinesia when the drug is withdrawn. In cases where the antischizophrenic medication continues for a prolonged time, the supersensitivity becomes so great that it overcompensates for the effects of the drug, causing the tardive dyskinesia to occur.

So far, so good. But consider this: why does a patient not develop "tardive schizophrenia"? If schizophrenia is alleviated by a blockade of dopaminergic neurons (perhaps in the nucleus accumbens or frontal cortex, rather than in the caudate nucleus), why is it that super-sensitivity to dopamine, and a resulting outbreak of schizophrenic symptoms, does not occur there? Clearly, we do not know enough about denervation supersensitivity in neurons, tardive dyskinesia, or the physiology of schizophrenia (or, more likely, all three).

This is a good place to mention a possible problem that is inherent in the current screening process used by pharmaceutical companies to identify new drugs that may have antischizophrenic effects (Baldessarini, 1979). The usual procedure is to make compounds whose molecular structure resembles that of known antischizophrenic drugs. The compounds are given to rats or mice, and those that produce motor disturbances (Parkinson-like symptoms) are investigated further. In general, motor disturbances and antischizophrenic effects go hand-in-hand, but if some compound had an antischizophrenic effect without motor side effects (which would be very desirable), it would not be discovered by this procedure.

AFFECTIVE PSYCHOSES

Description

Affect refers to feelings or emotions. If schizophrenia is characterized by thought disorders, the affective psychoses are characterized by disorders in feelings. Of course, feelings and emotions are essential parts of human existence; they represent our evaluation of the events in our lives. In a very real sense, feelings and emotions are what human life is all about.

The emotional state of most of us reflects what is happening; our feelings are tied to events in the real world, and usually they are the results of fair assessment of the significance of these events to our lives. But for some people, affect becomes divorced from reality. They possess feelings of extreme elation (mania) or despair (depression) that are not justified by what is happening to them. For example, depression that accompanies the loss of a loved one is normal; depression that becomes a way of life, and that will not respond to the sympathetic efforts of friends and relatives is pathological.

Some people suffer from alternating periods of mania and depression. They are said to have a **bipolar affective disorder.** This disorder afflicts approximately equal numbers of men and women. The mania can last as little as a few days or as much as several months, but usually takes a few weeks to run its course. The depression that follows generally lasts three times as long as the mania. Other people suffer from unremitting **unipolar depression.** This disorder strikes women two to three times more than men. Mania without periods of depression sometimes occurs, but it is rare.

The affective psychoses are very serious disorders, since a person who suffers from psychotic depression runs a significant risk of death by suicide. Psychotically depressed people usually feel extremely unworthy and have strong feelings of guilt. Often, they believe that their body is rotting away in retribution for sins that they have committed. They have very little energy, and move and talk very slowly, sometimes becoming almost torpid. At other times, they may move around restlessly and aimlessly. They may cry a lot. They lose their appetite for food and (especially) sex. They generally fall asleep readily, but awaken early, and find it impossible to get to sleep again. (In contrast, people with neurotic depression usually have trouble falling asleep, but do not awaken early.) They often become constipated, and secretion of saliva decreases.

The following interview illustrates a case of psychotic depression.

> Th.: Good morning, Mr. H., how are you today?
> Pt.: (Long pause—looks up and then head drops back down and stares at floor.)
> Th.: I said good morning, Mr. H. Wouldn't you like to tell me how you feel today?
> Pt.: (Pause—looks up again). . . I feel . . . terrible . . . simply terrible.
> Th.: What seems to be your trouble?
> Pt.: . . . There's just no way out of it . . . nothing but blind alleys . . . I have no appetite . . . nothing matters anymore . . . it's hopeless . . . everything is hopeless.
> Th.: Can you tell me how your trouble started?
> Pt.: I don't know . . . it seems like I have a lead weight in my stomach . . . I feel different . . . I am not like other people . . . my health is ruined . . . I wish I were dead.
> Th.: Your health is ruined?
> Pt.: . . . Yes, my brain is being eaten away. I shouldn't have done it . . . If I had any willpower I would kill myself . . . I don't deserve to live . . . I have ruined everything . . . and it's all my fault.
> Th.: It's all your fault?
> Pt.: Yes. . . I have been unfaithful to my wife and now I am being punished . . . my health is ruined . . . there's no use going on . . . (sigh) . . . I have ruined everything . . . my family . . . and now myself . . . I bring misfortune to everyone . . . I am a moral leper . . . a serpent in the Garden of Eden . . . why don't I die . . . why don't you give me a pill and end it all before I bring catastrophe on everyone . . . No one can help me . . . It's hopeless . . . I know that . . . it's hopeless. (From *Abnormal Psychology and Modern Life*, 5th Edition by James C. Coleman. Copyright © 1976 by Scott, Foresman and Co. Reprinted by permission.)

Episodes of mania are characterized by a sense of euphoria that does not seem to be justified by the circumstances. The diagnosis of mania is partly a matter of degree—one would not want to call moderate exuberance and a zest for life pathological. People with psychotic mania usually exhibit nonstop speech and motor activity. They flit from topic to topic, but without the severe disorganization that is seen in schizophrenia. They are usually full of their own importance, and become angry or defensive if they are contradicted. Often they go for long periods without sleep, working furiously on a project that may or may not be realistic.

The following interview illustrates a case of mania. The patient is a woman.

> Therapist: Well, you seem pretty happy today.
> Client: Happy! Happy! You certainly are a master of understatement, you rogue! (Shouting, literally jumping out of seat.) Why I'm ecstatic. I'm leaving for the West coast today, on my daughter's bicycle. Only 3100 miles. That's nothing, you know. I could probably walk, but I want to get there by next week. And along the way I plan to followup on my inventions of the past month, you know, stopping at the big plants along the way having lunch with the executives, maybe getting to know them a bit—you know, Doc, "know" in the biblical sense (leering at therapist seductively). Oh, God, how good it feels. It's almost like a nonstop orgasm. (Davison and Neale, 1974, p. 175).

The fact that mania is rarely seen without depression has suggested to some clinicians that the mania is a defense mechanism—an attempt to ward off a period of depression. They note that there is often something brittle and unnatural about the happiness of the manic phase. This is illustrated in the following description.

> He neglected his meals and rest hours, and was highly irregular, impulsive, and distractible in his adaptations to ward routine. Without apparent intent to be annoying or disturbing he sang, whistled, told pointless off-color stories, visited indiscriminately, and flirted crudely with the nurses and female patients. Superficially he apeared to be in high spirits, and yet one day when he was being gently chided over some particularly irresponsible act he suddenly slumped in a chair, covered his face with his hands, began sobbing, and cried, "For Pete's sake, doc, let me be. Can't you see that I've just got to act happy?" (Masserman, 1961, pp. 66–67.)

Heritability of Affective Psychoses

Although the heritability of the affective psychoses has not been studied as thoroughly as the heritability of schizophrenia has been, there is ample evidence that genetics plays a role in a person's susceptibility to these disorders. For example, Rosenthal (1971) found that close relatives of people who suffer from affective psychoses are ten times more likely to develop these disorders than people without afflicted relatives. Of course, this study does not prove that genetic mechanisms are operating; relatives have similar environments as well as similar genes. However, it was also observed that if one member of a set of monozygotic (identical) twins was afflicted with an affective disorder, the likelihood was 69 per cent that the other twin was similarly afflicted. In contrast, the likelihood was only 13 per cent for dizygotic (fraternal) twins, who are no more closely related than any pair of siblings (Gershon, Bunney, Leckman, Van Eerde-wegh, and DeBauche, 1976). Furthermore, high concordance for monozygotic twins appears to be the same whether the twins are raised together or apart (Price, 1968). The greater concordance for the presence of an affective disorder in monozygotic twins is strong evidence for the existence of a heritable trait. And if the affective psychoses are heritable, then they probably have a physiological basis.

Physiological Treatments for the Affective Psychoses

There are three treatments that have a profound effect on unipolar depression: electroconvulsive therapy (ECT), monamine oxidase inhibitors, and the tricyclic antidepressant drugs. In addition, bipolar affective disorders are effectively treated by lithium salts. The efficacy of these treatments provides additional evidence that the affective disorders have a physiological basis. Furthermore, the fact that lithium is very effective in treating bipolar affective disorders, but not unipolar depression, suggests that there is a fundamental difference between these pathologies.

A depressed patient does not immediately respond to treatment with monoamine oxidase inhibitors (like *iproniazid*) or one of the tricyclic antidepressant drugs (like *imipramine*); an improvement in symptoms is not usually seen before one to three weeks of drug treatment. In contrast, the effects of electroconvulsive shock are much more rapid. (I shall not describe the procedure here, since ECT was described in some detail in Chapter 18.) A few electroconvulsive treatments can often snap a person out of a deep depression. Although prolonged and excessive use of ECT can cause long-lasting impairments in memory (Squire, 1974), judicious use of ECT during the period before the antidepressant drugs have their effect has undoubtedly saved the lives of suicidal patients (Baldessarini, 1977).

The therapeutic effects of lithium (usually administered in the form of lithium carbonate) are very rapid. This drug is most effective in the treatment of the manic phase of a bipolar affective disorder, and once the mania is eliminated, the stage of depression usually does not follow. This suggests that depression is a reaction to mania, and not the reverse. A patient in the manic phase of a bipolar affective disorder usually responds as soon as the level of the lithium ion in the blood reaches a therapeutic level (Gerbino, Oleshansky, and Gershon, 1978). Lithium does not interfere with a person's sex drive, and leaves the patient able to feel and express joy and sadness to events in his or her life. Similarly, it does not impair intellectual process. Some patients have received this drug continuously for many years without any apparent ill effects (Fieve, 1979).

Lithium has received a bad name because lithium chloride, formerly used as a salt-substitute for patients with high blood pressure, was found to have toxic effects, and hence was removed from the market. However, monitoring of the blood levels of patients receiving lithium therapy can prevent toxic side effects.

THE PHARMACOLOGY OF ANTIDEPRESSANT TREATMENTS. The drug treatment for affective disorders prior to the 1950s was limited to amphetamine (for depression) or the barbiturates (for agitation). Neither of these drugs was very satisfactory. In the late 1940s, it was noted that some drugs that were used for the treatment of tuberculosis seemed to have a mood-elevating effect. It was subsequently found that a derivative of these drugs, iproniazid, was useful for psychotic depression (Crane, 1957). Iproniazid inhibits the activity of monoamine oxidase, and thus acts as an agonist for the monoamine neurotransmitters: dopamine, norepinephrine, and serotonin. Unfortunately, monoamine oxidase inhibitors can have harmful side effects. In some cases, the drugs can cause liver damage. A more common, and therefore more serious, effect has been called the *cheese effect*. Many foods (e.g., cheese, yogurt, wine, yeast breads, chocolate, and various fruits and nuts) contain *biogenic amines*—substances that are similar to the catecholamines in their effects on the nervous system. Normally, these amines are deactivated by monoamine oxidase that is present in the blood and in other tissues of the body. But if a person is being treated with an inhibitor of monoamine oxidase, he or she may suffer a serious sympathetic reaction after eating one of these foods. These reactions can raise blood pressure enough to produce intracranial bleeding or cardiovascular collapse.

Fortunately, another class of antidepressant drugs was soon discovered—the **tricyclic antidepressants.** (Figure 20.8 illustrates the molecular structure of imipramine, the first drug of this category to be discovered; you can see why it is called *tricyclic*.) (See **FIGURE 20.8.**) These drugs were found to inhibit the re-uptake mechanism in the terminal buttons of monoaminergic neurons. Thus, both the monoamine oxidase inhibitors and the tricyclic antidepressant drugs are monoaminergic agonists. Perhaps, then, depression is produced by insufficient activity of this class of neurons. You will recall from Chapter 17 that the catecholamines (especially dopamine) have been implicated in mechanisms of reinforcement. This may explain the

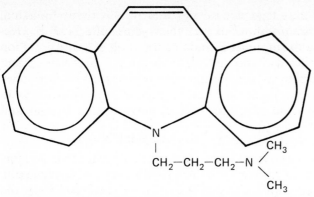

FIGURE 20.8   The molecular structure of a tricyclic antidepressant, imipramine.

inability of depressed patients to experience pleasure.

One class of tricyclic antidepressants (including imipramine) has a more pronounced effect on the re-uptake of serotonin. These drugs appear to have a sedating effect, whereas the other tricyclic antidepressants (including desipramine) have a more activating effect, presumably because they affect norepinephrine more than serotonin (Baldessarini, 1977). Thus, we might conclude that depression is caused by insufficient activity in monoaminergic synapses. Too little noradrenergic and/or dopaminergic activity results in retarded depression, and too little serotonergic activity results in agitated depression.

Some recently-discovered drugs cast doubt on the validity of these conclusions. These drugs, **iprindole** and **mianserin** (both of which are tricyclic in structure) have little or no effect on the re-uptake mechanisms of noradrenergic or serotonergic neurons. In fact, mianserin has been shown to block serotonergic receptors. However, both of these drugs have potent antidepressant effects. If they do not affect noradrenergic or serotonergic synapses, how do they alleviate the symptoms of depression?

There is independent evidence from two laboratories (Green and Maayani, 1977; Kanof and Greengard, 1978) that all of the clinically-effective tricyclic antidepressants are extremely potent blockers of histamine receptors. There is good evidence that histamine is a neurotransmitter in the brain (Schwartz, 1977). In fact, an ascending system of histaminergic neurons ascends through the medial forebrain bundle and projects diffusely to widespread regions of the forebrain. The anatomy at least supports the plausibility of a modulating effect of histaminergic neurons on mood. In addition, some (but not all) of the antischizophrenic drugs have therapeutic effects on depression; these, too, were found to block histamine receptors. As Kanof and Greengard point out, the only known common site of biochemical activity of all of these drugs is on histamine receptors.

It is also the case that some drugs that inhibit the re-uptake of norepinephrine or serotonin have no therapeutic effects on depression. To take one example, cocaine inhibits both NE and 5-HT re-uptake (Ross and Renyi, 1966, 1977) but, despite its euphoric effects in nondepressed people, this drug makes depressed patients feel worse when it is injected intravenously (Post, Kotin, and Goodwin, 1974). Cocaine has no effect on histamine receptors (Kanof and Greengard, 1978).

There is little to say about the pharmacological effects of ECT and lithium. Much has been hypothesized, but little has been found in the way of conclusive evidence. Electroconvulsive shock produces many alterations in the biochemical functions of the brain, so it will be difficult to isolate the ones that are related to its therapeutic effects on depression. Lithium has been implicated in the activity of catecholaminergic and cholinergic neurons, and in the concentration of amino acid neurotransmitters (Fieve, 1979). It has even been suggested that patients with affective disorders also have deficient lithium transport mechanisms in their cell membranes (Dorus, Pandey, Shaughnessy, Gaviria, Val, Ericksen, and Davis, 1979). We still lack enough data to develop a comprehensive theory.

Not all evidence supports the histamine hypothesis. Monoamine oxidase inhibitors are effective in treating depression, but they do not appear to affect histamine receptors (Kanof and Greengard, 1978). Do the MAO inhibitors and the tricyclic antidepressants act in pharmacologically different ways, or is the effect of the tricyclics on histamine receptors merely a red herring? It will take much more study to determine the ways in which antidepressant drugs exert their therapeutic effects.

Drugs that Produce Affective Disorders

One of the first drugs that was found to have an antischizophrenic effect was an alkaloid extract from *Rauwolfia serpentina,* a shrub of Southeast Asia. (Rauwolf was a seventeenth-century German botanist.) The alkaloid, which is now called **reserpine,** has been used in India for hundreds of years to treat snakebite, circulatory disorders, and insanity. Modern research has confirmed that it has both an antischizophrenic effect and a hypotensive effect (that is, it lowers blood pressure). This effect on blood pressure reduces its usefulness in the treatment of schizophrenia, but the drug is often used today to treat patients with high blood pressure.

Reserpine has a serious side effect; it can cause depression. In fact, in the early years of its use as a hypotensive agent up to 15 per cent of the people who received it became depressed (Sachar and Baron, 1979). Reserpine acts on the membrane of synaptic vesicles in the terminal buttons of monoaminergic neurons. It makes the membranes "leaky," so that the neurotransmitters are released from the vesicles and are destroyed by MAO. Thus, the drug serves as a potent NE, DA, and 5-HT antagonist.

The pharmacological and behavioral effects of reserpine complement the pharmacological and behavioral effects of MAO inhibitors and most of the tricyclic antidepressants. That is, a monoamine antagonist produces depression, whereas monoamine agonists alleviate it. However, we saw that histaminergic neurons have also been implicated; tricyclic antidepressants (including those with no effect on monoaminergic synapses) block histamine receptors, and cocaine, which is a potent monamine agonist, makes depressed patients feel even worse. It is probably safe to predict that the causes of the affective disorders will eventually be found to be complex.

Schizophrenia and the affective psychoses are disorders that disrupt the lives of many people. If it were not for drug treatments and ECT, many more people—and their families—would suffer. While it is true that all of the biological therapies we have now were discovered by accident, we can hope that research on these disorders will lead to a fundamental understanding of their physiological causes, and thus to more specific, and more effective, therapeutic agents.

AFTERWORD

In the preface I invited you to write to me so that we could discuss this book or issues raised (or not raised) here. I should like to extend this invitation again. My address is given in the preface. As I wrote this book, I had an invisible set of readers in mind; it would be very gratifying to hear from some of you.

KEY TERMS

acute schizophrenia p. 666
affect (*AFF ekt*) p. 682
affective psychosis p. 664
benzodiazepine (*ben zo dy AZ a peen*) p. 670
bipolar affective disorder p. 682
butyrophenone (*bew TEER oh fen OWN*) p. 671
catatonic schizophrenia p. 665
chlorpromazine p. 668
denervation supersensitivity p. 681
dihydroxyphenylacetic acid (DOPAC) p. 676
flupenthixol (*flew pen THIX all*) p. 670
haloperidol (*hal oh PAIR a dahl*) p. 671

hebephrenic schizophrenia (*heb a FREN ik*) p. 665
homovanillic acid (HVA) p. 675
hyperkinetic syndrome p. 679
imipramine (*im IP ra meen*) p. 685
iprindole (*IP rin dole*) p. 687
iproniazid (*ip roh NY a zid*) p. 685
isomer (*EYE so mur*) p. 670
major tranquilizer p. 669
mianserin (*my AN sur in*) p. 687
neurosis p. 664
paranoid schizophrenia p. 665
process schizophrenia p. 666

psychosis p. 664
reserpine (*ree SUR peen*) p. 688
schizophrenia p. 664
tardive dyskinesia p. 681

tricyclic antidepressant (*try SIK lik*) p. 686
unipolar depression p. 683
waxy flexibility p. 665

SUGGESTED READINGS

BALDESSARINI, R. J. *Chemotherapy in Psychiatry*. Cambridge, MA: Harvard University Press, 1977.

BERGSMA, D., and GOLDSTEIN, A. L. (eds.) *Neurochemical and Immunologic Components in Schizophrenia*. New York: Alan R. Liss, 1978.

WYNNE, L. C., CROMWELL, R. L., and MATTHYSSE, S. (eds.) *The Nature of Schizophrenia: New Approaches to Research and Treatment*. New York: Wiley and Sons, 1978.

The book by Baldessarini provides an excellent introduction to the use of psychotherapeutic drugs. The volume edited by Bergsma and Goldstein introduces a number of hypotheses that were not described here. The volume edited by Wynne, et al., covers behavioral as well as genetic and physiological investigations into the causes of schizophrenia. In addition to these books, you should look at recent volumes of *Annual Review of Medicine, Annual Review of Pharmacology and Toxicology*, and *Annual Review of Neuroscience*, which often carry reviews that are relevant to the topics of this chapter.

Glossary

Ablation: The intentional removal or destruction of portions of the central nervous system. Synonym: brain lesion.

Absorptive phase: The phase of metabolism during which nutrients are absorbed from the digestive system. Glucose and amino acids constitute the principal source of energy for cells during this phase. Stores of glycogen are increased and excess nutrients are stored in adipose tissue in the form of lipids.

Acetylcholine (ACh): A neurotransmitter found in the brain, spinal cord, ganglia of the autonomic nervous system, and postganglionic terminals of the parasympathetic division of the autonomic nervous system.

Acetylcholinesterase (AChE): The enzyme that destroys acetylcholine soon after it is liberated by the terminal buttons, thus terminating the postsynaptic potential.

ACh: See *acetylcholine.*

AChE: See *acetylcholinesterase.*

ACTH: See *adrenocorticotrophic hormone.*

Actin: Actin and *myosin* are the proteins that provide the physical basis for muscular contraction. See Figure 10.3.

Action potential: The brief electrical impulse that provides the basis for conduction of information along an axon. The action potential results from brief changes in membrane permeability to sodium and potassium ions. See Figure 3.12.

Activational effect: See *hormone.*

AD: See *androstenedione.*

Ad libitum: Literally, "to the desire." More generally, "as much as is wanted."

Adenine: One of the nucleotide bases of RNA and DNA.

Adenohypophysis: See *pituitary gland.*

Adenosine triphosphate (ATP): A molecule of prime importance to cellular energy metabolism: the conversion of ATP to ADP liberates energy. ATP can also be converted to *cyclic AMP,* which serves as intermediate messenger in the production of postsynaptic potentials by some neurotransmitters, and in the mediation of the effects of polypeptide hormones. See *adenyl cyclase* and Figure 5.10.

Adenyl cyclase: An enzyme that converts *ATP* to *cyclic AMP* when the postsynaptic receptor to which it is bound is stimulated by the appropriate substance. Important in mediating the intracellular effects of many neurotransmitters and polypeptide hormones. See *adenosine triphosphate* and Figure 5.10.

ADH: See *antidiuretic hormone.*

Adipose tissue: Fat tissue, composed of cells that can absorb nutrients from the blood and store them in the form of lipids during the *absorptive phase,* or release them in the form of fatty acids and ketones during the *fasting phase.*

Adipsia: Complete lack of drinking; can be produced by lesions of the lateral hypothalamus or region around the anteroventral third ventricle.

Adrenalin: See *epinephrine.*

Adrenocorticotrophic hormone (ACTH): A hormone produced and liberated by the anterior pituitary gland in response to ACTH releasing hormone, produced by the hypothalamus. ACTH stimulates the adrenal cortex to produce various corticosteroid hormones. ACTH also inhibits testosterone-dependent intermale aggression.

Affect: An emotional state of strong feelings, either positive or negative in nature.

Affective attack: A highly emotional attack of one animal upon another; can be elicited by electrical stimulation of certain regions of the brain. Contrasts with *quiet-biting attack.*

Agonist: Literally, a contestant. An agonistic drug facilitates the effects of a particular neurotransmitter on the postsynaptic cell. An agonistic muscle produces or facilitates a particular movement. Antonym: *antagonist.*

Aldosterone: A hormone of the adrenal cortex that causes the retention of sodium by the kidneys.

All-or-none law: Refers to the fact that once an action potential is triggered in an axon, it is propagated, without decrement, to the end of the fiber.

Alpha activity: Smooth electrical activity of 8 to 12 Hz, recorded from the brain. Alpha activity is generally associated with a state of relaxation.

Alpha-methyl-paratyrosine: See α-*methyl-paratyrosine.*

Alpha-methyltryptamine: A serotonin agonist: stimulates postsynaptic receptors.

Alpha motor neuron: A neuron whose cell body is located in the ventral horn of the spinal cord or in one of the motor nuclei of the cranial nerves. Stimulation of an alpha motor neuron results in contraction of the extrafusal muscle fibers upon which its terminal buttons synapse.

Alphafetoprotein: A protein found in the extracellular fluid of fetuses; binds with maternal estrogen, thus preventing masculinization of female fetuses.

Amino acid: A molecule that contains both an amino group and a carboxyl group. Amino acids are linked together by peptide bonds and serve as the constituents of proteins.

Amino group: NH_2; two molecules of hydrogen attached to a molecule of nitrogen.

Aminostatic theory: The theory that hunger and satiety depend on the level of amino acids in the blood plasma, low levels being associated with hunger.

Amitryptyline: A monoamine agonist: retards re-uptake by the terminal buttons.

Amphetamine: A *catecholamine* agonist: facilitates release, stimulates postsynaptic receptors (slightly), and retards re-uptake by the terminal buttons.

Amplifier: An electronic device that increases the amplitude of a small electrical signal.

AMPT: See α-*methyl-paratyrosine.*

Amygdala: The term commonly used for the amygdaloid complex, a set of nuclei located in the base of the temporal lobe. The amygdala is a part of the limbic system.

Analgesia: Lack of sensitivity to pain.

Androgen: A male sex steroid hormone. *Testosterone* is the principal mammalian androgen.

Androgenization: The process initiated by exposure of the cells of a developing animal to androgens. Exposure to androgens causes embryonic sex organs to develop as male, and produces certain changes in the brain. See *hormone.*

Androstenedione (AD): An androgen secreted by adrenal cortex of both males and females.

Angiotensin: See *renin.*

Angiotensinogen: See *renin.*

Anion: See *ion.*

Antagonist: An antagonistic muscle produces a movement contrary or opposite to the one being described. An antagonistic drug opposes or inhibits the effects of a particular neurotransmitter on the postsynaptic cell. Antonym: *agonist.*

Anterior: See Figure 6.1

Anterior pituitary gland: See *pituitary gland.*

Anterior thalamus: A group of three thalamic nuclei that receive fibers from the mammillary bodies and project fibers to the *cingulate gyrus,* and thus comprise a portion of *Papez's circuit.*

Anterograde amnesia: See *posttraumatic amnesia.*

Anterograde degeneration: Rapid degeneration of an axon distal to its point of damage. See Figure 7.6.

Antidiuretic hormone (ADH): A hormone secreted by the posterior pituitary gland that causes the kidneys to excrete a more concentrated urine, thus retaining water in the body.

Antipsychotic drug: A drug that reduces or eliminates the symptoms of psychosis. Antischizophrenic drugs appear to exert their effect by antagonizing dopaminergic synapses.

Aphagia: Complete lack of eating. Can be produced by lesions of the lateral hypothalamus.

Apomorphine: A dopamine agonist: stimulates postsynaptic receptors.

Arachnoid: The middle layer of the *meninges,*

between the outer *dura mater* and inner *pia mater.*

Arcuate fasciculus: A bundle of long associ-ation fibers that interconnect *Wernicke's area* on the left temporal lobe with *Broca's speech area* on the left frontal lobe. Damage to the arcuate fasci-culus results in *conduction aphasia.*

Arcuate nucleus: The hypothalamic nucleus that contains the cell bodies of the neurosecretory cells that produce the hypothalamic hormones.

Area postrema: A region of the medulla where the *blood-brain barrier* is weak. Systemic poisons can be detected there, and can initiate vomiting.

Aromatization: Conversion of a molecule into one that contains a benzene ring. Estradiol and estrone are aromatized forms of testosterone and androstenedione, respectively.

Arousal: A state of alertness characterized by pupillary dilation and *beta activity* of the EEG.

Astrocyte (astroglia): A glial cell that pro-vides support for neurons of the central nervous system. Astrocytes also participate in the forma-tion of scar tissue after injury to the brain or spinal cord.

ATP: See *adenosine triphosphate.*

Atropine: An acetylcholine antagonist: blocks postsynaptic receptors.

Auditory nerve: The auditory nerve has two principal branches. The cochlear nerve transmits auditory information and the vestibular nerve transmits information related to balance.

Autonomic nervous system (ANS): The por-tion of the peripheral nervous system that controls the body's vegetative function. The *sympathetic branch* mediates functions that accompany arousal, while the *parasympathetic branch* mediates func-tions that occur during a relaxed state.

Axoaxonic synapse: The synapse of a *terminal button* upon the axon of another neuron, near its terminals. These synapses mediate presynaptic in-hibition.

Axodendritic synapse: The synapse of a *ter-minal button* of the axon of one cell upon the den-drite of another cell.

Axon: A thin elongated process of a neuron that can transmit action potentials toward its ter-minal buttons, which synapse on other neurons, gland cells, or muscle cells.

Axon hillock: The initial part of the axon, at the junction of axon and soma. It is capable of pro-ducing an action potential, and generally has a slightly lower *threshold of excitation* than the rest of the axon.

Axosomatic synapse: The synapse of a *termi-nal button* of the axon of one neuron upon the mem-brane of the soma, or cell body, of another neuron.

Baroreceptor: A special receptor that trans-duces changes in barometric pressure (chiefly within the heart or blood vessels) into neural ac-tivity.

Basal ganglia: Caudate nucleus, globus pal-lidus, putamen, and amygdala. The first three are important parts of the *extrapyramidal motor system.*

Basilar artery: An artery found at the base of the brain, connecting the blood supplies of the *ver-tebral* and *carotid* arteries.

Basolateral group: The phylogenetically newer portion of the amygdaloid complex. See *amygdala.*

Beta activity: Irregular electrical activity of 13 to 30 Hz, recorded from the brain. Beta activity is generally associated with a state of arousal.

Bicuculline: A GABA antagonist: blocks post-synaptic receptors.

Bilateral: On both sides of the midline of the body.

Bipolar neuron: A neuron with only two processes—a dendritic process at one end and an axonal process at the other end. See Figure 2.8.

Blood-brain barrier: A barrier produced by the *astrocytes* and cells in the walls of the capillar-ies in the brain; this barrier permits passage of only certain substances.

Botulinum toxin: An acetylcholine antago-nist: prevents release by terminal buttons.

Bregma: The junction of the sagittal and co-ronal sutures of the skull. It is often used as a ref-erence point for stereotaxic brain surgery.

Broca's speech area: A region of frontal cor-tex, located at the base of the left precentral gyrus, that is necessary for normal speech production. Damage to this region results in Broca's aphasia, characterized by extreme difficulty in speech artic-ulation.

Cable properties: Passive conduction of elec-trical current, in a decremental fashion, down the length of an axon, similar to the way in which elec-trical current traverses a submarine cable.

Calcium pump: A metabolically active proc-ess in the cellular membrane that extrudes cal-cium, which enters the cell during an action po-tential. Found in terminal buttons and muscle fibers. The entry of calcium is necessary for release of neurotransmitters and contraction of muscle fibers.

Cannula: A small metal tube that may be inserted into the brain to permit introduction of chemicals or removal of fluid for analysis.

Carboxyl group: COOH; two atoms of oxygen and one atom of hydrogen bound to a single atom of carbon.

Carotid artery: An artery, the branches of which serve the rostral portions of the brain.

Cataplexy: A symptom of narcolepsy; complete paralysis that occurs during waking. Thought to be related to D sleep mechanisms.

Catecholamine: Dopamine and norepinephrine.

Catechol-o-methyl transferase (COMT): An enzyme that destroys dopamine and norepinephrine.

Cation: See *ion.*

Caudal: See Figure 6.1.

Caudate nucleus: A telencephalic nucleus; one of the basal ganglia. The caudate nucleus is principally involved with inhibitory control of movement.

Central canal: The narrow tube, filled with cerebrospinal fluid, that runs through the length of the spinal cord.

Central gray of the tegmentum: Stimulation of this brain region may result in aggressive behavior and also can produce analgesia.

Central tegmental tract: An alternate name for the *ventral noradrenergic* bundle, which carries fibers of noradrenergic neurons from regions of the medulla and from locus coeruleus and the subcoerulear area of the pons to the hypothalamus.

Cerebral aqueduct: A narrow tube interconnecting the third and fourth ventricles of the brain.

Cerebral cortex: The outermost layer of gray matter of the cerebral hemispheres.

Cerebrospinal fluid: A clear fluid, similar to blood plasma, that fills the ventricular system of the brain and spinal cord, and in which they float.

Cerveau isolé: A brainstem transection at the midcollicular level that results in a chronically comatose animal.

Chemoreceptor: A receptor that responds, by means of receptor potentials or neural impulses, to the presence of a particular chemical.

Chlorpromazine: A dopamine antagonist: blocks postsynaptic receptors. It is the most commonly-prescribed antischizophrenic drug.

Choroid plexus: Highly vascular tissue that protrudes into the ventricles and produces *cerebro-spinal fluid.*

Chromosome: Strand of DNA, with associated proteins, found in the nucleus. Carries genetic information.

Cingulate gyrus: A strip of limbic cortex lying along the lateral walls of the groove separating the cerebral hemispheres, just above the *corpus callosum.* See *Papez's circuit.*

Circumstriate belt: A region of visual association cortex; receives fibers from *striate cortex* and from the *superior colliculi,* and projects (in monkeys) to *inferotemporal cortex.*

CNS: Central nervous system; the brain and spinal cord.

Cochlea: The snail-shaped structure of the inner ear that contains the auditory transducing mechanisms.

Cochlear nerve: See *auditory nerve.*

Codon: The basic three-letter word of the genetic code. Each codon (represented by a sequence of three nucleotide bases on a strand of messenger RNA) specifies a particular amino acid that will be added to a *polypeptide* chain.

Commissure: A fiber bundle that interconnects corresponding regions on each side of the brain.

Conduction aphasia: Damage to the *arcuate fasciculus,* which interconnects *Wernicke's area* and *Broca's speech area,* results in the inability to repeat words that are heard, although they can usually be understood and responded to appropriately.

Cone: See *photoreceptor.*

Cosolidation: The process by which *short-term memories* are converted into *long-term memories.*

Contralateral: Residing in the side of the body opposite to the reference point.

Convergence: See Figure 10.10.

Coolidge effect: The restorative effect of introducing a new female sex partner to a male that has apparently become "exhausted" by sexual activity.

Coronal section: See Figure 6.2.

Corpus callosum: The largest commissure of the brain, interconnecting the areas of association cortex on each side of the brain.

Corpus luteum: After ovulation, the ovarian follicle develops into a corpus luteum, and secretes estrogen and progesterone.

Correctional mechanism: In a regulatory process the correctional mechanism is the one that is capable of changing the value of the system variable.

Corticomedial group: The phylogenetically older portion of the amygdaloid complex.

Cranial nerve: One of a set of twelve pairs of nerves that exit from the base of the brain.

Cranial nerve ganglion: See *ganglion.*

Cross section: See Figure 6.2.

CSF: See *cerebrospinal fluid.*

Curare: An acetylcholine antagonist: blocks nicotinic receptors and causes muscular paralysis.

Cyclic AMP: See *adenyl cyclase.*

Cyclic nucleotide: A compound such as cyclic AMP or cyclic GMP, important in mediating the intracellular effects of many neurotransmitters and polypeptide hormones. See *adenyl cyclase.*

Cytoplasm: The viscous semiliquid substance contained in the interior of a cell.

Decerebrate: An animal whose brainstem has been transected.

Decremental conduction: Conduction of a subthreshold stimulus along an axon, according to its *cable properties.*

Decussation: Crossing of a fiber to the other side of the brain.

Delta activity: Regular synchronous electrical activity of approximately 3 to 5 Hz, recorded from the brain. Delta activity is generally associated with slow wave sleep (*S sleep*).

Dendrite: Treelike process attached to the *soma* of a neuron, which receives *terminal buttons* from other neurons.

Dendritic spine: Small buds on the surface of a dendrite, on which synapse *terminal buttons* from other neurons.

Denervation supersensitivity: Increased sensitivity of neural postsynaptic membrane or *motor end plate* to the neurotransmitter: caused by damage to the afferent axons or long-term blockage of neurotransmitter release.

2-Deoxy-D-glucose (2-DG): A sugar that interferes with the metabolism of glucose.

Deoxyribonucleic acid (DNA): A long complex macromolecule consisting of two interconnected helical strands. Strands of DNA, along with their associated proteins, constitute the chromosomes that contain the genetic information of the animal.

Depolarization: Reduction (toward zero) of the membrane potential of a cell from its normal resting potential of approximately −70mV.

Desynchrony: Irregular electrical activity recorded from the brain, generally associated with periods of arousal. See *beta activity.*

Detector: In a regulatory process, a mechanism that signals when the system variable deviates from its set point.

DFP (diisopropylfluorophosphate): An acetylcholine agonist: destroys *acetylcholinesterase.*

2-DG See *2-deoxy-D-glucose.*

Diabetes mellitus: A disease that results from insufficient production of *insulin,* thus causing, in an untreated state, a high level of blood glucose.

Differentiation: The process by which cells in a developing organism begin to develop differentially, into specialized organs.

Diffusion: Movement of molecules from regions of high concentration to regions of low concentration.

Divergence: See Figure 10.10.

DNA: See *deoxyribonucleic acid.*

Dopamine (DA): A neurotransmitter; one of the catecholamines.

Dorsal: See Figure 6.1.

Dorsal root: See *spinal root.*

Dorsal root ganglion: See *ganglion.*

Dorsal tegmental bundle: An alternate name for the dorsal noradrenergic bundle, which carries fibers of noradrenergic neurons from locus coeruleus to various forebrain structures.

Drive-inducing effect: The hypothesized effect of rewarding electrical stimulation of the brain that provides the drive that motivates the animal to seek further stimulation.

D sleep: A period of desynchronized sleep, during which dreaming and rapid eye movements occur. Also called *REM sleep* for the rapid eye movements and *paradoxical sleep* for the fact that EEG activity resembles that of arousal.

Duodenum: The portion of the small intestine immediately adjacent to the stomach.

Dura mater: the outermost layer of the three *meninges.*

Dyslexia: An term that refers to a variety of reading disorders.

ECS: See *electroconvulsive shock.*

EEG: See *electroencephalogram.*

Effector: A muscle or gland, or one of the active cells of these organs.

Electroconvulsive shock (ECS): A brief electrical shock, applied to the head, that results in electrical seizure and convulsions. Used therapeutically to alleviate severe depression, and experimentally (in animals) to study the consolidation process.

Electrode: A conductive medium (generally made of metal) that can be used to apply electrical stimulation or record electrical potentials.

Electroencephalogram (EEG): Electrical brain potentials recorded by placing *electrodes* on or in the scalp or on the surface of the brain.

Electrolyte: An aqueous solution of a material that ionizes—namely, a soluble acid, base, or salt.

Electromyogram: Electrical potential recorded from an electrode placed on or in a muscle.

Electrostatic pressure: the attractive force between atomic particles charged with opposite signs, or the repulsive force between atomic particles charged with the same sign.

Encephale isolé: An animal whose central nervous system has been severed transversely between the brain and spinal cord, resulting in paralysis but possessing normal sleep-waking cycles.

Endocrine gland: A gland that liberates its secretions into the extracellular fluid around capillaries, and hence into the bloodstream.

Endorphin: A general term used to refer to all endogenous peptides that act as opiates, including beta-endorphin and the enkephalins.

Enzyme: A protein that facilitates a biochemical reaction without itself becoming part of the end product.

Ependyma: The layer of tissue around blood vessels and on the interior walls of the ventricular system of the brain.

Epinephrine: A hormone, secreted by the adrenal medulla, that produces physiological effects characteristic of the sympathetic division of the *autonomic nervous system.*

EPSP: See *postsynaptic potential.*

Equilibrium: A balance of forces, during which the system is not changing.

Eserine: An acetylcholine agonist: inactivates acetylcholinesterase.

Estradiol: The principal *estrogen* of many mammals, including humans.

Estrogen: A class of sex hormones that cause maturation of the female genitalia, growth of breast tissue, and development of other physical features characteristic of females. Estrogens are also necessary for normal sexual behavior of most mammals other than primates.

Estrous cycle: A cyclic change in the hormonal level and sexual receptivity of sub-primate mammals.

Estrus: that portion of the *estrous cycle* during which a female is sexually receptive.

Et al. (et alii): And others.

Evoked potential: A regular series of alterations in the slow electrical activity recorded from the central nervous system, produced by a sensory stimulus or an electrical shock to some part of the nervous system.

Excitatory postsynaptic potential: See *postsynaptic potential.*

Exocrine gland: A gland that liberates its secretions into a duct.

Extracellular thirst: See *volumetric thirst.*

Extrafusal muscle fiber: One of the muscle fibers that are responsible for the force exerted by a muscular contraction.

Extrapyramidal motor system: A complex system of structures in the brain, including the basal ganglia, pontine nuclei, cerebellum, parts of the reticular formation, and their connections with motor neurons of the spinal cord and cranial nerve nuclei.

False transmitter: An antagonist that works by blocking postsynaptic receptors without stimulating them, thus preventing the neurotransmitter from acting on them.

Fascia: A sheet of fibrous connective tissue encasing a muscle.

Fasting phase: The phase of metabolism during which nutrients are not available from the digestive system. Glucose, amino acids, fatty acids, and ketones are derived from glycogen, protein, and adipose tissue during this phase.

Fimbria: A fiber bundle that runs along the lateral surface of the hippocampal complex, connecting this structure with other regions of the forebrain, especially the hypothalamus. The fibers of the fimbria become the *fornix* as they course rostrally from the hippocampal complex.

Fissure: A major groove in the surface of the brain. A smaller groove is called a *sulcus.*

Fistula: An artificial opening or tube into a normal cavity of the body. For example, a gastric fistula permits introduction of substances into the stomach, or their removal from it.

FLA-63: A noradrenergic antagonist: inhibits dopamine-β-hydroxylase, thus preventing the conversion of dopamine into norepinephrine.

Follicle: A small secretory cavity. The ovarian follicle consists of epithelial cells surrounding an oocyte, which develops into an ovum.

Follicle-stimulating hormone: The hormone of the anterior pituitary gland that causes development of a follicle and maturation of its oocyte into an ovum.

Foramen: A normal passage that allows communication between two cavities of the body. The intervertebral foramen permits passage of the spinal nerves through the vertebral column. The foramina of Megendie and of Luschka permit the passage of CSF out of the fourth ventricle and into the subarachnoid space. The foramen of Monroe interconnects the lateral and third ventricles.

Forebrain: See Table 6.1.

Fornix: See *fimbria.*

Fourth ventricle: See *ventricle.*

Fovea: The region of the retina that mediates the most acute vision of birds and higher mammals. Color-sensitive cones constitute the only type of photoreceptor found in the fovea.

Foveal prestriate area: A region of the *circumstriate belt* of visual association cortex, located adjacent to primary visual cortex, mediating foveal vision.

Free fatty acid: A substance of importance to metabolism during the fasting phase. Fats can be broken down to free fatty acids and glycerol. Free fatty acids can be metabolized by most cells of the body. Their basic structure is an alkyl group (CH) attached to a carboxyl group (COOH).

Frontal section: See Figure 6.2

Fructose: Fruit sugar; a monosaccharide that can be used in metabolism. Since fructose cannot cross the blood-brain barrier, it cannot be utilized by the brain.

FSH: See *follicle-stimulating hormone.*

FTG: The *gigantocellular tegmental field* of the pons. Activity of neurons in this region was previously thought to be important in the production of D sleep.

Gamete: A mature reproductive cell; a sperm or ovum.

Gamma hydroxybutyrate: A dopamine antagonist: inhibits neurotransmitter release by the terminal buttons.

Gamma motor neuron: A lower motor neuron whose terminal buttons synapse upon *intrafusal muscle fibers.*

Ganglion: A collection of neural cell bodies, covered with connective tissue, located outside the central nervous system. *Autonomic ganglia* contain the cell bodies of postganglionic neurons of the sympathetic and parasympathetic branches of the *autonomic nervous system. Dorsal root ganglia (spinal nerve ganglia)* contain cell bodies of afferent spinal nerve neurons. *Cranial nerve ganglia* contain cell bodies of afferent cranial nerve neurons. The *basal ganglia* include the *amygdala, caudate nucleus, globus pallidus,* and *putamen;* in this case the term *ganglion* is a misnomer, since the basal ganglia are actually brain nuclei.

Gene: The functional unit of the chromosome, which directs synthesis of one or more proteins.

Generator potential: A graded electrical potential produced by the transducing action of a specialized neuron that serves as a *receptor cell.*

Gestagen: A group of hormones that promote and support pregnancy. Progesterone is the principal mammalian gestagen.

Gigantocellular tegmental field: See *FTG.*

Glia: The supportive cells of the central nervous system—the *astroglia, oligodendroglia,* and *microglia.*

Gliosis: The process by which dead neurons are destroyed and replaced with glial cells.

Globus pallidus: One of the basal ganglia (see *ganglion*); an excitatory structure of the *extrapyramidal motor system.*

Glucagon: A pancreatic hormone that promotes the conversion of liver *glycogen* into *glucose.*

Glucocorticoid: One of a group of hormones of the adrenal cortex that are important in protein and carbohydrate metabolism, secreted especially in times of stress.

Glucoprivation: A state in which the cells of the body are deprived of glucose; can occur when injections of insulin lower the blood glucose level drastically.

Glucose: A simple sugar, of great importance in metabolism. Glucose and *ketones* constitute the major source of energy for the brain.

Glucostatic theory: A theory that states that the level or availability of glucose in the blood determines whether an organism is hungry or sated.

Glutamic acid diethylester: A glutamic acid antagonist: blocks postsynaptic receptors.

Glutamic acid dimethylester: A glutamic acid agonist: blocks re-uptake by the terminal buttons.

Glycogen: A polysaccharide often referred to as animal starch. The hormone *glucagon* causes conversion of liver glycogen into glucose.

Glycogenolysis: The conversion of glycogen into glucose within a cell.

Gold thioglucose (GTG): A molecule consisting of glucose, to which are attached molecules of gold and sulfur. Administration of GTG results in brain damage, especially in the region of the *ventromedial nucleus of the hypothalamus (VMH).*

Golgi apparatus: A complex of parallel membranes in the cytoplasm that wraps the products of a secretory cell.

Golgi tendon organ: The receptive organ at the junction of the tendon and muscle that is sensitive to stretch.

Gonadotrophic hormone: A hormone of the anterior pituitary gland that has a stimulating effect on cells of the gonads. See *follicle-stimulating hormone* and *luteinizing hormone.*

Graded potential: A slow electrical potential in a neuron or a receptor cell (generator potential, PSP, or receptor potential).

Gradient: A grade or slope; more typically, a change in the amount or concentration of a substance with distance or with time.

Growth hormone (GH): also called *somatotrophic hormone (STH):* A hormone that is necessary for the normal growth of the body before adulthood; also causes the conversion of glycogen to glucose, and thus has an anti-insulin effect.

GTG: See *gold thioglucose.*

Guanine: One of the nucleotide bases of RNA and DNA.

Gyrus: A convolution of the cortex of the cerebral hemispheres, separated by *sulci* or *fissures.*

Hair cell: The receptive cell of the auditory or vestibular apparatus.

Hemicholinium: An acetylcholine antagonist: retards uptake of choline, and thus retards neurotransmitter synthesis.

Hepatic portal system: The system of blood vessels that drains the capillaries of the digestive system, travels to the liver, and divides again into capillaries.

Hindbrain: See Table 6.1.

Hippocampus: A forebrain structure of the temporal lobe, constituting an important part of the *limbic system* and of *Papez's circuit.*

Histology: The microscopic study of tissues of the body.

Histone protein. A basic, water-soluble protein that is found in the nucleus, attached to chromosomes; acts as an inhibitor of transcription, or production of *mRNA.*

Horizontal section: See Figure 6.2.

Hormone: A chemical substance liberated by an endocrine gland that has effects on target cells in other organs. *Organizational effects* of a hormone affect tissue differentiation and development; for example, androgens cause prenatal development of male genitalia. *Activational effects* of a hormone are those that occur in the fully developed organism; many of these depend upon the organism's prior exposure to the organizational effects of hormones.

Horseradish peroxidase: An enzyme extracted from the horseradish root. Can be made visible by special histological techniques. Since it is taken up by terminal buttons or by severed axons and is carried by axoplasmic transport, it is useful in anatomical studies.

6-Hydroxydopamine: A chemical that is selectively taken up by axons and terminals of noradrenergic or dopaminergic neurons and that acts as a poison, damaging or killing them.

Hyperphagia: Excessive intake of food.

Hyperpolarization: An increase in the membrane potential of a cell, relative to the normal resting potential. *Inhibitory postsynaptic potentials (IPSPs)* are hyperpolarizations.

Hypoglycemia: A low level of blood glucose.

Hypopolarization: See *depolarization.*

Hypothalamic hormone: A hormone produced by cells of the hypothalamus that affects the secretion and production of hormones of the anterior pituitary gland. The effects are excitatory in the case of releasing hormones and inhibitory in the case of inhibitory hormones.

Hypothalamic-hypophyseal portal system: A system of blood vessels that connects capillaries of the hypothalamus with capillaries of the anterior pituitary gland. Hypothalamic hormones travel to the anterior pituitary gland by means of this system.

Hypovolemia: See *volumetric thirst.*

Imipramine: A noradrenergic and serotonergic agonist: retards re-uptake of norepinephrine and serotonin by terminal buttons. One of the most commonly-used *tricyclic antidepressants.*

Immune system: The system by which the body protects itself from foreign proteins. In response to an infection the white blood cells produce antibodies that attack and destroy the foreign antigen.

Incus: One of the bones of the middle ear, shaped somewhat like an anvil. See Figure 8.12.

Inferior: See Figure 6.1.

Inferior colliculi: Protrusions on top of the midbrain that relay auditory information to the medial geniculate nucleus.

Inferotemporal cortex: In monkeys, the highest level of visual association cortex, located on the inferior surface of the temporal lobe.

Inhibitory postsynaptic potential: See *postsynaptic potential.*

Insulin: A pancreatic hormone that facilitates entry of glucose and amino acids into the cell, facilitates conversion of glucose into glycogen, and facilitates production of fats in adipose tissue.

Integration: The process by which inhibitory and excitatory postsynaptic potentials summate and control the rate of firing of a neuron.

Intermediate horn (of gray matter of spinal cord): Location of the cell bodies of *preganglionic* neurons of the sympathetic branch of the autonomic nervous system and of the sacral portion of the parasympathetic branch.

Interstitial: Refers to the space between the cells of the body.

Intrafusal muscle fiber: A muscle fiber that functions as a stretch receptor, arranged parallel to the *extrafusal muscle fibers,* thus detecting muscle length.

Intragastric fistula: A tube that can be used to introduce substances into the stomach, or to remove substances from it.

Intraoral fistula: A tube that can be used to introduce substances into the mouth.

Intraperitoneal (IP): Pertaining to the peritoneal cavity, the space surrounding the abdominal organs.

Intromission: Insertion of one part into another, especially of a penis into a vagina.

Ion: A charged molecule; *cations* are negatively charged and *anions* are positively charged.

IP: See *intraperitoneal.*

Iproniazid: A monoamine agonist: deactivates monoamine oxidase, and thus prevents destruction of extravesicular monoamines in the terminal buttons.

Ipsilateral: Located on the same side of the body as the point of reference.

IPSP: See *postsynaptic potential.*

Isolated cortical slab: A region of cortex surgically separated from surrounding tissue but with a relatively intact blood supply.

Iso-osmotic: Equal in osmotic pressure.

Isoproterenol: A noradrenergic agonist: stimulates postsynaptic beta receptors.

Isotonic: Equal in osmotic pressure to the contents of a cell. A cell placed in an isotonic solution neither gains nor loses water.

Ketone: An organic acid consisting of two hydrocarbon radicals attached to a carbonyl group (CO). Ketones are produced from the breakdown of fats, and can be utilized by the brain. (Often called ketone bodies.)

Krebs cycle (citric acid cycle, tricarboxylic acid cycle): A series of chemical reactions that involve oxidation of pyruvate. The Krebs cycle takes place on the cristae of the mitochrondria, and supplies the principal source of energy to the cell.

Lactate: Noun: a salt of lactic acid; produced when muscles metabolize glycogen in the absence of oxygen. Verb: to produce milk.

Latency: The time interval between an event and a succeeding event that is caused by it.

Lateral: See Figure 6.1.

Lateral geniculate nucleus: A group of cell bodies within the lateral geniculate body of the thalamus. The dorsal lateral geniculate nucleus receives fibers from the retina and projects fibers to primary visual cortex.

Lateral hypothalamus (LH): A region of the hypothalamus that contains cell bodies and diffuse fiber systems. The *LH syndrome* that follows destruction of the lateral hypothalamus is characterized by a relative lack of spontaneous movement, *adipsia,* and *aphagia,* from which the animal at least partially recovers.

Lateral lemniscus: A band of fibers running rostrally through the medulla and pons, which carries fibers of the auditory system.

Lateral ventricle: See *ventricle.*

L-DOPA: The levo isomeric form of dihydroxyphenylalanine; the precursor of the catecholamines, dopamine and norepinephrine. Often used to treat *Parkinson's disease* because of its effect as a dopamine agonist.

Lemniscal system: The somatosensory fibers of the lateral or trigeminal lemniscus, as contrasted with the extralemniscal system, a polysynaptic pathway that ascends through the reticular formation.

LH: See *lateral hypothalamus* or *luteinizing hormone.*

LH syndrome: See *lateral hypothalamus.*

Limbic system: A group of brain regions including the anterior thalamus, amygdala, hippocampus, limbic cortex, and parts of the hypothalamus, as well as their interconnecting fiber bundles.

Linear: Falling in a straight line. If points on a graph can be connected, at least roughly, by a straight line, they are said to describe a linear function.

Lipostatic theory: The theory that suggests that hunger and satiety depend on levels of lipids in the blood or stored in *adipose tissue.*

Locus coeruleus: A dark-colored group of noradrenergic cell bodies located near the rostral end of the floor of the fourth ventricle.

Long-term memory: Relatively stable memory as opposed to *short-term memory.*

Lower motor neuron: A neuron located in the *intermediate horn* or *ventral horn* of the gray matter of the spinal cord or in one of the motor nuclei of the cranial nerves, the axon of which synapses on muscle fibers.

Luteinizing hormone (LH): A hormone of the anterior pituitary gland that causes ovulation and development of the follicle into a *corpus luteum.*

Lysergic acid diethylamide (LSD): A serotonin antagonist. Acts as a hallucinogen.

Malleus: One of the bones of the middle ear, shaped somewhat like a hammer. See Figure 8.12.

Mammillary body: A protrusion of the bottom of the brain at the posterior end of the hypothalamus, containing the medial and lateral mammillary nuclei.

Mannose: A simple sugar that may be metabolized by all cells of the body.

Massa intermedia: A bridge of tissue across the third ventricle that connects the right and left portions of the thalamus.

Medial: See Figure 6.1.

Medial dorsal nucleus (dorsomedial nucleus): A large nucleus of the thalamus that projects to prefrontal cortex and has connections with temporal neocortex, other thalamic nuclei and the limbic system.

Medial forebrain bundle: A fiber bundle that runs in a rostral-caudal direction through the basal forebrain and lateral hypothalamus.

Medial geniculate nucleus: A group of cell bodies within the medial geniculate body of the thalamus; part of the auditory system.

Medial lemniscus: A fiber bundle that ascends rostrally through the *medulla* and *pons*, carrying fibers of the somatosensory system.

*Medulla oblongata (*usually *medulla):* The most caudal portion of the brain, immediately rostral to the spinal cord.

Meiosis: The process by which a cell divides to form *gametes* (sperm or ova).

Membrane: A structure consisting principally of lipid-like molecules that defines the outer boundaries of a cell, and also constitutes many of the cell organelles such as the Golgi apparatus.

Meninges (singular = *meninx):* The three layers of tissue that encase the central nervous system: the *dura mater, arachnoid,* and *pia mater.*

Mesencephalon: See Table 6.1.

Messenger ribonucleic acid: See *ribonucleic acid.*

Metabolism: The sum of all physical and chemical changes that take place in an organism, including all reactions that liberate energy. See *absorptive phase* and *fasting phase.*

Metencephalon: See Table 6.1.

α-*methyl-paratyrosine (AMPT):* A substance that interferes with the activity of tyrosine hydroxylase and thus prevents the synthesis of dopamine and norepinephrine.

Microelectrode: A very fine electrode, generally used to record activity of individual neurons.

Microglia: See *glia.*

Microphonic: Electrical activity recorded from the *cochlea* that corresponds to the sound vibrations received at the oval window.

Midbrain: See Table 6.1.

Mid-collicular transection: See *cerveau isolé.*

Midline nuclei: Thalamic nuclei that lie near the midline of the brain.

Midpontine transection (also called midpontine pretrigeminal transection): A transection through the middle of the *pons* that produces an animal showing ocular and electroencephalographic signs of wakefulness most of the time.

Midsagittal plane: The plane that divides the body in two symmetrical halves through the midline.

Mitochondrion: A cell organelle in which the chemical reactions of the *Krebs cycle* take place.

Mitosis: Duplication and division of a somatic cell into a pair of daughter cells.

Modality: See *sensory modality.*

Monoamine: A class of amines that includes indolamines (e.g., serotonin) and catecholamines (e.g., dopamine and norepinephrine).

Monoamine oxidase (MAO): A class of enzymes that destroy the monoamines: dopamine, norepinephrine, and serotonin.

Monosynaptic stretch reflex: A reflex consisting of sensory ending of the intrafusal muscle fiber, its afferent fiber, synapsing upon an alpha motor neuron, and the efferent fiber of this neuron, synapsing on its extrafusal muscle fibers in the same muscle. When a muscle is quickly stretched, the monosynaptic stretch reflex causes it to contract.

Morphology: Physical shape and structure.

Motor cortex: The precentral gyrus, which contains a considerable number of motor neurons.

Motor end plate: Region of the membrane of a muscle fiber upon which synapse terminal buttons of the efferent axon.

Motor neuron (motoneuron): A neuron, the stimulation of which results in contractions of muscle fibers.

Motor unit: A lower motor neuron and its associated muscle fibers.

mRNA: See *ribonucleic acid.*

Müllerian system: The embryonic precursors of the female internal sex organs.

Multipolar neuron: A neuron with a single axon and numerous dendritic processes originating from the somatic membrane.

Muscimol: A GABA agonist: stimulates postsynaptic receptors. Acts as a hallucinogen.

Muscle spindle: A sense organ found in muscles; an *intrafusal muscle fiber.*

Myelencephalon: See Table 6.1.

Myelin: A complex fat-like substance produced by the *oligodendroglia* in the central nervous system, and by the *Schwann cells* in the peripheral nervous system, that surrounds and insulates myelinated axons.

Myosin: *Actin* and *myosin* are the proteins that provide the physical basis for muscular contraction. See Figure 10.3.

Negative feedback: A process whereby the effect produced by an action serves to diminish or terminate that action. Regulatory systems are characterized by negative feedback loops.

Neostriatum: The caudate nucleus and putamen; the palostriatum includes the globus pallidus.

Nephrectomy: The removal of the kidneys.

Nernst equation: $V = k \log [x_o]/[x_i]$: where V = membrane potential in volts; k = constant, the value of which is determined by physical conditions such as temperature and by the valence of ion x; $[x_o]$ = concentration, in millimoles per liter, of ion x outside the cell; and $[x_i]$ = concentration of ion x inside the cell.

Neurin: A protein that coats the membrane of microtubules and the interior of the presynaptic membrane of terminal buttons; thought to interact with *stenin* in the release of transmitter substance.

Neurohypophysis: See *pituitary gland.*

Neuromodulator: A naturally-secreted substance that acts like a neurotransmitter, except that it is not restricted to the synaptic cleft, but diffuses through the interstitial fluid. Presumably, it activates receptors on neurons that are not located at synapses.

Neuromuscular junction: The synapse between terminal buttons of an axon and a muscle fiber.

Neurosecretory cell: A neuron that secretes a hormone or hormone-like substance into the interstitial fluid.

Neurotransmitter: Synonym for *transmitter substance.*

Nigrostriatal bundle: A bundle of axons, originating in the *substantia nigra,* terminating in the neostriatum (*caudate nucleus* and *putamen*).

Nissl substance: Cytoplasmic materal dyed by cell body stains (Nissl stains).

Node of Ranvier: A naked portion of a myelinated axon, between adjacent *oligodendroglia* or *Schwann cells.*

Nondecremental conduction: Conduction of an action potential in an all-or-none manner down an axon.

Nonhistone protein: An acidic protein that is found in the nucleus, attached to chromosomes; acts as a disinhibitor of transcription, or production of *mRNA.*

Noradrenalin: See *norepinephrine.*

Norepinephrine: A neurotransmitter found in the brain and in the terminal buttons of postganglionic fibers of the sympathetic branch of the autonomic nervous system.

Nucleolus: An organelle within the nucleus of a cell that produces the ribosomes.

Nucleotide cyclase: An enzyme that causes the conversion of a triphosphate nucleotide (ATP or GTP) to a cyclic monophosphate nucleotide (cyclic AMP or cyclic GMP). See *adenyl cyclase.*

Nucleus: 1. The central portion of an atom. 2. A spherical structure, enclosed by membrane, located in the cytoplasm of most cells, and containing the chromosomes. 3. A histologically identifiable group of neural cell bodies in the central nervous system.

Nucleus of the solitary tract: A nucleus of the medulla that receives information from visceral organs and from the gustatory system; appears to play a role in sleep.

Olfactory bulb: The protrusion at the end of the olfactory nerve; receives input from the olfactory receptors.

Oligodendroglia: A type of glial cell in the central nervous system that forms myelin sheaths.

Optic chiasm: A cross-shaped connection between the optic nerves, located below the base of the brain, just anterior to the pituitary gland.

Optic disk: See *optic nerve.*

Optic nerve: The second cranial nerve, carrying visual information from the retina to the brain. The *optic disk* is formed at the exit point of the fibers of the ganglion cells from the retina that form the optic nerve.

Organizational effect: See *hormone.*

Organ of Corti: The receptor organ situated on the basilar membrane of the inner ear.

Oscilloscope: A laboratory instrument capable of displaying a graph of voltage as a function time on the face of a cathode ray tube.

Osmometric thirst: Thirst produced by an increase in the osmotic pressure of the interstitial fluid relative to intracellular fluid, thus producing cellular dehydration.

Osmosis: Movement of ions through a semipermeable membrane, down their concentration gradient.

Oval window: An opening in the bone surrounding the *cochlea.* The base plate of the *stapes* presses against a membrane exposed by the oval window, and transmits sound vibrations into the fluid within the cochlea.

Oxytocin: A hormone of the posterior pituitary gland that strengthens uterine contractions and causes ejection of milk.

Papez's circuit: A neural circuit consisting of the mammillary bodies, anterior thalamus, cingulate cortex, hippocampus, and their interconnecting fibers.

Para-chlorophenylalanine (PCPA): A substance that blocks the action of the enzyme tryptophan hydroxylase and hence prevents synthesis of *serotonin* (5-HT).

Paradoxical sleep: See *D sleep.*

Parasympathetic division: See *autonomic nervous system.*

Paraventricular nucleus: A hypothalamic nucleus that contains cell bodies of neurons that produce oxytocin and transport it through their axons to the posterior pituitary gland.

Parkinson's disease: A neurological disease that is characterized by fine tremor, rigidity, and difficulty in movement. Caused by degeneration of the *nigrostriatal bundle.*

PCPA: See *para-chlorophenylalanine.*

Peptide bond: A bond between the amino group of one amino acid and the carboxyl group of its neighbor. Peptide bonds link amino acids to form *polypeptides* or *proteins.*

Permeability: The degree to which a membrane permits passage of a particular substance.

PGO spikes: Bursts of phasic electrical activity originating in the pons, followed by activity in the lateral geniculate nucleus and visual cortex; a characteristic of D sleep.

Phentolamine: A noradrenergic agonist: stimulates postsynaptic alpha receptors.

Phosphodiesterase: A class of enzymes that deactivate cyclic nucleotides.

Phosphorylation: The addition of phosphate to a protein, thus altering its physical characteristics.

Photoreceptor: The receptive cell of the retina, which transduces photic energy into electrical potentials. *Cones* are maximally sensitive to one of three different wave lengths of light, and hence encode color vision, whereas all *rods* are maximally sensitive to light of the same wave length, and hence do not encode color vision.

Pia mater: The meningeal layer adjacent to the surface of the brain.

Picrotoxin: A GABA antagonist: blocks postsynaptic receptors.

Pituitary gland: The "master endocrine gland" of the body, attached to the base of the brain. The *anterior pituitary gland (adenohypophysis)* secretes hormones in response to the hypothalamic hormones. The *posterior pituitary gland (neurohypophysis)* secretes oxytocin or antidiuretic hormone in response to stimulation from its neural input.

Plexus: A network formed by the junction of several adjacent nerves.

Polypeptide: A chain of amino acids joined together by peptide bonds. Proteins are long polypeptides.

Pons: The region of the brain rostral to the medulla and caudal to the midbrain.

Positive feedback: A process whereby the effect of an action serves to increase or prolong that action.

Posterior: See Figure 6.1.

Posterior pituitary gland: See *pituitary gland.*

Postganglionic: Referring to the neurons of the autonomic nervous system that synapse directly upon their target organ.

Postsynaptic: Referring to the cell upon which a terminal button synapses.

Postsynaptic potential: Alterations in the membrane potential of a postsynaptic neuron, produced by liberation of transmitter substance at the synapse. Excitatory postsynaptic potentials *(EPSPs)* are depolarizations and increase the probability of firing of the postsynaptic neuron. Inhibitory postsynaptic potentials *(IPSPs)* are hyperpolarizations, and decrease the probability of neural firing.

Posttraumatic amnesia (anterograde amnesia): Later amnesia for events that occur after some disturbance to the brain, such as head injury, electroconvulsive shock, or certain degenerative brain diseases.

Predatory attack: Attack of one animal directed at an individual of another species on which the attacking animal normally preys. When this behavior is elicited by electrical stimulation of the brain it is generally referred to as *quiet-biting attack.*

Preganglionic: Pertaining to the efferent neuron of the autonomic nervous system whose cell body is located in a cranial nerve nucleus or in the *intermediate horn* of the spinal gray matter, and whose terminal buttons synapse upon postganglionic neurons in the autonomic ganglia.

Preoptic area: An area of cell bodies (usually divided into the lateral and medial preoptic areas) just rostral to the hypothalamus. Some investigators refer to the preoptic area as a part of the hypothalamus, although embryologically they are derived from different tissue.

Presynaptic: Referring to a neuron that synapses upon another one, as opposed to *postsynaptic.*

Priming: A phenomenon often observed in studies that investigate the response of animals to rewarding brain stimulation. Priming refers to the fact that an animal that previously responded for brain stimulation will fail to do so until it is given a few "reminder shots" of brain stimulation. The analogy is to priming a pump that has run dry.

Primordial: In embryology, refers to the undeveloped early form of an organ.

Projection: The efferent connection between neurons in one specific region of the brain and those in another region.

Prolactin: A hormone of the anterior pituitary gland, necessary for production of milk and (in some sub-primate mammals) development of a *corpus luteum.* (Occasionally called *luteotrophic hormone.*)

Propranolol: A noradrenergic antagonist: blocks postsynaptic beta receptors.

n-Propylacetic acid: A GABA agonist; deactivates GABA-T, and thus prevents destruction of extravesicular GABA in the terminal buttons.

Protein: A long polypeptide that can serve in a structural capacity or as an enzyme.

Pulvinar: A region of the thalamus that receives fibers from, and projects fibers to, areas of visual cortex.

Putamen: One of the nuclei that constitute the basal ganglia. The putamen and *caudate nucleus* compose the neostriatum.

Pyramidal motor system: A system of long axons whose cell bodies reside in cortex, and whose terminals synapse upon neurons in motor nuclei of the cranial nerves and on interneurons or (in primates) lower motor neurons of the spinal cord.

Pyruvate: An intermediate product in the metabolism of carbohydrates and amino acids.

Quiet-biting attack: See *predatory attack.*

Raphe: A region of the medulla situated along the midline, just dorsal to the aqueduct of Sylvius.

Receptive field: That portion of the visual field in which the presentation of visual stimuli will produce an alteration in the firing rate of a neuron.

Receptor: A specialized type of cell that transduces physical stimuli into slow, graded *receptor potentials.*

Receptor potential: A slow, graded electrical potential produced by a receptor in response to a physical stimulus. Receptor potentials alter the firing rate of neurons upon which the receptors synapse.

Receptor site: The region of the membrane of a cell that is sensitive to a particular chemical, such as a neurotransmitter or hormone. When the appropriate chemical stimulates a receptor site, changes take place in the membrane or within the cell.

Reflex: A stereotyped glandular secretion or movement produced as the direct result of a stimulus.

REM sleep: See *D sleep.*

Renin: Sympathetic stimulation of the kidney, or reduction of its blood flow, results in liberation of renin, a hormone that causes conversion of *angiotensinogen* in the blood into *angiotensin.* Angiotensin produces thirst, constricts blood vessels (thus raising blood pressure), and stimulates the adrenal cortex to produce *aldosterone,* a hormone that stimulates the kidney to retain sodium.

Reserpine: A monoamine antagonist: makes synaptic vesicles leaky, so that the neurotransmitter (dopamine, norepinephrine, or serotonin) cannot be kept inside.

Reticular formation: A large network of neural tissue located in central regions of the brainstem, from the medulla to the diencephalon.

Retina: The neural tissue and photoreceptive cells located on the inner surface of the posterior portion of the eye.

Retrieval: The process by which memories stored in the brain are "recalled," so that they can affect the animal's behavior.

Retrograde amnesia: Amnesia for events that preceded some disturbance to the brain, such as head injury or electroconvulsive shock.

Retrograde degeneration: Gradual degeneration of an injured axon from the point of injury toward the *soma,* which itself often dies.

Re-uptake: The re-entry of a transmitter substance just liberated by a terminal button back through its membrane, thus terminating the *postsynaptic potential* that is induced in the postsynaptic neuron.

Reverberation: Circulating electrical activity maintained in a closed loop of neurons, which might constitute the physical basis for short-term memory.

Ribonucleic acid: A complex macromolecule composed of a sequence of nucleotide bases attached to a sugar-phosphate backbone. *Messenger RNA (mRNA)* delivers genetic information from a portion of a chromosome to a ribosome, where the appropriate molecules of *transfer RNA (tRNA)* assemble the appropriate *amino acids* to produce the *polypeptide* coded for by the active portion of the chromosome.

Ribosome: A cytoplasmic structure, made of protein, that serves as the site of production of proteins from *mRNA.*

RNA: See *ribonucleic acid.*

Rod: See *photoreceptor.*

Rostral: See Figure 6.1.

Round window: An opening in the bone surrounding the cochlea of the inner ear that permits vibrations to be transmitted, via the *oval window,* through the fluids and receptive tissue contained within the cochlea.

Sagittal section: See Figure 6.2.

Saltatory conduction: Conduction of action potentials by myelinated axons; the action potential "jumps" from one *node of Ranvier* to the next.

Satellite cell: A cell that serves to support neurons of the peripheral nervous system, such as the *Schwann cell* that provides the *myelin sheath.*

Satiety: Cessation of hunger produced by adequate and available supplies of nutrients.

Schwann cell: A cell in the peripheral nervous system that is wrapped around a myelinated axon, providing one segment of its myelin sheath.

Scotoma: A region of blindness within an otherwise-normal visual field, produced by localized damage somewhere in the visual system.

Semicircular canal: One of the three ringlike structures of the vestibular apparatus that transduce changes in head rotation into neural activity.

Sensory coding: Representation of sensory events in the form of neural activity.

Sensory cortex: Regions of cortex whose primary input is from one of the sensory systems.

Sensory modality: A particular form of sensory input, such as vision, audition, or olfaction.

Sensory transduction: The process by which sensory stimuli are transduced into slow, graded *generator potentials* or *receptor potentials.*

Septum: A portion of the limbic system, lying between the walls of the anterior portions of the lateral ventricles.

Serotonin: An alternate name for the neurotransmitter 5-hydroxytryptamine (5-HT); named because of its constricting effect on blood vessels.

Set point: The optimal value of the system variable in a regulatory mechanism. The set point for human body temperature, recorded orally, is approximately 37° C.

Short-term memory: Immediate memory for sensory events that may or may not be consolidated into *long-term memory.*

Single-unit: An individual neuron.

Smooth muscle: Nonstriated muscle innervated by the autonomic nervous system, found in the walls of blood vessels, in sphincters, within the eye, in the digestive system, and around hair follicles.

Sodium-potassium pump: A metabolically active process in the cellular membrane that extrudes sodium and transports potassium into the cell.

Solitary nucleus: A nucleus in the medulla that receives input from the gustatory system and sensory information from the viscera.

Soma: A cell body or, more generally, the body.

Somatosenses: Bodily sensation; sensitivity to such stimuli as touch, pain, and temperature.

Somatosensory cortex: The postcentral gyrus, which receives many projection fibers from the somatosensory system.

Somatotrophic hormone (STH): See *growth hormone.*

Species-typical behavior: A behavior that is typical of all or most members of a species of animal, especially a behavior that does not appear to have to be learned.

Spinal root: A bundle of axons surrounded by connective tissue that occurs in pairs, which fuse and form a spinal nerve. The *dorsal root* contains afferent fibers whereas the *ventral root* contains efferent fibers.

Spinal sympathetic ganglion: See *ganglion.*

Spiral ganglion: A group of small nodules located near the cochlea that contain the cell bodies of axons of the cochlear nerve.

S sleep: Slow wave sleep, or synchronized sleep, as opposed to *D sleep.*

Stapes: One of the bones of the inner ear, shaped somewhat like a stirrup. See Figure 8.12.

Stenin: A protein that coats the membrane of synaptic vesicles; thought to interact with *neurin* in the release of transmitter substance.

Steroid hormone: A hormone of low molecular weight, derived from cholesterol. Steroid hormones affect their target cells by attaching to receptors found within the cell.

Striate cortex: Primary visual cortex.

Stria terminalis: A long fiber bundle that connects portions of the amygdala with the hypothalamus.

Strychnine: A glycine antagonist: blocks postsynaptic receptors.

Subarachnoid space: The fluid-filled space between the *arachnoid space* and the *pia mater.*

Subcortical: Located within the brain, as opposed to being located on its cortical surface.

Substance P: A polypeptide that appears to act as a neurotransmitter for pain-sensitive neurons in the dorsal horn of the spinal cord.

Substantia gelatinosa: The tip of the dorsal horn of the spinal cord gray matter.

Substantia nigra: A darkly stained region of the pons, which communicates with the neostriatum via the *nigrostriatal bundle.*

Sulcus: A groove in the surface of the cerebral hemisphere, smaller than a *fissure.*

Superior: See Figure 6.1.

Superior colliculi: Protrusions on top of the midbrain; part of the visual system.

Supraoptic nucleus: A hypothalamic nucleus that contains cell bodies of neurons that produce *antidiuretic hormone* and transport it through their axons to the posterior pituitary gland.

Sympathetic chain: See *ganglion.*

Sympathetic division: See *autonomic nervous system.*

Synapse: A junction between the terminal button of an axon and the membrane of another neuron.

Synaptic vesicle: A small, hollow beadlike structure found in terminal buttons. Synaptic vesicles contain transmitter substance.

Synaptosome: Fragments prepared from homogenized neural tissue, consisting principally of sealed-off terminal buttons and a portion of the adjacent postsynaptic membrane.

Synchrony: High-voltage, low-frequency electroencephalographic activity, characteristic of *S sleep* or coma. During synchrony, neurons are presumably firing together in a regular fashion.

System variable: That which is controlled by a regulatory mechanism; for example, temperature in a heating system.

Tardive dyskinesia: A movement disorder that occasionally occurs after prolonged treatment with antischizophrenic medication, characterized by involuntary movements of the face and neck, sometimes interfering with speech.

Target cell: The type of cell that is directly affected by a hormone or nerve fiber.

Tectum: The roof of the midbrain, comprising the *inferior* and the *superior colliculi.*

Tegmentum: the portion of the midbrain beneath the *tectum,* containing the red nucleus and nuclei of various cranial nerves.

Telencephalon: See Table 6.1.

Tentorium: The tentlike fold of *dura mater* that separates the occipital lobe from the cerebellum.

Terminal button: The rounded swelling at the distal end of an axonic process that synapses upon another neuron, muscle fiber, or gland cell.

Testosterone: The principle *androgen* found in males.

Tetanus toxis: A GABA antagonist: inhibits neurotransmitter release.

Third ventricle: See *ventricle.*

Threshold of excitation: The value of the membrane potential that must be reached in order to produce an action potential.

Thymine: One of the nucleotide bases found in DNA. In RNA this base is replaced by *uracil.*

Transfer ribonucleic acid (tRNA): See *ribonucleic acid.*

Transmitter substance: A chemical that is liberated by the terminal buttons of an axon, which produces an EPSP or an IPSP in the membrane of the postsynaptic cell. Synonym: *neurotransmitter.*

Transverse section: See Figure 6.2.

Tricyclic antidepressant: A class of drugs that are used to treat psychotic depression, named for their molecular structure.

Trigeminal lemniscus: A bundle of fibers running parallel to the medial lemniscus, which conveys afferent fibers from the trigeminal nerve to the thalamus.

tRNA: See *ribonucleic acid.*

Trophic hormone: A hormone of the anterior pituitary gland that affects the hormonal secretion of another endocrine gland.

Unipolar neuron: A neuron with a long, continuous fiber that has dendritic processes on one end and axonal processes and terminal buttons on the other. The fiber connects with the *soma* of the neuron by means of a single, short process.

Uracil: One of the nucleotide bases of RNA. See *thymine.*

Urea: The by-product of metabolism of amino acids, excreted in the urine.

Vagus nerve: The largest of the cranial nerves, conveying efferent fibers of the parasympathetic nervous system to organs of the thoracic and abdominal cavities. The vagus nerve also carries nonpain sensory fibers from these organs to the brain.

Vena cava: The principal vein draining the blood supply of the body to the right atrium. The inferior vena cava drains the lower portion of the body, whereas the superior vena cava drains the upper portion of the body.

Ventral: See Figure 6.2.

Ventral amygdalofugal pathway: The diffuse system of fibers connecting portions of the *amygdala* with various forebrain structures.

Ventral horn (of gray matter of spinal cord): Location of the cell bodies of *alpha* and *gamma motor neurons* of the spinal cord.

Ventral noradrenergic bundle: A system of noradrenergic fibers, running through the hypothalamus, connecting brainstem and forebrain regions.

Ventral posterior nucleus: The thalamic nucleus that projects to somatosensory cortex.

Ventral root: See *spinal root:*

Ventricle: One of the hollow spaces within the brain, filled with cerebrospinal fluid, including the *lateral, third,* and *fourth ventricles.*

Ventromedial nucleus of the hypothalamus (VMH): A large nucleus of the hypothalamus located near the walls of the third ventricle.

Vertebral artery: An artery, the branches of which serve the posterior region of the brain and spinal cord.

Vestibular nerve: See *auditory nerve.*

VMH: See *ventromedial nucleus of the hypothalamus.*

Volumetric thirst (extracellular thirst): Thirst produced by *hypovolemia,* or reduction in the amount of extracellular fluid. Volumetric thirst is produced by baroreceptors in the right atrium of the heart, and by reduced blood flow from the kidneys.

Wernicke's area: A region of auditory association cortex on the left temporal lobe of humans, which is important in comprehension of words and production of meaningful speech. *Wernicke's aphasia,* which occurs as a result of damage to this area, results in fluent, but meaningless, speech.

Wolffian system: The embryonic precursors of the male internal sex organs.

References

ADAMS, R. D. The anatomy of memory mechanisms in the human brain. In *The Pathology of Memory*, edited by G. A. Talland and N. C. Waugh. New York: Academic Press, 1969.

AGHAJANIAN, G. K., BLOOM, F. E., and SHEARD, M. N. Electron microscopy of degeneration within the serotonin pathway of rat brain. *Brain Research*, 1969, *13*, 266–273.

AGNEW, H. W., JR., WEBB, W. B., and WILLIAMS, R. L. The effects of stage four sleep deprivation. *Electroencephalography and Clinical Neurophysiology*, 1964, *17*, 68–70.

AGNEW, H. W., JR., WEBB, W. B., and WILLIAMS, R. L. Comparison of stage four and 1-REM sleep deprivation. *Perceptual and Motor Skills*, 1967, *24*, 851–858.

AGRANOFF, B. W., BURRELL, H. R., DOKAS, L., and SPRINGER, A. D. Progress in biochemical approaches to learning and memory. In *Psychopharmacology: A Generation of Progress*, edited by M. A. Lipton, A. Di Masco, and K. F. Killam. New York: Raven Press, 1978.

AKIL, H. and LIEBESKIND, J. C. Monoaminergic mechanisms of stimulation-produced analgesia. *Brain Research*, 1975, *94*, 279–296.

AKIL, H. and MAYER, D. J. Antagonism of stimulation-produced analgesia by *p*-CPA, a serotonin synthesis inhibitor. *Brain Research*, 1972, *44*, 692–697.

AKIL, H., MAYER, D., and LIEBESKIND, J. C. Antagonism of stimulation-produced analgesia by Naloxone, a narcotic antagonist. *Science*, 1976, *191*, 961–962.

ALLISON, T. and CHICHETTI, D. Sleep in mammals: ecological and constitutional correlates. *Science*, 1976, *194*, 732–734.

AMENOMORI, Y., CHEN, C. L., and MEITES, J. Serum prolactin levels in rats during different reproductive states. *Endocrinology*, 1970, *86*, 506–510.

AMOORE, J. E. *Molecular Basis of Odor*. Springfield, Ill.: Charles C Thomas, 1970.

ANAND, B. K. and BROBECK, J. R. Hypothalamic control of food intake in rats and cats. *Yale Journal of Biology and Medicine*, 1951, *24*, 123–140.

ANDEN, N.-E., FUXE, K., HAMBERGER, B., and HÖKFELT, T. A quantitative study on the nigrostriatal dopamine neuron system in the rat. *Acta Physiologica Scandinavica*, 1966, *67*, 306–312.

ANDERSEN, V. O., BUCHMANN, B., and LENNOX-BUCHTHAL, M. A. Single cortical units with narrow spectral sensitivity in monkey (*Cercocebus torquatus atys*). *Vision Research*, 1962, *2*, 295–307.

ANDERSSON, B. The effect of injections of hypertonic NaCl solutions in different parts of the hypothalamus of goats. *Acta Physiologica Scandinavica*, 1953, *28*, 188–201.

ANDERSSON, B. Thirst and brain control of water balance. *American Scientist*, 1971, *59*, 408–415.

ANDERSSON, B. Regulation of water intake. *Physiological Review*, 1978, *58*, 582–603.

ANDERSSON, B., LEKSELL, L. G., and LISHAJKO, F. Perturbations in fluid balance induced by medially placed forebrain lesions. *Brain Research*, 1975, *99*, 261–275.

ANDREWS, W. H. H. and ORBACH, J. Sodium receptors activating some nerves of perfused rabbit livers. *American Journal of Physiology*, 1974, *227*, 1273–1275.

ANGELBORG, A. R., and ENGSTRÖM, H. The normal organ of Corti. In *Basic Mechanisms in Hearing*, edited by A. R. Møeller. New York: Academic Press, 1973.

ANGELERI, F., MARCHESI, G. F., and QUATTRINI, A. Effects of chronic thalamic lesions on the electrical activity of the neocortex and on sleep. *Archives Italiennes de Biologie*, 1969, *107*, 633–667.

ANGRIST, B., SATHANANTHAN, G., WILK, S., and GERSHON, S. Amphetamine psychosis: behavioural and biochemical aspects. *Journal of Psychiatric Research*, 1974, *11*, 13–24.

ANLEZARK, G. M., ARBUTHNOTT, G. W., CROW, T. S., ECCLESTON, D., and WALTER, D. S. Intracranial self-stimulation and noradrenaline metabolism in the cortex. *British Journal of Pharmacology*, 1973, *47*, 645P.

ANONYMOUS. Effects of sexual activity on beard growth in man. *Nature*, 1970, *226*, 867–870.

ANTELMAN, S. M., ROWLAND, N. E., and FISHER, A. E. Stress related recovery from lateral hypothalamic aphagia. *Brain Research*, 1975, *99*, 346–350.

ANTELMAN, S. M., SZECHTMAN, H., CHIN, P., and FISHER, A. D. Tail pinch-induced eating, gnawing and licking behavior in rats: Dependence on the nigrostriatal dopamine system. *Brain Research*, 1975, *99*, 319–377.

ARAI, Y., and GORSKI, R. A. Critical exposure time for androgenization of the rat hypothalamus determined by antiandrogen injection. *Proceedings of the Society for Experimental Biology and Medicine*, 1968, *127*, 590–593.

ASANUMA, H., and ROSÉN, I. Topographic organization of cortical efferent zones projecting to distal forelimb muscles in monkey. *Experimetal Brain Research*, 1972, *14*, 243–256.

ASCHOFF, J. Desynchronization and resynchronization of human circadian rhythms. *Aerospace Medicine*, 1969, *40*, 844–849.

ASCHOFF, J. Circadian rhythms: General features and endocrinological aspects. In *Endocrine Rhythms*, edited by D. T. Krieger. New York: Raven Press, 1979.

ASCHOFF, J. C., CONRAD, B., and KORNHUBER, H. H. Acquired pendular nystagmus in multiple sclerosis. *Proceedings of the Barany Society*, 1970, 127–132.

ASERINSKI, N. E. and KLEITMAN, N. Regularly occurring periods of eye motility and concomitant phenomena during sleep. *Science*, 1955, *118*, 273–274.

ASSCHER, A. W., and ANSON, S. G. A vascular permeability factor or renal origin. *Nature*, 1963, *198*, 1097–1099.

ATWEH, S. F. and KUHAR, M. J. Autoradiographic localization of opiate receptors in rat brain. I. Spinal cord and lower medulla. *Brain Research*, 1977, *124*, 53–67.

AUSSEL, C., URIEL, J., and MERCIER-BODARD, C. Rat alphafetoprotein: isolation, characterization, and estrogen-binding properties. *Biochimie*, 1973, *55*, 1431–1437.

BAEKELAND, F. Pentobarbital and dextroamphetamine sulfate: effects on the sleep cycle in man. *Psychopharmacologia*, 1967, *11*, 388–396.

BAKER, T. L. and MCGINTY, D. J. Reversal of cardiopulmonary failure during active sleep in hypoxic kittens: implications for sudden infant death. *Science*, 1977, *198*, 419–421.

BAKKE, J. L. A double-blind study of progestin-estrogen combination in the management of the menopause. *Pacific Medical-Surgical*, 1965, p. 200.

BALAGURA, S. *Hunger: A Biopsychological Analysis*. New York: Basic Books, 1973.

BALAGURA, S., and HOEBEL, B. G. Self-stimulation of the lateral hypothalamus modified by insulin and glucagon. *Physiology & Behavior*, 1967, *2*, 337–340.

BALASUBRAMANIAM, V., KANAKA, T. S., and RAMAMURTHI, B. Surgical treatment of hyperkinetic and behavior disorders. *International Surgery*, 1970, *54*, 18–23.

BALASUBRAMANIAM, V., KANAKA, T. S., RAMANUJAM, P. V., and RAMAMURTHI, B. Sedative Neurosurgery. *Journal of the Indian Medical Association*, 1969, *53*, 377–381.

BALDESSERINI, R. J. *Chemotherapy in Psychiatry*. Cambridge: Harvard University Press, 1977.

BALDESSERINI, R. J. The pathophysiological basis of tardive dyskinesia. *Trends in NeuroSciences*, 1979, *2*, 133–135.

BANCROFT, J. and SKAKKEBACK, N. Androgens and human sexual behavior. Paper read at Ciba Foundation Symposium No. 62, London, 1978.

BANCROFT, J. The relationship between hormones and sexual behavior in humans. In *Biological Determinants of Sexual Behaviour*, edited by J. B. Hutchinson. Chichester: John Wiley & Sons, 1978.

BANDLER, R. J. and FLYNN, J. P. Control of somatosensory fields for striking during hypothalamically elicited attack. *Brain Research*, 1972, *38*, 197–201

BARD, P. Studies in the cortical representation of somatic sensibility. *Harvey Lectures*, 1938, *33*, 143–169.

BARKLEY, M. S. Testosterone and the ontogeny of sexually dimorphic aggression in the mouse. Ph.D. Thesis, University of Connecticut, 1974.

BARRACLOUGH, C. A. Hormones and development. Modifications in the CNS regulation of reproduction after exposure of prepubertal rats to steroid hormones. *Recent Progress in Hormone Research*, 1966, *22*, 503–539.

BASBAUM, A. I., MARLEY, N. J. E., O'KEEFE, J., and CLANTON, C. H. Reversal of morphine and stimulus-produced analgesia by subtotal spinal cord lesions. *Pain*, 1977, *3*, 43–56.

BATESON, P. P. G., HORN, G., and MCCABE, B. J. Imprinting: the effect of partial ablation of the medial hyperstriatum ventrale of the chick. *Journal of Physiology (London)*, 1978, *285*, 23.

BATINI, C., MORUZZI, G., PALESTINI, M., ROSSI, G. F., and ZANCHETTI, A. Persistent patterns of wakefulness in the pretrigeminal midpontine preparation. *Science*, 1958, *128*, 30–32.

BATINI, C., MORUZZI, G., PALESTINI, M., ROSSI, G. F., and ZANCHETTI, A. Effect of complete pontine transections on the sleep-wakefulness rhythm: the midpontine pretrigeminal preparation. *Archives Italiennes de Biologie*, 1959, *97*, 1–12.

BATINI, C., PALESTINI, M., ROSSI, G. F., and ZANCHETTI, A. EEG activation patterns in the midpontine pretrigeminal cat following sensory deafferentation. *Archives Italiennes de Biologie*, 1959, *97*, 26–32.

BATSEL, H. L. Electroencephalographic synchronization and desynchronization in the chronic "cerveau isole" of the dog. *Electroencephalography and Clinical Neurophysiology*, 1960, *12*, 421–430.

BATSEL, H. L. Spontaneous desynchronization in the chronic cat "cerveau isole." *Archives Italiennes de Biologie*, 1964, *102*, 547–566.

BAUM, M. J., SLOB, A. K., DeJONG, F. H., and WESTBROEK, D. L. Persistence of sexual behavior in ovariectomized stumptail macaques following dexamethasone treatment or adrenalectomy. *Hormones and Behavior*, 1978, *11*, 323–347.

BAZETT, H. C., McGLONE, B., WILLIAMS, R. G., and LUFKIN, H. M. Sensation. I. Depth, distribution, and probable identification in the prepuce of sensory and end-organs concerned in sensations of temperature and touch: thermometric conductivity. *Archives of Neurology and Psychiatry (Chicago)*, 1932, *27*, 489–517.

BEACH, F. A. Effects of injury to the cerebral cortex upon sexually receptive behavior in the female cat. *Psychosomatic Medicine*, 1944, *6*, 40–55.

BEACH, F. A. Cerebral and hormonal control of reflexive mechanisms involved in copulatory behavior. *Physiological Review*, 1967, *47*, 289–316.

BEACH, F. A. Factors involved in the control of mounting behavior by female animals. In *Reproduction and Sexual Behavior*, edited by M. Diamond. Bloomington: University of Indiana Press, 1968.

BEACH, F. A. Coital behavior in dogs. VI. Long-term effects of castration upon mating in the male. *Journal of Comparative and Physiological Psychology*, 1970, *70*, 1–32.

BEACH, F. A., and WILSON, J. R. Effects of prolactin, progesterone, and estrogen on reactions of nonpregnant rats to foster young. *Psychological Reports*, 1963, *13*, 231–239.

BEALER, S. L., and SMITH, D. V. Multiple sensitivity to chemical stimuli in human taste papillae. *Physiology & Behavior*, 1975, *14*, 795–800.

BEAMER, W., BERMANT, G., and CLEGG, M. Copulatory behavior of the ram, *Ovis aries*. II: Factors affecting copulatory satiety. *Animal Behavior*, 1969, *17*, 706–711.

BEATTY, W. W., and SCHWARTZBAUM, J. S. Commonality and specificity of behavioral dysfunctions following septal and hippocampal lesions in rats. *Journal of Comparative and Physiological Psychology*, 1968, *66*, 60–68.

BEECHER, H. K. *Measurement of Subjective Responses: Quantitative Effects of Drugs*. New York: Oxford University Press, 1959.

BEEMAN, E. A. The effect of male hormone on aggressive behavior in mice. *Physiological Zoology*, 1947, *20*, 373–405.

BEHBEHANI, M. M. and FIELDS, H. L. Evidence that an excitatory connection between the periaqueductal gray and nucleus raphe magnus mediates stimulation produced analgesia. *Brain Research*, 1979, *170*, 85–93.

BEIDLER, L. M. Physiological properties of mammalian taste receptors. In *Taste and Smell in Vertebrates*, edited by G. E. W. Wolstenholme. London: J & A. Churchill, 1970.

BELLINGER, L. L., TRIETLEY, G. J., and BERNARDIS, L. L. Failure of portal glucose and adrenaline infusions or liver denervation to affect food intake in dogs. *Physiology & Behavior*, 1976, *16*, 299–304.

BELLUZZI, J. D., and STEIN, L. Enkephalin may mediate euphoria and drive-reduction reward. *Nature*, 1977, *266*, 556–558.

BENKERT, O. Studies on pituitary hormones in depression and sexual impotence. *Progress in Brain Research*, 1975, *42*, 25–36.

BERKUN, M. A., KESSEN, M. L., and MILLER, N. E. Hunger-reducing effects of food by stomach fistula versus food by mouth measured by a consummatory response. *Journal of Comparative and Physiological Psychology*, 1952, *45*, 550–554.

BERL, S., PUSZKIN, S., and NICKLAS, W. J. Actomyosin protein in brain. *Science*, 1973, *179*, 441–446.

BERLUCCHI, G., MAFFEI, L., MORUZZI, G., and STRATA, P. EEG and behavioral effects elicited by cooling of medulla and pons. *Archives Italiennes de Biologie*, 1964, *102*, 372–392.

BERMANT, G. Response latencies of female rats during sexual intercourse. *Science*, 1961, *133*, 1771–1773. (a)

BERMANT, G. Regulation of sexual contact by female rats. Unpublished Ph.D. dissertation, Harvard University, 1961. Cited by Bermant and Davidson (1974). (b)

BERMANT, G., and DAVIDSON, J. M. *Biological Bases of Sexual Behavior*. New York: Harper & Row, 1974.

BERMANT, G., and TAYLOR, L. Interactive effects of experience and olfactory bulb lesions in male rat copulation. *Physiology & Behavior*, 1969, *4*, 13–18.

BERNARD, C. *Lecons de Physiologie Expérimentale Appliquée à la Médicine faites au Collège de France*. Volume 2. Paris: Bailliere, 1856.

BINNS, E. *The Anatomy of Sleep; or the Art of Procuring Sound and Refreshing Slumber at Will*. London: John Churchill, 1852.

BISHOP, G. H. Responses to electrical stimulation of single sensory units of skin. *Journal of Neurophysiology*, 1943, *6*, 361–382.

BLAKEMORE, C., and CAMPBELL, F. W. On the existence of neurones in the human visual system selectively sensitive to the orientation and size of retinal images. *Journal of Physiology (London)*, 1969, *203*, 237–260.

BLASS, E. M., and EPSTEIN, A. N. A lateral preoptic osmosensitive zone for thirst. *Journal of Comparative and Physiological Psychology*, 1971, *76*, 378–394.

BLASS, E. M., and KRALY, F. S. Medial forebrain bundle lesions: Specific loss of feeding to decreased glucose utilization in rats. *Journal of Comparative and Physiological Psychology*, 1974, *86*, 679–692.

BLASS, E. M., NUSSBAUM, A. I., and HANSON, D. G. Septal hyperdipsia—specific enhancement of drinking to angiotensin in rats. *Journal of Comparative and Physiological Psychology*, 1974, *87*, 422–439.

BLAUSTEIN, J. D., and FEDER, H. H. Cytoplasmic progestin-receptors in guinea pig brain: characteristics and relationship to the induction of sexual behavior. *Brain Research*, 1979, *169*, 481–497.

BLAUSTEIN M. P., KENDRICK, N. C., FRIED, R. C., and RATZLAFF, R. W. Calcium metabolism at the mammalian presynaptic nerve terminal: lessons from the synaptosome. In *Society for Neuroscience Symposia, vol. II.*, edited by W. M. Cowan and J. A. Ferrendelli. Bethesda, Md.: Society for Neuroscience, 1977.

BLOCH, V., HENNEVIN, E., and LECONTE, P. Interaction between post-trial reticular stimulation and subsequent paradoxical sleep in memory consolidation processes. In *Neurobiology of Sleep and Memory*, edited by R. R. Drucker-Colin, and J. L. McGaugh. New York: Academic, 1977.

BLOCK, M., and ZUCKER, I. Circadian rhythms of rat locomotor activity after lesions of the midbrain raphe nuclei. *Journal of Comparative Physiology*, 1976, *109*, 235–247.

BLUMER, D. Hypersexual episodes in temporal lobe epilepsy. *American Journal of Psychiatry*, 1970, *126*, 1099–1106.

BOLLES, R. C. *Theory of Motivation*. New York: Harper & Row, 1975.

BONGIOVANNI, A. M., DiGEORGE, A. M., and GRUMBACH, M. M. Masculinization of the female infant associated with estrogenic therapy alone during gestation: four cases. *Journal of Clinical Endocrinology and Metabolism*, 1959, *19*, 1004–1011.

BONVALLET, M., and ALLEN, M. B., Jr. Prolonged spontaneous and evoked reticular activation following discrete bulbar lesions. *Electroencephalography and Clinical Neurophysiology*, 1963, *15*, 969–988.

BONVALLET, M., and SIGG, B. Etude électrophysiologique des afférénces vagales au neveau de leur pénétration dans le bulbe. *Journal of Physiology (Paris)*, 1958, *50*, 63–74.

BOOKIN, H. B. and PFEIFER, W. D. Adrenalectomy attenuates electroconvulsive shock-induced retrograde amnesia in rats. *Behavioral Biology*, 1978, *24*, 527–532.

BOWER, G. H., and MILLER, N. E. Rewarding and punishing effects from stimulating the same place in the rat's brain. *Journal of Comparative and Physiological Psychology*, 1958, *51*, 669–674.

BOWERS, M. B. Central dopamine turnover in schizophrenic syndromes. *Archives of General Psychiatry*, 1974, *31*, 50–54.

BRADY, J. V., and NAUTA, W. J. H. Subcortical mechanisms in emotional behavior: affective changes following septal forebrain lesions in the albino rat. *Journal of Comparative and Physiological Psychology*, 1953, *46*, 339–346.

BREGER, L., HUNTER, I., and LANE, R. W. The effects of stress on dreams. *Physiological Issues Monograph Number 27*. New York: International University Press, 1971.

BREMER, F. L'activité cérébrale au cours du sommeil et de la narcose. Contribution á l'étude du mécanisme du sommeil. *Bulletin de l'Académie Royale de Belgique*, 1937, *4*, 68–86.

BREMER, F. Preoptic hypnogenic focus and mesencephalic reticular formation. *Brain Research*, 1970, *21*, 132–134.

BREMER, F. A further study of the inhibitory processes induced by the activation of the preoptic hypnogenic structure. *Archives Italiennes de Biologie*, 1975, *113*, 79–88.

BRIDGES, R. S., ZARROW, M. X., and DENENBERG, V. H The role of neonatal androgen in the expression of hormonally induced maternal responsiveness in the adult rat. *Hormones and Behavior*, 1973, *4*, 315–322.

BROBECK, J. R., TEPPERMAN, J., and LONG, C. N. H. Effects of experimental obesity upon carbohydrate metabolism. *Yale Journal of Biology and Medicine*, 1943, *15*, 893–904.

BRODY, M. J., and JOHNSON, A. K. Role of the anteroventral third ventricle (AV3V) region in fluid and electrolyte balance, arterial pressure regulation, and hypertension. In *Frontiers in Neuroendocrinology*, edited by L. Martini and W. F. Ganong. New York: Raven Press, 1980.

BRONSON, F. H., and DESJARDINS, C. Aggressive behavior and seminal vesicle function in mice: differential sensitivity to androgen given neonatally. *Endocrinology*, 1969, *85*, 971–974.

BRONSON, F. H., and DESJARDINS, C. Neonatal androgen administration and adult aggressiveness in female mice. *General and Comparative Endocrinology*, 1970, *15*, 320–325.

BROOKS, D., and BIZZI, E. Brain stem electrical activity during sleep. *Archives Italiennes de Biologie*, 1963, *101*, 648–665.

BROOKS, V. B., and STONEY, S. D., JR. Motor mechanisms: The role of the pyramidal system in motor control. *Annual Review of Physiology*, 1971, *33*, 337–392.

BROUGHTON, R. J. Sleep disorders; disorders of arousal? *Science*, 1968, *159*, 1070–1078.

BROWMAN, L. G. Light in its relation to activity and estrous rhythms in the albino rat. *Journal of Experimental Zoology*, 1937, *75*, 375–388.

BUDDINGTON, R. W., KING, F. A., and ROBERTS, L. Emotionality and conditioned avoidance responding in the squirrel monkey following septal injury. *Psychonomic Science*, 1967, *8*, 195–196.

BUGGY, J., HOFFMAN, W. E., PHILLIPS, M. I., FISHER, A. E., and JOHNSON, A. K. Osmosensitivity of rat third ventricle and interactions with angiotensin. *American Journal of Physiology*, 1979, *236(1)*, R75–R82.

BUGGY, J. and JOHNSON, A. K. Preoptic-hypothalamic periventricular lesions: thirst deficits and hypernatremia. *American Journal of Physiology*, 1977, *233(1)*, R44–R52.

BULLOCK, N. P., and NEW, M. I. Testosterone and cortisol concentration in spermatic, adrenal and systemic venous blood in adult male guinea pigs. *Endocrinology*, 1971, *88*, 523–526.

BURNS, B. D. *The Mammalian Cerebral Cortex*. London: Arnold, 1958.

CAGGIULA, A. R. Analysis of the copulation-reward properties of posterior hypothalamic stimulation in male rats. *Journal of Comparative and Physiological Psychology*, 1970, *70*, 399–412.

CAGGIULA, A. R. Shock-elicited copulation and aggression in male rats. *Journal of Comparative and Physiological Psychology*, 1972, *83*, 393–407.

CAIN, D. P., and BINDRA, D. Responses of amygdala single units to odors in the rat. *Experimental Neurology*, 1972, *35*, 98–110.

CAIRNS, H., and MOSBERG, W. H. Colloid cyst of the third ventricle. *Surgery, Gynecology, and Obstetrics*, 1951, *92*, 545–570.

CAMPBELL, D. S., and DAVIS, J. D. Licking rate of rats is reduced by intraduodenal and intraportal glucose infusion. *Physiology & Behavior*, 1974, *12*, 357–365.

CANNON, W. B. The James-Lange theory of emotions: A critical examination and an alternative. *American Journal of Psychology*, 1927, *39*, 106–124.

CARLSON, H. E., GILLIN, J. C., GORDEN, P., and SNYDER, F. Absence of sleep-related growth hormone peaks in aged normal subjects and in acromegaly. *Journal of Clinical Endocrinology and Metabolism*, 1972, *34*, 1102–1105.

CARLSON, N. R., and THOMAS G. J. Maternal behavior of mice with limbic lesions. *Journal of Comparative and Physiological Psychology*, 1968, *66*, 731–737.

CARLSSON, A., and LINDQVIST, M. Effect of chlorpromazine or haloperidol on formation of 3-methoxytyramine and normetanephrine in mouse brain. *Acta Pharmacologica et Toxicologia*, 1963, *20*, 140–144.

CARROLL, D., LEWIS, S. A., and OSWALD, I. Effects of barbiturates on dream content. *Nature*, 1969, *223*, 865–866.

CARTER, C. S., CLEMENS, L. G., and HOEKEMA, D. J. Neonatal androgen and adult sexual behavior in the golden hamster. *Physiology & Behavior*, 1972, *9*, 89–95.

CARTWRIGHT, R. D., GORE, S., and WEINER, L. Paper presented at the meeting of the Association for the Psychophysiological Study of Sleep. San Diego, 1973. Cited by Greenberg and Pearlman, 1974.

CERLETTI, U. Electroshock therapy. In *The Great Psychodynamic Therapies in Psychiatry*, edited by F. Marti-Ibanez, A. M. Sackler, M. D. Sackler, and R. R. Sackler. New York: Hoeber-Harper, 1956.

CERLETTI, U., and BINI, L. Electric shock treatment. *Bollettino ed*

Atti della Accademia Medica di Roma, 1938, *64*, 36.

CHOROVER, S. L., and SCHILLER, P. H. Short-term retrograde amnesia in rats. *Journal of Comparative and Physiological Psychology*, 1965, *59*, 73–78.

CHOW, K. L. Effects of local electrographic after discharge on visual learning and retention in monkey. *Journal of Neurophysiology*, 1961, *24*, 391–400.

CLARK, A. W., HURLBUT, W. P., and MAURO, A. Changes in the fine structure of the neuromuscular junction of the frog caused by black widow spider venom. *Journal of Cell Biology*, 1972, *52*, 1–14.

CLARKE, P. G. H., and WHITTERIDGE, D. A comparison of stereoscopic mechanisms in cortical visual areas V1 and V2 of the cat. *Journal of Physiology, London*, 1978, *275*, 92–93.

CLAVIER, R. M., and FIBIGER, H. C. On the role of ascending catecholaminergic projections in intracranial self-stimulation of the substantia nigra. *Brain Research*, 1977, *131*, 271–286.

CLAVIER, R. M., FIBIGER, H. C., and PHILLIPS, A. G. Evidence that self-stimulation of the region of the locus coeruleus in rats does not depend upon noradrenergic projections to telencephalon. *Brain Research*, 1976, *113*, 71–81.

CLAVIER, R. M. and GERFEN, C. R. Neural inputs to the prefrontal agranular insular cortex in the rat—horseradish peroxidase study. *Brain Research Bulletin*, 1979, *4*, 347–353. (a)

CLAVIER, R. M. and GERFEN, C. R. Self-stimulation of the sulcal prefrontal cortex in the rat: Direct evidence for ascending dopaminergic mediation. *Neuroscience Letters*. 1979, *12*, 183–187. (b)

CLAVIER, R. M., and ROUTTENBERG, A. In search of reinforcement pathways: a neuroanatomical odyssey. In *Biology of Reinforcement: A Tribute to James Olds*. New York: Academic Press, 1980.

CLELAND, B. G., LEVICK, W. R., and WÄSSLE, H. Physiological identification of morphological class of cat retinal ganglion cells. *Journal of Physiology, London*, 1975a, *248*, 151–171.

CLEMENS, L. G. Influence of prenatal litter composition on mounting behavior of female rats. *American Zoologist*, 1971, *11*, 617–618.

CLEMENTE, C. D., and CHASE, M. H. Neurological substrates of aggressive behavior. *Annual Review of Physiology*, 1973, *35*, 329–356.

CLEMENTE, C. D., STERMAN, M. B., and WYRWICKA, W. Forebrain inhibitory mechanisms: Conditioning of basal forebrain induced EEG synchronization and sleep. *Experimental Neurology*, 1963, *7*, 404–417.

COBURN, P. C. and STRICKER, E. M. Osmoregulatory thirst in rats after lateral preoptic lesions. *Journal of Comparative and Physiological Psychology*, 1978, *92*, 350–361.

COHEN, H. D., ERVIN, F., and BARONDES, S. H. Puromycin and cycloheximide: Differential effects on hippocampal electrical activity. *Science*, 1966, *154*, 1557–1558.

COINDET, J., CHOVET, B., and MOURET, J. Effects of lesions of the suprachiasmatic nuclei on paradoxical sleep and slow wave sleep circadian rhythms in the rat. *Neuroscience Letters*, 1975, *1*, 243–247.

COLEMAN, J. C. *Abnormal Psychology and Modern Life*. Glenview, Ill.: Scott, Foresman, 1976.

COMARR, A. E. Sexual function among patients with spinal cord injury. *Urologia Internationalis*, 1970, *25*, 134–168.

CONNER, R. L., and LEVINE, S. Hormonal influences on aggressive behaviour. In *Aggressive Behaviour*, edited by S. Garattini and E. B. Sigg. New York: Wiley & Sons, 1969.

COONS, E. E., and CRUCE, J. A. F. Lateral hypothalamus: Food and current intensity in maintaining self-stimulation of hunger. *Science*, 1968, *159*, 1117–1119.

COOPER, B. R., KONKOL, R. J., and BREESE, G. R. Effects of catecholamine depleting drugs and *d*-amphetamine on self-stimulation of the substantia nigra and locus coeruleus. *The Journal of Pharmacology and Experimental Therapeutics*, 1978, *204*,

592–605.

COTES, M., CROW, T. J., JOHNSTONE, E. C., BARTLETT, W., and BOURNE, R. C. Neuroendocrine changes in acute schizophrenia as a function of clinical state and neuroleptic medication. *Psychological Medicine (London)*, 1978, *8(4)*, 657–665.

COULTER, J. D., LESTER, B. K., and WILLIAMS, H. L. Reserpine and sleep. *Psychopharmacologia*, 1971, *19*, 134–147.

COWEY, A., and WEISKRANTZ, L. Demonstration of cross-modal matching in rhesus monkeys, *Macaca mulatta. Neuropsychologia*, 1975, *13*, 117–120.

CRAGG, B. G. Centrifugal fibers to retina and olfactory bulb, and composition of supraoptic commissures in rabbit. *Experimental Neurology*, 1962, *5*, 406–427.

CRANE, G. E. Iproniazid (Marsilid) phosphate, a therapeutic agent for mental disorders and debilitating diseases. *Psychiatric Research Reports*, 1957, *8*, 142–152.

CREESE, I., BURT, D. R., and SNYDER, S. H. Dopamine receptor binding predicts clinical and pharmacological potencies of antischizophrenic drugs. *Science*, 1976, *192*, 481–483.

CRISWELL, H. E., and ROGERS, F. B. Narcotic analgesia: changes in neural activity recorded from periaqueductal gray matter of rat brain. *Society for Neuroscience Abstracts*, 1978, *4*, 458.

CRONHOLM, B. Post-ECT amnesias. In *The Pathology of Memory*, edited by G. A. Talland and N. C. Waugh. New York: Academic Press, 1969.

CROSBY, E. C., HUMPHREY, T., and LAUER, E. W. *Correlative Anatomy of the Nervous System*. New York: Macmillan, 1962.

CROW, T. J. A map of the rat mesencephalon for electrical self-stimulation. *Brain Research*, 1972, *36*, 265–273.

CROW, T. J., SPEAR, P. J., and ARBUTHNOTT, G. W. Intracranial self-stimulation with electrodes in the region of the locus coeruleus. *Brain Research*, 1972, *36*, 275–287.

DAVIDSON, J. M. Characteristics of sex behavior in male rats following castration. *Animal Behaviour*, 1966, *14*, 266–272.

DAVIS, J. D., and CAMPBELL, C. S. Peripheral control of meal size in the rat: effect of sham feeding on meal size and drinking rate. *Journal of Comparative and Physiological Psychology*, 1973, *83*, 379–387.

DAVIS, J. M. A two-factor theory of schizophrenia. *Journal of Psychiatric Research*, 1974, *11*, 25–30.

DAVIS, K. B. *Factors in the Sex Life of 2,200 Women*. New York: Harper & Row, 1929.

DAVIS, P. G., McEWEN, B. S., and PFAFF, D. W. Localized behavioral effects of tritiated estradiol implants in the ventromedial hypothalmus of female rats. *Endocrinology*, 1979, *104*, 898–903.

DAVSON, H. *The Physiology of the Eye, 3rd edition*. New York: Academic Press, 1972.

DAW, N. W. Colour-coded ganglion cells in the goldfish retina: Extension of their receptive fields by means of new stimuli. *Journal of Physiology (London)*, 1968, *197*, 567–592.

DELAY, J., and DENIKER, P. Le traitement des psychoses par une methode neurolytique derivée d'hibernothérapie; le 4560 RP utilisée seul un cure prolongée et continuée. *Comptes Rendus Congrès des Médecins Aliénistes et Neurologistes de France et des Pays de Langue Francaise*, 1952a, *50*, 497–502.

DELAY, J., and DENIKER, P. 38 cas des psychoses traitées par la cure prolongée et continuée de 4560 RP. *Comptes Rendus Congrès des Médecins Aliénistes et Neurologistes de France et des Pays de Langue Francaise*, 1952b, *50*, 503–513.

DELGADO, J. M. R. *Physical Control of the Mind*. New York: Harper & Row, 1969.

DeLONG, M. Motor functions of the basal ganglia: Single-unit activity during movement. In *The Neurosciences: Third Study Program*, edited by F. O. Schmitt and F. G. Worden. Cambridge: MIT Press, 1974.

DEMENT, W. The effect of dream deprivation. *Science*, 1960, *131*, 1705–1707.

DEMENT, W. C. *Some Must Watch While Some Must Sleep*. San Francisco: W. H. Freeman & Company, 1974.

DEMENT, W. C. The relevance of sleep pathologies to the function of sleep. In *The Functions of Sleep*, edited by R. Drucker-Colin, M. Shkurovich, and M. B. Sterman. New York: Academic, 1979.

DEMENT, W., FERGUSON, F., COHEN, H., and BARCHAS, J. Nonchemical methods and data using a biochemical model: The REM quanta. In *Psychochemical Research in Man*, edited by A. J. Mandell and M. P. Mandell. New York: Academic Press, 1969.

DEMENT, W., MITLER, M., and HENRIKSEN, S. Sleep changes during chronic administration of parachlorophenylalanine. *Revue Canadienne de Biologie*, 1972, *31*, 239–246.

DEMENT, W., and WOLPERT, E. Relation of eye movements, body motility, and external stimuli to dream content. *Journal of Experimental Psychology*, 1958, *55*, 543–553.

DENNY-BROWN, D. The general principles of motor integration. In *Handbook of Physiology. Section I: Neurophysiology*, edited by J. Field, H. W. Magoun, and V. E. Hall. Washington: American Physiological Society, 1960.

DESIRAJU, T., BANERJEE, M. G., and ANAND, B. K. Activity of single neurons in the hypothalamic feeding centers: Effect of 2-deoxy-D-glucose. *Physiology & Behavior*, 1968, *3*, 757–760.

DESISTO, M. J. and ZWEIG, M. Differentiation of hypothalamic feeding and killing. *Physiological Psychology*, 1974, *2*, 67–70.

DEUTSCH, J. A. The stomach in food satiation and the regulation of appetite. *Progress in Neurobiology*, 1978, *10*, 135–153.

DEUTSCH, J. A., and DiCARA, L. Hunger and extinction in intracranial self-stimulation. *Journal of Comparative and Physiological Psychology*, 1967, *63*, 344–347.

DEUTSCH, J. A., and HARDY, W. T. Cholecystokinin produces bait shyness in rats. *Nature*, 1977, *266*, 196.

DEUTSCH, J. A., and HOWARTH, C. I. Some tests of a theory of intracranial self-stimulation. *Psychological Review*, 1963, *70*, 444–460.

DEUTSCH, J. A., MOLINA, F., and PUERTO, A. Conditioned taste aversion caused by palatable nontoxic nutrients. *Behavioral Biology*, 1976, *16*, 161–174.

DEUTSCH, J. A., PUERTO, A., and WANG, M.-L. The pyloric sphincter and differential food preferences. *Behavioral Biology*, 1977, *19*, 543–547.

DEUTSCH, J. A., YOUNG, W. G., and KALOGERIS, T. J. The stomach signals satiety. *Science*, 1978, *201*, 165–167.

DeVALOIS, R. L. Behavioral and electrophysiological studies of primate vision. In *Contributions to Sensory Physiology, Vol. 1*, edited by W. D. Neff. New York: Academic Press, 1965.

DeVALOIS, R. L., ABRAMOV, I., and JACOBS, G. H. Analysis of response patterns of LGN cells. *Journal of the Optical Society of America*, 1966, *56*, 966–977.

DeVALOIS, R. L., ALBRECHT, D. G., and THORELL, L. Cortical cells: Bar detectors or spatial frequency filters? In *Frontiers in Visual Science*, edited by S. J. Cool and E. L. Smith. Berlin: Springer, 1978.

DeVALOIS, R. L. and DeVALOIS, K. K. Spatial vision. *Annual Review of Psychology*, 1980, *31*, 309–341.

DIAMOND, I. T., and NEFF, W. D. Ablation of temporal cortex and discrimination of auditory patterns. *Journal of Neurophysiology*, 1957, *20*, 300–315.

DOBELLE, W. H., MLADEJOVSKY, M. G., and GIRVIN, J. P. Artificial vision for the blind: Electrical stimulation of visual cortex offers hope for a functional prosthesis. *Science*, 1974, *183*, 440–444.

DOETSCH, G. S., and ERICKSON, R. P. Synaptic processing of taste quality information in the nucleus tractus solitarius of the rat. *Journal of Neurophysiology*, 1970, *33*, 490–507.

DORUS, E., PANDEY, G. N., SHAUGHNESSY, R., GAVIRIA, M., VAL, E., ERICKSEN, S., and DAVIS, J. M. Lithium transport across red cell membrane: a cell membrane abnormality in manic-depressive illness. *Science*, 1979, *205*, 932–933.

DOTT, N. M. Surgical aspects of the hypothalamus. In *The Hypothalamus: Morphological, Functional, and Clinical Aspects*, edited by W. E. Clark. Edinburgh: Oliver and Boyd, 1938.

DOUGHTY, C., BOOTH, J. E., McDONALD, P. G., and PARROTT, R. F. Effects of oestradiol-17, oestradiol benzoate, and synthetic oestrogen, RU2858 on sexual differentiation in the neonatal female rat. *Journal of Endocrinology*, 1975, *67*, 419–424.

DOWNER, J. L. deC. Interhemispheric integration in the visual system. In *Interhemispheric Relations and Cerebral Dominance*, edited by V. B. Mountcastle. Baltimore: Johns Hopkins University Press, 1962.

DREYFUS-BRISAC, C. Sleep ontogenesis in early human prematurity from 24 to 27 weeks of conceptual age. *Developmental Psychobiology*, 1968, *1*, 162–169.

DRUCKER-COLÍN, R. R., and SPANIS, C. W. Is there a sleep transmitter? *Progress in Neurobiology*, 1976, *6*, 1–22.

DRUCKER-COLÍN, R. R., SPANIS, C. W., COTMAN, C. W., and McGAUGH, J. L. Changes in protein level in perfusates of freely moving cats: relation to behavioral state. *Science*, 1975, *187*, 963–965.

DUA, S., and MACLEAN, P. D. Location for penile erection in medial frontal lobe. *American Journal of Physiology*, 1964, *207*, 1425–1434.

DUGGAN, A. W. Morphine, enkephalins and the spinal cord. In *Advances in Pain Research and Therapy*, Vol. 3, edited by J. J. Bonica, J. C. Liebeskind, and D. Albe-Fessard. New York: Raven Press, 1979.

DUNLEAVY, D. L., BREZINOVÁ, V., OSWALD, I., MACLEAN, A. W., and TINKER, M. Changes during weeks in effects of tricycle drugs on the human sleeping brain. *British Journal of Psychiatry*, 1972, *120*, 663–672.

DUNN, A. J. Neurochemistry of learning and memory: an evaluation of recent data. *Annual Review of Psychology*, 1980, *31*, 343–390.

DUNN, A. J., REES, H. D., and IUVONE, P. M. ACTH and the stress-induced changes of lysine incorporation into brain and liver protein. *Pharmacology, Biochemistry, and Behavior*, 1978, *8*, 455–465.

DZENDOLET, E. A. A structure common to sweet-evoking compounds. *Perception and Psychophysics*, 1968, *3*, 65–68.

ECCLES, J. C. *The Understanding of the Brain*. New York: McGraw-Hill, 1973.

EDWARDS, D. A. Mice: Fighting by neonatally androgenized females. *Science*, 1968, *161*, 1027–1028.

EGGER, M. D., and FLYNN, J. P. Effect of electrical stimulation of the amygdala on hypothalamically elicited attack behavior in cats. *Journal of Neurophysiology*, 1963, *26*, 705–720.

EGGER, M. D., and FLYNN, J. P. Further studies on the effects of amygdaloid stimulation and ablation on hypothalamically elicited attack behavior in cats. In *Progress in Brain Research, Volume 27*, edited by W. R. Adey and T. Tokizane. Amsterdam: Elsevier, 1967.

EICHENBAUM, H., QUENON, B. A., HEACOCK, A., and AGRANOFF, B. W. Differential behavioral and biochemical effects of regional injections of cyclohexamide into mouse brain. *Brain Research*, 1976, *101*, 171–176.

EINSTEIN, E. R. Basic protein of myelin and its role in experimental allergic encephalomyelitis and multiple sclerosis. In *Handbook of Neurochemistry, Volume 7*, edited by A. Lajtha. New York: Plenum Press, 1972.

ELDRED, E., and BUCHWALD, J. Central nervous system: Motor

mechanisms. *Annual Review of Physiology*, 1967, *29*, 573–606.

ELDREDGE, D. H., and MILLER, J. D. Physiology of hearing. *Annual Review of Physiology*, 1971, *33*, 281–310.

ELLISON, G. D., and FLYNN, J. P. Organized aggressive behavior in cats after surgical isolation of the hypothalamus. *Archives Italiennes de Biologie*, 1968, *106*, 1–20.

EL-WAKIL, F. W. Unpublished master's thesis. University of Massachusetts, Amherst, 1975.

ENGELMAN, K., LOVENBERG, W., and SJOERDSMA, A. Inhibition of serotonin synthesis by parachlorophenylalanine in patients with the carcinoid syndrome. *New England Journal of Medicine*, 1967, *277*, 1103–1108.

ENGSTRÖM, H., ADES, H. W., and HAWKINS, J. E. Cellular pattern, nerve structures, and fluid spaces of the organ of Corti. In *Contributions to Sensory Physiology*, edited by W. D. Neff. New York: Academic Press, 1965.

ENTINGH, D., DUNN, A., GLASSMAN, E., WILSON, J. E., HOGAN, E., and DAMSTRA, T. Biochemical approaches to the biology of memory. In *Handbook of Psychobiology*, edited by M. S. Gazzaniga and C. Blakemore. New York: Academic Press, 1975

EPSTEIN, A. N., FITZSIMONS, J. T., and ROLLS, B. J. Drinking induced by injection of angiotensin into the brain of the rat. *Journal of Physiology (London)*, 1970, *210*, 457–474.

EPSTEIN, A. N., and TEITELBAUM, P. Severe and persistent deficits in thirst produced by lateral hypothalamic damage. In *Thirst: Proceedings of the First International Symposium on Thirst in the Regulation of Body Water*, edited by M. J. Wayner. Oxford: Pergamon Press, 1964.

EPSTEIN, A. N., and TEITELBAUM, P. Specific loss of the hypoglycemic control of feeding in recovered lateral rats. *American Journal of Physiology*, 1967, *213*, 1159–1167.

ERHARDT, A. A., and MEYER-BAHLBURG, H. F. L. Psychosexual development: an examination of the role of prenatal hormones. *Ciba Foundation Symposium*, 1979, *62*, 41–57.

ERVIN, F. R. Discussion in workshop on regulation of behavior. In *Ethical Issues in Biology and Medicine*, edited by P. N. Williams. Cambridge, Mass.: Schenkman, 1973.

ETHOLM, B. Evoked responses in the inferior colliculus, medial geniculate body, and auditory cortex by single and double clicks in cats. *Acta Oto-Laryngologica (Stockholm)*, 1969, *67*, 319–325.

ETKIN, W. Reproductive behaviors. In *Social Behavior and Organization Among Vertebrates*, edited by W. Etkin. Chicago: University of Chicago Press, 1964.

EVARTS, E. V. Relation of discharge frequency to conduction velocity in pyramidal tract neurons. *Journal of Neurophysiology*, 1965, *28*, 215–228.

EVARTS, E. V. Relation of pyramidal tract activity to force exerted during voluntary movement. *Journal of Neurophysiology*, 1968, *31*, 14–27.

EVARTS, E. V. Activity of pyramidal tract neurons during postural fixation. *Journal of Neurophysiology*, 1969, *32*, 375–385.

EVARTS, E. V. Sensorimotor cortex activity associated with movements triggered by visual as compared to somesthetic inputs. In *The Neurosciences: Third Study Program*, edited by F. O. Schmitt and F. G. Worden. Cambridge: MIT Press, 1974.

EVERETT, J. W. and NICHOLS, D. C. The timing of ovulatory release of gonadotropin induced by estrogen in pseudopregnant and diestrous cyclic rats. *Anatomical Record*, 1968, *160*, 346.

EVERITT, B. J. A neuroanatomical approach to the study of monoamines and sexual behavior. In *The Biological Determinants of Sexual Behaviour*, edited by J. B. Hutchinson. Chichester: Wiley & Sons, 1978.

EVERITT, B. J., and HERBERT, J. The effects of implanting testosterone propionate into the central nervous system on the sexual behavior of adrenalectomized female rhesus monkeys. *Brain Re-*

search, 1975, 86, 109–120.

EVERITT, B. J., HERBERT, J., and HAMER, J. D. Sexual receptivity of bilaterally adrenalectomised female rhesus monkeys. *Physiology & Behavior*, 1972, 8, 409–415.

FALCK, B., HILLARP, N. Å. THIEME, G., and TORP, A. Fluorescence of catecholamines and related compounds condensed with formaldehyde. *Journal of Histochemistry and Cytochemistry*, 1962, 10, 348–364.

FANSELOW, M. S. Naloxone attenuates rat's preference for signaled shock. *Physiological Psychology*, 1979, 7, 70–74.

FELIX, D., and PHILLIPS, M. I. Angiotensin excitation of cells in the organum vasculosum laminar terminalis by microiotophoresis. *Federation Proceedings*, 1978, 38, 985.

FENCL, V., KOSKI, G., and PAPPENHEIMER, J. R. Factors in cerebrospinal fluid from goats that affect sleep and activity in rats. *Journal of Physiology (London)*, 1971, 216, 565–589.

FETZ, E. E. Personal communication. Cited by A. L. Towe, 1973.

FEX, J. Neural excitatory processes of the inner ear. In *Handbook of Sensory Physiology, Volume 5, Auditory System*, edited by H. H. Kornhuber. Berlin: Springer-Verlag, 1972.

FEX, J. Neuropharmacology and potentials of the inner ear. In *Basic Mechanisms in Hearing*, edited by A. R. Møeller. New York: Academic Press, 1973.

FIBIGER, H. C. Drugs and reinforcement mechanisms: a critical review of the catecholamine theory. *Annual Review of Pharmacology and Toxicology*, 1978, 18, 37–56.

FIBIGER, H. C., ZIS, A. P., and McGEER, E. G. Feeding and drinking deficits after 6-hydroxydopamine administration in the rat: Similarities to the lateral hypothalamic syndrome. *Brain Research*, 1973, 55, 135–148.

FIELD, P. M., and RAISMAN, G. Structural and functional investigations of a sexually dimorphic part of the rat preoptic area. In *Recent Studies of Hypothalamic Function. Symposium Proceedings*. Calgary, Alberta, May, 1973.

FIELDS, H. L., and BASBAUM, A. I. Brainstem control of spinal pain-transmission neurons. *Annual Review of Physiology*, 1978, 40, 217–248.

FIEVE, R. R. The clinical effects of lithium treatment. *Trends in NeuroSciences*, 1979, 2, 66–68.

FINKELSTEIN, J. W., ROFFWARG, H. P., BOYAR, R. M., KREAM, J., and HELLMAN, L. Age-related change in the twenty-four hour spontaneous secretion of growth hormone. *Journal of Clinical Endocrinology and Metabolism*, 1972, 35, 665–670.

FISHBEIN, W., and GUTWEIN, B. M. Paradoxical sleep and memory storage processes. *Behavioral Biology*, 1977, 19, 425–464.

FISHER, C. Dreaming and sexuality. In *Psychoanalysis: A General Psychology*, edited by L. Lowenstein, M. Newman, M. M. Schur, and A. Solnit. New York: International University Press, 1966.

FISHER, C., BYRNE, J., EDWARDS, A., and KAHN, E. A psychophysiological study of nightmares. *Journal of the American Psychoanalytic Association*, 1970, 18, 747–782.

FISHER, C., GROSS, J., and ZUCH, J. Cycle of penile erection synchronous with dreaming (REM) sleep. Preliminary report. *Archives of General Psychiatry*, 1965, 12, 29–45.

FITZSIMONS, J. T. Drinking by rats depleted of body fluid without increase in osmotic pressure. *Journal of Physiology (London)*, 1961, 159, 297–309.

FITZSIMONS, J. T. Thirst. *Physiological Reviews*, 1972, 52, 468–561.

FITZSIMONS, J. T., KUCHARCZYK, J., and RICHARDS, G. Systemic angiotensin-induced drinking in the dog: a physiological phenomenon. *Journal of Physiology (London)*, 1978, 276, 435–448.

FITZSIMONS, J. T., and LE MAGNEN, J. Eating as a regulatory control of drinking in the rat. *Journal of Comparative and Physiological Psychology*, 1969, 67, 273–283.

FITZSIMONS, J. T. and SIMONS, B. J. The effect on drinking in the rat of intravenous infusion of angiotensin, given alone or in combination with other stimuli of thirst. *Journal of Physiology (London)*, 1969, 203, 45–57.

FLEMING, A., and ROSENBLATT, J. S. Olfactory regulation of maternal behavior in rats: I. Olfactory bulbectomy in experienced and inexperienced females. *Journal of Comparative and Physiological Psychology*, 1974, 86, 221–232. (a)

FLEMING, A., and ROSENBLATT, J. S. Olfactory regulation of maternal behavior in rats: II. Effects of peripherally induced anosmia and lesions of the lateral olfactory tract in pup-induced virgins. *Journal of Comparative and Physiological Psychology*, 1974, 86, 233–246. (b)

FLOCK, A. Transducing mechanisms in the lateral line canal organ receptors. *Cold Spring Harbor Symposia on Quantitative Biology*, 1965, 30, 133–146.

FLYNN, J., VANEGAS, H., FOOTE, W., and EDWARDS, S. Neural mechanisms involved in a cat's attack on a rat. In *The Neural Control of Behavior*, edited by R. F. Whalen, M. Thompson, M. Verzeano, and N. Weinberger. New York: Academic Press, 1970.

FOURIEZOS, G., and WISE, R. A. Pimozide-induced extinction of intracranial self-stimulation: response patterns rule out motor or performance deficits. *Brain Research*, 1976, 103, 377–380.

FRAY, P. J., KOOB, G. F., and IVERSEN, S. D. Tail pinch versus brain stimulation: problems of comparison. Reply to letters by Perochio and Hendon, and Katz. *Science*, 1978, 201, 841–842.

FREDERICSON, E. The effects of food deprivation upon competitive and spontaneous combat in C57 black mice. *Journal of Psychology*, 1950, 29, 89–100.

FREEMON, F. R. *Sleep Research: A Critical Review*. Springfield, Ill.: Charles C Thomas, 1972.

FRIEDMAN, M. I. Hyperphagia in rats with experimental diabetes mellitus: a response to a decreased supply of utilizable fuels. *Journal of Comparative and Physiological Psychology*, 1978, 92, 109–117.

FRIEDMAN, M. I., and BRUNO, J. P. Exchange of water during lactation. *Science*, 1976, 191, 409–410.

FRIEDMAN, M. I., ROWLAND, N., SALLER, C., and STRICKER, E. M. Different receptors initiate adrenal secretion and hunger during hypoglycemia. *Neuroscience Abstracts*, 1976, 2, 299.

FRIEDMAN, M. I., and STRICKER, E. M. The physiological psychology of hunger; a physiological perspective. *Psychological Review*, 1976, 83, 409–431.

FROHMAN, L. A., and BERNARDIS, L. L. Growth hormone and insulin levels in weanling rats with ventromedial hypothalamic lesions. *Endocrinology*, 1968, 82, 1125–1132.

FROHMAN, L. A., GOLDMAN, J. K., and BERNARDIS, L. L. Metabolism of intravenously injected C-glucose in weanling rats with hypothalamic obesity. *Metabolism*, 1972, 21, 799–805.

FUJIMORI, M., and HIMWICH, H. E. Electroencephalographic analyses of amphetamine and its methoxy derivatives with reference to their sites of EEG alerting in the rabbit brain. *International Journal of Neuropharmacology*, 1969, 8, 601–615.

FUNAKOSHI, M. and NINOMIYA, Y. Neural code for taste quality in the thalamus of the dog. In *Food Intake and Chemical Senses*, edited by Y. Katsuki, M. Sateo, S. F. Takagi, and Y. Oomura. Tokyo: University of Tokyo Press, 1977.

GALABURDA, A., and KEMPER, T. L. Observations cited by Geschwind, 1979.

GALABURDA, A. M., LeMAY, M., KEMPER, T. L., and GESCHWIND, N. Right-left asymmetries in the brain. *Science*, 1978, 199, 852–856.

GALLISTEL, C. R. The incentive of brain stimulation reward. *Journal of Comparative and Physiological Psychology*, 1969, 69, 713–721.

GALLISTEL, C. R. Self-stimulation: The neurophysiology of reward and motivation. In *The Physiological Basis of Memory*, edited by J. A. Deutsch. New York: Academic Press, 1973.

GANDELMAN, R., ZARROW, M. X., DENENBERG, V. H., and MYERS, M. Olfactory bulb removal eliminates maternal behavior in the mouse. *Science*, 1971, *171*, 210–211.

GARCIA, J., ERVIN, F. R., YORKE, C. H. and KOELLING, R. A. Conditioning with delayed vitamin injections. *Science*, 1967, *155*, 716–718.

GARCIA, J., GREEN, K. F., and McGOWAN, B. K. X-ray as an olfactory stimulus. In *Taste and Olfaction*, edited by C. Pfaffman. New York: Rockefeller University Press, 1969.

GARCIA, J., KIMMELDORF, D. J., and KOELLING, R. A. Conditioned aversion to saccharin resulting from exposure to gamma radiation. *Science*, 1955, *122*, 157–158.

GARCIA, J., and KOELLING, R. A. Relation of cue to consequence in avoidance learning. *Psychonomic Science*, 1966, *4*, 123–124.

GARCIA-BENGOCHEA, F., DELATTORE, D., ESQUIVEL, O., VIETA, R., and CLAUDIO, F. The section of the fornix in the surgical treatment of certain epilepsies. *Transactions of the American Neurological Association*, 1954, *79*, 176–178.

GAW, A. C., CHANG, L. W., and SHAW, l. -C. Efficacy of acupuncture on osteoarthritic pain. *New England Journal of Medicine*, 1975, *293*, 375–378.

GAZZANIGA, M. S., and LeDOUX, J. E. *The Integrated Mind*. New York: Plenum Press, 1978.

GENOVESI, U., MORUZZI, G., PALESTINI, M., ROSSI, G. F., and ZANCHETTI, A. EEG and behavioral patterns following lesions of the mesencephalic reticular formation in chronic cats with implanted electrodes. *Abstracts of Communications of the Twentieth International Physiology Congress*, Brussels, 1956, 335–336.

GERBINO, L., OLESHANSKY, M., and GERSHON, S. Clinical use and mode of action of lithium. In *Psychopharmacology: A Generation of Progress*, edited by M. A. Lipton, A. DiMascio, and K. F. Killam. New York: Raven Press, 1978.

GERSHON, E. S., BUNNEY W. E. JR., LECKMAN, J., VAN EERDEWEGH, M., and DeBAUCHE, B. The inheritance of affective disorders: a review of data and hypotheses. *Behavior Genetics*, 1976, *6*, 227–261.

GESCHWIND, N. Disconnexion syndromes in animals and man. *Brain*, 1965, *88*, 237–294, 585–644.

GESCHWIND, N. Language and the brain. *Scientific American*, 1972, *226*, 76–83.

GESCHWIND, N. The apraxias: Neural mechanisms of disorders of learned movement. *American Scientist*, 1975, *63*, 188–195.

GESCHWIND, N. Specializations of the human brain. *Scientific American*, 1979, *241*, 180–201.

GESCHWIND, N., QUADFASEL, F. A., and SEGARRA, J. M. Isolation of the speech area. *Neuropsychologia*, 1968, *6*, 327–340.

GIACHETTI, I., MACLEOD, P., and LeMAGNEN, J. Influence des états de faim et de satieté sur les réponses du bulbe olfactif chez le rat. *Journal de Physiologie (Paris)*, 1970, *62 (Supplement 2)*, 280–281.

GIBBS, J., YOUNG, R. C., and SMITH, G. P. Cholecystokinin decreases food intake in rats. *Journal of Comparative and Physiological Psychology*, 1973, *84*, 488–495. (a)

GIBBS, J., YOUNG, R. C., and SMITH G. P. Cholecystokinin elicits satiety in rats with open gastric fistulas. *Nature*, 1973, *245*, 323–325. (b)

GILMAN, A. The relation between blood osmotic pressure, fluid distribution and voluntary water intake. *American Journal of Physiology*, 1937, *120*, 323–328.

GIRDEN, E., METTLER, F. A., FINCH, G., and CULLER, E. Conditioned responses in a decorticate dog to acoustic, thermal and tactile stimulation. *Journal of Comparative Psychology*, 1936, *27*, 367–385.

GISPEN, W. H., PERUMAL, R., WILSON, J. E., and GLASSMAN, E. Phosphorylation of proteins of synaptosome-enriched fractions of brain during short term training experience: the effects of various

behavioral treatments. *Behavioral Biology*, 1977, *21*, 358–363.

GLENDENNING, K. K. Effects of septal and amygdaloid lesions on social behavior of the cat. *Journal of Comparative and Physiological Psychology*, 1972, *80*, 199–207.

GODDARD, G. V. Development of epileptic seizures through brain stimulation at low intensity. *Nature*, 1967, *214*, 1020–1021.

GOLD, R. M. Hypothalamic obesity: The myth of the ventromedial nucleus. *Science*, 1973, *182*, 488–490.

GOLD, R. M., JONES, A. P., SAWCHENKO, P. E., and KAPATOS, G. Paraventricular area: critical focus of a longitudinal neurocircuitry mediating food intake. *Physiology & Behavior*, 1977, *18*, 1111–1119.

GOLDBERG, J. M., and FERNANDEZ, C. Vestibular mechanisms. *Annual Review of Physiology*, 1975, *37*, 129–162.

GOLDSTEIN, M. H., HALL, J. L., and BUTTERFIELD, B. O. Single-unit activity in the primary auditory cortex of unanesthetized cats. *Journal of the Acoustical Society of America*, 1968, *43*, 444–455.

GORSKI, R. A., and WAGNER, J. W. Gonadal activity and sexual differentiation of the hypothalamus. *Endocrinology*, 1965, *76*, 226–239.

GOTSICK, J. E., and MARSHALL, R. C. Time course of the septal rage syndrome. *Physiology & Behavior*, 1972, *9*, 685–687.

GOURAS, P. Identification of cone mechanisms in monkey ganglion cells. *Journal of Physiology (London)*, 1968, *199*, 533–538.

GOURAS, P., and KRÜGER, J. Similarities of color properties of neurons in areas 17, 18, and V4 of monkey cortex. Paper presented at the Seventh International Neurobiology Meeting, Gottingen, September, 1975.

GOY, R. W. Organizing effect of androgen on the behaviour of rhesus monkeys. In *Endocrinology of Human Behaviour*, edited by R. P. Michael. London: Oxford University Press, 1968.

GOY, R. W., and GOLDFOOT, D. A. Hormonal influences on sexually dimorphic behavior. In *Handbook of Physiology, Section 7, Volume 2, Part I*, edited by R. O. Green. Washington, D. C.: American Physiological Society, 1973.

GRANNEMAN, J. and FRIEDMAN, M. I. Hepatic control of feeding and gastric acid secretion: Inhibition mediated by the hepatic vagus. Paper presented at the Meeting of the Eastern Psychological Association, Philadelphia, 1978.

GRANNEMAN, J. and FRIEDMAN, M. I. Hepatic modulation of insulin-induced gastric acid secretion and EMG activity in rats. *American Journal of Physiology*, 1980, in press.

GRANT, E. C. G., and MEARS, E. Mental effects of oral contraceptives. *Lancet*, 1967, *2*, 945–946.

GREEN, J., CLEMENTE, C., and DeGROOT, J. Rhinencephalic lesions and behavior in cats. *Journal of Comparative Neurology*, 1957, *108*, 505–545.

GREEN, J. P., and MAAYANI, S. Tricyclic antidepressant drugs block histamine H2 receptor in brain. *Nature*, 1977, *269*, 163–165.

GREENBERG, R., PEARLMAN, C., FINGAR, R., KANTROWITZ, J., and KAWLICHE, S. The effects of dream deprivation: Implications for a theory of the psychological function of dreaming. *British Journal of Medical Psychology*, 1970, *43*, 1–11.

GREENBERG, R., PILLARD, R., and PEARLMAN, C. The effect of dream (stage REM) deprivation on adaptation to stress. *Psychosomatic Medicine*, 1972, *34*, 257–262.

GREENGARD, P. Phosphorylated proteins as physiological effectors. *Science*, 1978, *199*, 146–152.

GREENOUGH, W. T., CARTER, C. S., STEERMAN, C., and DeVOOGD, T. J. Sex differences in dendritic patterns in hamster preoptic area. *Brain Research*, 1977, *26*, 63–72.

GREISER, C., GREENBERG, R., and HARRISON, R. H. The adaptive function of sleep: The differential effects of sleep and dreaming on recall. *Journal of Abnormal Psychology*, 1972, *80*, 280–286.

GRIFFITH, J. D., CAVANAUGH, J., HELD, N. N., and OATES, J. A. Dextroamphetamine: evaluation of psychotomimetic properties in man. *Archives of General Psychiatry*, 1972, *26*, 97–100.

GROSS, C. G. Inferotemporal cortex and vision. In *Progress in Physiological Psychology*, edited by E. Stellar and J. M. Sprague. New York: Academic Press, 1973.

GROSS, C. G., ROCHA-MIRANDA, C. E., and BENDER, D. B. Visual properties of neurons in inferotemporal cortex of the macaque. *Journal of Neurophysiology*, 1972, *35*, 96–111.

GROSS, M. D. Violence associated with organic brain disease. In *Dynamics of Violence*, edited by J. Fawcett. Chicago: American Medical Association, 1971.

GROSSMAN, S. P. The biology of motivation. *Annual Review of Psychology*, 1979, *30*, 209–242.

GÜLDNER, F.-H. Synaptology of the rat suprachiasmatic nucleus. *Cell Tissue Research*, 1976, *165*, 509–544.

GÜLDNER, F.-H., and WOLFF, H. R. Dendrodendritic synapses in the suprachiasmatic nucleus of the rat hypothalamus. *Journal of Neurocytology*, 1974, *3*, 245–250.

GULEVICH, G., DEMENT, W. C., and JOHNSON, L. Psychiatric and EEG observations on a case of prolonged (264 hours) wakefulness. *Archives of General Psychiatry*, 1966, *15*, 29–35.

GULICK, W. L. *Hearing: Physiology and Psychophysics*. New York: Oxford University Press, 1971.

GUMULKA, W., SAMANIN, R., VALZELLI, L., and CONSOLOS, S. Behavioural and biochemical effects following the stimulation of the nucleus raphis dorsalis on rats. *Journal of Neurochemistry*, 1971, *18*, 533–534.

HAMMOND, P. H., MERTON, P. A., and SUTTON, G. G. Nervous gradation of muscular contraction. *British Medical Bulletin*, 1956, *12*, 214–218.

HAMMOND, W. A. *Sleep and Its Derangements*. Philadelphia: Lippincott, 1883.

HARDING, C. F., and LESHNER, A. I. The effects of adrenalectomy on the aggressiveness of differently housed mice. *Physiology & Behavior*, 1972, *8*, 437–440.

HARLOW, H. F. *Learning to Love*. New York: Ballantine, 1973.

HARMON, L. D., and JULESZ, B. Masking in visual recognition: Effects of two-dimensional filtered noise. *Science*, 1973, *180*, 1194–1197.

HARPER, A. E. Effects of dietary protein content and amino acid pattern on food intake and preference. In *Handbook of Physiology, Volume I*, edited by C. F. Code and W. Heidel. Washington, D. C.: American Physiological Society, 1967.

HARRIS, G. W., and JACOBSOHN, D. Functional grafts of the anterior pituitary gland. *Proceedings of the Royal Society, London*, Series B, 1951–1952, *139*, 263–267.

HART, B. Sexual reflexes and mating behavior in the male dog. *Journal of Comparative and Physiological Psychology*, 1967, *66*, 388–399.

HART, B. Gonadal hormones and sexual reflexes in the female rat. *Hormones and Behavior*, 1969, *1*, 65–71.

HART, B. L. Hormones, spinal reflexes, and sexual behaviour. In *Determinants of Sexual Behaviour*, edited by J. B. Hutchinson. Chichester: Wiley & Sons, 1978.

HART, R. D. A. Monthly rhythm of libido in married women. *British Medical Journal*, 1960, *1*, 1023–1024.

HARTLINE, H. K., and RATLIFF, F. Spatial inhibitory influences in the eye of the *Limulus*, and the mutual interaction of receptor units. *Journal of General Physiology*, 1958, *41*, 1049–1066.

HARTMANN, E. *The Biology of Dreaming*. Springfield, Ill.: C. C Thomas, 1967.

HARTMANN, E. L. *The Functions of Sleep*. New Haven: Yale University Press, 1973.

HATTON, G. I. Nucleus circularis: Is it an osmoreceptor in the brain? *Brain Research Bulletin*, 1976, *1*, 123–131.

HAWKE, C. C. Castration and sex crimes. *American Journal of Mental Deficiency*, 1950, *55*, 220–226.

HAWKINS, J. E. Cytoarchitectural basis of cochlear transducer. *Cold Springs Harbor Symposium on Quantitative Biology*, 1965, *30*, 147–157.

HAYWOOD, J., ROSE, S. P. R., and BATESON, P. P. G. Effects of an imprinting procedure on RNA polymerase activity in chick brain. *Nature*, 1970, *228*, 373–374.

HAYWOOD, J., ROSE, S. P. R., and BATESON, P. Changes in chick brain RNA polymerase associated with imprinting procedure. *Brain Research*, 1975, *92*, 227–235.

HEATH, R. G. Pleasure response of human subjects to direct stimulation of the brain: Physiologic and psychodynamic consideration. In *The Role of Pleasure in Behavior*, edited by R. G. Heath. New York: Harper & Row, 1964.

HEFFNER, R., and MASTERSON, B. Variation in form of the pyramidal tract and its relationship to digital dexterity. *Brain, Behavior, and Evolution*, 1975, *12*, 161–200.

HEIMER, L., and LARSSON, K. Impairment of mating behavior in male rats following lesions in the preoptic-anterior hypothalamic continuum. *Brain Research*, 1966/1967, *3*, 248–263.

HENDRICKSEN, S., DEMENT, W., and BARCHAS, J. The role of serotonin in the regulation of a phasic event of rapid eye movement sleep: The ponto-geniculo-occipital wave. *Advances in Biochemical Psychopharmacology*, 1974, *11*, 169–179.

HENN, F. A., HALJÄMAE, H., and HAMBERGER, A. Glial cell functions: active control of extracellular concentration. *Brain Research*, 1972, *43*, 437–443.

HENRISON, A. E., WAGONER, N., and COWAN, W. M. Autoradiographic and electron microscopic study of retino-hypothalamic connections. *Zeitschrift für Zellforschung und Mikroskopische Anatomie*, 1972, *125*, 1–26.

HENSEL, H. Thermoreceptors. *Annual Review of Physiology*, 1974, *36*, 233–249.

HENSEL, H., and KENSHALO, D. R. Warm receptors in the nasal region of cats. *Journal of Physiology (London)*, 1969, *204*, 99–112.

HERBERG, L. Seminal ejaculation following positively reinforcing electrical stimulation of the rat hypothalamus. *Journal of Comparative and Physiological Psychology*, 1963, *56*, 679–685.

HERBST, A. L., KURMAN, R. J., SCULLY, R. E., and POSKANZER, D. C. Clear-cut adenocarcinoma of the genital tract in young females. *Registry Report, New England Journal of Medicine*, 1972, *287*, 1259–1264.

HERZ, A., ALBUS, K., METYS, J. SCHUBERT, P., and TESCHEMACHER, H. On the central sites for the antinociceptive action of morphine and fentanyl. *Neuropharmacology*, 1970, *9*, 539–551.

HESS, W. R. Das Schlafsyndrom als Folge dienzephaler Reizung. *Helvetica Physiologica et Pharmacologica Acta*, 1944, *2*, 305–344.

HESTON, L. L. Psychiatric disorders in foster-home-reared children of schizophrenic mothers. *British Journal of Psychiatry*, 1966, *112*, 819–825.

HETHERINGTON, A. W., and RANSON, S. W. Experimental hypothalamohypophyseal obesity in the rat. *Proceedings of the Society for Experimental Biology and Medicine*, 1939, *41*, 465–466.

HEUSER, J. E. Synaptic vesicle exocytosis revealed in quick-frozen frog neuromuscular junctions treated with 4-aminopyridine and given a single electrical shock. In *Society for Neuroscience Symposia, Volume II*, edited by W. M. Cowan and J. A. Ferrendelli. Bethesda, Md.: Society for Neuroscience, 1977.

HEUSER, J. E., and REESE, T. S. Evidence for recycling of synaptic vesicle membrane during transmitter release at the frog neuromuscular function. *Journal of Cell Biology*, 1973, *57*, 315–344.

HEUSER, J. E., REESE, T. S., DENNIS, M. J., JAN, Y., JAN, L., and EVANS, L. Synaptic vesicle exocytosis captured by quick freezing and correlated with quantal transmitter release. *Journal of Cell Biology*, 1979, *81*, 275–300.

HIERONS, R., and SAUNDERS, M. Impotence in patients with temporal lobe lesions. *Lancet*, 1966, *2*, 761–764.

HILTON, S. M., and ZBROZYNA, A. W. Amygdaloid region for defense reactions and its afferent pathway to the brain stem. *Journal of Physiology (London)*, 1963, *165*, 160–173.

HIRSCH, H. V. B., and SPINELLI, D. N. Modification of the distribution of receptive field orientation in cats by selective visual exposure during development. *Experimental Brain Research*, 1971, *13*, 509–527.

HITT, J. C., HENDRICKS, S. E., GINSBERG, S. I., and LEWIS, J. H. Disruption of male, but not female, sexual behavior in rats by medial forebrain bundle lesions. *Journal of Comparative and Physiological Psychology*, 1970, *73*, 377–384.

HOBSON, J. A., McCARLEY, R. W., PIVIK, T., and FREEDMAN, R. Selective firing by cat pontine brain stem neurons in desynchronized sleep. *Journal of Neurophysiology*, 1974, *37*, 497–511.

HOEBEL, B. G. Inhibition and disinhibition of self-stimulation and feeding: Hypothalamic control and post-ingestional factors. *Journal of Comparative and Physiological Psychology*, 1968, *66*, 89–100.

HOEBEL, B. G., and TEITELBAUM, P. Weight regulation in normal and hypothalamic hyperphagic rats. *Journal of Comparative and Physiological Psychology*, 1966, *61*, 189–193.

HOHMAN, G. W. Some effects of spinal cord lesions on experienced emotional feelings. *Psychophysiology*, 1966, *3*, 143–156.

HÖKFELT, T., LJUNGDAHL, A., TERENIUS, L., ELDE, R., and NILSON, G. Immunohistochemical analysis of peptide pathways possibly related to pain and analgesia: enkephalin and substance P. *Proceedings of the National Academy of Science (USA)*, 1977, *74*, 3081–3085.

HOLADAY, J. W., WEI, E., LOH, H. H., and LI, C. H. Endorphins may function in heat adaptation. *Proceedings of the National Academy of Science (USA)*, 1978, *75*, 2923–2927.

HOLMES, G. The cerebellum of man. *Brain*, 1939, *62*, 21–30.

HOREL, J. A. The neuroanatomy of amnesia. A critique of the hippocampal memory hypothesis. *Brain*, 1978, *101*, 403–445.

HORN, G., McCABE, B. J.; and BATESON, P. P. G. An autoradiographic study of the chick brain after imprinting. *Brain Research*, 1979, *168*, 361–374.

HORN, G., ROSE, S. P. R., and BATESON, P. P. G. Experience and plasticity in the central nervous system. *Science*, 1973, *181*, 506–514.

HOROVITZ, A. P., PIALA, J. J., HIGH, J. P., BURKE, J. C., and LEAF, R. C. Effects of drugs on the mouse-killing (muricide) test and its relationships to amygdaloid function. *International Journal of Neuropharmacology*, 1966, *5*, 405–411.

HUBBELL, W. L., and BOWNDS, M. D. Visual transduction in vertebrate photoreceptors. *Annual Review of Neuroscience*, 1979, *2*, 17–34.

HUBEL, D. H., and WIESEL, T. N. Functional architecture of macaque monkey visual cortex. *Proceedings of the Royal Society of London*, 1977, *198*, 1–59.

HUBEL, D. H., and WIESEL, T. N. Brain mechanisms of vision. *Scientific American*, 1979, *241*, 150–162.

HUBEL, D. H., WIESEL, T. N., and STRYKER, M. P. Anatomical demonstration of orientation columns in macaque monkeys. *Journal of Comparative Neurology*, 1978, *177*, 361–380.

HUGHES, J., SMITH, T. W., KOSTERLITZ, H. W., FOTHERGILL, L. A., MORGAN, B. A., and MORRIS, H. R. Identification of two related pentapeptides from the brain with potent opiate agonist activity. *Nature*, 1975, *258*, 577–579.

HUNSICKER, J. P., and REID, L. D. "Priming effect" in conventionally reinforced rats. *Journal of Comparative and Physiological Psychology*, 1974, *87*, 618–621.

HUNSPERGER, R. In *The Sleeping Brain*, edited by M. H. Chase. Los Angeles: Brain Information Service/Brain Research Institute, UCLA, 1972.

HUNTER, R., LOGUE, V., and McMENEMY, W. H. Temporal lobe epilepsy supervening on longstanding transvestitism and fetishism. *Epilepsia*, 1963, *4*, 60–65.

HUTCHINSON, R. R., and RENFREW, J. W. Stalking attack and eating behavior elicited from the same sites in the hypothalamus. *Journal of Comparative and Physiological Psychology*, 1966, *61*, 300–367.

IBUKA, N. and KAWAMURA, H. Loss of circadian rhythm in sleep-wakefulness cycle in the rat by suprachiasmatic nucleus lesions. *Brain Research*, 1975, *96*, 76–81.

ILLIS, L. Changes in spinal cord synapses and a possible explanation for spinal shock. *Experimental Neurology*, 1963, *8*, 328–335.

INGELFINGER, F. J. The late effects of total and subtotal gastrectomy. *New England Journal of Medicine*, 1944, *231*, 321–327.

INGRAM, W. R., BARRIS, R. W., and RANSON, S. W. Catalepsy. An experimental study. *Archives of Neurology and Psychiatry (Chicago)*, 1936, *35*, 1175–1197.

ISSACSON, R. L. *The Limbic System*. New York: Plenum Press, 1974.

ITO, M. Neuronal events in the cerebellar flocculus associated with an adaptive modification of the vestibulo-ocular reflex of the rabbit. In *Control of Gaze by Brain Stem Neurons, Developments in Neuroscience*, edited by R. Baker and A. Berthoz. Amsterdam: Elsevier, 1977.

JACKLET, J. W. Neuronal circadian rhythms: phase shifting by protein synthesis inhibitor. *Science*, 1977, *198*, 69–71.

JACKLET, J. W. The cellular mechanisms of circadian clocks. *Trends in NeuroSciences*, 1978, *1*, 117–119.

JACOBS, B. L. and TRULSON, M. E. Dreams, hallucinations, and psychosis—the serotonin connection. *Trends in NeuroSciences*, 1979, *2*, 276–280.

JAMES, W. Principles of Psychology, New York: Henry Holt, 1890.

JAMES, W. H. The distribution of coitus within the human intermenstruum. *Journal of Biosocial Science*, 1971, *3*, 159–171.

JANOWITZ, H. D., and GROSSMAN, M. I. Some factors affecting the food intake of normal dogs and dogs with esophagostomy and gastric fistula. *American Journal of Physiology*, 1949, *159*, 143–148.

JANOWITZ, H. D. and HOLLANDER, F. Effect of prolonged intragastric feeding on oral ingestion. *Federation Proceedings*, 1953, *12*, 72.

JARVIK, M. E. Effects of chemical and physical treatments on learning and memory. *Annual Review of Psychology*, 1972, *23*, 457–486.

JOHNSON, A. K. Localization of angiotensin in sensitive areas for thirst within the rat brain. Paper presented at the meeting of the Eastern Psychological Association, Boston, April, 1972.

JOHNSON, A. K., and BUGGY, J. A critical analysis of the site of action for the dipsogenic effect of angiotensin II. In *International Symposium on the Central Actions of Angiotensin and Related Hormones*, edited by J. P. Buckley and C. Ferrario. Oxford: Pergamon Press, 1977.

JOHNSON, R. N., DeSISTO, M. J. and KOENIG, A. B. Social and developmental experience and interspecific aggression in rats. *Journal of Comparative and Physiological Psychology*, 1972, *79*, 237–242.

JOHNSTON, P. J., and DAVIDSON, J. M. Intracerebral androgens and sexual behavior in the male rat. *Hormones and Behavior*, 1972, *3*, 345–357.

JOHNSTONE, E. D., CROW, T. J., FRITH, C. D. CARNEY, M. W. P., and PRICE, J. S. Mechanism of the antipsychotic effect in the treatment of acute schizophrenia. *Lancet*, 1978, *1 (8069)*, 848–851.

JONES, B. Catecholamine-containing neurons in the brain stem of the cat and their role in waking. Unpublished master's thesis. Lyon: Tixier, 1969. Cited by Jouvet, 1972.

JONES, B. E., BOBILLIER, P., and JOUVET, M. Effets de la destruction des neurones contenant des catécholamines du mésencéphale sur le cycle veille-sommeils du chat. *Comptes Rendus de la Societe de Biologie, Paris*, 1969, *163*, 176–180.

JONES, B. E., HARPER, S. T., and HALARIS, A. E. Effects of locus coeruleus lesions upon cerebral monoamine content, sleep wakefulness states and the response to amphetamine in the cat. *Brain Research*, 1977, *124*, 473–496.

JOST, A. Embryonic sexual differentiation. In *Hermaphroditism, Genital Anomalies and Related Endocrine Disorders*, second edition. Baltimore: Williams & Wilkins, 1969.

JOUVET, M. Recherches sur les structures nerveuses et les mécanismes responsables des différentes phases du sommeil physiologique. *Archives Italiennes de Biologie*, 1962, *100*, 125–206.

JOUVET, M. Insomnia and decrease of cerebral 5-hydroxytryptamine after destruction of the raphe system in the cat. *Advances in Pharmacology*, 1968, *6*, 265–279.

JOUVET, M. The role of monoamines and acetylcholine-containing neurons in the regulation of the sleep-waking cycle. *Ergebnisse der Physiologie*, 1972, *64*, 166–307.

JOUVET, M., MICHEL, F., and MOUNIER, D. Comparative study of the "paradoxical phase" of sleep in cat and man (abstract). *Electroencephalography and Clinical Neurophysiology*, 1960, *160*, 1461–1465.

JOUVET-MOUNIER, D., ASTIC, L. and LACOTE, D. Ontogenesis of the states of sleep in rat, cat, and guinea pig during the first postnatal month. *Developmental Psychobiology*, 1970, *2*, 216–239.

KALRA, S. P., and SAWYER, C. H. Blockage of copulation-induced ovulation in the rat by anterior hypothalamic deafferentation. *Endocrinology*, 1970, *87*, 1124–1128.

KANE, F. J. Psychiatric reaction to oral contraceptives. *American Journal of Obstetrics and Gynecology*, 1968, *102*, 1053–1063.

KANOF, P. D., and GREENGARD, P. Brain histamine receptors as targets for antidepressant drugs. *Nature*, 1978, *272*, 329–333.

KAPATOS, G., and GOLD, R. M. Evidence for ascending noradrenergic mediation of hypothalamic hyperphagia. *Pharmacology, Biochemistry, and Behavior*, 1973, *1*, 81–87.

KARACAN, I., ROSENBLOOM, A. L., LONDONO, J. H., WILLIAMS, R. L., and SALIS, P. J. Growth hormone levels during morning and afternoon naps. *Behavioral Neuropsychiatry*, 1975, *6*, 67–70.

KARACAN, I., SALIS, P. J., and WILLIAMS, R. L. The role of the sleep laboratory in diagnosis and treatment of impotence. In *Sleep Disorders: Diagnosis and Treatment*, edited by R. J. Williams and I. Karacan. New York: Wiley & Sons, 1978.

KARLI, P. The Norway rat's killing response to the white mouse. *Behavior*, 1956, *10*, 81–103.

KARSCH, F. J., DIERSCHKE, D. J., and KNOBIL, E. Sexual differentiation of pituitary function: apparent difference between primates and rodents. *Science*, 1973, *179*, 484–486.

KASSIL, V. G., USGOLEV, A. M., and CHERNIGOVSKII, V. N. Regulation of selection and consumption of food and metabolism. *Progress in Physiological Sciences*, 1970, *1*, 387–404.

KEBABIAN, J. W., PETZOLD, G. L., and GREENGARD, P. Dopamine-sensitive adenylate cyclase in caudate nucleus of rat brain and its similarity to the "dopamine receptor." *Proceedings of the National Academy of Science (USA)*, 1972, *69*, 2145–2149.

KEELE, C. A. Measurement of responses to chemically induced pain. In *Touch, Heat, and Pain*, edited by A. V. S. deRenuck and J. Knight. Boston: Little, Brown, 1966.

KENT, E., and GROSSMAN, S. P. Evidence for a conflict interpre-

tation of anomalous effects of rewarding brain stimulation. *Journal of Comparative and Physiological Psychology*, 1969, *69*, 381–390.

KERTESZ, A. Anatomy of Jargon. In *Jargon Aphasia*, edited by J. Brown. New York: Academic Press, 1980.

KETY, S. S., ROSENTHAL, D., WENDER, P. H., and SCHULSINGER, K. F. The types and prevalence of mental illness in the biological and adoptive families of adopted schizophrenics. In *The Transmission of Schizophrenia*, edited by D. Rosenthal and S. S. Kety. New York: Pergamon Press, 1968.

KIANG, N. Y. -S. Discharge patterns of single fibers in the cat's auditory nerve. Cambridge: MIT Press, 1965.

KIM, Y. K., and UMBACH, W. Combined stereotaxic lesions for treatment of behavior disorders and severe pain. Paper presented at the Third World Congress of Psychosurgery. Cambridge, England, 1972. Cited by Valenstein, 1973.

KING, B. M., CARPENTER, R. G., STAMOUTSOS, B. A., FROHMAN, L. A., and GROSSMAN, S. P. Hyperphagia and obesity following ventromedial hypothalamic lesions in rats with subdiaphragmatic vagotomy. *Physiology & Behavior*, 1978, *20*, 643–651.

KLEITMAN, N. *Sleep and Wakefulness*, Second edition. Chicago: University of Chicago Press, 1963.

KLING, A., LANCASTER, J., and BENITONE, J. Amygdalectomy in the free-ranging vervet (*Cercopithecusalthiops*). *Journal of Psychiatric Research*, 1970, *7*, 191–199.

KLÜVER, H., and BUCY, P. C. Preliminary analysis of functions of the temporal lobes in monkeys. *Archives of Neurology and Psychiatry (Chicago)*, 1939, *42*, 979–1000.

KNOBIL, E. On the control of gonadotrophin secretion in the rhesus monkey. *Recent Progress in Hormone Research*, 1974, *30*, 1–46.

KOELLA, W. P. Serotonin—A hypnogenic transmitter and an antiawakening agent. *Advances in Biochemical Psychopharmacology*, 1974, *11*, 181–186.

KOLÁRSKÝ, A., FREUND, K., MACHEK, J., and POLÁK, O. Male sexual deviation. Association with early temporal lobe damage. *Archives of General Psychiatry*, 1967, *17*, 735–743.

KOMISARUK, B. R. Neural and hormonal interactions in the reproductive behavior of female rats. In *Reproductive Behavior*, edited by E. Montagna and W. A. Sadler. New York: Plenum Press, 1974.

KOMISARUK, B. R., ADLER, N. T., and HUTCHINSON, J. Genital sensory field: enlargement by estrogen treatment of female rats. *Science*, 1972, *178*, 1295–1298.

KOOB, G. F., FRAY, P. J., and IVERSEN, S. D. Tail-pinch stimulation: sufficient motivation for learning. *Science*, 1976, *194*, 637–639.

KORNHUBER, H. H. Cerebral cortex, cerebellum, and basal ganglia: An introduction to their motor functions. In *The Neurosciences: Third Study Program*, edited by F. O. Schmitt and F. G. Worden. Cambridge: MIT Press, 1974.

KOVNER, R., and STAMM, J. S. Disruption of short-term visual memory by electrical stimulation of inferotemporal cortex in the monkey. *Journal of Comparative and Physiological Psychology*, 1972, *81*, 163–172.

KRASNE, F. B. General disruption resulting from electrical stimulation of ventromedial hypothalamus. *Science*, 1962, 138, 822–823.

KRIECKHAUS, E. E., and WOLF, G. Acquisition of sodium by rats: Interaction of innate mechanisms and latent learning. *Journal of Comparative and Physiological Psychology*, 1968, *65*, 197–201.

KRINGLEN, E. Adult of offspring of two psychotic parents, with special reference to schizophrenia. In *The Nature of Schizophrenia*, edited by L. C. Wynne, R. L. Cromwell, and S. Matthysse. New York: Wiley & Sons, 1978.

KUCHARCZYK, J., and MOGENSON, G. J. Separate lateral hypothalamic pathways for extracellular and intracellular thirst. *American Journal of Physiology*, 1975, *228*, 295–301.

KUCHARCZYK, J., and MOGENSON, G. J. Specific deficits in regulatory drinking following electrolytic lesions of the lateral hypothalamus. *Experimental Neurology*, 1976, *53*, 371–385.

KUCHARCZYK, J., and MOGENSON, G. J. Effect of preoptic administration of angiotensin on lateral hypothalamic unit activity. *Physiology & Behavior*, 1977, *19*, 455–457.

KUFFLER, S. W. Discharge patterns and functional organization of mammalian retina. *Journal of Neurophysiology*, 1953, *16*, 37–68.

KUFFLER, S. W., and NICHOLS, J. G. *From Neuron to Brain*. Sunderland, Mass.: Sinauer Associates, 1976.

LABORIT, H. La thérapeutique neuro-végétate du choc et de la maladie post-traumatique. *Presse Medicale*, 1950, *58*, 138–140. Cited by Snyder, S. H. *Madness and the Brain*. New York: McGraw-Hill, 1974.

LAMOTTE, C., PERT, C. B., and SNYDER, S. H. Opiate receptor binding in primate spinal cord: distribution and changes after dorsal root section. *Brain Research*, 1976, *112*, 407–412.

LARSSON, K. Mating behavior in male rats after cerebral cortex ablation: II. Effects of lesions in the frontal lobes compared to lesions in the posterior half of the hemispheres. *Journal of Experimental Zoology*, 1964, *155*, 203–214.

LASCHET, U. Antiandrogen in the treatment of sex offenders: mode of action and therapeutic outcome. In *Contemporary Sexual Behavior: Critical Issues in the 1970's*, edited by J. Zubin and J. Money. Baltimore: Johns Hopkins Press, 1973.

LASEK, R. J., GAINER, H., and PRZYBYLSKI, R. J. Transfer of newly synthesized proteins from Schwann cells to the squid giant axon. *Proceedings of the National Academy of Sciences (USA)*, 1974, *71*, 1188–1192.

LASHLEY, K. In search of the engram. *Society of Experimental Biology*, 1950, Symposium 4, 454–482.

LAW, D. T., and MEAGHER, W. Hypothalamic lesions and sexual behavior in the female rat. *Science*, 1958, *128*, 1626–1627.

LAWRENCE, J. E. S. Science and sentiment: Overview of research on crowding and human behavior. *Psychological Bulletin*, 1974, *81*, 712–720.

LEFKOWITZ, M. A model of the glabrous skin of the fingertip. Unpublished Master's Thesis, Johns Hopkins University, Baltimore, 1979.

LEGRENDE, R., and PIERON, H. Recherches sur le besoin de sommeil consécutif a une veille prolongés. *Zietschrift für Allgemeine Physiologie*, 1913, *14*, 235–362.

LEMAGNEN, J. Hyperphagie, provoquée chez le rat blanc par altération du mécanisme de satiété péripherique. *Comptes Rendus des Sèances de la Sociètè de Biologie*, 1956, *150*, 32.

LEMAGNEN, J., and TALLEN, S. La periodicité spontanée de la prise d'aliments ad libitum du rat blanc. *Journal of Physiology (Paris)*, 1966, *58*, 323–349.

LENARD, H. G., and SCHULTE, F. J. Polygraph sleep study in cranipagus twins (where is the sleep transmitter?). *Journal of Neurology, Neurosurgery, and Psychiatry*, 1972, *35*, 756–762.

LESHNER, A. I., WALKER, W. A., JOHNSON, A. E., KELLING, J. S., KREISLER, S. J., and SVARE, B. B. Pituitary adrenocortical activity and intermale aggressiveness in isolated mice. *Physiology and Behavior*, 1973, *11*, 705–711.

LETTVIN, J. Y., and GESTELAND, R. C. Speculations on smell. *Cold Spring Harbor Symposium*, 1965, *30*, 217–225.

LEVINE, J. D., GORDON, N. C., and FIELDS, H. L. The role of endorphins in placebo analgesia. In *Advances in Pain Research and Therapy, Volume 3*, edited by J. J. Bonica, J. C. Liebeskind, and D. Albe-Fessard. New York: Raven Press, 1979.

LICKLIDER, J. C. R. Three auditory theories. In *Psychology: A Study of a Science, Volume I*, edited by I. S. Koch. New York: McGraw-Hill, 1959.

LIEBELT, R. A., BORDELON, C. B., and LIEBELT, A. G. The adipose tissue system and food intake. In *Progress in Physiological Psychology*, edited by E. Stellar and J. M. Sprague. New York: Academic Press, 1973.

LIEBERBURG, I. and MCEWEN, B. S. Brain cell nuclear retention of two testosterone metabolites, 5 α-dihydrotestosterone and estradiol-17 in adult rats. *Endocrinology*, 1977, *100*, 188–197.

LIEBERBURG, I., WALLACH, G., and MCEWEN, B. S. The effects of an inhibitor of aromatization (1,4,6 androstatariene-3,17-dione) and an anti-estrogen (C1628) on *in vivo* formed testosterone metabolites recovered from neonatal rat brain tissues and purified cell nuclei. Implications for sexual differentiation of the rat brain. *Brain Research*, 1977, *128*, 176–181.

LIEBLING, D. S., EISNER, J. D., GIBBS, J., and SMITH, G. P. Intestinal satiety in rats. *Journal of Comparative and Physiological Psychology*, 1975, *89*, 955–965.

LINDSLEY, D. B., SCHREINER, L. H., KNOWLES, W. B., and MAGOUN, H. W. Behavioral and EEG changes following chronic brain stem lesions in the cat. *Electroencephalography and Clinical Neurophysiology*, 1950, *2*, 483–498.

LIPPA, A. S., ANTELMAN, S. M., FISHER, A. E., and CANFIELD, D. R. Neurochemical mediation of reward: a significant role for dopamine? *Pharmacology, Biochemistry, and Behavior*, 1973, *1*, 23–28.

LISK, R. D., PRETLOW, R. A., and FRIEDMAN, S. Hormonal stimulation necessary for elicitation of maternal nest-building in the mouse *(Mus musculus)*. *Animal Behaviour*, 1969, 17, 730–737.

LJUNGBERG, T., and UNGERSTEDT, U. Sensory inattention produced by 6-hydroxydopamine-induced degeneration of ascending dopamine neurons in the brain. *Experimental Neurology*, 1976, *53*, 585–600.

LLOYD, D. P. C. The spinal mechanism of the pyramidal system in cats. *Journal of Neurophysiology*, 1941, *4*, 525–546.

LOEWENSTEIN, W. R., and MENDELSON, M. Components of receptor adaptation in a Pacinian corpuscle. *Journal of Physiology (London)* 1965, *177*, 377–397.

LOEWENSTEIN, W. R., and RATHKAMP, R. The sites for mechanoelectric conversion in a Pacinian corpuscle. *Journal of General Physiology*, 1958, *41*, 1245–1265.

LOTT, D. F., and FUCHS, S. S. Failure to induce retrieving by sensitization or the injection of prolactin. *Journal of Comparative and Physiological Psychology*, 1962, *55*, 1111–1113.

LUBAR, J. F., BOYCE, B. A., and SCHAEFER, C. S. Etiology of polydipsia and polyuria in rats with septal lesions. *Physiology & Behavior*, 1968, *3*, 289–292.

LUCERO, M. A. Lengthening of REM sleep duration consecutive to learning in the rat. *Brain Research*, 1970, *20*, 319–322.

LUND, J. S. Organization of neurons in the visual cortex, area 17, of the monkey *(Macaca mulatta)*. *Journal of Comparative Neurology*, 1973, *147*, 455–496.

LURIA, A. R. *Higher Cortical Functions in Man*, translated by B. Haigh. New York: Basic Books, 1966.

LURIA, A. R. The functional organization of the brain. *Scientific American*, 1970, *222*, 66–79.

MACDONNELL, M. F., and FLYNN, J. P. Control of sensory fields by stimulation of hypothalamus. *Science*, 1966, *152*, 1406–1408.

MACHLUS, B., ENTINGH, D., WILSON, J. E., and GLASSMAN, E. Brain phosphoproteins: The effect of various behaviors and reminding experiences on the incorporation of radioactive phosphate into nuclear proteins. *Behavioral Biology*, 1974, *10*, 63–73.

MACHLUS, B., WILSON, J. E., and GLASSMAN, E. Brain phosphoproteins: The effect of short experiences on the phosphorylation of nuclear proteins of rat brain. *Behavioral Biology*, 1974, *10*, 43–62.

MACKAY, E. M., CALLOWAY, J. W., and BARNES, R. H. Hyperalimentation in normal animals produced by protamine insulin. *Journal of Nutrition*, 1940, *20*, 59–66.

MACLEAN, P. D., DUA, S., and DENNISTON, R. H. Cerebral localization for scratching and seminal discharge. *Archives of Neurology*, 1963, *9*, 485–497.

MacLusky, N. J., and McEwen, B. S. Oestrogen modulates progestin receptor concentration in some brain regions, but not in others. *Nature*, 1978, *274*, 276–278.

Magnes, J., Moruzzi, G., and Pompeiano, O. Synchronization of the EEG produced by low-frequency electrical stimulation of the region of the solitary tract. *Archives Italiennes de Biologie*, 1961, *99*, 33–67.

Magni, F., Moruzzi, G., Rossi, G. F., and Zanchetti, A. EEG arousal following inactivation of the lower brain stem by selective injection of barbiturate into the vertebral circulation. *Archives Italiennes de Biologie*, 1959, *97*, 33–46.

Maier, S. F., and Jackson, R. L. Learned helplessness: all of us were right (and wrong): inescapable shock has multiple effects. In *Psychology of Learning and Motivation*, edited by G. H. Bower. New York: Academic Press, 1979.

Malsbury, C., and Pfaff, D. Supression of sexual receptivity in the hormone-primed female hamster by electrical stimulation of the medial preoptic area. *Abstracts, Society for Neuroscience*, 1973, 122.

Malvin, R. L., Mouw, D., and Vander, A. J. Angiotensin: physiological role in water-deprivation-induced thirst of rats. *Science*, 1977, *19*, 171–173.

Mann, F., Bowsher, D., Mumford, J., Lipton, S., and Miles, J. Treatment of intractable pain by acupuncture. *Lancet*, 1973, *2*, 57–60.

Margules, D. L. Noradrenergic rather than serotonergic basis of reward in the dorsal tegmentum. *Journal of Comparative and Physiological Psychology*, 1969, *67*, 32–35.

Mark, V. H., and Ervin, F. R. *Violence and the Brain*. New York: Harper & Row, 1970.

Mark, V. H., Ervin, F. R., and Yakovlev, P. I. The treatment of pain by stereotaxic methods. *Confina Neurologica*, 1962, *22*, 238–245.

Mark, V. H., Sweet, W. H., and Ervin, F. R. The effect of amygdalectomy on violent behavior in patients with temporal lobe epilepsy. In *Psychosurgery*, edited by E. Hitchcock, L. Laitinen, and K. Vernet. Springfield, Ill.: C. C Thomas, 1972.

Martin, G. E., and Bacino, C. B. Action of intrahypothalamically-injected β-endorphin on the body temperature of the rat. *Abstracts, Society for Neuroscience*, 1978, *4*, 411.

Martinez, C., and Bittner, J. J. A non-hypophyseal sex difference in estrous behavior of mice bearing pituitary grafts. *Proceedings of the Society for Experimental Biology and Medicine*, 1956, *91*, 506–509.

Masserman, J. H. *Principles of Dynamic Psychiatry*. Philadelphia: W. B. Saunders, 1961.

Mathews, D. and Money, J. Progestin-induced hermaphroditism: a follow-up report. In Eastern Conference of Reproductive Behavior, Madison, Wis., 1978. Program Abstract, p. 62.

Maxim, P. E. Behavioral effects of telestimulating hypothalamic reinforcement sites in freely moving rhesus monkeys. *Brain Research*, 1972, *42*, 243–262.

Maxim, P. E. Self-stimulation of a hypothalamic site in response to tension or fear. *Physiology & Behavior*, 1977, *18*, 197–201.

Mayer, D. J., and Liebeskind, J. C. Pain reduction by focal electrical stimulation of the brain: An anatomical and behavioral analysis. *Brain Research*, 1974, *68*, 73–93.

Mayer, D. J., Price, D. D., and Becker, D. P. Neurophysiological characterization of the anterolateral spinal cord neurons contributing to pain perception in man. *Pain*, 1975, *1*, 51–58.

Mayer, D. J., Price, D. D., Becker, D. P., and Young, H. F. Threshold for pain from anterolateral quadrant stimulation as a predictor of success of percutaneous cordotomy for relief of pain. *Journal of Neurosurgery*, 1975, *43*, 445–447.

Mayer, D. J., Price, D. D., Rafii, A., and Barber, J. Acupuncture hypalgesia: evidence for activation of a central control system as a mechanism of action. In *Advances in Pain Research and Therapy, Volume 1*, edited by J. J. Bonica and D. Albe-Fessard. New York: Raven Press, 1976, 751–754.

Mayer, J. Regulation of energy intake and the body weight: The glucostatic theory and the lipostatic hypothesis. *Annals of the New York Academy of Science*, 1955, *63*, 15–43.

McCauley, E., and Ehrhardt, A. A. Female sexual response: hormonal and behavioral interactions. *Primary Care*, 1976, *3*, 455.

McClintock, M. K. and Adler, N. T. The role of the female during copulation in wild and domestic Norway rats (*Rattus norvegicus*). *Behaviour*, 1978, *67*, 67–96.

McDonald, N. Living with schizophrenia. *Journal of the Canadian Medical Association*, 1960, *82*, 218–221.

McDonald, P. G., and Doughty, C. Effects of neonatal administration of different androgens in the female rat: correlation between aromatization and the induction of sterilization. *Journal of Endocrinology*, 1974, *61*, 95–103.

McEwen, B. S. Gonadal steroid receptors in neuroendocrine tissue. In *Hormone Receptors, Volume I*, edited by B. O'Malley and L. Birnbaumer. New York: Academic Press, 1978.

McEwen, B. S., Davis, P. G., Parsons, B., and Pfaff, D. W. The brain as a target for steroid hormone action. *Annual Review of Neuroscience*, 1979, *2*, 65–112.

McFayden, V. M., Oswald, I., and Lewis, S. A. Starvation and human slow wave sleep. *Journal of Applied Physiology*, 1973, *35*, 391–394.

McGinty, D., Epstein, A. N., and Teitelbaum, P. The contribution of oropharyngeal sensations to hypothalamic hyperphagia. *Animal Behaviour*, 1965, *13*, 413–418.

McGinty, D. J., Harper, R. M., and Fairbanks, M. K. Neuronal unit activity and the control of sleep states. In *Advances in Sleep Research*, edited by E. D. Weitzman. Flushing, New York: Spectrum Publications, 1974.

McGinty, D. J., and Sterman, M. B. Sleep suppression after basal forebrain lesions in the cat. *Science*, 1968, *160*, 1253–1255.

McGrath, M. J., and Cohen, D. B. REM sleep facilitation of adaptive waking behavior: a review of the literature. *Psychological Bulletin*, 1978, *85*, 24–57.

McKenna, T., McCarley, R. W., Amatruda, T., Black, D., and Hobson, J. A. Effects of carbachol at pontine sites yielding long duration desynchronized sleep episodes. In *Sleep Research*, Volume 3. Los Angeles: Brain Information Service/Brain Research Institute, UCLA, 1974, p. 39.

McKinley, M. J., Denton, D. A., and Weisinger, R. S. Sensors for antidiuresis and thirst-osmoreceptors or CSF sodium detectors. *Brain Research*, 1978, *141*, 89–103.

Meddis, R. *The Sleep Instinct*. London: Routledge and Kegan Paul, 1977.

Meddis, R., Pearson, A., and Langford, G. An extreme case of healthy insomnia. *Electroencephalography and Clinical Neurophysiology*, 1973, *35*, 213–214.

Mellinkoff, S. M., Frankland, M., Boyle, D., and Greipel, M. Relation between serum amino acid concentration and fluctuations in appetite. *Journal of Applied Psychology*, 1956, *8*, 535–538.

Mendelson, J. Role of hunger in T-maze learning for food by rats. *Journal of Comparative and Physiological Psychology*, 1966, *62*, 341–349.

Mendelson, J. Lateral hypothalamic stimulation in satiated rats: The rewarding effects of self-induced drinking. *Science*, 1967, *157*, 1077–1079.

Mennin, S. P., and Gorski, R. A. Effects of ovarian steroids on plasma LH in normal and persistent estrous adult female rats.

Endocrinology, 1975, *96*, 486–491.

MILEDI, R. Acetylcholine sensitivity of partially denervated frog muscles. *Journal of Physiology (London)*, 1959, *147*, 45–46P.

MILLER, G. A., and TAYLOR, W. G. The perception of repeated bursts of noise. *Journal of the Acoustical Society of America*, 1948, *20*, 171–182.

MILLER, N. E., BAILEY, C. J., and STEVENSON, J. A. F. Decreased hunger but increased food intake resulting from hypothalamic lesions. *Science*, 1950, *112*, 256–259.

MILLER, R. R., and SPRINGER, A. D. Amnesia consolidation and retrieval. *Psychological Review*, 1973, *80*, 69–70.

MILNER, B. Memory disturbance after bilateral hippocampal lesions. In Milner, P., and Glickman, S., *Cognitive Processes and the Brain*. Princeton: Van Nostrand, 1965.

MILNER, B. Memory and the temporal regions of the brain. In *Biology of Memory*, edited by K. H. Pribram and D. E. Broadbent. New York: Academic Press, 1970.

MILNER, B., BRANCH, C., and RASMUSSEN, T. Evidence for bilateral speech representation in some non-right-handers. *Transactions of the American Neurological Association*, 1966, *91*, 306–308.

MILNER, B., CORKIN, S., and TEUBER, H. -L. Further analysis of the hippocampal amnesic syndrome: 14-year follow-up study of H. M. *Neuropsychologia*, 1968, *6*, 317–338.

MISELIS, R. R., and EPSTEIN, A. N. Feeding induced by 2-deoxy-D-glucose injections into the lateral ventricle of the rat. *The Physiologist*, 1970, *13*, 262.

MISHKIN, M. Visual mechanisms beyond the striate cortex. In *Frontiers in Physiological Psychology*, edited by R. W. Russell. New York: Academic Press, 1966.

MITCHELL, W., FALCONER, M. A., and HILL, D. Epilepsy and fetishism relieved by temporal lobectomy. *Lancet*, 1954, *2*, 626–630.

MOGENSON, G. J. Hypothalamic limbic mechanisms in the control of water intake. In *The Neuropsychology of Thirst: New Findings and Advances in Concepts*, edited by A. N. Epstein, H. R. Kissileff, and E. Stellar. Washington: Winston, 1973.

MOLDOFSKY, H., and SCARISBRICK, P. Induction of neurasthenic musculoskeletal pain syndrome by selective sleep stage deprivation. *Psychosomatic Medicine*, 1976, *38*, 35–44.

MOLTZ, H., LUBIN, M., LEON, M., and NUMAN, M. Hormonal induction of maternal behavior in the ovariectomized nulliparous rat. *Physiology & Behavior*, 1970, *5*, 1373–1377.

MOLTZ, H., ROBBINS, D., and PARKS, M. Caesarian delivery and maternal behavior of primiparous and multiparous rats. *Journal of Comparative and Physiological Psychology*, 1966, *61*, 455–460.

MONCRIEFF, R. W. Olfactory adaptation and odour likeness. *Journal of Physiology (London)*, 1956, *133*, 301–316.

MONEY, J. Components of eroticism in man: Cognitional rehearsals. In *Recent advances in Biological Psychiatry*, edited by J. Wortis. New York: Grune and Stratton, 1960.

MONEY, J., and EHRHARDT, A. *Man & Woman, Boy & Girl*. Baltimore: The Johns Hopkins University Press, 1972.

MONNIER, M., DUDLER, L., GÄCHTER, R., MAIER, P. F., TOBLER, H. J. and SCHOENENBERGER, G. A. The delta sleep inducing peptide (DSIP). Comparative properties of the original and synthetic nonapeptide. *Experientia*, 1977, *33/34*, 548–552.

MONNIER, M., and HÖSLI, L. Dialysis of sleep and waking factors in blood of rabbit. *Science*, 1964, *146*, 796–798.

MONNIER, M., and HÖSLI, L. Humoral regulation of sleep and wakefulness by hypnogenic and activating dialysable factors. *Progress in Brain Research*, 1965, *18*, 118–123.

MONNIER, M., KOLLER, T., and GRABER, S. Humoral influences of induced sleep and arousal upon electrical brain activity of animals with crossed circulation. *Experimental Neurology*, 1963, *8*, 264–277.

MOORE, R. Y. Effects of some rhinencephalic lesions on retention of conditioned avoidance behavior in cats. *Journal of Comparative and Physiological Psychology*, 1964, *53*, 540–548.

MOORE, R. Y., and BLOOM, F. E. Central catecholamine neuron systems: anatomy and physiology of the dopamine systems. *Annual Review of Neuroscience*, 1978, *1*, 129–169.

MOORE, R. Y., and BLOOM, F. E. Central catecholamine neuron systems: anatomy and physiology of the norepinephrine and epinephrine systems. *Annual Review of Neuroscience*, 1979, *2*, 113–168.

MOORE, R. Y., and EICHLER, V. B. Loss of a circadian adrenal corticosterone rhythm following suprachiasmatic lesions in the rat. *Brain Research*, 1972, *42*, 201–206.

MOORE-EDE, M. C., LYDIC, C. A., CZEISLER, C. A., FULLER, C. A., and ALBERS, H. E. Structure of function of suprachiasmatic nuclei (SCN) in human and non-human primates. *Neuroscience Abstracts*, 1980, in press.

MORA, F., SANGUINETTI, A. M., ROLLS, E. T., and SHAW, S. G. Differential effects of self-stimulation and motor behavior produced by microintracranial injections of a dopamine-receptor blocking agent. *Neuroscience Letters*, 1975, *1*, 179–184.

MORGAN, C. T., and MORGAN, J. D. Studies in hunger: II. The relation of gastric denervation and dietary sugar to the effects of insulin upon food intake in the rat. *Journal of General Psychology*, 1940, *57*, 153–163.

MORGANE, J. P. Alterations in feeding and drinking of rats with lesions in the globi pallidi. *American Journal of Physiology*, 1961, *201*, 420–428.

MORISON, R. S., and DEMPSEY, E. W. A study of thalamo-cortical relations. *American Journal of Physiology*, 1942, *135*, 281–292.

MORTIMER, C. H., McNEILLY, A. S., FISHER, R. A., MURRAY, M. A. F., and BESSER, G. M. Gonadotrophin releasing hormone therapy in hypogonadal males with hypothalamic or pituitary dysfunction. *British Medical Journal*, 1974, *4*, 617–621.

MORUZZI, G. The sleep-waking cycle. *Ergebnisse der Physiologie*, 1972, *64*, 1–165.

MORUZZI, G., and MAGOUN, H. W. Brain stem reticular formation and activation of the EEG. *Electroencephalography and Clinical Neurophysiology*, 1949, *1*, 455–473.

MOTOKAWA, K., TAIRA, N., and OKUDA, J. Spectral responses of single units in the primate cortex. *Tohoku Journal of Experimental Medicine*, 1962, *78*, 320–337.

MOUNT, G. B., and HOEBEL, B. G. Lateral hypothalamic reward decreased by intragastric feeding: Self-determined "threshold" technique. *Psychosomatic Science*, 1967, *9*, 265–266.

MOUNTCASTLE, V. B. The problem of sensing and the neural coding of sensory events. In *The Neurosciences*, edited by G. Quarton, T. Melnechuk, and F. O. Schmitt. New York: Rockefeller University Press, 1967.

MOUNTCASTLE, V. B., and POWELL, T. P. S. Neural mechanisms subserving cutaneous sensibility, with special reference to the role of afferent inhibition in sensory perception and discrimination. *Bulletin of the Johns Hopkins Hospital*, 1959, *105*, 201–232.

MOYER, K. E. Kinds of aggression and their physiological basis. *Communications in Behavioral Biology*, 1968, *2*, 65–87.

MOYER, K. E. *The Psychobiology of Aggression*. New York: Harper & Row, 1976.

MOZELL, M. M. Evidence for a chromatographic model of olfaction. *Journal of General Physiology*, 1970, *55*, 46–63.

MUELLER, K., and HSIAO, S. Current status of cholescystokinin as a short-term satiety hormone. *Neuroscience and Biobehavioral Reviews*, 1978, *2*, 79–87.

MURRAY, R. G., and MURRAY, A. The anatomy and ultrastructure of taste endings. In *Taste and Smell in Vertebrates*, edited by G. E. W. Wolstenholme and J. Knight. London: J. & A. Churchill, 1970.

720

MYERS, R. D. An improved push pull cannula system for perfusing an isolated region of the brain. *Physiology & Behavior*, 1970, *5*, 243–246.

MYERSON, B. J., and MALMNÄS, C. -O. Brain monoamines and sexual behaviour. In *Biological Determinants of Sexual Behaviour*, edited by J. B. Hutchison. Chichester: Wiley & Sons, 1978.

NACHMAN, M. Taste preference for sodium salts by adrenalectomized rats. *Journal of Comparative and Physiological Psychology*, 1962, *55*, 1124–1129.

NACHMAN, M. Learned aversion to the taste of lithium chloride and generalization to other salts. *Journal of Comparative and Physiological Psychology*, 1963, *56*, 343–349.

NAFE, J. P., and WAGONER, K. S. The nature of pressure adaptation. *Journal of General Psychology*, 1941, *25*, 323–351.

NAFTOLIN, F., RYAN, K. J., DAVIES, I. J., REDDY, V. V., FLORES, F., PETRO, Z., and KUHN, M. The formation of estrogen by central neuroendocrine tissue. *Recent Progress in Hormone Research*, 1975, *31*, 295–315.

NAQUET, R., DENAVIT, M., and ALBE-FESSARD, D. Comparison entre le role du subthalamus et celui des différentes structures bulbomésencephaliques dans le maintien de la vigilance. *Electroencephalography and Clinical Neurophysiology*, 1966, *20*, 149–164.

NARABAYASHI, H. Stereotaxic amygdalectomy. In *The Neurobiology of the Amygdala*, edited by B. E. Eleftheriou. New York: Plenum Press, 1972.

NARABAYASHI, H., NAGAO, T., SAITO, Y., YOSHIDA, M., and NAGAHATA, M. Stereotaxic amygdalotomy for behavior disorders. *Archives of Neurology*, 1963, *9*, 1–16.

NAUTA, W. J. H. Hypothalamic regulation of sleep in rats. Experimental study. *Journal of Neurophysiology*, 1946, *9*, 285–316.

NICOLAIDIS, S. Réponses des unités osmosensibles hypothalamiques aux stimulations salienes et aqueuses de la langue. *Comptès Rendus Hebdomadaires des Sèances de l'Academie des Sciences. Sèries C.*, 1968, *267*, 2352–2355.

NIIJIMA, A. Afferent impulse discharge from glucoreceptors in the liver of the guinea pig. *Annals of the New York Academy of Science*, 1969, *157*, 690–700.

NIIJIMA, A. Coding mechanism of glucose sensitive afferent nerve fibers in the liver. *International Congress of Physiology of Food and Fluid Intake*, Paris, 1977.

NOIROT, E. Selective priming of maternal responses by auditory and olfactory cues from mouse pups. *Developmental Psychobiology*, 1970, *2*, 273–276.

NOIROT, E. Selective priming of maternal responses by auditory and olfactory cues from mouse pups. *Developmental Psychobiology*, 1972, *5*, 371–387.

NORMAN, R. J., BUCHWALD, J. S., and VILLABLANCA, J. R. Classical conditioning with auditory discrimination of the eye blink in decerebrate cats. *Science*, 1977, *196*, 551–553.

NOVIN, D., SANDERSON, J. D., and VANDERWEELE, D. A. The effect of isotonic glucose on eating as a function of feeding condition and infusion site. *Physiology & Behavior*, 1974, *13*, 3–7.

NOVIN, D., VANDERWEELE, D. A., and REZEK, M. Hepatic-portal 2-deoxy-D-glucose infusion causes eating: Evidence for peripheral glucoreceptors. *Science*, 1973, *181*, 858–860.

NOWLIS, G. H., and FRANK, M. Qualities in hamster taste: behavioral and neural evidence. In *Olfaction and Taste 6*, edited by J. LeMagnen and P. MacLeod. Washington, D.C.: Information Retrieval, 1977.

NUMAN, M. Medial preoptic area and maternal behavior in the female rat. *Journal of Comparative and Physiological Psychology*, 1974, *87*, 746–759.

NUNEZ, A. A., and CASATI, M. J. The role of efferent connections of the suprachiasmatic nucleus in the control of circadian rhythms. *Behavioral and Neural Biology*, 1979, *25*, 263–267.

OAKLEY, K., and TOATES, F. M. The passage of food through the gut of rats and its uptake of fluid. *Psychonomic Science*, 1969, *16*, 225–226.

O'KELLY, L. I. The psychophysiology of motivation. *Annual Review of Psychology*, 1963, *14*, 57–92.

OLDS, J. Commentary. In *Brain Stimulation and Motivation*, edited by E. S. Valenstein. Glenview, Ill.: Scott, Foresman, 1973.

OLDS, J. *Drives and Reinforcements*. New York: Raven Press, 1977.

OLDS, J., ALLAN, W. S., and BRIESE, E. Differentiation of hypothalamic drive and reward centers. *American Journal of Physiology*, 1971, *221*, 368–375.

OLDS, J., and MILNER, P. Positive reinforcement produced by electrical stimulation of septal area and other regions of rat brain. *Journal of Comparative and Physiological Psychology*, 1954, *47*, 419–427.

OLINS, D. E. and OLINS, A. L. Nucleosomes: The structural quantum in chromosomes. *American Scientist*, 1978, *66*, 704–711.

ORBACH, J., and ANDREWS, W. H. Stimulation of afferent nerve terminals in the perfused rabbit liver by sodium salts of some long chain fatty acids. *Quarterly Journal of Experimental Physiology*, 1973, *58*, 267–274.

OSWALD, I. Drugs and sleep. *Pharmacological Reviews*, 1968, *20*, 272–303.

OSWALD, I., ADAM, K., ALLEN, S., BURACK, R., SPENCE, M., and THACORE, V. Alpha adrenergic blocker, thymoxamine and mesoridazine, both increase human REM sleep duration. *Sleep Research*, 1974, *3*, 62.

OWEN, F., CROSS, A. J., CROW, T. J., LONGDEN, M., POULTER, M., and RILEY, G. J. Increased dopamine-receptor sensitivity in schizophrenia. *Lancet*, 1978, *2(8083)*, 223–226.

PANDYA, D. N., and KUYPERS, H. G. J. M. Corticocortical connections in the rhesus monkey. *Brain Research*, 1969, *13*, 13–36.

PANKSEPP, J. Drugs and stimulus-bound attack. *Physiology & Behavior*, 1971, *6*, 317–320. (a)

PANKSEPP, J. Aggression elicited by electrical stimulation of the hypothalamus in albino rats. *Physiology & Behavior*, 1971, *6*, 321–329. (b)

PANKSEPP, J., and TROWILL, J. A. Intraoral self-injection: II. The simulation of self-stimulation phenomena with a conventional reward. *Psychonomic Science*, 1967, *9*, 407–408.

PARMEGGIANI, P. L. Telencephalo-diencephalic aspects of sleep mechanisms. *Brain Research*, 1968, *7*, 350–359.

PARROTT, R. F. Aromatizable and 5 α-reduced androgens: differentiation between central and peripheral effects on male rat sexual behaviour. *Hormones and Behavior*, 1975, *6*, 99–108.

PATTERSON, K. E., and MARCEL, A. J. Aphasia, dyslexia, and the phonological coding of written words. *Quarterly Journal of Experimental Psychology*, 1977, *29*, 307–318.

PAUL, L., MILEY, W. M., and BAENNINGER, R. Mouse killing by rats: Roles of hunger and thirst in its initiation and maintenance. *Journal of Comparative and Physiological Psychology*, 1971, *76*, 242–249.

PAVLOV, I. P. *The Work of the Digestive Glands*. London: C. Griffin and Co., 1910.

PAVLOV, I. P. "Innere Hemmung" der bedingten Reflexe und der Schlaf—ein und derselbe Prozeß. *Skandinavisches Archiv für Physiologie*, 1923, *44*, 42–58.

PAXINOS, G., and BINDRA, D. Hypothalamic knife cuts: Effects on eating, drinking, irritability, aggression, and copulation in the male rat. *Journal of Comparative and Physiological Psychology*, 1972, *79*, 219–229.

PAXINOS, G., and BINDRA, D. Hypothalamic and midbrain neural pathways involved in eating, drinking, irritability, aggression, and copulation in rats. *Journal of Comparative and Physiological Psychology*, 1973, *82*, 1–14.

PECK, J. W. Discussion: Thirst(s) resulting from bodily water imbalances. In *The Neuropsychology of Thirst: New Findings and Advances in Concepts*, edited by A. N. Epstein, H. R. Kissileff, and E. Stellar. Washington: Winston, 1973.

PECK, J. W., and BLASS, E. M. Localization of thirst and antidiuretic osmoreceptors by intracranial injections in rats. *American Journal of Physiology*, 1975, *5*, 1501–1509.

PECK, J. W., and NOVIN, D. Evidence that osmoreceptors mediating drinking in rabbits are in the lateral preoptic region. *Journal of Comparative and Physiological Psychology*, 1971, *74*, 134–147.

PEÑALOZA-ROJAS, J., and RUSSEK, M. Anorexia induced by direct current blockade of the vagus nerve. *Nature*, 1963, *200*, 176.

PENFIELD, W., and JASPER, H. *Epilepsy and the Functional Anatomy of the Human Brain*. Boston: Little, Brown, 1954.

PENFIELD, W., and MATHIESON, G. Memory: autopsy findings and comments on the role of hippocampus in experiential recall. *Archives of Neurology*, 1974, *31*, 145–154.

PENFIELD, W., and MILNER, B. Memory deficit produced by bilateral lesions in the hippocampal zone. *American Medical Association Archives of Neurological Psychiatry*, 1958, *79*, 475–497.

PERKEL, D. H., and BULLOCK, T. H. Neural coding. *Neuroscience Research Progress Bulletin*, 1968, *6*, 221–347.

PERRY, T. L., HANSEN, S., and KLOSTER, M. Huntington's chorea: Deficiency of γ-aminobutyric acid in brain. *New England Journal of Medicine*, 1973, *288*, 337–342.

PERT, C. B., SNOWMAN, A. M., and SNYDER, S. H. Localization of opiate receptor binding in presynaptic membranes of rat brain. *Brain Research*, 1974, *70*, 184–188.

PETRIDES, M., and IVERSEN, S. D. Cross-modal matching and the primate frontal cortex. *Science*, 1976, *192*, 1023–1024.

PFAFF, D. W. Luteinizing hormone-releasing factor potentiates lordosis behavior in hypophysectomized female rats. *Science*, 1973, *182*, 1148–1149.

PFAFF, D. W., and KEINER, M. Atlas of estradiol-concentrating cells in the central nervous system of the female rat. *Journal of Comparative Neurology*, 1973, *151*, 121–158.

PFAFF, D. W., and ZIGMOND, R. E. Neonatal androgen effects on sexual and nonsexual behavior of adult rats tested under various hormone regimes. *Neuroendocrinology*, 1971, *7*, 129–145.

PFAFFMANN, C., FRANK, M., and NORGREN, R. Neural mechanisms and behavioral aspects of taste. *Annual Review of Psychology*, 1979, *30*, 283–325.

PFEIFFER, C. A. Sexual differences of the hypophysis and their determination by the gonads. *American Journal of Anatomy*, 1936, *58*, 195–226.

PHILLIPS, A. G. Enhancement and inhibition of olfactory bulb self-stimulation of odors. *Physiology & Behavior*, 1970, *5*, 1127–1131.

PHILLIPS, A. G., VAN DER KOOY, D., and FIBIGER, H. C. Maintenance of intracranial self-stimulation in hippocampus and olfactory bulb following regional depletion of noradrenaline. *Neuroscience Letters*, 1977, *4*, 77–84.

PHILLIPS, M. I. Angiotensin in the brain. *Neuroendocrinology*, 1978, *25*, 354–377.

PHILLIPS, M. I., and FELIX, D. Specific angiotensin II receptive neurons in the cat subfornical organ. *Brain Research*, 1976, *109*, 531–540.

PIÉRON, H. *Le Problème Physiologique de Sommeil*. Paris: Masson, 1913.

PITTENDRIGH, C. S. Circadian oscillations in cells and the circadian organization of multicellular systems. In *The Neurosciences: Third Study Program*, edited by F. O. Schmitt and F. G. Worden. Cambridge, Mass.: MIT Press, 1974.

PLAPINGER, L., and McEWEN, B. S. Gonadal steroid–brain interactions in sexual differentiation. In *Biological Determinants of Sexual Behaviour*, edited by J. B. Hutchinson. Chichester: Wiley & Sons, 1978.

POGGIO, G. F., and FISCHER, B. Binocular interaction and depth sensitivity in striate and prestriate cortex of behaving rhesus monkey. *Journal of Neurophysiology*, 1977, *40*, 1392–1405.

POLC, P., and MONNIER, M. An activating mechanism in the pontobulbar raphe system of the rabbit. *Journal of Pharmacology and Experimental Therapeutics*, 1966, *154*, 64–73.

POMPEIANO, O., and SWETT, J. E. EEG and behavioral manifestations of sleep induced by cutaneous nerve stimulation in normal cats. *Archives Italiennes de Biologie*, 1962, *100*, 311–342.

POMPEIANO, O., and SWETT, J. E. Actions of graded cutaneous and muscular afferent volleys on brain stem units in the decerebrate cerebellectomized cats. *Archives Italiennes de Biologie*, 1963, *101*, 552–583.

POSCHEL, B. P. H., and NINTEMAN, F. W. Psychotrophic drug effects on self-stimulation of the brain: a control of motor output. *Psychological Reports*. 1966, *19*, 79–82.

POST, R. M., FINK, E., CARPENTER, W. T., and GOODWIN, F. K. Cerebrospinal fluid amine metabolites in acute schizophrenia. *Archives of General Psychiatry*, 1975, *32*, 1063–1069.

POST, R. M., KOTIN, J., and GOODWIN, F. K. The effects of cocaine on depressed patients. *American Journal of Psychiatry*, 1974, *131*, 511–517.

POWERS, B., and VALENTSTEIN, E. S. Sexual receptivity: Facilitation by medial preoptic lesions in female rats. *Science*, 1972, *175*, 1003–1005.

POWERS, J. B. Hormonal control of sexual receptivity during the estrous cycle of the rat. *Physiology & Behavior*, 1970, *5*, 831–835.

POWERS, J. B. Facilitation of lordosis in ovariectomized rats by intracerebral progesterone implants. *Brain Research*, 1972, *48*, 311–325.

POWLEY, T. L. The ventromedial hypothalamic syndrome, satiety, and a cephalic phase hypothesis. *Psychological Review*, 1977, *84*, 89–126.

POWLEY, T. L., and KEESEY, R. E. Relationship of body weight to the lateral hypothalamic feeding syndrome. *Journal of Comparative and Physiological Psychology*, 1970, *70*, 25–36.

POWLEY, T. L., and OPSAHL, C. A. Ventromedial hypothalamic obesity abolished by subdiaphragmatic vagotomy. *American Journal of Physiology*, 1974, *226*, 25–33.

PRICE, D. D. and DUBNER, R. Neurons that subserve the sensory-discriminative aspects of pain. *Pain*, 1977, *3*, 307–338.

PRICE, J. The genetics of depressive behavior. *British Journal of Psychiatry*, 1968, *2*, 37–45.

PROUDFIT, H. K., and ANDERSON, E. G. Morphine analgesia blockade by raphe magnus lesions. *Brain Research*, 1975, *98*, 612–618.

PUERTO, A., DEUTSCH, J. A., MOLINA, F., and ROLL, P. L. Rapid discrimination of rewarding nutrient by the upper gastrointestinal tract. *Science*, 1976, *192*, 485–487. (a)

PUERTO, A., DEUTSCH, J. A., MOLINA, F., and ROLL, P. L. Rapid rewarding effects of intragastric injection. *Behavioral Biology*, 1976, *18*, 123–134. (b)

PUJOL, J. F., BUGUET, A., FROMENT, J. L., JONES, B., and JOUVET, M. The central metabolism of serotonin in the cat during insomnia: A neurophysiological and biochemical study after P-chlorophenylalanine or destruction of the raphe system. *Brain Research*, 1971, *29*, 195–212.

RAISMAN, G. Neuronal plasticity in the septal nuclei of the adult rat. *Brain Research*, 1969, *14*, 24–48.

RANDIC, M., and YU, H. H. Effects of 5-hydroxytryptamine and bradykinin in cat dorsal horn neurons activated by noxious stimuli. *Brain Research*, 1976, *111*, 197–203.

RAYBIN, J. B., and DETRE, T. P. Sleep disorder and symptomatology among medical and nursing students. *Comprehensive Psychiatry*, 1969, *10*, 452–467.

RAYNAUD, J.-P. Influence of rat estradiol binding plasma protein (EBP) on uterotrophic activity. *Steroids*, 1973, *21*, 249–258.

RAYNAUD, J.-P. Mercier-Bodard, C., and Baulieu, E. E. Rat estradiol binding plasma protein (EBP). *Steroids*, 1971, *18*, 767–788.

RECHTSCHAFFEN, A., WOLPERT, E. A., DEMENT, W. C., MITCHELL, S. A., and FISHER, C. Nocturnal sleep of narcoleptics. *Electroencephalography and Clinical Neurophysiology*, 1963, *15*, 599–609.

REED, D. J., and WOODBURY, D. M. Effects of hypertonic urea on cerebrospinal fluid pressure and brain volume. *Journal of Physiology (London)*, 1962, *164*, 252–264.

REES, H. D., BROGAN, L. L., ENTINGH, D. J., DUNN, A. J., SHINKMAN, P. G., DAMSTRA-ENTINGH, T., WILSON, J. E., and GLASSMAN, E. Effect of sensory stimulation on the uptake and incorporation of radioactive lysine into protein of mouse brain and liver. *Brain Research*, 1974, *68*, 143–156.

REES, H. D. and DUNN, A. J. The role of pituitary-adrenal system in the foot-shock–induced increase of [³H]lysine incorporation into mouse brain and liver protein. *Brain Research*, 1977, *120*, 317–325.

REYNOLDS, D. V. Surgery in the rat during electrical analgesia induced by focal brain stimulation. *Science*, 1969, *164*, 444–445.

RIBACK, C. E., and PETERS, A. An autoradiographic study of the projections from the lateral geniculate body of the rat. *Brain Research*, 1975, *92*, 341–368.

RICHARDS, W. Selective stereoblindness. In *Spatial Contrast*, edited by H. Spekreijse and L. H. van der Tweel. Amsterdam: North-Holland, 1977.

RICHTER, C. P. *Biological Clocks in Medicine and Psychiatry*. Springfield, Ill.: Thomas, 1965.

RICHTER, C. P. Sleep and activity: their relation to the 24-hour clock. *Proceedings of the Association for Research on Nervous and Mental Disorders*, 1967, *45*, 8–27.

RIDDLE, O., LAHR, E. L., and BATES, R. W. The role of hormones in the initiation of maternal behavior in rats. *American Journal of Physiology*, 1942, *137*, 299–317.

RIDEOUT, B. Non-REM sleep as a source of learning deficits induced by REM sleep deprivation. *Physiology & Behavior*, 1979, *22*, 1043–1047.

RIDLEY, R. M., and ETTLINGER, G. Impaired tactile learning and retention after removals of the second somatic sensory projection cortex (SII) in the monkey. *Brain Research*, 1976, *109*, 656–660.

RIDLEY, R. M., and ETTLINGER, G. Further evidence of impaired tactile learning after removals of the second somatic sensory projection cortex (SII) in the monkey. *Experimental Brain Research*, 1978, *31*, 475–488.

RINGLE, D. A., and HERNDON, B. L. Plasma dialysates from sleep deprived rabbits and their effect on the electrocorticogram of rats. *Pflügers Archives*, 1968, *303*, 344–349.

RITTER, S., and STEIN, L. Self-stimulation of noradrenergic cell group (A6) in locus coeruleus of rats. *Journal of Comparative and Physiological Psychology*, 1973, *85*, 443–452.

ROBBINS, M. J., and MEYER, D. R. Motivational control of retrograde amnesia. *Journal of Experimental Psychology*, 1970, *84*, 220–225.

ROBERTS, W. W. Rapid escape learning without avoidance learning motivated by hypothalamic stimulation in cats. *Journal of Comparative and Physiological Psychology*, 1958, *51*, 391–399. (a)

ROBERTS, W. W. Both rewarding and punishing effects from stimulation of posterior hypothalamus of cats with same electrode at same intensity. *Journal of Comparative and Physiological Psychology*, 1958, *51*, 400–407. (b)

ROBERTS, W. W., and KIESS, H. O. Motivational properties of hy-

pothalamic aggression in cats. *Journal of Comparative and Physiological Psychology*, 1964, *58*, 187–193.

ROBERTS, W. W., and ROBINSON, T. C. L. Relaxation and sleep induced by warming of preoptic region and anterior hypothalamus in cats. *Experimental Neurology*, 1969, *25*, 282–294.

ROBERTS, W. W., STEINBERG, M. L., and MEANS, L. W. Hypothalamic mechanisms for sexual, aggressive, and other motivational behaviors in the opossum, *Didelphis virginiana*. *Journal of Comparative and Physiological Psychology*, 1967, *64*, 1–15.

ROBINSON, T. E., and WHISHAW, I. Q. Effects of posterior hypothalamic lesions on voluntary behavior and hippocampal electroencephalograms in the rat. *Journal of Comparative and Physiological Psychology*, 1974, *86*, 768–786.

RODGERS, W. L. Specificity of specific hungers. *Journal of Comparative and Physiological Psychology*, 1967, *64*, 49–58.

RODIECK, R. W., and STONE, J. Response of cat retinal ganglion cells to moving visual patterns. *Journal of Neurophysiology*, 1965, *28*, 819–832.

ROEDER, F., ORTHNER, H., and MÜLLER, D. The stereotaxic treatment of pedophilic homosexuality and other sexual deviations. In *Psychosurgery*, edited by E. Hitchock, L. Laitinen, and K. Vaernet. Springfield, Ill.: C. C Thomas, 1972.

ROFFWARG, H. P., DEMENT, W. C., MUZIO, J. N., and FISHER, C. Dream imagery: Relationship to rapid eye movements of sleep. *Archives of General Psychiatry*, 1962, *7*, 235–258.

ROFFWARG, H. P., MUZIO, J. N., and DEMENT, W. C. Ontogenesic development of human sleep-dream cycle. *Science*, 1966, *152*, 604–619.

ROJAS-RAMIREZ, J. A., and DRUCKER-COLÍN, R. R. Phylogenetic correlations between sleep and memory. In *Neurobiology of Sleep and Memory*, edited by R. R. Drucker-Colín and J. L. McGaugh. New York: Academic Press, 1977.

ROLL, S. K. Intracranial self-stimulation and wakefulness: effect of manipulating ambient brain catecholamines. *Science*, 1970, *168*, 1370–1372.

ROLLS, E. T., ROLLS, B. J., KELLY, P. H., SHAW, S. G., WOOD, R. J., and DALE, R. The relative attenuation of self-stimulation, eating, and drinking produced by dopamine-receptor blockade. *Psychopharmacologia*, 1974, *38*, 219–230.

ROSE, J. E., BRUGGE, J. F., ANDERSON, D. J., and HIND, J. E. Phase-locked response to low-frequency tones in single auditory nerve fibers of the squirrel monkey. *Journal of Neurophysiology*, 1967, *30*, 769–793.

ROSE, R. M., BOURNE, P. G., POE, R. O., MOUGEY, E. H., COLLINS, D. R., and MASON, J. W. Androgen responses to stress: II. Excretion of testosterone, epitestosterone, androsterone, and etiocholanolone during basic combat training and under attack. *Pscyhosomatic Medicine*, 1969, *31*, 418–436.

ROSÉN, I., and ASANUMA, H. Peripheral inputs to the forelimb area of the monkey motor cortex: Input-output relations. *Experimental Brain Research*, 1972, *14*, 257–273.

ROSENBLATT, J. S. The development of maternal responsiveness in the rat. *Journal of Orthopsychiatry*, 1969, *39*, 36–56.

ROSENBLATT, J. S., and ARONSON, L. R. The influence of experience on the behavioural effects of androgen in prepubertally castrated male cats. *Animal Behaviour*, 1958, *6*, 171–182. (a)

ROSENBLATT, J. S., and ARONSON, L. R. The decline of sexual behavior in male cats after castration with special reference to the role of prior sexual experience. *Behaviour*, 1958, *12*, 285–338. (b)

ROSENTHAL, D. A program of research on heredity in schizophrenia. *Behavioral Science*, 1971, *16*, 191–201.

ROSENZWEIG, M. R. Evidence for anatomical and chemical changes in the brain during primary learning. In *Biology of Memory*, edited by K. H. Pribram and D. E. Broadbent. New York: Academic Press, 1970.

ROSENZWEIG, M. R. Conference summary, p. 593, in *Neural Mechanisms of Learning and Memory*, edited by M. R. Rosenzweig and E. L. Bennett. Cambridge, MA: MIT Press, 1976.

ROSS, J., CLAYBOUGH, C., CLEMENS, L. G., and GORSKI, R. A. Short latency induction of estrous behavior with intra-cerebral gonadal hormones in ovariectomized rats. *Endocrinology*, 1971, *81*, 32–38.

ROSS, S. B. and RENYI, A. L. Uptake of some tritiated sympathomimetic amines by mouse brain cortex slices in vitro. *Acta Pharmacologica et Toxicologica*, 1966, *24*, 297–309.

ROSS, S. B. and RENYI, A. L. Accumulation of tritiated 5-hydroxytryptamine in brain slices. *Life Sciences*, 1967, *6*, 1407–1415.

ROUTTENBERG, A., and LINDY, J. Effects of the availability of rewarding septal and hypothalamic stimulation on bar pressing for food under conditions of deprivation. *Journal of Comparative and Physiological Psychology*, 1965, *60*, 158–161.

ROUTTENBERG, A., and MALSBURY, C. Brainstem pathways of reward. *Journal of Comparative and Physiological Psychology*, 1969, *68*, 22–30.

ROWLAND, N. E., and ANTELMAN, S. M. Stress-induced hyperphagia and obesity in rats: A possible model for understanding human obesity. *Science*, 1976, *191*, 310–312.

ROZIN, P. Specific aversions and neophobia as a consequence of vitamin deficiency and/or poisoning in half-wild and domestic rats. *Journal of Comparative and Physiological Psychology*, 1968, *66*, 82–88.

ROZIN, P. The psychobiological approach to human memory. In *Neural Mechanisms of Learning and Memory*, edited by M. R. Rosenzweig and E. L. Bennet. Cambridge: The MIT Press, 1976.

ROZIN, P., and KALAT, J. W. Specific hungers and poison avoidance as adaptive specializations of learning. *Psychological Review*, 1971, *78*, 459–486.

ROZIN, P., PORITSKY, S., and SOTSKI, R. American children with reading problems can easily learn to read English represented by Chinese characters. *Science*, 1971, *171*, 1264–1267.

RUSSEK, M. Hepatic receptors and the neurophysiological mechanisms controlling feeding behavior. In *Neurosciences Research*, Volume 4, edited by S. Ehrenpreis. New York: Academic Press, 1971.

RUSSEK, M., and RACOTTA, R. A possible role of adrenaline and glucagon in the control of food intake. *Frontiers in Hormone Research*, 1980, in press.

RUSSEK, M., LORA-VILCHIS, M. C., and ISLAS-CHAIRES, M. Food intake inhibition elicited by intraportal glucose and adrenaline in dogs on a 22 hour-fasting/2 hour feeding schedule. *Physiology & Behavior*, 1980, *24*, 157–161.

RUSSELL, W. R., and NATHAN, P. W. Traumatic amnesia. *Brain*. 1946, 69, 280–300.

SABOURAND, O., GAGNEPAIN, J., CHATEL, M., and MERAULT, F. Un cas de lesions bilaterales de la convexité temporale: tentative de définition des symptoms. *Cortex*, 1976, *12*, 154–168.

SACHAR, E. J., and BARON, M. The biology of affective disorders. *Annual Review of Neuroscience*, 1979, *2*, 505–518.

SAGALES, T., and ERILL, S. Effects of central dopaminergic blockage with pimozide upon the EEG stages of sleep in man. *Psychopharmacologia*, 1975, *41*, 53–56.

SAR, M., and STUMPF, W. E. Androgen concentration in motor neurons of cranial nerves and spinal cord. *Science*, 1977, *197*, 77–79.

SASANUMA, S. Kana and kanji processing in Japanese aphasics. *Brain and Language*, 1975, *2*, 369–383.

SASSIN, J. F., PARKER, D. C., MACE, J. W., GOTLIN, R. W., JOHNSON, L. C., and ROSSMAN, L. G. Human growth hormone release: relation of slow-wave sleep and sleep-waking cycles. *Science*, 1969, *165*, 513–518.

SAWCHENKO, P. E., and FRIEDMAN, M. I. Sensory functions of the liver—a review. *American Journal of Physiology*, 1979, *236*, R5-R20.

SCHACHTER, S. Some extraordinary facts about obese humans and rats. *The American Psychologist*, 1971, *26*, 129–144.

SCHILLER, P. H., and MALPELI, J. G. Properties and tectal projections of monkey retinal ganglion cells. *Journal of Neurophysiology*, 1977, *40*, 428–445.

SCHEIBEL, M. E., and SCHEIBEL, A. B. On circuit patterns of the brain stem reticular core. *Annals of the New York Academy of Science*, 1961, *89*, 857–865.

SCHNEDORF, J. F., and IVY, A. C. An examination of hypnotoxin theory of sleep. *American Journal of Physiology*, 1939, *125*, 491–505.

SCHREINER, L., and KLING, A. Rhinencephalon and behavior. *American Journal of Physiology*, 1956, *184*, 486–490.

SCHWARTZ, J. -C. Histaminergic mechanisms in brain. *Annual Review of Pharmacology and Toxicology*, 1977, *17*, 325–339.

SCHWARTZ, W. J., and GAINER H. Suprachiasmatic nucleus: use of ^{14}C-labelled deoxyglucose uptake as a functional marker. *Science*, 1977, *197*, 1089–1091.

SCOVILLE, W. B. The limbic lobe in man. *Journal of Neurosurgery*, 1954, *11*, 64–66.

SCOVILLE, W. B., and MILNER, B. Loss of recent memory after bilateral hippocampal lesions. *Journal of Neurology, Neurosurgery, and Psychiatry*, 1957, *20*, 11–21.

SEKULER, R. Spatial Vision. *Annual Review of Psychology*, 1974, *25*, 195–232.

SEM-JACOBSEN, C. W. *Depth-electrographic Stimulation of the Human Brain and Behavior*. Springfield, Ill.: C C Thomas, 1968.

SEYLER, L. E., CANALIS, E., SPARE, S., and REICHLIN, S. Abnormal gonadotropin secretory responses to LRH in transsexual women after diethylstilbesterol priming. *Journal of Clinical Endocrinology and Metabolism*, 1978, *47*, 176–183.

SHAKOW, D. Segmental set: a theory of the formal psychological deficit in schizophrenia. *Archives of General Psychiatry*, 1962, *6*, 1–17.

SHALLICE, T., and WARRINGTON, E. K. Word recognition in a phonemic dyslexic patient. *Quarterly Journal of Experimental Psychology*, 1975, *27*, 187–199.

SHELLING, P., GANTEN, D., HECKL, R., HAYDUK, K., HUTCHINSON, J. S., SPONER, G., and GANTEN, U. On the origin of angiotensin-like peptides in cerebrospinal fluid. In *Central Actions of Angiotensin and Related Hormones*, edited by J. P. Buckley and C. Ferrario. New York: Pergamon Press, 1976.

SHERRINGTON, C. S. Experiments on the value of vascular and visceral factors for the genesis of emotion. *Proceedings of the Royal Society (London), Series B*, 1900, *66*, 390–403.

SIDMAN, M., STODDARD, L. T., and MOHR, J. P. Some additional quantitative observations of immediate memory in a patient with bilateral hippocampal lesions. *Neuropsychologia*, 1968, *6*, 245–254.

SIEGAL, A., and SKOG, D. Effect of electrical stimulation of the septum upon attack behavior elicited from the hypothalamus in the cat. *Brain Research*, 1970, *23*, 371–380.

SIEGEL, J. M. and McGINTY, D. J. Pontine reticular formation neurons: relationship of discharge to motor activity. *Science*, 1977, *196*, 678–680.

SIEGEL, J. M., McGINTY, D. J., and BREEDLOVE, S. M. Sleep and waking activity of pontine gigantocellular field neurons. *Experimental Neurology*, 1977, *56*, 553–573.

SIMPSON, B. A., and IVERSEN, S. D. Effects of substantia nigra lesions on the locomotor and stereotyped responses to amphetamine. *Nature*, 1971, *230*, 30–32.

SIMPSON, J. B., EPSTEIN, A. N., and CAMARDO, J. S., JR. The localization of dipsogenic receptors for angiotensin II in the subfornical organ. *Journal of Comparative and Physiological Psychology*, 1978, *92*, 581–608.

SIMPSON, J. B., MANGIAPANE, M. L., and DELLMAN, H. -D. Central receptor sites for angiotensin-induced drinking: a critical review. *Federation Proceedings*, 1978, *37*, 2676-2682.

SINCLAIR, D. C. *Cutaneous Sensation*. London: Oxford University Press, 1967.

SINGER, J. Hypothalamic control of male and female sexual behavior in female rats. *Journal of Comparative and Physiological Psychology*, 1968, *66*, 738–742.

SITARAM, V., NURNBERGER, J. I., JR., GERSHON, E. S., and GILLIN, J. C. Faster Cholinergic REM sleep induction in euthymic patients with primary affective illness. Science, 1980, 208, 200–202.

SITARAM, N., WYATT, R. J., DAWSON, S., and GILLIN, J. C. REM sleep induction by physostigmine infusion during sleep. *Science*, 1976, *191*, 1281–1283.

SLOTNICK, B. M. Disturbances of maternal behavior in the rat following lesions of the cingulate cortex. *Behaviour*, 1967, *29*, 203–236.

SLOTNICK, B. M., and McMULLEN, M. F. Intraspecific fighting in albino mice with septal forebrain lesions. *Physiology & Behavior*, 1972, *8*, 333–337.

SLOTNICK, B. M., McMULLEN, M. F., and FLEISCHER, S. Changes in emotionality following destruction of the septal area in albino mice. *Brain, Behavior, and Evolution*, 1974, *8*, 241–252.

SLOTNICK, B. M., and NIGROSH, B. J. Maternal behavior of mice with cingulate, cortical, amygdala, or septal lesions. *Journal of Comparative and Physiological Psychology*, 1975, *88*, 118–127.

SMITH, G. P., and EPSTEIN, A. N. Increased feeding in response to decreased glucose utilization in the rat and monkey. *American Journal of Physiology*, 1969, *217*, 1083–1087.

SMYTHE, G. E., and STERN, K. Tumours of the thalamus—a clinico-pathological study. *Brain*, 1938, *61*, 339–374.

SNYDER, S. H. *Madness and the Brain*. New York: McGraw-Hill, 1974.

SNYDER, S. H., BURT, D. R., and CREESE, I. Dopamine receptor of mammalian brain: direct demonstration of binding to agonist and antagonist states. *Neuroscience Symposia*, 1976, *1*, 28–49.

SNYDER, S. H., and CHILDERS, S. R. Opiate receptors and opioid peptides. *Annual Review of Neuroscience*, 1979, *2*, 35–64.

SOFRONIEW, M. V. and WEINDL, A. Projections from the parvocellular vasopressin and neurophysin containing neurons of the SCN. *American Journal of Anatomy*, 1978, *153*, 391–430.

SOLOFF, M. S., MORRISON, M. J., and SWARTZ, T. L. A comparison of the estrone-estradiol-binding proteins in the plasmas of prepubertal and pregnant rats. *Steroids*, 1972, *20*, 597–608.

SOLOMON, P. Insomnia. *New England Journal of Medicine*, 1956, *255*, 755–760.

SONDEREGGER, T. B. Intracranial stimulation and maternal behavior. *Proceedings of the Seventy-eighth Annual American Psychological Association Convention*, 1970.

SPIEGEL, E. A., WYCIS, H. T. ORCHINIK, L. W., and FREED, H. The thalamus and temporal orientation. *Science*, 1955, *121*, 771–772.

SPINELLI, D. H., and JENSEN, F. E. Plasticity: the mirror of experience. *Science*, 1979, *203*, 75–78.

SPINELLI, D. H., JENSEN, F. E., and DiPRISCO, G. V. Early experience effect on dendritic branching in normally reared kittens. *Experimental Neurology*, 1980, *62*, 1–11.

SPINELLI, D. H., PRIBRAM, K. H., and WEINGARTEN, M. Centrifugal optic nerve responses evoked by auditory and somatic stimulation. *Experimental Neurology*, 1965, *12*, 303–319.

SPITZ, C. J., GOLD, A. R., and ADAMS, D. B. Cognitive and hormonal factors affecting coital frequency. *Archives of Sexual Behavior*, 1975, *4*, 249–264.

SPOENDLIN, H. The innervation of the cochlear receptor. In *Basic Mechanisms in Hearing*, edited by A. R. Møeller. New York: Academic Press. 1973.

SPRAGUE, J. M., BERLUCCHI, G., and RIZZOLATTI, G. The role of the superior colliculus and pretectum in vision and visually guided behavior. In *Handbook of Sensory Physiology, Volume VII/3, Central Processing of Visual Information, Part B; Visual Centers in the Brain*, edited by R. Jung. Berlin: Springer, 1973.

SQUIRE, L. R. Stable impairment in remote memory following electroconvulsive therapy. *Neuropsychologia*, 1974, *13*, 51–58.

STEARNS, E. L., WINTER, J. S. D., and FAIMAN, C. Effects of coitus on gonadotropin, prolactin, and sex steroid levels in man. *Journal of Clinical Endocrinology and Metabolism*, 1973, *37*, 687–691.

STEBBINS, W. C., MILLER, J. M., JOHNSSON, L. -G., and HAWKINS, J. E. Ototoxic hearing loss and cochlear pathology in the monkey. *Annals of Otology, Rhinology, and Laryngology*, 1969, *78*, 1007–1026.

STEFFENS, A. B. Influence of reversible obesity on eating behavior, blood glucose, and insulin in the rat. *American Journal of Physiology*, 1975, *228*, 1738–1744.

STEIN, G. S., STEIN, J. S., and KLEINSMITH, L. J. Chromosomal proteins and gene regulation. *Scientific American*, 1975, *232*, 46–57.

STEIN, L. Self-stimulation of the brain and the central stimulant action of amphetamine. *Federation Proceedings*, 1964, *23*, 836–850.

STEIN, L. Chemistry of reward and punishment. In *Psychopharmacology: A Review of Progress*, edited by D. H. Efron. Washington, D. C.: Government Printing Office, 1968.

STEIN, L., and RAY, O. S. Brain stimulation reward "thresholds" self-determined in rat. *Psychopharmacologia*, 1960, *1*, 251–256.

STEINHAUSEN, W. Über den experimentallen Nachweis der Ablenkuns der Cupula terminalis in der intakten Bogengangsampulle des Labyrinths bei der thermischen und adaquäten rotatorischen Reizung. *Zeitschrift für hals-, nasen-und ohrenheilkunde*, 1931, *29*, 211–216.

STEPHAN, F. K., and NUNEZ, A. A. Elimination of circadian rhythms in drinking activity, sleep, and temperature by isolation of the suprachiasmatic nuclei. *Behavioral Biology*, 1977, *20*, 1–16.

STEPHAN, F. K., and ZUCKER, I. Rat drinking rhythms: central visual pathways and endocrine factors mediating responsiveness to environmental illumination. *Physiology & Behavior*, 1972, *8*, 315–326. (a)

STEPHAN, F. K., and ZUCKER, R. Circadian rhythms in drinking behavior and locomotor activity of rats are eliminated by hypothalamic lesion. *Proceedings of the National Academy of Science, USA*, 1972, *69*, 1583–1586. (b)

STEPHENS, D. B., and BALDWIN, B. A. The lack of effect of intrajugular or intraportal injections of glucose or amino acids on food intake in pigs. *Physiology & Behavior*, 1974, *12*, 923–929.

STERMAN, M. B., and CLEMENTE, C. D. Forebrain inhibitory mechanisms: Cortical synchronization induced by basal forebrain stimulation. *Experimental Neurology*, 1962, *6*, 91–102. (a)

STERMAN, M. B., and CLEMENTE, C. D. Forebrain inhibitory mechanisms: Sleep patterns induced by basal forebrain stimulation in the behaving cat. *Experimental Neurology*, 1962, *6*, 103–117. (b)

STERNBACH, R. A. *Pain: A Psychophysiological Analysis*. New York: Academic Press, 1968.

STETSON, R. H. The hair follicle and the sense of pressure. *Psychological Review Monographs*, 1923, *32*, 1–17.

STEVENS, S. S., and NEWMAN, E. B. Localization of actual sources of sound. *American Journal of Psychology*, 1936, *48*, 297–306.

STEWART, S. Long term memory for verbal and non-verbal material in Korsakoff patients. Unpublished Master's Thesis, University of Massachusetts, Amherst, 1977.

STONE, J. Morphology and physiology of the geniculo-cortical synapse in the cat: the question of parallel input to the striate cortex.

Investigative Ophthalmology, 1972, *11*, 338–346.

STOYVA, J., and METCALF, D. Sleep patterns following chronic exposure to cholinesterase-inhibiting organophosphate compounds. *Psychophysiology*, 1968, *5*, 206.

STURDEVANT, R. A., and GOETZ, H. Cholecystokinin both stimulates and inhibits human food intake. *Nature*, 1976, *261*, 713–715.

STRICKER, E. M. Osmoregulation and volume regulation in rats: Inhibition of hypovolemic thirst by water. *American Journal of Physiology*, 1969, *217*, 98–105.

STRICKER, E. M. Thirst, sodium appetite, and complementary physiological contributions to the regulation of intravascular fluid volume. In *The Neuropsychology of Thirst: New Findings and Advances in Concepts*, edited by A. N. Epstein, H. R. Kissileff, and E. Stellar. Washington: Winston, 1973.

STRICKER, E. M. The renin-angiotensin system and thirst: a reevaluation. II. Drinking elicited in rats by caval ligation or isoproterenol. *Journal of Comparative and Physiological Psychology*, 1977, *91*, 1220–1231.

STRICKER, E. M. The renin-angiotensin system and thirst: some unanswered questions. *Federations Proceedings*, 1978, *37*, 2704–2710.

STRICKER, E. M. Thirst and sodium appetite after colloid treatment in rats. *Journal of Comparative and Physiological Psychology*, in press, 1980. (a)

STRICKER, E. M. Thirst and sodium appetite after colloid treatment in rats: Effect of sodium deprivation. *Journal of Comparative and Physiological Psychology*, in press, 1980. (b)

STRICKER, E. M., COOPER, P. H., MARSHALL, J. F., and ZIGMOND, M. J. Acute homeostatic imbalances reinstate sensorimotor dysfunctions in rats with lateral hypothalamic lesions. *Journal of Comparative and Physiological Psychology*, 1979, *93*, 512–521.

STRICKER, E. M., FRIEDMAN, M. I., and ZIGMOND, M. J. Glucoregulatory feeding by rats after intraventricular 6-hydroxydopamine or lateral hypothalamic lesions. *Science*, 1975, *189*, 895–897.

STRICKER, E. M., ROWLAND, N., SALLER, C. F., and FRIEDMAN, M. I. Homeostasis during hypoglycemia: central control of adrenal secretion and peripheral control of feeding. *Science*, 1977, *196*, 79–81.

STRICKER, E. M., and WOLF, G. The effects of hypovolemia on drinking in rats with lateral hypothalamic damage. *Proceedings of the Society for Experimental Biology and Medicine*, 1967, *124*, 816–820.

STUMPF, W. E., and SAR, M. Autoradiographic localization of estrogen, androgen, progestin, and glucocorticosteroid in "target tissues" and "on-target tissues." In *Receptors and Mechanism of Action of Steroid Hormones*, edited by J. Pasqualini. New York: Marcel Dekker, 1976.

STURUP, G. K. Correctional treatment and the criminal sexual offender. *Canadian Journal of Correction*, 1961, *3*, 250–265.

SVARE, B. B., and GANDELMAN, R. Postpartum aggression in mice: experiential and environmental factors. *Hormones and Behavior*, 1973, *4*, 323–334.

SWANSON, H. H. Determination of the sex role in hamsters by the action of sex hormones in infancy. In *The Influence of Hormones on the Nervous System*, edited by D. H. Ford. New York: S. Karger, 1971.

SWANSON, L. W., and COWAN, W. M. Hippocampo-hypothalamic connections: origins in subicular cortex, not Ammon's horn. *Science*, 1975, *189*, 303–304 (a).

SWANSON, L. W., and COWAN, W. M. The efferent connections of the suprachiasmatic nucleus of the hypothalamus. *Journal of Comparative Neurology*, 1975, *160*, 1-12. (b)

SWANSON, L. W., KUCHARCZYK, J., and MOGENSON, G. J. Autoradiographic evidence for pathways from the medial preoptic area to the midbrain involved in the drinking response to angiotensin

II. *Journal of Comparative Neurology*, 1978, *178*, 645–660.

SWEET, C. P., and HOBSON, J. A. The effects of posterior hypothalamic lesions on behavioral and electrographic manifestations of sleep and waking in cats. *Archives Italiennes de Biologie*, 1968, *106*, 283–293.

SWEET, W. H. Participant in brain stimulation in behaving subjects. Neurosciences Research Program Workshop, 1966.

TAKAHASHI, Y. Growth hormone secretion related to the sleep waking rhythm. In *The Functions of Sleep*, edited by R. Drucker-Colín, M. Shkurovich, and M. B. Sterman. New York: Academic Press, 1979.

TANABE, T., IINO, M. OOSHIMA, Y., and TAKAGI, S. F. An olfactory area in the prefrontal lobe. *Brain Research*, 1974, *80*, 127–130.

TANABE, T., IINO, M., and TAKAGI, S. F. Discrimination of odors in olfactory bulb, pyriform-amygdaloid areas, and orbitofrontal cortex of the monkey. *Journal of Neurophysiology*, 1975, *38*, 1284–1296.

TEITELBAUM, P. The encephalization of hunger. In *Progress in Physiological Psychology*, Volume 4, edited by E. Stellar and J. M. Sprague. New York: Academic Press, 1971.

TEITELBAUM, P., and CYTAWA, J. Spreading depression and recovery from lateral hypothalamic damage. *Science*, 1965, *147*, 61–63.

TEITELBAUM, P., and EPSTEIN, A. N. The lateral hypothalamic syndrome: Recovery of feeding and drinking after lateral hypothalamic lesions. *Psychological Review*, 1962, *69*, 74–90.

TEITELBAUM, P., and STELLAR, E. Recovery from the failure to eat produced by hypothalamic lesions. *Science*, 1954, *120*, 894–895.

TENEN, S. S. Antagonism of the analgesia effect of morphine and other drugs by P-chlorophenylalanine, a serotonin depletor. *Psychopharmacologia (Berlin)*, 1968, *12*, 278–285.

TERENIUS, L. and WAHLSTRÖM, A. Morphine-like ligand for opiate receptors in human CSF. *Life Sciences*, 1975, *16*, 1759–1764.

TERKEL, J. Aspects of maternal behavior in the rat with special reference to humoral factors underlying maternal behavior at parturition. Unpublished Ph.D. dissertation. Rutgers University, 1970.

TERKEL, J. A. Chronic, cross-transfusion technique in freely behaving rats by use of a single heart catheter. *Journal of Applied Psychology*, 1972, *33*, 519–522.

TERKEL, J., and ROSENBLATT, J. S. Maternal behavior induced by maternal blood plasma injected into virgin rats. *Journal of Comparative and Physiological Psychology*, 1968, *65*, 479–482.

TIEFER, L., and JOHNSON, W. Neonatal androstenedione and adult sexual behavior in golden hamsters. *Journal of Comparative and Physiological Psychology*, 1975, *88*, 239–247.

TONNDORF, J., and KHANNA, S. M. Submicroscopic displacement amplitudes of the tympanic membrane (cat) measured by a laser interferometer. *Journal of the Acoustical Society of America*, 1968, *44*, 1546–1554.

TORAN-ALLERAND, C. D. Sex steroids and the development of the newborn mouse hypothalamus and preoptic area in vitro: implications for sexual differentiation. *Brain Research*, 1976, *106 (2)*, 407–412.

TOWE, A. L. Motor cortex and the pyramidal system. In *Efferent Organization and the Integration of Behavior*, edited by J. D. Maser. New York: Academic Press, 1973.

TOWE, A. L., and ZIMMERMAN. Unpublished observations. Cited by Towe, 1973.

TOWER, D. B. *Neurochemistry of Epilepsy*. Springfield, Ill.: C. C Thomas, 1960.

TSOU, K., and JANG, C. S. Studies on the site of analgesia action of morphine by intracerebral microinjection. *Scientia Sinica*, 1964, *13*, 1099–1109.

TUNTURI, A. R. A difference in the representation of auditory signals for the left and right ears in the isofrequency contours of right middle ectosylvian auditory cortex in the dog. *American Journal of Physiology*, 1952, *168*, 712–727.

UDRY, J. R., and MORRIS, N. M. Distribution of coitus in the menstrual cycle. *Nature*, 1968, *220*, 593–596.

UDRY, J. R., and MORRIS, N. M. Human sexual behaviour at different stages of the menstrual cycle. *Journal of Reproduction and Fertility*, 1977, *51*, 419.

UNGERER, A. Effets comparés de la puromycine et de *Datura stramonium* sur la retention d'un apprentissage instrumental chez la souris. *Comptes Rendus Academie Sciences, Serie D.* 1969, *269*, 910–913.

UNGERSTEDT, U. Stereotaxic mapping of the monoamine pathways in the rat. *Acta Physiologica Scandinavica*, 1971, *367*, 1–48.

UNGERSTEDT, U. and PYCOCK, C. Functional correlates of dopamine neurotransmission. *Bulletin der Schweizewrisschen Akademie der Medizinischen Wissenschaften*, 1974, *30*, 44–55.

URCA, G., and NAHIN, R. L. Morphine-induced multiple unit changes in analgesic and rewarding brain sites. *Pain Abstracts*, 1978, *1*, 261.

URSIN, H., and KAADA, B. R. Functional localization within the amygdaloid complex in the cat. *Electroencephalography and Clinical Neurology*, 1960, *12*, 1–20.

UTTAL, W. R. *The Psychobiology of Sensory Coding.* New York: Harper & Row, 1973.

VALENSTEIN, E. S. *Brain Control.* New York: Wiley & Sons, 1973.

VALENSTEIN, E. S., and BEER B. Continuous opportunity for reinforcing brain stimulation. *Journal of the Experimental Analysis of Behavior*, 1964, *7*, 183–184.

VALENSTEIN, E. S., COX, V. C., and KAKOLEWSKI, J. W. Polydipsia elicited by the synergistic action of a saccharin and glucose solution. *Science*, 1967, *157*, 522–554.

VALENSTEIN, E. S., and VALENSTEIN, T. Interaction of positive and negative reinforcing neural systems. *Science*, 1964, *145*, 1456–1458.

VALLBO, Å. B. Muscle spindle response at the onset of isometric voluntary contractions in man. Time differences between fusimotor and skeletomotor effects. *Journal of Physiology (London)*, 1971, *218*, 405–431.

VAN DIS, H., and LARSSON, K. Induction of sexual arousal in the castrated male rat by intracranial stimulation. *Physiology & Behavior*, 1971, *6*, 85–86.

VAN ESSEN, D. C. Visual areas of the mammalian cerebral cortex. *Annual Review of Neuroscience*, 1979, *2*, 227–263.

VAUGHAN, T., WYATT, R. J., and GREEN, R. Changes in REM sleep of chronically anxious depressed patients given alpha-methyl-paratyrosine (AMPT). *Psychophysiology*, 1972, *9*, 96.

VERNEY, E. B. The antidiuretic hormone and the factors which determine its release. *Proceedings of the Royal Society (London), Series B*, 1947, *135*, 25–106.

VERTES, R. P. Selective firing of rat pontine gigantocellular neurons during movement and REM sleep. *Brain Research*, 1977, *128*, 146–152.

VERZEANO, J., LAUFER, M., SPEAR, S., and MCDONALD, S. The activity of neuronal networks in the thalamus of the monkey. In *Biology of Memory*, edited by K. H. Pribram and D. E. Broadbent. New York: Academic Press, 1970.

VERZEANO, J., and NEGISHI, K. Neuronal activity in cortical and thalamic networks: A study with multiple electrodes. *Journal of General Physiology*, 1960, *43*, (6, part 2), 177–195.

VICTOR, M., ADAMS, R. D., and COLLINS, G. H. *The Wernicke-Korsakoff Syndrome.* Philadelphia: F. A. Davis, 1971.

VILBERG, T. R., and BEATTY, W. W. Behavioral changes following VMH lesions in rats with controlled insulin levels. *Pharmacology, Biochemistry, and Behavior*, 1975, *3*, 377–384.

VOCI, V. E., and CARLSON, N. R. Enhancement of maternal behavior and nest behavior following systemic and diencephalic administration of prolactin and progesterone in the mouse. *Journal of Comparative and Physiological Psychology*, 1973, *83*, 388–393.

VOGEL, G. W. A motivational function of REM sleep. In *The Functions of Sleep*, edited by R. Drucker-Colín and M. B. Sterman. New York: Academic Press, 1979.

VOGEL, G. W., AUGUSTINE, F., MCABEE, R., and THURMOND, A. New findings about how REM sleep deprivation improves depression. Paper presented to the Association for the Psychophysiological Study of Sleep, Houston, 1977. Cited by Vogel, 1979.

VOLICER, L., and LOEW, C. G. Penetration of angiotensin II into the brain. *Neuropharmacology*, 1971, *10*, 631–636.

VON BÉKÉSY, G. The vibration of the cochlear partition in anatomical preparations in the models of the inner ear. *Journal of the Acoustical Society of America*, 1949, *21*, 233–245.

VON BÉKÉSY, G. Sweetness produced electrically on the tongue and its relation to taste theories. *Journal of Applied Physiology*, 1964, *19*, 1105–1113.

VON MEDUNA, L. General discussion of the cardiazol therapy. *American Journal of Psychiatry (Supplement)*, 1938, *94*, 40–50.

WADE, G. N., and ZUCKER, I. Modulation of food intake and locomotion activity in female rats by diencephalic hormone implants. *Journal of Comparative and Physiological Psychology*, 1970, *72*, 328–336.

WALINDER, J., SKOTT, A., CARLSSON, A., and ROOS, B. Potentiation by metyrosine of thioridazine effects in chronic schizophrenics: a long term trial using double-blind cross-over technique. *Archives of General Psychiatry*, 1976, *33*, 501–505.

WALKER, W. A., and LESHNER, A. I. The role of the adrenals in aggression. *American Zoologist*, 1972, *12*, 652.

WALLACH, H., NEWMAN, E. G., and ROSENWEIG, M. R. The precedence effect in sound localization. *American Journal of Psychology*, 1949, *62*, 315–336.

WARD, A. A. The cingular gyrus: Area 24. *Journal of Neurophysiology*, 1948, *11*, 13–23.

WARD, I. Prenatal stress feminizes and demasculinizes the behavior of males. *Science*, 1972, *175*, 82–84.

WARD, OJEMAN, and CALVIN. Personal communication, cited by Towe, 1973.

WARRINGTON, E. K., and SHALLICE, T. Neuropsychological evidence of visual storage in short-term memory tests. *Quarterly Journal of Experimental Psychology*, 1972, *24*, 30–40.

WAXENBERG, S. E., DRELLICH, M. G., and SUTHERLAND, A. M. The role of hormones in human behavior: I. Changes in female sexuality after adrenalectomy. *Journal of Clinical Endocrinology and Metabolism*, 1959, *19*, 193–202.

WEBB, W. B. Paper presented at a symposium of the First International Congress of the Association for the Psychophysiological Study of Sleep, 1971, Bruges, Belgium.

WEBB, W. B. Sleep as an adaptive process. *Perceptual and Motor Skills*, 1974, *38*, 1023–1027.

WEBB, W. B. *Sleep: The Gentle Tyrant.* Englewood Cliffs, N. J.: Prentice-Hall, 1975.

WEBB, W. B. The sleep of conjoined twins. *Sleep*, 1978, *1*, 205–211.

WEGENER, G. Autoradiographische Untersuchungen über sesteigerte proteinsynthese in tectum opticum von Fröschen nach opticher Reizung. *Experimental Brain Research*, 1970, *10*, 363–379.

WEISKRANTZ, L., WARRINGTON, E. K., SANDERS, M. D., and MARSHALL, J. Visual capacity in the hemianopic field following a restricted occipital ablation. *Brain*, 1974, *97*, 708–728.

WERSÄLL, J., FLOCK, A., and LUNDQUIST, P. Structural basis for directional sensitivity in cochlear and vestibular sensory receptors. *Cold Spring Harbor Symposia on Quantitative Biology*, 1965,

30, 115–132.

WETZEL, M. C. Self-stimulation aftereffects and runway performance in the rat. *Journal of Comparative and Physiological Psychology*, 1963, *56*, 673–678.

WEVER, E. G. *Theory of Hearing*. New York: Wiley and Sons, 1949.

WEVER, E. G., and BRAY, C. W. Present possibilities for auditory theory. *Psychological Review*, 1930, *37*, 365–380.

WHITFIELD, I. C., and EVANS, E. F. Responses of auditory cortical neurons to stimuli of changing frequency. *Journal of Neurophysiology*, 1965, *28*, 655–672.

WIESNER, B. P., and SHEARD, N. *Maternal Behaviour in the Rat*. London: Oliver and Brody, 1933.

WILSKA, A. Eine Methode zur Bestimmung der Horschwellenamplituden der Tromenfells bei verscheideden Frequenzen. *Skandinavisches Archiv für Physiologie*, 1935, *72*, 161–165.

WILSON, J. G., HAMILTON, J. B., and YOUNG, W. C. Influence of age and presence of the ovaries on reproductive function in rats injected with androgens. *Endocrinology*, 1941, *29*, 784–789.

WISE, C. D., and STEIN, L. Facilitation of brain self-stimulation by central administration of norepinephrine. *Science*, 1969, *163*, 299–301.

WOLSTENCROFT, J. H. Reticulospinal neurones. *Journal of Physiology (London)*, 1964, *174*, 91–99.

WOODRUFF, M. L., and BOWNDS, M. D. Amplitude, kinetics, and reversibility of a light-induced decrease in guanosine 3',5'-cyclic monophosphate in isolated frog receptor membranes. *Journal of General Physiology*, 1979, *73*, 629–653.

WOODRUFF, M. L., BOWNDS, D., GREEN, S. H., MORRISEY, J. L., and SHEDLOVSKY, A. Guanosine 3',5'-cyclic monophosphate and the in vitro physiology of frog photoreceptor membranes. *Journal of General Physiology*, 1977, *69*, 667–679.

WRIGHT, R. H. Predicting olfactory quality from far infrared spectra. *Annals of the New York Academy of Sciences*, 1974, *237*, 129–136.

WYATT, R. J. The serotonin-catecholamine dream bicycle: a clinical study. *Biological Psychiatry*, 1972, *5*, 33–64.

WYNNE, L. C., CROMWELL, R. L., and MATTHYSSE, S. *The Nature of Schizophrenia*. New York: Wiley & Sons, 1978.

WYRWICKA, W., and DOBRZECKA, C. Relationship between feeding satiation centers of the hypothalamus. *Science*, 1960, *131*, 805–806.

YAKSH, T. L. Central nervous system sites mediating opiate analgesia. In *Advances in Pain Research and Therapy*, vol. 3, edited by J. J. Bonica, J. C. Liebeskind, and D. Albe-Fessard. New York: Raven Press, 1979.

YALON, I. D., GREEN, R., and FISK, F. Prenatal exposure to female hormones: effect on psychosexual development in boys. *Archives of General Psychiatry*, 1973, *28*, 554–561.

YOKOYAMA, A., HALASZ, B., and SAWYER, C. H. Effect of hypothalamic deafferentation on lactation in rats. *Proceeding of the Society for Experimental Biology and Medicine*, 1967, *125*, 623–626.

YOUNG, R. W. Visual cells. *Scientific American*, 1970, *223(4)*, 80–91.

YOUNG, W. C. The hormones and mating behavior. In *Sex and Internal Secretions*, third edition, edited by W. C. Young. Baltimore: Williams & Wilkins, 1961.

ZEIGLER, H. P., and KARTEN, H. S. Central trigeminal structures and the lateral hypothalamus syndrome in the rat. *Science*, 1974, *186*, 636–637.

ZWEIRS, H., VELDHUIS, H. J., SCHOTMAN, P., and GISPEN, W. H. ACTH, cyclic nucleotides, and brain phosphorylations in vitro. *Neurochemistry Research*, 1976, *1*, 669–677.

INDEX